L. E. J. BROUWER
COLLECTED WORKS
VOLUME 2

L. E. J. BROUWER (1881–1966)

L. E. J. BROUWER
COLLECTED WORKS

2

GEOMETRY, ANALYSIS, TOPOLOGY AND MECHANICS

EDITED BY
HANS FREUDENTHAL

1976
NORTH-HOLLAND PUBLISHING COMPANY – AMSTERDAM · OXFORD
AMERICAN ELSEVIER PUBLISHING COMPANY, INC. – NEW YORK

Library of Congress Catalog Card Number: 73–75529
ISBN North-Holland, Volume 2: 0 7204 2805 x
ISBN American Elsevier: 0 444 10643 x
Printed in The Netherlands

PUBLISHERS:
NORTH-HOLLAND PUBLISHING COMPANY – AMSTERDAM · OXFORD
SOLE DISTRIBUTORS FOR THE U.S.A. AND CANADA:
AMERICAN ELSEVIER PUBLISHING COMPANY, INC.
52 VANDERBILT AVENUE NEW YORK, N.Y. 10017

TABLE OF CONTENTS

See also the Bibliography, p. XVI

PREFACE

The second volume of Brouwer's Collected Works contains:

the greater part of his work in classical mathematics, written in English, French, and German, and a few translations from Dutch into English;

a few inedita (scraps and correspondence) in French and German, and translations of inedita from Dutch into English.

The inedita have come from Brouwer's estate and the *Universitäts-bibliothek Göttingen*. It is a pity that the greater part of Brouwer's papers seems to have been destroyed; perhaps a fire during the war in the cabin he inhabited has contributed to this irreparable loss. In August 1929 Brouwer had lost his scientific diary by the theft of his briefcase on the Brussels tramway. Nevertheless when visiting Brouwer's cabin in 1968, P. Alexandroff still found Brouwer's archive carefully watched over by Miss Jongejan. It is not clear what happened after her death. What is now left in Brouwer's estate, is in general incoherent and difficult to interpret. Brouwer's part of his correspondence with F. Klein and D. Hilbert is fairly well preserved.

The texts included in the present volume are arranged according to subject matter rather than in their temporal order. The greater part were copied from Brouwer's own reprints. Except for a few pencil writings all corrections and other notes in Brouwer's hand are visible in the copied text. Though the editor was not able to check or study the text systematically, a few errors were noticed and corrected.

Time did not allow the editor to tie lengthy historical investigations to the edition. His extensive, though unsystematic, comments draw on his previous knowledge of Brouwer's work in its historical frame. Lest it would be lost, some oral tradition was included into the commentary though nothing could be checked by testimony of others.

Brouwer's work, published during his life-time, is coded by means of the – approximate – year of first publication together with a discriminating capital letter; the alphabetic order of the letters does not necessarily reflect the temporal order of publication. A Dutch and an English (or German) version of the same paper are distinguished by a numeral 1 and 2, respectively. Inedita are coded by means of the capital Y together with a number.

References such as Brouwer 1904 A2, Alexandroff 1928 A refer to the Bibliography of Brouwer's work (pp. XVI–XXVII) and the Literature (pp. 689–706). respectively.

Numbers within ⟦ ⟧ in the margin of a piece of the text refer to the comment at the end of that piece.

Brackets ⟦ ⟧ within the text include additions; brackets ⟨ ⟩ within the text include parts that would better be deleted.

My collaborator Mrs. Th. Breughel–Vollgraff has again surpassed herself. She surprised me by reading better than I could, not only my own handwriting she is used to, but also Brouwer's, which was new to her, and by understanding better than I did, the implicit order in the most chaotic manuscript I ever produced. Without her assistance much of my work, and in particular the present one, would never have been published. My only worry is she is beyond my praise.

The Editor

THE LIFE OF L.E.J. BROUWER

(27 February 1881 – 2 December 1966)

by H. Freudenthal and A. Heyting*

Luitzen Egbertus Jan Brouwer was born at Overschie on 27th February, 1881. He attended the 'Hoogere Burgerschool' (modern secondary school) at Hoorn and Haarlem, and the 'gymnasium' at Haarlem. On the 27th September 1897 he registered at the University of Amsterdam to study mathematics and science. He completed his academic studies at the University of Amsterdam with the publication of his doctoral dissertation entitled 'Over de grondslagen der wiskunde' (The foundations of mathematics) on 19th February 1907 and was awarded the degree of Doctor of Mathematics and Sciences 'cum laude'. He took up the duties of 'privaatdocent' at the University of Amsterdam on 12th October 1909 with a public lecture 'Het wezen der Meetkunde' (the nature of geometry). In 1912 he was appointed 'Professor extraordinary' in set theory, function theory and axiomatics at the University of Amsterdam; on 14th October 1912 he delivered his inaugural address on 'Intuitionisme en Formalisme'. He was made Professor 'in ordinary' at the same University a year later and held the chair in set theory, function theory and axiomatics until his retirement in 1951. He was a member of the Natural Sciences division of the Royal Academy of Sciences. He was killed in a traffic accident on the 2nd December 1966.

It is most unlikely that in 1912 there were in any other universities chairs in set theory and axiomatics, the combination defined in Brouwer's mandate. They reflected the two main subjects of Brouwer's thesis and the two main lines along which his research had developed at that time: set theory with a strong emphasis on topology, and axiomatics as a part of the enquiry into the foundations of mathematics, referred to by Brouwer as 'intuitionism'.

Brouwer's publications, initially only in the Proceedings of the Royal Academy of Sciences, later on also in the *Mathematische Annalen*, of which he was a member of the Board of Editors from 1915 to 1928, started in 1904 with a kind of geometry which was cultivated at that period in the Netherlands. A paper on vector distributions in 1907, though still written in the classical style of differential geometry, can with hindsight be interpreted as a sign-post of topology.

Brouwer's contemporaries in 1907 may have been puzzled by his thesis and even to-day it still presents us with problems. On the one hand they knew him as a serious and promising mathematician, on the other hand, they were faced with

* Translated, with emendations, from *Levensbericht van L. E. J. Brouwer*, KNAW Jaarboek 1966–67, pp. 335–340.

a work that seemed to be philosophical in nature rather than mathematical in the technical sense. This impression, however, is quite unjustified. Brouwer's thesis is an account of his struggle with mathematical problems with which he was deeply involved, at a time when he had reconnoitred, though not yet conquered, the technical difficulties of the field. In Foundations of Mathematics Brouwer had already taken up a position in direct opposition to formalism and logicism; it would, however, be many years before he was to embark on the construction of an intuitionist mathematics. In the Foundations of Geometry he had favoured Helmholtz's philosophically-tinted analysis of the space problem in preference to Hilbert's more naive approach based on Euclid. Brouwer's approach, however, required much stronger technical tools: solving Helmholtz's space problem without the supposition of differentiability, or more generally, eliminating such suppositions from the theory of Lie groups, which was Hilbert's Fifth Problem at the Paris Congress of 1900. Brouwer's lecture at the International Congress in Rome in 1908 shows that these problems were very much on his mind when he was writing his thesis and that he knew more about them than is evident from his thesis – indeed, his Rome lecture deals with the topological foundation of Lie groups, a problem that was not in fact solved until the 1950's. Brouwer's publications on this subject in the 'Mathematische Annalen' in 1909–1910 only treated a small part of what was then mere conjecture and what much later was confirmed, but even this small piece looked at that stage like the result of an almost superhuman effort. The topological methods which Brouwer had adopted from others and then further developed, were inadequate to deal successfully with such problems; it is likely that Brouwer soon came to this conclusion himself. However, one beautiful result of these investigations is his Translation Theorem (1909), the proof of which topologists have again and again tried to simplify but so far only with limited success.

The topological results with which Brouwer's name will forever be associated were published in 1911–1912; it appears from datings, indirect references, and drafts among Brouwer's papers, that these results date back to 1910 or perhaps even 1909. Brouwer was an 'autodidact' in the sense that in his mathematical research he was not guided by a master. Dutch mathematicians at that time tended to work in very traditional areas, and the only evidence of Brouwer's personal contacts abroad before 1909 is his attendance and his contribution at the Congress in Rome in 1908. On the other hand, it is quite clear from his thesis that he was well-informed about contemporary research and controversy in mathematics; during his visit to Rome in 1908 he probably also became familiar with all the 'folklore' about it.

In 1910 Brouwer convinced the mathematical world of his penetrating mind and imagination: he constructed a surprising and paradoxical example of a curve which divides the plane into three domains while every one of its points is a boundary point of each of the three domains.

The sources of inspiration which led to his work in 1911–1912 are not difficult to find. After Cantor had proved the equipotency of all cartesian spaces, whatever their dimension, and Peano had constructed continuous transformations that increased dimension, Poincaré had formulated the problem of topological invariant definition of dimension. In spite of many attempts, the topological invariance of dimension was stil an unsolved problem. There is no doubt that this problem has been one of the chief sources of inspiration in Brouwer's work at that time. He was certainly aware of the urgency of the problem of the topological invariance of the n-dimensional domain, which was needed in Poincaré's continuity proof in the theory of automorphic functions (1883) and still at that time used in the theory of uniformisation, though it had never rigorously been justified. But most of all, Brouwer's own research on vector fields led the way to the topological work he was to publish in the years 1911–1912.

Seen in historical perspective, Brouwer's performance at that time, in particular his first great paper, on the invariance of dimension, looks like witchcraft. As his magic wand he used the seemingly simple theorem that under a continuous mapping of the unit cube into itself which displaces every point less than half a unit, the image has an interior point. He proved this theorem by simplicial approximation of the given mapping, a method which due to Brouwer now seems rather obvious but which was revolutionary in 1911. It is the first example of mixing up combinatorial and set theory methods, which later would lead to the most profound results in topology. Because of this discovery we may rightly claim that, notwithstanding all precursors, modern topology started with Brouwer.

It is no exaggeration to refer to Brouwer's new tools as a magic wand; the results indeed look like witchcraft, the invariance of dimension, the mapping degree, the singularities of spherical vector fields, which factually were his starting point, fixed point theorems, the invariance of domain, with applications to automorphic functions, the Jordan theorem for arbitrary dimension, the invariance of the closed curve and its connectivity, which in germ contains the homology theory of compacta. This indeed is a staggering series of old conjectures cometrue, and of new theorems and concepts which in due course were to be fundamental in topology, all produced in less than two years. In 1913 Brouwer added an intrinsic, inductive definition of dimension which was discovered anew by P. Urysohn (1922) and K. Menger (1924) and developed into the now classic theory of dimension. Up to 1920 there followed a number of papers on classification of mappings of surfaces. In 1939, a long time after Brouwer had said farewell to research in topology, he published a paper on triangulation of differentiable varieties. This publication suggests the existence of even more topological work among Brouwer's papers.*

* This hope has not materialised. We knew that much had been lost, but the estate contained less than could reasonably be expected. Brouwer used to carry all his papers with him when he travelled.

Brouwer made his first international public appearance during the Congress of Rome in 1908. His visit to Paris around New Year 1910 was the start of his contacts with H. Poincaré, J. Hadamard, and E. Borel. There modern topology was born, 1 January 1910, 6 rue de l'Abbée de l'Epée. In the summer of 1909 he seems to have met Hilbert, and from 1911 onwards he made regular visits to Göttingen. In 1911 he attended the *Deutsche Naturforscherversammlung*, and the Congress of *Deutsche Mathematiker-Vereinigung* in Karlsruhe. On this last occasion he had the honour of taking part in a select colloquium on automorphic functions. In 1912 he attended the International Congress in Cambridge. His contacts with A. Schoenflies date from 1908 when he discovered fundamental mistakes in Schoenflies' work. In 1911, when Schoenflies was preparing a revised edition of his *Bericht* 'Die Entwicklung der Lehre von den Punktmannigfaltigkeiten', Brouwer was asked to assist him. When this work appeared in 1913 it contained numerous contributions due to Brouwer's profound criticism.

On the eve of the first world war Brouwer ranked among the first mathematicians. The war paralysed international mathematical activity. Maybe, this lack of international contact was one of the reasons why Brouwer turned his attention away from topology which he had promoted and placed in the centre of mathematical interest. In the period 1925–1926 P. Alexandroff, K. Menger and L. Vietoris stayed with Brouwer. At that time he was no more actively working in topology, but his influence as a mentor can still be traced from the work of his collaborators.

Yet the influence exerted by the publications themselves was more important. His writings were known as difficult to read, pre-brouwerian methods still prevailed and some time would elapse before others fully comprehended his work and built on it. The first to do so were Erhard Schmidt and J. W. Alexander. Through Schmidt's lectures, Heinz Hopf became acquainted with Brouwer's work and, more than anybody else, carried on Brouwer's work directly and in a great style. The first-named of the two authors of this obituary, himself a student of Hopf, was introduced to Brouwer in 1927 when Brouwer gave a series of lectures at Berlin University; when he had completed his studies for the Ph.D. degree in 1930 he became one of Brouwer's assistants, together with W. Hurewicz, who had come to Amsterdam with Menger. Although Brouwer was no longer active in the field of topology, somehow his presence inspired his two assistants to continue his topological work. Hurewicz's pioneering research was done in Amsterdam at that time and presented by Brouwer to the Academy for publication in the Proceedings.

Brouwer's work in the Foundations of Mathematics is marked by the same

On a journey through Belgium in 1928 a pocket containing his scientific journal was stolen and never recovered. The other papers which he may have used for his 1939 publication were probably destroyed by fire in 1943. (Note added in 1972.) The material that has been spared is historically but not mathematically valuable. (Note added in 1975.) See Preface.

characteristics of originality and unbending rigour as his topological work. Here too Brouwer had his forerunners, especially among the French constructivists whose main spokesman was E. Borel.

Brouwer was the first to have the courage to deduce and accept all the consequences that follow from a rigorous constructive viewpoint. His most important ideas can already be found in his thesis of 1907, although they are still somewhat vaguely formulated. In the article 'The unreliability of the logical principles', which appeared in 1908 in Tijdschrift voor Wijsbegeerte, he expresses his most revolutionary conclusion: the logical principle of the excluded middle is not a reliable tool in mathematics. Against formalists and logicists, who had reduced mathematics to a game of symbols, Brouwer maintained that mathematics is an activity of constructing in the human mind and that to existence of mathematical objects independent of their mental construction no clear meaning can be given. Acceptance of this principle ultimately leads to the rejection of the principle of the excluded middle. To this consistently constructivist conception of mathematics he gave the name intuitionism, a term which has caused a great deal of misunderstanding.

During the first ten years following the publication of his dissertation Brouwer only occasionally expressed his views on these matters. He was probably entirely absorbed in his topological research. He did, however, say some years later that in his topological work he had tried to use only methods which he expected could be made constructive. It seems that the selfsame mental disposition drove Brouwer both towards constructive methods and a constructive philosophy. In 1918, he started a series of publications in which he systematically reconstructed mathematics in accordance with intuitionist principles. This reconstruction required a complete revision of all proofs; many results could not be retained and others only in a modified form. It is not surprising that only few mathematicians were prepared to accept such far-reaching consequences on the basis of philosophical considerations which to them seemed farfetched and ran counter to their usual train of thought. The interest of mathematicians and philosophers in Brouwer's work remained largely academic. Apart from his own students, there have been only a few mathematicians who worked and published along intuitionist lines; one of them was Hermann Weyl.

After 1927, Brouwer's publications became more and more polemical, some of them were published lectures in which he defended his views and developed his philosophy further.

Only during the last ten years has intuitionism been more universally recognised as important and exciting. Various mathematicians are now engaged in research into the formal suppositions in the practice of intuitionist mathematics. Already in his publications of 1907 and 1908 Brouwer had related his philosophy of mathematics with a philosophy of natural science. Later he was influenced by Mannoury, who introduced the Signific Movement to Holland and developed

'Significs' into an original philosophy. Even if Brouwer at first borrowed many ideas from Mannoury, his thinking soon moved in a direction diametrically opposite. Whereas Mannoury's philosophy is strongly social, Brouwer's philosophy centres round the individual and ultimately leads to solipsism; this is very clear from his address to the International Philosophical Congress of 1948. Even if Brouwer's purely philosophical writings are not of the same high standard as his contribution to mathematics and the foundations of mathematics, they are marked by the same unrelenting consistency in pursuing a chosen course.

We have tried to sketch the development of Brouwer's scientific activities and personality. We are well aware that this sketch is imperfect and far from complete. Without recourse to Brouwer's vast correspondence and unpublished papers, one can do no more than make a reasoned guess. With all the relevant information available one could certainly write a most fascinating account of the birth and rise of this great mathematician and remarkable man.

Abstraction together with a vivid geometrical intuition are the most typical characteristics of Brouwer's work. The unimpeachable preciseness of his argument is reflected in his style; his elaborate and precise formulation frightened and put off whole generations of mathematicians; it has now been universally accepted so that we can read Brouwer more easily than his contemporaries. The same impressive but exhausting style he used in his lectures and public addresses. Unfortunately, he rarely lectured on the mathematics he himself had helped create. He, therefore, had only a few direct followers.

Both authors of this obituary have in their studies been strongly influenced by Brouwer. His tall, lean and muscular figure, his sharp ascetic features, his soft but firm voice and his very personal handwriting remained even in his later years the physical reflection of a great mind which excelled in originality, depth and self-discipline and which in its pursuit of knowledge was not restricted by the narrow bounds of one subject and its methods.

BIBLIOGRAPHY

Luitzen Egbertus Jan Brouwer

1904 A1 Over de splitsing van de continue beweging om een vast punt O van R_4 in twee continue bewegingen om O van R_3's. KNAW Versl. 12, pp. 819–839.

1904 A2 On a decomposition of a continuous motion about a fixed point O of S_4 into two continuous motions about O of S_3's. KNAW Proc. 6, pp. 716–735.

1904 B1 Over symmetrische transformatie van R_4 in verband met R_r en R_l. KNAW Versl. 12, pp. 926–928.

1904 B2 On symmetric transformation of S_4 in connection with S_r and S_l. KNAW Proc. 6, pp. 785–787.

1904 C1 Algebraïsche afleiding van de splitsbaarheid der continue beweging om een vast punt van R_4 in die van twee R_3's. KNAW Versl. 12, pp. 941–947.

1904 C2 Algebraic deduction of the decomposability of the continuous motion about a fixed point of S_4 into those of two S_3's. KNAW Proc. 6, pp. 832–838.

1905 Leven, kunst en mystiek ⟦Life, Art and Mysticism⟧. Waltman, Delft 1905, 99 p.

1906 A1 Meerdimensionale vectordistributies. KNAW Versl. 15, pp. 14–26.

1906 A2 Polydimensional vectordistributions. KNAW Proc. 9, pp. 66–78.

1906 B1 Het krachtveld der niet-Euclidische negatief gekromde ruimten. KNAW Versl. 15, pp. 75–94.

1906 B2 The force field of the non-Euclidean spaces with negative curvature. KNAW Proc. 9, pp. 116–133.

1906 C1 Het krachtveld der niet-Euclidische positief gekromde ruimten. KNAW Versl. 15, pp. 293–310.

1906 C2 The force field of the non-Euclidean spaces with positive curvature. KNAW Proc. 9, pp. 250–266.

1907 Over de grondslagen der wiskunde ⟦On the foundations of mathematics⟧. Thesis, Amsterdam 1907, 183 p.

Die moeglichen Maechtigkeiten. Atti IV Congr. Intern. Mat. Roma* III, pp. 569–571.

1908 B Over de grondslagen der wiskunde ⟦On the foundations of mathematics⟧. Nieuw Arch. Wisk. (2) 8, pp. 326–328.

* "Bologna" as indicated in Vol. 1 is mistaken.

1908 C De onbetrouwbaarheid der logische principes ⟦The unreliability of the logical principles⟧. Tijdschrift voor Wijsbegeerte 2, pp. 152–158. Also in 1919 B.

1908 D1 Over differentiequotiënten en differentiaalquotiënten. KNAW Versl. 17, pp. 38–45.

1908 D2 About difference quotients and differential quotients. KNAW Proc. 11, pp. 59–66.

1909 A Het wezen der meetkunde ⟦The nature of geometry⟧. Amsterdam 1909, 24 p. ⟦Also in 1919 B.⟧

1909 B Die Theorie der endlichen continuierlichen Gruppen, unabhängig von den Axiomen von Lie. Atti IV Congr. Intern. Mat. Roma II, pp. 296–303.

1909 C Die Theorie der endlichen kontinuierlichen Gruppen, unabhängig von den Axiomen von Lie, I. MA 67, pp. 246–267. ⟦Berichtigung see 1910 G.⟧

1909 D Over de niet-Euclidische meetkunde. Nieuw Arch. Wisk. (2) 9, pp. 72–74. ⟦Not republished.⟧

1909 E Karakteriseering der Euclidische en niet-Euclidische bewegingsgroepen in R_n ⟦Characterization of the Euclidean and non-Euclidean motion groups in R_n⟧. Handelingen van het Nederlandsch Natuur- en Geneeskundig Congres 12, pp. 189–199.

1909 F1 Over één-éénduidige, continue transformaties van oppervlakken in zich zelf. KNAW Versl. 17, pp. 741–752.

1909 F2 Continuous one-one transformations of surfaces in themselves. KNAW Proc. 11, pp. 788–798.

1909 G1 Over continue vectordistributies op oppervlakken. KNAW Versl. 17, pp. 896–904.

1909 G2 On continuous vector distributions on surfaces. KNAW Proc. 11, pp. 850–858.

1909 H1 Over één-éénduidige, continue transformaties van oppervlakken in zichzelf, II. KNAW Versl. 18, pp. 106–117.

1909 H2 Continuous one-one transformations of surfaces in themselves, II. KNAW Proc. 12, pp. 286–297.

1910 A1 Over continue vectordistributies op oppervlakken, II. KNAW Versl. 18, pp. 702–721.

1910 A2 On continuous vector distributions on surfaces, II. KNAW Proc. 12, pp. 716–734.

1910 B1 Over de structuur der perfekte puntverzamelingen. KNAW Versl. 18, pp. 833–842.

1910 B2 On the structure of perfect sets of points. KNAW Proc. 12, pp. 785–794.

1910 C Zur Analysis Situs. MA 68, pp. 422–434

1910 D1 Over continue vector distributies op oppervlakken, III. KNAW Versl. 19, pp. 36–51.

1910 D2 On continuous vector distributions on surfaces, III. KNAW Proc. 12, pp. 171–186.

1910 E Beweis des Jordanschen Kurvensatzes. MA 69, pp. 169–175.

1910 F Über eineindeutige, stetige Transformationen von Flächen in sich. MA 69, pp. 176–180. ⟦Berichtigung see 1910 L, 1919 F.⟧

1910 G Berichtigung ⟦referring to 1909 C⟧. MA 69, p. 180.

1910 H Die Theorie der endlichen kontinuierlichen Gruppen, unabhängig von den Axiomen von Lie, II. MA 69, pp. 181–203.

1910 J Sur les continus irréductibles de M. Zoretti. Annales Ecole norm. sup. 27, pp. 565–566.

1910 L Berichtung ⟦referring to 1910 F⟧. MA 69, p. 592.

1911 A ⟦Review of:⟧ G. Mannoury, Methodologisches und Philosophisches zur Elementar-Mathematik (Haarlem 1909). Nieuw Arch. Wisk. (2) 9, pp. 199–201.

1911 B1 Over de structuur van perfecte puntverzamelingen, II. KNAW Versl. 19, pp. 1416–1426.

1911 B2 On the structure of perfect sets of points, II. KNAW Proc. 14, pp. 137–147.

1911 C Beweis der Invarianz der Dimensionenzahl. MA 70, pp. 161–165.

1911 D Über Abbildung von Mannigfaltigkeiten. MA 71, pp. 97–115. ⟦Berichtigung see 1911 M, 1921 E.⟧

1911 E Beweis der Invarianz des *n*-dimensionalen Gebiets. MA 71, pp. 305–313. ⟦Berichtigung see 1921 F.⟧

1911 F Beweis des Jordanschen Satzes für den *n*-dimensionalen Raum. MA 71, pp. 314–319.

1911 G Über Jordansche Mannigfaltigkeiten. MA 71, pp. 320–327. ⟦Bemerkung see 1911 N.⟧

1911 H1 Over één-éénduidige, continue transformaties van oppervlakken in zichzelf, III. KNAW Versl. 19, pp. 737–747.

1911 H2 Continuous one-one transformations of surfaces in themselves, III. KNAW Proc. 13, pp. 767–777.

1911 J1 Over één-éénduidige, continue transformaties van oppervlakken in zichzelf, IV. KNAW Versl. 20, pp. 24–34.

1911 J2 Continuous one-one transformations of surfaces in themselves, IV. KNAW Proc. 14, pp. 300–310.

1911 K Sur une théorie de la mesure. A propos d'un article de M. G. Combebiac. Enseignement Math. 13, pp. 377–380.

1911 L Sur le théorème de M. Jordan dans l'espace à *n* dimensions. C. R. Paris 153, pp. 542–543.

1911 M Berichtigung ⟦referring to 1911 D⟧. MA 71, p. 598.

1911 N Bemerkung ⟦referring to 1911 G⟧. MA 71, p. 598.

1912 A1 Intuitionisme en formalisme, Amsterdam 1912, 32 p. ⟦Also: Wiskundig tijdschrift 9 (1913), p. 180–211 and in 1919 B.⟧

1912 A2 Intuitionism and formalism ⟦translation of 1912 A1 by A. Dresden⟧. Bull. A. M. S. 20 (1913), pp. 81–96.

1912 B Beweis des ebenen Translationssatzes. MA 72, pp. 37–54.

1912 C Zur Invarianz des n-dimensionalen Gebiets. MA 72, pp. 55–56.

1912 D Über die topologischen Schwierigkeiten des Kontinuitätsbeweises der Existenztheoreme eindeutig umkehrbarer polymorpher Funktionen auf Riemannschen Flächen (Auszug aus einem Brief an R. Fricke). Nachr. Göttingen 1912, pp. 603–606.

1912 E1 Over omloopcoëfficiënten. KNAW Versl. 20, pp. 1049–1057.

1912 E2 On looping coefficients. KNAW Proc. 15, pp. 113–122.

1912 F Sur l'invariance de la courbe fermée. C. R. Paris 154, pp. 862–863.

1912 G Über die Singularitätenfreiheit der Modulmannigfaltigkeit. Nachr. Göttingen 1912, pp. 803–806.

1912 H Über den Kontinuitätsbeweis für das Fundamentaltheorem der automorphen Funktionen im Grenzkreisfalle. J. ber. D.M.V. 21, pp. 154–157.

1912 K1 Over één-éénduidige, continue transformaties van oppervlakken in zichzelf, V. KNAW Versl. 21, pp. 300–309.

1912 K2 Continuous one-one transformations of surfaces in themselves, V. KNAW Proc. 15, pp. 352–360. ⟦See 1920 G2.⟧

1912 L Beweis der Invarianz der geschlossenen Kurve. MA 72, pp. 422–425.

1912 M Sur la notion de 'classe' de transformations d'une multiplicité. Proc. V Intern. Congr. Math. Cambridge 1912, II, pp. 9–10 (1913).

1913 A Über den natürlichen Dimensionsbegriff. Journ. r.u. angew. Math. 142, p. 146–152. ⟦Berichtigung see 1923 D2, 1924 M.⟧

1913 B1 Eenige opmerkingen over het samenhangstype η. KNAW Versl. 21, pp. 1412–1419.

1913 B2 Some remarks on the coherence type η. KNAW Proc. 15, pp. 1256–1263.

1914 ⟦Review of:⟧ A. Schoenflies und H. Hahn, Die Entwickelung der Mengenlehre und ihrer Anwendungen (Leipzig und Berlin 1913). Jber. D.M.V. 23, pp. 78–83 kursiv.

1915 A1 Opmerking over inwendige grensverzamelingen. KNAW Versl. 23, pp. 1325–1326.

1915 A2 Remark on inner limiting sets. KNAW Proc. 18, pp. 48–49.

1915 B Over de loodrechte trajectoriën der baankrommen eener vlakke eenledige projectieve groep ⟦On the orthogonal trajectories of the orbits of a one parameter plane projective group⟧. Nieuw Arch. Wisk. (2) 11, pp. 265–290.

1916 A ⟦Review of:⟧ L. E. J. Brouwer, Eenige opmerkingen over het samenhangstype η. (Amst. Ak. Versl. 21, 1412–1419). Jahrbuch über die Fortschritte der Mathematik 44 (for 1913), p. 556.

1916 B ⟦Review of:⟧ E. Borel, Les ensembles de mesure nulle (S. M. F. Bull. 41, 1–19). Jahrbuch über die Fortschritte der Mathematik 44 (for 1913), pp. 556–557.

1917 A Addenda en corrigenda over de grondslagen der wiskunde ⟦Addenda and corrigenda on the foundations of mathematics⟧. KNAW Versl. 25, pp. 1418–1423.

1917 B1 Over lineaire inwendige grensverzamelingen. KNAW Versl. 25, pp. 1424–1426.

1917 B2 On linear inner limiting sets. KNAW Proc. 20, pp. 1192–1194.

1918 A (With H. P. J. Bloemers, H. Borel and F. van Eeden) Voorbereidend manifest ⟦Preparatory manifesto⟧ (with French, German and English translations). Meded. Internat. Inst. Wijsbegeerte 1, pp. 3–12. ⟦The main part of this manifesto is cited in 1946 A.⟧

1918 B Begründung der Mengenlehre unabhängig vom logischen Satz vom ausgeschlossenen Dritten. Erster Teil, Allgemeine Mengenlehre. KNAW Verh. 1ᵉ sectie 12 no. 5, 43 p.

1918 C Über die Erweiterung des Definitionsbereichs einer stetigen Funktion. MA 79, pp. 209–211. ⟦Bemerkung see 1919 E.⟧

1918 D Lebesguesches Mass und Analysis Situs. MA 79, pp. 212–222.

1918 E Beweging van een materieel punt op den bodem eener draaiende vaas onder den invloed der zwaartekracht ⟦Motion of a particle on the bottom of a rotating vessel under the influence of the gravitational force⟧. Nieuw Arch. Wisk. (2) 12, pp. 407–419.

1919 A Begründung der Mengenlehre unabhängig vom logischen Satz vom ausgeschlossenen Dritten. Zweiter Teil, Theorie der Punktmengen. KNAW Verh. 1ᵉ sectie 12 no. 7, 33 p.

1919 B Wiskunde, waarheid, werkelijkheid ⟦Mathematics, Truth, Reality⟧. ⟦Reprint of 1908 C, 1909 A, 1912 A1⟧. Groningen 1919, 12+24+29 p.

1919 C (With G. Mannoury, H. J. F. Borel and F. van Eeden) Signifisch taalonderzoek (Signifische Sprachforschung); Onderscheid der taaltrappen ten aanzien van de sociale verstandhouding (Vom Unterschied der Sprachstufen in Bezug auf die soziale Verständigung) ⟦with German and French translations⟧. Meded. Internat. Inst. Wijsbegeerte 2, pp. 5–32.

1919 D1 Intuitionistische Mengenlehre. Jber. D.M.V. 28, pp. 203–208. Also: KNAW Proc. 23 (1922), pp. 949–954.

1919 D2 Intuitionistische verzamelingsleer. KNAW Versl. 29 (1921), pp. 797–802.

1919 E Nachträgliche Bemerkung über die Erweiterung des Definitionsbereiches einer stetigen Funktion. ⟦referring to 1918 C⟧. MA 79, p. 403.

1919 F Berichtigung ⟦referring to 1910 F⟧. MA 79, p. 403.

1919 G Enumération des surfaces de Riemann régulières de genre un. C.R. Paris, 168, pp. 677–678, 832.

1919 H Enumération des groupes finis de transformations topologiques du tore. C.R. Paris 168, pp. 845–848, 1168; 169, p. 552.

1919 J Sur les points invariants des transformations topologiques des surfaces. C.R. Paris 168, pp. 1042–1044; 169, p. 552.

1919 K Sur la classification des ensembles fermés situés sur une surface. C.R. Paris 169, pp. 953–954. ⟦Not republished.⟧

1919 L1 Over één-éénduidige continue transformaties van oppervlakken in zichzelf, VI. KNAW Versl. 27, pp. 609–612.

1919 L2 Ueber eineindeutige, stetige Transformationen von Flächen in sich, VI. KNAW Proc. 21, pp. 707–710.

1919 M1 Opmerking over de vlakke translatiestelling. KNAW Versl. 27, pp. 840–841.

1919 M2 Remark on the plane translation theorem. KNAW Proc. 21, pp. 935–936.

1919 N1 Over topologische involuties. KNAW Versl. 27, pp. 1201–1203.

1919 N2 Ueber topologische Involutionen. KNAW Proc. 21, pp. 1143–1145.

1919 P1 Opsomming der periodieke transformaties van de torus. KNAW Versl. 27, pp. 1363–1367.

1919 P2 Aufzählung der periodischen Transformationen des Torus. KNAW Proc. 21, pp. 1352–1356.

1919 Q1 Opmerking over meervoudige integralen. KNAW Versl. 28, pp. 116–120.

1919 Q2 Remark on multiple integrals. KNAW Proc. 22, pp. 150–154.

1919 R Luchtvaart en photogrammetrie ⟦Aviation and Photographic survey⟧, I. Nieuw Tijdschrift voor Wiskunde 7, pp. 311–331. ⟦Not republished.⟧

1919 S Über die periodischen Transformationen der Kugel. MA 80, pp. 39–41.

1920 A1 Over de structuur van perfecte puntverzamelingen, III. KNAW Versl. 28, pp. 373–375.

1920 A2 Ueber die Struktur der perfekten Punktmengen, III. KNAW Proc. 22, pp. 471–474, 974.

1920 B1 Over één-éénduidige, continue transformaties van oppervlakken in zichzelf, VII. KNAW Versl. 28, pp. 1186–1190. ⟦Erroneously numbered VI.⟧

1920 B2 Ueber eineindeutige, stetige Transformationen von Flächen in sich, VII. KNAW Proc. 22, pp. 811–814. ⟦Erroneously numbered VI.⟧

1920 C Enumération des classes de transformations du plan projectif. C.R. Paris 170, pp. 834–835, 1295. ⟦Not republished.⟧

1920 D Enumération de classes de représentations d'une surface sur un autre surface. C.R. Paris 171, pp. 89–91, 830. ⟦Not republished.⟧

1920 E Über die Minimalzahl der Fixpunkte bei den Klassen von eindeutigen stetigen Transformationen der Ringflächen. MA 82, pp. 94–96.

1920 F Luchtvaart en photogrammetrie ⟦Aviation and photographic survey⟧, II. Nieuw Tijdschrift voor Wiskunde 8, pp. 300–307. ⟦Not republished.⟧

1920 G1 Over één-éénduidige, continue transformaties van oppervlakken in zichzelf, VIII. KNAW Versl. 29, pp. 640–642.

1920 G2 Über eineindeutige, stetige Transformationen von Flächen in sich, VIII. KNAW Proc. 23, pp. 232–234. ⟦Republished in part in 1912 K2.⟧ ⟦Erroneously numbered VII.⟧

1921 A Besitzt jede reelle Zahl eine Dezimalbruch-Entwickelung? KNAW Versl. 29, pp. 803–812. Also: KNAW Proc. 23, pp. 955–964. Also: MA 83, pp. 201–210.

1921 C Opmerking over de bepaling van alle complexe functies, voor welke $f(z) = \varphi(|z|)$ is. Christiaan Huygens 1, pp. 352–354. ⟦Not republished.⟧

1921 D Aufzählung der Abbildungsklassen endlichfach zusammenhängender Flächen. MA 82, pp. 280–286.

1921 E Berichtigung ⟦referring to 1911 D⟧. MA 82, p. 286.

1921 F Berichtigung ⟦referring to 1911 E⟧. MA 82, p. 286.

1923 A Begründung der Funktionenlehre unabhängig vom logischen Satz vom ausgeschlossenen Dritten. Erster Teil, Stetigkeit, Messbarkeit, Derivierbarkeit. KNAW Verh. 1e sectie 13 no. 2, 24 p.

1923 B1 Over de rol van het principium tertii exclusi in de wiskunde, in het bijzonder in de functietheorie. Wis- en natuurkundig tijdschrift 2, pp. 1–7.

1923 B2 Über die Bedeutung des Satzes vom ausgeschlossenen Dritten in der Mathematik, insbesondere in der Funktionentheorie. Journ. r.u. angew. Math. 154, pp. 1–7. ⟦English translation: On the significance of the principle of the excluded middle in mathematics, especially in function theory, in: J. van Heyenoort, From Frege to Gödel, Cambridge Mass. 1967, pp.334–341.⟧

1923 B3 Die Rolle des Satzes vom ausgeschlossenen Dritten in der Mathematik, J. ber. D.M.V. 33, p. 67 kursiv. ⟦Summary of 1923 B2.⟧

1923 C1 Intuitionistische splitsing van mathematische grondbegrippen. KNAW Versl. 32, pp. 877–880.

1923 C2 Intuitionistische Zerlegung mathematischer Grundbegriffe. J. ber. D.M.V. 33 (1925), pp. 251–256.

1923 D1 Over het natuurlijke dimensiebegrip. KNWA Versl. 32, pp. 881–886.

1923 D2 Ueber den natürlichen Dimensionsbegriff. KNAW Proc. 26, pp. 795–800. ⟦Revised edition of 1913 A.⟧

1924 A1 Over de toelating van oneindige waarden voor het functiebegrip. KNAW Versl. 33, p. 41.

1924 A2 Ueber die Zulassung unendlicher Werte für den Funktionsbegriff. KNAW Proc. 27, p. 248.

1924 B1 Perfecte puntverzamelingen met positief irrationale afstanden. KNAW Versl. 33, p. 81.

1924 B2 Perfect sets of points with positively-irrational distances. KNAW Proc. 27, p. 487.

1924 C1 (With B. de Loor) Intuitionistisch bewijs van de hoofdstelling der algebra. KNAW Versl. 33, pp. 82–84.

1924 C2 (With B. de Loor) Intuitionistischer Beweis des Fundamentalsatzes der Algebra. KNAW Proc. 27, pp. 186–188.

1924 D1 Bewijs dat iedere volle functie gelijkmatig continu is. KNAW Versl. 33, pp. 189–193.

1924 D2 Beweis, dass jede volle Funktion gleichmässig stetig ist. KNAW Proc. 27, pp. 189–193.

1924 E1 Intuitionistische aanvulling van de hoofdstelling der algebra. KNAW Versl. 33, pp. 459–462.

1924 E2 Intuitionistische Ergänzung des Fundamentalsatzes der Algebra. KNAW Proc. 27, pp. 631–634.

1924 F1 Bewijs van de onafhankelijkheid van de onttrekkingsrelatie van de versmeltingsrelatie. KNAW Versl. 33, pp. 479–480.

1924 F2 Zur intuitionistischen Zerlegung mathematischer Grundbegriffe. J. ber. D.M.V. 36 (1927), pp. 127–129.

1924 G1 Opmerkingen aangaande het bewijs der gelijkmatige continuiteit van volle functies. KNAW Versl. 33, pp. 646–648.

1924 G2 Bemerkungen zum Beweise der gleichmässigen Stetigkeit voller Funktionen. KNAW Proc. 27, pp. 644–646.

1924 H Zuschrift an den Herausgeber. J. ber. D.M.V. 33, p. 124 kursiv.

1924 J1 Opmerkingen over het natuurlijke dimensiebegrip. KNAW Versl. 33, pp. 476–478.

1924 J2 Bemerkungen zum natürlichen Dimensionsbegriff. KNAW Proc. 27, pp. 635–638.

1924 K1 Over de n-dimensionale simplex ster in R_n. KNAW Versl. 33, pp. 1008–1010.

1924 K2 On the n-dimensional simplex star in R_n. KNAW Proc. 27, pp. 778–780.

1924 L Zum natürlichen Dimensionsbegriff. MZ 21, pp. 312–314. ⟦Not republished.⟧

1924 M Berichtigung ⟦referring to 1913 A⟧. Journ. r.u. angew. Math. 153, p. 253.

1925 A Zur Begründung der intuitionistischen Mathematik, I. MA 93, pp. 244–257.

1925 B1 Intuitionistischer Beweis des Jordanschen Kurvensatzes. KNAW Proc. 28, pp. 503–508.

1925 B2 Intuitionistisch bewijs van de krommenstelling van Jordan. KNAW Versl. 34, p. 657. ⟦Summary in Dutch of 1925 B1⟧.

1925 C (With H. de Vries and R. Weitzenböck) Rapport over de verhandeling van P. Urysohn†: "Mémoire sur les multiplicités cantoriennes. Deuxième partie: Les lignes cantoriennes." KNAW Versl. 34, pp. 516–517. ⟦Not republished.⟧

1926 A Zur Begründung der intuitionistischen Mathematik, II. MA 95, pp. 453–472.

1926 B1 Intuitionistische invoering van het dimensiebegrip. KNAW Versl. 35, pp. 634–642.

1926 B2 Intuitionistische Einführung des Dimensionsbegriffes. KNAW Proc. 29, pp. 855–873.

1926 C1 De intuitionistische vorm van het theorema van Heine–Borel. KNAW Versl. 35, pp. 667–678.

1926 C2 Die intuitionistische Form des Heine–Borelschen Theorems. KNAW Proc. 29, pp. 866–867.

1926 D1 Over transformaties van projectieve ruimten. KNAW Versl. 35, pp. 643–644.

1926 D2 On transformations of projective spaces. KNAW Proc. 29, pp. 864–865.

1927 A Zur Begründung der intuitionistischen Mathematik, III. MA 96, pp. 451–488.

1927 B Über Definitionsbereiche von Funktionen. MA 97, pp. 60–75. ⟦English translation of § 1–3: On the domains of definition of functions, in: J. van Heyenoort, From Frege to Gödel, Cambridge Mass. 1967, pp. 446–463.⟧

1927 C Virtuelle Ordnung und unerweiterbare Ordnung. Journ. r.u. angew. Math. 157, pp. 255–257.

1928 A1 Intuitionistische Betrachtungen über den Formalismus. ⟦Summary in Dutch of 1928 A2⟧. KNAW Versl. 36, p. 1189.

1928 A2 Intuitionistische Betrachtungen über den Formalismus. KNAW Proc. 31, pp. 374–379. ⟦Also: Sitzber. Akad. Berlin 1928, pp. 48–52.⟧ ⟦English translation of § 1 in: J. van Heyenoort, From Frege to Gödel, Cambridge Mass. 1967, pp. 490–492.⟧

1928 B1 Beweis, dass jede Menge in einer individualisierten Menge enthalten ist. ⟦Summary in Dutch of 1928 B2.⟧ KNAW Versl. 36, p. 1189.

1928 B2 Beweis, dass jede Menge in einer individualisierten Menge enthalten ist. KNAW Proc. 31, pp. 380–381.

1928 C1 Zur Geschichtsschreibung der Dimensionstheorie. KNAW Versl. 37, p. 626.

1928 C2 Zur Geschichtsschreibung der Dimensionstheorie. KNAW Proc. 31, pp. 953–957.

1929 Mathematik, Wissenschaft und Sprache. Monatshefte Math. u. Phys. 36, pp. 153–164.

1930 A Die Struktur des Kontinuums. Wien 1930, 14 p.

1930 B ⟦Review of:⟧ A. Fraenkel, Zehn Vorlesungen über die Grundlegung der Mengenlehre. J. ber. D.M.V. 39, pp. 10–11 kursiv.

1931 Über freie Umschliessungen im Raume. KNAW Proc. 34, pp. 100–101.

1933 Willen, weten, spreken ⟦Volition, knowledge, language⟧. Euclides 9, pp. 177–193. ⟦Also: De uitdrukkingswijze der wetenschap, kennistheoretische voordrachten gehouden aan de Universiteit van Amsterdam (1932–1933), pp. 43–63.⟧

1937 (With F. van Eeden, J. van Ginneken and G. Mannoury) Signifische dialogen ⟦Signific Dialogues⟧. Synthese 2, pp. 168–174, 261–268, 316–324. ⟦Also: Utrecht 1939.⟧

1939 Zum Triangulationsproblem. KNAW Proc. 42, pp. 701–706 = Indag. Math. 1, pp. 248–253.

1942 A Zum freien Werden von Mengen und Funktionen. KNAW Proc. 45, pp. 322–323 = Indag. Math. 4, pp. 107–108.

1942 B Die repräsentierende Menge der stetigen Funktionen des Einheitskontinuums. KNAW Proc. 45, p. 443 = Indag. Math. 4, p. 154.

1942 C Beweis dass der Begriff der Menge höherer Ordnung nicht als Grundbegriff der intuitionistischen Mathematik in Betracht kommt. KNAW Proc. 45, pp. 791–793 = Indag. Math. 4, pp. 274–276.

1946 A Synopsis of the signific movement in the Netherlands. Synthese 5, pp. 201–208.

1946 B Address delivered on September 16th, 1946, on the conferment upon Professor G. Mannoury of the honorary degree of Doctor of Science. ⟦Dutch original: Jaarboek der Universiteit van Amsterdam 1946–1947, II.⟧

1947 Richtlijnen der intuitionistische wiskunde ⟦Guidelines of intuitionistic mathematics⟧. KNAW Proc. 50, p. 339 = Indag. Math. 9, p. 197.

1948 A Essentieel negative eigenschappen ⟦Essentially negative properties⟧. KNAW Proc. 51, pp. 963–964 = Indag. Math. 10, pp. 322–323.

1948 B Opmerkingen over het beginsel van het uitgesloten derde en over negatieve asserties ⟦Remarks on the principle of the excluded middle and on negative assertions⟧. KNAW Proc. 51, pp. 1239–1244 = Indag. Math. 10, pp. 383–387.

1948 C Consciousness, Philosophy and Mathematics. Proceedings of the 10th International Congress of Philosophy, Amsterdam 1948, III, pp. 1235–1249.

1949 A De non-aequivalentie van de constructieve en de negatieve orderelatie in het continuum ⟦The non-equivalence of the constructive and the negative order relation on the continuum⟧. KNAW Proc. 52, pp. 122–124 = Indag. Math. 11, pp. 37–39.

1949 B Contradictoriteit der elementaire meetkunde ⟦Contradictority of elementary geometry⟧. KNAW Proc. 52, pp. 315–316 = Indag. Math. 11, pp. 89–90.

1950 A Remarques sur la notion d'ordre. C.R. Paris 230, pp. 263–265.

1950 B Sur la possibilité d'ordonner le continu. C.R. Paris 230, pp. 349–350.

1950 C Discours final. Les méthodes formelles en axiomatique. Colloques C.N.R.S. Paris 1950, p. 75.

1951 On order in the continuum, and the relation of truth to non-contradictority. KNAW Proc. Ser. A 54, pp. 357–358 = Indag. Math. 13, pp. 357–358.

1952 A An intuitionist correction of the fixed-point theorem on the sphere. Proc. Roy. Soc. London Ser. A 213, pp. 1–2.

1952 B1 Historical background, principles and methods of intuitionism. South African J. Sci. 49, pp. 139–146.

1952 B2 Voorgeskiedenis, beginsels en metodes van die intuisionisme. Tydskrif vir Wetenskap en Kuns 1956, pp. 186–197.

1952 C Over accumulatiekernen van oneindige kernsoorten ⟦On accumulation cores of infinite core species⟧. KNAW Proc. Ser. A 55, pp. 439–441 = Indag. Math. 14, pp. 439–441.

1952 D Door klassieke theorema's gesignaleerde pinkernen die onvindbaar zijn ⟦Fixed cores which cannot be found, though they are claimed to exist by classical theorems⟧. KNAW Proc. Ser. A 55, pp. 443–445 = Indag. Math. 14, pp. 443–445.

1954 A Points and spaces. Canad. J. Math. 6, pp. 1–17.

1954 B Addenda en corrigenda over de rol van het principium tertii exclusi in de wiskunde ⟦Addenda and corrigenda on the role of the principium tertii exclusi in mathematics⟧. KNAW Proc. Ser. A 57, pp. 104–105 = Indag. Math. 16, pp. 104–105. ⟦English translation in: J. van Heyenoort, From Frege to Gödel, Cambridge Mass. 1967, pp. 341–342.⟧

1954 C Nadere addenda en corrigenda over de rol van het principium tertii exclusi in de wiskunde ⟦Further addenda and corrigenda on the role of the principium tertii exclusi in mathematics⟧. KNAW Proc. Ser. A 57, pp. 109–111 = Indag. Math. 16, pp. 109–111. ⟦English translation in: J. van Heyenoort, From Frege to Gödel, Cambridge Mass. 1967, pp. 342–345.⟧

1954 D Ordnungswechsel in Bezug auf eine coupierbare geschlossene stetige Kurve. KNAW Proc. Ser. A 57, pp. 112–113 = Indag. Math. 16, pp. 112–113.

1954 E Intuitionistische differentieerbaarheid ⟦Intuitionistic differentiability⟧. KNAW Proc. Ser. A 57, pp. 201–203 = Indag. Math. 16, pp. 201–203.

1954 F An example of contradictority in classical theory of functions. KNAW Proc. Ser. A 57, pp. 204–205 = Indag. Math. 16, pp. 204–205.

1955 The effect of intuitionism on classical algebra of logic. Proc. Roy Irish Acad. Sect. A 57, pp. 113–116.

Y1 Letter to O. Blumenthal, draft, 19–6–1911

Y2 Letter to O. Blumenthal, draft, 20–6–1911

Y3 Manuscript, 1910

Y4 Manuscript, 1910

Y5 Letter to O. Blumenthal, draft, 1911

Y6 Letter to F. Klein, draft, 1911

Y8 Letter to D. Hilbert, UB Göttingen, 28–10–1909

Y10 Letter to D. Hilbert, UB Göttingen, 14–5–1909

Y15 Letter to D. Hilbert, UB Göttingen, 15–10–1909

Y16 Letter to D. Hilbert, draft, 1–1–1910

Y17 Letter to J. Hadamard, draft, 4–1–1910

Y18 Letter to D. Hilbert, UB Göttingen, 18–3–1910

Y20 Letter to P. Koebe, UB Göttingen, 14–2–1912

Y22 Letter to A. Hurwitz, UB Göttingen, 13–12–1911

Y23 Letter to A. Hurwitz, UB Göttingen, 5–1–1912

Y43 Letter to F. Engel, copy, 21–1–1912

Y44 Letter to F. Engel, copy, 6–3–1912

Y45 Letter to F. Engel, copy, 29–3–1912

Y57 Letter to E. Borel, copy, 7–11–1913

Y58 Letter to H. Blaschke, copy, 19–11–1915

Y97 Letter to O. Blumenthal, original, 5–5–1914

Y98 Letter to G. Hamel, copy, 20–6–1914

CHAPTER 1

Non-euclidean spaces and integral theorems

1904 A2

Mathematics. — "*On a decomposition of a continuous motion about a fixed point O of S_4 into two continuous motions about O of S_3's*" by Mr. L. E. J. BROUWER, communicated by Prof. KORTEWEG.

(Communicated in the meeting of February 27, 1904). [[1]]

Two planes in S_4 making two equal angles of position are called mutually "equiangular to the right" if one is (with its normal plane) plane of rotation for an equiangular double rotation to the right about the other one and its normal plane.

We will call the sides of one and the same acute angle of position "corresponding vectors" through the point of intersection of two equiangular intersecting planes.

As is known a system of planes equiangular to the right or to the left is infinite of order two. Of course a determined equiangular system of planes to the right can have with a determined equiangular system of planes to the left not more than one pair of planes in common (two pairs of planes cannot intersect each other at the same time equiangularly to the right and to the left); but *one* pair of planes they always have in common. We will show how that common pair of planes can be found.

A pair of intersecting pairs of planes of both systems is of course easy to find. We lay through any vector OC the planes belonging to the two systems; their normal planes intersect each other in a second vector OD. Thus OCD is one plane of position of those two pairs of planes. In the second plane of position the four planes furnish four lines of intersection, let us say OH, OF, OK, OG, in such a way, that the considered pairs of planes must be OCH; ODK and OCF; ODG. The following figures are supposed to be situated in those two planes of position.

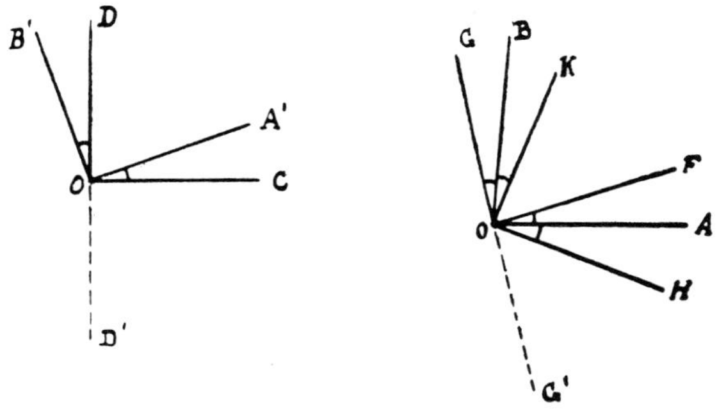

Fig. 1.

Let the pair of planes OCH; ODK belong to the given system equiangular to the right, and OCF; ODG to the given system equiangular to the left.

[[3]]

The vectors in the second plane of position are drawn in such a way that either

$$\begin{cases} OH \rightarrow OC \\ OK \rightarrow OD \end{cases} \quad \ldots \ldots \quad (1)$$

is an equiangular double rotation to the right, or that such is the case with

$$\begin{cases} OH \rightarrow OC \\ OK \rightarrow OD' \end{cases}$$

We shall suppose the first to be true (the reasoning is the same for the second case). Then

$$\begin{cases} OF \rightarrow OC \\ OG \rightarrow OD \end{cases}$$

is also an equiangular double rotation to the right; for, the planes OFC and OGD can be brought to coincide with these directions of rotation with the planes OHC and OKD, having the directions of rotation

$$\begin{cases} OH \rightarrow OC \\ OK \rightarrow OD \end{cases}$$

So

$$\begin{cases} OF \rightarrow OC \\ OG' \rightarrow OD \end{cases} \quad \ldots \ldots \ldots \quad (2)$$

is an equiangular double rotation to the left.

If farthermore OA and OB are bisectors of the angles HOF and KOG, and if we have made $\angle COA' = \angle DOB' = \angle HOA = \angle FOA = \angle KOB = \angle GOB$, then the pair of planes

$$\begin{cases} AOA' \\ BOB' \end{cases}$$

is a pair of planes of rotation as well for the equiangular double rotation to the right (1) as for the equiangular double rotation to the left (2). So it is the pair of common planes which was looked for of the two systems of planes.

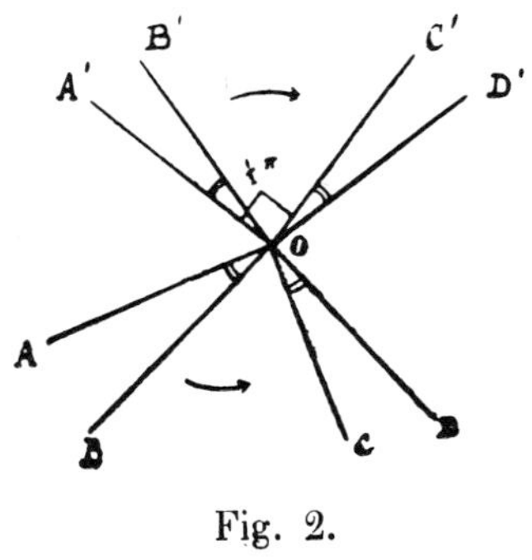

Fig. 2.

We shall think now that through two arbitrary vectors OA and OB two planes intersecting each other equiangularly to the left have been laid; we shall now consider more closely the position which two such planes have with respect to the plane OAB and its normal plane. We shall call the indicated equiangular planes to the left α and β; and indicate OAB by γ and its normal plane by δ. In fig. 2 the lines drawn upwards lie in δ and those drawn downwards in γ.

The plane OCC' is the plane of position of γ and α, intersected by γ in OF, by α in OC. Fig. 3 is supposed to lie in that plane of position. We have made fartheron in fig. 1 the angles $A'OB'$, $C''OD'$, COD equal to $\angle AOB = \varphi$, and the directions of rotation indicated in those planes belong to a double rotation to the right. Fig. 4 is supposed to lie in the plane ODD', and the lines OG and OG' are drawn in it in such a way, that $ODGD'G' \cong OCFC''F'$.

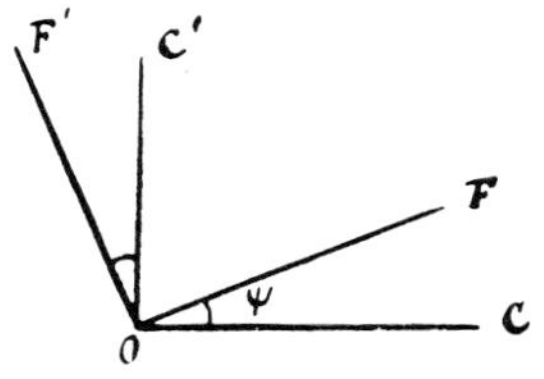

Fig. 3.

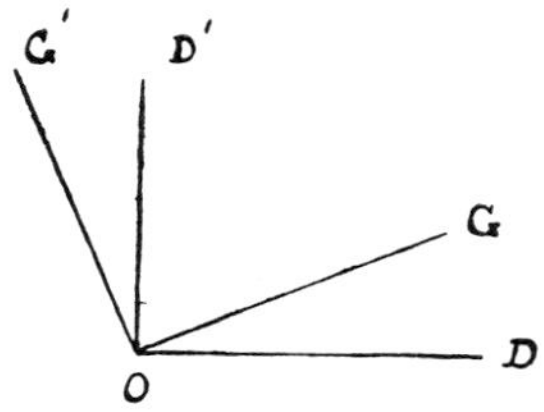

Fig. 4.

We shall consider the plane BOG more closely. Let us project OB and OG both on α, then it is not difficult to see that the execution of those two operations, each of which is threedimensional, gives as a result two lines OH and OK, mutually perpendicular (see fig. 5, supposed to lie in α).

The projecting planes are successively: OBF' (fig. 6) and OGA' (fig. 7).

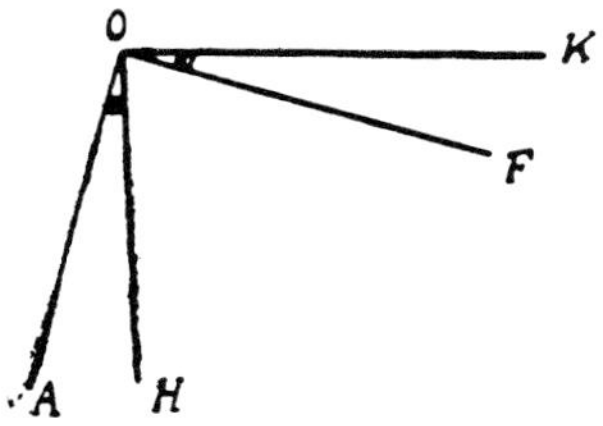

Fig. 5.

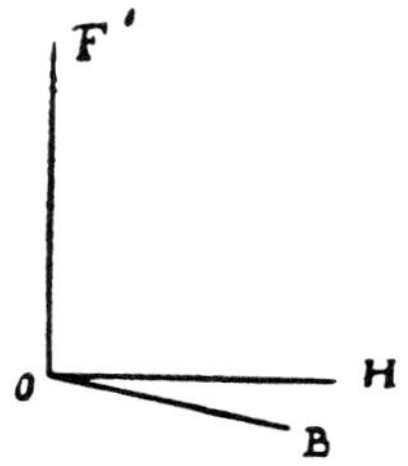

Fig. 6.

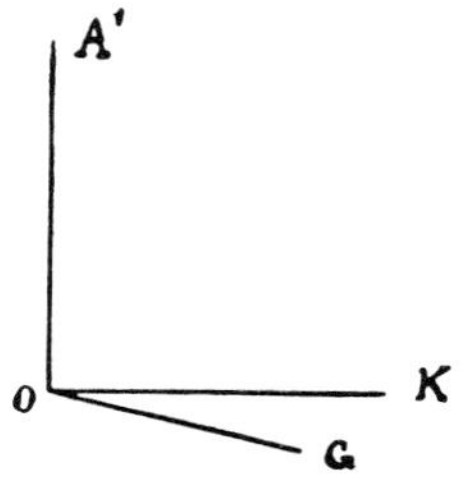

Fig. 7

We shall directly see that OB and OF' are situated on different sides of OH, and OG and OA' on different sides of OK, and that $\angle HOB = \angle KOG$, if we suppose ourselves to be successively in the threedimensional spaces, in which the projecting takes place.

So we see that the plane BOG has two mutually perpendicular vectors, making equal angles with α and projecting itself on α according to two perpendicular vectors namely OB and OG, projecting themselves according to OH and OK: the characteristic of equiangular intersection.

Let us still examine of which kind that equiangular intersection is; we shall then perceive that on account of OB being transferred into OG by the equiangular double rotation

$$OF'' \rightarrow OA'$$
$$OH \rightarrow OK$$

and this being of the same kind as

$$OF'' \rightarrow OA'$$
$$OA \rightarrow OF$$

which in its turn can be made to coincide with

$$OC' \rightarrow OA'$$
$$OA \rightarrow OC$$

by a single rotation about the plane OAA', the kind of equiangular intersection is the same as the kind of the double rotation

$$OC' \rightarrow OA'$$
$$OA \rightarrow OC$$

which is to the left according to the data.

So the plane OBG is identical with the plane β, for through OB only *one* plane equiangular to the left with α can pass.

If we now introduce the notation "$\begin{pmatrix} ab \\ cd \end{pmatrix}$ equiangular to the right" indicating if $abcd$ denote four vectors through O, that the planes (ab) and (cd) are equiangular to the right and that the same double equiangular rotation to the right transferring a into b, also transfers c into d, then

$$\begin{pmatrix} OA, OB \\ OF, OG \end{pmatrix}$$

is equiangular to the right and the corresponding equiangular double rotation to the right transfers α into β. In other words we have proved the

Theorem 1. If $\begin{pmatrix} ab \\ cd \end{pmatrix}$ is equiangular to the right, then $\begin{pmatrix} ac \\ bd \end{pmatrix}$ is equiangular to the left; or in other words though less significant:

By an equiangular double rotation to the right any plane passes into one equiangular to it to the left.

If we suppose three vectors abc (whose position of course determines the position of S_4) to have come after some equiangular double rotations to the right into the position def, then $\begin{pmatrix} ab \\ de \end{pmatrix}$ is equiangular to the left and $\begin{pmatrix} ac \\ df \end{pmatrix}$ equiangular to the left; so $\begin{pmatrix} ad \\ be \end{pmatrix}$ equiangular to the right and $\begin{pmatrix} ad \\ cf \end{pmatrix}$ equiangular to the right, so finally

$$\begin{pmatrix} ad \\ be \\ cf \end{pmatrix}$$

equiangular to the right; in other words the final position would have been obtainable out of the initial position by a single equiangular double rotation to the right; with which is proved:

Theorem 2. Equiangular double rotations of the same kind form a group.

Let us suppose given two equiangular systems to the right and two vectors OA and OB through each of which we bring the planes belonging to both systems; then the equiangular double rotation to the left, transferring OA into OB, will transfer at the same time the angle of position formed in OA into the one formed in OB, thus:

Theorem 3. Two equiangular systems to the right form in each vector the same angle, which can be called the angle of the two systems.

The obtained results we shall verify by deducing analytically theorems 1 and 3, which deduction will also throw some more light.

Suppose a rectangular system of coordinates to be given in such a way that

$$\begin{pmatrix} OX_2, & OX_3 \\ OX_1, & OX_4 \end{pmatrix}$$

is equiangular to the right. The same then holds good for

$$\begin{pmatrix} OX_3, & OX_1 \\ OX_2, & OX_4 \end{pmatrix} \text{ and } \begin{pmatrix} OX_1, & OX_2 \\ OX_3, & OX_4 \end{pmatrix}.$$

A vector α through O we can determine by its four cosines of direction $\alpha_1, \alpha_2, \alpha_3, \alpha_4$..

A plane passing through the vectors α and β with direction of rotation from α to β, is determined by its six coefficients of position (i. e. projections of a vector unity) $\lambda_{23}, \lambda_{31}, \lambda_{12}. \lambda_{14}, \lambda_{24}, \lambda_{34}$, which are defined by the following relations, if we represent $\alpha_2\beta_3-\alpha_3\beta_2$ by ξ_{23}:

$$\lambda_{23} = \frac{\xi_{23}}{+\sqrt{\xi_{23}^2+\xi_{31}^2+\xi_{12}^2+\xi_{14}^2+\xi_{24}^2+\xi_{34}^2}}, \text{ etc.}$$

We must take the positive sign in the denominator, for λ_{23} must be positive, when the projection of α on OX_2X_3 to that of β on OX_2X_3 rotates through an angle less than π in the same way as OX_2 to OX_3; and in that same case ξ_{23} gives us a positive value. If we now represent that positive root of the denominator by K, then

$$\lambda_{23} = \frac{\xi_{23}}{K}; \ \lambda_{31} = \frac{\xi_{31}}{K}, \text{ etc.}$$

An equiangular double rotation to the right can be given by the system of equiangular planes of rotation to the right with direction and the angle of rotation.

For all those planes of rotation

$$\lambda_{23} + \lambda_{14}$$
$$\lambda_{31} + \lambda_{24}$$
$$\lambda_{12} + \lambda_{34}$$

have the same values. These three values a_1, a_2, a_3, besides the cosinus of the angle of rotation a_4, we can take as determining quantities of the equiangular double rotation to the right. A rotation $< 2\pi$ is unequivocally determined by that (for, whether the angle of rotation is $\gtrless \pi$, which is left undecided by the value of the cosinus, follows from the sign of the a_1, a_2, a_3).

Suppose an arbitrary vector α to be transferred by the rotation into β, then it holds good for each pair of vectors $\alpha\beta$ that:

$$\alpha_2\beta_3 - \alpha_3\beta_2 + \alpha_1\beta_4 - \alpha_4\beta_1 = K \cdot a_1$$
$$\alpha_3\beta_1 - \alpha_1\beta_3 + \alpha_2\beta_4 - \alpha_4\beta_2 = K \cdot a_2$$
$$\alpha_1\beta_2 - \alpha_2\beta_1 + \alpha_3\beta_4 - \alpha_4\beta_3 = K \cdot a_3$$
$$\alpha_1\beta_1 + \alpha_2\beta_2 + \alpha_3\beta_3 + \alpha_4\beta_4 = a_4.$$

If however we consider that $K = +\sqrt{\sin^2\vartheta}$, if ϑ is the angle of rotation, then K proves to be a constant for all pairs of vectors so that we may regard $K.a_1$, $K.a_2$, $K.a_3$ and a_4 as determining quantities of the double rotation which we shall call $\pi_1, \pi_2, \pi_3, \pi_4$; and we shall write the relations:

$$\left.\begin{array}{r} -\alpha_4\beta_1 - \alpha_3\beta_2 + \alpha_2\beta_3 + \alpha_1\beta_4 = \pi_1 \\ \alpha_3\beta_1 - \alpha_4\beta_2 - \alpha_1\beta_3 + \alpha_2\beta_4 = \pi_2 \\ -\alpha_2\beta_1 + \alpha_1\beta_2 - \alpha_4\beta_3 + \alpha_3\beta_4 = \pi_3 \\ \alpha_1\beta_1 + \alpha_2\beta_2 + \alpha_3\beta_3 + \alpha_4\beta_4 = \pi_4 \end{array}\right\} \quad . \quad . \quad . \quad . \quad . \quad (H)$$

in which we have at the same time arranged the first members according to $\beta_1, \beta_2, \beta_3, \beta_4$. We now perceive:

$$\pi_1^2 + \pi_2^2 + \pi_3^2 + \pi_4^2 = K^2\{(\lambda_{23}^2 + \lambda_{31}^2 + \lambda_{12}^2 + \lambda_{14}^2 + \lambda_{24}^2 + \lambda_{34}^2) + $$
$$+ 2(\lambda_{23}\lambda_{14} + \lambda_{31}\lambda_{24} + \lambda_{12}\lambda_{34})\} + a_4^2.$$
$$= K^2(1 + 2 \times 0) + a_4^2$$
$$= \sin^2\vartheta + \cos^2\vartheta$$
$$= 1.$$

So we can regard $\pi_1, \pi_2, \pi_3, \pi_4$ as cosines of direction of a vector through O in S_4 and we can represent an equiangular double rotation

[2] to the right by a vector / through O in S_4, which determines it

/ without length

[8]

unequivocally. ~~(A vector *without* length; lateron we shall determine it, likewise unequivocally, by a vector *with* length, in S_3).~~ [3]

It we farthermore consider the determinant on the coefficients of $\beta_1, \beta_2, \beta_3, \beta_4$ in (H) it proves to satisfy all the conditions of an orthogonal transformation.

We call that transformation with that determinant

$$\begin{array}{cccc} -\alpha_4 & -\alpha_3 & +\alpha_2 & +\alpha_1 \\ +\alpha_3 & -\alpha_4 & -\alpha_1 & +\alpha_2 \\ -\alpha_2 & +\alpha_1 & -\alpha_4 & +\alpha_3 \\ +\alpha_1 & +\alpha_2 & +\alpha_3 & +\alpha_4 \end{array}$$

the $(+r)$ α-transformation; it appears in the relations (H), if the first members are arranged according to the cosines of direction of the final position of the rotating vector. If they are arranged according to the cosines of direction of the initial position the determinant of the coefficients becomes

$$\begin{array}{cccc} \beta_4 & \beta_3 & -\beta_2 & -\beta_1 \\ -\beta_3 & \beta_4 & \beta_1 & -\beta_2 \\ \beta_2 & -\beta_1 & \beta_4 & -\beta_3 \\ \beta_1 & \beta_2 & \beta_3 & \beta_4 , \end{array}$$

which we shall call the $(-r)$ β-transformation.

Quite analogous to this we have for equiangular double rotations to the left $(\varrho_1\ \varrho_2\ \varrho_3\ \varrho_4)$ bilinear homogeneous equations between the cosines of direction of initial and final position of a vector, let us say α and β, which arranged according to the β's, give as determinant of the coefficients

$$\begin{array}{cccc} \alpha_4 & -\alpha_3 & \alpha_2 & -\alpha_1 \\ \alpha_3 & \alpha_4 & -\alpha_1 & -\alpha_2 \\ -\alpha_2 & \alpha_1 & \alpha_4 & -\alpha_3 \\ \alpha_1 & \alpha_2 & \alpha_3 & \alpha_4 , \end{array}$$

the $(+l)$ α-transformation and arranged according to the α's

$$\begin{array}{cccc} -\beta_4 & \beta_3 & -\beta_2 & \beta_1 \\ -\beta_3 & -\beta_4 & \beta_1 & \beta_2 \\ \beta_2 & -\beta_1 & -\beta_4 & \beta_3 \\ \beta_1 & \beta_2 & \beta_3 & \beta_4 , \end{array}$$

the $(-l)$ β-transformation.

We can now state the following:

If the equiangular double rotation to the right $(\pi_1\ \pi_2\ \pi_3\ \pi_4)$ transfers the vector $(\alpha_1\ \alpha_2\ \alpha_3\ \alpha_4)$ into $(\beta_1\ \beta_2\ \beta_3\ \beta_4)$ then the $(+r)$ α-transformation transfers the vector $(\pi_1\ \pi_1\ \pi_3\ \pi_4)$ into $(\beta_1\ \beta_2\ \beta_3\ \beta_4)$

and the $(-r)$ β-transformation transfers the vector $(\pi_1\ \pi_2\ \pi_3\ \pi_4)$
[[4]] into $(\alpha_1\ \alpha_2\ \alpha_3\ \alpha_4)$.

Analogous to this:

If the equiangular double rotation to the left $(\varrho_1\ \varrho_2\ \varrho_3\ \varrho_4)$ transfers the vector $(\alpha_1\ \alpha_2\ \alpha_3\ \alpha_4)$ into $(\beta_1\ \beta_2\ \beta_3\ \beta_4)$, then the $(+l)$ α-transformation transfers the vector $(\varrho_1\ \varrho_2\ \varrho_3\ \varrho_4)$ into $(\beta_1\ \beta_2\ \beta_3\ \beta_4)$

and the $(-l)$ β-transformation transfers the vector $(\varrho_1\ \varrho_2\ \varrho_3\ \varrho_4)$ into $(\alpha_1\ \alpha_2\ \alpha_3\ \alpha_4)$.

Let us now suppose that S_4 has first an equiangular double rotation to the right (π) transferring an arbitrary vector α into β'; then an equiangular double rotation to the left (ϱ), transferring β' into γ, then we can write:

$$\pi = [(+r)\ \alpha]\ \beta' \qquad \varrho = [(-l)\ \gamma]\ \beta'$$

$$\pi = \{[(+r)\ \alpha] \quad [(-l)\ \gamma]\}\ \varrho$$

where the form between $\{\ \}$ denotes the determinant of transformation having as first row the sum of the products of the terms of the first row of $[(+r)\ \alpha]$ with the corresponding ones of respectively the first, second, third and fourth of $[(-l)\ \gamma]$, whilst the second etc. row in a corresponding manner is deduced out of the second etc. row of $[(+r)\ \alpha]$ (all this in the way of forming a product of determinants).

If S_4 has first an equiangular double rotation to the left (ϱ) transferring α into β'' and then an equiangular double rotation to the right (π'), transferring β'' into γ, we have:

$$\varrho = [(+l)\ \alpha]\ \beta'' \qquad \pi' = [(-r)\ \gamma]\ \beta''$$

$$\pi' = \{[(-r)\ \gamma] \cdot [(+l)\ \alpha]\}\ \varrho.$$

But now

$$[(+r)\ \alpha] \cdot [(-l)\ \gamma] = [(-r)\ \gamma] \cdot [(+l)\ \alpha]$$

which can be at once verified, so:

$$\pi' = \pi.$$

Thus:

If S_4 is allowed successively an equiangular double rotation to the right (π) and one to the left (ϱ) the order of the two rotations may be interchanged. For, in both cases an initial position of a vector α gives the same final position γ.

And if we regard the quadruple $\alpha, \beta', \beta'', \gamma$, then $\begin{pmatrix} \alpha & \beta' \\ \beta'' & \gamma \end{pmatrix}$ is equiangular to the right, according to the rotation (π) and $\begin{pmatrix} \alpha & \beta'' \\ \beta' & \gamma \end{pmatrix}$ equiangular to the left according to the rotation (ϱ); by which we assuredly once more have proved theorem 1.

[[10]]

Let us farthermore suppose (σ) and (τ) to be two equiangular double rotations to the right, transferring a given vector ε the former into ζ, the latter into η. Then

$$\sigma = [(+r)\,\varepsilon]\,\zeta \qquad \tau = [(+r)\,\varepsilon]\,\eta.$$

The same orthogonal transformation transferring σ into ζ, transfers τ into η, so that the angle between ζ and η is the angle between the vectors σ and τ independent of the initial position ε. As a special case theorem 3 is included in this, for the case that the two double rotations take place about an angle $\frac{1}{2}\pi$; for then the angle between ζ and η is the angle of the two planes of rotation through ε, proving to be independent of ε.

We have still to mention that theorem 1 in the second form is entirely included in the applications of the biquaternions on S_4 as given by Dr. W. A. WIJTHOFF in his dissertation: "De Biquaternion als bewerking in de ruimte van vier afmetingen" (the biquaternion as an operation in fourdimensional space). For an equiangular double rotation to the right is represented by $Q \cdot \varepsilon_1 + \varepsilon_2$ (p. 127) where Q represents a certain quaternion with norm unity.

This applied to an arbitrary double vector

$$\alpha_1\varepsilon_1 + \alpha_2\varepsilon_2,$$

changes it into

$$Q\alpha_1\varepsilon_1 + \alpha_2\varepsilon_2,$$

so it leaves the isosceles part of that double vector to the left unchanged and so also the equiangular system of planes to the left to which it belongs. This holds good for an arbitrary double vector, so particularly for a plane.

Finally theorem 3 can be proved as follows:

If φ and ψ are the acute angles of position of two planes, then if we represent the coefficients of position respectively by λ's and μ's:

$$\cos\varphi\,\cos\psi = \lambda_{23}\,\mu_{23} + \lambda_{31}\,\mu_{31} + \lambda_{12}\,\mu_{12} + \lambda_{14}\,\mu_{14} + \lambda_{24}\,\mu_{24} + \lambda_{34}\,\mu_{34} =$$
$$= \Sigma\,(\lambda_{23} + \lambda_{14})\,(\mu_{23} + \mu_{14}) - \Sigma\,(\lambda_{23}\,\mu_{14} + \lambda_{14}\,\mu_{23}).$$

For intersecting planes with angle of position φ:

$$\cos\varphi = \Sigma\,(\lambda_{23} + \lambda_{14})\,(\mu_{23} + \mu_{14}) = \Sigma\,(\lambda_{23} - \lambda_{14})\,(\mu_{23} - \mu_{14}).$$

So for two intersecting planes, belonging to two definite equiangular systems to the right or two to the left

$$\lambda'_{23} + \lambda'_{14} = \lambda''_{23} + \lambda''_{14} \quad \text{and} \quad \mu'_{23} + \mu'_{14} = \mu''_{23} + \mu''_{14}$$

resp. $\lambda'_{23} - \lambda'_{14} = \lambda''_{23} - \lambda''_{14}$ and $\mu'_{23} - \mu'_{14} = \mu''_{23} - \mu''_{14}$ and $\cos\varphi' = \cos\varphi''$.

We shall now resume our geometrical reasoning dropped after theorem 3. Let us take through O a definite vector OW in S_4 but not movable with S_4 and let us represent each system of planes equiangular to the right by the line of intersection of the plane through OW belonging to it with $S_3 \perp OW$. That S_3 is thus, entirely filled with these representing lines which are in (1,1)-correspondence with the represented systems of planes.

We shall call that $S_3 \perp OW$ regarded as a complex of the rays representing the equiangular to the right systems of planes, "the representing S_3 to the right of S_4" or shorter "the S_r of S_4". In the same way we form the "S_l of S_4". Each pair of planes in S_4 is then unequivocally determined by its representants in S_r and S_l and reversely the pair of planes determines unequivocally its representants.

Theorem 4. An equiangular double rotation to the right of the S_4 about a certain equiangular system to the right which double rotation leaves according to theorem 1 the position of S_l unchanged, gives a rotation of S_r about the representant of the system of the planes of rotation over an angle equal to double the angle, over which the equiangular double rotation to the right of S_4 takes place.

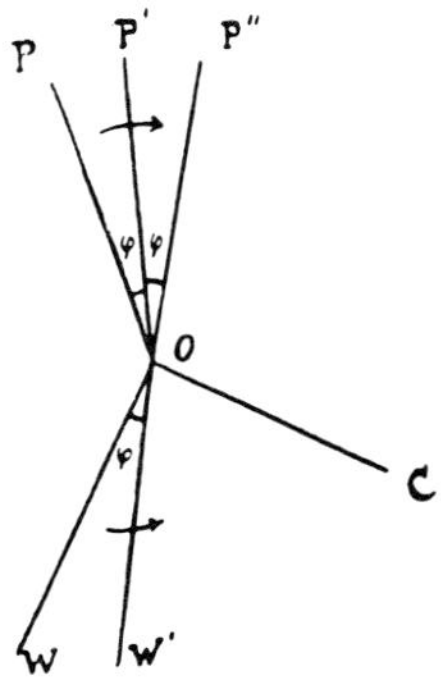

Fig. 8.

Proof. In the first place ensues from theorem 3 that the representants in S_r keep making mutually the same angles; so S_r has a "motion as a solid". We suppose through OW to be laid its plane of rotation α in S_4 and its normal plane β. In fig. 8 we suppose the lines tending downward to lie in α and those tending upward to lie in β whilst the indicated directions of rotation are those of the equiangular double rotation to the right which we consider. Angle WOC is made equal to $\frac{1}{2}\pi$. Then the S_r is the S_3 through OC and β. Let OP be an arbitrary vector in β and φ the angle, over which the equiangular double rotation to the right takes place. The double rotation leaves OC unchanged as representant of the equiangular system to the right on $(\alpha\beta)$. Moreover it transfers OW into OW' and OP into OP'. If we then still make $\angle P''OP'$ equal to $\angle P'OP$ we have:

$$\begin{pmatrix} OP, & OP' \\ OW, & OW' \end{pmatrix} \text{ equiangular to the right, thus:}$$

$$\begin{pmatrix} OP', & OP \\ OW, & OW' \end{pmatrix} \text{ equiangular to the left, or also:}$$

$$\begin{pmatrix} OP'', & OP' \\ OW, & OW' \end{pmatrix} \text{ equiangular to the left, so at last}$$

$$\begin{pmatrix} OP'', & OW \\ OP', & OW' \end{pmatrix} \text{ equiangular to the right.}$$

The plane POW giving OP as representant of its equiangular system to the right before its double rotation, gives after that rotation (transferred to $P'OW'$) as representant OP'' making an angle 2φ with OP; so an arbitrary vector OP in $S_r \perp OC$ (the invariable vector) rotates about OC over an angle 2φ, with which the theorem is proved.

We can now say: For an S_4 moving as a solid about a fixed point the position is at every moment determined by its "position to the right" (the position of the S_r moving as a solid about a fixed point) and its "position to the left" (the position of S_l). For, if of two positions the pairs of planes through O coincide, then this is the case too for all planes, thus for all rays too.

N.B. We can remark by the way, that in this way we have proved quite synthetically that two positions of S_4 have a common pair of planes, namely that pair, which has as representants the axis of rotation of the two positions to the right and that of the two positions to the left; so, taking into consideration that also the common fixed point is always there (having as projections on the positions of planes remained invariable the centres of rotation of the projections of S_4 on it), that the double rotation is the most general displacement of S_4. However, until now we have occupied ourselves only and wish to keep doing so with the motions of S_4, where always the same point O is in rest.

For a continuous motion of S_4 the position and the condition of motion are at every moment determined by S_r and S_l; so the motion of S_4 is quite determined by the motions of S_r and S_l; and at every moment the resulting displacement of S_4 is quite determined by that of S_r and of S_l, independent of the order of succession or combining of the two latter; they have no influence upon each other. We can regard a motion of S_4 as a sum of two entirely heterogeneous things, i. e. which cannot influence each other in any way or be transformed into each other.

We can proceed another step by indicating not only by a line in S_r a system of equiangular planes of rotation to the right, but also by a line vector along it an equiangular velocity of rotation to the right, that line vector being equal in size to the double velocity of rotation of the double rotation and directed along the vector moving with S_4 in the direction of OW. Then equal and opposite velocities of rotations to the right of S_4 are indicated by equal and opposite vectors in S_r.

Let OP_r be such an indicating vector and OQ_r and OS_r two mutually perpendicular vectors in the plane erected perpendicularly to OP_r in S_r in such a way that

$$\begin{pmatrix} OP_r, OW \\ OQ_r, OS_r \end{pmatrix}$$

is equiangular to the right, then to the equiangular double rotation to the right of S_4, indicated by OP_r corresponds the rotation of S_r about OP_r equal to the length of OP_r in the direction of $OQ_r \rightarrow OS_r$.

Let OP'_r be another indicating vector and let us determine in an analogous way OQ'_r and OS'_r, then the orthogonal systems of vectors

$$OWP_r\, Q_r\, S_r \text{ and}$$

$$OWP'_r\, Q'_r\, S'_r$$

are congruent, can thus be made to coincide by a rotation of S_4, with OW, thus S_r too, remaining in its place, so that the indicating vector OP_r by a motion of S_r in itself can be made to coincide with the indicating vector OP'_r in such a way, that at the same time the directions of rotation of S_r belonging to it coincide in the normal planes. But then an indicating vector in S_r represents that velocity of rotation entirely in the way usual in space of three dimensions, as also by its length it indicates the size of the velocity of rotation of S_r belonging to it; if namely we endeavour to regard the definition of that usual mode of representation entirely apart from the notion "motion with or against the hands" which is lacking in S_4; and say simply after having taken in that space a system of coordinates $OXYZ$: the vector of rotation will be erected to that side of the plane of rotation, that for the plane of rotation being by motion inside the space made to coincide with the plane of XY in such a way, that the direction of rotation runs from OX to OY, the vector of rotation is directed along the positive axis of Z.

For us that system of coordinates in S_r: OX_r, OY_r, OZ_r has been chosen in such a way that with OW it forms a system of coordinates in S_4, for which

$$\begin{pmatrix} OY_r, & OZ_r \\ OX_r, & OW \end{pmatrix}, \quad \begin{pmatrix} OZ_r, & OX_r \\ OY_r, & OW \end{pmatrix}, \quad \begin{pmatrix} OX_r, & OY_r \\ OZ_r, & OW \end{pmatrix}$$

are equiangular to the right.

A vector along OX_r then indicates a rotation of S_r in the sense of $OY_r \rightarrow OZ_r$.

Entirely analogous reasonings hold good for S_l. It being however more profitable to be able to say also for S_l: a vector along OX_l represents a velocity of rotation of S_l in the sense of $OY_l \rightarrow OZ_l$, we must modify the preceding either by choosing the system $OX_l\, Y_l\, Z_l\, W$ in such a way that

$$\begin{pmatrix} OY_l, & OZ_l \\ OX_l, & OW \end{pmatrix}$$

is no more equiangular to the right, but to the left, or if we suppose

$$\begin{pmatrix} OY_l, & OZ_l \\ OX_l, & OW \end{pmatrix}$$

to be also equiangular to the right, we must take as indicating vector in S_l that vector in the direction of which OW would move together with S_4, not the one moving together with S_4 in the direction of OW. We shall do the latter. The advantage of this choice will be evident from what follows.

We have still to remark, that if only the position of S_r and S_l is determined, the position of S_4 ensues from it not in one, but in two ways; for, a position of S_4 gives no other positions of S_r and S_l as its "opposite position" for which all vectors are reversed; that opposite position can be obtained by an arbitrary equiangular double rotation over an angle π; S_r and S_l then rotate 2π and are again in their former position.

But a continuous motion of S_4 out of a given initial position is unequivocally determined by the given continuous movements of S_r and S_l out of the corresponding initial positions. So we shall have completely answered the question how a solid S_4 moves under the action of determined forces if we can point out how S_r and S_l move under those actions; in other words if we can point out "the cones of rotation in the solid and in space".

APPLICATION. The Euler motion in S_4.

The equations of motion for this have been given for the first time by FRAHM in the "Mathematische Annalen" Band 8, 1874 p. 35. [5]
The system of coordinates $OX_1\ X_2\ X_3\ X_4$ in the solid we shall choose in such a way that the products of inertia disappear. We shall suppose

$$\begin{pmatrix} OX_2, & OX_3 \\ OX_1, & OX_4 \end{pmatrix}$$

to be equiangular to the right.

And we choose $OX_r\ Y_r\ Z_r$ and $OX_l\ Y_l\ Z_l$ in such a way that:

$$\begin{pmatrix} OX_1, & OX_2, & OX_3, & OX_4 \\ OX_r, & OY_r, & OZ_r, & OW \end{pmatrix}$$

is equiangular to the right and

$$\begin{pmatrix} OX_1, & OX_2, & OX_3, & OX_4 \\ OX_l, & OY_l, & OZ_l, & OW \end{pmatrix}$$

equiangular to the left, (from which ensues as a matter of fact that also

$$\begin{pmatrix} OX_r, & OY_r \\ OZ_r, & OW \end{pmatrix} \text{ and } \begin{pmatrix} OX_l, & OY_l \\ OZ_l, & OW \end{pmatrix}$$

are equiangular to the right).

The systems $OX_r\ Y_r\ Z_r$ and $OX_l\ Y_l\ Z_l$ execute the motions of S_r and S_l which are to be considered.

Let us farther call ${}_2\omega_3$, ${}_3\omega_1$, ${}_1\omega_2$, ${}_1\omega_4$, ${}_2\omega_4$, ${}_3\omega_4$ the components of the velocities of rotation according to the system $OX_1\ X_2\ X_3\ X_4$; and φ_1, φ_2, φ_3 the components of the velocities of rotation of S_r according to OX_r, OY_r, OZ_r, likewise ψ_1, ψ_2, ψ_3 the components of the velocities of rotation of S_l according to OX_l, OY_l, OZ_l. Then we know the components of velocity of rotation to the right $\frac{1}{2}({}_2\omega_3 + {}_1\omega_4)$ according to $\begin{matrix} OX_2 \rightarrow OX_3 \\ OX_1 \rightarrow OX_4 \end{matrix}$ or according to $\begin{matrix} OY_r \rightarrow OZ_r \\ OX_r \rightarrow OW \end{matrix}$ and analogues, and likewise the components of velocity of rotation to the left $\frac{1}{2}({}_2\omega_3 - {}_1\omega_4)$ according to $\begin{matrix} OX_2 \rightarrow OX_3 \\ OX_4 \rightarrow OX_1 \end{matrix}$ or according to $\begin{matrix} OY_l \rightarrow OZ_l \\ OW \rightarrow OX_l \end{matrix}$ and analogues.

Therefore:

$$\begin{aligned} \varphi_1 &= {}_2\omega_3 + {}_1\omega_4 & \psi_1 &= {}_2\omega_3 - {}_1\omega_4 \\ \varphi_2 &= {}_3\omega_1 + {}_2\omega_4 & \psi_2 &= {}_3\omega_1 - {}_2\omega_4 \\ \varphi_1 &= {}_1\omega_2 + {}_3\omega_4 & \psi_3 &= {}_1\omega_2 - {}_3\omega_4. \end{aligned}$$

The Euler equations of motion "in the solid" (giving the opposite of the apparent motion of the surrounding space) follow more simply than according to FRAHM out of the vector formula

fluxion of moment of motion = moment of force — rotation $\times$ moment of motion

which is easy to understand for a three dimensional space as well as for a four dimensional one,

(and where the sign $\times$ indicates the vector product)

For, "in the space" the fluxion of the moment of motion = moment of force; but of this for the position in solid has already been marked the fluxion wanted to keep the position constant in the solid, i. e. the fluxion which corresponds to the rotation of the moment-vector about the rotationvector and this is = rotation $\times$ moment of motion.

Let us call the squares of inertia $\Sigma m x_1^2$ etc. P_1 etc. and let us put

$$P_1 + P_2 + P_3 + P_4 = R$$
$$-P_1 + P_2 + P_3 - P_4 = A_1$$
$$P_1 - P_2 + P_3 - P_4 = A_2$$
$$P_1 + P_2 - P_3 - P_4 = A_3.$$

Then we can write the rotationvector in the form:

$${}_2\omega_3\, i + {}_3\omega_1\, j + {}_1\omega_2\, k + h\,({}_1\omega_4\, i + {}_2\omega_4\, j + {}_3\omega_4\, k)$$

or in the form

$$\varepsilon_1\,(\varphi_1\, i + \varphi_2\, j + \varphi_3\, k) + \varepsilon_2\,(\psi_1\, i + \psi_2\, j + \psi_3\, k).$$

The notations h and ε are taken from the above-named dissertation of Dr. W. A. WIJTHOFF; h is defined on page 67; ε_1 and ε_2 on page 78. [[6]]

The moment of motion becomes

$$(P_2 + P_3)\,{}_2\omega_3\, i + (P_3 + P_1)\,{}_3\omega_1\, j + (P_1 + P_2)\,{}_1\omega_2\, k +$$
$$+ h\,\{(P_1 + P_4)\,{}_1\omega_4\, i + (P_2 + P_4)\,{}_2\omega_4\, j + (P_3 + P_4)\,{}_3\omega_4\, k\}$$

or in an other form

$$\tfrac{1}{2}\,\varepsilon_1\,\{(R\varphi_1 + A_1\,\psi_1)\, i + (R\varphi_2 + A_2\,\psi_2)\, j + (R\varphi_3 + A_3\,\psi_3)\, k\} +$$
$$+ \tfrac{1}{2}\,\varepsilon_2\,\{(R\psi_1 + A_1\,\varphi_1)\, i + (R\psi_2 + A_2\,\varphi_2)\, j + (R\psi_3 + A_3\,\varphi_3)\, k\}.$$

If φ and ψ represent the rotation vectors in R_r and R_l, we can write the rotation:

$$\varepsilon_1\,\varphi + \varepsilon_2\,\psi,$$

and the moment

$$\tfrac{1}{2}\,R\,(\varepsilon_1\,\varphi + \varepsilon_2\,\psi) + (\tfrac{1}{2}\,\varepsilon_1 . (A)\,\psi + \tfrac{1}{2}\,\varepsilon_2 . (A)\,\varphi),$$

where the notation $(A)\varphi$ means: $A_1\,\varphi_1\, i + A_2\,\varphi_2\, j + A_3\,\varphi_3\, k$.

The first and strongest of these terms falls along the rotationvector; for a body with equal squares of inertia it is the only one; the second, which, together with the A's, becomes stronger as the body is more asymmetric, we might call the "crossed moment" because its right part is caused by the left part of the rotation and inversely.

Let us put finally the moment of force in the form $\varepsilon_1\,\mu + \varepsilon_2\,\nu$, where μ and ν are threedimensional vectors; then the above given formula of the vector can be broken up into the six following components, given successively by the coefficients of $\varepsilon_1\, i$, $\varepsilon_2\, i$, $\varepsilon_1\, j$, $\varepsilon_2\, j$, $\varepsilon_1\, k$, $\varepsilon_2\, k$.

$$R\varphi_1 + A_1\,\dot{\psi}_1 = A_2\,\psi_2\,\varphi_3 - A_3\,\psi_3\,\varphi_2 + 2\mu_1$$
$$R\psi_1 + A_1\,\dot{\varphi}_1 = A_2\,\varphi_2\,\psi_3 - A_3\,\varphi_3\,\psi_2 + 2\nu_1$$
$$R\varphi_2 + A_2\,\dot{\psi}_2 = A_3\,\psi_3\,\varphi_1 - A_1\,\psi_1\,\varphi_3 + 2\mu_2$$
$$R\psi_2 + A_2\,\dot{\varphi}_2 = A_3\,\varphi_3\,\psi_1 - A_1\,\varphi_1\,\psi_3 + 2\nu_2$$
$$R\varphi_3 + A_3\,\dot{\psi}_3 = A_1\,\psi_1\,\varphi_2 - A_2\,\psi_2\,\varphi_1 + 2\mu_3$$
$$R\psi_3 + A_3\,\dot{\varphi}_3 = A_1\,\varphi_1\,\psi_2 - A_2\,\varphi_2\,\psi_1 + 2\nu_3.$$

If we put $R^2 - A_p{}^2 = a_p$ and $RA_p + A_q A_r = b_p$, and if we represent the vector $a_1 \varphi_1 i + a_2 \varphi_2 j + a_3 \varphi_3 k$, by $(a)\varphi$, we can write the six equations of motion:

$$(a)\dot{\varphi} = V.(b)\psi.\varphi + 2R\mu - 2A\nu$$

$$(a)\dot{\psi} = V.(b)\varphi.\psi + 2R\nu - 2A\mu.$$

For absence of external forces:

$$\left.\begin{aligned}(a)\dot{\varphi} &= V.(b)\psi.\varphi \\ (a)\dot{\psi} &= V.(b)\varphi.\psi\end{aligned}\right\} \quad . \; . \; . \; . \; . \; . \quad (h)$$

In this form we can directly read:

1st. If in the initial position $\varphi = \psi$, then φ remains equal to ψ, i. e. if the initial rotation of S_4 is a rotation // to a principal space of inertia, then the motion remains a rotation // to that space. The equations of motion for that case can be reduced to a system to be treated as the well known Euler motion in S_3 when the forces are missing.

2nd. For unequal A's "invariable rotating" is only possible under the following two conditions which are each in itself sufficient:

a. for φ and ψ both directed along one and the same axis of coordinate (X-, Y- or Z-axis of the representing spaces) i. e. for a double rotation of S_4 about a pair of principal planes of inertia;

b. for $\varphi = 0$ or $\psi = 0$, i. e. for an equiangular double rotation of S_4.

[[7]] It has been pointed out by KÖTTER (see "Berliner Berichte" 1891, p. 47), how a system of equations analogous with what was given, can be integrated. (The problem treated there is the motion of a solid in a liquid). According to the method followed by him the six components of rotation can be expressed explicitly by hyper-elliptic functions in the time. If however we have $\varphi_1, \varphi_2, \varphi_3, \psi_1, \psi_2, \psi_3$ expressed in the time, we have the "cones in the solid" for S_r and S_l. To deduce from these the "cones in space" we set about as follows. We notice that the moment of motion to the right $R\varphi + (A)\psi$ in S_r remains invariable "in space" (in S_l that vector changes of course its direction, but there $R\psi + (A)\varphi$ remains invariable); calling the two spheric coordinates ("polar distance" and "length") of φ with respect to $R\varphi + (A)\psi$ during the motion of φ in space ϑ and χ and remarking that each element of the "cone in space" at the moment of contact coincides with the corresponding element of the "cone in the solid", we can express ϑ, $\dot{\vartheta}$ and χ in the time, with which the

[[18]]

"cone in space" for S_r is found. Analogously the "cone in space" is determined for S_l with respect to $R\psi + (A)\varphi$.

We shall just show that as soon as two of the squares of inertia P become equal, which means the same as two A's becoming equal we can do with usual elliptic functions only.

For instance let $A_2 = A_3$, thus also $a_2 = a_3 = {}_2a_3$; $b_2 = b_3 = {}_2b_3$. Let us call ${}_2\varphi_3$ and ${}_2\psi_3$ the value of the decomponents of φ and ψ in the YZ-plane, F_φ and F_ψ the anomalies of those decomponents (counted from Y to Z), and F the difference of anomaly of ${}_2w_3$ and ${}_2\gamma_3$; then the equations (h) become for this case:

$$a_1\dot\varphi_1 = -\,{}_2b_3\,{}_2\varphi_3\,{}_2\psi_3 \sin F \qquad a_1\dot\psi_1 = {}_2b_3\,{}_2\varphi_3\,{}_2\psi_3 \sin F$$

$$ {}_2a_3\,{}_2\dot\varphi_3 = {}_2b_3\,\varphi_1\,{}_2\psi_3 \sin F \qquad {}_2a_3\,{}_2\dot\psi_3 = -\,{}_2b_3\,{}_2\varphi_3\,\psi_1 \sin F$$

$$\dot F_\varphi = \frac{b_1}{{}_2a_3}\psi_1 - \frac{{}_2b_3}{{}_2a_3}\varphi_1 \cos F\,\frac{{}_2\psi_3}{{}_2\varphi_3} \qquad \dot F_\psi = \frac{b_1}{{}_2a_3}\varphi_1 - \frac{{}_2b_3}{{}_2a_3}\psi_1 \cos F\,\frac{{}_2\gamma_3}{{}_2\psi_3}$$

$$\dot F = \frac{b_1}{{}_2a_3}(\varphi_1 - \psi_1) + \frac{{}_2b_3}{{}_2a_3}\cos F\left(\varphi_1\frac{{}_2\psi_3}{{}_2\varphi_3} - \psi_1\frac{{}_2\varphi_3}{{}_2\psi_3}\right)$$

from which we deduce four integral relations between

φ_1, ${}_2\varphi_3$, ψ_1, ${}_2\psi_3$ and F, namely

$$\varphi_1 + \psi_1 = \frac{c_1}{\sqrt{a_1}} \quad \ldots\ldots\ldots\ldots (I)$$

$${}_2a_3\,{}_2\varphi_3^{\,2} + a_1\varphi_1^2 = c_2^2 \quad . . (II), \qquad {}_2a_3\,{}_2\psi_3^{\,2} + a_1\psi_1^2 = c_3^2 \quad . . (III)$$

$${}_2a_3\,{}_2b_3\,{}_2\varphi_3\,{}_2\psi_3 \cos F + a_1 b_1 \varphi_1 \psi_1 = c_4 \quad \ldots\ldots (IV)$$

If we put in these

$${}_2\varphi_3 = \frac{c_2}{\sqrt{{}_2a_3}}\cos\eta, \qquad \varphi_1 = \frac{c_2}{\sqrt{a_1}}\sin\eta,$$

$${}_2\psi_3 = \frac{c_3}{\sqrt{{}_2a_3}}\cos\zeta, \qquad \psi_1 = \frac{c_3}{\sqrt{a_1}}\sin\zeta,$$

we have the differential eqations

$$\dot\eta = -\frac{{}_2b_3}{{}_2a_3\sqrt{a_1}}c_3\cos\zeta\sin F, \quad \dot\zeta = \frac{{}_2b_3}{{}_2a_3\sqrt{a_1}}c_2\cos\eta\sin F$$

with the two integrals:

$$c_2\sin\eta + c_3\sin\zeta = c_1,$$

$${}_2b_3\cos\eta\cos\zeta\cos F + b_1\sin\eta\sin\zeta = \frac{c_4}{c_2c_3},$$

so that after elimination we obtain the following differential equation in η:

$$\begin{aligned}{}_2a_3^{\,2}\,a_1 c_2^2 . \cos^2\eta . \dot\eta^2 = &({}_2b_3^{\,2} - b_1^2)\,c_2^4 \sin^4\eta \\ &- 2c_1({}_2b_3^{\,2} - b_1^2)\,c_2^3\sin^3\eta \\ &+ \{c_1^2({}_2b_3^{\,2} - b_1^2) - (c_2^2 + c_3^2)\,{}_2b_3^{\,2} - 2c_4 . b_1\}\,c_2^2\sin^2\eta \\ &+ 2c_1(c_2^2 . {}_2b_3^{\,2} + c_4 . b_1)\,c_2\sin\eta \\ &+ c_2^2(c_3^2 - c_1^2)\,{}_2b_3^{\,2} - c_4^2\end{aligned}$$

or $c_2 \sin \eta \, (= \sqrt{a_1} \cdot \dot{\varphi}_1)$ put equal to u:

$_2a_3^{\,2} \, a_1 \, \dot{u}^2 = F_4(u)$, in which we can easily verify that $F_4(u)$ has two real roots between $-c_2$ and $+c_2$ (the two other roots are real outside those limits or imaginary according to $_2b_3^{\,2} - b_1^{\,2}$ being $\gtrless$ than 0); those two roots indicate the limits between which η swings to and fro according to a course indicated by elliptic functions. For the case $_2b_3^{\,2} > b_1^{\,2}$ for instance, thus for four real roots $u_1 < u_2 < u_3 < u_4$, that course becomes:

$$u = u_2 + \frac{_1u_2 \cdot {_2u_3} \, sn^2}{_1u_3 - {_2u_3} \, sn^2}, \quad \text{where } sn = sn \left\{ \tfrac{1}{2} t \sqrt{\frac{_2u_4 \cdot {_1u_3} \, (_2b_3^{\,2} - b_1^{\,2})}{_2a_3^{\,2} \, a_1}} \right\},$$

and

$$_pu_q = u_q - u_p.$$

Farthermore:

$$\dot{F}_\varphi = \frac{b_1}{_2a_3} \psi_1 - \frac{_2b_3}{_2a_3} \varphi_1 \frac{_2\varphi_3 \cdot {_2\psi_3} \cos F}{_2\varphi_3^{\,2}}$$

where the second member is a rational function of φ_1 (ψ_1 can be rationally expressed in φ_1 according to (I), $_2\varphi_3^{\,2}$ according to (II), $_2\varphi_3 \, {_2\psi_3} \cos F$ according to (IV)), so that F_φ too can be expressed in t by elliptic functions and by that the entire "cone in the solid"; and further according to the above method also "the cone in space".

The following special cases can very easily be traced to the end.

1st The four squares of inertia are equal two by two. This case is obtained by putting $A_2 = A_3 = 0$.

Then

$$a_1 = R^2 - A_1^{\,2} \qquad b_1 = RA_1$$
$$_2a_3 = R^2 \qquad _2b_3 = 0.$$

And the equations of motion pass into:

$$a_1 \dot{\varphi}_1 = 0 \qquad a_1 \dot{\psi}_1 = 0$$
$$_2a_3 \, {_2\dot{\varphi}_3} = 0 \qquad _2a_3 \, {_2\dot{\psi}_3} = 0$$
$$\dot{F}_\varphi = \frac{b_1}{_2a_3} \psi_1 \qquad \dot{F}_\psi = \frac{b_1}{_2a_3} \varphi_1$$
$$\dot{F} = \frac{b_1}{_2a_3} (\varphi_1 - \psi_1)$$

from which we directly read, φ_1, $_2\varphi_3$, ψ_1 and $_2\psi_3$ remaining constant, that the "cone in the solid" for S_r and for S_l is a cone of revolution with the X axis as axis of revolution. Farthermore the moment

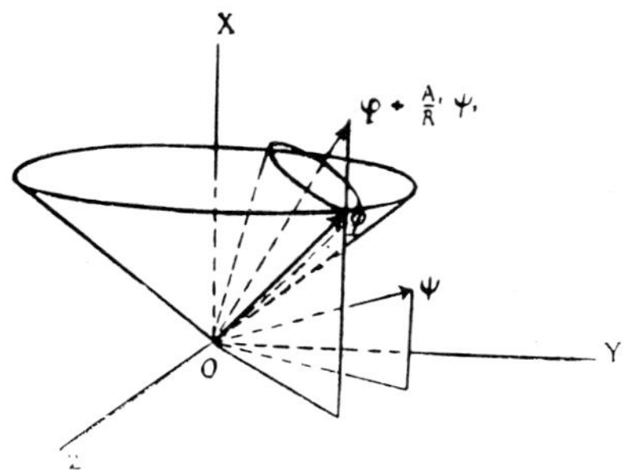

Fig. 9.

$R\varphi + A_1\psi_1$ to the right lying in the meridian plane of φ remains in S_r invariable. Thus "in space" that meridian plane rotates about the vector $R\varphi + A_1\psi_1$, by which the "cone in space" is known, and likewise proves to be a cone of revolution. Analogous for S_l. Fig. 9 shows the two cones of rotation in S_r. The outer cone is the moving one.

2nd. Three of the squares of inertia are equal and unequal to the fourth. We take the axis of the unequal one as X_4 axis in S_4. Then $A_1 = A_2 = A_3 = A$; $a_1 = a_2 = a_3 = a$; $b_1 = b_2 = b_3 = b$; and the equations (h) pass into

$$\dot{\varphi} = \frac{b}{a} V \psi . \varphi$$

$$\dot{\psi} = \frac{b}{a} V \varphi . \psi$$

therefore $\dot{\varphi}$ and $\dot{\psi}$ are both perpendicular to φ and to ψ, whilst $\dot{\varphi} + \dot{\psi} = 0$, so $\varphi + \psi$ is constant and φ and ψ are each for itself constant in absolute value, so that they both rotate about their sum ("in space" that vector of the sum has in general quite a different position for S_r than for S_l) by which the two "cones in the solid" are determined. "Invariable rotating" of S_4 we have here wherever φ and ψ, regardless of their value, coincide. To find the "cone in space" for S_r, we notice the invariability in S_r of $R\varphi + A\psi$. "In space" φ rotates about $R\varphi + A\psi$, for the angle between those two vectors remains constant. In S_l rotates analogously "in space" ψ about $R\psi + A\varphi$. Fig. 10 represents the two cones of rotations in S_r. (Here too the outer cone is the moving one).

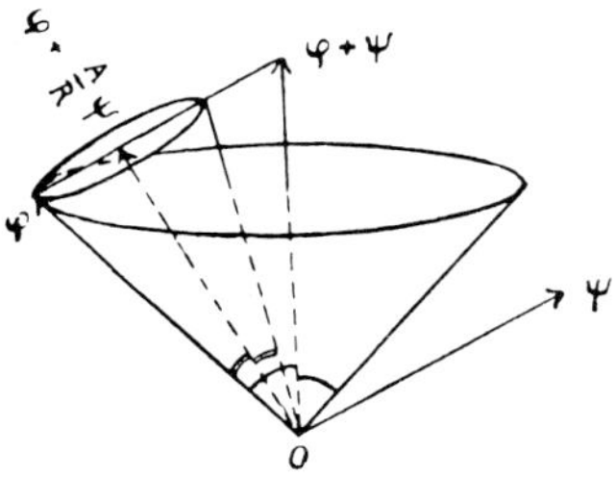

Fig. 10.

We remind the readers once more, that where we bring φ and ψ, as far as their positions in the solid are concerned, into relationship with each other, we must of course in our mind make the positions to the right and the left that is of the systems of coordinates $OX_r Y_r Z_r$ and $OX_l Y_l Z_l$ to coincide with each other, so that but one system of coordinates $OXYZ$ is left (for instance for the equations

(h) this must be noticed); but in space (i. e. the $S_3 \perp OW$) those systems of coordinates $OX_r\ Y_r\ Z_r$ and $OX_l\ Y_l\ Z_l$ have at each moment very definite positions differing from one another.

NOTES

⟦1⟧ The papers Brouwer 1904 A2, 1904 B2, 1904 C2 are closely related. Their subject is rotations and reflexions in 4-space. Fundamental concepts are: equiangular 'double' rotation to the right and to the left, equiangular pairs of planes, symmetries. The main result is the decomposability of any 'double' rotation into a product of a right and a left equiangular one, and the involved decomposition of the rotation group of 4-space into two groups which are isomorphic with that of 3-space.

Rotations with a fixed point O in 4-space leave two orthogonal planes invariant; in the general case these planes are uniquely determined – this explains the term 'double rotation'. A rotation is called equiangular if the angle between the original and the image vector is constant. There are two systems of equiangular rotations – the rotation senses in the invariant orthogonal planes together produce one of the two orientations of 4-space. Among the rotations of 4-space the equiangular rotations are those for which the invariant planes are not uniquely determined.

A pair of planes in 4-space meeting in one point O is characterised by two position angles. If the two angles happen to be equal, the pair is called equiangular. The orientation of 4-space makes it possible to distinguish right and left equiangularity. Brouwers' wording of the definition of right and left equiangularity says that one of the planes is invariant under a right or left equiangular rotation with the other as an axis.

The most efficient way to deal with these concepts is to use quaternions. The right and left equiangular rotations are represented by the left and right quaternion multiplications respectively; according to Cayley, 1855, the general rotation in 4-space can be written in the form $x \rightarrow axb$. From the point of view of elliptic 3-space A. Cayley's result plays a part in W. K. Clifford, 1873 (equiangular planes correspond to Clifford parallels). Clifford's paper inspired F. Klein, 1890, which is a clear and rather complete treatment of the related questions. Se also E. Study, 1902. The representation of rotations of 3-space by $x \rightarrow axa^{-1}$ had been known as long ago as 1843 (W. R. Hamilton, 1848 and A. Cayley, 1845).

There can be no doubt that in 1904 Brower knew about quaternions – in fact he cited W. A. Wythoff, 1898, who used them in an essential way and mentioned Clifford, 1873. Perhaps Brouwer intentionally avoided quaternions because he

⟦22⟧

aimed at a synthetic approach or, at least, an algebraic one in which no axis played a special part, but it is also likely that he was unfamiliar with the use of quaternions and had no knowledge of the relevant literature; indeed he may even have overlooked connexions which look trivial to us. The aftermath of Brouwer's first paper indicates that the use of quaternions and the relevant literature were not yet familiar at that time:

E. Jahnke replied to Brouwer 1904 A2, with a communication which was accepted by the Academy and printed, Jahnke 1904. Its text was as follows:

> Die genannte Notiz steht in Zusammenhang mit Untersuchungen von FERDINAND CASPARY und mit Arbeiten, die ich in den Jahren 1896–1901 veröffentlicht habe. Da Herr BROUWER hierauf keinen Bezug nimmt, erlaube ich mir folgendes zu bemerken: Probleme aus der Theorie der Thetafunktionen einerseits und aus der Mechanik andererseits haben mich dazu geführt, die Drehung im R_4 zu zwei Drehungen im R_3 in Beziehung zu setzen. Die Relationen zwischen den Elementen der vierdimensionalen Drehung und den Elementen der beiden zugehörigen dreidimensionalen Drehungen habe ich in den Sitzungsberichten der Berliner Akademie vom 30. Juli 1896 und im Journal für die reine und angewandte Mathematik Bd. **118**, S. 225, 1897 ⟦E. Jahnke, 1897⟧ explizite angegeben. Insbesondere habe ich gefunden, dass sich die Geschwindigkeitskomponenten der ersteren Drehung aus den Geschwindigkeitskomponenten der beiden anderen in einfacher Weise zusammensetzen (vgl. auch meinen Vortrag auf der Naturforscherversammlung in Hamburg 1901: "Ueber Drehungen im vierdimensionalen Raum").
>
> Herr BROUWER gelangt in seiner Abhandlung ebenfalls zu diesen Resultaten, wenn auch auf verschiedenem, nämlich auf geometrischem Wege, während ich den algebraischen Weg gegangen bin. Herr BROUWER kommt zu einer *Zerlegung* (splitsing) der vierdimensionalen Drehung in zwei dreidimensionale, während ich mich des Ausdrucks *Zuordnung* bediene.
>
> *Berlin*, den 28. März 1904.

Jahnke's communication shows that its author was no better acquainted with the relevant literature, and that he did not grasp the contents of Brouwer 1904 A2, either. The orthogonal transformations of 4-space that E. Jahnke had found in Caspary 1883, and reconsidered, look in quaternion form as $x \to a\bar{x}$. Jahnke mistook them for rotations. Since conjugation interchanges left and right multiplications, every rotation of 4-space can be obtained as a product of two transformations of this kind (instead of a left and a right multiplication). This is the theorem Jahnke claimed against Brouwer on the strength of Jahnke, 1896, 1897, 1902, though in fact it is at least Caspary's result, if it is not older.

Caspary's and Jahnke's context was theta functions, where Caspary's decomposition is quite natural; the decomposition of the 4-rotation group into two groups isomorphic with the 3-rotation group is natural in geometry. Though Jahnke cited geometers, he did not grasp the essence of Brouwer's geometric approach. Though the cited Hamilton, he was equally little acquainted with quaternions. He did not grasp either how the rotation group of 3-space entered into Brouwer's approach.

In 1904 C2 Brouwer answered Jahnke's claims. This paper shows complete insight though even now neither quaternions nor other relevant literature were mentioned. An unpublished sharp reaction on Jahnke's communication is found in Brouwer's papers.

If specialization is thought to be a new development in mathematics, it may come as a surprise to learn that 70 years ago mathematics was sharply divided according to such topics as theta functions, four-dimensional space, elliptic geometry, and quaternions, and so were the mathematicians.

⟦2⟧, ⟦3⟧ The emendations are in Brouwer's hand.

⟦3⟧ Brouwer's own copy of the Dutch text contains the remark (translated): Not correct; the vector in R_4 represents a finite rotation; that in R_3 a rotation velocity.

⟦4⟧ Brouwer's own copy of the Dutch text contains the remark (translated): What is the relation between the corresponding three-dimensional length vectors? (There are three of them.)

⟦5⟧ W. Frahm, 1874.

⟦6⟧ W. A. Wythoff, 1898.

⟦7⟧ F. Kötter, 1895.

1904 B2

Mathematics. — "*On symmetric transformation of S_4 in connection with S_r and S_l.*" By Mr. L. E. J. BROUWER. (Communicated by Prof. D. J. KORTEWEG). [[1]]

Let us for the present occupy ourselves with a particular case of symmetric transformation — the *reflection*, and let us investigate its influence on S_r and S_l. As S_r and S_l are independent of the choice of a system of axes, we make a suitable choice by selecting the X_4 axis along the axis of reflection. Let us call $\alpha_1, \alpha_2, \alpha_3, \alpha_4$ the cosines of direction of a vector before the reflection; $\beta_1, \beta_2, \beta_3, \beta_4$ those after it; let us moreover represent $\alpha_2' \alpha_3'' - \alpha_3' \alpha_2''$ etc. by ξ_{23} etc. and $\beta_2' \beta_3'' - \beta_3' \beta_2''$ etc. by χ_{23} etc. and let us call λ_{23} etc. the coefficients of position of a plane with sense of rotation included before the reflection and μ_{23} etc. those after it. Then: [[2]]

$$\alpha_1 = \beta_1$$
$$\alpha_2 = \beta_2$$
$$\alpha_3 = \beta_3$$
$$\alpha_4 = -\beta_4$$

$$\xi_{23} = \chi_{23} \qquad \xi_{14} = -\chi_{14}$$
$$\xi_{31} = \chi_{31} \qquad \xi_{24} = -\chi_{24}$$
$$\xi_{12} = \chi_{12} \qquad \xi_{34} = -\chi_{34}$$

$$\sqrt{\xi^2_{23}+\xi^2_{31}+\xi^2_{12}+\xi^2_{14}+\xi^2_{24}+\xi^2_{34}} = \sqrt{\chi^2_{23}+\chi^2_{31}+\chi^2_{12}+\chi^2_{14}+\chi^2_{24}+\chi^2_{34}}.$$

So also:

$$\lambda_{23} = \mu_{23} \qquad \lambda_{14} = -\mu_{14}$$
$$\lambda_{31} = \mu_{31} \qquad \lambda_{24} = -\mu_{24}$$
$$\lambda_{12} = \mu_{12} \qquad \lambda_{34} = -\mu_{34}$$

[[25]]

or :

$$\left.\begin{aligned} \lambda_{23} + \lambda_{14} &= \mu_{23} - \mu_{14} & \qquad \lambda_{23} - \lambda_{14} &= \mu_{23} + \mu_{14} \\ \lambda_{31} + \lambda_{24} &= \mu_{31} - \mu_{24} & \qquad \lambda_{31} - \lambda_{24} &= \mu_{31} + \mu_{24} \\ \lambda_{12} + \lambda_{34} &= \mu_{12} - \mu_{34} & \qquad \lambda_{12} - \lambda_{34} &= \mu_{12} + \mu_{34} \end{aligned}\right\} \quad . \qquad (a)$$

Now however

$$\begin{gathered} \lambda_{23} + \lambda_{14} \\ \lambda_{31} + \lambda_{24} \\ \lambda_{12} + \lambda_{34} \end{gathered}$$

are the cosines of direction of the representant of the system of planes equiangular to the right with λ with respect to a system
[3] of coordinates $OX_r\,Y_r\,Z_r$ taken in S_r, as that was defined (These Proceedings Febr. 1904, page 729).

And likewise

$$\begin{gathered} \lambda_{23} - \lambda_{14} \\ \lambda_{31} - \lambda_{24} \\ \lambda_{12} - \lambda_{34} \end{gathered}$$

are the cosines of direction of the representant of the system of planes equiangular to the left with λ with respect to a system of coordinates $OX_l\,Y_l Z_l$ taken in an analogous manner in S_l.

So from the formulae (a) ensues that the effect of a reflection is what we might call a *reciprocal interchange of S_r and S_l*, i.e. a suchlike interchange that every vector of S_l takes the place of that vector of S, which has substituted itself for it.

But now an arbitrary symmetric transformation of S_4 can be replaced by a reflection preceded or followed by a double rotation; which is represented by a reciprocal interchange of S_r and S_l preceded or followed by a rotation of S_r and one of S_l; therefore:

The arbitrary symmetric transformation of S_4 is represented by an *interchange of S_r and S_l in arbitrary positions.*

Let us now consider that for such an arbitrary interchange of S_r and S_l a system of coordinates α of S_r is placed on a system of coordinates β of S_l whilst that system of coordinates β of S_l itself is placed on a system γ of S_r; then we can replace the interchange by a "reciprocal interchange" placing α on β and β on α, followed by a rotation of S_r, placing α on γ, or also by a rotation of S_r, placing α on γ, followed by a reciprocal interchange, placing γ on β and β on γ.

Consequently we have proved :

"An arbitrary symmetric transformation of S_4 can be replaced by a reflection preceded or followed by a double rotation equiangular to the right and likewise of course by a reflection preceded or followed by a double rotation equiangular to the left."

The plane of rotation of the equiangular double rotation passing through the axis of reflection remains for both parts of the transformation in an unaltered position; it undergoes by the double rotation a congruent transformation and by the reflection a symmetric one.

The plane of rotation of the equiangular double rotation situated in the space perpendicular to the axis of reflection remains also for both parts of the transformation in an unaltered position; it is not transformed at the reflection and undergoes by the double rotation a congruent transformation.

Those two planes of rotation are perpendicular to each other, so that geometrically the wellknown property is proved:

"For symmetric transformation of S_4 about a fixed point one pair of planes remains at its place; and one plane of it is transformed congruently, the other symmetrically."

NOTES

⟦1⟧ The Dutch version was communicated at the meeting of 23 April 1904. The paper extends 1904A2.
⟦2⟧ Instead of 'for the present' read 'first'.
⟦3⟧ Brouwer 1904A2.

1904 C2

Mathematics. — "*Algebraic deduction of the decomposability of the continuous motion about a fixed point of S_4 into those of two S_3's*". By Mr. L. E. J. BROUWER. (Communicated by
⟦1⟧ Prof. KORTEWEG).

As the position of S_4 is determined with respect to a fixed system of axes by *six* independent variables and that of S_3 with respect to a fixed system of axes by *three* independent variables we understand at once that in an infinite number of ways two S_3 motions can be *coordinated* to an S_4 motion, so that position and velocities of S_4 are determined by position and velocities of the two S_3's. On such a
⟦2⟧ coordination Mr. JAHNKE has been engaged in the papers mentioned above and has deduced the relations between positions and velocities of S_4 and the two S_3's. Interpreted geometrically his coordination amounts to the following: Let us suppose in S_4 a fixed system of axes X_1 X_2 X_3 X_4, and a movable one Y_1 Y_2 Y_3 Y_4; let us consider the part equiangular to the right of the double rotation, which transfers X_1 X_2 X_3 X_4 into Y_1 Y_2 Y_3 Y_4; let us add to it an equal equiangular double rotation to the left (namely equal with respect to the system of axes X_1 X_2 X_3 X_4; only with respect to a definite system of axes can we call an equiangular double rotation to the right and one to the left equal); the resulting rotation becomes a single

⟦28⟧

rotation parallel to the space X_1 X_2 X_3 which would transfer the system of axes X_1 X_2 X_3 into an other Z_1 Z_2 Z_3. Thus to each position Y with respect to X_1 X_2 X_3 X_4 answers a position Z with respect to X_1 X_2 X_3, and by interchanging right and left, in an analogous manner a position U with respect to X_1 X_2 X_3; and we may consider the positions Z and U as coordinated to the position Y.

Not immediately to be seen are the two following properties of the S_4 motion geometrically deduced in what was communicated in the February meeting.

1st. The continuous motion of S_4 can be *decomposed*, that is: independent of the choice of a system of axes two definite threefoldly infinite motion groups exist in S_4 in such a way that an arbitrary motion can be composed out of two motions each of which belongs to one of the groups mentioned.

2nd. The continuous motion of S_4 can be *decomposed into two S_3 motions*, that is, two twodimensional manifoldnesses (namely [[3]]
those of the systems of planes equiangular to the right and to the left) exist in S_4 in such a way that each of the motion groups mentioned transforms the elements of one of them into each other and leaves the other unchanged; to which furthermore we can allow twodimensional Euclidean stars to answer in such a way that to congruent combinations in one of the manifoldnesses congruent combinations of the Euclidean stars correspond, that to the corresponding motion group of S_4 answers the motion group of the Euclidean star movable as a solid and that to congruent combinations in the motion group of S_4 answer congruent combinations in the motion group of the Euclidean star movable as a solid: reason why we may call the two twodimensional manifoldnesses *twodimensional Euclidean stars* and the motion groups of S_4 transforming them *Euclidean threedimensional motion groups about a fixed point.*

We shall now see how we can arrive algebraically at both results.

Mr. Jahnke takes from Caspary the so called "Elementary trans- [[4]]
formation" (see a.o. Jahresbericht der Deutschen Mathematiker-
Vereiniging XI, 4, 1902, p. 180 and F. Caspary, *Zur Theorie der [[5]]
Thêtafunctionen mit zwei Argumenten*, Crelle's Journal, vol. 94, page 75), which has the property that an arbitrary congruent transformation of S_4 can be replaced by two successive elementary transformations. The name "Elementary rotation" (Elementardrehung) of Mr. Jahnke seems to me less fortunate, because it is a symmetric transformation, not a rotation. The real meaning of the "Elementary transformation" will be made clear furtheron. For the present we remind the readers
of its determinant type (see Jahresbericht, l. c., page 180). [[6]]

$$\left.\begin{matrix} \pi_1 & -\pi_2 & \pi_3 & \pi_4 \\ \pi_2 & \pi_1 & -\pi_4 & \pi_3 \\ -\pi_3 & \pi_4 & \pi_1 & \pi_2 \\ \pi_4 & \pi_3 & \pi_2 & -\pi_1 \end{matrix}\right| , \quad \ldots \ldots \quad (I)$$

and we notice that it *does not represent a group* and *does not possess any threedimensional properties* (it *does* after composition with itself, compare for instance the theorem of Mr. JAHNKE, Jahres-
[[6]] bericht, l. c., page 182: "Jede endliche Drehung im R_3 lässt sich als eine Zusammensetzung aus einer Elementardrehung im R_4 mit sich selbst auffassen"[1]); which operation is for the rest bound to a once chosen system of axes).

We shall now deduce two different determinant types likewise determined by a system of cosines of direction $\pi_1, \pi_2, \pi_3, \pi_4$ which *do represent a group* and *have threedimensional properties*. Those will be the determinants of equiangular double rotation to the right and to the left.

[[7]] Let us solve the α's out of the equations (H) (see Proceedings of March 19th, 1904, page 721); then we have

$$\left.\begin{aligned} \alpha_1 &= \pi_4\beta_1 + \pi_3\beta_2 - \pi_2\beta_3 + \pi_1\beta_4 \\ \alpha_2 &= \pi_4\beta_2 - \pi_3\beta_1 + \pi_2\beta_4 + \pi_1\beta_3 \\ \alpha_3 &= \pi_4\beta_3 + \pi_3\beta_4 + \pi_2\beta_1 - \pi_1\beta_2 \\ \alpha_4 &= \pi_4\beta_4 - \pi_3\beta_3 - \pi_2\beta_2 - \pi_1\beta_1 \end{aligned}\right\} \quad \ldots \quad (a).$$

Thus the determinant type of the equiangular double rotation to the right is

$$\left.\begin{matrix} \pi_4 & \pi_3 & -\pi_2 & \pi_1 \\ -\pi_3 & \pi_4 & \pi_1 & \pi_2 \\ \pi_2 & -\pi_1 & \pi_4 & \pi_3 \\ -\pi_1 & -\pi_2 & -\pi_3 & \pi_4 \end{matrix}\right| . \quad \ldots \ldots \quad (II)$$

Directly can be verified that this determinant type forms a group.

Likewise we deduce for the equiangular double rotation to the left $(\pi_1\,\pi_2\,\pi_3\,\pi_4)$, transferring the vector $(\alpha_1\,\alpha_2\,\alpha_3\,\alpha_4)$ into $(\beta_1\,\beta_2\,\beta_3\,\beta_4)$, the relations:

$$\left.\begin{aligned} \alpha_1 &= -\pi_4\beta_1 - \pi_3\beta_2 + \pi_2\beta_3 + \pi_1\beta_4 \\ \alpha_2 &= \pi_3\beta_1 - \pi_4\beta_2 - \pi_1\beta_3 + \pi_2\beta_4 \\ \alpha_3 &= -\pi_2\beta_1 + \pi_1\beta_2 - \pi_4\beta_3 + \pi_3\beta_4 \\ \alpha_4 &= -\pi_1\beta_1 - \pi_2\beta_2 - \pi_3\beta_3 - \pi_4\beta_4 \end{aligned}\right\} , \quad \ldots \quad (b)$$

[1]) "Each finite rotation in S_3 can be regarded as a composition of an elementary rotation in S_4 with itself."

from which ensues the determinant type for equiangular double rotation to the left:

$$\left.\begin{matrix} -\pi_4 & -\pi_3 & \pi_2 & \pi_1 \\ \pi_3 & -\pi_4 & -\pi_1 & \pi_2 \\ -\pi_2 & \pi_1 & -\pi_4 & \pi_3 \\ -\pi_1 & -\pi_2 & -\pi_3 & -\pi_4 \end{matrix}\right\}, \quad . \; . \; . \; . \; . \; . \qquad \text{(III)}$$

and for this too the property of a group can be verified.

If we call (I') the determinant type formed by interchange of the rows and columns of (I), we can remark:

If we reverse the signs in the bottom row of type (II) type (I') appears.

If we reverse the signs in the last column of type (III) type (I') appears.

If we ask ourselves whether each arbitrary congruent transformation can be replaced by the succession of a transformation (III) and a transformation (II) the answer must be affirmative; for we shall have but to take those transformations (III) and (II) belonging to the transformations (I') which when successively applied transfer the given initial position into the given final one. (For those two ways only the intermediate positions will differ in as far as they will be each other's reflection with regard to their X_4-axis.)

This is the algebraic proof of 1°.

At the same time it has become evident that the meaning of the type (I') is an arbitrary equiangular double rotation to the right preceded by a reflection according to the X_4-axis (that is the X_4-axis of the initial position) or an arbitrary equiangular double rotation to the left followed by a reflection according to the X_4-axis (that is the X_4-axis of the final position), and that the meaning of the type (I) is an arbitrary equiangular double rotation to the right followed by a reflection according to the X_4-axis or an arbitrary equiangular double rotation to the left preceded by a reflection according to the X_4-axis.

Thus according to a preceding communication made in this meeting
of the Academy (see page 785) it has been proved that the types (I) ⟦8⟧
and (I') represent the most general symmetric transformation of S_4, of which the determinant type has been simplified only by particular choice of the system of coordinates.

We shall now give a proof for 2°.

Out of the relations (a) for equiangular double rotation to the right we deduce, representing for shortness' sake $\alpha'_2\,\alpha''_3 - \alpha'_3\,\alpha''_2$ etc. by ξ_{23} etc; $\beta'_2\,\beta''_3 - \beta'_3\,\beta''_2$ etc. by χ_{23} etc.:

$$\xi_{23}=(\pi_1^2+\pi_4^2)\chi_{23}+(\pi_1\pi_2+\pi_3\pi_4)\chi_{31}+(\pi_3\pi_1-\pi_2\pi_4)\chi_{12}-(\pi_2^2+\pi_3^2)\chi_{14}+$$
$$+(\pi_1\pi_2+\pi_3\pi_4)\chi_{24}+(\pi_3\pi_1-\pi_2\pi_4)\chi_{34}$$

$$\xi_{31}=(\pi_1\pi_2-\pi_3\pi_4)(\chi_{23}+\chi_{14})+(\pi_2^2+\pi_4^2)\chi_{31}-(\pi_3^2+\pi_1^2)\chi_{24}+$$
$$+(\pi_2\pi_3+\pi_1\pi_4)(\chi_{12}+\chi_{34})$$

$$\xi_{12}=\pi_3\pi_1+\pi_2\pi_4)(\chi_{23}+\chi_{14})+(\pi_2\pi_3-\pi_1\pi_4)(\chi_{31}+\chi_{24})+$$
$$+(\pi_3^2+\pi_4^2)\chi_{12}-(\pi_1^2+\pi_2^2)\chi_{34}.$$

$$\chi_{14}=-(\pi_2^2+\pi_3^2)\chi_{23}+(\pi_1^2+\pi_4^2)\chi_{14}+(\pi_1\pi_2+\pi_3\pi_4)(\chi_{31}+\chi_{24})+$$
$$+(\pi_3\pi_1-\pi_2\pi_4)(\chi_{12}+\chi_{34})$$

$$\chi_{24}=(\pi_1\pi_2-\pi_3\pi_4)(\chi_{23}+\chi_{14})-(\pi_3^2+\pi_1^2)\chi_{31}+(\pi_2^2+\pi_4^2)\chi_{24}+$$
$$+(\pi_2\pi_3+\pi_1\pi_4)(\chi_{12}+\chi_{34})$$

$$\xi_{34}=(\pi_3\pi_1+\pi_2\pi_4)(\chi_{23}+\chi_{14})+(\pi_2\pi_3-\pi_1\pi_4)(\chi_{31}+\chi_{24})-$$
$$-(\pi_1^2+\pi_2^2)\chi_{12}+(\pi_3^2+\pi_4^2)\chi_{34},$$

from which ensues:

$$\xi_{23}+\xi_{14}=(\pi_1^2+\pi_4^2-\pi_2^2-\pi_3^2)(\chi_{23}+\chi_{14})+2(\pi_1\pi_2+\pi_3\pi_4)(\chi_{31}+\chi_{24})+$$
$$+2(\pi_3\pi_1-\pi_2\pi_4)(\chi_{12}+\chi_{34})$$

$$\xi_{31}+\xi_{24}=2(\pi_1\pi_2-\pi_3\pi_4)(\chi_{23}+\chi_{14})+(\pi_2^2+\pi_4^2-\pi_3^2-\pi_1^2)(\chi_{31}+\chi_{24})+$$
$$+2(\pi_2\pi_3+\pi_1\pi_4)(\chi_{12}+\chi_{34})$$

$$\xi_{12}+\xi_{34}=2(\pi_3\pi_1+\pi_2\pi_4)(\chi_{23}+\chi_{14})+2(\pi_2\pi_3-\pi_1\pi_4)(\chi_{31}+\chi_{24})+$$
$$+(\pi_3^2+\pi_4^2-\pi_1^2-\pi_2^2)(\chi_{12}+\chi_{34})$$

$$\xi_{23}-\xi_{14}=\chi_{23}-\chi_{14}$$
$$\xi_{31}-\xi_{24}=\chi_{31}-\chi_{24}$$
$$\xi_{12}-\xi_{34}=\chi_{12}-\chi_{34}.$$

So also if we call λ_{23} etc. the coefficients of position of a plane *before* the equiangular double rotation to the right and μ_{23} etc. the coefficients of position after it:

$$\lambda_{23}+\lambda_{14}=(\pi_1^2+\pi_4^2-\pi_2^2-\pi_3^2)(\mu_{23}+\mu_{14})+2(\pi_1\pi_2+\pi_3\pi_4)(\pi_{31}+\mu_{24})+$$
$$+2(\pi_3\pi_1-\pi_2\pi_4)(\mu_{12}+\mu_{34})$$

$$\lambda_{31}+\lambda_{24}=2(\pi_1\pi_2-\pi_3\pi_4)(\mu_{23}+\mu_{14})+(\pi_2^2+\pi_4^2-\pi_3^2-\pi_1^2)(\mu_{31}+\mu_{24})+$$
$$+2(\pi_2\pi_3+\pi_1\pi_4)(\mu_{12}+\mu_{34})$$

$$\lambda_{12}+\lambda_{34}=2(\pi_3\pi_1+\pi_2\pi_4)(\mu_{23}+\mu_{14})+2(\pi_2\pi_3-\pi_1\pi_4)(\mu_{31}+\mu_{24})+$$
$$+(\pi_3^2+\pi_4^2-\pi_1^2-\pi_2^2)(\mu_{12}+\mu_{34})$$

$$\lambda_{23}-\lambda_{14}=\mu_{23}-\mu_{14}$$
$$\lambda_{31}-\lambda_{24}=\mu_{31}-\mu_{24}$$
$$\lambda_{12}-\lambda_{34}=\mu_{12}-\mu_{3}$$

In an analogous way we deduce from the relations (b) for an equiangular double rotation to the left the following relations between the coefficients of position of a plane before and after the rotation:

$$\lambda_{23} + \lambda_{14} = \mu_{23} + \mu_{14}$$
$$\lambda_{31} + \lambda_{24} = \mu_{31} + \mu_{24}$$
$$\lambda_{12} + \lambda_{34} = \mu_{12} + \mu_{34}$$

$$\lambda_{23} - \lambda_{14} = (\pi_1^2 + \pi_4^2 - \pi_2^2 - \pi_3^2)(\mu_{23} - \mu_{14}) + 2(\pi_1\pi_2 + \pi_3\pi_4)(\mu_{31} - \mu_{24}) + \\ + 2(\pi_3\pi_1 - \pi_2\pi_4)(\mu_{12} - \mu_{34})$$

$$\lambda_{31} - \lambda_{24} = 2(\pi_1\pi_2 - \pi_3\pi_4)(\mu_{23} - \mu_{14}) + (\pi_2^2 + \pi_4^2 - \tau_3^2 - \pi_1^2)(\mu_{31} - \mu_{24}) + \\ + 2(\pi_2\pi_3 + \pi_1\pi_4)(\mu_{12} - \mu_{34})$$

$$\lambda_{12} - \lambda_{34} = 2(\pi_3\pi_1 + \pi_2\pi_4)(\mu_{23} - \mu_{14}) + 2(\pi_2\pi_3 - \pi_1\pi_4)(\mu_{31} - \mu_{24}) + \\ + (\pi_3^2 + \pi_4^2 - \pi_1^2 - \pi_2^2)(\mu_{12} - \mu_{34}).$$

As now $\Sigma\,(\lambda_{23} + \lambda_{14})^2 = 1$ and $\Sigma\,(\lambda_{23} - \lambda_{14})^2 = 1$ and the determinant

$$\left.\begin{matrix} \pi_1^2 + \pi_4^2 - \pi_2^2 - \pi_3^2 & 2(\pi_1\pi_2 + \pi_3\pi_4) & 2(\pi_3\pi_1 - \pi_2\pi_4) \\ 2(\pi_1\pi_2 - \pi_3\pi_4) & \pi_2^2 + \pi_4^2 - \pi_3^2 - \pi_1^2 & 2(\pi_2\pi_3 + \pi_1\pi_4) \\ 2(\pi_3\pi_1 + \pi_2\pi_4) & 2(\pi_2\pi_3 - \pi_1\pi_4) & \pi_3^2 + \pi_4^2 - \pi_1^2 - \pi_2^2 \end{matrix}\right| \quad (IV)$$

represents the general congruent threedimensional transformation about a fixed point expressed in the four parameters of homogeneity, we can regard the motion group with the determinant type (II) as a congruent motion group of the twodimensional Euclidean star of the $(\lambda_{bc} + \lambda_{a4})$'s and the motion group with the determinant type (III) as a congruent motion group of the twodimensional Euclidean star of the $(\lambda\ \ - \lambda_{a4})$'s; namely according to the determinant type (IV) about an axis with cosines of direction

$$\frac{\pi_1}{\sqrt{1 - \pi_4^2}},\ \frac{\pi_2}{\sqrt{1 - \pi_4^2}},\ \frac{\pi_3}{\sqrt{1 - \pi_4^2}}$$

over an angle equal to $2\ arc\ cos\ \pi_4$.

Let us call the S_3 of the $(\lambda_{bc} + \lambda_{a4})$'s "the representing space to the right" or the S_r of S_4 and the S_3 of the $(\lambda_{bc} - \lambda_{a4})$'s the "representing space to the left" of the S_l of S_4, then we find that to two equiangular double rotations to the right (left) $(\pi_1'\ \pi_2'\ \pi_3'\ \pi_4')$ and $(\pi_1''\ \pi_2''\ \pi_3''\ \pi_4'')$ of S_4 whose angles of rotation are $arc\ cos\ \pi_4'$ and $arc\ cos\ \pi_4''$ and whose systems of planes of rotation make an angle with each other equal to

$$arc\ cos\ \frac{\pi_1'\pi_1'' + \pi_2'\pi_2'' + \pi_3'\pi_3''}{\sqrt{1 - \pi_4'^2}\ .\ \sqrt{1 - \pi_4''^2}},$$ (see Proceedings, March, 1904, page 724) [7]

correspond two rotations of S_r (S_l) over angles $2\ arc\ cos\ \pi'_4$ and $2\ arc\ cos\ \pi_4''$, whose axes make an angle equal to $arc\ cos\ \dfrac{\pi_1'\pi_1'' + \pi_2'\pi_2'' + \pi_3'\pi_3''}{\sqrt{1 - \pi_4'^2}\ .\ \sqrt{1 - \pi_4''^2}}$

with each other. So to congruent combinations in the group of the equiangular double rotations to the right (left) in S_4 correspond congruent combinations in the motion group of S_r (S_l). As moreover the

$$\begin{array}{c} \lambda_{23} + \lambda_{14} \\ \lambda_{31} + \lambda_{24} \\ \lambda_{12} + \lambda_{34} \end{array}$$

of a plane are the cosines of direction of the representant of the system equiangular to the right with that plane with respect to the
[[7]] system of axes $OX_r\,Y_r\,Z_r$ (defined Proceedings March 1904, p. 728), and likewise

$$\begin{array}{c} \lambda_{23} - \lambda_{14} \\ \lambda_{31} - \lambda_{24} \\ \lambda_{12} - \lambda_{34} \end{array}$$

the cosines of direction of the representant of the system equiangular to the left with that plane with respect to the system of axes $OX_l\,Y_l\,Z_l$ (defined in the same place) the S_r and S_l introduced just now prove to be identical with those introduced here formerly (see
[[7]] Proceedings March, 1904, p. 725) so that they represent not only by their motions the equiangular motion groups of S_4 to the right and to the left, but also by their vectors the systems of planes equiangular to the right and to the left (with direction of rotation) of S_4 and so that the angle of the representing vectors is the angle of the systems of planes themselves.

So also to congruent combinations in the twodimensional manifoldness formed by the equiangular systems of planes to the right (left) correspond congruent combinations in S_r (S_l). This is an algebraic proof for 2nd. to its full extent.

This deduction has at the same time made clear the meaning of the four parameters of homogeneity for the general congruent threedimensional transformation about a fixed point, namely the cosines of direction of the vector indicating the corresponding equiangular double rotation to the right (left) of an S_4 of which this S_3 is the S_r (S_l) and the system of axes in S_3 the system $OX_r\,Y_r\,Z_r$ ($OX_l\,Y_l\,Z_l$).

[[34]]

NOTES

⟦1⟧ The Dutch version was communicated at the meeting of 23 April 1904. The paper extends 1904 A2, 1904 B2. Cp. in particular 1904 A2 ⟦1⟧.

⟦2⟧ E. Jahnke 1904. Cp. 1904 A2 ⟦1⟧.

⟦3⟧ 'manifoldness' – a literal translation of German 'Mannigfaltigkeit' or French 'multiplicité' – is better replaced by 'variety' or 'manifold'.

⟦4⟧ E. Jahnke 1896, 1897, 1902.

⟦5⟧ F. Caspary 1883.

⟦6⟧ E. Jahnke 1902.

⟦7⟧ Brouwer 1904 A2.

⟦8⟧ Brouwer 1904 B2.

1906 A2 **Mathematics.** — "*Polydimensional Vectordistributions*". [1]) By L. E. J.
[[1]] BROUWER. (Communicated by Prof. D. J. KORTEWEG.)

Let us call the plane space in which to operate S_n; we suppose in it a rectangular system of coordinates in which a C_p represents a coordinatespace of p dimensions. Let a ${}^pX$-distribution be given in S_n; i.e. let in each point of S_n a p-dimensional system of vectors be given. By $X_{\alpha_1\alpha_2\ldots\alpha_p}$ we understand the vector component parallel to C_p indicated by the indices, whilst as positive sense is assumed the one corresponding to the indicatrix indicated by the sequence of the indices. By interchanging two of the indices the sense of the indicatrix changes, hence the sign of the vectorcomponent.

Theorem 1. The integral of ${}^pX$ in S_n over an arbitrary curved bilateral closed S_p is equal to the integral of ${}^{p+1}Y$ over an arbitrary curved S_{p+1}, enclosed by S_p as a boundary, in which ${}^{p+1}Y$ is determined by

[[2]]
$$Y_{\alpha_1\alpha_2\ldots\alpha_p\alpha_{p+1}} = \sum_{\alpha_{q_1}=\alpha_1,\alpha_2,\ldots\alpha_{p+1}} \frac{\partial X_{\alpha_{q_2}\alpha_{q_3}\ldots\alpha_{q_{p+1}}}}{\partial x_{\alpha_{q_1}}},$$

where for each of the terms of the second member the indicatrix $(a_{q_1}a_{q_2}\ldots a_{q_p}a_{q_{p+1}})$ has the same sense as $(a_1\,a_2\ldots a_{p+1})$. We call the vector Y the *first derivative* of ${}^pX$.

Proof. We suppose the limited space S_{p+1} to be provided with curvilinear coordinates $u_1\ldots u_{p+1}$ determined as intersection of curved C_p's, i. e. curved coordinatespaces of p-dimensions. We suppose the system of curvilinear coordinates to be inside the boundary without singularities and the boundary with respect to those coordinates to be everywhere convex.

The integral element of ${}^{p+1}Y$ becomes when expressed in differential quotients of ${}^pX$:

$$\sum_{Y_{\alpha_1..\alpha_{p+1}}} \sum_{\alpha_{q_1}=\alpha_1\ldots\alpha_{p+1}} \frac{\partial X_{\alpha_{q_2}\ldots\alpha_{q_{p+1}}}}{\partial x_{\alpha_{q_1}}} \begin{vmatrix} \dfrac{\partial x_{\alpha_1}}{\partial u_1} & \cdots\cdots & \dfrac{\partial x_{\alpha_{p+1}}}{\partial u_1} \\ \vdots & & \vdots \\ \dfrac{\partial x_{\alpha_1}}{\partial u_{p+1}} & \cdots\cdots & \dfrac{\partial x_{\alpha_{p+1}}}{\partial u_{p+1}} \end{vmatrix} du_1\ldots du_{p+1}.$$

1) The Dutch original contains a few errors (see Erratum at the end of Verslagen 31 Juni 1906), which have been rectified in this translation.

[[36]]

We now unite all terms containing one of the components of ${}^pX$, e.g. $X_{123\ldots p}$. We then find:

$$\frac{\partial X_{123\ldots p}}{\partial x_{p+1}} \begin{vmatrix} \frac{\partial x_{p+1}}{\partial u_1} & \frac{\partial x_1}{\partial u_1} & \cdots\cdots & \frac{\partial x_p}{\partial u_1} \\ \vdots & \vdots & & \vdots \\ \frac{\partial x_{p+1}}{\partial u_{p+1}} & \frac{\partial x_1}{\partial u_{p+1}} & \cdots\cdots & \frac{\partial x_p}{\partial u_{p+1}} \end{vmatrix} du_1 \ldots du_{p+1} +$$

$$+ \frac{\partial X_{123\ldots p}}{\partial x_{p+2}} \begin{vmatrix} \frac{\partial x_{p+2}}{\partial u_1} & \frac{\partial x_1}{\partial u_1} & \cdots\cdots & \frac{\partial x_p}{\partial u_1} \\ \vdots & \vdots & & \vdots \\ \frac{\partial x_{p+2}}{\partial u_{p+1}} & \frac{\partial x_1}{\partial u_{p+1}} & \cdots\cdots & \frac{\partial x_p}{\partial u_{p+1}} \end{vmatrix} du_1 \ldots du_{p+1} +$$

$$+ \ldots (n-p \text{ terms}).$$

If we add to these the following terms with the value 0:

$$\frac{\partial X_{123\ldots p}}{\partial x_1} \begin{vmatrix} \frac{\partial x_1}{\partial u_1} & \frac{\partial x_1}{\partial u_1} & \cdots\cdots & \frac{\partial x_p}{\partial u_1} \\ \vdots & \vdots & & \vdots \\ \frac{\partial x_1}{\partial u_{p+1}} & \frac{\partial x_1}{\partial u_{p+1}} & \cdots\cdots & \frac{\partial x_p}{\partial u_{p+1}} \end{vmatrix} du_1 \ldots du_{p+1} +$$

$$+ \frac{\partial X_{123\ldots p}}{\partial x_2} \begin{vmatrix} \frac{\partial x_2}{\partial u_1} & \frac{\partial x_1}{\partial u_1} & \cdots\cdots & \frac{\partial x_p}{\partial u_1} \\ \vdots & \vdots & & \vdots \\ \frac{\partial x_2}{\partial u_{p+1}} & \frac{\partial x_1}{\partial u_{p+1}} & \cdots\cdots & \frac{\partial x_p}{\partial u_{p+1}} \end{vmatrix} du_1 \ldots du_{p+1} +$$

$$+ \ldots (p \text{ terms}),$$

the n-terms can be summed up as:

$$\begin{vmatrix} \frac{\partial X_{123\ldots p}}{\partial u_1} du_1 & \frac{\partial x_1}{\partial u_1} du_1 & \cdots\cdots & \frac{\partial x_p}{\partial u_1} du_1 \\ \vdots & \vdots & & \vdots \\ \frac{\partial X_{123\ldots p}}{\partial u_{p+1}} du_{p+1} & \frac{\partial x_1}{\partial u_{p+1}} du_{p+1} & \cdots\cdots & \frac{\partial x_p}{\partial u_{p+1}} du_{p+1} \end{vmatrix}$$

Let us suppose this determinant to be developed according to the first column, let us then integrate partially each of the terms of the development according to the differential quotient of $X_{123\ldots p}$, appearing in it; there will remain under the $(p+1)$-fold integration sign $p(p+1)$ terms neutralizing each other two by two. Thus for instance:

$$du_1 \ldots du_{p+1} \left| \begin{array}{c|ccc} & \frac{\partial^2 x_1}{\partial u_1 \partial u_{p+1}} & \cdots\cdots & \frac{\partial^2 x_p}{\partial u_1 \partial u_{p+1}} \\ & \frac{\partial x_1}{\partial u_2} & \cdots\cdots & \frac{\partial x_p}{\partial u_2} \\ & \vdots & & \vdots \\ & \frac{\partial x_1}{\partial u_p} & \cdots\cdots & \frac{\partial x_p}{\partial u_p} \\ \hline X_{123\ldots p} & & & \end{array} \right|$$

$$\text{and } du_1 \ldots du_{p+1} \left| \begin{array}{c|ccc} X_{123\ldots p} & & & \\ \hline & \frac{\partial x_1}{\partial u_2} & \cdots\cdots & \frac{\partial x_p}{\partial u_2} \\ & \vdots & & \vdots \\ & \frac{\partial x_1}{\partial u_p} & \cdots\cdots & \frac{\partial x_p}{\partial u_p} \\ & \frac{\partial^2 x_1}{\partial u_1 \partial u_{p+1}} & \cdots\cdots & \frac{\partial^2 x_p}{\partial u_1 \partial u_{p+1}} \end{array} \right| ,$$

as they transform themselves into one another by interchangement of two rows of the matrix-determinant.

So the p-fold integral remains only, giving under the integration sign

$$X_{123\ldots p} \left| \begin{array}{cccc} \pm 1 & \frac{\partial x_1}{\partial u_1} du_1 & \cdots\cdots & \frac{\partial x_p}{\partial u_1} du_1 \\ \vdots & \vdots & & \vdots \\ \pm 1 & \frac{\partial x_1}{\partial u_{p+1}} du_{p+1} & \cdots & \frac{\partial x_p}{\partial u_{p+1}} du_{p+1} \end{array} \right|$$

to be integrated over the boundary, whilst in a definite point of that boundary the hth term of the first column gets the sign $+$ when for the coordinate u_p the point lies on the positive side of the boundary.

Let us now find the integral of $X_{123\ldots p}$ over the boundary and let us for the moment suppose ourselves on the part of it lying for all

u's on the positive side. The indicatrix is in the sense $u_1\, u_2 \ldots u_{p+1}$ and if we integrate $X_{123\ldots p}$ successively over the components of the elements of boundary according to the curved C_p's we find:

$$\sum \int X_{123\ldots p} \begin{vmatrix} \frac{\partial x_1}{\partial u_{\alpha_1}} du_{\alpha_1} & \cdots\cdots & \frac{\partial x_p}{\partial u_{\alpha_1}} du_{\alpha_1} \\ \vdots & & \vdots \\ \frac{\partial x_1}{\partial u_{\alpha_p}} du_{\alpha_p} & \cdots\cdots & \frac{\partial x_p}{\partial u_{\alpha_p}} du_{\alpha_p} \end{vmatrix},$$

where $(\alpha_{p+1}\, \alpha_1 \ldots \alpha_p) = (1\,2\,3 \ldots p\,(p+1))$; so that we can write as well

$$\sum_{q=1,2\ldots(p+1)} \int \left| X_{123\ldots p} \begin{vmatrix} \frac{\partial x_1}{\partial u_1} du_1 & \cdots\cdots & \frac{\partial x_p}{\partial u_1} du_1 \\ \vdots & & \vdots \\ \frac{\partial x_1}{\partial u_{q-1}} du_{q-1} & \cdots & \frac{\partial x_p}{\partial u_{q-1}} du_{q-1} \\ \hline \frac{\partial x_1}{\partial u_{q+1}} du_{q+1} & \cdots & \frac{\partial x_p}{\partial u_{q+1}} du_{q+1} \\ \vdots & & \vdots \\ \frac{\partial x_1}{\partial u_{p+1}} du_{p+1} & \cdots & \frac{\partial x_p}{\partial u_{p+1}} du_{p+1} \end{vmatrix} \right| .$$

$$\text{or: } \int X_{123\ldots p} \begin{vmatrix} 1 & \frac{\partial x_1}{\partial u_1} du_1 & \cdots\cdots & \frac{\partial x_p}{\partial u_1} du_1 \\ \vdots & \vdots & & \vdots \\ 1 & \frac{\partial x_1}{\partial u_{p+1}} du_{p+1} & \cdots & \frac{\partial x_p}{\partial u_{p+1}} du_{p+1} \end{vmatrix} .$$

If we now move to other parts of the boundary we shall continually see, where we pass a limit of projection with respect to one of the coordinates u, the projection of the indicatrix on the relative curved C_p change in sense.

So in an arbitrary point of the boundary the integral is found in the same way as on the entirely positive side; we shall find only, that for each coordinate u_q for which we are on the negative side, the corresponding term under the sign Σ will have to be taken negatively, by which we shall have shown the equality of the p-fold integral of ${}^{p}X$ over the boundary and the $(p+1)$ fold integral of ${}^{p+1}Y$ over the bounded S_{p+1}.

We can also imagine the scalar values of ${}^{p}X$ set off along the normal-S_{n-p}'s. As such the integral over an arbitrary curved bilateral closed S_{n-p} can be reduced to an $(n-p+1)$-dimensional vector over a curved S_{n-p+1}, bounded by S_{n-p}. If again we set off the scalar values of that vector along its normal-S_{p-1}, the vector ${}^{p-1}Z$ appears, which we shall call the *second derivative* of ${}^{p}X$. For the component vectors of ${}^{p-1}Z$ we find:

$$Z_{\alpha_1 \ldots \alpha_{p-1}} = \sum_{\alpha_q = \alpha_p, \alpha_{p+1} \ldots \alpha_n} - \frac{\partial X_{\alpha_q \alpha_1 \ldots \alpha_{p-1}}}{\partial x_{\alpha_q}}.$$

The particularity may appear that one of the derivatives becomes 0. If the first derivative of an ${}^{m}X$ is zero we shall speak of an ${}_{m-1}^{\;m}X$, if the second is zero of an ${}_{m+1}^{\;m}X$.

Theorem 2. The first derivative of a ${}^{p}X$ is a ${}^{p+1}_{\;\;p}X$, the second a ${}^{p-1}_{\;\;p}X$; in other words the process of the first derivation as well as that of the second applied twice in succession gives zero.

The demonstration is simple analytically, but also geometrically the theorem is proved as follows:

Find the integral of the first derivative of ${}^{p}X$ over a closed S_{p+1}, then we can substitute for the addition given by an S_{p+1} element the integral of ${}^{p}X$ along the bounding S_p of that element. Along the entire S_{p+1} each element of those S_p boundaries is counted twice with opposite indicatrix, so that the integral must vanish.

The analogous property for the second derivative is apparent, when we evaluate the integral of the normalvector over a closed S_{n-p+1}.

By *total derivative* we shall understand the sum of the first and
[[3]] second derivatives and we shall represent the operation of total derivation by ∇.

Theorem 3. $\nabla^2 = -\sum_{h=1}^{h=n} \frac{\partial^2}{\partial x_h^2}$.

Proof. In the first place it is clear from theorem 2 that the vector ∇^2 is again a ${}^{p}X$. Let us find its component $X_{12\ldots p}$.

The first derivative supplies the following terms

[[40]]

$$T_1 = -\sum_{q=p+1}^{q=n} \frac{\partial Y_{q1\dots p}}{\partial x_q},$$

where

$$Y_{q1\dots p} = \sum_{u=1}^{u=p} \pm \frac{\partial X_{q12\dots(u-1)(u+1)\dots p}}{\partial x_u}$$

$$\left(+ \text{ sign for } (uq\,12\,..\,(u-1)\,(u+1)\,\dots\,p) = (q\,1\,\dots\,p)\right)$$

$$+ \frac{\partial X_{12\dots p}}{\partial x_q}$$

So

$$T_1 = \sum_{u=1}^{u=p} \sum_{q=p+1}^{q=n} \mp \frac{\partial^2 X_{q12\dots(u-1)(u+1)\dots p}}{\partial x_u\, \partial x_q}$$

$$\left(- \text{ sign for } (uq\,12\,\dots(u-1)(u+1)\dots p) = (q\,1\,\dots\,p)\right)$$

$$- \sum_{q=p+1}^{q=n} \frac{\partial^2 X_{12\dots p}}{\partial x_q^2}.$$

The second derivative supplies the terms

$$T_2 = \sum_{u=1}^{u=p} \pm \frac{\partial Z_{12\dots(u-1)(u+1)\dots p}}{\partial x_u}$$

$$\Big(+ \text{ sign for } (u\,12\,\dots\,(u-1)\,(u+1)\,\dots\,p) = (12\,\dots\,p)$$

$$\text{or for } (qu\,1\,\dots\,(u-1)\,(u+1)\,\dots\,p) = (q\,12\,\dots\,p)\Big),$$

where $Z_{12\dots(u-1)(u+1)\dots p} = \sum_{q=p+1}^{q=n} - \frac{\partial X_{q12..(u-1)(u+1)\dots p}}{\partial x_q} \mp$

$$\mp \frac{\partial X_{12\,..p}}{\partial x_u} \left(- \text{ sign for } (u\,12\,\dots\,(u-1)\,(u+1)\,\dots\,p) = (12\,\dots\,p)\right).$$

So $T_2 = \sum_{u=1}^{u=p} \sum_{q=p+1}^{q=n} \mp \frac{\partial^2 X_{q12\,..(u-1)(u+1)\,..\,p}}{\partial x_u \partial x_q}$

$$\left(- \text{ sign for } (qu1\,..\,(u-1)(u+1)\,..\,p) = (q12\,..\,p)\right)$$

$$- \sum_{u=1}^{u=p} \frac{\partial^2 X_{12\,..\,p}}{\partial x_u^2}.$$

The terms under the sign $\Sigma\,\Sigma$ of T_1 are annulled by those of T_2, so that only

$$-\sum_{h=1}^{h=n} \frac{\partial^2 X_{12..p}}{\partial x_h{}^2}$$

is left.

Corollary. If a vectordistribution ${}^pV$ is given, then the vector-distribution $\int \frac{V dv}{k_n(n-2)r^{n-2}}$, integrated over the entire space, has for second derivative V. (if $k_n r^{n-1}$ expresses the surface of the ${}^{n-1}$-sphere in S_n).

The theorem also holds for a distribution of sums of vectors of various numbers of dimensions, e.g. quaternions.

We shall say that a vectordistribution has the *potential property* when its scalar values satisfy the demands of vanishing at infinity, which must be put to a scalar potential function in S_n. [1]) And in the following we shall suppose that the vectordistribution from which we start possesses the potential property. Then holds good:

Theorem 4. A vectordistribution V is determined by its total derivative of the second order.

For, each of the scalar values of V is uniformly determined by the scalar values of $\nabla^2 V$, from which it is derived by the operation $\int \frac{dv}{k_n(n-2)r^{n-2}}$.

Theorem 5. A vectordistribution is determined uniformly by its total derivative of the first order.

For, from the first total derivative follows the second, from which according to the preceding theorem the vector itself.

[[4]] We shall say that a vectordistribution has the *field property*, if the scalar values of the total derivative of the first order satisfy the demands which must be put to an agens distribution of a scalar potential function in S_n. And in the following we shall suppose that the vectordistribution under consideration possesses the field property. Then we have:

Theorem 6. Each vectordistribution is to be regarded as a total

---

[1]) Generally the condition is put: the function must become infinitesimal of order $n-2$ with respect to the reciprocal value of the distance from the origin. We can, however, prove, that the being infinitesimal only is sufficient.

derivative, in other words each vectordistribution has a potential and that potential is uniformly determined by it.

Proof. Let V be the given distribution, then

$$P = \int \frac{\nabla V \,.\, dv}{k_n(n-2)r^{n-2}}$$

is its potential. For $\nabla^2 P = \nabla V$, or $\nabla(\nabla P) = \nabla V$, or $\nabla P = V$. Farther follows out of the field property of V, that P is uniformly determined as ∇^{-2} of ∇V, so as ∇ of V. So P has clearly the potential property; it need, however, not have the field property.

N.B. A distribution not to be regarded here, because it has not the field property, though it has the potential property, is e. g. the fictitious force field of a single agens point in S_2. For, here we have not a potential vanishing at infinity — and as such determined uniformly. The magnetic field in S_2 has field property and also all the fields of a single agens point in S_3 and higher spaces.

Let us call $\overset{1}{\nabla} V$ the first derivative of ${}^pV$ and $\overset{2}{\nabla} V$ the second; we can then break up ${}^pV$ into

$$\nabla \int \frac{\overset{1}{\nabla} V \,.\, dv}{k_n\,(n-2)r^{n-2}} = \nabla P_1 = \overset{2}{\nabla} P_1 = {}^{\;\;p}_{p+1}V.$$

and

$$\nabla \int \frac{\overset{2}{\nabla} V \,.\, dv}{k_n\,(n-2)r^{n-2}} = \nabla P_2 = \overset{1}{\nabla} P_2 = {}^{\;\;p}_{p-1}V.$$

From the preceding follows immediately:

Theorem 7. Each ${}^{\;\;p}_{p-1}V$ has as potential a ${}^{p-1}_{\;\;p}V$. Each ${}^{\;\;p}_{p+1}V$ has as potential a ${}^{p+1}_{\;\;p}V$.

We can indicate of the ${}^{\;\;p}_{p+1}V$ the elementary distribution, i. e. that particular ${}^{\;\;p}_{p+1}V$ of which the arbitrary S_n integral must be taken to obtain the most general ${}^{\;\;p}_{p+1}V$.

For, the general ${}^{\;\;p}_{p+1}V$ is $\overset{2}{\nabla}$ of the general ${}^{p+1}V$, so it is the general S_n integral of the $\overset{2}{\nabla}$ of an isolated $(p+1)$-dimensional vector, which, as is easily seen geometrically, consists of equal ^{p}vectors in the surface of a ^{p}sphere with infinitesimal radius described round the point of the given isolated vector in the R_{p+1} of the vector.

In like manner the general ${}^{\;\;p}_{p-1}V$ is the $\overset{1}{\nabla}$ of the general ${}^{p-1}V$, so it is the general S_n integral of the $\overset{1}{\nabla}$ of an isolated ${}^{p-1}$vector, consisting of equal ^{p}vectors normal to the surface of an ${}^{n-p}$sphere with infinitesimal radius described round the point of the given isolated vector in the R_{n-p+1}, normal to that vector.

From this follows:

Theorem 8. The general ${}^{p}V$ is an arbitrary integral of elementary fields E_1 and E_2, where:

$$E_1 = \nabla \cdot \int \frac{\overset{p-1}{{}_p Z}\, dv}{k_n(n-2)r^{n-2}}, \text{ where } \overset{p-1}{{}_p Z} \text{ consists of the } {}^{p-1}\text{vectors in the}$$

surface of an infinitesinal ${}^{p-1}$sphere Sp_z, ... (1)

$$E_2 = \nabla \cdot \int \frac{\overset{p+1}{{}_p Y}\, dv}{k_n(n-2)r^{n-2}}, \text{ where } \overset{p+1}{{}_p Y} \text{ consists of the } {}^{p+1}\text{vectors normal}$$

to the surface of an infinitesinal ${}^{n-p-1}$sphere Sp_y. . . . (2)

For the rest the fields E_1 and E_2 must be of a perfectly identical structure at finite distance from their origin; for two fields E_1 and E_2 with the same origin must be able to be summed up to an isolated ^{p}vector in that point.

We can call the spheres Sp_y and Sp_z with their indicatrices the *elementary vortex systems* Vo_y and Vo_z. A field is then uniformly determined by its elementary vortex systems and can be regarded as caused by those vortex systems.

We shall now apply the theory to some examples.

The force field in S_2.

The field E_1. The elementary sphere Sp_z becomes here two points lying quite close to each other, the vortex system Vo_z passes into two equal and opposite scalar values placed in those two points. It furnishes a scalar potential $\frac{\cos \varphi}{r}$ in which φ denotes the angle of the radiusvector with the S_1 of Vo_z, i. e. the line connecting the two points. The elementary field is the (first) derivative of the potential (the gradient); it is the field of an agens double point in two dimensions.

The field E_2. The elementary sphere Sp_y again consists of two points lying in close vicinity, the elementary vortex system Vo_y has in those two points two equal and opposite planivectors. The planivector potential (determined by a scalar value) here again becomes $\frac{\cos \varphi}{r}$; so the field itself is obtained by allowing all the vectors of a field E_1 to rotate 90°. As on the other hand it has to be of an identical structure to E_1 outside the origin we may call the field E_1 resp. E_2 "dual to itself".

In our space the field E_1 can be realized as that of a plane, infinitely long and narrow magnetic band with poles along the edges; the field E_2 as that of two infinitely long parallel straight electric currents, close together and directed oppositely.

The planivector (vortex) field in S_4.

The field E_1. The elementary sphere Sp_z is a circlet, the elementary vortex system Vo_z a current along it. It furnishes a linevector potential $= \frac{\sin \varphi}{r^3}$ directed along the circles which project themselves on the plane of Vo_z as circles concentric to Vo_z, and where φ is the angle of the radiusvector with the normal plane of Vo_z. The field is the first derivative (rotation) of this potential.

The field E_2. The elementary sphere Sp_y is again a circlet, the elementary vortex system Vo_y assumes in the points of that circlet equal ^{3}vectors normal to it. The ${}^3V$-potential consists of the ${}^3V$'s normal to the potential vectors of a field E_1; the field E_2 is thus obtained by taking the normal planes of all planivectors of a field E_1. As on the other hand E_1 and E_2 are of the same identical structure outside the origin, we can say here again, that the field E_1 resp. E_2 is dual to itself.

So we can regard the vortex field in S_4 as caused by elementary circular currents of two kinds; two equal currents of a different kind cause vortex fields of equal structure, but one field is perfectly normal to the other.

So if of a field the two generating systems of currents are identical, it consists of isosceles double-vortices.

The force field in S_3.

The field E_1. Vo_z gives a double point, causing a scalar potential $\frac{\cos \varphi}{r^2}$, where φ is the angle of the radiusvector with the axis of the double point; the derivative (gradient) gives the wellknown field of an elementary magnet.

The field E_2. Vo_y consists of equal planivectors normal to a small circular current. If we represent the planivector potential by the linevector normal to it, we shall find for that linevector $\frac{\sin \varphi}{r^2}$ directed along the circles, which project themselves on the plane of Vo_z as circles concentric to Vo_z, and where φ is the angle of the radius-vector with the normal on the circular current. The field E_2 is the second derivative of the planivector potential, i.e. the rotation of the normal linevector.

According to what was derived before the field E_2 of a small circular current is outside the origin equal to the field E_1 of an elementary magnet normal to the current.

In this way we have deduced the principle that an arbitrary force field can be regarded as generated by elementary magnets and elementary circuits. A finite continuous agglomeration of elementary magnets furnishes a system of finite magnets; a finite continuous
[5] agglomeration of elementary circuits furnishes a system of finite closed currents, i.e. of finite dimensions; the linear length of the separate currents may be infinite.

Of course according to theorem 6 we can also construct the scalar potential out of that of single agens points $\left(\frac{1}{4\pi}\times\right.$ the second derivative of the field), and the vector potential out of that of rectilinear elements of current (perpendicular to $\frac{1}{4\pi}\times$ the first derivative of the field), but the fictitious "field of a rectilinear element of current" has everywhere rotation, so it is the real field of a rather complicated distribution of current. A field having as its only current a rectilinear element of current, is not only physically but also mathematically impossible. A field of a single agens point though physically perhaps equally impossible, is mathematically just possible in the Euclidean space in consequence of its infinite dimensions, as the field of a magnet of which one pole is removed at infinite distance.

In hyperbolic space also the field of a single agens point is possible for the same reason, but in elliptic and in spherical space being finite it has become as impossible as the field of a rectilinear
[6] element of current. The way in which Schering (Göttinger Nachr. 1870, 1873; compare also Fresdorf Diss. Göttingen 1873; Opitz Diss. Göttingen 1881) and Killing (Crelle's Journ. 1885) construct the potential of elliptic space, starting from the supposition that as unity of field must be possible the field of a single agens point,.
[7] leads to absurd consequences, to which Klein (Vorlesungen über Nicht-Euklidische Geometrie) has referred, without, however, proposing an improvement. To construct the potential of the elliptic and spherical spaces nothing but the field of a double point must be assumed as unity of field, which would lead us too far in this

paper but will be treated more in details in a following communication.

With the force field in S_3 the vortex field in S_3 dual to it has been treated at the same time. It is an integral of vortex fields as they run round the force lines of an elementary magnet and as they run round the induction lines of an elementary circuit.

The force field in S_n .

The field E_1. Vo_z again gives a double point, which furnishes a scalar potential $\frac{\cos\varphi}{r^{n-1}}$, where φ is the angle between radiusvector and axis of the double point; its gradient gives what we might call the field of an elementary magnet in S_n.

The field E_2. Vo_y consists of equal planivectors normal to a small ${}^{n-2}$sphere Sp_y. To find the planivector potential in a point P, we call the perpendicular to the S_{n-1} in which Sp_y is lying OL, and the plane LOP the "meridian plane" of P; we call φ the angle LOP and OQ the perpendicular to OL drawn in the meridian plane. We then see that all planivectors of Vo_y have in common with that meridian plane the direction OL, so they can be decomposed each into two components, one lying in the meridian plane and the other cutting that meridian plane at right angles. The latter components, when divided by the n — 2nd power of their distance to P, and placed in P, neutralize each other two by two; and the former consist of pairs of equal and opposite planivectors directed parallel to the meridian plane and at infinitely small distance from each other according to the direction OQ. These cause in P a planivector potential lying in the meridian plane $= c\,\frac{\sin\varphi}{r^{n-1}}$. The field E_2 is of this potential the $\overline{\nabla} = \overline{\nabla^{2}}$, and outside the origin is identical to the field of an elementary magnet along OL.

The force field in S_n can be regarded as if caused 1st. by magnets, 2nd. by vortex systems consisting of the plane vortices erected normal to a small ${}^{n-2}$sphere. We can also take as the cause the spheres themselves with their indicatrices and say that the field is formed by magnets and vortex spheres of n—2 dimensions (as in S_3 the cause is found in the closed electric current instead of in the vortices round about it).

Here also fields of a single plane vortex element are impossible. Yet we can speak of the fictitious "field of a single vortex" although that really has a vortex i.e. a rotation vector everywhere in space. We can say namely:

If of a force field in each point the divergence (a scalar) and the rotation (a planivector) are given, then it is the ∇ of a potential: $\int \frac{div.\, dv}{k_n(n-2)r^{n-2}} + \int \frac{rot.\, dv}{k_n(n-2)r^{n-2}}$; this formula takes the field as an integral of fictitious fields of agens points and of single vortices.

NOTES

⟦1⟧ The Dutch version was communicated at the meeting of 26 May 1906. The English version appeared in the Proceedings of the meeting of 30 June 1906. On the same subject: Brouwer 1906 B2, 1906 C2, 1919 Q2. Compared with the Dutch version the English version contains a large number of corrections.
⟦2⟧ This definition is unusual. It has been improved in Brouwer 1906 B2, p. 116–117, footnote 1. A non-metric form of this Theorem is given in Brouwer 1919 Q2, p. 150.
⟦3⟧ A complement to this definition is given in Brouwer 1919 Q2, p. 154.
⟦4⟧ Compare the definition in Brouwer 1906 B2, p. 118, l. 4.
⟦5⟧ According to Brouwer 1919 Q2, p. 150, footnote 1, 'system of finite closed currents' is to be replaced by 'finite sourceless current system'.
⟦6⟧ E. Schering 1873, Fresdorf 1873, P. Opitz 1881, W. Killing 1885.
⟦7⟧ This very probably refers to one of Klein's *Autographierte Vorlesungshefte*. Cp. F. Klein 1894, 1923 B. See also F. Klein 1890, pp. 557–558 = Abh. I, pp. 366–367.

Mathematics. — *"Remark on Multiple Integrals."* By Prof. L. E. J. Brouwer. 1919 Q2

(Communicated in the meeting of June 28, 1919).

The object of this communication is to make two remarks in conjunction with the first part of my paper: *"Polydimensional Vectordistributions"* [1]) presented at the meeting of May 26, 1906.

§ 1.

I. The proof of the generalisation of Stokes's theorem given l.c. pp. 66—70 provides this generalisation not only in the Euclidean but also in the following ametric form:

Theorem. *In the n-dimensional space* $(x_1, \ldots x_n)$ *let the* $(p-1)$*-tuple integral*

$$\int \Sigma F_{\alpha_1 \ldots \alpha_{p-1}} (x_1, \ldots x_n)\, dx_{\alpha_1} \ldots dx_{\alpha_{p-1}} \quad . \quad . \quad . \quad . \quad (1)$$

be given, where the F's are continuous and finitely and continuously differentiable; consider also the p-tuple integral

$$\int \Sigma f_{\alpha_1 \ldots \alpha_p} (x_1, \ldots x_n)\, dx_{\alpha_1} \ldots dx_{\alpha_p}, \quad . \quad . \quad . \quad . \quad . \quad (2)$$

where

$$f_{\alpha_1 \ldots \alpha_p} = \sum_{\nu=1}^{p} \frac{\partial F_{j_\nu}}{\partial x_{\alpha_\nu}}$$

(*indicatrix* $j_\nu \alpha_\nu$ *aeq. indicatrix* $\alpha_1 \ldots \alpha_p$).

Then, if the two-sided p-dimensional fragment [2]) *G is bounded by the two-sided* $(p-1)$*-dimensional closed space g, the indicatrices of G and g corresponding and both G and g possessing a continuously varying plane tangent space, the value of* (1) *over g is equal to the value of* (2) *over G.*

---

[1]) See Vol. IX, pp. 66—78; we take the definitions modified in accordance with note [1]) on p. 116 l. c. I take this opportunity of pointing out that on p. 76 l.c. lines 13 and 14 "finite sourceless current system" should be read instead of "system of finite closed currents". [[1]]

[2]) Math. Annalen 71, p. 306. [[2]]

Of this theorem, which was enunciated by POINCARÉ[1]) in 1899 already, without proof however, and in a form expressing the rule of signs in a less simple manner, I shall here give the proof anew, editing it somewhat more precisely than in my quoted paper.

II. In the n-dimensional space $(x_1, \dots x_n)$, which we shall denote by S, let the p-tuple integral

$$\int \Sigma\, \varphi_{\alpha_1 \dots \alpha_p} (x_1, \dots x_n)\, dx_{\alpha_1} \dots dx_{\alpha_p} \quad \dots \dots \quad (3)$$

be given, where the φ's are continuous.

Let Q be a two-sided p-dimensional net fragment[2]) provided with an indicatrix and situated in S, σ a base simplex of Q with the vertex indicatrix $A_1 A_2 \dots A_p A_{p+1}$, A an arbitrary point of σ, ${}_A\varphi_{\alpha_1 \dots \alpha_p}$ the value of $\varphi_{\alpha_1 \dots \alpha_p}$ at A, ${}_\nu x_\mu$ the value of x_μ at A_ν. For every σ we determine the value of

$${}_\sigma\varphi \equiv \Sigma\, {}_A\varphi_{\alpha_1 \dots \alpha_p} \cdot {}_\sigma i_{\alpha_1 \dots \alpha_p},$$

where

$${}_\sigma i_{\alpha_1 \dots \alpha_p} \equiv \frac{1}{p\,!} \begin{vmatrix} {}_1x_{\alpha_1} & \dots & {}_{p+1}x_{\alpha_1} \\ \cdot & \cdot\ \cdot\ \cdot & \cdot \\ {}_1x_{\alpha_p} & \dots & {}_{p+1}x_{\alpha_p} \\ 1 & \dots & 1 \end{vmatrix},$$

and where, for different terms under the Σ sign A may be chosen differently; and we sum ${}_\sigma\varphi$ over the different base simplexes of Q. The upper and lower limit between which this latter sum varies on account of the free choice of the points A, we call *the upper and lower value of* (3) *over* Q.

If we now subject Q to a sequence of indefinitely condensing simplicial divisions which give rise to a sequence Q', Q'', ... of net fragments covering Q, then, as ν increases indefinitely, the upper and lower value of (3) over $Q^{(\nu)}$ converge to the same limit, which we call *the value of* (3) *over* Q.

Let F be a two-sided p-dimensional fragment provided with an

⟦3⟧ 1) Les méthodes nouvelles de la mécanique céleste III, p. 10. The significance of the rule of signs here formulated, is apparent only after comparing former publications of the same author from the Acta Mathematica and the Journal de l'École Polytechnique in which the equivalence of the identically vanishing of (2) and the vanishing of (1) over every g was pronounced.

⟦2⟧ 2) Math. Annalen 71, p. 316.

indicatrix and situated in S, f a sequence of indefinitely condensing simplicial approximations P', P'', ... of F corresponding to a category ψ of simplicial divisions. If the values of (3) over P', P'', ... converge to a limit which is independent of the choice of f so far as it is left free by ψ, then we call this limit *the value of* (3) *over* F *for* ψ.

III. We shall now occupy ourselves with the value of (1) over the boundary β of a p-dimensional simplex σ provided with an indicatrix and situated in S. In doing so we take it that the indicatrices of β and σ *correspond*, that is to say, the indicatrix of an arbitrary $(p-1)$-dimensional side of σ is obtained by placing the vertex of σ which does not belong to this side last in the indicatrix of σ and subsequently omitting it. We begin by confining ourselves to the contribution of the single term

$$\int F_{\alpha_1 \ldots \alpha_{p-1}} dx_{\alpha_1} \ldots dx_{\alpha_{p-1}}$$

to the value of (1) over β. By a suitable simplicial division ζ of the space $(x_{\alpha_1}, \ldots x_{\alpha_{p-1}})$ we determine a simplicial division of β, whose base simplexes correspond in pairs to those of $(x_{\alpha_1}, \ldots x_{\alpha_{p-1}})$. The family of those $(n-p+1)$-dimensional spaces within which $x_{\alpha_1}, \ldots x_{\alpha_{p-1}}$ are constant, cuts the plane p-dimensional space in which σ is contained, in a family of straight lines which connect pairs of corresponding base simplexes of β into p-dimensional truncated simplicial prisms. If e_1 and e_2 are a pair of corresponding base simplexes of β, d the concomitant truncated simplicial prism, l a line segment having components $r_{\alpha_p}, \ldots r_{\alpha_n}$ which leads from a point of e_1 to the corresponding point of e_2, then the contribution of the term $\int F_{\alpha_1 \ldots \alpha_{p-1}} dx_{\alpha_1} \ldots dx_{\alpha_{p-1}}$ to the value of (1) over e_1 and e_2 becomes

$$- {}_{e_1}i_{\alpha_1 \ldots \alpha_{p-1}} \cdot \sum_{\nu=p}^{n} r_{\alpha_\nu} \cdot {}_A\left\{\frac{\partial F_{\alpha_1 \ldots \alpha_{p-1}}}{\partial x_{\alpha_\nu}}\right\} + \varepsilon,$$

where A denotes a point of σ which may be different for the different terms under the Σ sign, and ε becomes indefinitely small with respect to ${}_{e_1}i_{\alpha_1 \ldots \alpha_{p-1}}$ for indefinite condensation of ζ.

Now let $B_1 B_2 \ldots B_p$ be a vertex indicatrix of e_1 and ${}_\nu x_\mu$ the value of x_μ at B_ν, then the value of $r_{\alpha_\nu} \cdot {}_{e_1}i_{\alpha_1 \ldots \alpha_{p-1}}$ can be expressed as

$$\frac{1}{(p-1)!}\begin{vmatrix} {}_1x_{\alpha_1} - {}_px_{\alpha_1} & \cdots & {}_{p-1}x_{\alpha_1} - {}_px_{\alpha_1} & 0 \\ \cdot & \cdot & \cdot & \cdot \\ {}_1x_{\alpha_{p-1}} - {}_px_{\alpha_{p-1}} & \cdots & {}_{p-1}x_{\alpha_{p-1}} - {}_px_{\alpha_{p-1}} & 0 \\ {}_1x_{\alpha_\nu} - {}_px_{\alpha_\nu} & \cdots & {}_{p-1}x_{\alpha_\nu} - {}_px_{\alpha_\nu} & r_{\alpha_\nu} \end{vmatrix},$$

thus also as

$$-\, d\dot{i}_{\alpha_1 \ldots \alpha_{p-1}\alpha_\nu} + \varepsilon,$$

so that the contribution of the term $\int F_{\alpha_1 \ldots \alpha_{p-1}} dx_{\alpha_1} \ldots dx_{\alpha_{p-1}}$ to the value of (1) over e_1 and e_2 can be expressed as

$$\varepsilon + \sum_{\nu=p}^{n} \left\{ \frac{\partial F_{\alpha_1 \ldots \alpha_{p-1}}}{\partial x_{\alpha_\nu}} \right\}_A d\dot{i}_{\alpha_1 \ldots \alpha_{p-1}\alpha_\nu},$$

and the value of (1) over β is obtained in the form:

$$\sum_{\alpha_1 \ldots \alpha_p} \sum_{\nu=1}^{p} \left\{ \frac{\partial F_{j_\nu}}{\partial x_{\alpha_\nu}} \right\}_A \sigma i_{\alpha_1 \ldots \alpha_p}$$

(*indicatrix* $j_\nu \alpha_\nu$ *aeq. indicatrix* $\alpha_1 \ldots \alpha_p$),

where A represents a point of σ which may be different for different terms under the $\Sigma\Sigma$ sign.

Hence it follows immediately, that, if Q is a two-sided p-dimensional net fragment situated in S, and R denotes the boundary of Q, while the indicatrices of Q and R correspond, then *the value of* (1) *over* R *is equal to the value of* (2) *over* Q.

IV. To complete the proof of the theorem formulated in I, we consider a category ψ of simplicial divisions of g such that the aggregate of the base sides possesses for ψ uniformly continuously varying plane tangent spaces, and the ratio of the volume of a base simplex to the $(p-1)^{\text{th}}$ power of its greatest coordinate fluctuation does not fall below a certain minimum for ψ. Let $g', g'', \ldots$ be a sequence of indefinitely condensing simplicial approximations of g corresponding to ψ. If, on $g^{(\mu)}$ we construct an approximating simplicial representation $g^{(\nu\mu)}$ of $g^{(\nu)}$, then, by choosing both μ and ν above a suitable limit, we can, in virtue of III, see to it that the values of (1) over $g^{(\nu\mu)}$ and $g^{(\nu)}$ differ from each other by as little as we please, whilst $g^{(\mu)}$ is covered by $g^{(\nu\mu)}$ with degree one, so that the values of (1) over $g^{(\mu)}$ and $g^{(\nu\mu)}$ are equal. Thus there exists a *value of* (1) *over* g *for* ψ which, naturally, does not change if, instead of ψ, some other category of the same kind is chosen.

Now let χ be a category of simplicial divisions of G analogous to ψ. The resulting simplicial divisions of g belong to a category ψ of the kind just described. Let $G', G'', \ldots$ be a sequence of indefinitely condensing simplicial approximations of G corresponding to χ, then, at the same time, there is hereby determined a sequence g', g'', ... of indefinitely condensing simplicial approximations of g corresponding to ψ. Since, in virtue of III, the value of (1) over $g^{(\nu)}$ is equal to the value of (2) over $G^{(\nu)}$, *there exists, just as there does a value of* (1) *over* g *for* ψ, *a value of* (2) *over* G *for* χ, *both values being equal, and not changing if some other category of the same kind is chosen in the place of either* ψ *or* χ.

§ 2.

In introducing l.c. p. 70 the notion of a second derivative, we have omitted to give the definition of the underlying concept of *normality* of an S_p provided with an indicatrix and an S_{n-p} provided with an indicatrix which are perpendicular to each other in an S_n provided with an indicatrix. This definition we shall here give.

Let T be the point of intersection of S_p and S_{n-p}, $\alpha_1 \ldots \alpha_p T$ the indicatrix of S_p and $\beta_1 \ldots \beta_{n-p} T$ the indicatrix of S_{n-p}; we call S_p *normal to* S_{n-p} and S_{n-p} *postnormal to* S_p, if $\alpha_1 \ldots \alpha_p T \beta_{n-p} \ldots \beta_1$ is an indicatrix of S_n.

Thus, for some values of n the concepts *normal* and *postnormal* are equivalent, for other values not equivalent.

Furthermore we call a p-dimensional vector system V *normal* to an $(n-p)$-dimensional vector system W at the same point, and W *postnormal* to V, if, with respect to a rectangular system of coordinates the components of V are respectively normal to and of equal scalar values as the components of W.

In this terminology, the second derivative of the vector distribution X is the normal distribution of the first derivative of the postnormal distribution of ${}^{P}X$.

NOTES

⟦1⟧ Brouwer 1906 A2.
⟦2⟧ Brouwer 1911 E.
⟦3⟧ H. Poincaré 1899.

1906 B2 **Mathematics.** — "*The force field of the non-Euclidean spaces with negative curvature*". By Mr. L. E. J. Brouwer. (Communicated by Prof. D. J. Korteweg).

⟦1⟧

A. *The hyperbolic* Sp_3.

I. Let us suppose a rectangular system of coordinates to be placed thus that $ds = \sqrt{A^2\,du^2 + B^2\,dv^2 + C^2\,dw^2}$, and let us assume a line-vector distribution X with components X_u, X_v, X_w, then the integral of X along a closed curve is equal to that of the planivector Y over an arbitrary surface bounded by it; here the components of Y are determined by:

$$Y_u = \frac{1}{BC}\left\{\frac{\partial (X_v B)}{\partial w} - \frac{\partial (X_w C)}{\partial v}\right\}, \text{ etc.}$$

For, if we assume on the bounded surface curvilinear coordinates ξ and η, with respect to which the boundary is convex, the surface integral is

$$\int \sum \left(\frac{\partial v}{\partial \xi} \cdot \frac{\partial w}{\partial \eta} - \frac{\partial v}{\partial \eta} \cdot \frac{\partial w}{\partial \xi}\right)\left(\frac{\partial (X_v B)}{\partial w} - \frac{\partial (X_w C)}{\partial v}\right) d\xi\, d\eta.$$

Joining in this relation the terms containing $X_w\, C$ and adding and subtracting $\frac{\partial (X_w C)}{\partial w} \cdot \frac{\partial w}{\partial \xi} \cdot \frac{\partial w}{\partial \eta}$ we obtain:

$$\int d\xi\, d\eta \left\{\frac{\partial (X_w C)}{\partial \eta} \cdot \frac{\partial w}{\partial \xi} - \frac{\partial (X_w C)}{\partial \xi} \cdot \frac{\partial w}{\partial \eta}\right\}.$$

Integrating this partially, the first term with respect to η, the second to ξ, we shall get $\int X_w\, C\, dw$ along the boundary, giving with the integrals $\int X_v\, B\, dv$ and $\int X_u\, A\, du$ analogous to them the line integral of X along the boundary.

⟦2⟧ In accordance with the terminology given before (see Proceedings of this Meeting p. 66—78) [1]) we call the planivector Y the first derivative of X.

[1]) The method given there derived from the indicatrix of a convex boundary that for the bounded space by front-position of a point of the interior; and the method understood by the vector $X_{pqr\ldots}$ a vector with indicatrix $opqr\ldots$. We can however determine the indicatrix of the bounded space also by post-position of a point of the interior with respect to the indicatrix of the boundary; and moreover assign to the vector $X_{pqr\ldots}$ the indicatrix $pqr..o$. We then find:

⟦54⟧

Analogously we find quite simply as second derivative the scalar:

$$Z = \frac{1}{ABC} \sum \frac{\partial \{X_u . BC\}}{\partial u}.$$

According to the usual way of expressing, the first derivative is the rotation vector and the second the divergency.

II. If X is to be a ${}^1_2X$, i. e. a second derivative of a planivector Ξ, we must have:

$$X_u = \frac{1}{BC} \left\{ \frac{\partial (\Xi_v B)}{\partial w} - \frac{\partial (\Xi_w C)}{\partial v} \right\}, \text{ etc.,}$$

and it is easy to see that for this is necessary and sufficient

$$Z = 0.$$

III. If X is to be a ${}^1_0X$, i. e. a first derivative (gradient) of a scalar distribution φ, we must have:

$$X_u = -\frac{\partial \varphi}{A \partial u} \quad , \quad X_v = -\frac{\partial \varphi}{B \partial v} \quad , \quad X_w = -\frac{\partial \varphi}{C \partial w}$$

and it is easy to see, that to this end it will be necessary and sufficient that

$$Y = 0.$$

IV. It is easy to indicate (comp. SCHERING, Göttinger Nachrichten, 1870) the ${}^1_0X$, of which the divergency is an isolated scalar value in the origin. [[3]]

It is directed according to the radius vector and is equal to:

$$\frac{1}{sinh^2 r},$$

when we put the space constant $= 1$ [2]).

$$Y_{\alpha_1 \alpha_2 \ldots \alpha_p \alpha_{p+1}} = \sum_{\alpha_{q_{p+1}} = \alpha_1 \ldots \alpha_{p+1}} \frac{\partial X_{\alpha_{q_1} \ldots \alpha_{q_p}}}{\partial x_{\alpha_{q_{p+1}}}} ;$$

$$Z_{\alpha_1 \ldots \alpha_{p-1}} = - \sum_{\alpha_q = \alpha_p \ldots \alpha_n} \frac{\partial X_{\alpha_1 \ldots \alpha_{p-1} \alpha_q}}{\partial x_{\alpha_q}} .$$

These last definitions include the well known *divergency* of a vector, and the *gradient* of a potential also as regards the sign; hence in the following we shall start from it and we have taken from this the extension to non-Euclidean spaces.

2) For another space constant we have but to substitute in the following formulae $\frac{r}{R}$ for r.

It is the first derivative of a scalar distribution:

$$-1 + \coth r,$$

and has in the origin an isolated divergency of 4π.

V. In future we shall suppose that X has the *field property* and shall understand by it, that it vanishes at infinity in such a manner that in the direction of the radius vector it becomes of lower order than $\frac{1}{r}$ and in the direction perpendicular to the radius vector of lower order than e^{-r}.

For a ${}^{1}_{0}X$ this means that it is derived from a scalar distribution, having the *potential property*, i.e. the property of vanishing at infinity.

Now the theorem of GREEN holds for two scalar distributions (comp.
[[4]] FRESDORF, diss. Göttingen, 1873):

$$\int \varphi \frac{\partial \psi}{\partial \nu} dO - \int \varphi \nabla^2 \psi . d\tau = \int \psi \frac{\partial \varphi}{\partial \nu} dO - \int \psi \nabla^2 \varphi . d\tau$$

$$\left(= \int S\{grad.\, \varphi,\, grad.\, \psi\}\, dt \right).$$

If now φ and ψ both vanish at infinity whilst at the same time $\lim. \varphi \psi e^{2r} = 0$, then the surface integrals disappear, when we apply the theorem of GREEN to a sphere with infinite radius and

$$\int \varphi . \nabla^2 \psi . d\tau = \int \psi . \nabla^2 \varphi . d\tau,$$

integrated over the whole space, is left.

Let us now take an arbitrary potential function for φ and $-1 + \coth r$ for ψ, where r represents the distance to a point P taken arbitrarily, then these functions will satisfy the conditions of vanishing at infinity and $\lim. \varphi \psi e^{2r} = 0$, so that we find:

$$4\pi . \varphi_P = \int (-1 + \coth r) \nabla^2 \varphi . d\tau.$$

So, if we put $-1 + \coth r \equiv F_1(r)$, we have:

$$ {}^{1}_{0}X = \nabla_1 \int \frac{\nabla_2\, {}^{1}_{0}X}{4\pi} F_1(r)\, d\tau. \quad . \quad . \quad . \quad . \quad . \quad . \quad (I)$$

VI. We now see that there is no vector distribution with the field property, which has in finite nowhere rotation and nowhere divergency. For, such a vector distribution would have to have a potential, having nowhere rotation, but that potential would have to be everywhere 0 according to the formula, so also its derived vector.

[[56]]

From this ensues: a vector field is determined uniformly by its rotation and its divergency.

VII. So, if we can indicate elementary distributions of divergency and of rotation, the corresponding vector fields are elementary fields, i. e. the arbitrary vector field is an arbitrary space-integral of such fields.

For such elementary fields we find thus analogously as in a Euclidean space (l. c. p. 74 seq.): [[2]]

1. a field E_1, of which the second derivative consists of two equal and opposite scalar values, close to each other.

2. a field E_2, of which the first derivative consists of equal planivectors in the points of a small circular current and perpendicular to that same current.

At finite distance from their origin the fields E_1 and E_2 are here again of the same identical structure.

VIII. To indicate the field E_1 we take a system of spherical coordinates and the double point in the origin along the axis of the system. Then the field E_1 is the derivative of a potential:

$$\frac{\cos \varphi}{\sinh^2 r}.$$

It can be regarded as the sum of two fictitious "fields of a single agenspoint", formed as a derivative of a potential $-1 + \coth r$, which have however in reality still complementary agens at infinity.

IX. The field E_2 of a small circular current lying in the equator plane in the origin is outside the origin identical to the above field E_1. Every line of force however, is now a closed vector circuit with a line integral of 4π along itself. We shall find of this field E_2 a planivector potential, lying in the meridian plane and independent of the azimuth.

In order to find this in a point P with a radius vector r and spherical polar distance φ we have but to divide the total current between the meridian plane of P and a following meridian plane with difference of azimuth $d\vartheta$, passing between P and the positive axis of revolution, by the element of the parallel circle through P over $d\vartheta$. For, if ds is an arbitrary line element through P in the meridian plane making with the direction of force an angle F, if dh is the element of the parallel circle, Σ the above mentioned current and H the vector potential under consideration, we find:

$$d\Sigma = dh \,.\, X ds \sin F,$$

whilst the condition for H is:

$$d\,(Hdh) = dh\;ds\;X\;sin\;F.$$

So we have but to take $\frac{\Sigma}{dh}$ for H.

To find Σ we integrate the current of force within the meridian zone through the spherical surface through P. The force component perpendicular to that spherical surface is $2\cos\varphi\frac{\cosh r}{\sinh^3 r}$, therefore:

$$\Sigma = \int_0^{\varphi} 2\cos\varphi\frac{\cosh r}{\sinh^3 r}\,.\,\sinh r\,d\varphi\,.\,\sinh r\sin\varphi\,d\vartheta = d\vartheta\,\coth r\,.\,\sin^2\varphi.$$

So:

$$H = \frac{\Sigma}{dh} = \frac{\Sigma}{\sinh r\sin\varphi d\vartheta} = \frac{\cosh r}{\sinh^2 r}\sin\varphi.$$

X. From this ensues, that if two arbitrary vectors of strength unity are given in different points along whose connecting line we apply a third vector $= \frac{\cosh r}{\sinh^2 r}$, the volume product of these three vectors, i. e. the volume of the parallelepipedon having these vectors as edges taken with proper sign, represents the linevector potential according to the first (second) vector, caused by an elementary magnet with moment unity according to the second (first) vector.

To find that volume product, we have first to transfer the two
[[5]] given vectors to a elfsame point of their connecting line, each one parallel to itself, i. e. in the plane which it determines with that connecting line, along which the transference is done, and maintaining the same angle with that connecting line.

The volume product $\psi(S_1, S_2)$ is a symmetric function of the two vectors unity of which we know that with integration of S_1 along a closed curve s_1 it represents the current of force of a magnet unity according to S_2 through s_1, in other words the negative reciprocal energy of a magnet unity in the direction of S_2 and a magnetic scale with intensity unity within s_1, in other words the force in the direction of S_2 by a magnetic scale with intensity unity within s_1, in other words the force in the direction of S_2 by a current with intensity unity along s_1. So we can regard $\psi(S_1, S_2)$ as a force in the direction of S_2 by an element of current unity in the direction of S_1.

With this we have found for the force of an element of current with intensity unity in the origin in the direction of the axis of the system of coordinates:

[[58]]

$$\frac{cosh r}{sinh^2 r} sin \varphi,$$

directed perpendicular to the meridian plane.

XI. For the fictitious field of an element of current (having meanwhile everywhere current, i. e. rotation) introduced in this way we shall find a linevector potential V, everywhere "parallel" (see above under § X) to the element of current and the scalar value of which is a function of r only.

Let us call that scalar value U, and let us regard a small elementary rectangle in the meridian plane bounded by radii vectores from the origin and by circles round the origin, then the line integral of V round that rectangle is:

$$-\frac{\partial}{\partial r}\{U sin \varphi \, sinh \, r \, d\varphi\} dr - \frac{\partial}{\partial \varphi}\{U cos \varphi \, dr\} d\varphi.$$

This must be equal to the current of force through the small rectangle:

$$\frac{cosh \, r}{sinh^2 r} sin \varphi \, . \, sinh \, r \, d\varphi . \, dr,$$

from which we derive the following differential equation of U with respect to r:

$$U - \frac{\partial}{\partial r}\{U sinh \, r\} = coth \, r,$$

the solution of which is:

$$U = cosech \, r - \tfrac{1}{2} r \, sech^2 \tfrac{1}{2} r + c \, . \, sech^2 \tfrac{1}{2} r.$$

Let us take $c = 0$, we shall then find as vector potential V of an element of current unity E:

$$cosech \, r - \tfrac{1}{2} r \, sech^2 \tfrac{1}{2} r \equiv F_2(r),$$

directed parallel to E.

Let us now apply in an arbitrary point of space a vector G, then the vector V has the property that, when integrated in G along an elementary circuit whose plane is perpendicular to G, it indicates the force in the direction of G, caused by the element of current E, or likewise the vector potential in the direction of E caused by an elementary magnet with intensity unity in the direction of G.

So, if we call of two vectors unity E and F the potential $\chi(E, F)$, the symmetric function $F_2(r$ $cos \varphi$, where r represents the distance of the points of application of the two vectors and φ their angle after parallel transference to a selfsame point of their connecting line, we know that this function χ gives, by integration of e.g. E over a closed curve e not only the negative energy of a magnetic scale with intensity unity bounded by e in the field of an element of current unity F, but also the component along F of the vector potential caused by a current unity along e.

From this ensues for the vector V of an element of current, that when the element of current is integrated to a closed current it becomes the vector potential of that current determined uniformly on account of its flux property.

So really *the* vector potential of a ${}^{1}_{2}X$, i. e. of a field of currents is obtained as an integral of the vectors V of the elements of current.

XII. We can now write that in an arbitrary point:

$$ {}^{1}_{2}X = \overset{2}{\nabla} \int \frac{\overset{1}{\nabla}\, {}^{1}_{2}X}{4\pi} F_2(r)\, d\tau, \quad \ldots\ldots \quad (H) $$

where we first transfer in a parallel manner the vector elements of the integral to the point under consideration and then sum up.

Let us now consider an arbitrary force field as if caused by its two derivatives (the magnets and currents), we can then represent to ourselves, that both derivatives, propagating themselves according to a function of the distance vanishing at infinity, generate the potential of the field.

The field X is namely the total derivative of the potential:

$$ \int \frac{\overset{2}{\nabla} X}{4\pi} F_1(r)\, d\tau + \int \frac{\overset{1}{\nabla} X}{4\pi} F_2(r)\, d\tau. $$

The extinguishment of the scalar potential is greater than that of the vector potential; for, the former becomes at great distances of order e^{-2r}, the latter of order re^{-r}. Farther the latter proves not to decrease continuously from ∞ to 0, but at the outset it passes quickly through 0 to negative, it then reaches a negative maximum and then according to an extinguishment re^{-r} it tends as a negative (i. e. directed oppositely to the generating element of current) vector to zero.

XIII. The particularity found in Euclidean spaces, that $F_1(r) = F_2(r) = \frac{1}{r}$, is founded upon this, that in Euclidean spaces the operation of twice total derivation is found to be alike for scalar

[[2]] distributions and vector distributions of any dimensions (l.c. p. 70).

Not so in non-Euclidean spaces; e. g. in the hyperbolic Sp_3 we find for the ∇^2 of a scalar distribution u in an arbitrary point when choosing that point as centre of a system of RIEMANN normal coordinates

$$\left(\text{i. e. a system such that } ds = \frac{\sqrt{dx^2 + dy^2 + dz^2}}{1 - \frac{x^2 + y^2 + z^2}{4}}\right)$$

$$\nabla^2 u = -\left(\frac{\partial^2 u}{\partial x^2} + \frac{\partial^2 u}{\partial y^2} + \frac{\partial^2 u}{\partial z^2}\right),$$

but as ∇^2 of a vector distribution with components X, Y and Z we find for the x-component X_{∇^2}:

$$X_{\nabla^2} = -\left(2\,X + \frac{\partial^2 X}{\partial x^2} + \frac{\partial^2 X}{\partial y^2} + \frac{\partial^2 X}{\partial z^2}\right).$$

The hyperbolic Sp_2.

I. As first derivative Y of a vector distribution X we find a planivector determined by a scalar value:

$$\frac{1}{AB}\left\{\frac{\partial\,(X_u A)}{\partial v} - \frac{\partial\,(X_v B)}{\partial u}\right\}.$$

As second derivative Z we find the scalar:

$$\frac{1}{AB}\left\{\frac{\partial\,(X_u B)}{\partial u} + \frac{\partial\,(X_v A)}{\partial v}\right\}.$$

II. If X is to be a ${}_2^1X$, i. e. a second derivative of a planivector with scalar value ψ we must have:

$$X_u = -\frac{\partial\psi}{B\partial v} \quad ; \quad X_v = \frac{\partial\psi}{A\partial u},$$

to which end is necessary and sufficient: $Z = 0$.

If X is to be a ${}_0^1X$, i. e. a first derivative of a scalar φ we must have:

$$X_u = -\frac{\partial\varphi}{A\partial u} \quad ; \quad X_v = -\frac{\partial\varphi}{B\partial v},$$

to which end is necessary and sufficient: $Y = 0$.

III. The ${}_0^1X$, of which the divergency is an isolated scalar value in the origin, becomes a vector distribution in the direction of the radius vector:

$$\frac{1}{\sinh r}.$$

It is the first derivative of a scalar distribution:

$$l\,\coth \tfrac{1}{2}\,r.$$

The divergency in the origin of this field is 2π.

The scalar distribution $l \coth \frac{1}{2} r$ has thus the potential property. (This was not the case for the field of a single agens point in the Euclidean Sp_2).

IV. In the following we presuppose again for the given vector distribution the field property (which remains equally defined for 2 and for n dimensions as for 3 dimensions); no vector field is possible that has nowhere rotation and nowhere divergency; so each vector field is determined by its rotation and its divergency and we have first of all for a gradient distribution:

$$ {}^1_0X = \overline{\nabla}^1 \int \frac{\overline{\nabla}^2\, {}^1_0X}{2\pi}\, l \coth \tfrac{1}{2} r\, d\tau, $$

$$ {}^1_0X = \overline{\nabla}^1 \int \frac{\overline{\nabla}^2\, {}^1_0X}{2\pi}\, F_1(r)\, d\tau \quad . \; . \; . \; . \; . \; . \quad (I) $$

V. For the field E_1 of an agens double point we find the gradient of the potential:

$$ \frac{\cos \varphi}{\sinh r}. $$

It can be broken up into "fields of a single agens point" formed as a derivative of a potential $l \coth \frac{1}{2} r$.

VI. Identical outside the origin to the above field E_1 is the field E_2 of a double point of rotation, whose axis is perpendicular to the axis of the agens double point of the field E_1. For that field E_2 we find as scalar value of the planivector potential in a point P the total current of force between P and the axis of the agens double point, that is:

$$ \sin \varphi \coth r. $$

So if are given a vector unity V and a scalar unity S and if we apply along their connecting line a vector $\coth r$, the volume product ψ of V, S and the vector along the connecting line is the scalar value of the planivector potential in S by a magnet unity in the direction of V.

Of ψ we know that when summing up S out of a positive scalar unity S_2 and a negative S_1 it represents the current of force of a magnet unity in the direction of V passing between S_1 and S_2, in other words the negative reciprocal energy of a magnet unity in the direction of V and a magnetic strip $S_1 S_2$ with intensity unity, in other words the force in the direction of V by a couple of rotation $S_1 \rightarrow S_2$. So we can regard ψ as the force in the direction of V by an isolated rotation in S. So that we must take as fictitious "force field of an element of rotation unity"

$$coth\ r,$$

directed perpendicularly to the radius vector. In reality, however, this force field has rotation everywhere in Sp_2.

VII. Let us now find the scalar value U, function of r, which we must assign to a planivector potential, that the "field of an element of rotation unity" be its second derivative. We must have:

$$-\frac{dU}{dr} = coth\ r.$$

$$U = l\ cosech\ r.$$

And we find for an arbitrary ${}_2^1X$:

$${}_2^1X = \overline{\nabla}^2 \int \frac{\overline{\nabla}^1\ {}_2^1X}{2\pi}\ l\ cosech\ r\ d\tau,$$

$${}_2^1X = \overline{\nabla}^2 \int \frac{\overline{\nabla}^1\ {}_2^1X}{2\pi}\ F_2\ (r)\ d\tau. \quad . \quad . \quad . \quad . \quad . \quad (II)$$

And an arbitrary vector field X is the total derivative of the potential

$$\int \frac{\overline{\nabla}^2 X}{2\pi}\ F_1\ (r)\ d\tau + \int \frac{\overline{\nabla}^1 X}{2\pi}\ F_2\ (r)\ d\tau.$$

VIII. We may now wonder that here in Sp_2 we do not find F_1 and F_2 to be identical, as the two derivatives and the two potentials of a vectordistribution are perfectly dually related to each other in the hyperbolic Sp_2 as well as in the Euclidean Sp_2. The difference, however, is in the principle of the field property, which postulates a vanishing at infinity for the scalar potential, not for the planivector potential; and from the preceding the latter appears not to vanish, so with the postulation of the field property the duality is broken.

But on the other hand that postulation in Sp_2 lacks the reasonable basis which it possesses in spaces of more dimensions. For, when putting it we remember the condition that the total energy of a field may not become infinite. As soon as we have in the infinity of Sp_n forces of order e^{-r}, this furnishes in a spherical layer with thickness dr and infinite radius described round the origin as centre an energy of order $e^{-2r} \times e^{(n-1)r}\ dr = e^{(n-3)r}\ dr$; which for $n \geqq 3$ would give when integrated with respect to r an infinite energy at infinity of Sp_n. So for $n \geqq 3$ are excluded by the field property only vector distributions which cannot have physical meaning.

For $n = 2$ however the postulation lacks its right of existence; more sense has the condition (equivalent for $n > 2$ to the field property) that for given rotation and divergency the vector distribution must have a minimum energy. Under these conditions we shall once more consider the field and we shall find back there too the duality with regard to both derivatives and both potentials.

IX. Let us consider first of all distributions with divergency only and let us find the potential function giving a minimum energy for given ∇^2.

We consider the hyperbolical Sp_2 as a conform representation of a part of a Euclidean Sp_2 bounded by a circle; if we then apply in corresponding points of the representation the same potential, we retain equal energies and equal divergencies in corresponding plane elements. So the problem runs:

Which potential gives within a given curve (in this case a circle) in the Euclidean Sp_2 under given divergency distribution a minimum energy?

According to the theorem of GREEN we have for this:

$$\tfrac{1}{2}\int \delta\Sigma \left(\frac{\partial u}{\partial x}\right)^2 d\tau = \int \Sigma \frac{\partial u}{\partial x} \cdot \frac{\partial \delta u}{\partial x} \cdot d\tau = \int u \cdot \frac{\partial . \delta u}{\partial \nu} \cdot dO - \int u \nabla^2 \delta u \, . \, d\tau,$$

so that, as $\nabla^2 \delta u$ is 0 everywhere within the boundary curve, the necessary and sufficient condition for the vanishing of the variation of the energy is:

$$u = 0, \text{ along the boundary curve.}$$

For the general vector distribution with divergency only in the hyperbolical Sp_2 we thus find under the condition of minimum energy also, that the potential at infinity must be 0. So we find it, just as under the postulation of the field property, composed of fields E_1, derived from a potential $\dfrac{\cos\varphi}{\sinh r}$.

The lines of force of this field E_1 have the equation,

$$\sin\varphi \coth r = c.$$

Only a part of the lines of force (in the Euclidean plane all of them) form a loop; the other pass into infinity. None of the equipotential lines, however, pass into infinity; they are closed and are all enclosed by the circle at infinity as the line of 0-potential.

The same holds for the arbitrary ${}^{1}_{0}X$; of the lines of force one part goes to infinity; the potential lines however are closed.

X. If we now have to find the field with rotation only, giving for given rotation distribution a minimum energy, it follows from a consideration of the rotation as divergency of the normal vector, that the scalar value of the planivector potential at infinity must be 0, and the general ${}^{1}_{2}X$ is composed of fields E_2, derived from a planivector potential $\frac{\sin \varphi}{\sinh r}$ (whilst we found under the postulation of the field property $\sin \varphi \ \coth r$).

In contrast to higher hyperbolical spaces and to any Euclidean and elliptic spaces the fields E_1 and E_2 cannot be summed up here to a single isolated vector.

For this field E_2 and likewise for the arbitrary ${}^{1}_{2}X$ the lines of force (at the same time planivector potential lines) are closed curves.

XI. We have now found

$$ {}^{1}_{0}X = \overline{\nabla}^{1} \int \frac{\overline{\nabla}^{2}\, {}^{1}_{0}X}{2\pi} \, l \coth \tfrac{1}{2} r \, d\tau, $$

$$ {}^{1}_{2}X = \overline{\nabla}^{2} \int \frac{\overline{\nabla}^{1}\, {}^{1}_{2}X}{2\pi} \, l \coth \tfrac{1}{2} r \, d\tau. $$

And from this ensues that also the general vector distribution X having under given rotation and divergency a minimum energy is equal to:

$$ X_{div.} + X_{rot.} = \overline{\nabla}^{1} \int \frac{\overline{\nabla}^{2} X}{2\pi} \, l \coth \tfrac{1}{2} r \, d\tau + \overline{\nabla}^{2} \int \frac{\overline{\nabla}^{1} X}{2\pi} \, l \coth \tfrac{1}{2} r \, d\tau. $$

For, if V is an arbitrary distribution without divergency and without rotation in finite, it is derived from a scalar potential function, so it has (according to § VIII) no reciprocal energy with $X_{div.}$; neither (as according to § IX all lines of force of $X_{rot.}$ are closed curves and a flux of exclusively closed vector tubes has no reciprocal energy with a gradient distribution) with $X_{rot.}$; so that the energy of $X_{div.} + X_{rot.} + V$ is larger than that of $X_{div.} + X_{rot.}$.

So finally we have for the general vector distribution of minimum energy X:

$$ X = \nabla \int \frac{\Delta X}{2\pi} \, . \, l \coth \tfrac{1}{2} r \, d\tau. $$

C. *The hyperbolic Sp_n.*

I. Let us suppose a system of rectangular coordinates, so that

$$ds = \sqrt{A_1 u_1^2 + \dots A_n u_n^2},$$

and let us suppose a linevector distribution X with components $X_1 \dots X_n$, then the integral of X along a closed curve is equal to that of a planivector Y over an arbitrary surface bounded by it, in which the components of Y are determined by:

$$Y_{\alpha_1 \alpha_2} = \frac{1}{A_{\alpha_1} A_{\alpha_2}} \left\{ \frac{\partial (X_{\alpha_2} A_{\alpha_2})}{\partial x_{\alpha_1}} - \frac{\partial (X_{\alpha_1} A_{\alpha_1})}{\partial x_{\alpha_2}} \right\}.$$

Y is the *first derivative* or *rotation* of X.

Further the starting vector current of X over a closed curved Sp_{n-1} is equal to the integral of the scalar Z over the bounded volume of that Sp_{n-1}; here

$$Z = \frac{1}{A_1 \dots A_n} \sum \frac{\partial (X_{\alpha_1} \,.\, A_{\alpha_2} \dots A_{\alpha_n})}{\partial x_{\alpha_1}}$$

Z is the *second derivative* or *divergency* of X.

II. If X is to be a ${}_2^1X$, i. e. a second derivative of a planivector Ξ, we must have:

$$X_{\alpha_1} = \frac{1}{A_{\alpha_2} \dots A_{\alpha_n}} \sum \frac{\partial \,.\, (\Xi_{\alpha_1 \alpha_{q_1}} A_{\alpha_{q_2}} \dots A_{\alpha_{q_{n-1}}})}{\partial_{\alpha_{q_1}}}.$$

The necessary and sufficient condition for this is:

$$Z = 0.$$

If X is to be a ${}_0^1X$, i. e. a first derivative of a scalar φ, we must have:

$$X_\alpha = -\frac{\partial \varphi}{A_\alpha \partial x_\alpha}.$$

The necessary and sufficient condition for this is:

$$Y = 0.$$

III. The ${}_0^1X$, which has as divergency an isolated scalar value in
[[6]] the origin (comp. Opitz., Diss. Göttingen, 1881), is directed along the radius vector, and if we put the space constant equal to 1 is equal to

$$\frac{1}{\sinh^{n-1} r}.$$

It is the first derivative of a scalar distribution

$$\int_r^\infty \frac{dr}{\sinh^{n-1} r} \equiv w_n(r),$$

and it has in the origin an isolated divergency of k_n (if $k_n r^{n-1}$ expresses the spherical surface of the Euclidean space Sp_n).

IV. For two scalar distributions φ and ψ the theorem of GREEN holds (comp. Opitz., l.c.):

$$\int \varphi \frac{\partial \psi}{\partial \nu} \cdot dO_{n-1} - \int \varphi \nabla^2 \psi \cdot d\tau_n = \int \psi \frac{\partial \varphi}{\partial \nu} \cdot dO_{n-1} - \int \psi \nabla^2 \varphi \cdot d\tau_n$$

$$\left(= \int S(\nabla \varphi, \nabla \psi) \cdot d\tau_n \right).$$

If at infinity φ and ψ both become 0 whilst at the same time

$$\lim \varphi \psi e^{(n-1)r} = 0,$$

then for an ${}^{n-1}$sphere with infinite radius the surface integrals disappear and we have left

$$\int \varphi \cdot \nabla^2 \psi \cdot d\tau_n = \int \psi \cdot \nabla^2 \varphi \cdot d\tau_n,$$

integrated over the whole space.

If here we take an arbitrary potential function for φ and $w_n(r)$ for ψ, where r represents the distance to an arbitrarily chosen point P — these functions satisfying together the conditions of the formula — we have:

$$k_n \varphi_P = \int w_n(r) \cdot \nabla^2 \varphi \cdot d\tau_n.$$

If thus we postulate for the vector distributions under consideration the *field property* (which remains defined just as for Sp_3) we have, if we put $w_n(r) \equiv F_1(r)$, for an arbitrary ${}_0^1X$:

$$ {}_0^1X = \overline{\nabla}^1 \int \frac{\overline{\nabla}^2\, {}_0^1X}{k_n} F_1(r)\, d\tau ; \quad . \quad . \quad . \quad . \quad . \quad (I)$$

from which we deduce (compare A § VI) that there is no vector field which has in finite nowhere rotation nor divergency; so that a vector field is uniformly determined by its rotation and its divergency.

V. So a vectorfield is an arbitrary integral of:

1. Fields E_1, of which the second derivative consists of two equal and opposite scalar values close to each other.

2. Fields E_2, of which the first derivative consists of planivectors distributed regularly in the points of a small ${}^{n-2}$sphere and perpendicular to that ${}^{n-2}$sphere.

At finite distance from their origin the fields E_1 and E_2 are of identical structure.

VI. In order to indicate the field E_1 we assume a spherical system of coordinates [1]) and the double point in the origin along the first axis of the system. Then the field E_1 is the derivative of a potential:

$$\frac{\cos \varphi}{\sinh^{n-1} r}.$$

The lines of force of this field run in the meridian plane. It can be regarded as the sum of two fictitious "fields of a single agens-point" constructed as derivative of a potential $w_n(r)$ to which, however, must be assigned still complementary agens at infinity.

VII. The field E_2 of a small vortex-${}^{n-2}$sphere according to the space perpendicular to the axis of the double point just considered is identical outside the origin to the field E_1. Each line of force is now however a closed vector tube with a line integral k_n along itself. We shall find for this field E_2 a planivector potential H, lying in the meridian plane and dependent only on r and φ. It appears then simply that this H is a ${}^2_1X$.

Let ε be an $(n-2)$-dimensional element in the $n-2$ coordinates existing besides r and φ, then it defines for each r and φ an element on the surface of an ${}^{n-2}$-sphere of a size $dh = c\varepsilon \sinh^{n-2} r \sin^{n-2} \varphi$, and for the entire Sp_n what may be called a "meridian zone".

We then obtain for the current of force Σ, passing inside a meridian zone between the axis of the system and a point P with coordinates r and φ, if ds represents an arbitrary line element through P in the meridian plane under an angle F with the direction of force:

$$d\Sigma = dh \,.\, X\, ds \sin F,$$

whilst we can easily find as necessary and sufficient condition for H:

$$d\,(Hdh) = dh \,.\, ds \,.\, X \sin F\,;$$

so we have but to take $\dfrac{\Sigma}{dh}$ for H.

---

[1]) By this we understand in Sp_n a system which with the aid of a rectangular system of numbered axes determines a point by 1. r, its distance to the origin, 2. φ, the angle of the radius vector with X_1, 3. the angle of the projection of the radius vector on the coordinate space $X_2 \ldots X_n$ with X_2, 4. the angle of the projection of the last projection on the coordinate space $X_3 \ldots X_n$ with X_3; etc. The plane through the X_1-direction and the radius vector we call the *meridian plane.*

[[68]]

To find Σ we integrate the current of force inside the meridian zone passing through the ${}^{n-1}$spherical surface through P between the axis of the system and P. As we have $(n-1)\cos\varphi\dfrac{\cosh r}{\sinh^{n} r}$ for the force component perpendicular to that spherical surface we find:

$$\Sigma=\int_{0}^{\varphi}(n-1)\cos\varphi\frac{\cosh r}{\sinh^{n} r}\,.\sinh r\,d\varphi\,.\,c\varepsilon\sinh^{n-2} r\sin^{n-2}\varphi=c\varepsilon\sin^{n-1}\varphi\coth r.$$

$$H=\frac{\Sigma}{dh}=\frac{\cosh r}{\sinh^{n-1} r}\sin\varphi.$$

VIII. If thus are given in different points a line vector L unity and an ${}^{n-2}$vector W unity and if we put along their connecting line a line vector $\dfrac{\cosh r}{\sinh^{n-1} r}$, then the volume product ψ of L, W and the vector along the connecting line is the ${}^{n-2}$vector potential in the direction of W caused by an elementary magnet with moment unity in the direction of L.

We know of $\psi(L, W)$ that with integration of W along a closed curved Sp_{n-2} Q it represents the current of force of a magnet unity in the direction of L through Q, in other words the negative reciprocal energy of a magnet unity in the direction of L and a magnetic ${}^{n-1}$scale with intensity unity, bounded by Q, in other words the force in the direction of L by a magnetic ${}^{n-1}$scale bounded by Q, in other words the force in the direction of L by a vortex system, regularly distributed over Q and perpendicular to Q. So we can regard $\psi(L, W)$ as the force in the direction of L by a vortex unity, perpendicular to W. With this we have found for the force of a plane vortex with intensity unity in the origin:

$$\frac{\cosh r}{\sinh^{n-1} r}\sin\varphi,$$

directed parallel to the operating vortex element and perpendicular to the "meridian plane", if now we understand by that plane the projecting plane on the vortex element; whilst φ is here the angle of the radiusvector with the Sp_{n-2} perpendicular to the vortex element.

IX. For the fictitious field of a vortex element in the origin introduced in this way (which meanwhile has vorticity everywhere in space) ⟦5⟧
we shall find a planivector potential, directed everywhere "parallel" to the vortex element and of which the scalar value U is a function of r only.

Let us suppose a point to be determined by its azimuth parallel to the vortex element and then farther in the Sp^{n-1} of constant azimuth by a system of spherical coordinates, of which we take the first axis in the "meridian plane" (see above under § VIII), and in the plane of the vortex element, the second in the meridian plane perpendicular to the first, and the rest arbitrarily; let us understand meanwhile by φ here the angle of the radius vector with the Sp_{n-2}, perpendicular to the vortex element; let further ε be an $(n-3)$-dimensional element in the $n-3$ last coordinates, then this defines for each r and φ an element on the surface of an $^{n-3}$sphere, of a size

$$dk = c\varepsilon \sinh^{n-3} r \cos^{n-3} \varphi .$$

We then consider a small elementary rectangle in the meridian plane bounded by radii vectores out of the origin and circles about the origin and a Sp_{n-1} element consisting of the elements dk erected in each point of this small elementary rectangle. Applying to this Sp_{n-1}-element the reduction of an $(n-2)$-fold integral along the boundary to a $(n-1)$-fold integral over the volume according to the definition of second derivative, we find:

$$-\frac{\partial}{\partial\varphi}\{U \cos\varphi \,.\, dr \,.\, c\varepsilon \sinh^{n-3} r \cos^{n-3}\varphi\}\, d\varphi -$$

$$-\frac{\partial}{\partial r}\{U \sin\varphi \,.\, \sinh r\, d\varphi \,.\, c\varepsilon \sinh^{n-3} r \cos^{n-3}\varphi\}\, dr =$$

$$= c\varepsilon \sinh^{n-3} r \cos^{n-3}\varphi \;.\; \sinh r\, d\varphi \,.\, dr \,.\, \frac{\cosh r}{\sinh^{n-1} r} \sin\varphi .$$

$$(n-2)\, U - \frac{dU}{dr} \sinh r - (n-2)\, U \cosh r = \frac{\cosh r}{\sinh^{n-2} r} .$$

$$\frac{dU}{dr} + (n-2) \tanh \tfrac{1}{2} r \,.\, U = -\frac{\cosh r}{\sinh^{n-1} r} .$$

The solution of this equation is:

$$U = -\frac{1}{2^{n-3}} \,.\, \cosh^{-2(n-2)} \tfrac{1}{2} r \,.\int \coth^{n-3} \tfrac{1}{2} r \,.\, d\, \tfrac{1}{2} r + \frac{1}{(n-2) \sinh^{n-2} r} .$$

So we find as planivector potential V of a plane vortex:

$$\frac{1}{(n-2) \sinh^{n-2} r} - \frac{1}{2^{n-3} \cosh^{2(n-2)} \frac{1}{2} r} \int \coth^{n-3} \tfrac{1}{2} r \,.\, d\, \tfrac{1}{2} r \equiv F_2 (r),$$

directed parallel to that plane vortex.

Let us now call E the $^{n-2}$vector, perpendicular to the plane vortex, the field of which we have examined, and let us also set off the vector potential V as an $^{n-2}$vector; let us then bring in an arbitrary point of space a line vector G; then the vector V has the property that when integrated in G along a small curved closed Sp_{n-2} in a Sp_{n-1} perpendicular to G, it indicates the force in the direction of G caused by the current element E, or also the vector potential in the direction of E, caused by an elementary magnet with intensity unity in the direction of G.

Let us now call the potential $\chi(E, F)$ of two ${}^{n-2}$vectors unity E, F the symmetric function $F_2(r) \cos\varphi$, where r represents the distance of the points of application of both vectors and φ their angle after parallel transference to one and the same point of their con- [5]
necting line, then we know that this function χ gives, when e. g. E is integrated over a closed curved Sp_{n-2} which we shall call e, not only the negative energy of a magnetic ${}^{n-1}$scale with intensity unity bounded by e in the field of a vortex unity perpendicular to F but also the component along F of the vector potential caused by a system of vortices about e with intensity unity.

From this ensues again for the vector potential V of a vortex element, that when the vortex element is integrated to a system of vortices about a closed curved Sp_{n-2} it becomes the vector potential determined according to § VII of that vortex Sp_{n-2}; so that the vector potential of an arbitrary ${}^{1}_{2}X$ is obtained as integral of the vectors V of its vortex elements, in other words:

$$ {}^{1}_{2}X = \overset{2}{\nabla} \int \frac{\overset{1}{\nabla}\, {}^{1}_{2}X}{k_n'} F_2(r)\, d\tau, \quad . \; . \; . \; . \; . \quad (II) $$

where for each point the vector elements of the integral are first brought over to that point parallel to themselves and there are summed up.

X. So let us consider an arbitrary force field as if caused by its two derivatives (the magnets and the vortex systems), we can then imagine that both derivatives are propagated through the space according to a function of the distance vanishing at infinity, causing thereby the potential of the field.

For, the field X is the total derivative of the potential:

$$ \int \frac{\overset{2}{\nabla} X}{k_n} F_1(r)\, d\tau + \int \frac{\overset{1}{\nabla} X}{k_n} F_2(r)\, d\tau. $$

The extinguishment of the scalar potential is the stronger, as it is at great distances of order $e^{-(n-1)r}$, the vector potential only of order $re^{-(n-2)r}$

NOTES

⟦1⟧ The Dutch version was communicated at the meeting of 30 June 1906. The paper extends Brouwer 1906 A2.
⟦2⟧ Brouwer 1906 A2.
⟦3⟧ E. Schering 1870.
⟦4⟧ Fresdorf 1873.
⟦5⟧ A definition of parallel displacement in non-Euclidean space. There are pencil strokes at just these three places, probably in Brouwer's hand.
⟦6⟧ P. Opitz 1881.

 1

Mathematics. — "*The force field of the non-Euclidean spaces with positive curvature*" by Mr. L. E. J. BROUWER. (Communicated by Prof. D. J. KORTEWEG). 1906 C2 [[1]]

(Communicated in the meeting of September 29, 1906).

D[1]). *The spherical Sp_2*.

I. The theorems under C § I and II hold invariably for the spherical and elliptical Sp_n's. But on account of the finiteness of these spaces we need not postulate a limiting field property for the following developments. We shall first consider the spherical spaces. [[2]]

Firstly we remark for the general linevector distribution of the spherical Sp_n that the total sum of the divergency is 0; for the outgoing vectorcurrents out of the different space-elements destroy each other. This proves already that as elementary ${}_0^1X$ we can but take the field of a double point.

SCHERING (Göttinger Nachrichten 1873), and KILLING (Crelle's Journal, 1885) give as elementary gradient field the derivative of the potential function $\int_r^{1/2\pi} \frac{dr}{\sin^{n-1} r} \equiv v_n(r)$. [2]) [[3]]

But the derivative of this field consists of two equal and opposite divergencies in two opposite points; and it is clear that an arbitrary integral of such fields always keeps equal and opposite divergencies in the opposite points, so it cannot furnish the general divergency-distribution limited only to a total divergency sum $= 0$.

II. If we apply for a spherical Sp_n the theorem of GREEN to the whole space (i. e. to the two halves, in which it is divided by an arbitrary closed Sp_{n-1}, together), doing this particularly for a scalar function φ which we presuppose to have nowhere divergency and a scalar function having only in two arbitrary points P_1 and P_2 equal and opposite divergencies and nowhere else (such functions we shall deduce in the following), we then find

$$\varphi_{P_1} - \varphi_{P_2} = 0,$$

i. o. w. φ is a constant, the points P_1 and P_2 being taken arbitrarily.

1) A, B and C refer to: "The force field of the non-Euclidean spaces with negative curvature". (See these Proceedings, June 30, 1906). [[2]]

2) We put the space constant $= 1$, just as we did in the hyperbolic spaces.

So there is no ${}_0^1X$ possible with nowhere divergency, thus no ${}^1X$ having nowhere rotation and nowhere divergency, and from this ensues:

A linevector distribution in a spherical Sp_n is determined uniformly by its rotation and its divergency.

III. The general vector distribution in a spherical Sp_n must thus be obtainable again as an arbitrary integral of:

1. fields E_1, whose second derivative consists of two equal and opposite scalar values close to each other.

2. fields E_2, whose first derivative consists of planivectors distributed regularly in the points of a small ${}^{n-2}$sphere and perpendicular to that ${}^{n-2}$sphere.

At finite distance from their origin the fields E_1 and E_2 have an identical structure.

IV. For the spherical Sp_2 there exists a simple way to find the field E_1 namely conform representation by stereographic projection of a Euclidean plane with a doublepoint potential, which double point is situated in the tangential point of the sphere and the plane. If we introduce on both surfaces as coordinates the distance to the double point and the angle of the radiusvector with the doublepoint-axis — in the plane ϱ and φ, on the sphere r and φ — we have:

$$\tfrac{1}{2}\varrho = \tan\tfrac{1}{2}r.$$

The potential in the plane: $\dfrac{\cos\varphi}{\varrho}$ becomes on the sphere:

$$\tfrac{1}{2}\cos\varphi\cot\tfrac{1}{2}r.$$

This potential shows nothing particular in the centre of projection on the sphere, so it is really the potential to be found of a single double point, the field E_1. (If we place in the opposite point of the double point an other double point in such a way that the unequal poles correspond as opposite points, we find as potential $\tfrac{1}{2}\cos\varphi\,(\cot\tfrac{1}{2}r+\tan\tfrac{1}{2}r)=\dfrac{\cos\varphi}{\sin r}$, which is the Schering potential of a double point).

V. Here too we can meanwhile break up the field of a double point into two fictitious "fields of a single agens point"; for this we have but to take $\int_r^\pi \tfrac{1}{2}\cot\tfrac{1}{2}r\,dr = -\,l\sin\tfrac{1}{2}r \equiv F(r)$; so that for an arbitrary gradient distribution holds

$$ {}^{1}_{0}X = \overline{\nabla}^{1} \int \frac{\overline{\nabla}^{2}\, {}^{1}_{0}X}{2\pi} F(r)\, d\tau \,. \quad \ldots \ldots \quad (I) $$

The "field of a single agens point" has however divergencies everywhere on the sphere.

VI. Out of the field E_1 we deduce in an analogous way as under B § VI the field E_2 of a rotation double point normal to the agens-doublepoint of the field E_1. As scalar value of the planivector potential we find there:

$$ \tfrac{1}{2} \sin \varphi \cot \tfrac{1}{2} r, $$

as we had to expect, completely dual to the scalar potential of the field E_1.

As fictitious force field of a unity-rotationelement we deduce out of this (in the manner of B § VI):

$$ \tfrac{1}{2} \cot \tfrac{1}{2} r, $$

directed normally to the radiusvector. For the rest this force field has rotation everywhere in Sp_2.

VII. Out of this we find (comp. under B § VII) for the scalar value of the planivector potential of a rotation-element:

$$ \int_r^{\pi} \tfrac{1}{2} \cot \tfrac{1}{2} r \, dr \equiv F(r), $$

so that for an arbitrary ${}^{1}_{2}X$:

$$ {}^{1}_{2}X = \overline{\nabla}^{2} \int \frac{\overline{\nabla}^{1}\, {}^{1}_{2}X}{2\pi} F(r)\, d\tau. \quad \ldots \ldots \quad (II) $$

And an arbitrary vectorfield is the ∇ of a potential:

$$ \int \frac{\nabla X}{2\pi} F(r)\, d\tau. $$

E. The spherical Sp_3.

I. The purpose is in the first place to find E_1; we shall compose it of some singular potential functions with simple divergency distributions, and which are easy to construct.

Let us suppose a principal ²sphere B with poles P_1 and P_2, and on B a principal circle C with poles Q_1 and Q_2 determining on B meridian circles M cutting C in points H.

We can construct in the first place out of the SCHERING potential the potential of two double points, in P_1 and P_2, the positive poles of which are both directed towards Q_1 (so that in opposite points unequal poles correspond). Let us determine a point S of the hypersphere by the distance $PS = r$ and $\angle QPS = \varphi$ (where for P and Q the index 1 or 2 must be taken according to S lying with P_1 or with P_2 on the same side of B), then this potential (α) becomes

$$\pm \frac{\cos \varphi}{\sin^2 r},$$

where the sign $+$ ($-$) must be taken for the half hyperspheres between P_1 (P_2) and B.

This field has no other divergency but that of the double points P_1 and P_2.

If we now reverse the sign of the potential in the half hypersphere on the side of P_2, we obtain the potential (β):

$$\frac{\cos \varphi}{\sin^2 r}.$$

The divergency of this consists in the first place of two double points, *one* directed in P_1 towards Q_1 and *one* directed in P_2 towards Q_2 (so that now in two opposite points equal poles correspond); and then of a magnetic scale (indeed a potential discontinuity) in sphere B varying in intensity according to $\cos \varphi$.

II. By the side of this we wish to find a potential, the divergency of which consists of only such a magnetic scale in sphere B with an intensity proportional to $\cos \varphi$. Now a field of a magnetic scale in B with an intensity varying according to an other zonal spherical harmonic, is easy to find. Let us namely take in each "meridian sphere" PQH as potential of a point S the angle $PHS = \frac{1}{2}\pi - \angle QHS$ (P and Q to be provided with indices in the way indicated above according to the place of S) $= \tan^{-1}\{\cos \varphi \tan r\}$, then we have such a potential: in the hypersphere it is a zonal spherical harmonic about PQ as axis; on the sphere B it has its only divergency in the shape of a magnetic scale, the intensity of which varies according to a zonal spherical harmonic with pole Q.

Let us now take in turns all the points of the sphere B as pole Q' of such a potential function, and let us integrate all those potentials over the solid angle about P each potential being multiplied by $\cos Q'Q$, then according to a wellknown theorem on spherical harmonics the integral is a zonal harmonic of form $\cos \varphi f(r)$, where $f(r) = \int \cos \varphi \, . \tan^{-1}\{\cos \varphi \tan r\} \, d\omega$, ($d\omega$ representing the element of the solid angle about P), whilst this integral field has as only divergency a magnetic scale in B with intensity proportional to $\cos\varphi$.

Effecting the integration we obtain:

$$f(r) = 2\pi \int_0^{\pi} \sin\varphi \cos\varphi \tan^{-1}\{\cos\varphi \tan r\}\, d\varphi.$$

$$f(r) = 2\pi \left\{ -\cot r + \frac{r}{\sin^2 r} \right\},$$

and for the corresponding potential function (γ) we find:

$$2\pi \cos\varphi \left\{ -\cot r + \frac{r}{\sin^2 r} \right\}.$$

III. If we take the difference of the field (β) multiplied by $\frac{1}{2}$ and the field (γ) multiplied by $\frac{1}{2\pi^2}$ the magnetic scale in B disappears and we have left the field (δ):

$$\frac{\cos\varphi}{\pi} \left\{ \frac{\frac{\pi}{2} - r}{\sin^2 r} + \cot r \right\},$$

which field has as only divergency two double points in P_1 and P_2 of which in the opposite points equal poles correspond.

The sum of this field (δ) and the field (α) multiplied by $\frac{1}{2}$ must now give a field having as divergency a single double point with unity-moment in P_1, i. o. w. the field E_1.

We therefore find on the half hypersphere between P_1 and B:

$$\frac{1}{\pi} \cos\varphi \left\{ \frac{\pi - r}{\sin^2 r} + \cot r \right\}$$

and on the half hypersphere between P_2 and B:

$$\frac{1}{\pi} \cos\varphi \left\{ \frac{-r}{\sin^2 r} + \cot r \right\},$$

or if we define on both halves the coordinates r and φ according to P_1 and $P_1\,Q_1$ we obtain the following expression holding for both halves:

$$\frac{1}{\pi} \cos\varphi \left\{ \frac{\pi - r}{\sin^2 r} + \cot r \right\} \equiv \psi(r) \cos\varphi.$$

IV. To break up this field into two fictitious "fields of a single agens point" (having however divergency along the whole hypersphere) we take for the latter $\int_r^{\pi} \psi(r)\, dr \equiv F_1(r)$.

Then for an arbitrary gradient distribution holds:

$$ {}_0^1X = \overline{\nabla}^1 \int \frac{\overline{\nabla}^2\, {}_0^1X}{4\pi} F_1(r)\, d\tau \quad . \; . \; . \; . \; . \; . \; . \qquad (I)$$

V: The field E_1 of a circular current according to the equator plane in the origin, is identical outside the origin to the above field E_1; but now each force line is closed, and has a line integral of 4π along itself.

According to the method of A § IX we find of this field E_2 the planivector potential H in the meridian plane and independent of the azimuth.

We find when writing $\pi - r = \beta$:

$$\Sigma = \frac{1}{\pi} \sin^2\varphi\, (1 + \beta \cot r)\, d\vartheta.$$

So

$$H = \frac{1}{\pi} \sin\varphi \frac{1 + \beta \cot r}{\sin r},$$

vanishing along all principal circles in the opposite point.

From which we deduce for the force of an element of current with unity-intensity in the origin directed according to the axis of the spherical system of coordinates:

$$\frac{1}{\pi} \sin\varphi \frac{1 + \beta \cot r}{\sin r},$$

directed normally to the meridianplane.

VI. From this we deduce as in A § XI a vector potential V of

[[4]] an element of current parallel to that element of current and a function of r only. For the scalar value U of that vector potential we have the differential equation:

$$-\frac{\partial}{\partial r}\left\{U \sin\varphi \sin r\, d\varphi\right\} dr - \frac{\partial}{\partial\varphi}\left\{U \cos\varphi\, dr\right\} d\varphi =$$

$$= \frac{1}{\pi} \sin\varphi \frac{1 + \beta \cot r}{\sin r} \,.\, dr \,.\, \sin r\, d\varphi.$$

Or:

$$U - \frac{\partial}{\partial r}\left\{U \sin r\right\} = \frac{1}{\pi}(1 + \beta \cot r),$$

of which the solution is

$$U = \frac{c}{\cos^2 \frac{1}{2} r} + \frac{1}{\pi}\left\{\frac{\frac{1}{4}\beta^2}{\cos^2 \frac{1}{2} r} - \frac{\beta}{\sin r}\right\}.$$

We choose $c = 0$, and we find as vector potential V of a unity-element of current:

$$\frac{1}{\pi}\left\{\frac{\frac{1}{4}\beta^2}{\cos^2\frac{1}{2}r} - \frac{\beta}{\sin r}\right\} \equiv F_2(r).$$

directed parallel to the element of current. The function $F_2(r)$ vanishes in the opposite point.

For an arbitrary flux now holds:

$$ {}_2^1X = \overset{2}{\nabla}\int \frac{\overset{1}{\nabla}\, {}_2^1X}{4\pi} F_2(r)\, d\tau. \quad . \quad . \quad . \quad . \quad . \quad (II)$$

And finally the arbitrary vector field X is the ∇ of the potential:

$$\int \frac{\overset{2}{\nabla}X}{4\pi} F_1(r)\, d\tau + \int \frac{\overset{1}{\nabla}X}{4\pi} F_2(r)\, d\tau.$$

F. *The spherical Sp_n.*

I. To find the field E_1 we set to work in an analogous way as for the spherical Sp_3. The principal sphere B becomes here a ${}^{n-1}$sphere B; the principal circle C of the points H a principal ${}^{n-2}$sphere C of the points H.

For the potential (α) is found:

$$\pm\frac{\cos\varphi}{\sin^{n-1} r};$$

for the potential (β):

$$\frac{\cos\varphi}{\sin^{n-1} r};$$

this field (β) has in the sphere B a magnetic ${}^{n-1}$scale.

The potential (γ) is integrated out of fields $\tan^{-1}\{\cos\varphi \tan r\}$ according to $\cos\varphi$, the first zonal ${}^{n-1}$spherical harmonic on B. This integration furnishes when dw represents the element of the n-dimensional solid angle about P:

$$\cos\varphi\, f(r),$$

where:

$$f(r) = \int \cos\varphi \tan^{-1}\left\{\cos\varphi \tan r\right\} dw = k_{n-1}\int_0^\pi \sin^{n-2}\varphi \cos\varphi \tan^{-1}\left\{\cos\varphi \tan r\right\} d\varphi =$$

$$= \frac{k_{n-1}}{n-1}\int_0^\pi \sin^n\varphi \frac{\tan r\, d\varphi}{1 + \tan^2 r \cos^2\varphi}$$

(k_n defined as under C § III).

Putting under the sign of the integral a factor $\sin^2\varphi \tan^2 r$ outside the brackets and, by regarding that factor as $\frac{1}{\cos^2 r} - (1 + \cos^2\varphi \tan^2 r)$, writing the integral as sum of two integrals to the former of which the same division in two is applied, etc., we find, if we write

$$\int_0^\pi \sin^h r \, dr = S_h :$$

$$\frac{(n-1) f(r)}{k_{n-1}} \sin^{n-1} r = -\sin^{n-2} r \cos r \, S_{n-2} - \sin^{n-4} r \cos r \, S_{n-4} \ldots .$$

$$\ldots . - \sin^2 r \cos r \, S_2 + \pi (1 - \cos r)$$

(for n even)

$$= -\sin^{n-2} r \cos r \, S_{n-2} - \sin^{n-4} r \cos r \, S_{n-4} \ldots .$$

$$\ldots . - \sin r \cos r \, S_1 + 2 r$$

(for n odd)

$$= \pi . \frac{(n-1)(n-3)\ldots .}{(n-2)(n-4)\ldots .} \int_0^r \sin^{n-1} r \, dr = (n-1) S_{n-2} \int_0^r \sin^{n-1} r \, dr,$$

(for n even)

$$= 2 . \frac{(n-1)(n-3)\ldots .}{(n-2)(n-4)\ldots .} \int_0^r \sin^{n-1} r \, dr = (n-1) S_{n-2} \int_0^r \sin^{n-1} r \, dr.$$

(for n odd)

If we write ξ_n for $2 . \pi . 2 . \pi . 2 \ldots .$, to n factors, we have

$$k_n = \frac{\xi_n}{(n-2)(n-4)\ldots .}, \quad \text{and} \quad \frac{k_{n+1}}{k_n} = S_{n-1}.$$

Therefore:

$$f(r) \sin^{n-1} r = k_n \int_0^r \sin^{n-1} r \, dr ,$$

and the potential (γ) becomes:

$$k_n \frac{\cos\varphi}{\sin^{n-1} r} \int_0^r \sin^{n-1} r \, dr.$$

II. We find the field (δ) by taking difference of field (β) multiplied by $\frac{1}{2}$ and field (γ) by $\frac{1}{k_n S_{n-1}} = \frac{1}{k_{n+1}}$, i. e.

$$\frac{\cos\varphi}{\sin^{n-1}r}\cdot\frac{\frac{1}{2}S_{n-1}-\int_0^r \sin^{n-1}r\,dr}{S_{n-1}}=\frac{\cos\varphi}{\sin^{n-1}r}\cdot\frac{\int_r^{\frac{1}{2}\pi}\sin^{n-1}r\,dr}{S_{n-1}}.$$

This field has as only divergency two double points, in P_1 and P_2, of which equal poles correspond in the opposite points. The field E_1 is then obtained by adding to it the field (α) multiplied by $\frac{1}{2}$. We find on the half ^{n}sphere between P_1 and B:

$$\frac{1}{S_{n-1}}\cdot\frac{\cos\varphi}{\sin^{n-1}r}\int_r^{\pi}\sin^{n-1}r\,dr.$$

On the half ^{n}sphere between P_2 and B:

$$-\frac{1}{S_{n-1}}\cdot\frac{\cos\varphi}{\sin^{n-1}r}\int_0^{r}\sin^{n-1}r\,dr.$$

Or, if we define on both halves the coordinates r and φ according to P_1 and P_1Q_1, we arrive at the expression holding for both halves:

$$\frac{1}{S_{n-1}}\cdot\frac{\cos\varphi}{\sin^{n-1}r}\cdot\int_r^{\pi}\sin^{n-1}r\,dr\equiv\psi_n(r\,\iota)\cos\varphi.$$

III. For the potential of the fictitious "field of a single agens point" we find:

$$\int_r^{\pi}\psi_n(r)\,dr\equiv F_1(r).$$

And for the arbitrary gradient distribution holds:

$$ {}^1_0X=\overline{\nabla}^{1}\!\!/\int\frac{\overline{\nabla}^{2}\!\!/\,{}^1_0X}{k_n}F_1(r)\,d\tau \quad . \quad . \quad . \quad . \quad . \quad . \quad (I)$$

Of the divergency distribution of $F_1(r)$ in points of a general position we know that, taken for two completely arbitrary centra (fictitious agens points) with opposite sign and then summed up, it furnishes 0: so on one side that distribution is independent of the position of the centre and on the other side it lies geometrically equivalent with respect to all points; so it is a constant. But if the function $F_1(r)$ has constant divergency in points of general position it satisfies a differential equation putting the divergency constant. In this is therefore a second means to determine the function F_1 and out of this the field E_1.

The differential equation becomes:

$$\frac{d}{dr}\left\{ \sin^{n-1} r \,.\, \frac{dF_1}{dr} \right\} = c \sin^{n-1} r \,. \quad . \quad . \quad . \quad . \quad (H)$$

$$\sin^{n-1} r \,.\, \frac{dF_1}{dr} = c \int \sin^{n-1} r \, dr.$$

$$\frac{dF_1}{dr} = c \,.\, \frac{\int \sin^{n-1} r \, dr}{\sin^{n-1} r}.$$

If the field E_1 is to be composed out of the function $F_1(r)$ then the opposite point of the centre may not have a finite outgoing vector current; we therefore put $\int \sin^{n-1} r \, dr = 0$, so that we get

$$\frac{dF_1}{dr} = - \frac{c}{\sin^{n-1} r} \int_r^{\pi} \sin^{n-1} r \, dr,$$

which corresponds to the above result.

IV. The field E_2 of a small vortex ${}^{n-2}$sphere according to Sp_{n-1}, perpendicular to the axis of the just considered double point, is identical to that field E_1 outside the origin; but now each force line is closed and has a line integral k_n along itself.

According to the method of C § VII we shall find of this field E_2 the planivector potential H, lying in the meridian plane and dependent only on r and φ ;so that it is a ${}^{2}_{1}X$. We find :

$$dh = c\varepsilon \sin^{n-2} r \sin^{n-2} \varphi.$$

Force in r-direction :

$$(n-1) \cos \varphi \left\{ \frac{1}{(n-1)\, S_{n-1}} + \frac{\cot r}{S_{n-1}} \,.\, \frac{\int_r^{\pi} \sin^{n-1} r \, dr}{\sin^{n-1} r} \right\} \equiv (n-1) \cos \varphi \,.\, \omega_n (r.)$$

$$\Sigma = \int_0^{\varphi} (n-1) \cos \varphi \, \omega_n (r) \,.\, c\varepsilon \sin^{n-2} r \sin^{n-2} \varphi \,.\, \sin r \, d\varphi =$$

$$= \omega_n r \,.\, c\varepsilon \sin^{n-1} r \sin^{n-1} \varphi.$$

$$H = \frac{\Sigma}{dh} = \omega_n (r) \sin r \sin \varphi \equiv \chi_n (r) \sin \varphi.$$

From this ensues for the force of a plane vortex element with unity-intensity in the origin :

$$\chi_n (r) \sin \varphi,$$

directed parallel to the acting vortex element and projecting itself on that plane according to the tangent to a concentric circle; whilst φ is the angle of the radiusvector with the Sp_{n-2} perpendicular to the vortex element.

V. In the same way as in C § IX we deduce from this the planivector potential V of a vortex element directed everywhere parallel to the vortex element and of which the scalar value is a function of r only. That scalar value U of that vector potential is here determined by the differential equation: [[4]]

$$-\frac{\partial}{\partial\varphi}\left\{U\cos\varphi\,.\,dr\,.\,c\varepsilon\sin^{n-3}r\cos^{n-3}\varphi\right\}d\varphi -$$

$$-\frac{\partial}{\partial r}\left\{U\sin\varphi\,.\,\sin r\,d\varphi\,.\,c\varepsilon\sin^{n-3}r\cos^{n-3}\varphi\right\}dr =$$

$$= \chi_n(r)\sin\varphi\,.\,\sin r\,d\varphi\,.\,dr\,.\,c\varepsilon\sin^{n-3}r\cos^{n-3}\varphi.$$

$$(n-2)\,U - \frac{dU}{dr}\sin r - (n-2)\,U\cos r = \chi_n(r)\sin r.$$

$$\frac{dU}{dr} - (n-2)\,U\,tg\,\tfrac{1}{2}r = -\chi_n(r).$$

$$U = \frac{1}{\cos^{2(n-2)}\frac{1}{2}r}\cdot\int_r^{\pi}\cos^{2(n-2)}\tfrac{1}{2}r\,.\,\chi_n(r)\,dr,$$

a function vanishing in the opposite point, which we put $\equiv F_2(r)$.
We then find for an arbitrary flux:

$$\tfrac{1}{2}X = \overline{\nabla}^{2}\int\frac{\overline{\nabla}^{1}\,\tfrac{1}{2}X}{k_n}F_2(r)\,d\tau \quad \ldots\ldots \quad (II)$$

And taking an arbitrary vector field to be caused by its two derivatives (the magnets and the vortex systems) propagating themselves through space as a potential according to a function of the distance vanishing in the opposite point, we find:

$$X = \nabla\left\{\int\frac{\overline{\nabla}^{2}X}{k_n}F_1(r)\,d\tau + \int\frac{\overline{\nabla}^{1}X}{k_n}F_2(r)\,d\tau\right\}.$$

G. *The Elliptic Sp_n.*

Also for the elliptic Sp_n the derivative of an arbitrary linevector distribution is an integral of elementary vortex systems Vo_y and Vo_z, which are respectively the first and the second derivative of an isolated line vector. For elementary ${}_0^1X$ we shall thus have to put the field of a divergency double point.

The Schering elementary potential $\int_r^{1/2\pi} \frac{dr}{\sin^{n-1} r} \equiv v_n(r)$ is here a plurivalent function (comp. KLEIN, Vorlesungen über Nicht-Euklidische Geometrie II, p. 208, 209); it must thus be regarded as senseless.

[[5]] II. The unilateral elliptic Sp_n is enclosed by a plane Sp_{n-1}, regarded twice with opposite normal direction, as a bilateral singly connected Sp_n-segment by a bilateral closed Sp_{n-1}. If we apply to the Sp_n enclosed in this way the theorem of GREEN for a scalar function φ having nowhere divergency, and for one having in two arbitrary points P_1 and P_2 equal and opposite divergencies and fartheron nowhere (such a function will prove to exist in the following), we shall find:

$$\varphi_{P_2} - \varphi_{P_1} = 0,$$

i. o. w. φ is a constant, the points P_1 and P_2 being arbitrarily chosen.

So no ${}_0^1X$ is possible having nowhere divergency, so no ${}^1X$ having nowhere rotation and nowhere divergency; and from this ensues:

A linevector distribution in an elliptical Sp_n is uniformly determined by its rotation and its divergency.

III. So we consider:

1. the field E_1, with as second derivative two equal and opposite scalar values quite close together.

2. the field E_2 with as first derivative planivectors regularly distributed in the points of a small ${}^{n-2}$sphere and perpendicular to that small ${}^{n-2}$sphere.

At finite distance from their origin the fields E_1 and E_2 are of identical structure.

IV. To find the potential of the field E_1 we shall represent it uni-bivalently on the spherical Sp_n; the representation will have as divergency two doublepoints in opposite points, where equal poles correspond as opposite points; it will thus be the field (σ), deduced under F § II, multiplied by 2:

[[84]]

$$\frac{\cos\varphi}{\sin^{n-1} r} \cdot \frac{\int_r^{1/2\pi} \sin^{n-1} r \, dr}{\frac{1}{2} S_{n-1}} \equiv \lambda_n(r) \cos\varphi.$$

In the field corresponding to this in the elliptic space, all force lines move from the positive to the negative pole of the double point; a part cuts the pole Sp_{n-1} of the origin: these force lines are unilateral in the meridian plane; the remaining do not cut it; these are bilateral in the meridian plane.

The two boundary force lines forming together a double point in the pole Sp_{n-1}, have the equation:

$$\sin^{n-1}\varphi \left\{ \sin^{n-1} r + (n-1) \cot r \int_r^{1/2\pi} \sin^{n-1} r \, dr \right\} = \pm 1.$$

The Sp_{n-1} of zero potential consists of the pole Sp_{n-1} and the equator Sp_{n-1} of the double point; its line of intersection with the meridian plane has a double point in the force lines doublepoint. All potential curves in the meridian plane are bilateral.

V. For the fictitious "field of a single agens point" the potential is $\int \lambda_n(r)\, dr$. It is rational to let it become 0 in the pole Sp_{n-1}; so we find:

$$\int_r^{1/2\pi} \lambda_n(r)\, dr \equiv F_1(r),$$

and for the arbitrary gradient distribution holds:

$$\overset{1}{_0}X = \overline{\sqrt[1]{}} \int \frac{\overline{\sqrt[2]{}}\ \overset{1}{_0}X}{k_n} F_1(r)\, d\tau \quad . \quad . \quad . \quad . \quad . \quad . \quad (I)$$

We could also have found $F_1(r)$ out of the differential equation (H) of F § III, which it must satisfy on the same grounds as have been asserted there. For the elliptic Sp_n also we find:

$$\frac{dF_1}{dr} = c \, . \, \frac{\int \sin^{n-1} r \, dr}{\sin^{n-1} r}.$$

But here in the pole Sp_{n-1}, lying symmetrically with respect to the centre of the field, the force, thus $\int \sin^{n-1} r \, dr$ must be 0; so that we find:

$$\frac{dF_1}{dr} = -\frac{c}{\sin^{n-1} r}\int_r^{1/2\pi} \sin^{n-1} r\, dr.$$

VI. In the usual way we deduce the $\overset{2}{{}_1X}$, which is planivector potential of the field E_2.

$$dh = c\varepsilon \sin^{n-2} r \sin^{n-2} \varphi.$$

Force in r-direction:

$$(n-1)\cos\varphi \left\{\frac{2}{(n-1)S_{n-1}} + \frac{2\cot r}{S_{n-1}} \cdot \frac{\int_r^{1/2\pi} \sin^{n-1} r\, dr}{\sin^{n-1} r}\right\} \equiv (n-1)\cos\varphi \,.\, \mu_n(r).$$

$$\Sigma = \int_0^{\varphi} (n-1)\cos\varphi \,.\, \mu_n(r) \,.\, c\varepsilon \sin^{n-2} r \sin^{n-2}\varphi \,.\, \sin r\, d\varphi =$$

$$= \mu_n(r) \,.\, c\varepsilon \sin^{n-1} r \sin^{n-1}\varphi.$$

$$H = \frac{\Sigma}{dh} = \mu_n(r) \sin r \sin\varphi \equiv \varkappa_n(r) \sin\varphi.$$

From which ensues for the force of a plane vortex element with unity-intensity in the origin:

$$\varkappa_n(r) \sin\varphi,$$

directed parallel to the acting vortex element and projecting itself on its plane according to the tangent to a concentric circle; φ is here the angle of the radiusvector with the Sp_{n-2} perpendicular to the vortex element.

[[4]] VII. Here too a planivector potential of a vortex element can be deduced, but we cannot speak of a direction propagated parallel to itself, that direction not being uniformly determined in elliptic space; after a circuit along a straight line it is transferred into the symmetrical position with respect to the normal plane on the straight line.

But we can obtain a vector potential determined uniformly, by taking that of two antipodic vortex elements in the spherical Sp_n (in their ^{2}sphere the two indicatrices are then oppositely directed).

The vector potential in a point of the elliptic Sp_n then lies in the space through that point and the vortex element; if we regard the plane of the element as equator plane in that space then the planivector potential V is normal to the meridian plane: it consists of:

1. a component U_1 normal to the radiusvector, according to the formula:

$$\frac{U_1}{\cos\varphi} = \frac{1}{\cos^{2(n-2)}\frac{1}{2}r}\int_r^{\pi}\cos^{2(n-2)}\tfrac{1}{2}r\,\chi_n(r)\,dr +$$

$$+\frac{1}{\sin^{2(n-2)}\frac{1}{2}r}\int_{\pi-r}^{\pi}\cos^{2(n-2)}\tfrac{1}{2}r\,.\,\chi_n(r)\,dr.$$

2. a component U_2 through the radiusvector, according to the formula:

$$\frac{U_2}{\sin\varphi} = \frac{1}{\cos^{2(n-2)}\frac{1}{2}r}\int_r^{\pi}\cos^{2(n-2)}\tfrac{1}{2}r\,\chi_n(r)\,dr -$$

$$-\frac{1}{\sin^{2(n-2)}\frac{1}{2}r}\int_{\pi-r}^{\pi}\cos^{2(n-2)}\tfrac{1}{2}r\,.\,\chi_n(r)\,dr.$$

If we regard this planivector potential as function of the vortex element and the coordinates with respect to the vortex element and represent that function by G_2, then

$$\tfrac{1}{2}X = \overline{\nabla}^{2}\int\frac{G_2\left\{\overline{\nabla}^{1}\,\tfrac{1}{2}X,\,r,\,\varphi\right\}}{k_n}\,d\tau. \quad . \quad . \quad . \quad . \quad (II)$$

holds for an arbitrary flux in the elliptic Sp_n.

And regarding an arbitrary vector field as caused by the two derivatives (the magnets and the vortex systems) propagating themselves through the space to a potential, we write:

$$X = \nabla\int\frac{\overline{\nabla}^{2}X\,.\,F_1(r) + G_2\left(\overline{\nabla}^{1}X,\,r,\,\varphi\right)}{k_n}\,d\tau.$$

VIII. In particular for the elliptic Sp_3 the results are:

Potential of an agens double point:

$$\frac{\cos\varphi}{\sin^2 r}\cdot\frac{\int_r^{\frac{1}{2}\pi}\sin^2 r\,dr}{\frac{1}{2}S_2} = \frac{2\cos\varphi}{\pi}\cdot\left\{\frac{(\frac{1}{2}\pi - r)}{\sin^2 r} + \cot r\right\},$$

or if we put $\frac{1}{2}\pi - r = \gamma$:

$$\frac{2\cos\varphi}{\pi}\cdot\left\{\frac{\gamma}{\sin^2 r} + \cot r\right\}.$$

Equation of the boundary lines of force:

$$\sin^2\varphi\,(1 + \gamma \cot r) = \pm 1.$$

Potential of a single agens point:

$$\frac{2}{\pi} \cdot \gamma \cdot \cot r.$$

Vector potential of an elementary circular current:

$$\frac{2}{\pi} \sin\varphi \cdot \frac{1 + \gamma \cot r}{\sin r}.$$

So also force of an element of current:

$$\frac{2}{\pi} \sin\varphi \cdot \frac{1 + \gamma \cot r}{\sin r}.$$

Linevector potential of an element of current:

according to the radiusvector: $\dfrac{\cos\varphi}{\pi}\left\{\dfrac{\frac{1}{4}\beta^2}{\cos^2\frac{1}{2}r} - \dfrac{\pi}{\sin r} + \dfrac{\frac{1}{4}r^2}{\sin^2\frac{1}{2}r}\right\}.$

normal to the radiusvector: $\dfrac{\sin\varphi}{\pi}\left\{\dfrac{\frac{1}{4}\beta^2}{\cos^2\frac{1}{2}r} + \dfrac{2r - \pi}{\sin r} - \dfrac{\frac{1}{4}r^2}{\sin^2\frac{1}{2}r}\right\}.$

IX. For the elliptic plane we find:

Potential of an agens double point:

$$\cos\varphi \cot r.$$

Equation of the boundary lines of force:

$$\sin\varphi = \pm \sin r, \quad \text{or} \quad \varphi = \begin{cases} r \\ \pi - r \end{cases}.$$

Potential of a single agens point:

$$- l \sin r.$$

Scalar value of the planivector potential of a double point of rotation:

$$\frac{\sin\varphi}{\sin r}.$$

Thus also force of a rotation element:

$$\frac{\sin\varphi}{\sin r}.$$

Planivector potential of a rotation element:

$$l \cot \tfrac{1}{2} r.$$

We notice that the duality of both potentials and both derivatives existing for the spherical Sp_2, has disappeared again in these results. The reason of this is that for the representation on the sphere a divergency in the elliptic plane becomes two equal divergencies in opposite points with equal signs; a rotation two equal rotations in opposite points with different signs; for the latter we do not find the analogous potential as for the former; the latter can be found here according to the Schering potential formula.

With this is connected immediately that in the elliptic plane the field of a single rotation (in contrast to that of a single divergency) has as such possibility of existence, so it can be regarded as unity of field. That field consists of forces touching concentric circles and great $\frac{1}{\sin r}$.

P o s t s c r i p t. In the formula for vector fields in hyperbolic spaces:

$$Pot.\, X = \int \frac{\overline{\nabla}^{2} X}{k_n} F_1(r)\, d\tau + \int \frac{\overline{\nabla}^{1} X}{k_n} F_2(r)\, d\tau$$

nothing for the moment results from the deduction but that to $\overline{\nabla}^{2} X$ and $\overline{\nabla}^{1} X$ also must be counted the contributions furnished by infinity. From the field property ensues, however, immediately that the effect of these contributions disappears in finite, so that under the integral sign we have but to read $\overline{\nabla}^{2} X$ and $\overline{\nabla}^{1} X$ in finite.

For the $\overline{\nabla}^{1}$ at infinity pro surface-unity of the infinitely great sphere is $<$ order e^{-r}; the potential-effect of this in finite becomes $<$ order $re^{-(n-2)r} \times e^{-r} = re^{-(n-1)r}$; so the force-effect $<$ order $e^{-(n-1)r}$; so the force-effect of the entire infinitely great spherical surface is infinitesimal.

And the $\overline{\nabla}^{2}$ at infinity pro surface-unity is $<$ order $\frac{1}{r}$; it furnishes a potential-effect in finite $<$ order $e^{-(n-1)r} \cdot \frac{1}{r}$, thus a force-effect $<$ order $e^{-(n-1)r} \cdot \frac{1}{r^2}$; so the force-effect, caused by the infinite, remains $<$ order $\frac{1}{r^2}$.

The reasoning does not hold for the force field of the hyperbolical Sp_2 in the second interpretation (see under B § VIII), but it is in the nature of that interpretation itself that the derivatives at infinity are indicated as such, therefore also counted.

NOTES

⟦1⟧ The English version appeared in the Proceedings of the meeting of October 27, 1906. The paper extends Brouwer 1906 A2, 1906 B2.

⟦2⟧ Brouwer 1906 B2.

⟦3⟧ E. Schering 1873, W. Killing 1885.

⟦4⟧ Again pencil strokes at three places where parallel displacement of vectors is mentioned. See 1906 B2 ⟦5⟧.

⟦5⟧ The displacement of 'unilateral' from 'the' to 'a' is in Brouwer's hand. See also the Erratum in Brouwer 1909 H2, p. 297.

CHAPTER 2

Lie groups

Mathematics. — "*About difference quotients and differential quotients*". By Dr. L. E. J. BROUWER (Communicated by Prof. D. J. KORTEWEG). 1908 D2 [[1]]

(Communicated in the meeting of May 30, 1908).

Different investigations have been made which are very completely summed up in the work of DINI: "Grundlagen für eine Theorie der Functionen einer veränderlichen reellen Grösse" Chapt. XI and XII, on the connection between difference quotients and differential quotients, particularly on the necessary and satisfactory properties which the difference quotients must possess in order that there be a differential quotient. One however always regards in the first place these different difference quotients in one and the same point x_0 together, forming as a function of the increase of x the *derivatory function in* x_0. The existence of a differential quotient means then, that that derivatory function has a single limiting point in x_0 i.o.w. that in x_0 the right as well as the left *derivatory oscillation* is equal to zero. [[2]]

Other conditions for the existence of a differential quotient are found when in the first place the difference quotient for constant x-increase Δ is regarded as a function of x and then the set or these functions for varying Δ is investigated. Let $f(x)$ be the given function which we suppose to be finite and continuous and let $\varphi_\Delta(x)$ be the difference quotient for a constant x-increase Δ. The different functions $\varphi_\Delta(x)$ form an infinite set of functions, in which each function is continuous. We shall occupy ourselves with the case that the set is *uniformly continuous*, i. e. that for any quantity ε, however small it may be, a quantity σ can be pointed out so that in any interval of the size of σ not one of the functions of the set has oscillations larger than ε. Concerning infinite uniformly continuous sets of functions there is a theorem that if they are limited (i.e., if a maximum value and a minimum value can be given between which all functions move) they possess *at least one* continuous limiting curve, to which uniform convergence takes place [1]).

We shall prove, that for the set of functions of the difference quotients of a finite continuous function, if it be uniformly continuous, follows in the first place the limitedness and furtheron for indefinite decrease of the x-increase the existence of only *one* limiting curve, so that holds:

Theorem 1. If a finite continuous function $f(x)$ has a uniformly continuous set of difference quotients, then it possesses a finite continuous differential quotient [2]).

To prove this we call ${}_{\delta}\varepsilon_{\Delta}(x)$ the size of the region of oscillation between x and $x+\Delta$ of the difference quotient for an x-increase δ. If we allow δ to assume successively all positive values, then it follows from the supposed uniform continuity, that Δ can always be chosen so small as to keep all values ${}_{\delta}\varepsilon_{\Delta}(x)$ below a certain amount as small as one cares to make it. If we thus call $\varepsilon_{\Delta}(x)$ the maximum of the values ${}_{\delta}\varepsilon_{\Delta}(x)$ for definite x and Δ, then $\varepsilon_{\Delta}(x)$ tends with Δ uniformly to 0.

We have fartheron if $\frac{p}{n}$ is a proper fraction:

$$\varphi_{\Delta}(x)=\frac{1}{n}\varphi_{\frac{\Delta}{n}}(x)+\frac{1}{n}\varphi_{\frac{\Delta}{n}}\left(x+\frac{1}{n}\Delta\right)+\dots+\frac{1}{n}\varphi_{\frac{\Delta}{n}}\left(x+\frac{n-1}{n}\Delta\right)\quad(1)$$

$$\varphi_{\frac{p}{n}\Delta}(x)=\frac{1}{p}\varphi_{\frac{\Delta}{n}}(x)+\frac{1}{p}\varphi_{\frac{\Delta}{n}}\left(x+\frac{1}{n}\Delta\right)+\dots+\frac{1}{p}\varphi_{\frac{\Delta}{n}}\left(x+\frac{p-1}{n}\Delta\right).\quad(2)$$

If we break up each of the n terms of the second member of (1) into p equal parts and each of the p terms of the second member of (2) into n equal parts, then the difference of those two second members can be divided into pn terms, each remaining in absolute value smaller than $\frac{1}{pn}\cdot\varepsilon_{\Delta}(x)$, so that the difference of $\varphi_{\Delta}(x)$ and $\varphi_{\frac{p}{n}\Delta}(x)$ remains smaller than $\varepsilon_{\Delta}(x)$ in absolute value.

⟦3⟧ 1) Compare ARZELÀ, "*Sulle funzioni di linee*", Memorie della Accademia di Bologna, serie 5, V, page 225.

2) We suppose the function to be given in a certain interval of values of the independent variable x.

So if we regard for any definite x all difference quotients the x-increases of which are equal to proper fractions of Δ, then the amount $\tau_\Delta(x)$ of their region of oscillation is smaller than $2\varepsilon_\Delta(x)$. The same holds for the region of oscillation of *all* difference quotients for definite x with x-increases smaller than Δ, because these can be approximated by the preceding on account of the continuity of f.

So if we allow Δ to decrease indefinitely, then also $\tau_\Delta(x)$ decreases indefinitely; as furthermore when Δ becomes smaller, each following region of oscillation is a part of the preceding, the limit of the region of oscillation is for each x a single definite value, to which uniform convergence takes place, which is the limit of the difference quotients, the *differential quotient.*

That this (forward) differential quotient cannot show any discontinuities, is evident as follows: If there were a discontinuity, then there would be a quantity σ which could be overstepped there for any interval by the oscillations of the differential quotient; but if the value of the differential quotient differs in two points more than σ, then there is also a difference quotient, the values of which in those two points differ more than σ; so there would be for each interval, which contains the indicated discontinuity, a difference quotient with a region of oscillation larger than σ, i. o. w. the functional set of the difference quotients would not be uniformly continuous.

Out of the continuity of the forward differential quotient follows at the same time that the forward and the backward differential quotient are equal.

Analogously it is evident that also a point at infinity in the differential quotient would disturb the uniform continuity of the difference quotients; in this is at the same time included the limitedness of the difference quotients, for they would otherwise on account of the finiteness of f be able to tend to infinity only for indefinitely decreasing x-increase, but that would furnish an infinity point in the differential quotient.

Theorem 2. Of a function with finite continuous differential quotient the difference quotients are uniformly continuous.

Let namely ε be a definite quantity, to be taken as small as one likes. Now we may have each x included by an interval i in such a way, that the oscillations of the differential quotient within each of those intervals remain smaller than $^1/_2\,\varepsilon$. On account of the uniform convergence, evident from the formula $\varphi_\Delta(x) = f'(x + \vartheta\Delta)$, a Δ' can be pointed out in such a way that all φ_Δ for which $\Delta < \Delta'$ differ from the differential quotient less than $^1/_4\,\varepsilon$ for any x, thus have their oscillations below ε in the intervals mentioned. On account of the uniform continuity of f we may furthermore have each x included by an interval i' chosen in such a way that for all $\Delta \geqq \Delta'$ the corresponding φ_Δ have within those intervals oscillations below ε only; to that end we have but to choose i' in such a way that the oscillations of f remain within the intervals below $^1/_2\, \varepsilon\, \Delta'$. If thus i'' is the smaller of the two quantities i and i', each x can be included by an interval i'' in such a way, that the oscillations of *all* difference quotients within it remain below ε, with which we have proved the uniform continuity of the difference quotients.

Theorem 3. If there is among the difference quotients of a finite continuous function a uniform continuous fundamental series with indefinitely decreasing x-increases, there exists a finite continuous differential quotient.

Let namely $\varphi_{\Delta'}(x), \varphi_{\Delta''}(x), \ldots$ be the fundamental series of functions under consideration, then for any quantity ε we can point out a quantity σ in such a way that $\varphi_{\Delta^{(\nu)}}(x+h) - \varphi_{\Delta^{(\nu)}}(x) < \varepsilon$ for any x, any $h < \sigma$ and any ν. If now the set of *all* difference quotients were not uniformly continuous, it would have to occur that for a certain Δ° not belonging to the fundamental series we should have $\varphi_{\Delta^\circ}(x+h) - \varphi_{\Delta^\circ}(x) > \varepsilon$. If we now approximate Δ° by a series $\alpha_1\Delta', \alpha_2\Delta'', \ldots$, where the α's represent integers, in such a way that $\alpha_p\,\Delta^{(p)} < \Delta^\circ < (\alpha_p+1)\Delta^{(p)}$, then also $\varphi_{\Delta^\circ}(x+h) - \varphi_{\Delta^\circ}(x)$ is approximated by $\varphi_{\alpha_p\Delta^{(p)}}(x+h) - \varphi_{\alpha_p\Delta^{(p)}}(x)$, which last expression always remains $< \varepsilon$ however large p may become, so that $\varphi_{\Delta^\circ}(x+h) - \varphi_{\Delta^\circ}(x)$ cannot be $> \varepsilon$, so the set of *all* difference quotients is uniformly continuous, and there is a finite continuous differential quotient.

Theorem 1 is applied when building up the theory of continuous groups out of the theory of sets, (where one remains independent of LIE's postulates), in a certain region finite and continuous functions of one or more variables occurring there, whose difference quotients are in a certain system of coordinates linear functions of the original functions. [1]) As on account of the finiteness of the original functions there cannot be a region within which any quantity could be overstepped everywhere by one and the same difference quotient, the

[[4]] [1]) Comp. L. E. J. BROUWER, "*Die Theorie der endlichen continuierlichen Gruppen unabhängig von den Axiomen von* LIE", Atti del IV° Congresso Internazionale dei Matematici. It is the differentiability in one and the same system of coordinates of all the functions, which express the different infinitesimal transformations of a group, which is proved in this way.

coefficients of the above mentioned linear functions remain within finite limits, the system of the difference quotients is uniformly continuous, and the differential quotients exist.

Theorem 4. If the conditions of theorem 1 are satisfied and if the system of all second difference quotients (of which each is determined by two independent x-increases) forms a uniformly continuous system, then there exists a finite continuous "second differential quotient", which at the same time is the only limit of the above set of functions when both x-increases decrease indefinitely, and the differential quotient of the (first) differential quotient.

To prove this we call $\varepsilon'_\Delta(x)$ the maximum size of the regions of oscillation of the different second difference quotients between x and $x + \Delta$; then again $\varepsilon'_\Delta(x)$ tends with Δ uniformly to zero.

If we represent the difference quotient of $\varphi_{\Delta_1}(x)$ for an x-increase Δ_2 by $\varphi_{\Delta_1\Delta_2}(x)$ and if $\frac{p_1}{n_1}$ and $\frac{p_2}{n_2}$ are proper fractions then we have:

$$\varphi_{\Delta_1\Delta_2}(x) = \frac{1}{n_1 n_2}\sum_{k_1=0}^{n_1-1}\sum_{k_2=0}^{n_2-1}\varphi_{\frac{\Delta_1}{n_1}\frac{\Delta_2}{n_2}}\left(x + k_1\frac{\Delta_1}{n_1} + k_2\frac{\Delta_2}{n_2}\right). \quad . \quad (1)$$

$$\varphi_{\frac{p_1\Delta_1}{n_1}\frac{p_2\Delta_2}{n_2}}(x) = \frac{1}{p_1 p_2}\sum_{k_1=0}^{p_1-1}\sum_{k_2=0}^{p_2-1}\varphi_{\frac{\Delta_1}{n_1}\frac{\Delta_2}{n_2}}\left(x + k_1\frac{\Delta_1}{n_1} + k_2\frac{\Delta_2}{n_2}\right). \quad . \quad (2)$$

If we break up each of the $n_1 n_2$ terms of the second member of (1) into $p_1 p_2$ equal parts and each of the $p_1 p_2$ terms of the second member of (2) into $n_1 n_2$ equal parts, then the difference or those two members breaks up into $p_1 p_2\ n_1 n_1$ terms, each of which remaining in absolute value smaller than $\frac{1}{p_1 p_2\, n_1 n_2}\,\varepsilon'_{\Delta_1+\Delta_2}(x)$, so that the difference of $\varphi_{\Delta_1\Delta_2}(x)$ and $\varphi_{\frac{p_1\Delta_1}{n_1}\frac{p_2\Delta_2}{n_2}}(x)$ remains in absolute value smaller than $\varepsilon'_{\Delta_1+\Delta_2}(x)$.

So if we consider for any definite x all difference quotients whose x-increases are equal to proper fractions of Δ_1 and Δ_2, then the size $\tau_{\Delta_1\Delta_2}(x)$ of their region of oscillation is smaller than $2\varepsilon'_{\Delta_1+\Delta_2}(x)$, from which we deduce as above in the proof of theorem 1 the existence of one single limit, to which the convergence is uniform and which is finite and continuous.

If we now regard the difference quotient with x-increase Δ_2, on one hand for all φ_Δ's, whose Δ is smaller than Δ_1, and on the other hand for the (first) differential quotient, then the former all differ less than $\varepsilon'_{\Delta_1+\Delta_2}(x)$ from the limiting function just deduced, so also the latter, which can be approximated by them. This holds independently of Δ_1;

the difference for x-increase Δ_2 of the (first) differential quotient can therefore not differ more than $\varepsilon'_{\Delta_2}(x)$ from the just deduced limiting function which is thus differential quotient of the (first) differential quotient i.e. second differential quotient.

Theorem 5. If a function possesses a finite continuous second differential quotient, then the system of the first and second difference quotients is uniformly continuous.

To find namely an interval size i'' which keeps the oscillations of all second difference quotients everywhere $< \varepsilon$, we first take the interval size i, which keeps the oscillations of the second differential quotient everywhere $< \frac{1}{2}\varepsilon$; then a Δ'_1 and a Δ'_2 in such a way, that all $\varphi_{\Delta_1 \Delta_2}$, for which $\Delta_1 < \Delta'_1$ and $\Delta_2 < \Delta'_2$, differ along the whole course less than $^1/_4\varepsilon$ from the second differential quotient [1]; finally an interval size i' which keeps the oscillations of the function f everywhere $< {}^1/_4\, \varepsilon\, \Delta'_1\, \Delta'_2$. For i'' we take the smaller of the two quantities i and i'.

Theorem 6. If there is among the second difference quotients of a finite continuous function with finite continuous differential quotient a uniformly continuous fundamental series, in which the two x-increases decrease indefinitely, then there exists a finite continuous second differential quotient.

Let namely $\varphi_{\Delta_1' \Delta_2'}(x)$, $\varphi_{\Delta_1'' \Delta_2''}(x)$... be the indicated fundamental series, then for any quantity ε a quantity σ can be pointed out in such a way, that $\varphi_{\Delta_1^{(\nu)} \Delta_2^{(\nu)}}(x+h) - \varphi_{\Delta_1^{(\nu)} \Delta_2^{(\nu)}}(x) < \varepsilon$ for any x, any $h < \sigma$ and any ν. If now the set of *all* second difference quotients were not uniformly continuous, then it would be possible for a certain Δ_1° and Δ_2° not belonging to the fundamental series, that $\varphi_{\Delta_1^\circ \Delta_2^\circ}(x+h) - \varphi_{\Delta_1^\circ \Delta_2^\circ}(x) > \varepsilon$. Let us now approximate Δ_1° by means of a series $\alpha_1 \Delta_1'$, $\alpha_2 \Delta_1''$, ... and Δ_2° by means of a series $\beta_1 \Delta_2'$, $\beta_2 \Delta_2''$, ..., where the α's and β's represent integers, in such a way that

$$\alpha_p \Delta_1^{(p)} < \Delta_1^\circ < (\alpha_p + 1)\, \Delta_1^{(p)} \text{ and } \beta_p \Delta_2^{(p)} < \Delta_2^\circ < (\beta_p + 1)\, \Delta_2^{(p)},$$

[1]) The uniform convergence of all difference quotients is evident from that of the difference quotients, for which $\Delta_1 = \Delta_2$ (out of these the other can be approximated in the manner indicated in the proof of theorem 6); the latter is evident by developing the terms of $f(x+2\Delta) - 2f(x+\Delta) + f(x)$ according to TAYLOR's series, in which we make the second differential quotient form the restterm; the terms preceding this restterm then destroy each other.

then also $\varphi_{\Delta_1^\circ\ \Delta}(x+h) - \varphi_{\Delta_1^\circ\ \Delta_2^\circ}(x)$ is approximated by

$$\varphi_{\alpha_v \Delta_1^{(p)}, \beta_p \Delta_2^{(p)}}(x+h) - \varphi_{\alpha_p \Delta_1^{(p)}, \beta_p \Delta_2^{(p)}}(x),$$

which last expression remains $< \varepsilon$, however great p may become, so that the first can neither be $> \varepsilon$,; so the set of *all* second difference quotients is uniformly continuous and there is a finite continuous second differential quotient.

Theorem 7. If there is among the second difference quotients of a finite continuous function a uniformly continuous fundamental series, in which both x-increases decrease indefinitely, the function possesses finite and continuous first and second differential quotients.

For, according to the above given proof of theorem 6 the whole system of the second difference quotients proves to be uniformly continuous, and out of the above given proof of theorem 4 this system proves to possess for indefinite decrease of the two x-increases one single finite continuous limiting function $f''(x)$ to which they converge uniformly. Let τ' be the maximum deviation from this limiting function of the second difference quotients, whose x-increases are smaller than Δ'_1 and Δ'_2, and let us regard the system ζ of all $\varphi_\Delta(x)$ whose $\Delta < \Delta'_1$, then all difference quotients with x-increase $< \Delta'_2$ of the system ζ lie between $f''(x) + \tau'$ and $f''(x) - \tau'$, from which may be deduced easily, that the system ζ is uniformly continuous, so that now first according to the proof of theorem 1 a finite continuous first differential quotient exists and then according to the proof of theorem 4 a finite continuous second differential quotient.

Analogous to the preceding are the proofs of the following more general theorems:

Theorem 8. If there is among the nth difference quotients of a finite continuous function a uniformly continuous fundamental series, in which all x-increases decrease indefinitely, then the function possesses finite and continuous first, second, up to the nth differential quotients; each pth differential quotient is here first the only limit for indefinitely decreasing x-increases of the pth difference quotients to which limit a uniform convergence takes place, and then differential quotient of the $(p-1)^{\text{st}}$ differential quotient.

Theorem 9. If a function possesses a finite continuous n^{th} differential quotient, then the system of the first, second, up to the nth difference quotients is uniformly continuous [1]).

Theorem 10. If $n_1, n_2, n_3 \ldots$ is an infinite series of increasing integers and if of a finite continuous function the systems of the n_1^{st}, of the $n_2^{nd} \ldots$ difference quotients are uniformly continuous, then the function has all differential quotients and these are all finite and continuous.

Theorem 11. A finite continuous function of several variables, among whose difference quotients of the nth order there is for each kind a uniformly continuous fundamental series in which the increases of the independent variables decrease indefinitely, possesses all differential quotients up to the nth order; these finite and continuous differential quotients are first each other's differential quotients in the manner expressed by their form, where the order of succession of the differentiations proves to be irrelevant, and then each differential quotient is the only limit of the corresponding difference quotients for indefinitely decreasing increases of the independent variables, to which limit a uniform convergence takes place.

Theorem 12. If a function of several variables possesses all kinds of nth differential quotients and if these are finite and continuous, then the system of the 1st, 2nd up to the nth difference quotients is uniformly continuous. [2])

Finally the observation, that what was treated here leads infinite differentiability back to continuity in a more extensive sense, and in this way may somewhat explain, that for so long all finite continuous functions were supposed to be infinitely differentiable, and may somewhat justify that so many wish to limit themselves in natural science to infinitely differentiable functions.

---

1) To prove the uniform convergence of the nth difference quotients for equal independent x-increases we break off just as for theorem 5 the Taylor development at the nth term and we apply the formula:

$$n^a - \binom{1}{n}(n-1)^a + \binom{2}{n}(n-2)^a \ldots\ldots = 0$$

(n and a integers; $a < n$).

2) To prove the uniform convergence of the nth difference quotients with equal independent increases Δ (these Δ indefinitely decreasing) we develop the elements of such a difference quotient according to TAYLOR's series, in which we make the nth differential quotients form the restterms. The terms preceding the restterms then fall out and of the restterms one kind converges uniformly to the differential quotient corresponding to the difference quotient considered and the other converge uniformly to zero.

NOTES

⟦1⟧ The English version appeared in the Proceedings of the meetings of 30 May and 27 June 1908.

It is strange that the concept of uniform differentiability, though more relevant, never succeeded in superseding the classical one of continuous differentiability. It is a characteristic feature that Brouwer hit on this concept in investigations which would be classified as differential topology today. See applications in Brouwer 1909 B.

⟦2⟧ U. Dini 1892.

⟦3⟧ C. Arzelà 1895.

⟦4⟧ Brouwer 1909 B.

Y 8 Letter to Hilbert

⟦Univ. Bibl. Göttingen⟧

Amsterdam, 28 Oktober 1909

⟦1⟧ Sehr geehrter Herr Geheimrat,

Darf ich gleich anfangen, Ihnen meine Bemerkungen zu Ihren Grundlagen der
⟦2⟧ Geometrie mitzuteilen? Die Voraussetzung der Umkehrbarkeit der Transforma-
tionen spielt in die Ueberlegungen der §§3 und 6 hinein, und es sind in der Beweis-
⟦3⟧ führung einige Ergänzungen anzubringen, damit sie von dieser Voraussetzung
unabhängig werden. Diese Ergänzungen hatte ich mir in folgender Weise gedacht:

Zu §1. Der Kreis $\mathfrak{k}$ wird hier so bestimmt, dass die Punkte innerhalb $\mathfrak{k}$ bei beliebigen Drehungen um M innerhalb $\mathfrak{K}$ bleiben.

Ueberdies sollte man nun mit Rücksicht auf §3 $\mathfrak{k}$ noch die Bedingung auferlegen, dass auch die Punkte ausserhalb $\mathfrak{K}$ bei beliebigen Drehungen um M ausserhalb $\mathfrak{k}$ bleiben. Dazu braucht dieser §1 nur unwesentlich abgeändert zu werden.

Zu §2. Zunächst kann man hier die Definition des Kreises k analog vervollständigen, wie die Definition von $\mathfrak{k}$ im §1.

Dann aber sollte hier, wieder mit Rücksicht auf §3, ausdrücklich bewiesen werden, dass die *bedeckte Punktmenge* für die Drehungen um M invariant ist. Für nicht paarweise inverse Transformationen ist ja a priori *nur* evident, dass diese Menge durch jede Drehung um M in eine Teilmenge von sich übergeht. Der Beweis kann wie folgt geführt werden:

Die bedeckte Punktmenge besteht aus einer *Gebietsmenge* g (d.h. eine Menge mit der Eigenschaft, dass, wenn ein Punkt P zu ihr gehört, auch das Innere eines gewissen um P geschlagenen Kreises zu ihr gehört), und aus der Grenze G von g. Ginge nun g durch eine Drehung Δ über in ein Teilgebiet g' von g, so sei r ein Teilgebiet von g, welches gänzlich ausserhalb g' liegt. Die sukzessiven Wiederholungen von Δ führen sodann r über in eine Reihe von Gebieten $r_1, r_2, r_3, \cdots$, welche einerseits ausserhalb von einander, andererseits innerhalb $\mathfrak{k}$ liegen; die grösste in r_n enthaltene Kreisfläche konvergiert somit gegen Null.

Sei nun μ ein in r liegender Kreis, dessen Inneres zu r gehört, π sein Mittelpunkt, und bezeichnen wir mit Δ_n die n-malige Wiederholung von Δ, mit π_n den Punkt, in den π durch Δ_n übergeht, und mit μ_n die geschlossene Jordansche Kurve, in welche μ durch Δ_n übergeht. Wir können dann eine Fundamentalreihe π_{α_1}, π_{α_2}, π_{α_3}, $\cdots$ auserwählen, welche nur einen einzigen Grenzpunkt π_ω besitzt. Sei nun P_{α_n} der π_{α_n} *am nächsten liegende* Punkt auf $\mu_{\alpha n}$, so hat die Reihe P_{α_1}, P_{α_2}, P_{α_3}, $\cdots$ ebenfalls π_ω als ihren einzigen Grenzpunkt. Sei Q_{α_n} der Punkt auf μ, welcher durch Δ_{α_n} in P_{α_n} übergeht. Dann können wir unter den Q_{α_n} eine Fundamentalreihe $Q_{\beta_1}, Q_{\beta_2}, Q_{\beta_3}, \cdots$ auserwählen, welche gegen einen einzigen Grenzpunkt Q_ω konvergiert. Die Drehungen Δ_{β_n} führen dann der Reihe nach sowohl π, wie Punkte in beliebiger Nähe von Q_ω über in beliebige Nähe von π_ω. Es gäbe also eine Drehung um M, welche sowohl π wie Q_ω in π_ω überführte, was ein Widerspruch ist.

Zu §3. Nach den vorangehenden Aenderungen in den §§1 u. 2 trifft nun der erste Absatz des §3, welche[[r]] in der gegenwärtigen Form die Umkehrbarkeit der Transformationen voraussetzt, auch ohne diese Annahme zu.

Indes kommen die letzten sechs Zeilen des §3, welche sich ebenfalls auf die Umkehrbarkeit stützen, zunächst in Fortfall, was aber durch die folgende Bemerkung zu §6 wieder ins reine gebracht wird.

Zu §6. Die Beweisführung dieses § stützt sich stillschweigend auf den folgenden Hauptsatz:

Gegen einen willkürlichen Punkt K des wahren Kreises κ konvergiert eine Fundamentalreihe von Punkten $K_1, K_2, \cdots$ des wahren Kreises in solcher Weise, dass es Drehungen um M gibt, welche K der Reihe nach in $K_1, K_2, \cdots$ überführen.

Diesen Satz wollen wir ausführlich beweisen, und dazu zunächst zwei Hilfssätze formulieren:

Hilfssatz 1. Eine Fundamentalreihe $K_1, K_2, K_3, \cdots$ von Punkten des wahren Kreises κ mit der Eigenschaft, dass jedesmal K_{n-1} und K_{n+1} durch K_n und einen festen Punkt K_∞ getrennt werden, besitzt nur *eine* Verdichtungsstelle.

Beweis. Kann z.B. nach dem zweitletzten Absatz von §6 geführt werden.

Hilfssatz 2. Es sei δ eine Drehung um M, welche den Punkt A des wahren Kreises κ in A', und den Punkt B des wahren Kreises κ in B' überführt, während A, A' und B verschieden sind, so kann weder B' mit B zusammenfallen, noch können A und B' durch A' und B getrennt werden. (In diesem Hilfssatz ist der Satz des §14 enthalten).

Beweis. Bezeichnen wir wieder mit δ_n die n-fache Wiederholung von δ, und mit $A^{(n)}$ den Punkt, in den A durch δ_n wird übergeführt. Wenn nun entweder B' mit B zusammenfiele, oder A und B' durch A' und B getrennt würden, so würden jedesmal $A^{(n-1)}$ und $A^{(n+1)}$ durch $A^{(n)}$ und B getrennt, und besässe mithin nach Hilfssatz 1 die Reihe des $A^{(n)}$ nur *eine* Verdichtungsstelle A_ω. Durch die sukzessiven δ_n würde dann das Punktepaar $A\ A'$ in beliebige Nähe von A_ω geführt werden, was nach Axiom III auf einen Widerspruch führt.

Wir kommen zum Beweise des Hauptsatzes und bemerken, dass es auf dem wahren Kreise κ einen bestimmten Punkt $_IK$ gibt, mit dessen Hilfe der wahre Kreis definiert wird, nämlich als die Menge der Punkte, in welche $_IK$ durch Drehungen um M übergehen kann. Nach den Axiomen II u. III gibt es eine Verdichtungsstelle K^* von Punkten von κ, welche ebenfalls zu κ gehört. Wir unterscheiden nun zwei Fälle.

Erstens nehmen wir an, dass es eine Drehung um M gibt, welche $_IK$ in K^* überführt, aber nach einer endlichen Zahl von Wiederholungen wieder in die Anfangslage $_IK$ zurückführt. Dann ist auch $_IK$ eine Verdichtungsstelle von zu κ gehörenden Punkten. Sei $_IK_1$, $_IK_2$, $\cdots$ eine Fundamentalreihe von Punkten von κ, welche gegen $_IK$ konvergiert, so gibt es eine Fundamentalreihe von Drehungen um M, welche der Reihe nach $_IK$ in $_IK_1$, $_IK_2$, $\cdots$ überführen. Seien τ_1, τ_2, τ_3, $\cdots$ diese Transformationen, K ein willkürlicher Punkt von κ, und K_1, K_2, K_3, $\cdots$ die Punkte, in welche K durch τ_1, τ_2, τ_3, $\cdots$ übergeführt wird. Hätten dann K_1, K_2, K_3, $\cdots$ eine Verdichtungsstelle V, welche von K verschieden wäre, so gäbe es nach Axiom III eine Drehung um M, welche $_IK$ festliesse, und K in V überführte, was nach Hilfssatz 2 unmöglich ist. Die Reihe K_1, K_2, K_3, $\cdots$ konvergiert somit gegen K und nur gegen K; zugleich gibt es Drehungen um M, welche K in K_1, K_2, $\cdots$ überführen; der Hauptsatz ist mithin für den vorausgesetzten ersten Fall bewiesen. *Zweitens* dürfen wir nun annehmen, dass es eine Drehung Δ um M gibt, dessen ⟦sic⟧ n-fache Wiederholung Δ_n für *keines* ⟦sic⟧ n den Punkt $_IK$ in seine Anfangslage zurückführt. Die Punkte R_n, in welche $_I$ durch die Δ_n übergeführt wird, sind dann alle verschieden. Unter den Punkten R_n wählen wir nun eine Fundamentalreihe von *Punktepaaren* $(R_{\alpha_n}, R_{\alpha'_n})$ aus in solcher Weise, dass jedesmal $\alpha'_n > \alpha_n$, und dass die Reihe nur eine einzige Verdichtungsstelle S besitzt. Es sei nun K ein willkürlicher Punkt von κ. Wir bezeichnen $\alpha'_n - \alpha_n$ mit β_n, und den Punkt, in den K durch Δ_{β_n} übergeführt wird mit K_{β_n}. Nach Axiom III und Hilfssatz 2 zeigen wir nun ebenso wie für den ersten Fall, dass die K_{β_n} nur den Punkt K als Verdichtungsstelle besitzen, womit auch für den zweiten Fall der Hauptsatz bewiesen ist, und mit ihm der Satz des §6.

Es bleiben jetzt nur noch ein Paar ⟦sic⟧ kleinere Bemerkungen:

Zu §15. Der Satz dieses § wird auf die Definition des wahren Kreises gegründet. Das ist aber nur erlaubt, wenn die Umkehrbarkeit vorausgesetzt wird. Indes wissen wir hier schon, dass der wahre Kreis κ eine geschlossene Jordansche Kurve ist, und dass bei den Drehungen um M die Stelle eines willkürlichen Punktes von κ eine stetige Funktion der Stelle des als beweglich gedachten Punktes $_IK$ ist. Hieraus kann der Satz des §15 dann gleich gefolgert werden.

⟦4⟧ *Zu §19 bis 21.* Hier scheint mir die Darstellung nicht ganz rein. Zunächst wird im Beweise des §19 die Ebene zerlegt in ein System von einander umschliessenden geschlossenen Jordanschen Kurven. Dies ist zwar möglich für die Zahlenebene; für die Ebene der Geometrie darf es aber nicht vorausgesetzt werden; wir wissen ja noch nicht, ob die Ebene der Geometrie einfach zusammenhängend ist.

Zweitens wird der Satz des §19 nur aufgestellt für einen wahren Kreis, in dessen Innerem M liegt; im letzten Absatze von §21 wird er aber angewandt für einen wahren Kreis, der eine geschlossene Jordansche Kurve ist, von der noch nicht feststeht, dass M in ihrem Inneren liegt.

Aus diesen Gründen hatte ich mir den Gedankengang der §§ 19–21 in folgender Weise abgeändert:

Wir beweisen folgenden Hauptsatz:

"*Es kann nicht in beliebiger Nähe von κ Punkte geben, denen die Eigenschaft* *abgeht, dass ihre Stellen bei Drehungen um M durch eindeutige und stetige Funktionen von ihren Anfangsstellen und ω (mit der kleinsten Periode 2π in ω) ausgedrückt werden, und dass die dabei beschriebenen wahren Kreise M in ihrem Inneren enthalten.*"

Zunächst nämlich kann es nach dem ersten Teile des Beweises von §19 nicht in beliebiger Nähe von κ Punkte geben, welche nicht zugleich mit κ festbleiben.

Ebensowenig können nach dem Beweise des §21 in beliebiger Nähe von κ Punkte liegen, deren Stellen bei Drehungen um M nicht stetige Funktionen von ihren Anfangsstellen und ω wären. Jeder Punkt B in hinreichender Nähe von κ beschreibt mithin für eine kleinste Periode $\frac{2\pi}{n}$ in ω eine geschlossene Jordansche Kurve, welche mit B gleichmässig gegen κ konvergiert.

Gäbe es nun weiter in beliebiger Nähe von κ solche Punkte, für welche $n > 1$ wäre, so konvergierte gegen einen gewissen Punkt P von κ eine Fundamentalreihe von Punkten $P_1, P_2, \cdots$, welche je für eine gewisse zwischen $\frac{1}{2}\pi$ und $\frac{3}{2}\pi$ liegende Zunahme von ω, der Reihe nach mit $\zeta_1, \zeta_2, \zeta_3, \cdots$ zu bezeichnen, in ihre Anfangslagen zurückkehrten. Aus dieser Reihe $P_1, P_2, \cdots$ wählen wir nun eine Reihe $P_{\alpha_1}, P_{\alpha_2}, \cdots$ aus, für welche die entsprechenden $\zeta_{\alpha_1}, \zeta_{\alpha_2}, \zeta_{\alpha_3}, \cdots$ gegen einen einzigen Limeswert ζ konvergieren. Durch diese Zunahme $\zeta_{\alpha_1}, \zeta_{\alpha_2}, \zeta_{\alpha_3}, \cdots$ von ω würde dann P in eine Punktreihe $S_{\alpha_1}, S_{\alpha_2}, S_{\alpha_3}, \cdots$ übergeführt werden, welche gegen einen einzigen von P verschiedenen Grenzpunkt S konvergierte. Nach Axiom III gäbe es somit eine Drehung um M, welche P einerseits fest liesse, andererseits in S überführt, was ein Widerspruch ist.

Gäbe es schliesslich in beliebiger Nähe von κ wahre Kreise κ', welche M *nicht* in ihrem Inneren enthielten, d.h. mit κ ein den unendlich fernen Punkt enthaltendes Ringgebiet R' bestimmten, so seien P und Q zwei Punkte von κ, welche um π verschiedenen Werten von ω entsprechen: Aus jedem dieser Punkte ziehen wir die kürzeste Strecke PP' resp. QQ' nach κ'. Wenn dann κ' gegen κ konvergiert, wird auch der Parameterunterschied der Punkte P' und Q' auf κ' π beliebig nahe kommen. Seien P'' und Q'' die letzten Kreuzungspunkte von PP' resp. QQ' mit κ, so wird auch der Parameterunterschied von P'' und Q'' auf κ π beliebig nahe kommen, während die Strecken $P''P'$ und $Q''Q'$ das Ringgebiet R' in zwei von den geschlossenen Jordanschen Kurven λ_1 und λ_2 begrenzte Teilgebiete zerlegen. Sei dann H ein willkürlicher Punkt der Zahlenebene in endlicher Entfernung von κ, so lässt er sich mit dem unendlich fernen Punkte verbinden durch eine Jordansche Kurve j, welche κ entweder nicht, oder nur in einem nicht zu λ_1 gehörigen Punkte trifft. Ebensowenig kann dann die Kurve λ_1, wenn κ' in hinreichende Nähe von κ gelangt ist, diesen Weg treffen. Das Innengebiet I_1 von λ_1 kann also schliesslich nur Punkte in beliebiger Nähe von κ enthalten, und analoges gilt für das Innengebiet I_2 von λ_2. Zudem liegen I_1 und I_2 ausserhalb von einander.

Die Strecken $P''P'$ und $Q''Q'$ gehören sowohl zur Grenze von I_1, wie von I_2, und zu jedem nicht mit P'' oder P' zusammenfallenden Punkte der Strecke $P''P'$ gibt es eine Umgebung, deren Punkte entweder zu Strecke $P''P'$ oder zu I_1 oder zu

I_2 gehören, während analoges gilt für die Strecke $Q''Q'$. Wir bilden nun die Vereinigungsmenge von den Strecken $P''P'$ und $Q''Q'$ und den Gebieten I_1 und I_2; sie bildet nach dem vorhergehenden ein einziges Gebiet, dessen Grenze genau aus den Kurven κ und κ' besteht, also mit dem von diesen beiden Kurven bestimmten Ringgebiete R' identisch ist. Dieses Ringgebiet konvergiert somit mit κ' gegen κ, und kann den unendlich fernen Punkt nicht enthalten, so dass der Hauptsatz in seinem vollen Umfange bewiesen ist.

Nennen wir die im Haupsatze ausgedrückte Eigenschaft eines wahren Kreises, die kleinste Periode 2π in ω zu besitzen, und M in seinem Inneren zu enthalten, die Haupteigenschaft, so wissen wir jetzt, dass es zu jedem wahren Kreise κ_1, der die Haupteigenschaft besitzt, einen ihn umschliessenden wahren Kreis κ_2 gibt, der, ebenso wie alle zwischen κ_1 und κ_2 enthaltenen wahren Kreise, gleichfalls die Haupteigenschaft besitzt.

Wir können mithin eine Fundamentalreihe $\kappa_1, \kappa_2, \kappa_3, \cdots$ von wahren Kreisen bilden, deren jeder alle ihm vorangehenden umschliesst, und welche alle die Haupteigenschaft besitzen. Wir nehmen nun an, dass es einen Punkt D der Geometrie gibt, welcher Grenzpunkt dieser Fundamentalreihe ist, und mithin ausserhalb aller κ_n liegt.

Dann ist seine Stelle in der Zahlenebene als Funktion von der Anfangsstelle und ω zunächst eindeutig, und nach §21 auch stetig. Die kleinste Periode ist wieder von der Form $\frac{2\pi}{n}$, und dass $n = 1$ ist, folgern wir nach dem ersten Teile des Beweises von §19. Schliesslich zeigen wir genau wie oben, mittels des Ringgebietes, dass der von D beschriebene wahre Kreis M in seinem Inneren enthält.

Bezeichnen wir nun diesen wahren Kreis mit κ_ω, so gibt es wieder einen $\kappa_\omega$umschliessenden wahren Kreis $\kappa_{\omega+1}$, der ebenso wie alle zwischen κ_ω und $\kappa_{\omega+1}$ liegenden wahren Kreise, die Haupteigenschaft besitzt. Dieses Verfahren können wir unbegrenzt fortsetzen, bei einer gewissen Zahl der zweiten Zahlenklasse muss es aber sein Ende erreichen. Jedem wahren Kreis κ_α entspricht nämlich ein von κ_α und $\kappa_{\alpha+1}$ begrenztes Ringgebiet, und die Zahlenebene, also auch die geometrische Ebene, kann nur eine abzählbare Zahl von ausserhalb von einander liegenden Gebieten enthalten.

Hiermit ist der Satz des §20 in seinem vollen Umfange bewiesen.

Zu §42. Formal fehlt hier der Nachweis, dass wenn das Euklidische Axiom für *einen* Punkt bezüglich *einer* Geraden zutrifft, es für *jeden* Punkt bezüglich *jeder* Geraden zutrifft. Da ein solcher Nachweis aber sehr einfach geführt werden kann, haben Sie ihn wohl absichtlich fortgelassen, wie dieser §42 auch sonst ganz knapp gehalten ist.

Wenn ich in die Dünen komme, denke ich immer an unsere schönen Ausflüge, und für mich so lehrreiche und anregende Unterhaltungen. Auch hier war der Herbst wunderschön: bis Mitte Oktober habe ich noch einige Male am Strande baden können. Ihre Nachricht, dass das Knie der werten Frau Geheimrat noch immer nicht wiederhergestellt ist, tat uns sehr leid; der einzige Trost ist, dass es jedenfalls einmal ausgeheilt sein wird, und dass es bei ihrer gesunden Konstitution nicht langsamer gehen wird, wie bei einem anderen Menschen. Wir wünschen Ihnen beiden damit das beste; und hoffen auf ein Wiedersehen in voller Gesundheit. Meine Frau will noch etwas hinzuschreiben. ⟦5⟧

Ihr herzlich ergebener
L E J Brouwer

NOTES

⟦1⟧ Brouwer's correspondence with Hilbert, or at least what is left of it, starts on 14 May 1909. It includes a collection of 27 letters and postcards from Brouwer to Hilbert in the Niedersächsische Staats- und Universitätsbibliothek in Göttingen, the draft or copy of a letter of Brouwer to Hilbert in Brouwer's papers, here published as Y16, and Brouwer's copy of pages 5–7 of the present letter as well as a few related remarks in Dutch. Sometime between 25 May 1909 and 15 October 1909 the Hilbert and Brouwer families seem to have met for the first time. Brouwer's addresses develop from *Sehr geehrter Herr Geheimrat* via *Lieber Herr Geheimrat* to *Lieber Herr Hilbert*; Brouwer's style becomes more and more cordial but always expresses veneration for Hilbert. Up to the last postcard of the collection (20 August 1919) nothing foreshadows the future quarrel.

Brouwer's papers include a copy of a part of this letter (from ⟦4⟧ onwards, except the last, personal, paragraph), one page of the draft, and a few fragments in Dutch.

⟦2⟧ D. Hilbert, 1902. The paper was included as *Anhang IV* in D. Hilbert, 1909, and following editions. The third edition of 1909 contains changes by which part of Brouwer's objections are avoided, but it is not clear whether they were made in response to Brouwer's criticism or originated independently. It is strange that in a letter to Hilbert of 16 June 1913 (Univ. Bibl. Göttingen) Brouwer wrote: 'Neulich las ich, dass von Ihren Grundlagen der Geometrie eine 4. Auflage erscheinen soll. Ist in ihr den Bemerkungen Rechnung getragen, die ich Ihnen im Herbste 1909 zu Anhang IV (d.h. zur Arbeit aus Annalen 56) mitteilte? Ich wäre jedenfalls gern bereit, zur Richtigstellung der bezüglichen Absätze behilflich zu sein, falls Sie es wünschten und das Imprimatur noch nicht erteilt ist.

⟦3⟧ In Hilbert 1909 inverses are expressly postulated. At that time assuming or omitting the postulate of the inverse was regarded as an important feature.

⟦4⟧ It seems that the first of these two objections, on the topology of the plane, has not been accounted for in later editions. The second, on tacitly extending theorems about true circles that contain M in their interior, to true circles in general, is avoided in Hilbert 1909 by simply mentioning at the beginning of §19 that the properties in question can be proved for arbitrary true circles in the same way, but it is not clear whether this has been added in order to meet Brouwer's objection.
⟦5⟧ Mrs. Brouwer's addition is lacking.

L. E. J. BROUWER 1909 B

DIE THEORIE DER ENDLICHEN CONTINUIERLICHEN GRUPPEN, UNABHÄNGIG VON DEN AXIOMEN VON LIE ⟦1⟧

Die Frage nach einem Aufbaue der LIE'schen Gruppentheorie unabhängig von den LIE'schen Axiomen ist mehrere Male aufgeworfen worden, z. B. von HILBERT in seinem auf dem Pariser Congress gehaltenen Vortrage: « Mathematische Probleme » (1) ⟦2⟧
und von POINCARÉ in seinem zur dritten Verteilung des LOBATCHEFSKY-Preises aus- ⟦3⟧
gebrachten Gutachten über die Arbeiten von HILBERT (2), welche die Grundlagen
der Geometrie betreffen. Letzterem war nämlich inzwischen (3) die Bestimmung der ⟦4⟧
ebenen Bewegungsgruppen gelungen durch ein System von Axiomen, frei von den LIE'schen Voraussetzungen über die Existenz von gewissen Differentialquotienten der die Gruppe definierenden Funktionen, welche Differentialquotienten bei der LIE'schen Methode notwendig sind, um die im Mittelpunkte dieser Untersuchungen stehenden Infinitesimaltransformationen, welche durch continuierliche und differenzierbare Funktionen gemessen werden, herzuleiten.

Eine allgemeine Formulierung des seiner Beschränkungen entledigten LIE'schen Problems: « Alle endlichen continuierlichen Gruppen der n-dimensionalen Mannigfaltigkeit zu bestimmen », wird indes von HILBERT nicht angegeben, und man sieht nicht gleich, wie sich seine für die Bewegungsgruppen der Ebene aufgestellten Axiome zu solchen, die für alle endlichen continuierlichen Gruppen gelten, verallgemeinern lassen. Ich habe nun diese Fragestellung festgelegt durch folgende Definitionen:

Unter einer *n-dimensionalen Mannigfaltigkeit* verstehe ich ein zusammenhän- ⟦5⟧
gendes endliches Gebiet des n-dimensionalen Zahlenraumes, c. q. modifiziert durch Hinzufügung der Grenze, paarweise Identifizierung von in ihr zu bildenden, auf einander applizierbaren Gebietsmengen (4) in der Weise, dass die Innenseite der einen auf die Aussenseite der anderen appliziert und mit ihr identifiziert wird, weiter paarweise Identifizierung von den Grenzpunkten der genannten Gebiete und Iden-

(1) Abgedruckt in den Göttinger Nachrichten, 1900.

(2) Kasan Bulletin. Série 2. XIV. No. 1. Section I.

(3) Mathematische Annalen 56.

(4) d. h. Teilmengen der Grenze, mit der Eigenschaft, dass, wenn ein Punkt P zu ihr gehört, auch der innerhalb einer gewissen im n-dimensionalen Raume un P beschriebenen Kugel liegende Teil der Grenze zu ihr gehört.

tifizierung von anderen auf einander applizierbaren abgeschlossenen Teilmengen der Grenze, und schliesslich wieder Fortlassung der Grenze ausser den paarweise zusammen mit Innen- und Aussenseite identifizierten zusammenhängenden Gebieten.

Unter einer *endlichen continuierlichen Gruppe*, insbesondere unter einer p-dimensionalen (¹) continuierlichen Gruppe einer n-dimensionalen Mannigfaltigkeit, als *Transformationsmannigfaltigkeit* zu bezeichnen, verstehe ich eine Gruppe von stetigen bi-uniformen und paarweise inversen Transformationen der Punkte der Mannigfaltigkeit in sich derart, dass sich die verschiedenen Transformationen der Gruppe stetig und bi-uniform auf die Punkte einer p-dimensionalen Mannigfaltigkeit, der wir den Namen *Parametermannigfaltigkeit* beilegen wollen, abbilden lassen (²). Wir bemerken gleich, dass diese Parametermannigfaltigkeit auch selber einer p-dimensionalen continuierlichen Gruppe unterliegt (³).

Die Gesamtheit der Punkte, in welche ein bestimmter Punkt der T. M. (Transformationsmannigfaltigkeit) durch die Gruppe übergehen kann, hat die Eigenschaft, dass jeder Punkt des Systems alle anderen Punkte des Systems, aber nur diese, durch die Gruppe erreichen kann. Ich nenne diese Punktmenge einen *Transformationsbereich*, und bemerke über sie zunächst, dass sie jedenfalls zusammenhängend, und wenn man die Grenztransformationen hinzunimmt, abgeschlossen ist. Beide Eigenschaften bleiben bestehen, wenn man nur denjenigen Teil des T. B. (Transformationsbereiches) ins Auge fasst, der einer beliebigen Umgebung mit Grenze in der P. M. (Parametermannigfaltigkeit) entspricht. Ein solcher Teil eines T. B. ist also entweder eine zusammenhängende perfekte Menge, oder der ganze T. B. besteht aus einem isolierten Punkt.

Natürlich bildet die P. M. für die Gruppe nur einen einzelnen T. B.

Halten wir einen oder mehrere Punkte der T. M. fest, so bilden die jetzt noch möglichen Transformationen eine Untergruppe mit paarweise inversen Transformationen, die wir *reduzierte Gruppe* nennen wollen. Scheiden wir in der P. M. mit ihrer Grenze resp. in einem beliebigen zusammenhängenden Gebiete mit Grenze der P. M. die Punkte aus, die dieser Untergruppe entsprechen, so bilden sie eine abgeschlossene Menge. Zusammenhängend braucht sie nicht zu sein (⁴). Die verschiedenen

(¹) Ob p für jede Gruppe eine Invariante ist, steht aus, solange die « Nichtapplizierbarkeit »
[[6]] zweier Räume, deren Dimensionenzahl verschieden, unbewiesen ist.

(²) Diese Definition soll meinen, dass die Stellen der Punkte der T. M. ausserhalb der Grenze der P. M. gleichmässig stetige Funktionen des repräsentierenden Punktes der P. M. sein sollen. Hieraus ergibt sich aber keineswegs die Gleichmässigkeit dieser Stetigkeit, wenn in der P. M. der Grenze unbeschränkt genähert wird.

(³) Durch die Abbildung der Gruppe auf die P. M. ist zugleich eine Abbildung der Grenztransformationen der Gruppe auf die Grenze der P. M. bestimmt. Um diese in der T. M. zu deuten, kann man genötigt sein, auch in der T. M. die Grenze hinzuzunehmen. Die Grenztransformationen sind sicher uniform, nicht aber stetig und biuniform.

(⁴) Man betrachte z. B. die Gruppe der Euklidischen Bewegungen der elliptischen Ebene, der man als P. M. den im vierdimensionalen Raume liegenden quadratischen Raum $x^2 + y^2 = r^2$ zuordnen kann. Hält man hier in der T. M. einen Punkt der invarianten geraden Linie fest, so werden in der P. M. zwei parallele Ebenen ausgeschieden.

[[110]]

innerhalb der P. M. zusammenhängenden Bestandteile der reduzierten Gruppe sind aber jedenfalls bi-uniform und stetig auf einander abzubilden, und es gibt eine beschränktere Untergruppe, die jeden dieser Bestandteile in sich transformiert, und deren T. B. zusammenhängend sind.

Wir wenden uns jetzt zur Bestimmung der endlichen continuierlichen Gruppen der eindimensionalen Mannigfaltigkeit. Erst die Bestimmung aller dieser Gruppen öffnet den Weg zur Lösung des analogen Problems in mehrdimensionalen Räumen. Es hat sich mir dabei bis jetzt herausgestellt, dass ausser nur Gruppen trivialer Art ekartierenden Restrictionen wenigstens in der zweidimensionalen Mannigfaltigkeit noch keine endlichen continuierlichen Gruppen, die von den LIE'schen wesentlich verschieden sind, existieren (1). Für höhere Räume habe ich diese Untersuchungen noch nicht abschliessen können. Dazu wird nämlich erfordert eine zur Zeit noch nicht vorliegende Erweiterung von den Sätzen über Analysis Situs der Ebene, die SCHOENFLIES (2) in den letzten Jahren veröffentlicht hat, auf drei und mehr Di- [[7]]
mensionen. Nur die Gruppen der zweidimensionalen Mannigfaltigkeiten lassen sich bestimmen mittelst der bisherigen SCHOENFLIES'schen Resultate, und ich glaube dass damit zum ersten Male eine Anwendung dieser Resultate gegeben wird.

Für die Gruppen der eindimensionalen Mannigfaltigkeit zeige ich zunächst, dass sowohl für die ursprüngliche Gruppe (3), die wir Hauptgruppe nennen könnten, wie für die verschiedenen reduzierten Gruppen, die T. B. entweder Systeme von isolierten Punkten, oder Systeme von Segmenten ohne ihre Endpunkte sind. Sodann enthält jede reduzierte Gruppe, der in der P. M. noch zusammenhängende Mengen entsprechen, noch in Segmenten bewegliche Punkte, und kann weiter reduziert werden durch Festhaltung eines solchen Punktes. Sei P ein solcher Punkt, P_ν eine willkürliche Stelle von P in seinem T. B., μ_ν die entsprechende Menge in der P. M., so zeigen wir zunächst, dass, wenn P_ω Grenzpunkt von $P_1, P_2, \ldots$, und Q_ω ein Punkt von μ_ω, es in $\mu_1, \mu_2, \ldots$ der Reihe nach Punkte $Q_1, Q_2, \ldots$ gibt, die gegen Q_ω convergieren, und hieraus weiter, dass, wenn μ_ν aus verschiedenen zusammenhängenden Bestandteilen ${}_hZ_\nu$ besteht, diese sich für jedes Segment des T. B. von P zusammensetzen zu ebenso vielen zusammenhängenden Mengen ${}_hY$, deren je zwei von einander isoliert sind, während in jedem ${}_hY$ die ${}_hZ_\nu$ bi-uniformes und stetiges Bild der P_ν sind.

Lassen wir jetzt von den bisher festgehaltenen Punkten einen, den wir R nennen wollen, frei, so setzen sich in derselben Weise für jedes Segment des T. B. von R die ${}_hY$ zusammen zu zusammenhängenden ${}_hX$, in jeder von welchen die ${}_hY_\tau$ bi-uniformes und stetiges Bild der R_τ sind, und es kann gezeigt werden, dass das ganze

(1) Auch wenn man nur von Infinitesimaltransformationen generierte Gruppen in Betracht zieht, geht dies über die LIE'schen Voraussetzungen hinaus, nicht nur insoweit man diese Infinitesimaltransformationen nicht als durch differenzierbare Funktionen bestimmt voraussetzt, sondern auch hierin, dass man kein Coordinatensystem voraussetzt, in dem alle die Gruppe erzeugenden Infinitesimaltransformationen als solche zu lesen sind.

(2) Mathematische Annalen 59, 62; Bericht über die Mengenlehre II. [[7]]

(3) Für diese kann freilich, wenn die T. M. ein geschlossenes Continuum ist, dieses auch als ein einziger T. B. auftreten.

[[111]]

jetzt erhaltene System der Z nicht nur bi-uniformes, sondern auch stetiges Bild der von den Stellen von P und R gebildeten zweidimensionalen Mannigfaltigkeit ist.

So kann man fortfahren, und schliesslich zeigen, dass wenn Z eine nicht weiter ausdehnbare zusammenhängende Teilmenge der P. M., welche nach Festhaltung von einem nach dem andern n je zuvor stetig beweglichen Punkten übrig bleibt, die zusammenhängende Umgebung eines willkürlichen Punktes in der P. M. sich aus einem gewöhnlichen n-dimensionalen Raume von Mengen Z zusammensetzt.

Hieraus folgern wir gleich, dass die sich zu einer Menge μ zusammensetzenden Mengen Z isoliert sind. Sonst nämlich gäbe es in beliebiger Nähe eines Punktes der P. M. andere Punkte, die nur auf endlichem Wege zusammenhängend erreichbar wären. Diese zusammenhängende Erreichbarkeit einer zweiten Menge Z derselben Menge μ ist sogar nur dann möglich, wenn wir als T. B. ein geschlossenes Continuum haben, nämlich durch einen vollen Umlauf dieses Continuums. In jedem anderen Falle enthält jedes μ sicher nur ein einziges Z.

Jetzt beweisen wir, dass es immer möglich ist, die nach einander festgehaltenen Punkte derart zu wählen, dass nach Festhaltung einer endlichen Zahl von je zuvor noch stetig beweglichen Punkten keine zusammenhängende Teilmenge der P. M. mehr zur Verfügung steht, und dies gelingt, indem man die Punkte in der Weise einen nach dem andern festhält, dass sie, wenn man unbeschränkt fortfahren könnte, die T. M. überall dicht überdecken würden, und von diesem die Unmöglichkeit zeigt. Man kann somit durch Festhaltung einer endlichen Zahl von Punkten erreichen, dass die T. B. nur noch aus isolierten Punkten bestehen. Ist n die Zahl dieser Punkte, so ist die Gruppe n-dimensional; ob n eine Invariante, ist noch unsicher.

Sodann studieren wir die reduzierte Gruppe, welche durch Festhaltung noch eines Punktes wird festgesetzt. Dieser Gruppe entsprechen in der P. M. isolierte einfache Kurven, während die zusammenhängende, die Identität enthaltende Untergruppe jede dieser Kurven in sich, und in der T. M. jedes Segment eines T. B. in sich transformiert. Die Analyse dieser letzteren Gruppe lehrt alsbald, dass sie in jedem von ihr auf der T. M. bestimmten Segmente eine überall dichte Skala vom Ordnungstypus η zu definieren gestattet, in der sie die gewöhnliche Additionsgruppe ist. Wir können somit den Satz aussprechen:

Jede endlich continuierliche Gruppe der eindimensionalen Mannigfaltigkeit enthält eine reduzierte Gruppe, deren zusammenhängender, die Identität enthaltender, Bestandteil von einer Infinitesimaltransformation erzeugt wird.

Um systematisch die Gruppen der eindimensionalen Mannigfaltigkeit zu bestimmen, stellen wir folgende Definition auf:

« Eine endliche continuierliche Gruppe der eindimensionalen Mannigfaltigkeit heisse p-gliedrig, wenn es möglich ist, durch Festhaltung von einem nach dem andern p je zuvor noch stetig beweglichen Punkten, zu erreichen, dass in der P. M. keine zusammenhängende Menge mehr zur Verfügung steht ».

Es zeigt sich, dass für jede Gruppe p eine Invariante ist.

Die Bestimmung der *eingliedrigen* Gruppen hat jetzt keine Schwierigkeit mehr, weil diese ihre eigenen letzten reduzierten Untergruppen sind, und wir solche letzte reduzierte Gruppen schon studiert haben. Indem wir die beiden zu unterscheidenden eindimensionalen Raumformen berücksichtigen, finden wir folgendes Resultat:

Durch eine eingliedrige Gruppe des offenen Continuums wird eine endliche oder abzählbare Menge von Segmenten bestimmt; jedes dieser Segmente wird von einer durch die Gruppe bestimmten überall dichten Skala von $-\infty$ bis $+\infty$ gemessen, und auf jeder dieser Skalen ist die Gruppe die Additionsgruppe der reellen Zahlen. Die nicht zu diesen Segmenten gehörigen Punkte bleiben bei der Gruppe invariant.

Eine eingliedrige Gruppe des geschlossenen Continuums ist entweder eine Rotationsgruppe, oder wird durch Ausscheidung eines festen Punktes zu einer solchen des offenen Continuums.

Wir gehen über zur Analyse der *zweigliedrigen* Gruppen, und fangen an, von ihren Infinitesimaltransformationen zwei solche zu bestrachten, für die ein Transformationssegment der einen eines der anderen teilweise überdeckt. Wir nennen die von diesen Infinitesimaltransformationen erzeugten eingliedrigen Gruppen g und g_1, und zeigen zunächst, dass in dem genannten Segmente von g nur *ein* fester Punkt von g_1, den wir B nennen wollen, liegen kann. Sodann wählen wir die Skala, welche in dem Segmente g zur Additionsgruppe macht, als eine X-Coordinatenaxe mit B zu ihrem Anfangspunkte, und construieren die Kurve $y = f_h(x)$, wo zu jedem x die Schiebung des entsprechenden Punktes durch die Transformation h aus g_1 als y genommen wird. Für variierendes h liegen die verschiedenen Kurven $y = f_h(x)$ gänzlich ausserhalb von einander mit Ausnahme des einzigen gemeinschaftlichen Punktes B.

Aus den Eigenschaften der zweigliedrigen Gruppe leiten wir her, dass jede Kurve

$$y = f_h(x + c) - f_h(c)$$

zum Büschel $y = f_h(x)$ gehört.

Man kann jetzt weiter zeigen, dass die Kurven $y = f_h(x)$ entweder gerade Linien sind, oder ihre Differenzenquotienten alle zusammen mit x steigen, oder aber alle zusammen mit x abnehmen, und hieraus kann gefolgert werden, dass die Kurven in jedem ihrer Punkte sowohl einen oberen wie einen unteren Differentialquotienten besitzen.

Die oberen und die unteren Differentialquotienten könnten noch verschieden sein, aber indem man bemerkt, dass der Unterschied der Transformationen aa_1 und $a_1 a$ auch eine Transformation der Hauptgruppe sein muss, zeigt man, dass die Kurven $y = f_h(x)$ keinen Knick aufweisen können.

Nachdem dies erreicht, kann man zeigen, dass die Gruppe g_1 auf der Additionsskala der Gruppe g von Infinitesimaltransformationen erzeugt wird, und dass die diese Infinitesimaltransformationen messenden Funktionen von x stetig sind; so dass wir jede Schiebung $u(x)$ der Hauptgruppe schreiben können:

$$u(x) = u(x_1) \left\{ \frac{\psi(x) + a}{\psi(x_1) + a} + \eta_u(x) \right\}$$

wo $\psi(x)$ eine stetige Funktion ist, und η_u mit u verschwindet.

Indem man von neuem die Gruppeneigenschaft heranzieht, findet man, dass jeder Differenzenquotient der Funktion $\psi(x)$ die Form $a\psi(x)+b$ haben muss, also ist die Menge der Differenzenquotienten gleichmässig stetig, und hieraus folgt ([1]) die Differenzierbarkeit der Funktion $\psi(x)$, in der wir somit die Differentialgleichung

$$\frac{d\psi}{dx}=a\psi+b$$

aufstellen können, aus der man schliesslich zu folgendem Resultat über die Gestalt der Gruppe in der Transformationsmannigfaltigkeit gelangt:

Durch eine zweigliedrige Gruppe des offenen oder geschlossenen Continuums wird eine endlich oder abzählbare Menge von Segmenten bestimmt; jedes dieser Segmente wird von einer durch die Gruppe bestimmten Skala gemessen von $-\infty$ nach $+\infty$, derart, dass auf jeder dieser Skalen die Gruppe entweder jene der Addition und Multiplikation der reellen Zahlen, oder jene der Addition der reellen Zahlen ist. Die nicht zu diesen Segmenten gehörigen Punkte bleiben bei der Gruppe invariant ([2]).

Die *dreigliedrigen* Gruppen der eindimensionalen Mannigfaltigkeit werden gefunden aus der Bemerkung, dass die eingliedrige Untergruppe, die übrig bleibt, wenn man zwei nicht zusammen fest bleibenden Punkte P und Q festhält, sich ergänzen lassen muss, sowohl zu einer zweigliedrigen Untergruppe für P fest und Q stetig beweglich, als zu einer zweigliedrigen Untergruppe für P stetig beweglich und Q fest. Dies gibt eine Beziehung zwischen den Skalen der beiden zweigliedrigen Untergruppen; setzt man noch beide Skalen in Beziehung zu der einer dritten zweigliedrigen Untergruppe, so gelangt man zu folgendem Resultat:

Durch eine dreigliedrige Gruppe des offenen Continuums wird eine endlich oder abzählbare Menge von Segmenten bestimmt; in jedem dieser Segmente ist die Gruppe entweder eine periodisch-projektive Gruppe mit unendlich vielen Perioden; oder eine Gruppe von Addition und Multiplikation, oder eine Additionsgruppe. Die nicht zu diesen Segmenten gehörigen Punkte bleiben bei der Gruppe invariant.

Eine dreigliedrige Gruppe des geschlossenen Continuums wird entweder durch Ausscheidung eines invarianten Punktes zu einer solchen des offenen Continuums, oder ist eine rotatorisch- projektive Gruppe, oder aber eine einfache projektive Gruppe.

Die Umöglichkeit einer viergliedrigen Gruppe zeigt sich durch die Bemerkung, dass in der zweigliedrigen Untergruppe, die durch Festhaltung zweier nicht zusammen

[[8]] ([1]) Verslag Akademie Amsterdam, XVII, S. 38.

([2]) Wir sehen also, dass bei der axiomatischen Definition der Rechnungsoperationen auf dem Continuum nach Aufstellung der commutativen und assoziativen Eigenschaften der Addition und Multiplikation, nicht noch die volle distributive Eigenschaft zur Bestimmung der Operationen notwendig ist. Diese nämlich sagt aus: $\varphi_\alpha\{f_\beta\}=f_{\varphi_\alpha(\beta)}\{\varphi_\alpha\}$; während die Forderung $\varphi_\alpha\{f_\beta\}=f_\gamma\{\varphi_\delta\}$ genügt.

[[114]]

fest bleibenden Punkte P und Q entsteht, P und Q einerseits die gleiche Rolle, und andererseits eine verschiedene Rolle spielen müssten. Und wenn viergliedrige Gruppen unmöglich sind, können ebensowenig mehr-als viergliedrige auftreten.

Nach der Bestimmung der zweigliedrigen Gruppen hätte man für die höheren Gruppen auch wie folgt fortfahren können, wobei man auf die LIE'sche Theorie zurückgreift:

Zunächst kann man beweisen, dass in einer gewissen Umgebung der Identität jeder Zustand aus jedem anderen erreicht werden kann durch successive Ausführung einer endlichen Zahl von Transformationen τ, die jede ein solches System σ von Punkten festlassen, dass nach Festhaltung *eines* weiteren Punktes die Gruppe feststehen würde. Nun haben wir aber gesehen, dass die Transformationen τ von Infinitesimaltransformationen generiert sind, und dass zwei dieser Infinitesimaltransformationen, die nur in einem ihrer festgehaltenen Punkte verschieden sind, sich beide auf einer selben Skala als Infinitesimaltransformationen mit differenzierbaren bestimmenden Funktionen lesen lassen. Es gibt also auch eine Skala, in der alle Transformationen τ, also auch die Hauptgruppe selbst, sich lesen lassen als von differenzierbaren Infinitesimaltransformationen generiert. Dann ist aber nach LIE (1) die Gruppe ähnlich mit einer von seinen analytischen Gruppen.

Indes tritt auf diesem Wege die Struktur der höheren Gruppen nur in einer gewissen Umgebung, nicht in der ganzen Transformationsmannigfaltigkeit deutlich hervor.

Um die Gruppen der zweidimensionalen Mannigfaltigkeit in Angriff zu nehmen, haben wir zunächst die Transformationsbereiche der Hauptgruppe und der reduzierten Gruppen zu bestimmen, und auf Grund des JORDAN'schen Kurvensatzes und dessen SCHOENFLIES'scher Umkehrung gelingt es zu zeigen, dass sie sich zusammensetzen aus einer endlichen oder abzählbaren Menge entweder von zusammenhängenden Gebieten, oder von isolierten einfachen Kurvenbogen oder einfachen geschlossenen Kurven, oder aber von isolierten Punkten.

Sodann gelingt es, wenngleich nicht so einfach wie beim Analogon in der eindimensionalen Mannigfaltigkeit, folgenden Satz zu beweisen:

« Hat man nach einander $m+n$ Punkte festgehalten, waren vor ihrer Festhaltung m dieser Punkte in Gebieten, und n in einfachen Kurven beweglich, und bleibt dann schliesslich in der Parametermannigfaltigkeit eine Menge μ, bestehend aus den grössten zusammenhängenden Teilmengen Z_ν (ν variabel), so setzt sich in einer gewissen Umgebung der Identität die Parametermannigfaltigkeit zusammen aus gewissen grössten zusammenhängenden Teilmengen J_ν. Jedes J_ν ist ein gewöhnlicher $(2m+n)$-dimensionaler Raum im entsprechenden Z_ν. Weiter gehört zu jedem Z_ω, das Grenzmenge der Reihe $Z_1, Z_2, \ldots$, ein J_ω, das Grenzmenge der Reihe $J_1, J_2, \ldots$ ist ».

Auch lässt sich beweisen, obgleich wieder umständlicher als für die eindimensionale Mannigfaltigkeit, dass man stets durch Festhaltung einer endlichen Zahl von Punkten die Gruppe festsetzen kann, und kann man also übergehen zur Untersuchung der *letzen reduzierten Gruppe*, die durch Festhaltung noch eines Punktes nur

(1) Theorie der Transformationsgruppen III. S. 365-369. [9]

noch isolierte Punkte in der Parametermannigfaltigkeit übrig lässt. Die zusammenhängenden Bestandteile der P. M. dieser Gruppe sind entweder einfache Kurven oder zweidimensionale Gebiete. In ersten Falle findet man leicht entweder die eingliedrige Translationsgruppe mit im allgemeinen offenen Kurven als Transformationsbereichen, die für alle zweidimensionalen Mannigfaltigkeiten auftreten kann, und bei der die Distribution der festbleibenden Mengen noch sehr verschiedenartig sein kann, oder die eingliedrige Rotationsgruppe mit einander umschliessenden geschlossenen Kurven als Transformationsbereichen, die immer alle zusammen durchrotiert werden.

Im zweiten Falle ist die Existenz der Infinitesimaltransformationen nicht so leicht nachzuweisen. Man kann aber zunächst durch Heranziehung der Parametergruppe $T'_a = T_x^{-1} T_a T_x$ zeigen, dass jede Transformation sich halbieren lässt, und sodann durch ziemlich schwierige Betrachtungen, welche sich in ausgiebigster Weise auf die SCHOENFLIES'sche Theorie der Analysis Situs stützen, dass diese von einer Transformation und ihren unbeschränkt fortgesetzten Halbierungen erzeugte Gruppe eine Translations- oder Rotationsgruppe ist, wie wir oben discutierten. Dann aber ist es leicht, alle möglichen zweidimensionalen Gruppen aufzuzählen, und wir finden nur solche, die in einer gewissen Umgebung die Struktur der LIE'schen Gruppen aufweisen.

Bei der weiteren Analyse nennen wir eine Gruppe der zweidimensionalen Mannigfaltigkeit $(2n + m)$-*gliedrig*, wenn sie festgesetzt werden kann durch Festhaltung von einem nach dem andern $n + m$ Punkten, deren im Augenblicke ihrer Festhaltung n in Gebieten, m auf Kurven beweglich waren. Die Zahl $2n + m$ zeigt sich für jede Gruppe invariant.

Nachdem die ein-und zweigliedrigen Gruppen, und somit auch die zusammenhängenden Bestandteile der letzten reduzierten Gruppe einer willkürlichen Gruppe, gefunden sind, werden die höheren Gruppen ziemlich leicht durch geometrische Ueberlegungen und Zurückgreifen auf die Resultate für die eindimensionale Mannigfaltigkeit gefunden. Indes würde die Mitteilung der Einzelheiten jetzt zu weit führen. Nur bemerke ich, dass die Construction eines invarianten projektiven Kurvensystems, wie es für jede primitive Gruppe, mit Ausnahme der conformen Gruppe, existiert, eine vornehme Rolle spielt; und dass, so bald bis einschliesslich die viergliedrigen Gruppen gefunden sind, man für die höheren Gruppen auf die LIE'schen Resultate zurückgreifen kann. Diese Erleichterung tritt hier aus analogem Grunde auf, wie bei der eindimensionalen Mannigfaltigkeit für höhere als zweigliedrige Gruppen; aber um eine Einsicht in die Struktur der Gruppen über die ganze Mannigfaltigkeit, nicht nur in einer gewissen Umgebung, zu gewinnen, tut man auch hier besser, von dieser Erleichterung keinen Gebrauch zu machen.

[[116]]

NOTES

⟦1⟧ Hilbert's fifth Paris problem (Hilbert 1900) proposed the elimination of differentiability assumptions from Lie groups. When conceiving his thesis (Brouwer 1908 B), Brouwer must have been confronted with this problem, at least in the weaker form of eliminating differentiability assumptions from the Helmholtz–Lie space problem – in this weaker form it had also been tackled by Hilbert 1902, for two dimensions.

Brouwer published his results on this subject in the present paper, which is a communication made at the 1908 Congress of Rome, and in two papers 1909 C, 1910 H. Further, 1911 F is methodically related to these studies. Brouwer's plane translation theorem (1912 B) is a byproduct of these investigations.

Brouwer solved Hilbert's fifth problem for groups acting on a one-dimensional variety. For groups acting on a two-dimensional variety he brought it up to a point where he 'possessed the tools for enumerating the groups of the two-dimensional varieties' and promised a future communication in which this plan would be carried out. It is quite certain that Brouwer possessed more extensive results. According to an oral communication he knew about the difficulties caused in three dimensions by such pathologies as revealed later by L. Antoine's point set (L. Antoine 1921) and J. W. Alexander's horned sphere (J. W. Alexander 1924).

After Brouwer, the first progress on Hilbert's fifth problem was made by J. v. Neumann, 1927, (linear groups), the next, thanks to Haar measure, by J. v. Neumann, 1933, (compact groups). When the analytical method using Haar measure appeared to fail in the case of noncompact groups, the problem was tackled anew by the more direct methods Brouwer had tried. A number of mathematicians contributed to the solution which finally was reached by A. M. Gleason, 1952, Montgomery and Zippin 1952, and H. Yamabe, 1953. See also Montgomery and Zippin 1955. On the history of Hilbert's fifth problem, see H. Freudenthal, 1968, Aleksandrov 1969 B, 101–115.

The problem of eliminating differentiability assumptions from the Helmholtz–Lie space problem was satisfactorily resolved by J. Tits 1953 A, 1953 B, On the history of the Helmholtz–Lie space problem, see H. Freudenthal, 1965.

⟦2⟧ D. Hilbert 1900.

⟦3⟧ H. Poincaré 1904.

⟦4⟧ D. Hilbert 1902.

⟦5⟧ A modified definition in Brouwer 1909 C, p. 247, footnote **) (see also 1910 G, and 1911 D, p. 98, footnote *)).

⟦6⟧ Proved in Brouwer, 1911 C.

⟦7⟧ A. Schoenflies 1904, 1906 A, 1908 A.

⟦8⟧ Brouwer 1908 D2.

⟦9⟧ S. Lie 1893.

1909 C

Die Theorie der endlichen kontinuierlichen Gruppen, unabhängig von den Axiomen von Lie.*)

⟦1⟧ (Erste Mitteilung.)

Von

L. E. J. Brouwer in Amsterdam.

§ 1.

Allgemeines.

Die Forderung eines Aufbaues der Lieschen Gruppentheorie, unabhängig von den Lieschen Voraussetzungen, wurde in den letzten Jahren mehrmals erhoben, so von Hilbert in seinem auf dem Pariser Kongreß
⟦2⟧ gehaltenen Vortrage „Mathematische Probleme"**) und von Poincaré in seinem zur dritten Verteilung des Lobatchefsky-Preises ausgebrachten Gut-
⟦3⟧ achten über die Arbeiten von Hilbert***), welche die Grundlagen der Geometrie betreffen. Letzterem war nämlich inzwischen†) die Bestimmung
⟦4⟧ der ebenen Bewegungsgruppen durch ein System von Axiomen gelungen, ohne Benutzung der Lieschen Voraussetzungen über die Existenz von gewissen Differentialquotienten der die Gruppe definierenden Funktionen, Voraussetzungen, die bei der Lieschen Methode notwendig sind, um die im Mittelpunkte dieser Untersuchungen stehenden Infinitesimaltransformationen, welche durch kontinuierliche und differenzierbare Funktionen gemessen werden, herzuleiten.

Eine allgemeine Formulierung des seiner Beschränkungen entledigten Lieschen Problems: *„Alle endlichen kontinuierlichen Gruppen der n-dimensionalen Mannigfaltigkeit zu bestimmen"*, wird indes von Hilbert nicht angegeben, und man sieht nicht gleich, wie sich seine für die Bewegungsgruppen der Ebene aufgestellten Axiome zu solchen, die für alle endlichen kontinuierlichen Gruppen gelten sollen, erweitern lassen. Ich habe nun die Liesche Fragestellung festgelegt durch folgende Definitionen, wobei

---

*) Über diese Untersuchungen wurde vom Verfasser auf dem IV^{ten} intern. Mathematiker-Kongreß in Rom ein referierender Vortrag gehalten.

**) Auch abgedruckt in Göttinger Nachrichten 1900.

***) Kasan, Bulletin, Série 2, XIV, N°. 1, Section I.

†) Mathematische Annalen Bd. 56.

ich mich mehr an Hilberts Pariser Vortrag als an seine „Grundlagen der Geometrie" anschließe.

Unter einer *n-dimensionalen Mannigfaltigkeit* wollen wir im folgenden verstehen das eineindeutige und stetige Abbild eines endlichen zusammenhängenden Gebietes des n-dimensionalen Zahlenraumes, nur modifiziert durch Hinzufügung der Grenze, paarweise Identifizierung von in ihr zu bildenden, aufeinander eineindeutig und stetig abbildbaren Gebietsmengen*) in der Weise, daß die Innenseite der einen auf die Außenseite der anderen eineindeutig und stetig abgebildet und mit ihr identifiziert wird, und schließlich wieder Fortlassung der Grenze außer den bei jener Identifizierung verwendeten Teilen.**)

Unter einer *endlichen kontinuierlichen Gruppe*, insbesondere unter einer ⟦5⟧
*p-dimensionalen****) *kontinuierlichen Gruppe einer* (als *„Transformationsmannigfaltigkeit"* zu bezeichnenden) *n-dimensionalen Mannigfaltigkeit* verstehen wir eine Gruppe von paarweise inversen†) und einschließlich der eventuellen Grenze eineindeutigen und stetigen Transformationen der Punkte der Mannigfaltigkeit in sich derart, daß diese Transformationen das für jede abgeschlossene Teilmenge gleichmäßig stetige und eineindeutige Abbild einer p-dimensionalen Mannigfaltigkeit, der wir den Namen *„Parametermannigfaltigkeit"* beilegen wollen, bilden.††)

Wir bemerken gleich, daß diese Parametermannigfaltigkeit auch selber einer p-dimensionalen kontinuierlichen Gruppe unterliegt.

Die Gesamtheit der Punkte, in welche ein bestimmter Punkt der T. M. (Transformationsmannigfaltigkeit) durch die Gruppe übergehen kann, hat die Eigenschaft, daß jeder Punkt der Gesamtheit alle anderen Punkte

---

*) D. h. Teilmengen der Grenze, mit der Eigenschaft, daß, wenn ein Punkt P zu ihr gehört, auch der innerhalb einer gewissen im n-dimensionalen Raume um P beschriebenen Kugel liegende Teil der Grenze zu ihr gehört.

**) Dies soll hier (in Gegensatz zu meinem römischen Vortrage, wo ich ein wenig ⟦6⟧
anders an die Sache herangegangen bin) so verstanden werden, daß auf die fortgelassene Grenze weder die Abbildung selber, noch ihre Stetigkeit sich erweitern zu lassen brauchen.

***) Ob p für jede Gruppe eine Invariante ist, ist unentschieden, solange die Unmöglichkeit der eineindeutigen und stetigen Abbildung zweier Räume von verschiedener Dimensionenzahl unbewiesen ist. ⟦7⟧

†) Dies ist keine wesentliche Beschränkung. Man kann, wie ich glaube, zeigen, daß, wären die Transformationen nicht paarweise invers, dies durch eine einfache Erweiterung der Gruppe zu erreichen wäre.

††) Existieren Grenztransformationen, so kann man diese der Gruppe, und in dieser Weise der Parametermannigfaltigkeit eine Grenze hinzufügen. Für die Grenztransformationen bleiben die Stetigkeit und Eineindeutigkeit nicht erhalten, und ebensowenig die Gleichmäßigkeit der stetigen Änderung der Transformationsmannigfaltigkeit, wenn man in der Parametermannigfaltigkeit sich der Grenze unbeschränkt nähert.

⟦119⟧

der Gesamtheit, aber nur diese, durch die Gruppe erreichen kann. Wir wollen eine solche Punktmenge einen *„Transformationsbereich“* nennen, und bemerken über sie zunächst, daß sie jedenfalls zusamenhängend ist, und daß jede Teilmenge von ihr, welche einem beliebigen geschränkten*) zusammenhängenden Gebiete der P. M. (Parametermannigfaltigkeit) mit seiner Grenze zusammen entspricht, zusammenhängend und abgeschlossen ist. Eine solche Teilmenge eines T. B. (Transformationsbereiches) ist also entweder eine zusammenhängende perfekte Menge, oder der ganze T. B. enthält nur einen isolierten Punkt.

Die P. M. bildet für die Gruppe natürlich nur einen einzelnen T. B.

Halten wir einen oder mehrere Punkte der T. M. fest, so bilden die jetzt noch möglichen Transformationen eine Untergruppe mit paarweise inversen Transformationen, die wir *„festgelegte Untergruppe“* nennen wollen.

Heben wir in einem beliebigen geschränkten zusammenhängenden Gebiete der P. M. mit seiner Grenze zusammen (oder auch eventuell in der ganzen P. M., nachdem sie zuvor durch ihre Grenze abgeschlossen ist) die einer festgelegten Untergruppe entsprechenden Punkte heraus, so bilden sie eine abgeschlossene Menge.

Zusammenhängend braucht diese Menge nicht zu sein.**) Die verschiedenen innerhalb der P. M. zusammenhängenden Bestandteile der festgelegten Untergruppe sind aber jedenfalls eineindeutiges und stetiges Bild voneinander, und es gibt eine beschränktere Untergruppe, die jeden dieser Bestandteile in sich transformiert, und deren T. B. zusammenhängend sind. Letztere Untergruppe nennen wir *„reduzierte Untergruppe“*.

Nach diesen allgemeinen Auseinandersetzungen wenden wir uns zur Bestimmung aller endlichen kontinuierlichen Gruppen der eindimensionalen Mannigfaltigkeit. Erst die Bestimmung aller dieser Gruppen öffnet den Weg zur Lösung des analogen Problems in mehrdimensionalen Räumen. Es hat sich mir dabei bis jetzt herausgestellt, daß wenigstens in der zweidimensionalen Mannigfaltigkeit noch keine endlichen kontinuierlichen Gruppen, die von den Lieschen wesentlich verschieden sind, existieren. Für höhere Räume habe ich für diese Untersuchungen noch nicht so abschließende Resultate erreichen können, und glaube auch, daß dies erst möglich sein wird durch eine zur Zeit noch nicht vorliegende Erweiterung der

---

[[8]] *) Schoenflies, Bericht über die Lehre von den Punktmannigfaltigkeiten, Bd. II, S. 97, Note [3].

**) Man betrachte z. B. die Gruppe der Euklidischen Bewegungen der projektiven Ebene, der man als P. M. den im vierdimensionalen Raume liegenden quadratischen Raume $x^2 + y^2 = r^2$ zuordnen kann. Hält man hier in der T. M. einen Punkt der invarianten geraden Linie fest, so werden in der P. M. zwei parallele Ebenen herausgehoben.

[[120]]

Sätze über Analysis situs der Ebene, die Schoenflies*) in den letzten Jahren veröffentlicht hat, auf drei und mehr Dimensionen. Nur die Gruppen der zweidimensionalen Mannigfaltigkeiten lassen sich mittels der bisherigen Schoenfliesschen Resultate bestimmen, und ich glaube, daß damit zum erstenmal eine Anwendung jener Resultate gegeben wird.

§ 2.

Die Transformationsbereiche und reduzierten Gruppen der Gruppen der eindimensionalen Mannigfaltigkeit.

Als eindimensionale T. M. können nur zwei Raumformen, nämlich die offene und die geschlossene Linie, auftreten. (Die offene Linie kann an ihren beiden Enden je entweder keine Grenze oder eine Grenze, deren Bau noch sehr verschiedenartig sein kann, aufweisen.) Die Anordnungsbeziehungen der Punkte der Linie untereinander bleiben bei der Gruppe ungeändert.

Wir konstruieren in der P. M. eine Kugel K um die Identität und betrachten den innerhalb K gelegenen Teil T des T. B. eines Punktes P für die Gruppe selbst oder für eine reduzierte Untergruppe H. Wir behaupten, daß, wenn in T $P_1, P_2, P_3, \cdots$ gegen P_ω konvergieren, es eine Folge von Transformationen der Gruppe gibt, welche der Reihe nach $P_1, P_2, \cdots$ in P_ω überführen, und gegen die identische Transformation konvergieren.

Sei nämlich μ_ν die Teilmenge von H in der P. M., welche der Stelle P_ν von P entspricht. Sei nun Q_ω ein Punkt von μ_ω innerhalb K, und $Q_1', Q_2', Q_3', \cdots$ der Reihe nach Punkte von $\mu_1, \mu_2, \mu_3, \cdots$ innerhalb K. Für jeden Grenzpunkt der Fundamentalreihe Q' liegt P im Grenzpunkte von $P_1, P_2, \cdots$, d. h. in P_ω. Also liegt jeder dieser Grenzpunkte in μ_ω, und für jedes n giebt es einen dieser Grenzpunkte ${}_nQ_\omega'$ derart, daß $\varrho(Q_n', {}_nQ_\omega') < \varepsilon_n'$, wo ε_n' für hinreichend großes n unter jede Grenze herabsinken muß. Dieselbe Transformation aber, welche ${}_nQ_\omega'$ in Q_n' überführt, führt Q_ω in Q_n über, und Q_n liegt in μ_n derart, daß $\varrho(Q_n', {}_nQ_\omega') < \varepsilon_n$, wo wieder ε_n für hinreichend großes n unter jede [[10]]
Grenze herabsinkt. Q_ω ist also Grenzpunkt von $Q_1, Q_2, \cdots$, welche Punkte der Reihe nach in $\mu_1, \mu_2, \cdots$ liegen, und die Folge der Transformationen, welche der Reihe nach $Q_1, Q_2, \cdots$ in Q_ω überführen, konvergiert gegen die identische Transformation. Zugleich sehen wir, daß gegen jede abgeschlossene Teilmenge von μ_ω abgeschlossene Teilmengen von $\mu_1, \mu_2, \cdots$ gleichmäßig konvergieren.

Es sind jetzt zwei Fälle möglich:

---

*) Mathematische Annalen Bd. 58, 59, 62; Bericht über die Mengenlehre II. [[9]]

1⁰. T ist in einem Segmente der T. M. überall dicht. In diesem Falle hat P für die Gruppe H ein Segment*), und für die entsprechende festgelegte Untergruppe eine endliche oder abzählbare Menge von Segmenten zum Transformationsbereich.

2⁰. T ist nirgends dicht. In diesem Falle ist der Transformationsbereich für die Gruppe H ein isolierter Punkt, und für die entsprechende festgelegte Untergruppe eine endliche oder abzählbare Punktmenge.

Eine reduzierte Untergruppe enthält dann und nur dann eine noch weiter reduzierte Untergruppe, wenn für sie noch Segmente als Transformationsbereiche existieren. Dies wird wenigstens solange der Fall sein, als ihr in der P. M. noch zusammenhängende Mengen entsprechen.**)

Wir betrachten jetzt eine festgelegte Untergruppe, für welche ein Punkt P als T. B. eine Reihe von Segmenten besitzt, und nennen λ_a die der Stelle P_a von P entsprechende Teilmenge der P. M. Sei P_1 die Stelle von P für die Identität, P_2 ein Punkt im selben Segmente des T. B. von P, sei mit P_{1F} ein willkürlicher zwischen P_1 und P_2 liegender Punkt bezeichnet, und sei P_2 so nahe an P_1 gewählt, daß es möglich ist, durch Transformationen $< \eta_1$***) P_1 sowohl in P_2, wie in jeden Punkt P_{1F} überzuführen. Sei Z_1 die zusammenhängende, die Identität enthaltende Teilmenge von λ_1, welche durch die eben genannten Transformationen in Z_2 resp. Z_{1F} übergeführt wird. Bezeichnen wir mit $E(Z_1, Z_2)$ das Minimum der Transformationen, durch welche Z_1 in Z_2 oder Z_2 in Z_1 übergeführt werden kann, so haben wir:

$$E(Z_1, Z_2) < \eta_1 \quad \text{und} \quad E(Z_1, Z_{1F}) < \eta_1.$$

Bezeichnen wir weiter mit ${}_{h_1}I_1$ die Menge jener zusammenhängenden Teilmengen von λ_1, welche sich aus Z_1 innerhalb λ_1 schrittweise erreichen lassen nur mittels Transformationen, welche die Größe h_1 nicht übersteigen, und ihrer Inversen. Will man diese Teilmengen aus Z_1 durch je *eine* Transformation erreichen, so braucht man dazu im allgemeinen größere Transformationen, aber die Grenze k_1, welche *diese* Transformationen nicht zu übersteigen brauchen, konvergiert mit h_1 gegen Null.

Die Menge ${}_{h_1}I_1$ hat mit ihrer Komplementärmenge in λ_1 ein $E > h_1$, und wird von den oben genannten, Z_1 in Z_2 resp. Z_{1F} überführenden Transformationen $< \eta_1$, welche ein gewisses endliches Gebiet der P. M.

*) Ist H die ursprüngliche Gruppe, so kann freilich in einer geschlossenen T. M. diese auch selbst als einziger T. B. auftreten.

**) Sonst nämlich wäre kein anderer Zustand der T. M. zusammenhängend erreichbar, was andererseits gerade vorausgesetzt wird.

***) D. h. deren repräsentierende Punkte in der P. M. weniger als η_1 von der Identität entfernt sind.

nicht überschreiten, transformiert in ${}_{h_1}I_2$ resp. ${}_{h_1}I_{1F}$; diese Teilmengen von λ_2 resp. λ_{1F} weisen alle mit ihren entsprechenden Komplementärmengen ein $E > h_1'$ auf, wo h_1' eine gewisse endliche Größe ist, welche mit h_1, aber auch nur mit h_1, gegen Null konvergiert.

Nennen wir $\eta_1\eta_1$ die obere Grenze der Transformationen, welche sich aus zwei Transformationen $< \eta_1$ oder inversen von solchen zusammensetzen lassen, und setzen wir η_1 so klein gewählt voraus, daß $\eta_1\eta_1 < h_1'$, so wird ${}_{h_1}I_1$ auch von allen anderen Transformationen $< \eta_1$, welche P_1 in P_2 resp. P_{1F} überführen, in dieselben Mengen ${}_{h_1}I_2$ resp. ${}_{h_1}I_{1F}$ übergeführt.

Wir wählen jetzt zwischen P_1 und P_2 weitere Punkte $P_{11}, P_{12}, P_{13}, \cdots$ derart, daß es möglich ist, jedes P_{1n}, mittels Transformationen $< \eta_2$, sowohl in $P_{1,n+1}$, wie in P_{1nF} (d. h. einen willkürlichen Punkt zwischen P_{1n} und $P_{1,n+1}$) überzuführen. Es ist dann möglich, am Segmente P_1P_2 entlang eine solche Reihe $Z_{10}, Z_{10F}, Z_{11}, Z_{11F}, Z_{12}, Z_{12F}, \cdots$ auszuwählen, daß

$$E(Z_{1n}, Z_{1,n+1}) < \eta_2 \quad \text{und} \quad E(Z_{1n}, Z_{1nF}) < \eta_2.$$

Ist $\eta_2 < \eta_1$, so wird jede Reihe von Transformationen, welche Z_1 sukzessive in diese verschiedenen Z überführt, ${}_{h_1}I_1$ überführen in ${}_{h_1}I_{10F}, {}_{h_1}I_{11}, \cdots$, welche mit den oben gefundenen ${}_{h_1}I_{1F}$ identisch sind. Die Transformationen dieser Reihe können mithin alle $< \eta_1 k_1$, also innerhalb eines gewissen endlichen Gebietes der P. M. gewählt werden.

Eine Menge ${}_{h_2}I_1$ wird von diesen Transformationen übergeführt in ${}_{h_2}I_{10F}, {}_{h_2}I_{11}, \cdots$, welche, falls $h_2 < h_1$, der Reihe nach Teilmengen von ${}_{h_1}I_{10F}, {}_{h_1}I_{11} \cdots$ sind, und jede mit ihrer entsprechenden Komplementärmenge ein $E > h_2'$ aufweisen, wo h_2' eine gewisse endliche Größe ist, welche mit h_2, aber auch nur mit h_2, gegen Null konvergiert.

Wir wählen jetzt noch η_2 so klein, daß $\eta_2\eta_2 < h_2'$: dann wird ${}_{h_2}I_1$ auch bei anderer den Bedingungen genügender Wahl von $Z_{10F}, Z_{11}, \cdots$ stets in dieselben Mengen ${}_{h_2}I_{10F}, {}_{h_2}I_{11}, \cdots$ übergeführt werden.

So fortfahrend, können wir, $h_1, h_2, h_3, \cdots$ unbegrenzt abnehmen lassend, durch immer neue Zwischenfügungen von Punkten $P_{mnl\ldots}$, für jedes h_n immer wieder eine ganz bestimmte Reihe von Abbildungen von ${}_{h_n}I_1$ am Segmente P_1P_2 entlang konstruieren, und jedes folgende ${}_{h_n}I_1$ gibt eine Reihe von Abbildungen, welche innerhalb der Abbildungen der vorangehenden Reihe liegen.

Schließlich bleibt als Limes eine ganz bestimmte Reihe von Abbildungen von Z_1.

Wenn man aber ein Teilsegment von P_1P_2 um einen seiner Punkte P_w zusammenschrumpfen läßt, werden die Minimaltransformationen, welche Z_w in die übrigen Limesabbildungen des Segmentes überführen können, schließlich unter jede Größe der Reihe $\eta_1, \eta_2, \eta_3, \cdots$ herabsinken. Die Limesreihe der Z bildet also eine Gesamtheit, welche zusammenhängend ist, und in der die Z eineindeutiges und stetiges Bild der entsprechenden Stellen von P sind. Die Limesreihe läßt sich über das ganze Segment des Transformationsbereiches von P ausdehnen, bleibt nach wie vor eineindeutiges und stetiges Abbild der entsprechenden Stellen von P, und bildet schließlich eine nicht weiter ausdehnbare zusammenhängende Teilmenge der Mannigfaltigkeit ν, welche in der P. M. der von uns in Betracht gezogenen festgelegten Untergruppe entspricht. Und allgemein erzeugt jedes Segment des Transformationsbereiches von P aus jeder zusammenhängenden Teilmenge Z eines λ eine zusammenhängende Teilmenge Y von ν, während keine zwei dieser Mengen Y innerhalb der P. M. miteinander zusammenhängen.

Wir setzen jetzt voraus, daß die Gruppe ν entstanden ist durch Festhaltung von der Reihe nach $p-1$ Punkten, welche je vor ihrer Festhaltung noch über Segmente beweglich waren. Machen wir sodann den letzten Punkt dieser Reihe, den wir R nennen, frei, halten aber die übrigen fest wie zuvor, so erzeugen wir die festgelegte Untergruppe $\varkappa$, in der jede zusammenhängende Teilmenge X von einer zusammenhängenden Teilmenge Y aus ν, durch Bewegung von R an einem der Segmente seines Transformationsbereiches entlang, erzeugt wird. Die verschiedenen Z, aus welchen sich ein X zusammensetzt, sind natürlich eineindeutiges Bild der entsprechenden Stellen von P und R; wir behaupten aber, daß dieses Bild auch stetig ist.

Denn wenn zunächst P und R nicht beide je gegen eine einzige bestimmte Stelle konvergieren, so kann auch Z nicht gegen ein einziges Limes-Z konvergieren; und wenn P und R beide je gegen eine einzige bestimmte Stelle konvergieren, so kann Z nur gegen das diesen Stellen entsprechende Z konvergieren; gäbe es nämlich noch ein zweites Limes-Z, so müßten P und R auch gegen die letzterem Z entsprechenden Stellen konvergieren.

Indem wir in dieser Weise fortfahren, finden wir, wenn wir nach Freilassung von noch q weiteren Punkten bis zur ursprünglichen Gruppe aufgestiegen sind, daß in der P. M. die mit der Identität zusammenhängende Menge wenigstens innerhalb einer gewissen Umgebung der Identität (diese Einschränkung ist notwendig wegen der Möglichkeit des geschlossenen Transformationsbereiches für die ursprüngliche Gruppe selber) sich aus einem gewöhnlichen, durch die Stellen der Punkte P, R, usw. bestimmten, $(q+2)$-dimensionalen Raume von Mengen Z zusammensetzt.

Hieraus folgern wir sogleich, daß jede Teilmenge Z einer Menge λ innerhalb λ isoliert liegt. Sonst nämlich gäbe es in beliebiger Nähe

[[124]]

eines Punktes der P. M. andere Punkte, welche innerhalb der P. M. nur auf endlichem Wege zusammenhängend erreichbar wären. Dieser Zusammenhang zweier Mengen Z eines und desselben λ innerhalb der P. M. ist sogar auf unserem jetzigen Standpunkte nur noch dann möglich, wenn wir in der ursprünglichen Gruppe einen geschlossenen T. B. haben, nämlich durch einen vollen Umlauf dieses T. B. Gibt es in der ursprünglichen Gruppe nur offene T. B., so enthält jedes λ nur ein einziges Z.

Wir behaupten jetzt weiter, daß es immer möglich ist, die nacheinander festgehaltenen Punkte derart zu wählen, daß nach Festhaltung einer endlichen Zahl von Punkten keine zusammenhängende Teilmenge der P. M. mehr zur Verfügung steht.

Im entgegengesetzten Falle könnten wir nämlich eine unendliche Reihe von festgehaltenen Punkten in ihrer abzählbaren Folge allmählich eine überall dichte Skala in der T. M. konstruieren lassen, und die Punkte dieser Folge der Reihe nach in der Gruppe festhalten. Die Gesamtheit der nach jeder neuen Festhaltung noch möglichen reduzierten Untergruppen konvergiert sodann gegen die Identität.*) Durch Festhaltung einer endlichen Zahl von Punkten kann man somit erreichen, daß in der P. M. als reduzierte Untergruppe nur noch eine gewisse im endlichen liegende Punktmenge π, welche sich der Grenze der P. M. nicht unbeschränkt nähert, zur Verfügung steht. Gäbe es jetzt in der T. M. noch T. B., welche Segmente enthalten, so hätte π einen Grenzpunkt K, welcher einen inneren Punkt eines solchen Segmentes in einen seiner Endpunkte überführte. Damit träte aber dieser Punkt aus seinem T. B., was für eine Transformation der Gruppe nicht möglich ist; K müßte also der Grenze der P. M. angehören, aber wir haben auch dies als unmöglich erkannt.

Somit kann durch Festhaltung einer endlichen Zahl von Punkten erreicht werden, daß die T. B. in der P. M. keine zusammenhängende Teilmenge mehr enthalten, also aus isolierten Punkten bestehen. Ist dies durch Festhaltung von der Reihe nach n zuvor noch in Segmenten beweglichen Punkten erreicht, so nennen wir die Gruppe *n-gliedrig* und

*) Nennen wir nämlich Z_n die nach Festhaltung von n Punkten übrig bleibende reduzierte Untergruppe, und β das von zwei in der P. M. um die Identität beschriebenen Kugeln begrenzte abgeschlossene Gebiet, so kann nicht jedes Z_n innerhalb β eine Teilmenge besitzen. Sonst nämlich gäbe es wegen der Abgeschlossenheit aller dieser Teilmengen wenigstens *einen* Punkt innerhalb β, welcher allen Z_n gemeinsam wäre, für den also alle Punkte der T. M. dieselben Stellen, wie für die Identität, einnehmen würden, was unmöglich ist, da dieser Punkt andererseits in der P. M. in endlicher Entfernung von der Identität liegen würde.

[[125]]

behaupten, daß eine n-gliedrige Gruppe nicht zugleich $(n-a)$-gliedrig sein kann. Seien nämlich $P_1, P_2, \cdots, P_n$ die Punkte, deren Festhaltung die n-gliedrige Gruppe festsetzt, und sei Q ein für die Gruppe nicht invarianter Punkt. Wird die Gruppe festgelegt durch Festhaltung zunächst von Q und sodann von verschiedenen zuvor noch beweglichen Punkten P in willkürlicher Reihenfolge, so sind wenigstens $n-1$ der Punkte P erforderlich, ehe alle Punkte P und damit die Gruppe festgelegt sind. Denn wenn schon $m(<n-1)$ Punkte P genügten, wäre die reduzierte Untergruppe, welche der Festhaltung dieser m Punkte entspricht, einerseits eindimensional, und andererseits $(n-m)$-dimensional, was unmöglich ist. Sei nun $Q_1, Q_2, \cdots, Q_{n'}$ eine Reihe von Punkten, welche die Gruppe n'-gliedrig festsetzt. Wir haben gesehen, daß nach Festhaltung nur von Q_1 noch wenigstens $n-1$ Punkte P notwendig sind, um durch ihre Festhaltung die Gruppe festzulegen, und folgern hieraus durch eine analoge Beweisführung, daß nach Festhaltung von Q_2 und Q_1 noch wenigstens $n-2$ Punkte P erfordert werden, und durch genügende Wiederholung, daß, wenn $n' < n$, nach Festhaltung aller Punkte Q noch wenigstens $n-n'$ Punkte P festgehalten werden müßten, die Gruppe also durch Festhaltung der Q allein nicht festgelegt werden könnte. Also kann n' nicht $<n$ sein, aber nun auch umgekehrt nicht $n<n'$, und *n ist eine Invariante der Gruppe.*

Die P. M. einer n-gliedrigen Gruppe ist sicher n-dimensional; sie könnte aber zugleich m-dimensional $(m \neq n)$ sein.

§ 3.

Die Infinitesimaltransformationen der Gruppen der eindimensionalen Mannigfaltigkeit.

Wir betrachten jetzt von einer mehrgliedrigen Gruppe diejenige festgelegte Untergruppe, welche nach Festhaltung noch *eines* weiteren Punktes in der P. M. als T. B. für die Identität nur isolierte Punkte übrig läßt. Diese Gruppe enthält in der P. M. ein System von eindimensionalen Mannigfaltigkeiten; in einer von diesen, nämlich in der entsprechenden reduzierten Untergruppe, liegt die Identität. Wir untersuchen jetzt, wie diese reduzierte Untergruppe jene sie selbst repräsentierende eindimensionale Mannigfaltigkeit $\mathfrak{z}$ in sich transformiert.

Wir legen der Identität den Namen „0" oder „Nullpunkt" bei, wählen eine willkürliche Transformation, welche den Nullpunkt in den Punkt „a" überführt, und nennen diese Transformation „$+a$". Ebenso bezeichne im allgemeinen „$+k$" diejenige Transformation, welche den Nullpunkt in den Punkt „k" überführt.

[[126]]

Sei „h“ der Punkt, in den die Transformation „$+a$“ den Punkt „a“ überführt, so liegt, wenn a rechts vom Nullpunkt, auch h rechts von a. Weil $h = a + a$ und $+h = +a + a$, setzen wir $h = 2a$, d. h. wir nennen „h“ mit anderen Worten „$2a$“.

Ebenso nennen wir „$3a$“ den Punkt, in den $2a$ von der Transformation $+a$ übergeführt wird, „$4a$“ den, in welchen $3a$ von ihr übergeführt wird, usw.

Analog nennen wir „$-a$“ die Inverse von der Transformation „$+a$“, und desgleichen den Punkt, in welchen der Nullpunkt von dieser Inversen übergeführt wird, setzen

$$-a - a \equiv -2a, \quad -2a - a \equiv -3a, \quad \text{usw.}$$

Die in dieser Weise bestimmte Punktreihe vom Ordnungstypus $^*\omega + \omega$ kann weder rechts noch links innerhalb der betrachteten eindimensionalen Mannigfaltigkeit einen Grenzpunkt besitzen. Sonst nämlich wäre die von einem solchen Grenzpunkte repräsentierte Transformation nicht eineindeutig.

Weiter kann die Transformation $+a$ keinen Punkt von $\mathfrak{z}$ festlassen. Denn einen Punkt zwischen na und $(n+1)a$ führt sie über in einen Punkt zwischen $(n+1)a$ und $(n+2)a$. Also erhalten auch alle Punkte eine Verrückung nach derselben Seite.

Zwischen den Punkten 0 und a liegt sicher ein Punkt b derart, daß die Transformation $+b$ den Punkt b in den Punkt a überführt, m. a. W. derart, daß $a = 2b$. Es kann aber nicht zwei Punkte dieser Art, etwa b und b', geben, denn wenn b' rechts von b, liegt auch $b' + b'$ rechts von $b' + b$, und $b' + b$ wieder rechts von $b + b$. Wir setzen $b \equiv \frac{1}{2}a$, und haben:

$$na = 2n \cdot \frac{1}{2}a.$$

Zugleich ist jetzt zwischen jedem na und $(n+1)a$ ein Punkt

$$na + \frac{1}{2}a = \frac{1}{2}a + na = (2n+1) \cdot \frac{1}{2}a$$

bestimmt, während für jeden Ausdruck $p + q + r + \cdots$, wo $p, q, r, \cdots$ Vielfache von $\frac{1}{2}a$ bezeichnen, die assoziative und die kommutative Eigenschaft zutreffen.

In derselben Weise fortfahrend bestimmen wir $c \equiv \frac{1}{2}b$, und zugleich alle Punkte $nc \equiv n \cdot \frac{1}{2}b$; sodann $\frac{1}{2}c$, und zugleich alle Punkte $n \cdot \frac{1}{2}c$ usw., und erzeugen so schließlich innerhalb $\mathfrak{z}$ einen Ordnungstypus η.

Dieser Ordnungstypus liegt in $\mathfrak{z}$ überall dicht. Denn es liegt wenigstens ein Grenzpunkt z. B. zwischen 0 und a; gäbe es nun in $\mathfrak{z}$ ein freies Intervall, so könnte dieses durch eine eigentliche Transformation (d. h. eine, welche keine Grenztransformation wäre) ganz in den ebengenannten Grenzpunkt übergeführt werden.

[[127]]

Weil nun aber die Transformationen in der T. M. eineindeutiges und stetiges Abbild von $\mathfrak{z}$ sind, so erzeugen dieselben Transformationen, welche $\mathfrak{z}$ überall dicht überdecken, auch in jedem Segmente, welches T. B. in der T. M. ist, eine überall dichte Skala.

Setzen wir $a = 1$, $2a = 2$, $\frac{1}{2}a = \frac{1}{2}$ usw., so bestimmen wir in dieser Weise eine eineindeutige und stetige Abbildung jeder dieser Skalen auf das System der positiven und negativen ganzen Zahlen und Dualbrüche, während dabei jeder T. B. selber eineindeutiges und stetiges Bild des Systems der reellen Zahlen wird. Das Abbild der T. B. in den reellen Zahlen unterliegt durch unsere reduzierte Untergruppe der *Gruppe der Addition.*

Wir wollen nun jede endliche kontinuierliche Gruppe, welche eineindeutiges und für jede abgeschlossene Teilmenge gleichmäßig stetiges Abbild der Additionsgruppe der reellen Zahlen ist, bezeichnen als „von einer Infinitesimaltransformation erzeugt", und können mithin den Satz aussprechen:

Jede endliche kontinuierliche Gruppe der eindimensionalen Mannigfaltigkeit enthält eine letzte reduzierte Untergruppe, welche von einer Infinitesimaltransformation erzeugt wird.

§ 4.

Die eingliedrigen Gruppen der eindimensionalen Mannigfaltigkeit.

Sei zunächst die T. M. offen. In diesem Falle gilt für die eingliedrigen Gruppen dasselbe, was wir im vorigen Paragraphen für die letzten reduzierten Untergruppen mehrgliedriger Gruppen hergeleitet haben. Wir können mithin folgenden Satz aufstellen:

Durch eine eingliedrige Gruppe der offenen eindimensionalen Mannigfaltigkeit wird in dieser eine endliche oder abzählbare Menge von Segmenten bestimmt; jedes dieser Segmente wird von einer durch die Gruppe bestimmten überall dichten Skala von $-\infty$ bis $+\infty$ gemessen, und in jeder dieser Skalen ist die Gruppe die Additionsgruppe der reellen Zahlen. Die nicht zu diesen Segmenten gehörigen Punkte bleiben bei der Gruppe invariant.

Ist aber die T. M. geschlossen, so müssen wir zwei Fälle unterscheiden.

Entweder es gibt einen Punkt P, der nach einem vollen Umlauf in seine alte Stelle P_0 wiederkehren kann. In diesem Falle konstruieren wir in analoger Weise, wie oben für die letzten reduzierten Untergruppen ausgeführt ist, eine Skala, der wir die Transformation, welche P aus P_0 nach einem einzigen Umlauf in bestimmtem Sinne wieder in P_0 zurückführt, und welche wir mit „$+ 2\pi$" bezeichnen, zugrunde legen. Diese Skala läuft unendlich viele Male um die T. M. herum, und jedem Punkte werden unendlich viele reelle Zahlen zugeordnet, von denen je eine zwischen $2n\pi$ und $2(n+1)\pi$ für jedes ganzzahlige n liegt; insbesondere sind dem Punkte P_0 alle Zahlen $2n\pi$ zugeordnet. Da nun aber jede Transformation der Gruppe jedem Punkte der T. M. nur eine einzige bestimmte Verrückung gibt, muß jede Teilskala von $2n\pi$ bis $2(n+1)\pi$ sich decken mit der Teilskala von Null bis 2π, und die Gruppe ist in dieser Skala die Rotationsgruppe.

Oder es gibt keinen solchen Punkt P. In diesem Falle hat die Gruppe in der T. M. wenigstens einen festen Punkt, und außerhalb dieses Punktes gilt für sie dasselbe, was für die Gruppen der offenen T. M. hergeleitet ist.

Also gilt der Satz:

Eine eingliedrige Gruppe der geschlossenen eindimensionalen Mannigfaltigkeit ist entweder eine Rotationsgruppe, oder wird durch Ausscheidung eines festen Punktes zu einer solchen der offenen eindimensionalen Mannigfaltigkeit.

Wir bemerken noch, daß die offene eindimensionale Mannigfaltigkeit zu beiden Seiten noch sehr verschiedenartig gebaute Grenzmengen („Enden“) aufweisen kann; daß in ihnen aber als T. B. für die Gruppe außer festen Punkten nur eindimensionale Mannigfaltigkeiten auftreten können.

§ 5.

Die zweigliedrigen Gruppen der eindimensionalen Mannigfaltigkeit.

Unter einer *eigentlichen n-gliedrigen Gruppe* verstehen wir eine solche n-gliedrige Gruppe, deren n die Gruppe festlegende Punkte in einem und demselben T. B. der Gruppe gewählt werden können.

Im entgegengesetzten Falle sprechen wir von einer *uneigentlichen n-gliedrigen Gruppe.* Wenn bei einer solchen Gruppe von n die Gruppe festlegenden Punkten höchstens m in einem und demselben T. B. gewählt werden können, so ist die Struktur der Gruppe auf die der eigentlichen ein-, zwei-, usw. bis einschließlich m-gliedrigen Gruppen zurückzuführen. Wir beschäftigen uns deshalb im folgenden nur mit eigentlichen n-gliedrigen Gruppen, und zwar zunächst mit eigentlichen zweigliedrigen Gruppen.

Wir bezeichnen eine solche Gruppe mit g, und bemerken zunächst, daß einzelne, aber nicht alle, Transformationsbereiche eingliedrig transformiert werden können. Sei also α ein T. B., der zweigliedrig transformiert wird. Der Festhaltung eines Punktes A dieses T. B. entspricht eine von einer Infinitesimaltransformation erzeugte letzte reduzierte Untergruppe g_1.

Bleiben für diese Gruppe g_1 mehrere Punkte von α fest, und ist A' einer von ihnen, so bleibt für die der Festhaltung von A' entsprechende letzte reduzierte Untergruppe g_1' auch A fest. Denn bliebe A beweglich, so würde g_1' durch Festhaltung von A festgelegt werden, und für festes A und A', m. a. W. für festes A wäre keine stetige Bewegung mehr möglich, was wider die Voraussetzung ist.

Da nun das System der für die Gruppe g_1 in α festbleibenden Punkte eine transitive Gruppe von eineindeutigen und stetigen Transformationen zulassen muß, so sind diese Punkte untereinander isoliert. Wenn weiter rechts von einem dieser Punkte A_1 ein zweiter Punkt A_2 liegt, so liegt rechts von A_2 ein dritter usw.; das System enthält also entweder nur einen einzigen Punkt, oder sein Ordnungstypus ist $^*\omega + \omega$ für ein offenes α, resp. endlich für ein geschlossenes α.

Im letzteren Falle können zwischen zwei aufeinander folgenden Punkten A_1 und A_2 nicht zwei Punkte eines anderen solchen Systems liegen. Sonst nämlich existierten zwischen diesen letzten zwei Punkten wieder zwei eines dritten Systems usw., und es wäre möglich, in einem gewissen Punkte von α zwei Punkte eines selben Systems von zusammen festbleibenden Punkten nur durch Transformationen innerhalb einer abgeschlossenen Teilmenge der P. M. sich unbeschränkt nähern und schließlich zusammenfallen zu lassen.

Umgekehrt ist es auch nicht möglich, daß zwischen A_1 und A_2 kein einziger Punkt des anderen Systems liegen sollte. Es liegt also zwischen je zwei aufeinanderfolgenden Punkten eines Systems je ein Punkt eines willkürlichen anderen Systems.

Sei β einer der T. B. von g_1 in α, B und C Punkte von β, und g_2 resp. g_3 die reduzierte Untergruppe, welche der Festhaltung von B resp. C entspricht, und welche nach dem vorigen keinen weiteren festen Punkt innerhalb β enthält. Sodann führt jede Transformation aus g, also auch jede Transformation aus g_1, welche B in C überführt, g_2 in g_3 über.

Wir bilden jetzt das Segment β mit der von g_1 in ihm konstruierten Skala auf die X-Achse eines Koordinatensystems einer Ebene ab, wobei wir den Koordinatenanfangspunkt dem Punkte B entsprechen lassen, und konstruieren zu jeder Transformation h aus g_2 die Kurve $y = f_h(x)$, wo zu jedem x die durch h bewirkte Verrückung des entsprechenden Punktes der X-Achse als y genommen wird. Die erhaltene Kurve nennen wir die *Inkrementkurve* der Transformation h.

Nach den Entwicklungen des § 4 liegen die verschiedenen Kurven $y = f_h(x)$ mit Ausnahme des einzigen gemeinschaftlichen Punktes B gänzlich außerhalb voneinander und von der die identische Transformation repräsentierenden X-Achse.

Ist c die Abszisse von C, so führt eine Translation „$+ c$" die Gruppe g_2 in g_3 über. Das System $y = f_h(x - c)$ bildet somit die Inkrementkurven der Gruppe g_3.

[[130]]

Jede Transformation aus g, welche C invariant läßt, und sich innerhalb g bei festem C stetig aus der Identität erreichen läßt, muß zu g_3 gehören. Solche Transformationen sind aber folgende:

$$x' = x + f_h(x) - f_h(c).$$

Also gehört die Kurvenschar

$$y = f_h(x) - f_h(c)$$

zur Schar

$$y = f_h(x - c),$$

und die Schar

$$y = f_h(x + c) - f_h(c)$$

zur Schar

$$y = f_h(x).$$

Jede Funktion $f_h(x)$ muß also entweder fortwährend steigen, oder fortwährend abnehmen. Sonst nämlich gäbe es eine Kurve

$$y = f_h(x + c) - f_h(c),$$

welche die X-Achse noch in anderen Punkten als B treffen würde.

Wir bezeichnen mit ${}_\Delta\varphi_h(x)$ die Zunahme der Ordinate der Kurve $y = f_h(x)$ zwischen den Abszissen x und $x + \Delta$, und nehmen einen Augenblick an, daß ${}_\Delta\varphi_{h_1}(x)$ zwei gleiche Werte hätte, für $x = x_1$ und $x = x_1 + p$. In jenem Falle hätten wir in der Kurvenschar $y = f_h(x)$, welche ja auch die Kurven $y = f_h(x + k) - f_h(k)$ enthält, zwei Kurven, welche einander sowohl für $x = 0$ wie für $x = \Delta$ träfen. Da nun aber die verschiedenen Kurven $y = f_h(x)$ mit Ausnahme des einzigen gemeinschaftlichen Punktes B gänzlich außerhalb voneinander liegen, ist das vorausgesetzte Verhalten nur dann möglich, wenn die Kurven

$$y = f_{h_1}(x + x_1) - f_{h_1}(x_1) \quad \text{und} \quad y = f_{h_1}(x + x_1 + p) - f_{h_1}(x_1 + p)$$

dieselben sind, m. a. W. wenn die Kurve $y = f_{h_1}(x)$ eine homothetisch-periodische Gestalt mit der Periode p besitzt.

Dies aber schließt ein, daß ${}_p\varphi_{h_1}(x_1 + t) = {}_p\varphi_{h_1}(x_1)$, unabhängig von t, so daß die Kurve $y = f_{h_1}(x)$ nun sogar eine homothetisch-periodische Gestalt mit jeder willkürlichen Periode t besitzen muß, was nur möglich ist, wenn sie eine gerade Linie ist.

Es hat sich also herausgestellt, daß, wenn $y = f_{h_1}(x)$ keine gerade Linie ist, wenigstens ${}_\Delta\varphi_{h_1}(x)$ nicht zweimal denselben Wert erhalten kann. Für jedes bestimmte Δ muß dieser Wert also mit x entweder fortwährend steigen, oder fortwährend abnehmen, was nur möglich ist, wenn er entweder für alle Δ fortwährend steigt, oder für alle Δ fortwährend abnimmt.

Wir setzen beispielsweise das erste voraus, und bezeichnen jetzt die Differenzenquotienten von $y = f_{h_1}(x)$ zwischen den Abszissen Null und Δ_1,

Δ_1 und $2\Delta_1$, $2\Delta_1$ und $3\Delta_1$, usw. der Reihe nach mit ${}_1Z$, ${}_2Z$, ${}_3Z$ usw., jene zwischen Null und $\frac{1}{2}\Delta_1$, $\frac{1}{2}\Delta_1$ und Δ_1, Δ_1 und $1\frac{1}{2}\Delta_1$, usw. mit ${}_1Z_0$, ${}_1Z_1$, ${}_2Z_0$ usw.; jene zwischen Null und $\frac{1}{4}\Delta_1$, $\frac{1}{4}\Delta_1$ und $\frac{1}{2}\Delta_1$, $\frac{1}{2}\Delta_1$ und $\frac{3}{4}\Delta_1$, usw. mit ${}_1Z_{00}$, ${}_1Z_{01}$, ${}_1Z_{10}$ usw., und fahren in dieser Weise fort. Gehören sodann ${}_aZ_p$ und ${}_aZ_q$ zu aneinander grenzenden, gleich großen Intervallen (derart daß ${}_aZ_q$ rechts von ${}_aZ_p$ liegt), so besitzen ${}_aZ_q$, ${}_aZ_{q0}$, ${}_aZ_{q00}$ usw. abnehmende Werte, welche aber oberhalb des Wertes von ${}_aZ_p$ bleiben, und ${}_aZ_p$, ${}_aZ_{p1}$, ${}_aZ_{p11}$ usw. haben zunehmende Werte, welche aber unterhalb des Wertes von ${}_aZ_q$ bleiben. Erstere Reihe hat somit einen unteren, letztere einen oberen Limeswert, und man zeigt leicht, daß nun auch alle Differenzenquotienten für unbeschränkt abnehmendes Δ gegen diese Werte konvergieren, m. a. W. die Kurve $y = f_{h_1}(x)$ hat in jedem Punkte, dessen Abszisse zur überall dichten Skala $\frac{m}{2^n}\Delta_1$ gehört, sowohl einen rechtsseitigen wie einen linksseitigen Differentialquotienten. Dasselbe sehen wir nun aber auch leicht ein für einen willkürlichen Punkt P der Kurve, wenn wir die Differenzenquotienten zwischen P und den Punkten einer Fundamentalreihe der oben genannten überall dichten Menge, welche P zu ihrem Grenzpunkte hat, in Betracht ziehen. (Letztere Differenzenquotienten lassen sich nämlich mit Hilfe der Z approximieren.)

Unser weiteres Ziel ist, zu zeigen, daß die rechtsseitigen und die linksseitigen Differentialquotienten einer Kurve $y = f_h(x)$ einander gleich sind, oder, was auf dasselbe hinauskommt, daß der rechtsseitige Differentialquotient eine stetige Funktion von x ist.

Die Punkte x, welche bei der Transformation h einer Verschiebung von der Größe $f_h(x)$ unterliegen, werden durch Überführung dieser Transformation in die zur Inkrementkurve $y = f_h(x + \Delta)$ gehörige weiter geschoben über eine Strecke ${}_\Delta\varphi_h(x)$. Letztere Verschiebung, zusammengesetzt mit einer ihr folgenden Translation δ, gibt eine Verrückung $\delta + {}_\Delta\varphi_h(x)$; zusammengesetzt mit einer ihr vorangehenden Translation δ, eine Verrückung $\delta + {}_\Delta\varphi_h(x + \delta')$, wo δ', mit x variabel, das Abszisseninkrement bezeichnet, welches durch die in der Kurve $y = f_h(x)$ zugehörige Ordinatenzunahme zu δ ergänzt wird.

Der Unterschied dieser beiden resultierenden Verrückungen stellt eine Verrückung $\sigma(x) \equiv {}_\Delta\varphi_h(x + \delta') - {}_\Delta\varphi_h(x)$ dar.

Besäße nun die Kurve $y = f_h(x)$ in P einen Knick, so sei Q ein Punkt der Kurve links von P; wir ziehen dann folgende Werte von x in Betracht:

x_1, die Abszisse von Q,
x_2, derart, daß $x_2 + \delta'(x_2) = x_3$,
x_3, die Abszisse von P.

Sei k der Unterschied zwischen dem rechtsseitigen und dem linksseitigen Differentialquotienten in P; sei weiter $\Delta < \delta'(x_2)$ gewählt, $x_1 < x_2$, und Δ und δ beide so klein, daß sowohl in P, wie in Q:

$$\frac{{}_\Delta\varphi_h(x+\delta')}{\Delta} - \frac{{}_\Delta\varphi_h(x)}{\Delta} < k.$$

Sodann würde die Inkrementfunktion $\sigma(x)$ für $x = x_1$ kleiner als $k\Delta$, für $x = x_2$ größer als $k\Delta$, und für $x = x_3$ wieder kleiner als $k\Delta$ sein, was unmöglich ist, da eine Inkrementfunktion mit x entweder fortwährend steigen, oder fortwährend abnehmen muß. Die Kurve $y = f_h(x)$ kann somit keinen Knick aufweisen.

Wir behaupten jetzt, daß die Gruppe g_2 auf der Skala der Gruppe g_1 durch eine Infinitesimaltransformation erzeugt wird, d. h. daß in dieser Skala der Quotient zweier unbeschränkt abnehmender Inkrementfunktionen von g_2 sich unbeschränkt einer Konstanten nähert.

Werden nämlich die Punkte x_0 durch die Inkrementkurve $y = f_h(x_0)$ in die Punkte x und durch die Inkrementkurve $y = f_h(x_0+\delta) - f_h(\delta)$ in die Punkte x' übergeführt, so bedeutet die Überführung von x in x' eine Verschiebung:

$$\Delta x = {}_\delta\varphi_h(x_0) - {}_\delta\varphi_h(0) = \delta \cdot \{f_h'(x_0) + \varepsilon_\delta(x_0) - f_h'(0) - \varepsilon_\delta(0)\},$$

wo die Funktion ε_δ mit δ verschwindet.

Wir können also schreiben:

$$\Delta x = \left\{\frac{\psi(x)}{\psi(x_1)} + \eta_\Delta(x)\right\} \cdot \Delta x_1,$$

wo x_1 eine willkürlich bestimmte Abszisse bezeichnet, und die Funktion η_Δ mit Δ verschwindet, während die Stetigkeit und gebietsweise Endlichkeit von $f_h'(x)$, sowie ihre fortwährende Steigung (resp. Abnahme) mit x mitbringt, daß auch $\psi(x)$ stetig und gebietsweise endlich ist, und mit x fortwährend steigt (resp. abnimmt).

Zugleich stellt sich heraus, daß die Verschiebung einer willkürlichen Transformation aus g sich schreiben läßt:

$$u(x) = u(x_1) \cdot \left\{\frac{\psi(x)+a}{\psi(x_1)+a} + \eta_u(x)\right\},$$

wo die Funktion η_u mit u verschwindet.

Wir können jetzt auch gleich zeigen, daß die Funktion $\psi(x)$, welche die Infinitesimaltransformation definiert, ihrerseits wieder differenzierbar ist. Die Gleichungen der Inkrementkurven von g_2 lassen sich nämlich schreiben:

$$y = \varepsilon \cdot \{\psi(x) + \varepsilon'(x)\},$$

wo die Funktion $\varepsilon'(x)$ mit ε verschwindet.

Zu diesen Inkrementkurven gehören aber alle Kurven:

$$y = \varepsilon \cdot \{ \psi(x+\delta) - \psi(\delta) + \varepsilon'(x+\delta) - \varepsilon'(\delta) \},$$

was nur möglich ist, wenn

$$\psi(x+\delta) - \psi(\delta) \equiv c \cdot \psi(x),$$

m. a. W. wenn

$$\frac{\psi(x+\delta) - \psi(x)}{\delta} \equiv a\psi(x) + b.$$

Jeder Differenzenquotient von $\psi(x)$ hat somit die Form $a\psi(x) + b$. Da weiter $\psi(x)$ gebietsweise endlich ist, ist es unmöglich, daß bei unbegrenzter Abnahme des x-Inkrementes schließlich in einem gewissen Gebiete jede Größe von einem und demselben Differenzenquotienten für jeden Wert von x überstiegen werden könnte, daß also einer der Koeffizienten a und b jede Größe übersteigen könnte.

Die Koeffizienten a und b liegen somit für die verschiedenen Differenzenquotienten zwischen endlichen Grenzen, also ist die Funktionsmenge der Differenzenquotienten gebietsweise gleichmäßig stetig, und die Funktion $\psi(x)$ ist differenzierbar.*)

Zur Bestimmung von ψ existiert mithin folgende Differentialgleichung:

$$\frac{d\psi}{dx} = a\psi + b.$$

Sei in dieser Gleichung zunächst $a = 0$, so haben wir:

$$\psi = bx,$$

und die endliche Gleichung der Gruppe g_2 wird:

$$\frac{x'}{x} = t.$$

Wir finden mithin für g_2 die Multiplikationsgruppe mit B als Nullpunkt in der Skala, welche g_1 als Additionsgruppe definiert. Wir bemerken, daß diese Lösung den Fall enthält, in dem die Kurven $y = f_h(x)$ gerade Linien sind.

Sei jetzt $a \neq 0$, so haben wir:

$$\psi = \frac{b e^{ax} - b}{a},$$

und die endliche Gleichung der Gruppe g_2 wird:

$$\frac{1 - e^{-ax'}}{1 - e^{-ax}} = t,$$

[[11]] *) Vergl. Verslag Akademie Amsterdam, Mei 1908, p. 38.

[[134]]

oder wenn wir $1 - e^{-ax} = \xi$ setzen:

$$\frac{\xi'}{\xi} = t.$$

Wir finden also für g_2 die Multiplikationsgruppe mit B als Nullpunkt in der Skala von e^{-ax}, d. h. in einer Skala, in der schon g_1 als eine Multiplikationsgruppe, nur mit anderem Nullpunkt als g_2, auftritt.

Im ersten Falle ist der Transformationsbereich β von g_1 identisch mit dem Transformationsbereiche α von g. Da dieser Bereich offen ist, ist die festgelegte Untergruppe für festes A mit g_1 identisch. Die Gruppe g ist in dem Bereiche α auf einer gewissen Skala, welche α von $-\infty$ bis $+\infty$ mißt, die Gruppe der Addition und Multiplikation.

Im zweiten Falle kann β nicht ein ganzer T. B. von g sein, weil die Gruppe g_2 es nicht in sich transformiert. Umso weniger kann β eine ganze geschlossene T. M. ausfüllen, seine Endpunkte A_1 und A_2, deren einer, z. B. A_2, gemeinsamer fester Punkt für g_1 und g_2 ist*), sind also jedenfalls verschieden. Sei A_2 rechts von A_1, so hat g_2 links von B und von A_1 einen weiteren festen Punkt B_1, der nach links höchstens so weit fortrückt, daß er nach einem einzigen Umlauf in A_2 fällt. Wie aber g_2 einen der beiden festen Punkte A_1 und A_2 mit g_1 gemeinsam hat, so hat g_1 einen der beiden festen Punkte B_1 und B, welcher nur B_1 sein kann, mit g_2 gemeinsam. Das Segment $B_1A_1BA_2$ ist also gemeinsamer Transformationsbereich für g_1 und g_2, und hieraus folgern wir, indem wir das oben über die Verteilung der innerhalb α zusammen fest bleibenden Punkte hergeleitete beachten, daß dieses Segment $B_1A_1BA_2$ mit α zusammenfällt. Es wird von $-\infty$ bis $+\infty$ gemessen durch eine Skala, in der g_1 und g_2 beide nur Ähnlichkeitstransformationen enthalten, und welche die einzig mögliche Skala mit dieser Eigenschaft ist. In ihr ist g auch hier wieder die Gruppe der Addition und Multiplikation.

Wir können jetzt folgenden Satz aussprechen:

Durch eine eigentliche zweigliedrige Gruppe der offenen oder geschlossenen eindimensionalen Mannigfaltigkeit wird in dieser eine endliche oder abzählbare Menge von Segmenten bestimmt; jedes dieser Segmente wird von einer durch die Gruppe bestimmten überall dichten Skala von $-\infty$ bis $+\infty$ gemessen, und in jeder dieser Skalen ist die Gruppe entweder die der Addition

*) Dieser Punkt A_2 (und analog der weiter unten genannte Punkt B_1) kann übrigens, wenn die T. M. sich auf dieser Seite nicht weiter als β fortsetzt, entweder ganz wegfallen, oder durch eines der Enden der offenen T. M. ersetzt werden müssen. Diese Enden, welche sehr verschiedenartig gebaut sein können, besitzen übrigens für g_2 und jede Gruppe g_3 dieselben T. B. (wie leicht zu beweisen ist), so daß auch für die Gruppe g die T. B. in den Enden nur feste Punkte oder eindimensionale Mannigfaltigkeiten sein können.

und Multiplikation der reellen Zahlen), oder die der Addition der reellen Zahlen. Die nicht zu diesen Segmenten gehörigen Punkte bleiben bei der Gruppe invariant.*

§ 6.

Die dreigliedrigen Gruppen der eindimensionalen Mannigfaltigkeit.

Sei α ein offener Transformationsbereich, in dem die eigentliche dreigliedrige Gruppe, welche wir g nennen wollen, wirklich genau dreigliedrig ist. (Es ist natürlich möglich, daß ein Teil der Transformationsbereiche nur ein- oder zweigliedrig transformiert wird.) Von den Systemen von in α zusammen fest bleibenden Punkten zeigen wir genau dieselben Eigenschaften wie im § 5, insbesondere daß zwischen je zwei aufeinander folgenden Punkten eines Systems je ein Punkt eines willkürlichen anderen Systems liegt. Seien $P_1, P_2, \cdots$ und $Q_1, Q_2, \cdots$ zwei solche Systeme in α, und sei $Q_1 P_1 Q_2 P_2$ eine Folge von Punkten, zwischen welchen keine weiteren Punkte der Systeme liegen.

Der Festhaltung von P_1 und Q_2 (wobei zugleich alle Punkte P und Q festbleiben) entspricht in der P. M. eine eingliedrige reduzierte Untergruppe g_0, welche sich sowohl zu einer zweigliedrigen Gruppe g_1, welche nur die P festläßt, als zu einer zweigliedrigen Gruppe g_2, welche nur die Q festläßt, ergänzen lassen muß.

Messen wir die Additionsskala von g_2 durch eine Koordinate x, welche in P_1 ihren Nullpunkt hat, und die Additionsskala von g_1 durch eine Koordinate y, welche in Q_2 ihren Nullpunkt hat. Die Gruppe g_0 muß sodann zwischen P_1 und Q_2 sowohl in x wie in y eine Multiplikationsgruppe sein.

Also ist zwischen P_1 und Q_2:

$$y = -x^{-c}.$$

Die Transformationen von g_1 werden also zwischen P_1 und Q_2 dargestellt durch die Gleichung:

$$\frac{1}{x'^c} = \frac{r}{x^c} + s.$$

---

*) Will man also die Grundoperationen der Arithmetik der Geometrie analog axiomatisch definieren, so ist nach Voraussetzung der eindimensionalen T. M. für das Verhältnis zwischen Addition und Multiplikation nicht die volle distributive Eigenschaft zur Bestimmung der Operationen notwendig. Diese nämlich sagt aus:

$$\varphi_\alpha \{ f_\beta \} = f_{\varphi_\alpha(\beta)} \{ \varphi_\alpha \},$$

während die Forderung

$$\varphi_\alpha \{ f_\beta \} = f_\gamma \{ \varphi_\delta \}$$

genügt.

Sei R ein Punkt zwischen P_1 und Q_2, und g_3 die zweigliedrige reduzierte Untergruppe, welche R festläßt, so werden, wenn in der x-Skala $P_1R = a$, die Transformationen von g_3 zwischen P_1 und Q_2 dargestellt durch die Gleichung:

$$\frac{1}{(x'-a)^c} = \frac{r}{(x-a)^c} + s. \tag{1}$$

Andererseits wird diejenige eingliedrige reduzierte Untergruppe, welche P_1 und R zusammen festläßt, dargestellt durch die Gleichung:

$$\frac{1}{x'^c} - \frac{1}{a^c} = t\left\{\frac{1}{x^c} - \frac{1}{a^c}\right\}. \tag{2}$$

Wenn nun aber (2) sich als besonderer Fall von (1) auffassen lassen soll, so muß $c = 1$ sein, wodurch andererseits eine dreigliedrige Gruppe auch wirklich zustande kommt als eine *periodisch-projektive Gruppe,* in der stets eine abzählbar unendliche Reihe von Punkten vom Ordnungstypus $^*\omega + \omega$ in α zusammen fest wird. (Alle diese Systeme von zusammen fest bleibenden Punkten sind derselben Additionstransformation in α gegenüber invariant.)

Wir können jetzt folgenden Satz aussprechen:

Durch eine eigentliche dreigliedrige Gruppe der offenen eindimensionalen Mannigfaltigkeit wird eine endliche oder abzählbare Menge von Segmenten bestimmt: in jedem dieser Segmente ist die Gruppe entweder eine periodisch-projektive Gruppe mit unendlich vielen Perioden, oder eine Gruppe von Addition und Multiplikation, oder eine Additionsgruppe. Die nicht zu diesen Segmenten gehörigen Punkte bleiben bei der Gruppe invariant.

Das vorhergehende wird für einen geschlossenen T. B. nur insoweit modifiziert, daß die Systeme von zusammen fest bleibenden Punkten hier entweder je einen einzigen Punkt oder je eine endliche Punktmenge bilden.

Im ersten Falle finden wir als einzig mögliche dreigliedrige Gruppe die *einfache projektive Gruppe;* im zweiten Falle ist die Gruppe eine *rotatorisch-projektive Gruppe,* bei der eine gewisse Rotationstransformation alle Systeme von zusammen fest bleibenden Punkten invariant läßt.

Mithin gilt der Satz:

Eine eigentliche dreigliedrige Gruppe der geschlossenen eindimensionalen Mannigfaltigkeit wird entweder durch Ausscheidung eines invarianten Punktes zu einer solchen der offenen eindimensionalen Mannigfaltigkeit, oder ist eine rotatorisch-projektive Gruppe, oder aber eine einfache projektive Gruppe.

§ 7.

Die vier- und mehrgliedrigen Gruppen der eindimensionalen Mannigfaltigkeit.

Eine eigentliche viergliedrige Gruppe der eindimensionalen Mannigfaltigkeit kann nicht existieren.

Für eine solche Gruppe hätten nämlich die Systeme von zusammen fest bleibenden Punkten eines bestimmten T. B. die im § 5 hergeleiteten Eigenschaften. Weiter entspräche der Festhaltung von zwei nicht zusammen fest bleibenden Punkten P und Q eines T. B., in dem die Gruppe viergliedrig wäre, eine zweigliedrige reduzierte Untergruppe g_0.

Wird nur P festgehalten, so bleibt eine dreigliedrige reduzierte Untergruppe übrig, in der P Endpunkt eines offenen T. B. ist, welcher dreigliedrig transformiert wird. Bei weiterer Festhaltung von Q wird in diesem T. B. ein mit Q zusammen fest bleibender Punkt Q' festgehalten, und mit ihm (der obigen Analyse der dreigliedrigen Gruppen gemäß) ein Ordnungstypus $^*\omega + \omega$ von Punkten, welcher P als Häufungsstelle besitzt. Das System der für g_0 fest bleibenden Punkte hätte also P zur Häufungsstelle, und Q nicht zur Häufungsstelle.

Aber in analoger Weise kann man zeigen, daß Q eine Häufungsstelle der für g_0 fest bleibenden Punkte sein muß, während P nicht eine solche sein kann. Die Voraussetzung einer eigentlichen viergliedrigen Gruppe führt somit auf einen Widerspruch.

Und ebensowenig können jetzt natürlich eigentliche mehr als viergliedrige Gruppen existieren. Dagegen gibt es uneigentliche n-gliedrige Gruppen für jedes endliche n.

§ 8.

Schlußbemerkung zu den Gruppen der eindimensionalen Mannigfaltigkeit.

Die obige Analyse hätte in ihren letzten Teilen unter Zurückgreifung auf die Liesche Theorie ein wenig anders geführt werden können:

Seien für eine n-gliedrige Gruppe n Punkte $P_1, P_2, \cdots, P_n$ gegeben, deren Festhaltung die Gruppe festlegt. Man kann sodann jeden Punkt der P. M. innerhalb der genannten Umgebung von der Identität aus stetig erreichen durch n stetige Transformationen τ, deren jede einen der genannten n Punkte in die geforderte Endstelle überführt, und die übrigen festläßt.

Wir haben aber im § 3 gesehen, daß diese Transformationen τ von Infinitesimaltransformationen erzeugt werden, und im § 5, daß zwei solche

[[138]]

Infinitesimaltransformationen, welche von ihren $n-1$ definierenden festen Punkten $n-2$ gemeinsam haben, beide auf derselben überall dichten Messungsskala durch differenzierbare Funktionen gemessen erscheinen. Es gibt also auch eine Messungsskala, in der alle n Infinitesimaltransformationen, welche die Transformationen τ erzeugen, durch n differenzierbare Funktionen gemessen werden, die, wie leicht einzusehen, linear unabhängig sind. Eine von solchen Infinitesimaltransformationen erzeugte Gruppe muß aber nach Lie*) ähnlich sein mit einer der von ihm aufgezählten analytischen Gruppen.

Wir haben indes oben den direkten Weg zur Bestimmung der drei- und mehrgliedrigen Gruppen vorgezogen, weil dabei ihre Struktur nicht nur in einer gewissen Umgebung, sondern in der ganzen T. M. deutlich hervortritt.

---

*) Theorie der Transformationsgruppen III, p. 365—369. ⟦12⟧

NOTES

⟦1⟧ Together with 1910 H this is an elaboration of 1909 B. See 1909 B ⟦1⟧. Corrections to the present paper are given in 1910 G.
⟦2⟧ D. Hilbert 1900.
⟦3⟧ H. Poincaré 1904.
⟦4⟧ D. Hilbert 1902.
⟦5⟧ More precisely in 1910 G. See also 1911 D, p. 98, footnote*).
⟦6⟧ Brouwer 1909 B.
⟦7⟧ Proved in Brouwer 1911 C.
⟦8⟧ A. Schoenflies, 1908 A.
⟦9⟧ A. Schoenflies 1904, 1906 A, 1908 A.
⟦10⟧ Correction $\rho(Q_n, Q_\omega)$ instead of $\rho(Q'_n, {}_nQ'_\omega)$ according to Brouwer 1910G.
⟦11⟧ Brouwer 1908 D2.
⟦12⟧ S. Lie 1893.

MA 69, 180

1910G

Berichtigung

[[1]] zu dem Aufsatze von L. E. J. Brouwer: „Die Theorie der endlichen kontinuierlichen Gruppen, unabhängig von den Axiomen von Lie (Erste Mitteilung)". Math. Ann. Bd. 67, S. 246—267.

S. 247 wird zur Herstellung der n-dimensionalen Mannigfaltigkeit ausgegangen von einem endlichen Gebiete des n-dimensionalen Zahlenraumes, in dessen Grenze $\mathfrak{G}$ gewisse Gebiete γ derart paarweise eineindeutig und stetig aufeinander abgebildet und identifiziert werden, daß sie auf Gebiete des n-dimensionalen Zahlenraumes eineindeutig und stetig abbildbare Umgebungen bekommen, wonach sie zur Mannigfaltigkeit hinzugerechnet werden.

Diese Definition ist dahin zu präzisieren, daß die Abbildung und Identifizierung sich auch auf die Grenzen der Gebiete γ erstrecken soll, und daß jeder Punkt von $\mathfrak{G}$, welcher dabei eine auf ein Gebiet des n-dimensionalen Zahlenraumes eineindeutig und stetig abbildbare Umgebung bekommt, zur Mannigfaltigkeit hinzuzurechnen ist.

S. 249, Z. 8 v. u. lese man $\varrho(Q_n, Q_\omega)$ statt $\varrho(Q'_n, {}_nQ'_\omega)$.

NOTE

[[1]] Brouwer 1909C.

Letter to F. Engel, copy Y 43

Amsterdam, 21.1.12

Overtoom 565.

Sehr geehrter Herr Professor [[1]]

In Ihrem auf S. 194 von Bd. 40 der "Fortschritte der Mathematik" befindlichen Referate erheben Sie gegen die betreffende Arbeit von mir Einwände zweierlei Art. Erstens sind Sie der Meinung, dass ich dem Probleme von vorn herein zu grosse Beschränkungen auferlegt habe, zweitens finden Sie auch nach Akzeptierung dieser Beschränkungen meine Darstellung nicht ganz lückenlos. [[2]]

Hinsichtlich des ersten Punktes wäre ich Ihnen sehr dankbar, wenn Sie die Güte haben wollten, mir genau mitzuteilen, welche ausgedehntere Fragestellung Sie dabei im Auge haben; es ist mir nämlich aus Ihren Andeutungen nicht gelungen, darüber zur vollen Klarheit zu kommen.

Der Wortlaut Ihres Referates würde etwa auf folgende von Ihnen gewünschte Voraussetzungen deuten:

„Eine n-dimensionale Mannigfaltigkeit μ unterliege einer kontinuierlichen, die Identität mitenthaltenden *Schar* von Transformationen, welche in der Umgebung der Identität *erstens* die Gruppeneigenschaft besitzt und *zweitens* sich durch p reelle Parameter eineindeutig und stetig darstellen lässt."

Diese Voraussetzungen wären aber sicher ungenügend, denn zu jeder p-dimensionalen Gruppe g kann man in mannigfacher Weise eine p-dimensionale Schar von Transformationen konstruieren, welche in einer gewissen Umgebung der Identität mit g identisch ist, ausserhalb dieser Umgebung aber weder mit g identisch ist noch überhaupt die Gruppeneigenschaft besitzt. Bei der mengentheoretischen Fassung des Problems kommt ja die bei Voraussetzung einer analytischen Abhängigkeit der Transformationen von p Parametern bestehende Möglichkeit, aus Eigenschaften in der Nähe der Identität auf Eigenschaften in beliebiger Entfernung von der Identität zu schliessen, in Fortfall.

Die Voraussetzungen müssten also jedenfalls in folgender, engerer Form angesetzt werden:

„Eine n-dimensionale Mannigfaltigkeit μ unterliege einer kontinuierlichen, die Identität mitenthaltenden *Gruppe* von Transformationen, welche sich in der Umgebung der Identität durch p reelle Parameter eineindeutig und stetig darstellen lässt."

Aus diesen Voraussetzungen lässt sich aber unmittelbar folgern, dass die ganze Gruppe sich eineindeutig und stetig auf eine p-dimensionale „Parametermannigfaltigkeit" abbilden lässt, und dass die Gruppe aus paarweise inversen

Transformationen besteht. Letztere Eigenschaft zieht ihrerseits nach sich, das in der n-dimensionalen Mannigfaltigkeit μ die Transformationen der Gruppe *überall eineindeutig* sind, womit wir die Voraussetzungen meiner Annalenarbeit alle zurückbekommen haben.

Vielleicht aber habe ich im vorstehenden die Gedanken Ihres Referates durchaus falsch interpretiert? Ist mir ja z.B. der Sinn Ihrer Worte: „unnötige Beschränkung, dass die Gruppe abgeschlossen sein soll", völlig dunkel geblieben.

Was den zweiten Punkt Ihrer Kritik, nämlich die Unvollständigkeit meiner Darstellung betreifft, so führen Sie davon zwei Beispiele an:

1) Gewisse Unklarheiten in der im §1 enthaltenen Formulierung der Voraussetzungen. Ich möchte gern vernehmen, welche Dunkelheiten oder Unbestimmtheiten Sie hier gefunden haben, ich glaubte nämlich jeder Forderung der mengentheoretischen Schärfe und Strenge genügt zu haben.

⟦3⟧ 2) Das „liegt sicher" auf S. 255, Z. 20. Dies ist aber in der That ganz selbstverständlich. Wenn nämlich ein Punkt x sich stetig von 0 nach a bewegt, so bewegt der Punkt $2x$ sich in derselben Richtung stetig von 0 nach $2a$. Im Augenblick, in dem der Punkt $2x$ den Punkt a erreicht, befindet sich der Punkt x im gesuchten Punkte b.

Ich würde grossen Wert darauf legen, mit einem Gruppentheoretiker Ihrer
⟦4⟧ Autorität über das obenstehende zur Uebereinstimmung zu gelangen.

Mit ausgezeichneter Hochachtung
bin ich

Ihr ergebenster

L E J Brouwer.

NOTES

⟦1⟧ The correspondence between Brouwer and F. Engel was sparked off by F. Engel's review, F. Engel 1911, of Brouwer 1909 C. It is a discussion between people living in different worlds: Engel, the co-author of Lie's great treatise, who could not grasp a group except in its analytic setting, and Brouwer, who had shaken off the algorithmic yoke and from his conceptual viewpoint could not comprehend his correspondent's difficulties. Manifolds and one-to-one mappings as substrate and action of Lie-groups instead of cartesian space and many-valued mappings was indeed a great step forward, though for older contemporaries of Brouwer it was too much.

What is left of the correspondence are three letters by Brouwer Y43, Y44, Y45, and three by Engel. A note by Brouwer on Engel's second letter indicates that Brouwer gave an immediate, though provisional, answer, perhaps on a postcard, to Engel's first letter by asking Engel a question; no copy of this communication exists.

⟦2⟧ F. Engel 1911.

⟦3⟧ Even after this explication Engel did not admit the triviality of the conclusion.

⟦4⟧ Engel answered with two letters, probably separated by a short communication of Brouwer's (see ⟦1⟧):

Letter of F. Engel to Brouwer

Greifswald 28.1.1912.

Sehr geehrter Herr Dr!
Es ist mir sehr erfreulich, dass Sie durch Ihren Brief vom 21. sich unmittelbar mit mir in Verbindung setzen. Schade nur, dass sich alle solche Dinge auf schriftlichem Wege so schlecht verhandlen lassen. Mündlich würden wir uns viel leichter verständigen können.

Zunächst sollte es mir leid thun, wenn Sie aus meiner Besprechung irgend ein abfälliges Urteil über Ihre Arbeit herausgelesen hätten. Ich habe nichts Derartiges beabsichtigt, sondern "mit Bewunderung zweifelnd, mit Zweifel bewundernd" so geurteilt, wie man nach Lessings Skala dem Meister gegenüber urteilen soll. Damit will sie ich ⟦sic⟧ aber freilich noch nicht als Meister schlechthin anerkennen, sondern nur meine allgemeine Schätzung für Ihre Arbeiten zum Ausdruck bringen, obwohl ich sie andererseits in der Mengenlehre, deren Anwendung mir gar nicht liegt, unbedingt als meinen Meister anerkenne.

Dass die auf S. 247 angegebenen Voraussetzungen an Klarheit der Formulierung sehr zu wünschen übrig lassen, daran muss ich festhalten. Vielleicht erscheinen sie einem eingefleischten Mengentheoretiker klar, aber ich muss sagen: "die Ausdrücke des Systems klingen unbeschnittenen Ohren dunkel."

Auf S. 247, Z. 3–11 haben Sie so viel in einen Satz zusammengepackt, dass schon dadurch die Möglichkeit klar zu sein ausgeschlossen ist, für mich wenigstens. Auch Z. 17–19 ist viel zu knapp gefasst, um klar zu sein.

Was nun die anderen Punkte angeht, so kann ich mich nachwievor nicht damit befreunden,

d.30-1-1912.

dass Sie von vornherein die Transformationen in der ganzen Mannigfaltigkeit als eindeutig umkehrbar voraussetzen. Ich kann nicht einsehen, dass der Fall, wo man eine Gruppe von analytischen mehrdeutigen Transformationen hat, so ohne Weiteres auf den Fall der Eindeutigkeit zurückgeführt werden kann.

Wenn ich nun vorschlug, die Eineindeutigkeit der Transformationen zunächst nur in der Umgebung eines Punktes und die Gruppeneigenschaft zunächst nur in der Umgebung der id. Trf. vorauszusetzen, so meinte ich natürlich: Ich habe eine Schaar von Trff., die in der Umgebung eines Punktes eindeutig umkehrbar sind, diese Schaar enthält die id. Trf. und zwei Trf. der Schaar, die in einer gewissen Umgebung der id. Trf. liegen, liefern wieder eine Trf. der Schaar[1]). Selbstverständlich sollten die Trf. der Schaar, wenn man sie beliebig oft hinter einander ausführt, eine Gruppe erzeugen, von der nur vorauszusetzen wäre, dass sie in der Umgebung der id. Trf. keine andere Trf. enthält, als die der ursprünglichen Schaar.

[1]) Paarweise inverse Trff. würde ich auch der Einfachheit wegen voraussetzen.

Auf diese Weise wird nicht nur die ursprüngliche Schaar von Trff. fortgesetzt, sondern man erhält zugleich auch die ursprünglichen Transformationen ausserhalb des Bereiches definirt, in dem sie ursprünglich definirt waren. Man erhält also eine Fortsetzung der Transformationen auch dann, wenn man von der analytischen Fortsetzung keinen Gebrauch machen kann. Darin zeigt sich eben die ungeheure Macht des Gruppenbegriffs.

Ob nun aus diesen Voraussetzungen die folgen, die Sie gemacht haben, das kann ich nicht übersehen, und ich getraue mich nicht, darüber etwas zu sagen. Sollte es der Fall sein und sollten Sie selbst schon sich das zurecht gelegt haben, so würde ich es offen gestanden sehr missbilligen, dass Sie es in Ihrer Arbeit nicht gesagt haben. Dann würden Sie es eben unterlassen haben, die gemachten Voraussetzungen auf möglichst einfache und natürliche zurückzuführen, oder Sie hätten die wichtige Eigenschaft des Gruppenbegriffs, dass er selbst ein Princip zur analytischen Fortsetzung liefert nicht in ihrer wahren Bedeutung durchschaut.

Ich hoffe, dass es Ihnen nunmehr klar ist, was ich gemeint habe, als ich sagte man solle die Gruppe nicht von vornherein als abgeschlossen voraussetzen; man soll sich eben nicht die ganze Gruppe gegeben denken sondern nur ein Stück von ihr. Bei allen Anwendungen ist man ja sowieso in diesem Falle.

Dass das "liegt" auf S. 255 so ganz selbstverständlich ist, kann ich immer noch nicht einsehen. Ich sehe nicht, dass es ohne Weiteres auf den Satz hinauskommt, dass eine stetige Funktion, die zwei Werthe annimmt, auch jeden Zwischenwerth annimmt. Wie man diesen Satz doch auch erst beweist, so scheint mir auch hier ein Beweis erforderlich oder die Zurückführung auf einen bewiesenen Satz. Ich kann mich ja täuschen, aber ich habe das Gefühl, dass ein Beweis nöthig ist.

Uebrigens kann ich nicht umhin, zu gestehen, dass Sie durch Ihre Arbeit und Ihren Brief mir, vielleicht ohne es zu wollen, zu einer neuen Einsicht verholfen haben. Es erscheint mir nämlich jetzt, auch wenn man analytische Gruppen betrachtet, nicht mehr als zweckmässig, die auftretenden Gleichungen auf dem gewöhnlichen Wege analytisch fortzusetzen, sondern man soll [1]) nur die analytischen Fortsetzungen betrachten, die sich aus der Anwendung des Gruppenbegriffs ergeben. Man wird auf diese Weise unter Umständen nicht das ganze Gebiet erschöpfen, über das sich die auftretenden Funktionen analytisch fortsetzen lassen, aber man wird dafür aus dem Gebiet, in dem die Trff. der Gruppe alle eindeutig umkehrbar sind, nicht herauskommen.

31.1.1912.

Dieser Brief ist in mehreren Absätzen geschrieben, denn abgesehen davon, dass ich zur Zeit als Dekan ziemlich viel zu thun habe, bin ich auch noch zum Geschworenen ausgelost worden und dadurch ist meine ohnedies knappe Zeit gerade jetzt noch mehr beschränkt.

[1]) wenigstens bei transitiven Gruppen

Zum Schluss möchte ich doch noch erwähnen, dass ich, wenn ich ganz offen sein soll, den Gewinn derartiger Untersuchungen nicht ganz der darauf verwandten Mühe und dem dazu erforderlichen Aufwande von Scharfsinn entsprechend finden kann. Sie mögen mich deswegen immerhin als einen Ketzer ansehen. Um so mehr aber würde ich mich freuen, wenn Sie nun auch über die Gruppentheorie selbst arbeiten würden, denn da ist noch sehr viel zu thun.

Mit vorzüglicher Hochachtung

Ihr sehr ergebener

F. Engel

Letter of F. Engel to Brouwer

Greifswald 4.2.1912.

Sehr geehrter Herr Dr.!

Am Besten wird Ihnen, denke ich, ein Beispiel klar machen, was ich meine.

Ich betrachte eine eingliedrige Gruppe, die von einer mehrdeutigen inf. Trf. erzeugt wird. Haben wir z.B. die inf. Trf.

$$x' = x + \tfrac{1}{2}\sqrt{x}\delta t,$$

so erhalten wir die eingliedrige Gruppe von zweideutigen Trf.

$$\sqrt{x'} = \sqrt{x} + t$$

oder:

$$x' = x + 2t\sqrt{x} + t^2.$$

Hier ist jede Transformation zweideutig, fällt aber mit ihrer inversen Trf. zusammen oder vielleicht besser: sie geht durch analyt. Fortsetzung in ihre inverse Trf. über.

7.2.1912.

Es sei nun $x_0 \neq 0$ und $\sqrt{x_0}$ ein bestimmter der beiden Wurzelwerthe. Wir gehen von der Schaar der Transformationen:

$$x' = x + 2t \cdot \sqrt{x_0}\left(1 + \frac{x - x_0}{x_0}\right)^{\frac{1}{2}} + t^2$$

aus, wo $(1+\alpha)^{\frac{1}{2}} = 1 + \frac{1}{2}\alpha + \cdots$ ist. Diese Schaar ist definiert für jedes t und für $|x - x_0| < |x_0|$. Ihre Trff. haben die Form:

$$x' - (t + \sqrt{x_0})^2 = (x - x_0)\left(1 + \frac{t}{\sqrt{x_0}}\right) + \cdots,$$

wo die weggelassenen Glieder von zweiter und höherer Ordnung in $x - x_0$ sind.

Bezeichnen wir die Trff. dieser Schaar mit S_t, so definiert nun die Trf. $S_t^{-1} \cdot S_{\tau+t}$ die Art, wie der Punkt, in den x_0 bei S_τ übergeht, bei der Transformation S_t transformirt wird.

Ich wähle $\tau = -2\sqrt{x_0}$, dann ist:

$$S_{-2\sqrt{x_0}} : x' - x_0 = -(x - x_0) + \cdots,$$

also führt diese Trf. den Punkt x_0 wieder in x_0 über. Aber es ist:

$$(S_{-2\sqrt{x_0}})^{-1}: x' - x_0 = -(x - x_0) + \cdots$$

$$S_{t-2\sqrt{x_0}}: x'' - (t - \sqrt{x_0})^2 = (x' - x_0)\left(\frac{t}{\sqrt{x_0}} - 1\right) + \cdots$$

also ergiebt sich:

$$(S_{-2\sqrt{x_0}})^{-1} \cdot S_{t-2\sqrt{x_0}}:$$

$$x'' - (t - \sqrt{x_0})^2 = (x' - x_0)\left(1 - \frac{t}{\sqrt{x_0}}\right) + \cdots$$

d.h. diese Trf. ist nicht S_t sondern S_{-t}. Man erhält aber durch gruppentheoretische Fortsetzung aus S_t die inverse Trf. S_{-t}, geradeso wie wenn man die Trf. S_t analytisch forsetzt und einmal um den Nullpunkt herum geht.

Hiermit scheint doch deutlich bewiesen, dass die Voraussetzung der Eindeutigkeit der Trff. in der Umgebung eines Punktes und die Annahme, dass die Gruppe zu jeder dieser Trff. auch die inverse in der Umgebung dieses Punktes enthält, keineswegs die Eindeutigkeit, geschweige denn die eindeutige Umkehrbarkeit der Trff. in dem Gebiete nach sich zieht, das jener Punkt bei den Trff. der Gruppe erreicht.

Auch wenn die inf. Trf. eindeutig ist, können doch die Trff. der eingliedrigen Gruppe mehrdeutig sein, z.B. $x' = x + e^{-x}\delta t$ gibt

$$x' = \lg(e^x + t).$$

Doch genug für diesmal. Mit den besten Grüssen

Ihr ganz ergebener
F. Engel

In margine on p. 1: Zu meinem Schrecken sehe ich eben, dass nur die Hälfte meines Briefes abgegangen ist.

The postmark on the envelope is 7 February 1912.

On p. 1 Brouwer's remark: Antwort auf meine Frage, ob nach Herrn E. die Eineindeutigkeit nicht eine Folge der paarweisen Inversivität sei.

Letter to F. Engel, copy Y 44

Amsterdam, 6.3.12

Overtoom 565.

Sehr geehrter Herr Professor

Durch vielerlei Tätigkeit absorbiert, war es mir bis heute nicht möglich, Ihnen ⟦1⟧
auf Ihre beiden ausführlichen Briefe, welche mir Ihre Gedanken nun ganz klar gemacht haben, zu antworten. Ich gehe jetzt auf alle Punkte meines ersten Briefes nochmal ein.

Das in Ihrem Referate gebrauchte Wort "Unklarheit" ist nach Ihrer näheren Erklärung im Sinne von "Schwerverständlichkeit für den Uneingeweihten" aufzufassen. Ueber eine solche *subjektive* Meinung lässt sich natürlich nicht streiten, aber mancher Leser wird gegen Ihre Absicht aus Ihren Worten den Eindruck bekommen haben, als hätte ich meine Grundbegriffe mit ungenügender Schärfe formuliert, was ein sehr *objektiver* Fehler sein würde, den ich noch immer aufs entschiedenste zurückweisen muss (Die der zweiten Mitteilung vorausgeschickte ⟦2⟧
"Berichtigung" in Bd. 69 beabsichtigt nur, gewisse bei endlichen kontinuierlichen Gruppen nicht auftretende singuläre Zusammenhangsverhältnisse dennoch nicht von vornherein auszuschliessen)

Das "liegt sicher" auf S. 255 kommt, wie Sie jetzt selber wohl einsehen, wirklich ⟦3⟧
ohne weiteres auf den Satz hinaus, dass jede stetige Funktion („stetige monotone Funktion" würde für diesen Fall sogar genügen), welche zwei Werte annimmt, auch jeden Zwischenwert annimmt; an einen so trivialen Satz braucht man die Annalenleser doch wohl kaum zu erinnern. Auch hier wird indes, wie ich glaube, mancher Leser Ihres Referates den Eindruck bekommen haben, als enthielte meine Arbeit *mehrere objektive Lücken*, von denen Sie die ebengenannte nur als Beispiel hervorgehoben hätten.

Mithin würden Sie mir eine grosse Freude machen, wenn Sie Sich entschliessen konnten, im voraussichtlich bevorstehenden Referate über meine zweite Mitteilung ⟦4⟧
eine mich rehabilitierende Bemerkung einzuschalten.

Was nun den inneren Grund meiner allgemeinen Voraussetzungen angeht, ich glaube ihn am vollständigsten aufzudecken durch die folgenden Betrachtungen:

Eine Punktmenge m heisse *verkettet*, wenn nach irgend einem Gesetze gewissen ⟦5⟧
zu m gehörigen unendlichen Punktfolgen f gewisse ebenfalls zu m gehörige, als *Grenzpunkte* von f zu bezeichnende Punkte p_f derart zugeordnet sind, dass es zu jedem Punkte p_f immer eine *Teilfolge* von f gibt, welche nur diesen einzigen Punkt als Grenzpunkt besitzt, und dass die Grenzpunkte einer Teilfolge von f eine Teilmenge der Grenzpunkte von f bilden während schliesslich folgende Eigen-

schaft besteht: wenn α_μ der einzige Grenzpunkt der Folge $\alpha_{\mu\nu}$ ist, und α der einzige Grenzpunkt der Folge α_μ, so enthält jedes $\alpha_{\mu\nu}$ ein solches Endsegment $\alpha_{\mu\pi}$, dass die Folge dieser Endsegmente *nur* α als Grenzpunkt besitzt.

Eine nicht verkettete Punktmenge, d.h. eine Punktmenge ohne Grenzpunkte heisse *diskret*.

Unter einer *Umgebung* des zu m gehörigen Punktes p verstehen wir eine solche Teilmenge von m, welche von jeder den Punkt p als Grenzpunkt besitzenden zu m gehörigen Punktfolge unendlich viele Punkte enthält.

(Konstruieren wir nun um jeden Punkt p von m eine Umgebung u_p, und wählen wir zwei Punkte p_1 und p_2 von m beliebig aus. Wenn es, unabhängig von der Wahl der u_p und von p_1 und p_2, möglich ist, zwischen p_1 und p_2 eine solche *endliche* Punktfolge von m einzuschalten, das je zwei aufeinanderfolgende Punkte dieser Folge einer und derselben u_p angehören, so heisse m eine *zusammenhängende*
⟦6⟧ *Punktmenge*.)

Eine Abbildung einer Punktmenge m auf eine Punktmenge m' heisse *stetig*, wenn jedem Grenzpunkte einer Punktfolge von m ein Grenzpunkt der korrespondierenden Punktfolge von m' entspricht.

Die Punktmenge m heisse *homogen*, wenn zu jeder Umgebung u_p eines willkürlichen Punktes p von m jeder andere Punkt von m eine Umgebung besitzt, welche sich eineindeutig und stetig auf u_p abbilden lässt.

Eine willkürliche Punktmenge m unterliege nun einer willkürlichen die Identität mitenthaltenden Gruppe γ von paarweise inversen (ein- oder mehrdeutigen) Transformationen. Wir überlagern dann m mit einer solchen Punktmenge μ, dass je zwei zusammenfallende Punkte von m dann und nur dann auch in μ als identisch betrachtet werden, wenn jede Transformation von γ sie wieder in zwei zusammenfallende Punkte von m überführt. Weiter werde in μ der Punkt π dann und nur dann als Grenzpunkt der Folge $\mathcal{F}$ betrachtet, wenn erstens der π in m entsprechende Punkt p Grenzpunkt ist der $\mathcal{F}$ in m entsprechenden Folge f, und zweitens diese Beziehung zwischen p und f bei einer willkürlichen Transformation der Gruppe γ bestehen bleibt. Schliesslich werde in γ die Transformation τ dann und nur dann als Grenzelement der unendlichen Transformationsfolge φ betrachtet, wenn jeder willkürliche Punkt von μ durch τ in einen solchen Punkt π und durch φ in eine solche Punktfolge $\mathcal{F}$ übergeführt wird, dass π Grenzpunkt von $\mathcal{F}$ ist.

Alsdann sind sowohl die „Transformationsmannigfaltigkeit" μ wie die „Parametermannigfaltigkeit" γ *homogene Punktmengen*, und in bezug auf μ sind die Transformationen von γ nicht nur paarweise invers, sondern auch *eineindeutig und stetig*, während die Punktmenge m nunmehr als eine *Faltung* (d.h. als ein eindeutiges – nicht eineindeutiges – und stetiges Bild) der Punktmenge μ erscheint.

Mithin kann jede Gruppe von (ein- oder mehrdeutigen) paarweise inversen Transformationen einer willkürlichen Punktmenge *aus einer homogenen Gruppe von eineindeutigen und stetigen, paarweise inversen Transformationen einer homo-*

genen Transformationsmannigfaltigkeit (eine Gruppe der letzteren Art werden wir, wenn sowohl die T.M. wie die P.M. abgeschlossen [1]) sind, als eine *kanonische Gruppe* bezeichnen) *durch Faltung der Transformationsmannigfaltigkeit erhalten werden.* ⟦7⟧

Nun sind bis heute nur folgende Typen von abgeschlossenen homogenen Punktmengen bekannt (und wahrscheinlich existieren auch keine andere):

a) diskrete Punktmengen.
b) endlichvieldimensionale Mannigfaltigkeiten R_n nach meiner Definition.
c) abzählbarunendlichvieldimensionale Mannigfaltigkeiten R_ω (vgl. die diesbezüglichen Arbeiten von Fréchet). ⟦8⟧
d) Punktmengen vom Ordnungstypus ζ der zusammenhangslosen, nirgends dichten, perfekten Punktmengen des R_n.
e) aus Mengen der vier vorstehenden Arten gebildete „*Produktmengen*" (z.B. ⟦9⟧ eine diskrete Menge von Ordnungstypen ζ von dreidimensionalen Räumen).

Und die allgemeinste kanonische Gruppe, für welche sowohl die T.M. wie die P.M. zum Typus e) gehören, lässt sich in einfacher Weise zusammensetzen aus kanonischen Gruppen, für welche sowohl die T.M. wie die P.M. zu einem der Typen a), b), c), d) gehören, welche mithin als *Primgruppen* bezeichnet werden können.

Beispiele von Primgruppen sind die endlichen Substitutionsgruppen (P.M. und T.M. vom Typus a)), die Fuchs'schen und Klein'schen Gruppen (T.M. vom Typus b), P.M. vom Typus a)), die endlichen kontinuierlichen Gruppen nach der Definition meiner Annalenarbeit (P.M. und T.M. vom Typus b)), die unendlichen kontinuierlichen Gruppen (T.M. vom Typus b), P.M. vom Typus c)), die ζ-Gruppen, auf welche ich in den Amsterdamer Proceedings vom April 1910 hin- ⟦10⟧ gewiesen habe (P.M. und T.M. vom Typus d)).

In völligem Einklang mit dem Obenstehenden entsteht das Beispiel Ihres letzten Briefes (die Gruppe $\sqrt{x'} = \sqrt{x}+t$) aus der Translationsgruppe der Ebene, d.h. aus einer endlichen kontinuierlichen Gruppe nach meiner Definition, *durch Faltung* (in diesem speziellen Falle durch zwei-eindeutige Abbildung) *der Transformationsmannigfaltigkeit.*

Die Art dieser Faltung unterliegt indes keinerlei Beschränkung; weil sie völlig ⟦7⟧ willkürlich ist, kann man sie insbesondere auch so kompliziert gestalten, dass die Gruppe sich in keiner Weise mehr durch analytische Formeln ausdrücken lässt. Die zugehörige kanonische Gruppe wird dadurch aber nicht beeinflusst; sie bleibt eine endliche kontinuierliche Gruppe nach meiner Definition.

Ich würde mich sehr freuen, durch das Obenstehende Ihnen klar gemacht zu haben, warum die in meiner Annalenarbeit formulierte Umgränzung des Pro-

[1]) wir nennen eine Punktmenge *abgeschlossen*, wenn zu jedem ihrer Punkte eine Umgebung existiert, in welcher jede Fundamentalreihe einen ebenfalls zur Punktmenge gehörigen Grenzpunkt aufweist.

blems: „alle endlichen kontinuierlichen Gruppen zu bestimmen" meiner Ansicht nach vollkommen naturgemäss ist, und nicht die geringste künstliche Einschränkung enthält.

⟦11⟧ Mit den besten Grüssen

Ihr ganz ergebener

L E J Brouwer.

NOTES

⟦1⟧ See Y43 ⟦4⟧.
⟦2⟧ Brouwer 1910 G.
⟦3⟧ Brouwer 1909 C.
⟦4⟧ Brouwer 1910 H.
⟦5⟧ The following shows Brouwer's interest in general topology. See also Brouwer 1913 B2. The axiomatic is not entirely sufficient. This must have sounded like Chinese to poor Engel!
⟦6⟧ This is an interesting version of the definition of connectedness. It is, in an even more explicit form than in Brouwer 1911 E (see ⟦12⟧) the so-called Lennes–Hausdorff concept.
⟦7⟧ Of course this does not solve all problems. It would be a meaningful question to ask for a definition of what is a decent *Faltung*.
⟦8⟧ M. Fréchet 1906.
⟦9⟧ Solenoids are, of course, lacking in this list.
⟦10⟧ 1910 B2, ⟦13⟧.
⟦11⟧ Engel answered:

Letter of F. Engel to Brouwer

Greifswald 26.3.12.

Sehr geehrter Herr Dr.!

Für Ihren Brief vom 6. d.M. danke ich Ihnen bestens. Ich hätte längst gern geantwortet, aber trotz der Ferien hatte ich immer Abhaltungen.

In dem Referat über Ihre 2. Mitteilung, das ich gerade jetzt mache, gebe ich eine solche Erklärung, wie Sie sie wünschen. Ich hoffe, dass Sie damit zufrieden sein werden.

Die Entwickelungen in Ihrem Briefe haben meinen ganzen Beifall, und ich gebe gerne zu, dass auf diese Weise Ihre Umgränzung des Problems durchaus natürlich erscheint. Aber ich vermisse in Ihren beiden Arbeiten jede Hindeutung darauf, dass damit auch ein viel allgemeineres Problem erledigt ist, und eine solche Hinweisung erscheint mir doch durchaus nöthig, denn welcher Leser wird sich das selbst zurechtlegen können?

Andererseits übersehe ich doch noch nicht, dass damit auch der Fall vollständig erledigt ist, dass von der Gruppe bloss ein Stück gegeben ist, in der Umgebung der id. Trf. u. in der Umg. eines Punktes. Denn bei den Gruppen, die man wirklich trifft, ist eigentlich immer von vornherein nur ein solches Stück bekannt.

Ferner kann auch eine Gruppe, die man in ihrer ganzen Ausdehnung kennt, in einer Form vorliegen, dass, wie mir scheint, Schwierigkeiten entstehen.

Schreibt man z.B. die allgemeine projektive Gruppe homogen und mit kanonischen Parametern [1]), so sind die Koefficienten beständig konv. Potenzreihen der Parameter, aber ∞ viele Parametersysteme liefern dieselbe Transformation. Auf so etwas muss man doch eigentlich immer gefasst sein. Können Sie nun eine solche Parametermannigfaltigkeit auch immer durch eine solche ersetzen, in der die Beziehung zwischen den Punkten und den Transformationen eineindeutig ist? Mich schaudert vor der Allgemeinheit solcher Betrachtungen und ich kann mich der Befürchtung nicht erwehren, dass es doch nicht ausführbar ist, alle Möglichkeiten zu erschöpfen.

Jedenfalls habe ich sehr den Wunsch, mich einmal mündlich mit Ihnen aussprechen zu können.

Mit den besten Grüsen

Ihr ergebener

F. Engel

[1]) Ich würde Ihnen sehr dankbar sein, wenn Sie das Wort "kanonisch" nicht noch in einer neuen Bedeutung gebrauchen würden. Dass kann Verwirrung anrichten.

Y 45 Letter to F. Engel, copy

Blaricum, 29.3.12.

[[1]] Sehr geehrter Herr Professor

[[2]] Für die in Ihrem Briefe enthaltene Zusage betreffs Ihres Referates über meine 2. Mitteilung bringe ich Ihnen meinen verbindlichsten Dank.

[[3]] Was Ihr Beispiel einer p-dimensionalen P.M. (Parametermannigfaltigkeit) γ, in welcher unendlich viele Punkte derselben Transformation entsprechen, betrifft, so folgt aus meinem vorigen Briefe, dass diese P.M. γ von meinem Standpunkte aus nicht die wahre P.M. ist, sondern erst durch Identifizierung aller derselben Transformation entsprechenden Punkte je zu einem einzigen Punkt, in die wahre, ebenfalls p-dimensionale P.M. γ', welche zu jeder Transformation nur einen einzigen Punkt enthält, übergeht. Natürlich wird im allgemeinen γ' einen ganz anderen Zusammenhang wie γ aufweisen; insbesondere wird, wenn γ den einfachen Zusammenhang des p-dimensionalen *Zahlenraumes* besitzt, für γ' diese Eigenschaft im allgemeinen verloren gehen, sodass man für rechnerische Zwecke oft genötigt sein wird, zu γ zurückzukehren.

Ich komme jetzt zum von Ihnen hervorgehobenen Falle, dass anfangs nur ein n-dimensionales Raumstück τ in der Umgebung eines Punktes P, welches einer p-imensionalen, in der Umgebung der Identität liegenden und diese mitenthaltenden Menge π von eineindeutigen und stetigen, paarweise inversen Transformationen unterliegt, gegeben ist.

Soll von der Erzeugung einer Gruppe aus diesem Systeme (τ,π) die Rede sein können, so muss natürlich gleichzeitig irgend eine Verfahren F angegeben sein, mittels dessen die Transformation von π auch für alle diejenigen Punkte, in welche die Punkte von τ durch beliebige Wiederholung von π übergehen, einen Sinn erhalten, welches Verfahren F für durch Potenzreihen gegebene Transformationen natürlich in ihrer analytischen Fortsetzung bestehen wird.

Dies vorausgesetzt, unterliegt aber die durch beliebige Wiederholung von π aus τ erzeugte Punktmenge m auf jeden Fall einer aus diesen Wiederholungen von π bestehenden *Gruppe* γ von paarweise inversen (ein- oder mehrdeutigen) Transformationen, welche Gruppe für die durch „Entfaltung" von m entstehende
[[4]] *homogene* Punktmenge μ (vgl. meinen vorigen Brief) eine „Primgruppe" ist, *und wir werden dann und nur dann sagen, das „das System* (τ, π) *eine endliche, kontinuierliche Gruppe bestimmt"*, *wenn* μ *in einer gewissen Umgebung von P mit* τ, *und* γ *in einer gewissen Umgebung der Identität mit* π *identisch ist.* (Bei völlig willkürlicher Annahme von τ und π werden nämlich im allgemeinen für μ und γ

Mannigfaltigkeiten von *höherer* Dimensionenzahl als τ resp. π, meistens sogar von abzählbar-unendlicher Dimensionenzahl gefunden.)

Hierbei stellt sich zugleich heraus, das von einem Systeme (τ, μ), von dem man weiss, dass es eine endliche, kontinuierliche Gruppe bestimmt, hiermit das Verfahren F vollkommen festgelegt ist, nämlich notwendig in „gruppentheoretischer Fortsetzung" bestehen muss. ⟦5⟧

Mit den besten Grüssen und auch meinerseits hoffend, Sie in Bälde einmal perönlich zu treffen

Ihr ergebenster

L E J Brouwer.

NOTES

⟦1⟧ See Y44 ⟦10⟧.

⟦2⟧ Brouwer 1910 H, reviewed by F. Engel 1913.
A copy of this review is among Brouwer's papers.

⟦3⟧ Y44.

⟦4⟧ Y44.

⟦5⟧ This is, properly speaking a circumvention of Engel's question. E. Cartan 1930 gave an example of a local Lie group of local mappings that does not extend globally. Brouwer may have guessed that such things existed. Anyway he should have said that the possibility of a global extension was not self-evident.

1910 H

Die Theorie der endlichen kontinuierlichen Gruppen, unabhängig von den Axiomen von Lie.

(Zweite Mitteilung.)

Von

L. E. J. Brouwer in Amsterdam.

---

§ 1.

Die Transformationsbereiche der Gruppen der zweidimensionalen Mannigfaltigkeiten.

⟦1⟧ Wir konstruieren, wie im § 2 der ersten Mitteilung,*) in der P. M. eine Kugel K um die Identität, und betrachten denjenigen Teil T des T. B. eines Punktes P für die Gruppe selbst oder für eine reduzierte Untergruppe H, welcher dem Innengebiete von K entspricht. Überdies setzen wir K so klein voraus, daß T in der T. M. innerhalb einer gewissen Kreisfläche F liegt. Genau wie im § 2 der ersten Mitteilung zeigen wir dann, daß, wenn $P_1, P_2, P_3, \cdots$ in T gegen P_ω konvergieren, es eine Folge von Transformationen der Gruppe gibt, welche der Reihe nach $P_1, P_2, \cdots$ in P_ω überführen und gegen die identische Transformation konvergieren.

Ist weiter μ_ν diejenige Teilmenge von H in der P. M., welche der Stelle P_ν von P entspricht, so konvergieren gegen jede abgeschlossene Teilmenge von μ_ω gleichmäßig abgeschlossene Teilmengen von $\mu_1, \mu_2, \mu_3, \cdots$

Wir unterscheiden jetzt folgende drei Fälle:

1. *T enthält keinen zusammenhängenden Bestandteil.* Dann ist P ein für die Gruppe H fester Punkt.

2. *T ist nirgends dicht, enthält aber zusammenhängende Bestandteile.*

Sei z ein solcher Bestandteil, so setzen wir erstens voraus, daß die Komplementärmenge von z ein einziges Gebiet ist. Dann wählen wir in F einen Punkt außerhalb z, den wir innerhalb F mit z verbinden durch zwei einfache Kurvenbogen a und b, welche einander nicht, und z in A und B treffen, und in dieser Weise mit z ein Gebiet G bestimmen, welches in F enthalten ist. Sei u der mit a und b den äußeren Rand von G bildende Teilkurvenbogen von z, und G' das Komplementärgebiet von G.

*) Math. Ann. Bd. 67, S. 246.

⟦156⟧

Gäbe es nun Teilmengen von z, welche mit u in einem anderen Punkte als A und B zusammenhingen, so gäbe es, außerhalb von A und B, in beliebiger Nähe eines in u liegenden gemeinschaftlichen Grenzpunktes von G und G' Punkte von z, welche nicht gemeinschaftliche Grenzpunkte von G und G' wären.

Es könnten also durch beliebig kleine Transformationen aus einem Punkte von u, welcher gemeinschaftlicher Grenzpunkt von G und G' außerhalb von A und B wäre, andere Punkte von z, welche nicht gemeinschaftliche Grenzpunkte von G und G' wären, erreicht werden; dies ist aber unmöglich.

Gäbe es weiter Punkte von u, welche für G nicht erreichbar*) wären, so könnten sie durch beliebig kleine Transformationen in für G erreichbare Punkte übergeführt werden, was ebenfalls unmöglich ist. In dieser Weise stellt sich heraus, daß jeder Punkt von z für das Komplementärgebiet allseitig erreichbar ist.

Also ist sowohl u wie z ein *einfacher Kurvenbogen.*

Indem wir nun weiter beachten, daß jeder Punkt von H aus jedem andern Punkte von H innerhalb H zusammenhängend erreichbar ist, während man dabei innerhalb einer endlichen Zahl von Analogen der Kugel K bleiben kann, folgern wir, daß der ganze T. B. von P für die Gruppe H sich zusammensetzt aus einem Ordnungstypus $^*\omega + \omega$ von aneinander schließenden einfachen Kurvenbogen derart, daß entweder jede endliche ununterbrochene Folge aus diesem Ordnungstypus wieder ein einfacher Kurvenbogen ist, oder alle Kurvenbogen Teilmengen einer gewissen einfachen geschlossenen Kurve sind.

Im ersten Falle dürfen wir nicht schließen, daß nun auch der T. B. innerhalb einer gewissen Umgebung eines willkürlichen zu ihm gehörigen Punktes nur einen einfachen Kurvenbogen besitzt; vielmehr darf hier eine Gestalt, wie die einmal beiläufig von Lorentz**) angedeutete Kurve besitzt, nicht ausgeschlossen werden.***)

Wir setzen nun zweitens voraus, daß die Komplementärmenge von z mehrere Gebiete enthält. Die oben angewendete Methode bringt dann

*) Schoenflies, Bericht über die Lehre von den Punktmannigfaltigkeiten, Bd. II, S. 126. [[2]]

**) Encyklopädie der Math. Wiss. V 2, S. 120, 121. Die Kurve läßt sich auch leicht in der Weise abändern, daß sie nirgends dicht wird. [[3]]

***) Für T sind wir indes sicher, daß sie nicht in beliebiger Nähe von z weitere Teilkurvenbogen besitzen kann. Denn zunächst könnten diese nur auf einer der beiden Seiten von z auftreten. Hätte man nun aber ein z' *links* von z und $\varrho(z, z') < \varepsilon$, so könnte man z' in z überführen durch eine Transformation $< \varepsilon'$; weil aber z' *rechts* von sich weitere Teilkurvenbogen im Abstande $< \varepsilon$ besitzen würde, so besäße z auch solche rechts von sich im Abstande $< \varepsilon''$, wo ε'' mit ε unter jede Grenze herabsinkt.

heraus, daß jeder Punkt von z gemeinschaftlicher Grenzpunkt von allen durch z begrenzten Gebieten ist, und für alle diese Gebiete allseitig erreichbar sein muß. Also ist z eine *einfache geschlossene Kurve.*

3. *T ist in einem Gebiete der T. M. überall dicht.* Dann hat P für die Gruppe H ein Gebiet, und für die entprechende festgelegte Untergruppe eine endliche oder abzählbare Gebietsmenge zum Transformationsbereich.

§ 2.

Die Zusammensetzung der Hauptgruppe aus den reduzierten Gruppen.

Wir denken der Reihe nach eine endliche Zahl von je zuvor noch stetig beweglichen Punkten festgehalten, und setzen voraus, daß die hierdurch bestimmte festgelegte Untergruppe v immer noch einen stetig beweglichen Punkt P besitzt.

Sei zunächst P nur auf einer Kurvenmenge k beweglich, so zeigen wir analog wie im § 2 der ersten Mitteilung, daß, wenn wir wieder mit Z_{α_1} eine zusammenhängende Teilmenge der der Stelle P_{α_1} von P auf k entsprechenden festgelegten Untergruppe λ_{α_1} bezeichnen, ein Teilkurvenbogen von k aus jedem Z eine zusammenhängende Teilmenge Y von v erzeugt, in der die Z eineindeutiges und stetiges Bild der entsprechenden Stellen von P sind, während je zwei aus verschiedenen Mengen Z_{α_1} erzeugte Mengen Y miteinander *nicht* zusammenhängen.

Sei weiter P in einer Gebietsmenge beweglich, so konstruieren wir in einem Teilquadrate q ein rechtwinkliges Koordinatensystem, wählen zu den Punkten einer Linie $y = c$ ein System von zusammenhängenden Teilmengen Z_{uc}, welche eineindeutiges stetiges Bild der Punkte (u, c) sind, und analog zu den Punkten *jeder* Linie $x = c'$ ein analoges System von Mengen $Z_{c'v}$, zu denen die Menge $Z_{c'c}$ des ersten Systems gehört. Wir konstruieren in dieser Weise eine zusammenhängende Teilmenge Y von v, und behaupten, daß *alle* sie zusammensetzenden Mengen Z eineindeutiges stetiges Bild der entsprechenden Punkte von q sind.

Nehmen wir nämlich einen Augenblick an, daß dies nicht der Fall wäre, so gäbe es für $P(u, v)$ beliebig nahe an $P_1(u_1, v_1)$ entsprechende Mengen Z_{uv} in unbeschränkter Nähe einer mit $Z_{u_1 v_1}$ *nicht* identischen Menge $Z'_{u_1 v_1}$, und wir könnten die Menge $\lambda_{u_1 v_1}$ zerlegen in zwei Teilmengen $T_{u_1 v_1}$ und $T'_{u_1 v_1}$ derart, daß $E(T_{u_1 v_1}, T'_{u_1 v_1})$*) einen endlichen Wert besitzt, während $Z_{u_1 v_1}$ zu $T_{u_1 v_1}$ und $Z'_{u_1 v_1}$ zu $T'_{u_1 v_1}$ gehört.

---

*) D. h. das Minimum der Transformationen, durch welche ein Punkt von $T_{u_1 v_1}$ in einen Punkt von $T'_{u_1 v_1}$, oder ein Punkt von $T'_{u_1 v_1}$ in einen Punkt von $T_{u_1 v_1}$ übergeführt werden kann.

Wir können sodann in der bekannten Weise an der Linie $x = u_1$ entlang zwischen den Punkten (u_1, v_1) und (u_1, c) eine Folge von Mengen $Z'_{u_1 v}$ konstruieren, welche $Z'_{u_1 v_1}$ enthält und eineindeutiges stetiges Bild der Punkte (u_1, v) ist; die Mengen $T_{u_1 v_1}$ und $T'_{u_1 v_1}$ bestimmen dabei je eine Folge von Abbildungen $T_{u_1 v}$ resp. $T'_{u_1 v}$, und die dabei auftretenden $E\,(T_{u_1 v}, T'_{u_1 v})$ besitzen ein endliches Minimum.

Wählt man δ hinreichend klein, so muß, unabhängig von v, entweder $E(Z_{u_1+\delta, v}, T_{u_1 v})$ oder $E(Z_{u_1+\delta, v}, T'_{u_1 v})$ unter jede Grenze herabsinken.

Nun ist es aber unmöglich, daß eine an der Linie $x = u_1 + \delta$ entlang konstruierte, stetige Folge von Mengen Z in eine dem Werte von δ entsprechende Nähe zunächst von $T_{u_1 c}$ und sodann von $T'_{u_1 v_1}$ gelangt, eben wegen des endlichen Minimums der $E\,(T_{u_1 v}, T'_{u_1 v})$. Unsere Annahme hat sich also als unmöglich erwiesen.

Zwei im Quadrate q aus verschiedenen Mengen $Z_{u_1 v_1}$ erzeugte Mengen Y hängen miteinander *nicht* zusammen. Wenn man sie aber über ein volles Gebiet des T. B. von P erweitert, so können die Zusammenhangsverhältnisse desselben ermöglichen, daß mehrere Mengen Y sich zu einer einzigen zusammenhängenden Teilmenge η von ν zusammenfügen.

Machen wir jetzt von denjenigen Punkten, welche durch ihre Festhaltung die Gruppe ν bestimmt haben, den zuletzt festgehaltenen, den wir R nennen wollen, frei, lassen aber die übrigen fest wie zuvor, so kommt eine festgelegte Untergruppe ϱ heraus, für welche R wieder entweder in einer Kurvenmenge oder in einer Gebietsmenge beweglich ist. Wir behandeln zunächst den ersten Fall.

Jeder Teilkurvenbogen s des T. B. von R erzeugt dann aus einer zusammenhängenden Teilmenge Y von ν eine zusammenhängende Teilmenge von ϱ. Wir behaupten nun, daß man in jeder dieser Mengen Y eine Menge Z auserwählen kann derart, daß diese Mengen Z eineindeutiges stetiges Bild der entsprechenden Punkte von s sind.

Seien nämlich nämlich ${}_1B$ und ${}_2B$ zwei Punkte von s, und bezeichnen wir mit ${}_1B_\mathcal{F}$ einen willkürlichen Punkt von s zwischen ${}_1B$ und ${}_2B$. Wir wählen ${}_1B$ und ${}_2B$ so nahe aneinander, daß man, nachdem ${}_1Z$ willkürlich angenommen ist, für jeden Punkt ${}_1B_\mathcal{F}$ eine Menge Z wählen kann, derart, daß

(1) $$E\,({}_1Z, {}_1Z_\mathcal{F}) < \varepsilon_0 .$$

Sodann wählen wir zwischen ${}_1B$ und ${}_2B$ eine endliche Zahl von Punkten ${}_1B_1, {}_1B_2, {}_1B_3, \cdots$ in solcher Weise, daß für jede der Formel (1) entsprechende Wahl der ${}_1Z_p$ in ${}_1Y_{p+1}$ eine Menge ${}_1Z'_{p+1}$ gewählt werden kann, für welche $E\,({}_1Z_p, {}_1Z'_{p+1}) < \varepsilon_1$, und in jedem ${}_1Y_{p\mathcal{F}}$ eine Menge

${}_1Z'_{p\mathfrak{f}}$, für welche $E\left({}_1Z_p, {}_1Z'_{p\mathfrak{f}}\right) < \varepsilon_1$. In dieser Weise wählt man nun, nachdem einmal die ${}_1Z_p$ angenommen sind, zu jedem von ihnen ein ${}_1Z'_{p+1}$.

Weiter bestimmen wir in jedem ${}_1Y_{p+1}$ die Mengen ${}_1Z'_{p+1}, {}_1Z''_{p+1}, \cdots, {}_1Z^{(h)}_{p+1}, {}_1Z_{p+1}$ in solcher Weise, daß

$$E\left({}_1Z^{(q)}_{p+1}, {}_1Z^{(q+1)}_{p+1}\right) < \varepsilon_1 \text{ und } E\left({}_1Z^{(h)}_{p+1}, {}_1Z_{p+1}\right) < \varepsilon_1,$$

während zugleich

$$E\left({}_1Z^{(q)}_{p+1}, {}_1Z_{p+1}\right) < f(\varepsilon_0),$$

wo $f(\varepsilon_0)$ mit ε_0 unter jede Grenze herabsinkt.

Schließlich nehmen wir die Punkte ${}^1_1B_p, {}^2_1B_p, \cdots, {}^h_1B_p$ in der angegebenen Reihenfolge auf s zwischen ${}_1B_p$ und ${}_1B_{p+1}$ derart an, daß es möglich ist, die Mengen ${}^q_1Z_p$ so zu bestimmen, daß

$$E\left({}^q_1Z_p, {}_1Z^{(q)}_{p+1}\right) < \varepsilon_1.$$

Sodann gelten für die Reihe ${}_1Z_p, {}^1_1Z_p, {}^2_1Z_p, \cdots, {}^h_1Z_p, {}_1Z_{p+1}$ die Formeln:

$$E\left({}_1Z_p, {}^1_1Z_p\right) < f(\varepsilon_1),\ E\left({}^q_1Z_p, {}^{q+1}_{\ \ 1}Z_p\right) < f(\varepsilon_1),\ E\left({}^h_1Z_p, {}_1Z_{p+1}\right) < f(\varepsilon_1),$$
$$E\left({}_1Z_p, {}^q_1Z_p\right) < f(\varepsilon_0),$$

wo $f(\varepsilon_0)$ und $f(\varepsilon_1)$ mit ε_0 resp. ε_1 unter jede Grenze herabsinken.

Wir erhalten in dieser Weise zwischen ${}_1Z$ und ${}_2Z$ eine endliche Reihe von Mengen Z, welche, wenn man ihr ${}_1Z$ und ${}_2Z$ hinzurechnet, die Eigenschaft besitzt, daß je zwei Elemente ein $E < f(\varepsilon_0)$ und je zwei aufeinanderfolgende Elemente ein $E < f(\varepsilon_1)$ bestimmen.

Zwischen je zwei aufeinanderfolgende Elemente dieser Reihe können wir eine weitere Reihe einschalten mit der Eigenschaft, daß in ihr je zwei Elemente ein $E < f(\varepsilon_1)$, und je zwei aufeinanderfolgende Elemente ein $E < f(\varepsilon_2)$ bestimmen.

In dieser Weise fortfahrend, gelangen wir, indem wir die Zahlen $\varepsilon_0, \varepsilon_1, \varepsilon_2, \varepsilon_3, \cdots$ gegen Null konvergieren lassen, zu einer in s überall dichten Punktmenge, und zu einer entsprechenden Menge von Mengen Z, welche ihr eineindeutiges stetiges Bild ist; zugleich ist nun zu *jedem* Punkte von s eine sich mit diesem Punkte stetig ändernde Menge Z bestimmt. Die Transformationen, welche diese Mengen Z ineinander überführen, erzeugen zugleich eine eindimensionale Mannigfaltigkeit von Mengen Y, und, je nachdem Y ein- oder zweidimensional in den Z war, eine zwei- oder dreidimensionale Mannigfaltigkeit von Mengen Z, welche zugleich eine innerhalb einer gewissen Umgebung nicht weiter ausdehnbare zusammenhängende Teilmenge ϑ von ϱ ist.

Wir gehen zum zweiten Falle über und nehmen an, daß R für die Gruppe ϱ in einer Gebietsmenge beweglich ist. Alsdann erzeugt jedes

Teilquadrat q dieses T. B. aus einer Menge Y eine zusammenhängende Teilmenge von ϱ.

Wir wollen zeigen, daß man auch hier wieder in jedem Y ein Z auswählen kann derart, daß diese Z eineindeutiges stetiges Bild der entsprechenden Punkte von q sind. Dazu wollen wir in einem Rechtecke innerhalb q ein rechtwinkliges Netz konstruieren, und zu jedem seiner Kreuzungspunkte eine solche Menge Z, daß je zwei von ihnen ein $E < f(\varepsilon_0)$, und je zwei in einer horizontalen oder vertikalen Querlinie des Netzes aufeinanderfolgende ein $E < f(\varepsilon_1)$ bestimmen.

Um dies zu erreichen, gehen wir aus von einem Rechtecke $\underset{1}{{}_1B}\,\underset{2}{{}_1B}\,\underset{1}{{}_2B}\,\underset{2}{{}_2B}$ mit den horizontalen Seiten $\underset{1}{{}_1B}\,\underset{2}{{}_1B}$ und $\underset{1}{{}_2B}\,\underset{2}{{}_2B}$, welches die Eigenschaft besitzt, daß man, nachdem $\underset{1}{{}_1Z}$ willkürlich angenommen ist, zu jedem Punkte $\underset{1\;\mathcal{F}}{{}_1B_{\mathcal{F}}}$ (d. h. zu jedem inneren Punkte des Rechteckes) ein solches $\underset{1\;\mathcal{F}}{{}_1Z_{\mathcal{F}}}$ wählen kann, daß $E(\underset{1}{{}_1Z}, \underset{1\;\mathcal{F}}{{}_1Z_{\mathcal{F}}}) < \varepsilon_0$.

Sodann bestimmen wir zwischen $\underset{1}{{}_1B}$ und $\underset{2}{{}_1B}$ solche Punkte $\underset{1\;1}{{}_1B}$, $\underset{1\;2}{{}_1B}$, $\cdots$, zwischen $\underset{1}{{}_1B}$ und $\underset{1}{{}_2B}$ solche Punkte $\underset{1}{{}_1B_1}$, $\underset{1}{{}_1B_2}$, $\cdots$, und zu jedem Punkte $\underset{1\;p}{{}_1B_q}$ ein solches $\underset{1\;p}{{}_1Z_q}$, daß zunächst in jedem $\underset{1\;p\mathcal{F}}{{}_1Y_{q\mathcal{F}}}$ ein solches $\underset{1\;p\mathcal{F}}{{}_1Z'_{q\mathcal{F}}}$ und in jedem $\underset{1\;p}{{}_1Y_{q+1}}$ ein solches $\underset{1\;p}{{}_1Z'_{q+1}}$ bestimmt werden kann, daß $E(\underset{1\;p}{{}_1Z_q}, \underset{1\;p\mathcal{F}}{{}_1Z'_{q\mathcal{F}}}) < \varepsilon_1$ und $E(\underset{1\;p}{{}_1Z_q}, \underset{1\;p}{{}_1Z'_{q+1}}) < \varepsilon_1$, daß aber zweitens

$$E(\underset{1}{{}_1Z}, \underset{1\;p}{{}_1Z_q}) < f(\varepsilon_0),\; E(\underset{1}{{}_1Z_q}, \underset{1}{{}_1Z_{q+1}}) < f(\varepsilon_1),$$

$$E(\underset{1\;p}{{}_1Z}, \underset{1\;p+1}{{}_1Z}) < f(\varepsilon_1),$$

$$E(\underset{1\;p}{{}_1Z_q}, \underset{1\;p+1}{{}_1Z_q}) < f(\varepsilon_1).$$

Wenn dann zu jedem $\underset{1\;p}{{}_1Z_q}$ ein entsprechendes $\underset{1\;p}{{}_1Z'_{q+1}}$ gewählt wird, dann gelten zugleich die Formeln

$$E(\underset{1\;p}{{}_1Z'_q}, \underset{1\;p'}{{}_1Z'_q}) < f(\varepsilon_0),\; E(\underset{1\;p}{{}_1Z'_q}, \underset{1\;p+1}{{}_1Z'_q}) < f(\varepsilon_1).$$

Transformationen $\tau_1, \tau_2, \tau_3, \cdots$, welche $\underset{1\;1}{{}_1Z_q}$, $\underset{1\;2}{{}_1Z_q}$, $\underset{1\;3}{{}_1Z_q}$, $\cdots$ in $\underset{1}{{}_1Z_q}$ überführen, führen $\underset{1\;1}{{}_1Z'_q}$, $\underset{1\;2}{{}_1Z'_q}$, $\cdots$ über in $\underset{1}{{}^1_1Z_q}$, $\underset{1}{{}^2_1Z_q}$, $\underset{1}{{}^3_1Z_q}$, $\cdots$ Letztere Reihe

[[161]]

liegt ganz in einer und derselben Menge $\underset{1}{{}_1Y_q}$, und auch für sie gelten die Relationen:

$$E\left(\underset{1}{{}_1^pZ_q},\ \underset{1}{{}_1^{p'}Z_q}\right) < f(\varepsilon_0),\quad E\left(\underset{1}{{}_1^pZ_q},\ \underset{1}{{}_1^{p+1}Z_q}\right) < f(\varepsilon_1).$$

Weil nun aber die Menge $\underset{1}{{}_1Y_q}$ eine gewöhnliche ein- oder zweidimensionale Mannigfaltigkeit in den Z ist, kann man in ihr zwischen die Kette $\underset{1}{{}_1^1Z_q}, \underset{1}{{}_1^2Z_q}, \cdots$ und die Menge $\underset{1}{{}_1Z_q}$ eine gewisse Zahl u_q von solchen weiteren Ketten $\underset{1}{{}_1^1Z_q^h}, \underset{1}{{}_1^2Z_q^h}, \underset{1}{{}_1^3Z_q^h}, \cdots$ einschalten, daß

$$E\left(\underset{1}{{}_1^pZ_q^h},\ \underset{1}{{}_1^{p+1}Z_q^h}\right) < f(\varepsilon_1),\quad E\left(\underset{1}{{}_1^pZ_q^h},\ \underset{1}{{}_1^pZ_q^{h+1}}\right) < f(\varepsilon_1),$$

$$E\left(\underset{1}{{}_1^pZ_q^h},\ \underset{1}{{}_1^{p'}Z_q^{h'}}\right) < f(\varepsilon_0).$$

Wenden wir nun wieder auf jede Menge $\underset{1}{{}_1^pZ_q^h}$ die Transformation τ_p^{-1} an, so finden wir zwischen den beiden Ketten:

$$\underset{1}{{}_1Z_q},\ \underset{1\;1}{{}_1Z'_q},\ \underset{1\;2}{{}_1Z'_q},\ \underset{1\;3}{{}_1Z'_q},\ \cdots$$

und

$$\underset{1}{{}_1Z_q},\ \underset{1\;1}{{}_1Z_q},\ \underset{1\;2}{{}_1Z_q},\ \underset{1\;3}{{}_1Z_q},\ \cdots$$

eine Zahl u_q von Ketten

$$\underset{1}{{}_1Z_q},\ \underset{1\;1}{{}_1Z_q^h},\ \underset{1\;2}{{}_1Z_q^h},\ \underset{1\;3}{{}_1Z_q^h},\ \cdots$$

eingeschaltet, und auch für diese gelten die Formeln:

$$E\left(\underset{1\;p}{{}_1Z_q^h},\ \underset{1\;p+1}{{}_1Z_q^h}\right) < f(\varepsilon_1),\quad E\left(\underset{1\;p}{{}_1Z_q^h},\ \underset{1\;p}{{}_1Z_q^{h+1}}\right) < f(\varepsilon_1),$$

$$E\left(\underset{1\;p}{{}_1Z_q^h},\ \underset{1\;p'}{{}_1Z_q^{h'}}\right) < f(\varepsilon_0).$$

Zwischen $\underset{1}{{}_1B_{q-1}}$ und $\underset{1}{{}_1B_q}$ bringen wir nun durch Punkte $\underset{1}{{}_1B_q^{(0)}}, \underset{1}{{}_1B_q^{(1)}}, \underset{1}{{}_1B_q^{(2)}}, \cdots u_q + 1$ weitere horizontale Querschnitte des Rechteckes an, welche von den alten Vertikalschnitten getroffen werden in horizontalen Punktreihen:

$$\underset{1}{{}_1B_q^{(h)}},\ \underset{1\;1}{{}_1B_q^{(h)}},\ \underset{1\;2}{{}_1B_q^{(h)}},\ \cdots$$

Diese neuen Horizontalschnitte wählen wir in hinreichender Nähe desjenigen, welcher $\underset{1}{{}_1B_q}$ enthält, damit wir Mengen $\underset{1\;p}{{}_1Z_q^{(h)}}$ bestimmen können, welche den Formeln

$$E\left(\underset{1}{{}_1Z_q^{(h)}},\ \underset{1}{{}_1Z_q}\right) < \varepsilon_1,\quad E\left(\underset{1\;p}{{}_1Z_q^{(h)}},\ \underset{1\;p}{{}_1Z_q^h}\right) < \varepsilon_1,\quad E\left(\underset{1\;p}{{}_1Z_q^{(0)}},\ \underset{1\;p}{{}_1Z'_q}\right) < \varepsilon_1$$

genügen.

[[162]]

In den Punkten ${}_1\underset{1\;\;p}{B_q}$ zusammen mit den Punkten ${}_1\underset{1\;\;p}{B_q^{(h)}}$ besitzen wir nunmehr die Kreuzungspunkte des gesuchten Netzes, wenn man diesen Punkten noch die gefundenen Mengen ${}_1\underset{1\;\;p}{Z_q}$ und ${}_1\underset{1\;\;p}{Z_q^{(h)}}$ entsprechen läßt.

Innerhalb jedes Elementarrechteckes dieses Netzes konstruieren wir sodann ein weiteres Netz, das nach Ersetzung von ε_0 durch ε_1 und von ε_1 durch ε_2 analoge Eigenschaften besitzt.

So fahren wir fort, und gelangen, indem wir $\varepsilon_0, \varepsilon_1, \varepsilon_2, \cdots$ gegen Null konvergieren lassen, zu einer in q überall dichten Punktmenge; den Punkten dieser Menge, also nun auch allen Punkten von q, entspricht eine Menge von Mengen Z, welche ihr eineindeutiges stetiges Bild ist, während die Transformationen, welche diese Mengen Z ineinander überführen, eine zweidimensionale Mannigfaltigkeit von Mengen Y erzeugen, welche, je nachdem Y ein- oder zweidimensional in den Z ist, eine drei- oder vierdimensionale Mannigfaltigkeit von Mengen Z, und zugleich eine innerhalb einer gewissen Umgebung nicht weiter ausdehnbare, zusammenhängende Teilmenge ϑ von ϱ ist.

Ziehen wir ein volles Teilgebiet des T. B. von R für die Gruppe ϱ in Betracht, so kann sein Zusammenhang ermöglichen, daß jedem seiner Punkte mehrere, durch nicht kontrahierbare Umläufe getrennte Mengen Z, und diesen entsprechende Mengen Y, zugeordnet sind. Diese Mengen Z und Y bleiben, unabhängig von ihrer Vieldeutigkeit, für jedes Teilrechteck eineindeutiges stetiges Bild voneinander.

Machen wir jetzt wieder von denjenigen Punkten, welche durch ihre Festhaltung die Gruppe ϱ bestimmt haben, den zuletzt festgehaltenen, U, frei, so gibt es für die jetzt zugelassene festgelegte Untergruppe eine innerhalb einer gewissen Umgebung nicht weiter ausdehnbare zusammenhängende Teilmenge v, welche einem Teilbogen resp. Teilrechtecke des T. B. von U entspricht, dessen Punkten eineindeutig und stetig Mengen Z, und diese enthaltende Mengen ϑ zugeordnet sind. Die Menge v ist eine ein- oder zweidimensionale Mannigfaltigkeit von Mengen ϑ, und eine drei-, vier-, fünf-, oder sechsdimensionale Mannigfaltigkeit von Mengen Z.

Sind wir, in dieser Weise fortfahrend, bis zur Freilassung des anfangs zuerst festgehaltenen Punktes F aufgestiegen, so gelangen wir zu einer gewöhnlichen Mannigfaltigkeit von gewisser Dimensionenzahl in den Z, welche innerhalb einer gewissen Umgebung die ganze P. M. enthält; also ist auch die ganze P. M. eine solche Mannigfaltigkeit von Mengen Z.

Aus dieser Rekonstruktion der P. M. aus Mengen Z folgern wir noch, daß, wenn es in beliebiger Nähe eines Punktes W einer Menge Z_W andere Punkte der entsprechenden festgelegten Untergruppe λ_W gäbe, welche nicht aus W innerhalb Z_W auf beliebig kurzem Wege zusammenhängend erreichbar wären, dieselbe Eigenschaft für die W enthaltenden Mengen Y_W, ϑ_W, v_W usw. bestehen würde, also auch für die P. M. selber, was absurd ist. *Die Teilmengen Z einer Menge λ liegen also in der P. M. außerhalb der Grenze der P. M. isoliert,* und hieraus zeigt sich, daß wir bei der obigen Rekonstruktion statt Teilkurvenbogen oder Teilrechtecken auch die vollen zusammenhängenden Bestandteile der T. B. benutzen dürfen.

Schließlich bemerken wir, daß, wenn die P. M. eine p-dimensionale Mannigfaltigkeit von durch Festhaltung der Punkte $P_1, P_2, \cdots, P_m$ bestimmten Mengen Z ist, das Stellensystem, wo die Punkte P festgehalten sind, für eine gewisse Umgebung der Identität eine solche eineindeutige Funktion der entsprechenden Menge Z ist, welche auf der Seite des Stellensystems, also wegen der Eineindeutigkeit auch auf der Seite der Menge Z, stetig ist. Und hieraus folgern wir, daß diese Stellensysteme eine gewöhnliche p-dimensionale Mannigfaltigkeit bilden.

§ 3.

Die Invarianz der Gliederzahl.

Wir wollen weiter zeigen, daß man immer durch Festhaltung einer endlichen Zahl von Punkten erreichen kann, daß die entsprechende festgelegte Untergruppe keinen zusammenhängenden Bestandteil mehr enthält.

Im entgegengesetzten Falle könnten wir nämlich der Reihe nach eine Fundamentalreihe von Punkten festhalten, welche in der T. M. überall dicht liegt. Wir folgern dann genau wie im § 2 der ersten Mitteilung, daß die nach jeder neuen Festhaltung noch mögliche reduzierte Untergruppe gegen die Identität konvergiert, und hieraus weiter, daß man durch Festhaltung einer endlichen Zahl von Punkten bewirken kann, daß in einer gewissen abgeschlossenen Teilmenge der T. M. die T. B. einerseits eine Breite unterhalb einer beliebig klein angenommenen Größe ε besitzen, und andererseits abgeschlossen sind.

Diese T. B. können somit nur einfache geschlossene Kurven sein.

Hält man in einer solchen Kurve einen Punkt fest, so bleiben alle Punkte fest; sonst nämlich würden wieder nicht abgeschlossene T. B. auftreten; und hieraus schließen wir, daß die in Betracht gezogene reduzierte Untergruppe in jedem ihrer T. B. eine gewöhnliche Rotationsgruppe bestimmt.

Sei γ eine dieser einfachen geschlossenen Kurven, und P ein Punkt auf ihr. Weil die reduzierte Untergruppe abgeschlossen ist, müssen, wenn

[[164]]

ein Punkt P' außerhalb γ sich P unbeschränkt nähert, alle Punkte des T. B. γ' von P' gleichmäßig gegen γ konvergieren, und dabei die ganze Menge γ approximieren. Aus demselben Grunde muß P' schließlich die Eigenschaft erhalten, mit P zusammen festzubleiben, bzw. seinen T.B. zu durchlaufen.

Betrachten wir nun die Orte von P und P' bei dieser Rotation als Funktionen des Rotationsargumentes, so konvergiert der Ort von P' gleichmäßig gegen den von P; also muß γ' schließlich γ umschließen.

In derselben Weise zeigen wir, daß außerhalb von γ' wieder eine einfache geschlossene Kurve γ'' existiert, welche γ' umschließt, T. B. für die reduzierte Gruppe ist, und mit γ und γ' zusammen durchlaufen wird. Diese Reihe

$$\gamma, \gamma', \gamma'', \gamma''', \cdots$$

läßt sich fortsetzen bis zu jedem Index der ersten, und, weil es zu $\gamma^{(\omega)}$ wieder ein $\gamma^{(\omega+1)}$ gibt, auch der zweiten Zahlenklasse. Die Breite dieser Kurven muß dabei aber sicher einmal die Größe ε übersteigen, was die Unmöglichkeit unserer Annahme zeigt.

Man kann somit durch Festhaltung einer endlichen Zahl von Punkten eine festgelegte Untergruppe bestimmen, welche in der P. M. nur noch aus isolierten Punkten besteht.

Ist dies erreicht durch Festhaltung von der Reihe nach $n_1 + n_2$ Punkten, und waren davon gerade vor ihrer Festhaltung n_1 in Gebieten, n_2 nur in Kurven beweglich, so nennen wir die Gruppe $(2n_1 + n_2)$-*gliedrig*, und werden zeigen, daß die Zahl $n = 2n_1 + n_2$ eine *Invariante der Gruppe* ist.

Zunächst beweisen wir, daß die Zahl n sich nicht ändern kann, wenn man *dieselben* Punkte in anderer Reihenfolge festhält, und dabei eventuell einen Punkt, welcher, wenn die Reihe an ihn kommt, schon feststeht, *nichts* zur Zahl n beitragen läßt; dabei genügt es zu zeigen, daß eine Vertauschung von zwei in der Reihe *aufeinander folgenden* Punkten die Zahl n nicht ändert. Dies gelingt nun in folgender Weise:

Sei W die reduzierte Untergruppe *vor* der Festhaltung der beiden Punkte, Z die reduzierte Untergruppe *nach* dieser Festhaltung, so würde, wenn die Zahl n sich durch die Verwechselung änderte, W sowohl p-dimensionaler, wie q-dimensionaler Raum von Mengen Z sein, wo $p < q$, und $p < 4$. Dies ist aber nach dem Beweise von Lüroth*) gerade für $p < 4$ sicher unmöglich. [[4]]

Sei ferner die Gruppe ein erstes Mal mittels der Punkte $P_1, P_2, \cdots, P_m$ festgesetzt, ein zweites Mal durch Festhaltung zunächst von einem neuen

*) Math. Annalen Bd. 63, S. 222—238.

Punkte Q_1, und sodann von so vielen Punkten P, als wir noch brauchen. Seien diese Punkte $P_{h_1}, P_{h_2}, \cdots, P_{h_\tau}$.

Die so gefundene Gliederzahl n' ist dieselbe, welche herauskommt, wenn wir zunächst die Punkte $P_{h_1}, \cdots, P_{h_\tau}$, und erst dann Q_1 festhalten. Hierbei aber trägt einerseits Festhaltung von $P_{h_1}, \cdots, P_{h_\tau}$, je nachdem Q_1 gerade vor seiner Festhaltung in Gebieten oder in Kurven beweglich ist, $n' - 2$ oder $n' - 1$ zur Gliederzahl n' bei, und es zeigt sich andererseits, daß die der Festhaltung von $P_{h_1}, \cdots, P_{h_\tau}$ entsprechende reduzierte Untergruppe eine gewöhnliche zwei- resp. eindimensionale Punktmannigfaltigkeit ist. Also auch wenn man nach dieser Festhaltung die endgültige Festlegung der Gruppe mittels anderer Punkte, z. B. Punkte P, zustande bringt — und im letzteren Falle gebraucht man für das ganze Verfahren wieder nur Punkte P — findet man die Gliederzahl $(n' - 2) + 2$ resp. $(n' - 1) + 1$, also n'. Und da man andererseits auch n finden muß, so ist $n' = n$.

Man darf also die Festlegung bewerkstelligen mittels Punkte, welche unter den verschiedenen P und Q_1 willkürlich gewählt werden, und wird immer dieselbe Gliederzahl n finden.

Wir fügen nun einen zweiten Punkt Q_2 hinzu, und zeigen in derselben Weise, daß man die Festlegung bewerkstelligen darf mittels Punkte, welche unter den verschiedenen P, Q_1 und Q_2 willkürlich gewählt werden, und immer dieselbe Gliederzahl finden wird.

In dieser Weise fortfahrend, kann man allmählich so viele neue Punkte Q hinzufügen, welche zur Festsetzung benutzt werden dürfen, daß schließlich die Punkte Q allein dazu ausreichen, und wird immer dieselbe Gliederzahl n herausbringen. Da andererseits die Punkte Q willkürlich gewählt sind, ist hiermit unser Satz bewiesen.

§ 4.

Die eingliedrigen letzten reduzierten Untergruppen.

Wir betrachten jetzt eine solche reduzierte Untergruppe, welche nach Festhaltung noch eines auf einer Kurve beweglichen Punktes nur noch isolierte Punkte in der P. M. enthält. Dieser Gruppe entspricht nach dem vorhergehenden in der P. M. entweder eine einfache offene Kurve, welche nur in der Grenze der P. M. Enden besitzen kann,*) oder eine einfache geschlossene Kurve.

Im ersten Falle zeigen wir, genau wie im § 3 der ersten Mitteilung, daß die Gruppe eineindeutiges und für jede abgeschlossene Teilmenge gleichmäßig stetiges Bild der Additionsgruppe der reellen Zahlen ist. Die zusammenhängenden T. B. in der T. M. sind dabei entweder einfache

[[3]] *) Lorentzsche Kurven sind hier ausgeschlossen.

geschlossene Kurven, oder einfache Kurvenbogen, oder einfache offene Kurven, welche auf einer oder auf beiden Seiten in der T. M. oder ihrer Grenze zusammenhängende Enden besitzen können, oder schließlich Lorentzsche Kurven.

Die eventuellen Enden in der Grenze der T. M. können entweder fest sein, oder T. B. aufweisen von derselben Art, wie die in der T. M. selber liegenden.*)

Im zweiten Falle zeigen wir genau wie in § 4 der ersten Mitteilung, daß die Gruppe eineindeutiges und für jede abgeschlossene Teilmenge gleichmäßig stetiges Bild der Rotationsgruppe des Kreises ist. Die T. B. in der T. M. sind jetzt alle einander umschließende, einfache, geschlossene Kurven, welche dieselbe Umlaufperiode besitzen.

Geht man von einer solchen Kurve aus, und betrachtet auf jeder Seite von ihr die an sie anschließende stetige Folge der geschlossenen T. B., so können diese beiden Folgen entweder ineinander übergehen, oder sich noch jede für sich in dreierlei Weise verhalten, nämlich erstens sich in einen Punkt zusammenziehen, zweitens in einer unilateralen einfachen geschlossenen Kurve ihren Abschluß finden, und drittens innerhalb der T. B. kein Ende besitzen.

Hieraus folgern wir, daß im Falle einer geschlossenen eingliedrigen letzten reduzierten Untergruppe die T. M. folgende sieben Gestalten besitzen kann:

a) die zweiseitigen: Cartesische Ebene, Zylinder, zweiseitige Ringfläche, Kugel.

b) die einseitigen: elliptische Ebene, Komplementärmenge einer Kreisfläche in der elliptischen Ebene, einseitige Ringfläche.

Bezeichnen wir nun eine endliche kontinuierliche Gruppe, welche eineindeutiges und für jede abgeschlossene Teilmenge gleichmäßig stetiges

---

*) Besitzt die P. M. der reduzierten Gruppe auf einer resp. beiden Seiten ein Ende, so müssen auch alle zusammenhängenden T. B. der T. M. entweder geschlossen sein, oder auf einer resp. beiden Seiten ein Ende besitzen. Näheres läßt sich schließen, wenn die P. M. der reduzierten Gruppe nicht nur ein Ende besitzt, sondern dazu in diesem Ende eine einfache geschlossene Kurve als T. B. In diesem Falle müssen alle Enden innerhalb der T. M. einfache geschlossene Kurven sein, die ihrerseits wieder T. B. der Gruppe sind. Letztere können, weil eine geschlossene Kurve von Grenztransformationen existiert, sich einander nicht unbeschränkt nähern, ohne daß sie schließlich einander umschließen und dieselbe Umlaufperiode bekommen. Sie fügen sich also zusammen zu einer abzählbaren Menge von Ringgebieten, welche jedes für sich eine feste Umlaufperiode besitzen, während die Umlaufperioden der verschiedenen Ringgebiete untereinander kommensurabel sind, da sie alle in der Periode der geschlossenen Kurve von Grenztransformationen enthalten sein müssen.

Die T. B. in der unmittelbaren Nähe eines solchen Ringgebietes sind einfache offene Kurven, welche sich ihm spiralförmig anschmiegen.

Abbild entweder der Additionsgruppe der reellen Zahlen oder der Rotationsgruppe des Kreises ist, als „von einer Infinitesimaltransformation erzeugt“, so formulieren wir folgenden Satz:

Jede eingliedrige letzte reduzierte Untergruppe einer Gruppe einer zweidimensionalen Mannigfaltigkeit wird von einer Infinitesimaltransformation erzeugt.

§ 5.

Die zweigliedrigen letzten reduzierten Untergruppen; die Gestalt der Fläche F.

Wir studieren weiter eine solche reduzierte Untergruppe, für welche die zusammenhängenden T. B. in der T. M. sämtlich Gebiete sind, während die Festhaltung eines Punktes davon die Gruppe festlegt. Einer solchen Gruppe entspricht nach dem vorhergehenden in der P. M. eine Fläche, die wir F nennen wollen, und die nur in der Grenze der P. M. eine Grenze besitzen kann. Wir wollen die Gestalt dieser Fläche im Sinne der Analysis Situs untersuchen.

Enthält sie nicht kontrahierbare geschlossene Kurven, so kann es deren nicht zwei voneinander unabhängige geben, welche einander nicht treffen. Denn, wenn c und c' zwei solche Kurven wären, so würde ein voller Umlauf von c' eine Transformation bedeuten, welche c in ihre alte Lage zurückführen müßte; andererseits aber sehen wir, daß, wenn ein bestimmter Punkt P von c dabei seine ursprüngliche Lage wieder erreicht hat, die Kurve c in eine Lage gelangt sein müßte, welche nicht bei Festhaltung von P in ihre ursprüngliche Lage stetig übergeführt werden könnte.

Hieraus folgern wir leicht, daß nur die sieben am Schlusse des § 4 aufgezählten Gestalten möglich sind, und wir behaupten weiter, daß von diesen noch die drei einseitigen in Fortfall kommen.

Denn wenn man in einer solchen Fläche F nach einem Umlaufe, welcher den Umlaufssinn ändert, einen Punkt P in seine ursprüngliche Lage zurückgeführt hätte, müßten auch alle Punkte von F ihre ursprünglichen Lagen wieder eingenommen haben, was, da der Umlaufssinn in P sich geändert hat, nicht zutrifft.

Schließlich kann auch die Kugel nicht auftreten, weil eine willkürliche Transformation der Gruppe F sie ohne Fixpunkt transformieren müßte, was unmöglich ist.*)

Die Fläche F unterliegt erstens einer zweigliedrigen Gruppe, für

---

[[5]] *) Vgl. „*Über eineindeutige, stetige Transformationen von Flächen in sich*“, Math. Annalen Bd. 69, S. 176; Amsterdamer Berichte, holländische Ausgabe XVII 2, S. 741; englische Ausgabe XI 2, S. 788.

[[168]]

welche sie selbst die P. M. ist, aber zweitens auch einer *Parametergruppe*, nach welcher die Transformation T_x von der Transformation T_a in $T_x' = T_a^{-1} T_x T_a$ übergeführt wird.

Die Transformationen dieser Parametergruppe sind paarweise invers, und jedenfalls außerhalb der Grenze eineindeutig und stetig; von der Fläche F sind sie aber zwar eindeutiges und für jede abgeschlossene Teilmenge gleichmäßig stetiges, nicht aber notwendig eineindeutiges Bild. Dennoch besitzen die T. B. in F für die Parametergruppe genau dieselben Eigenschaften, wie die T. B. in der T. M. für eine willkürliche reduzierte Untergruppe; die Herleitung des § 1 bleibt hier nämlich ungeändert bestehen.

Einem für die Parametergruppe invarianten Punkte σ der Fläche F entspricht eine mit allen übrigen vertauschbare Transformation σ.

Die Menge der σ enthält immer wenigstens einen Punkt, nämlich die Identität.

Bezeichnen wir mit s eine willkürliche, *nicht* mit jeder andern vertauschbare Transformation, so dürfen wir schreiben:

$$\sigma\sigma = \sigma, \qquad \sigma s = s\sigma = s.$$

Die Menge der σ bildet also eine Untergruppe der Gruppe F; weiter ist sie außerhalb der Grenze von F abgeschlossen.

Nach der Methode des § 1 können wir nun folgern, daß die Menge der σ entweder keinen zusammenhängenden Bestandteil enthält, oder aus isolierten, einfachen, offenen oder geschlossenen Kurven, oder aus Gebieten besteht.

Jeden dieser drei Fälle werden wir zur Herleitung der Infinitesimaltransformationen der Gruppe F einzeln diskutieren.

§ 6.

Erster Fall: Die Menge der σ ist punktartig.

In diesem Falle gibt es unter den Transformationsbereichen der übrigen Punkte von F für die Parametergruppe sicher eine Kurve.

Sonst nämlich gäbe es außerhalb der Menge der σ nur einen einzigen T. B., das Komplementärgebiet G der σ. Läßt man dann mittels einer Fundamentalreihe von Transformationen der Parametergruppe einen willkürlichen Punkt von G gegen die Identität konvergieren, so muß auch jede zusammenhängende perfekte Menge μ in G, welche diesen Punkt enthält, gleichmäßig gegen die Identität konvergieren. Sonst nämlich besäße die Fundamentalreihe der Lagen von μ einen Limespunkt L in G, und deshalb auch die Anfangslage von μ einen Punkt M, den in der Parametergruppe eine Limestransformation der ebengenannten Fundamentalreihe in L überführte. Diese Limestransformation wäre sodann eine eigentliche Transformation der Gruppe F, und keine Grenztransformation, während andererseits nur eine Grenztransformation in der Parametergruppe eine Transformation, welche nicht die identische ist, in die identische überführen kann.

Indem wir nun für die Menge μ eine geschlossene Kurve um die Identität wählen, sehen wir, daß die Menge der σ keine weiteren Punkte, als nur die Identität enthalten kann.

Weil weiter sicher Grenztransformationen existieren, kann die Fläche F nur noch eine Cartesische Ebene oder ein Zylinder sein. Letztere Gestalt kommt aber in Fortfall, denn man müßte in ihr mittels der Parametergruppe eine nicht kontrahierbare geschlossene Kurve gleichmäßig gegen die Identität konvergieren lassen können, was unmöglich ist. Ist aber die Fläche F eine Cartesische Ebene, so ist der T. B. für die Parametergruppe ein Zylinder, dessen beide Ränder der Rand der Fläche F und die Identität sind; dieser Zylinder muß sich also, um *alle* Transformationen der Gruppe F in der Parametergruppe operieren zu lassen, $^*\omega + \omega$ Male abwickeln auf die Fläche F, und dementsprechend müßte eine Transformation τ existieren, welche nicht die identische wäre, und welche in der Parametergruppe, nicht aber in der Gruppe F, alle Punkte in ihre ursprüngliche Lage zurückführte. Diese Transformation τ müßte sodann zur Menge der σ gehören, aber diese Menge enthält nur die Identität; also hat unsere Annahme auf einen Widerspruch geführt.

Wir besitzen also sicher eine einfache Kurve c, welche T. B. für die Parametergruppe ist. Halten wir in ihr einen Punkt P an der Stelle P' fest, so wird dadurch in der Gruppe F eine stetige Untergruppe u' bestimmt; die T. B. in der T. M. für diese Gruppe u' können nur einfache Kurven sein, und in der Fläche F entspricht ihr eine eindimensionale Mannigfaltigkeit $\varkappa'$ (welche nicht eine Lorentzsche Kurve sein kann).

Auf jeder der Kurven $\varkappa'$ kann man mittels Überlegungen wie in § 2 einen Punkt p' so bestimmen, daß diese Punkte zusammen eine einfache Kurve π bilden, welche jede Kurve $\varkappa'$ nur in einem Punkte trifft. Die Transformationen, welche den Schnittpunkt von $\varkappa_1'$ und π in die übrigen Punkte von π überführen, erzeugen aus jedem Punkte von $\varkappa_1'$ eine weitere einfache Kurve π'. Sowohl die Schar der π', wie die invariante Schar der $\varkappa'$ erstrecken sich über die ganze Fläche F, und innerhalb jedes von ihnen gebildeten Viereckes sind sie eineindeutiges stetiges Bild der horizontalen und vertikalen Querschnitte eines gewöhnlichen Rechteckes.

Je nachdem die Kurven $\varkappa'$ und π' offen oder geschlossen sind, ist die Fläche F eine Cartesische Ebene, oder ein Zylinder, oder eine zweiseitige Ringfläche.

[[170]]

Bei der wirklichen Aufzählung der Gruppen werden wir aber sehen, daß im Falle dieses Paragraphen nur die Gestalt der Cartesischen Ebene auftreten kann. Dennoch lassen wir in den folgenden Überlegungen lieber alle drei Gestalten zu, damit sie sich auf § 8 ungeändert übertragen lassen.

Die eindimensionale Mannigfaltigkeit der Kurven $\varkappa'$ wird von der Gruppe F entweder ein- oder zweigliedrig transformiert. Im letzteren Falle muß die Fläche F zugleich P. M. dieser zweigliedrigen Gruppe der eindimensionalen Mannigfaltigkeit sein. *Mithin wird dann jede endliche Transformation von einer Infinitesimaltransformation erzeugt, und genügen zwei willkürliche dieser Infinitesimaltransformationen, um die ganze Gruppe F zu erzeugen.*

Im ersteren Falle gibt es zu einer willkürlichen Transformation T, welche $\varkappa_1'$ in $\varkappa_2'$ überführt, sicher eine weitere Transformation λ, welche $\varkappa_1'$ in $\varkappa_3'$ und $\varkappa_3'$ in $\varkappa_2'$ überführt. Wenn nun auch $\lambda\lambda$ nicht gleich T ist, so kann man jedenfalls λ stetig variieren in solcher Weise, daß sie immer noch $\varkappa_1'$ in $\varkappa_3'$ überführt; auch wird $\lambda\lambda$ dann immer $\varkappa_1'$ in $\varkappa_2'$ überführen, und muß, wenn die stetige Änderung von λ hinreichend weit fortgesetzt wird, schließlich in T übergehen. Dann ist also $\lambda\lambda = T$; dasselbe soll die Formel $\lambda = \frac{1}{2} T$ besagen, oder auch die Ausdrucksweise, daß λ eine *Halbierung* von T ist. *Jede Transformation der Gruppe F ist also halbierbar.*

Konstruieren wir nun zu T Transformationen $\frac{1}{2} T$, $\frac{1}{4} T$, $\frac{1}{8} T$, usw.

Diese Transformationen $\frac{1}{2^n} T$ sind, wenn die Kurven $\varkappa'$ entweder geschlossen sind, oder untereinander einen zyklischen Ordnungstypus bilden, nicht eindeutig bestimmt; man kann aber leicht solche Festsetzungen treffen, daß wenigstens die Kurve $\varkappa'$, in der die Halbierung liegt, eindeutig bestimmt ist, und weiter für Transformationen in genügender Nähe der die Identität enthaltenden Kurve $\varkappa'$ volle Eindeutigkeit erhalten. Dann konvergiert sicher $\frac{1}{2^n} T$ gegen Null. Wir werden aber weiter zeigen, daß auch die Transformation $\frac{f}{2^n} T$ zugleich mit der Zahl $\frac{f}{2^n}$ gegen Null konvergiert.

Zunächst bemerken wir, daß, wenn T die Identität J in den Punkt P, welcher innerhalb einer gewissen Umgebung von J liegt, überführt, man jeden in obiger Weise konstruierten Punkt $\frac{a}{2^n} T (a < 2^n)$ mit $\frac{a+1}{2^n} T$ verbinden kann durch einen einfachen Kurvenbogen, welcher keinen weiteren Punkt der von den Punkten $\frac{a}{2^n} T$ und $\frac{a+1}{2^n} T$ bestimmten Kurven $\varkappa'$ enthält und bei unbeschränkter Zunahme von n gegen Null konvergiert. Der aus diesen Kurvenbogen gebildete, J und P verbindende Weg ist sodann, unabhängig von n, für festes J und festes P im Sinne der Analysis Situs äquivalent mit dem kürzesten Wege zwischen J und P.

Wir wollen nun annehmen, daß $\frac{f}{2^n}T$ *nicht* mit der Zahl $\frac{f}{2^n}$ gegen Null konvergiert, und setzen dabei zunächst voraus, daß die Kurven $\varkappa'$ geschlossen sind. Es gibt dann eine Transformation $T_1\left(=\frac{1}{2^k}T\right) < \varepsilon$, und eine Transformation $T_2 = \frac{h}{2^m}T_1\left(\frac{h}{2^m} < 1\right) > a$, wo wir ε beliebig klein und a oberhalb einer gewissen endlichen Grenze voraussetzen dürfen. Wir wiederholen nun die Transformation T_2 n Male, nämlich so lange, bis die Zahl $\frac{nh}{2^m} > 1$ ist. Darauf wenden wir die Transformation T_1^{-1} an; das Produkt aller angewandten Transformationen ist dann äquivalent mit $\frac{r}{2^m}T_1\left(\frac{r}{2^m} < 1\right)$. Jetzt wenden wir wieder n' Male T_2 an, nämlich bis $\frac{n'h+r}{2^m} > 1$ ist, subtrahieren wieder T_1 und fahren so fort, bis wir schließlich die Transformation $\frac{1}{2^m}T_1$ erreichen.

Wir sind dann immer im selben Sinne in einer Entfernung $< \varepsilon_1$ an $\varkappa'_J$ entlang gelaufen und haben dabei, indem wir jeden vollen Umlauf mitrechnen, einen gewissen Kurvenbogen $> a$ in $\varkappa'_J$ zurückgelegt.

Eine Wiederholung von $\frac{1}{2^m}T_1$, bis T_1 erreicht wird, kann diesen Gang an $\varkappa'_J$ entlang nur noch verlängern. Nennen wir also P_1 den Punkt, in den J von T_1 übergeführt wird, so kann der von den Punkten $\frac{a}{2^m}T_1$ (m fest; $a \leqq 2^m$) im obigen Sinne bestimmte Weg niemals für festes J und festes P_1 äquivalent sein mit dem kürzesten Wege zwischen J und P_1, und wir sind auf einen Widerspruch gekommen.

Wenn zweitens die Kurven $\varkappa'$ offen sind, so gibt es jedenfalls ein gewisses endliches Gebiet um die Identität, innerhalb dessen, bei Anwendung auf einen willkürlichen Punkt, T_1 und ihre Inverse $< \varepsilon'$, und T_2 und ihre Inverse $> a'$ bleiben. Hieraus folgern wir aber, daß, wenn bei den oben geschilderten Wiederholungen zunächst von T_2 und sodann von $\frac{1}{2^m}T_1$ dieses Gebiet einmal verlassen wird, es auf immer verlassen wird, was wieder die Unmöglichkeit der Annahme zeigt.

Die Transformationen $\frac{a}{2^n}T$ (a und n willkürliche ganze Zahlen) bilden also nebst ihren Grenztransformationen für jedes abgeschlossene Zahlensegment einen einfachen Kurvenbogen, und die Transformationen an diesem

[[172]]

Kurvenbogen und seinen Fortsetzungen auf beiden Seiten (welche in einer Ringfläche auch eine Lorentzsche Kurve bilden können) entlang *werden von einer Infinitesimaltransformation erzeugt.*

Man kann auch hier wieder zeigen, daß *zwei willkürliche dieser Infinitesimaltransformationen* φ_1 *und* φ_2 *hinreichen, um die ganze Gruppe* F *zu erzeugen.*

Seien nämlich c_1 und c_2 die Bahnkurven von φ_1 und φ_2 durch die Identität J. Ein endlicher, J enthaltender Teilbogen von c_1 trifft dann alle Bahnkurven von φ_2, welche durch eine gewisse Umgebung O von J hindurchgehen. Also lassen alle Punkte von O, aber deshalb nun auch alle Punkte von F, sich aus J mittels φ_1 und φ_2 erreichen.

§ 7.

Zweiter Fall: Die Menge der σ ist linienartig.

Die Menge der σ enthält in diesem Falle eine isolierte einfache Kurve $\varkappa$ durch die Identität, welche eine invariante Untergruppe der Gruppe F repräsentiert. Jede Transformation der Gruppe F führt diese Kurve über entweder in sich selbst oder in eine Kurve $\varkappa'$, welche gänzlich außerhalb $\varkappa$ liegt. Die Fläche F wird in eine für die Gruppe F invariante Schar von gänzlich getrennt liegenden einfachen Kurven $\varkappa'$ zerlegt.

Die Überlegungen des § 6 bleiben nun weiter ungeändert in Kraft. Wenn wir aber das Schlußresultat erreicht haben, sehen wir, daß der Fall dieses Paragraphen unmöglich ist. Weil nämlich die Infinitesimaltransformation φ_1, welche die Untergruppe $\varkappa$ erzeugt, und eine willkürliche andere Infinitesimaltransformation φ_2 hinreichen, um jede Transformation der Gruppe F zu erzeugen, während φ_1 und φ_2 miteinander vertauschbar sind, so sind *alle* Transformationen der Gruppe, welche ja als Produkte von aus φ_1 und aus φ_2 erzeugten anzusehen sind, miteinander vertauschbar, d. h. wir sind im Falle des nächsten Paragraphen.

§ 8.

Dritter Fall: Die Menge der σ ist flächenartig.

In diesem Falle kann ein von den σ eingenommenes Gebiet G keine Grenze innerhalb der Fläche F besitzen. Denn ein Punkt P einer solchen Grenze würde ebenfalls zur Menge der σ gehören, und in beliebiger Nähe von ihm gäbe es Punkte, welche zur Menge der s gehörten; es gäbe also auch beliebig kleine Transformationen s, und diese müßten das Gebiet G überführen in ein Gebiet ganz außerhalb von G, was unmöglich ist.

Das Gebiet G überdeckt mithin die ganze Fläche F, und *je zwei Transformationen der Gruppe F sind vertauschbar.*

Bezeichnen wir nun weiter eine willkürliche halbierbare Transformation mit τ, eine willkürliche nicht halbierbare mit t, so dürfen wir schreiben:

$$\tau\tau = \tau, \quad \tau t = t\tau = t.$$

Die Menge der τ ist eine die Identität enthaltende Untergruppe der Gruppe F; außerdem ist sie zusammenhängend, und außerhalb der Grenze von F abgeschlossen.

Nach der Methode des § 1 läßt sich nun weiter zeigen, daß sie entweder ein Gebiet ist oder eine einfache Kurve, welche nicht eine Lorentzsche Kurve sein kann.

Letzterer Fall kann aber gleich ausgeschlossen werden, denn wenn die Menge der τ eine einfache Kurve β wäre, so gäbe es beliebig kleine nicht halbierbare Transformationen t, welche β in eine einfache Kurve β' außerhalb von β, z. B. rechts von β, überführten; tt führte sodann β über in β'', rechts von β, also ebenfalls außerhalb β. Andererseits ist aber tt halbierbar, müßte also geradezu zu β gehören.

Die Menge der τ ist also jedenfalls ein Gebiet; von diesem aber zeigt man an genau derselben Weise, wie oben von dem von der Menge der σ gebildeten, daß es keine Grenze innerhalb der Fläche F aufweisen kann, also die ganze Fläche F überdeckt. *Mithin sind wieder alle Transformationen der Gruppe F halbierbar.*

Und in derselben Weise zeigen wir, daß es für jedes n zu jeder Transformation T eine Transformation $\frac{T}{n}$ gibt.

Wir unterscheiden nun folgende drei Fälle, von denen wir die ersten beiden auf den dritten zurückführen:

1. *Die Fläche F ist eine Cartesische Ebene.* Wir konstruieren zu einer willkürlichen Transformation T das *Transformationsfeld**), welches, da T keinen Punkt invariant läßt, in solcher Weise gebildet werden kann, daß es die Identität J enthält und begrenzt wird von zwei einfachen Kurven, welche einander nicht treffen, und sich beiderseits bis zum Unendlichen erstrecken.**)

Betrachten wir dann je zwei Punkte von F, welche durch T ineinander übergeführt werden, als identisch, so wird aus F eine Zylinderfläche F', und weil die Transformation T bei der Gruppe F invariant ist,

---

*) Vgl. Math. Annalen Bd. 69, S. 178; Amsterdamer Berichte, holl. Ausg. XVIII 1,
[[5]] S. 107; engl. Ausg. XII 1, S. 286.

**) Vgl. Math. Ann. Bd. 69, S. 179; Amster Ber., holl. Ausg. XVIII 1, S. 117;
[[5]] engl. Ausg. XII 1, S. 297

[[174]]

unterliegt auch F' einer Gruppe von eineindeutigen, stetigen und paarweise inversen Transformationen, welche sich, da jede von ihnen von T in sich selbst übergeführt wird, eineindeutig und für jede abgeschlossene Teilmenge gleichmäßig stetig auf F' selbst abbilden lassen.

Hiermit ist der Fall 1 auf 3 zurückgeführt.

2. *Die Fläche F ist eine zweiseitige Ringfläche.* Wir können dann jeden ihrer Punkte $^*\omega + \omega$ Male vervielfältigen, und dieselbe Vervielfältigung auch für jede der von ihnen repräsentierten Transformationen erreichen, indem wir für diese nicht nur eine Endlage, sondern zugleich einen stetigen Weg, auf dem die Endlage aus der Identität erreicht ist, berücksichtigen, wobei wir uns beschränken auf Unterschiede der Wege bezüglich *einer* der beiden voneinander unabhängigen, nicht kontrahierbaren geschlossenen Kurven der Ringfläche.

Aus der Fläche F wird dann eine Zylinderfläche F', welche einer zweigliedrigen Gruppe mit paarweise vertauschbaren Transformationen, für die sie selbst die P. M. ist, unterliegt, und hiermit ist auch der Fall 2 auf 3 zurückgeführt.

3. *Die Fläche F ist eine Zylinderfläche.* Sie besitzt dann zwei getrennte Ränder. Wenn wir eine Fundamentalreihe f von Transformationen gegen einen dieser Ränder konvergieren (bzw. ins Unendliche rücken) lassen, muß durch sie jede zusammenhängende perfekte Menge μ von F gleichmäßig gegen diesen Rand konvergieren (bzw. gleichmäßig auf der Seite dieses Randes ins Unendliche rücken). Denn wenn die Fundamentalreihe der Lagen von μ einen Limespunkt L innerhalb F besäße, gäbe es auch in der Anfangslage von μ einen Punkt M, den eine Limestransformation von f in L überführte, was unmöglich ist, da die Punkte, in welche M von f übergeführt wird, nur gegen einen Rand konvergieren können.

Seien nun ϱ und ϱ' die beiden Ränder, so können wir in F durch die Identität eine geschlossene Kurve c legen, welche diese beiden Ränder trennt. Sie teilt F in zwei Teilgebiete, nämlich G auf der Seite von ϱ, und G' auf der Seite von ϱ'. Man kann sodann immer den Rand ϱ mit einer solchen geschlossenen Kurve γ umgeben, daß jede zwischen ϱ und γ enthaltene Transformation G in ein beliebig weit gegen ϱ fortgerücktes Teilgebiet überführt. Analog kann man ϱ' mit einer geschlossenen Kurve γ' umgeben, welche die analoge Eigenschaft in bezug auf G' besitzt.

Wir behaupten nun, daß es unmöglich ist, daß Wiederholungen derselben Transformation x die Identität beliebig weit sowohl gegen ϱ, wie gegen ϱ' fortrücken lassen.

Wenn nämlich das Gebiet G von px in ein Teilgebiet transformiert wird, so wird es von jeder Transformation $(p+n)x$ in ein Gebiet transformiert, das gänzlich in der Summe des Gebietes G und derjenigen Gebiete, in welche G durch x, $2x$, $3x$, $\cdots$, $(n-1)x$ übergeführt wird, enthalten sein muß.

Sei also T eine willkürliche Transformation der Gruppe F, und sei ϱ' der Rand, gegen den ihre Wiederholungen nicht unbeschränkt weit fortrücken, so kann man die geschlossene Kurve γ' in solcher Weise wählen, daß für jede zwischen γ' und ϱ' enthaltene Transformation T' die Unmöglichkeit stattfindet, daß aT in beliebige Nähe von bT' gelangen könnte (a und b willkürliche positive ganze Zahlen).

Wir sehen also, daß γ' derart gewählt werden kann, daß die Menge $\frac{a}{b}T$ $\left(\frac{a}{b}\right.$ eine willkürliche positive rationale Zahl$\left.\right)$ nicht zwischen ϱ' und γ' eindringen kann.

In analoger Weise können wir zeigen, daß wir zu T γ und γ' *beide* in solcher Weise wählen können, daß die Menge $\frac{1}{b}T$ (b eine willkürliche positive ganze Zahl) *weder* zwischen ϱ und γ, *noch* zwischen ϱ' und γ' eindringen kann.

Wir betrachten nun im weiteren eine solche Transformation T, welche keine Wiederholung besitzt, die die identische Transformation ist. (Dieser Bedingung genügen nach dem vorhergehenden alle Transformationen, die hinreichend weit gegen einen der Ränder fortgerückt sind.) Gibt es dann zu einer gewissen positiven rationalen Zahl $\frac{a}{b}$ mehrere Transformationen $\frac{a}{b}T$, so unterscheiden sich je zwei von ihnen um eine Transformation $\frac{1}{b}J$, wo J die identische Transformation bedeutet.

Wir bezeichnen weiterhin $\frac{a}{b}T^{-1}$ auch mit $-\frac{a}{b}T$, und unterscheiden zwei Fälle, je nachdem die Menge $z \cdot T$ (z eine willkürliche positive rationale Zahl) in gewissen Gebieten überall dicht ist oder nicht.*)

Erster Fall. *Die Menge $z \cdot T$ enthält überall dichte Bestandteile.* Ist sie dann um den Punkt $p \cdot T$ (p ein spezieller Wert von z) überall dicht, so ist sie es auch um einen Punkt $(p+z_1)T$. Dies leuchtet ein durch Anwendung der Transformation $z_1 \cdot T$ auf den Punkt $p \cdot T$ und seine Umgebung, welche dabei wieder in eine Teilmenge von $z \cdot T$ übergeht.

Wäre nun h eine solche positive Zahl, daß für $z_1 > h$ die Menge $z \cdot T$ um $z_1 \cdot T$ überall dicht ist, für $z_1 < h$ aber nicht, so würde die Menge $(z-\varepsilon_1) \cdot T$ (ε_1 eine bestimmte, beliebig klein gewählte, positive rationale Zahl) um $z_1 \cdot T$ dann und nur dann überall dicht liegen, wenn $z_1 > h - \varepsilon_1$.

*) Daß, wenn man die eventuelle Vielwertigkeit von $\frac{a}{b}T$ in Betracht zieht, beide Fälle vorkommen können, zeigt schon die Bewegungsgruppe eines Kreiszylinders in sich im Euklidischen Raume.

Die Menge $-\varepsilon \cdot T$ (ε eine willkürliche, zwischen 0 und ε_1 variierende, positive rationale Zahl) müßte also überall dichte Bestandteile enthalten. Dies ist aber, weil andererseits die Menge $y \cdot T$ (y eine willkürliche, zwischen 0 und h variierende, positive rationale Zahl) keinen überall dichten Bestandteil enthalten darf, ein Widerspruch.

Die Menge $z \cdot T$ muß also um jeden ihrer Punkte überall dicht sein, und zugleich muß die Menge $w \cdot T$ (w eine willkürliche positive oder negative rationale Zahl) in F überall dicht sein, weil sie nämlich eine Untergruppe mit paarweise inversen Transformationen ist, und überall dichte Bestandteile besitzt. Dann muß aber auch für jedes ε_1 die Menge $-\varepsilon \cdot T$ überall dichte Bestandteile außerhalb der von $z \cdot T$ bedeckten Gebietsmenge $\mathfrak{F}$ besitzen. Sie muß sogar gänzlich außerhalb von $\mathfrak{F}$ liegen; sonst nämlich gäbe es eine Transformation $-\varepsilon_2 \cdot T$, welche einerseits die von $z \cdot T$ bedeckte Gebietsmenge in eine sie enthaltende überführen müßte, andererseits aber approximiert werden würde von Transformationen, welche sie in eine Teilmenge von sich überführten. Also ist auch die Menge $-\varepsilon \cdot T$ um jeden ihrer Punkte überall dicht.

Der Rand ϱ' wird von der Menge $z \cdot T$ nicht erreicht; mithin wird ϱ', nicht aber ϱ, von der Menge $-z \cdot T$ erreicht. Dann aber gehört zu $\mathfrak{F}$ ein gewisses Ringgebiet, in dessen Grenze ϱ und außerdem ein nicht zusammenziehbarer Rand g als Teile enthalten sind. Bezeichnen wir mit g_h den Rand, in den g von einer Transformation $h \cdot T$ (h eine positive oder negative rationale Zahl) übergeführt wird, so kann g_h für positives h nicht zwischen g und ϱ', für negatives h nicht zwischen g und ϱ eindringen, und konvergiert g_h für gegen Null konvergierende Werte von h gleichmäßig gegen g.

Gibt man also h alle rationalen Werte, so bilden die g_h nebst ihren Grenzmengen g_l die Elemente einer gewöhnlichen eindimensionalen Mannigfaltigkeit, welche die ganze Fläche F überdeckt, und von allen Transformationen der Gruppe $w \cdot T$, also auch von allen Transformationen der Gruppe F in sich transformiert wird und zwar eingliedrig.

Die stetige Untergruppe von F, welche jede Menge g_h in sich transformiert, wird in F auf Grund der Überlegungen der §§ 1 und 2 von einer eindimensionalen Punktmannigfaltigkeit durch die Identität repräsentiert; ihre T. B. in F sind getrennt liegende einfache Kurven, welche eine die ganze Fläche F überdeckende invariante Schar bilden. Diese Schar spielt nun genau dieselbe Rolle, wie die Schar der Kurven $\varkappa'$ im § 6, und nach den Methoden jenes Paragraphen zeigen wir wieder, *daß jede endliche Transformation der Gruppe F von einer Infinitesimaltransformation erzeugt wird, und je zwei dieser Infinitesimaltransformationen hinreichen, um die ganze Gruppe F zu erzeugen.*

Zweiter Fall. *Die Menge $z \cdot T$ ist nirgends dicht.* Dann ist auch die Menge $w \cdot T$ nirgends dicht. Letztere Menge muß sich aber jedenfalls aus zusammenhängenden Bestandteilen zusammensetzen. Sonst nämlich könnten wir einen gewissen Punkt $w_1 \cdot T$ mit P, den Punkt $(w_1 + 1) \cdot T$ mit Q, und eine gewisse geschlossene Kurve, welche P in ihrem Inneren, Q in ihrem Äußeren enthält, mit j bezeichnend, eine solche Größe ε bestimmen, daß es unmöglich wäre, eine solche Reihe von zu $w \cdot T$ gehörigen Punkten $P, P', P'' \cdots P^{(n)}$ anzugeben, daß $\varrho(P^{(k)}, P^{(k+1)}) < \varepsilon$ wäre, und $P^{(n)}$ außerhalb von j läge.

Nun muß aber die Menge $\frac{1}{2^n} T$, weil sie zwischen zwei Kurven γ und γ' eingeschlossen ist, mindestens einen Grenzpunkt innerhalb F besitzen; es gibt also sicher eine Transformation $\frac{a}{2^n} T$ (a ungerade), welche innerhalb j jeden Punkt um weniger als ε verrückt. Aus demselben Grunde gibt es eine Transformation $\frac{p}{a^m} T$ (p und a relativ prim) mit dieser Eigenschaft.

Man kann nun $\frac{1}{2^n a^m} T$, also auch T selbst, ersetzen durch eine Folge von Transformationen $\frac{a}{2^n} T$ und $\frac{p}{a^m} T$. Eine solche Folge, auf P angewandt, oder ein gewisser Anfangsteil von ihr besäße aber gerade die Eigenschaft, welche wir, wenn $w \cdot T$ keinen zusammenhängenden Bestandteil enthielte, als unmöglich erkannt haben.

Fügen wir der Gruppe $w \cdot T$ ihre Grenzpunkte hinzu, so erhalten wir eine Untergruppe mit paarweise inversen Transformationen der Gruppe F. Diese muß eine isolierte einfache Kurve $\varkappa$ durch die Identität enthalten, welche von der Gruppe F übergeführt wird in eine invariante Schar von getrennt liegenden, die ganze Fläche F überdeckenden, einfachen Kurven $\varkappa'$, auf welche sich wieder die Überlegungen des § 6 anwenden lassen.

Mithin ist folgender Satz allgemein bewiesen:

Jede Transformation einer zweigliedrigen letzten reduzierten Untergruppe einer Gruppe einer zweidimensionalen Mannigfaltigkeit wird von einer Infinitesimaltransformation erzeugt; irgend zwei Infinitesimaltransformationen erzeugen die ganze Untergruppe; die Untergruppe ist imprimitiv.

In diesem Satze, kombiniert mit dem Resultate des § 4, besitzen wir die Mittel, die Gruppen der zweidimensionalen Mannigfaltigkeiten der
[[6]] Reihe nach aufzuzählen, was ich in der nächsten Mitteilung auszuführen beabsichtige.

NOTES

⟦1⟧ Brouwer 1909 C. This second part was promised to Hilbert on 26 July 1909 for August 1919

⟦2⟧ A. Schoenflies 1908 A.

⟦3⟧ H. A. Lorentz 1904, p. 120–121, footnote 61. Simple curves which are dense in a domain.

⟦4⟧ J. Lüroth 1906 proved this special case of the invariance of dimension.

⟦5⟧ Brouwer 1910 F, 1909 F1, 1909 F2.

⟦6⟧ The publication has not been continued. See 1909 B ⟦1⟧.

1911 K

SUR UNE THÉORIE DE LA MESURE

⟦1⟧

A propos d'un article de M. G. Combebiac.

⟦2⟧ Dans son étude sur une théorie de la mesure publiée dans l'*Enseignement mathématique* du 15 mars 1910, M. G. Combebiac considère les fonctions F (x, y) possédant les propriétés suivantes :

1° F (x, y) est continue et croissante comme fonction de y, continue et décroissante comme fonction de x ; il s'ensuit qu'elle est encore continue comme fonction des deux variables x et y.

2° Les valeurs de F (x, y) et de F (x, z) déterminent la valeur de F (y, z).

En supposant de plus que la fonction F (x, y) possède des dérivées premières continues, M. Combebiac établit qu'elle peut se mettre sous la forme

$$\Phi\left\{ f(y) - f(x) \right\},$$

où Φ et f sont des fonctions continues, croissant avec leur argument.

Je me propose de démontrer ici, comme M. Combebiac le présume, que ce résultat est indépendant de l'existence des dérivées de F (x, y).

1. — Si nous ne considérons des valeurs de x que celles qui sont comprises dans un certain intervalle i_1, et des valeurs de y que celles qui sont comprises dans un certain intervalle i_2, x est une fonction continue de F et de y, croissante comme fonction de y, décroissante comme fonction de F.

En effet, si cette fonction n'était pas continue, on pourrait déterminer une telle suite de valeurs x', x'', x''', ... possédant une seule valeur limite x_l, et une telle suite de valeurs y', y'', y''', ... possédant une seule valeur limite y_l, que

$$\lim F(x^{(n)}, y^{(n)}) = F(x_a, y_l),$$

ou x_a serait une valeur différente de x_l, ce qui est absurde, puisque d'autre part $\lim F(x^{(n)}, y^{(n)})$ doit être égale à $F(x_l, y_l)$.

Cette propriété établie, choisissons deux nombres arbitraires a et b. Il existe un intervalle i_b contenant b, tel que, le nombre β étant arbitrairement choisi dans i_b, on peut déterminer un nombre α satisfaisant l'égalité

$$F(\alpha, \beta) = F(a, b) = \gamma. \tag{1}$$

Soit γ' un nombre variable différant suffisamment peu de γ, et tendant vers γ. Il détermine un nombre β' tendant vers β, et un nombre b' tendant vers b, tels que

$$F(\alpha, \beta') = F(a, b') = \gamma'. \tag{2}$$

Des égalités (1) et (2) nous concluons

$$F(\beta, \beta') = F(b, b'),$$

ou en passant à la limite

$$F(\beta, \beta) = F(b, b)$$

C'est dire que le nombre arbitraire b est contenu dans un intervalle i_b, dans lequel $F(x, x)$ est une constante. *Donc* $F(x, x)$ *est une constante dans tout le continu numérique.* Désignons cette constante par j.

2. — Choisissons arbitrairement deux nombres d_0 et d_1, et déterminons une série de nombres

$$\ldots, d_{-2}, d_{-1}, d_0, d_1, d_2, \ldots$$

se succédant dans leur succession naturelle, et satisfaisant la relation

$$F(d_n, d_{n+1}) = F(d_0, d_1) = \nu.$$

Cette série, prolongée autant que possible de chaque côté, où d'ailleurs elle peut être trouvée finie ou infinie, sera désignée par σ.

Entre d_0 et d_1 il existe un nombre $d_{\frac{1}{2}}$, défini univoquement par la relation

$$F(d_0, d_{\frac{1}{2}}) = F(d_{\frac{1}{2}}, d_1) .$$

Les nombres d_0 et $d_{\frac{1}{2}}$ définissent une série σ' :

$$\ldots, d_{-\frac{3}{2}}, d_{-1}, d_{-\frac{1}{2}}, d_0, d_{\frac{1}{2}}, d_1, d_{\frac{3}{2}}, \ldots$$

contenant les éléments de la série σ, et telle qu'on a pour chaque $d_{\frac{n}{2}}$:

$$F(d_{\frac{n}{2}}, d_{\frac{n+1}{2}}) = F(d_0, d_{\frac{1}{2}}) = \nu' .$$

Si σ ne possède pas de premier élément, σ' n'en possède pas non plus. Si, au contraire, σ possède un premier élément, c'est ou le premier, ou le second élément de σ'.

Si σ ne possède pas de dernier élément, σ' n'en possède pas non plus. Si, au contraire, σ possède un dernier élément, c'est ou le dernier, ou l'avant-dernier élément de σ'.

En opérant sur σ' comme sur σ, on obtient une série σ'' :

$$\ldots, d_{-\frac{3}{2}}, d_{-\frac{5}{4}}, d_{-1}, d_{-\frac{3}{4}}, d_{-\frac{1}{2}}, d_{-\frac{1}{4}}, d_0, d_{\frac{1}{4}}, d_{\frac{1}{2}}, \ldots$$

contenant les éléments de σ', et telle qu'on a pour chaque $d_{\frac{n}{4}}$:

$$F(d_{\frac{n}{4}}, d_{\frac{n+1}{4}}) = F(d_0, d_{\frac{1}{4}}) = \nu'' .$$

En répétant la même opération un nombre infini de fois, on obtient *un ensemble e, composé des nombres* $d_{\frac{n}{2^m}}$ *appartenant à l'ensemble des séries* $\sigma^{(m)}$.

3. — L'ensemble e possédant au moins une valeur limite finie, on peut faire tendre une suite d'intervalles $(d_{\frac{n}{2^m}}, d_{\frac{n+1}{2^m}})$ vers une seule valeur limite finie. Par conséquent $\lim \nu^{(m)} = j$, et toute suite d'intervalles $(d_{\frac{n}{2^m}}, d_{\frac{n+1}{2^m}})$, dont chaque terme fait partie du terme précédent, tend vers une seule valeur limite finie.

De plus, l'ensemble e ne peut pas posséder de limite supérieure l_s, puisque celle-ci entraînerait l'existence d'un nombre l'_s supérieur à l_s et d'un entier positif p tels que $F(l_s, l'_s) = \nu^{(p)}$, de sorte que l'_s serait, comme l_s, point limite de l'ensemble e.

Donc *l'ensemble e est partout dense dans le continu numérique*.

4. — Nous définissons une fonction $f(x)$ de la manière suivante :

Si x est un nombre $d_{\frac{n}{2^m}}$ de l'ensemble e, $f(x)$ sera égal à l'indice $\frac{n}{2^m}$. Si, au contraire, x n'appartient pas à e, toute suite de nombres appartenant à e et tendant vers x, aura la même valeur pour limite des indices, et c'est cette valeur limite que nous assignerons à $f(x)$.

Alors $f(x)$ *est une fonction continue et croissante de* x.

Par conséquent $F(x, y)$ *est une fonction continue de* $f(y)$ *et de* $f(x)$, *croissante comme fonction de* $f(y)$, *décroissante comme fonction de* $f(x)$.

5. — Soient x_1, y_1, x_2, y_2 quatre nombres arbitraires satisfaisant la relation

$$f(y_2) - f(x_2) = f(y_1) - f(x_1) \ . \qquad (3)$$

Soient $x'_1, x''_1, x'''_1, \ldots$; $x'_2, x''_2, x'''_2, \ldots$; $y'_1, y''_1, y'''_1, \ldots$ des suites de nombres appartenant à e, et tendant la première vers x_1, la seconde vers x_2, la troisième vers y_1.

Déterminons $y_2^{(r)}$ de manière que

$$f(y_2^{(r)}) - f(x_2^{(r)}) = f(y_1^{(r)}) - f(x_1^{(r)}) \ . \qquad (4)$$

Alors $y_2^{(r)}$ appartient à e, et la relation (4) entraîne celle-ci :

$$F(x_2^{(r)}, y_2^{(r)}) = F(x_1^{(r)}, y_1^{(r)}) \ .$$

Comme d'autre part la série $y'_2, y''_2, y'''_2, \ldots$ tend vers y_2, on a, en passant à la limite :

$$F(x_2, y_2) = F(x_1, y_1) \ . \qquad (5)$$

Par conséquent l'égalité (3) entraîne l'égalité (5) ; c'est dire que $F(x, y)$ est une fonction de $f(y) - f(x)$ seulement.

En combinant ce résultat avec la propriété déduite dans le § précédent, nous concluons :

$F(x, y)$ *est une fonction continue et croissante de* $f(y) - f(x)$.

C. Q. F. D. L.-E.-J. Brouwer (Amsterdam). [[3]]

NOTES

⟦1⟧ Related in method to 1909 C.
⟦2⟧ G. Combebiac 1910.
⟦3⟧ Brouwer's paper is followed in *Enseignement Math.* by G. Combebiac 1911.

Characterization of the Euclidean and non-Euclidean motion groups in R_n 1909 E

In the development of the non-Euclidean geometries two phases must be distinguished: ⟦1⟧

1° The *construction* of geometries satisfying all ordinary axioms with the exception of the axiom of parallel lines, in other words, the *mathematical creation* of n-dimensional non-Euclidean geometries.

2° The *reduction of the characterization* of these geometries, that is, the process of designing the most restricted system of postulates satisfied by the three geometries under consideration and by no others.

The first process was completed long ago. In 1854, *Riemann* [1]) specified three kinds of arc elements which possess completely free mobility in R_n. For one of them *Klein* [2]) showed two forms of space which possess this free mobility not only for bounded figures but also as a whole. For each of the other two there is only one single spatial form. We therefore find that since 1872 the Euclidean, the hyperbolic, the spherical and the elliptic geometries have existed side by side as well-defined mathematical theories. In defining them, one can start from Riemann's arc element as well as from Cayley's [3]) absolute metric.

The second phase is still topical and will probably remain topical for a long time. A satisfactory system which in its formulations only does not go beyond point set theory was worked out by Hilbert albeit only for plane geometry [4]). An analogous system for R_n still requires – though without any a priori motivation – the existence of certain differential quotients of certain functions.

In this context Riemann [5]) was the first to express the theorem that the *transitive transformation groups in R_n which leave a homogeneous essentially positive quadratic differential form invariant while transforming most generally the line elements at an arbitrary fixed point, are similar to a Euclidean, elliptic or hyperbolic motion group.*

A first proof of this theorem resulted from Christoffel's [6]) and Lipschitz's [7]) investigations by which the homogeneous quadratic differential forms in general were classified according to the transformation groups they admit.

[1]) Habilitationsschrift 1854; published in Göttinger Abhandlungen 1868. ⟦2⟧

[2]) Math. Ann. 6. ⟦3⟧

[3]) 'Sixth memoir upon quantics', Phil. Trans. 149. ⟦4⟧

[4]) Math. Ann. 56. ⟦5⟧

[5]) l.c.; Riemann does not speak of groups, he uses a different language. ⟦2⟧

[6]) J. f. Math. 70. ⟦6⟧

[7]) ibid. 70, 71, 72. In the last treatise it is proved that apart from the Euclidean, the hyperbolic and elliptic arc elements no other arc elements *of constant curvature* are possible. The proof of this fact was simplified by Schur (Math. Ann. 27) and Bianchi (Rend. Linc. 1898). The proof for $n = 2$ had already been given by Minding (J. f. Math. 19). ⟦7⟧

In a different way, i.e. by applying the general theory of continuous groups [8]) the same result was obtained by Lie [9]) and Killing [10]).

The following extension of Riemann's theorem is due to Killing: If one adds to Riemann's suppositions – which as the theorem itself are intended to apply only to certain finite neighbourhoods – that *the transformation group transforms* R_n *as a whole into itself, then there are only four spatial forms possible: the Euclidean,*
⟦10⟧ *the hyperbolic, the spherical and the elliptic space.* [11])

We will give a simple geometric proof of this extension of Riemann's theorem [12]) completely different from existing proofs; it introduces a minimum of suppositions for the functions which define the arc element.

We write the quadratic differential form as

$$\sum f_{pq}(x_1, \ldots, x_n)\,\mathrm{d}x_p\,\mathrm{d}x_q \text{ }^{13})$$

and suppose that the functions f_{pq} are finite and continuous, and have finite and continuous first and second differential quotients.

The '*distance element*'

$$\sqrt{\sum f_{pq}(x_1, \ldots, x_n)\,\mathrm{d}x_p\,\mathrm{d}x_q}$$

determines for every point and for every direction through that point *one geodesic*
⟦11⟧ of which the equation can be written down (see for example Schur (l.c. p. 541)), and which possesses continuous first, second and third differential quotients [14]).

[8]) It should be noted that in the theory of the (real) continuous groups the existence of the first and of the second differential quotients of the transformation functions is supposed; this cannot simply be considered to be equivalent with the analogous suppositions concerning the differentiability of the coefficients of the quadratic differential form as introduced in our text.

⟦8⟧ [9]) Arch. for Math. og Naturvid. X. Later (Leipziger Ber. 1890) Lie succeeded in eliminating several suppositions included in the group-theoretical version of Riemann's theorem.

⟦9⟧ [10]) J. f. Math. 109.

⟦10⟧ [11]) 'Einführung in die Grundlagen der Geometrie' I. Paderborn 1893, p. 313.

[12]) In fact under the same conditions with respect to the arc element, the proof of the non-extended theorem is parallel to that of the extended theorem. We then prove first the existence of a finite domain in which no pair of geodesics intersect twice. The geometry for the plane then follows by infinitesimal
⟦10⟧ arguments, for example according to Killing ('Einführung', pp. 86–89; we must then justify the various cases of intersection and non-intersection as well as the right angles and differential quotients; but this is easily done). For the derivation of the geometry of R_n from this plane geometry we then follow the procedure as described in this text.

[13]) Since in general the connection of R_n must be supposed to be arbitrary we imagine it to be built up of coordinate parallelepipeds overlapping like roof tiles in such a way that two arbitrary points can be joined by continuous transitions within the same parallelepiped, together with a finite number of moves at a certain point from one of the parallelepipeds covering the point to another.

[14]) Finiteness of the second and third differential quotient follows from their representation by fractions whose numerators are finite and denominators cannot vanish because of the definite character of the quadratic differential form.

If we let a fundamental sequence of directions at a point P converge to a limit direction then the corresponding geodesics, extended to a length a, converge uniformly to the geodesic belonging to the limit direction similarly extended to the length a.

We first consider the case $n = 2$ and mark a length λ on all geodesics departing from a point P; we call the endpoints obtained in this way the λ-endpoints of P.

We now let λ increase from 0 and investigate whether there can be a λ-endpoint which coincides either with another λ-endpoint or with a λ'-endpoint obtained earlier ($\lambda' < \lambda$).

Below a certain maximum value of λ this cannot happen except in the following way:

From P two geodesics depart to meet again for the first time at Q, enclosing a two-sided simply connected domain. Let the two geodesics be γ and γ' and their lengths between P and Q be l and l' respectively. If we suppose that $l' < l$, we can use our transformation group to interchange P and Q in such a way that γ remains invariant, γ' becomes γ'' on the other side of γ, and γ' and γ'' are disjoint between P and Q, *have the same length* l', and enclose a two-sided simply connected domain.

Coincidence of a λ-endpoint with a λ'-endpoint ($\lambda' < \lambda$) is therefore impossible, unless two λ'-endpoints had already coincided earlier. Let us now investigate this possibility.

Let α' be the angle of departure of two geodesics whose λ'-endpoints coincide. If this λ' can be made arbitrarily small, no finite $\lim \alpha'$ is possible since for an angle of departure above a finite magnitude no intersection at indefinitely small distances is possible because of the uniform continuity of the directions along the geodesics. On the other hand, if there exists a $\lim \alpha' = 0$ then there are also cases of finite $\lim \alpha'$. Therefore, λ' cannot be arbitrarily small for coinciding λ'-end- [[12]]
points. We can then certainly indicate a finite λ_g such that for $\lambda < \lambda_g$ there is neither coincidence of the two λ-endpoints nor coincidence of a λ-endpoint and a λ'-endpoint ($\lambda' < \lambda$).

The segments of the geodesics departing from P are therefore completely disjoint before the λ_g-endpoints and form a simply connected domain G which contains all geodesic circles of centre P and radius $< \lambda_g$ and whose boundary is the geodesic circle of radius λ_g.

From this it follows immediately that a λ_g-endpoint cannot coincide with a λ-endpoint for which $\lambda < \lambda_g$; if it coincides with another λ_g-endpoint we meet with one of the following two cases:

A. All λ_g-endpoints coincide; the transformation-R_2 is then immediately completed as a simply connected, two-sided, closed surface.

B. One complete turn of the initial direction at P causes the λ_g-endpoints to describe a simple, closed curve κ an integral number of times. Let Q be an arbitrary point of this curve; starting from a fixed geodesic, we follow the pencil of geodesics from P until we reach again a geodesic PQ. Then we have described a simply connected domain, bounded by a simple closed curve consisting of κ and the two geodesics PQ. In R_2, however, no more than *two* such disjoint domains can be connected along κ. For every complete turn of κ the initial direction at P travels through an angle π and the two geodesics PQ form one single geodesic. [15]) Our transformation-R_2 is therefore completed once more, but this time as a simply connected, one-sided, closed surface.

Let us finally consider the case that all λ_g-endpoints are different: the existence of a λ_u-endpoint coinciding with a λ'_u-endpoint ($\lambda'_u < \lambda_u$) at an indefinitely small distance above λ_g is then impossible. If it did exist, λ_u and λ'_u would both converge to λ_g and since the domain G is simply connected, the geodesics at P, extended to their λ_u-endpoints, would finally lie inside a certain simply connected domain and, by our earlier proof which remains valid, such λ_u-endpoints which approach the λ_g circle indefinitely, would coincide for *equal* values λ_u. For each of these values λ_u, the λ_u-endpoints, as a function of the initial direction, would show an angular period α_u of the initial direction, which is an integral fraction of 2π.

However, as λ_u approaches λ_g $\lim \alpha_u$ cannot vanish; otherwise the geodesic circles of radius λ_u would become indefinitely small and approach a single point which would appear as a geodesic circle of radius λ_g; this contradicts our supposition.

But the existence of a finite limit of α_u is equally impossible, because this limit would be an angular period for λ_g-endpoints, and this again contradicts our supposition.

Therefore, no maximum value can exist for which all λ-endpoints are different, whereas a minimum value of λ, for which not all λ-endpoints are different, leads to the cases A and B described above. Therefore, apart from these two cases, there is only one possibility left.

C. The geodesics leaving one and the same point remain otherwise disjoint and are of infinite length. The transformation R_2 in this case becomes a simply connected, two-sided and infinite two-dimensional domain.

In case A every point has an anti-point. R_2 is in this way divided into pairs of points. If we introduce in this R'_2 of pairs of points the same arc element as is given in the R_2 of single points (where the arc element will be the same for anti-points if these are covered by equal coordinate parallelograms), R'_2 is in the case B. Conversely, we may reduce case B to case A by a suitable duplication of R_2. We can therefore consider case A and case B together as one '*first geometry*'; especially considering case B we can state the following axioms:

1) *Two different points determine one and only one 'straight line' which is a simple closed curve through these points.*

[15]) They must also form an angle π at Q; otherwise two geodesics whose initial directions at P formed an angle $\gtrless \pi$ would meet at their λ_g-endpoints.

2) *The transformation-R_2 is a simply connected one-sided closed surface with unilateral straight lines.*
3) *All straight lines and pencils of straight lines are made measurable by finite scales so that any two points have a 'distance' and any two straight lines an 'angle', both determined up to a constant of periodicity; this constant of periodicity is the same for all straight lines and again the same for all pencils.*
4) *Of two triangles which have two sides and the included angle equal, all elements are equal.*

From these axioms elliptic geometry can be developed synthetically without any infinitesimal considerations.[16])
For the '*second geometry*', peculiar to case C, the following axioms hold:
1. *Two different points determine one and only one 'straight line', which is a simple open curve through those points.*
2. *The transformation-R_2 is a simply connected two-sided domain, divided into two subdomains by any straight line.*
3. *All straight lines are made measurable by infinite scales and all pencils of straight lines by finite scales so that every two points possess a uniquely determined 'distance' and every two intersecting lines an 'angle' which is determined up to a constant of periodicity which is the same for all pencils.*
4. *Of two triangles which have two sides equal and the included angle – in the same sense of succession – all elements are equal.*

From these axioms we arrive by synthetic methods either at Euclidean or hyperbolic geometry [17]).

We now move to the case $n > 2$. Again we let a bundle of geodesics leave a point P, and we investigate whether it is possible for two of these geodesics γ and γ', whose initial directions form an angle $\phi \gtrless \pi$, to intersect again at a point Q after completing the lengths l and l' respectively.

In that case we can – by means of our transformation group – move γ' continuously while keeping γ invariant. In this way we obtain from γ' a set of geodesics which form with γ' all possible angles below a certain maximum, all passing through P and Q, and all of length l' between P and Q. It follows that all geodesics from P, after completing a length l', will reach Q.

If there exist different lengths after which all geodesics from P meet again in one and the same point, then there must of course exist a minimum length μ, and with it a point P′ where all μ-endpoints coincide. However, the relation between P and P′ is then reciprocal and the geodesics between P and P′, which are other-

[16]) Cp. e.g. Hessenberg in Math. Ann. 61, p. 173. [[13]]

[17]) Cp. Hilbert in Math. Ann. 56, p. 421; one should then insert a fairly simple proof that the validity [[14]]
of the Euclidean axiom for one point with respect to one line implies general validity.

We may, of course, after laying down the axioms, use infinitesimal arguments which will then on the basis of these axioms lead to the geometry not just of a limited domain but of the whole of R_2.

wise disjoint, already complete the transformation-R_n into a closed R_n with the connectedness of the sphere.

If intersection of two geodesics from P only takes place for $\phi = \pi$, then these two geodesics form a single geodesic, and the geodesics through P are closed curves which have no points in common except P and whose transformation space in this way becomes a closed R_n with the connectedness of a projective space.

If neither of these cases occurs (and as for R_2 each of these two cases can be reduced to the other by a similar argument), then the geodesics leaving point P are otherwise completely disjoint and are of infinite length; their transformation space becomes a simply connected, two-sided, infinite n-dimensional domain.

In all cases the bundle of geodesics leaving P determines at every point (except at the anti-point of P, if it exists) one single geodesic with first differential quotients, which (as the second and third differential quotients) are continuous functions of the coordinates.

By variational calculus it can further be easily confirmed that from our suppositions regarding the f_{pq} the following properties follow:

Every geodesic is also a *shortest* line, at least up to a certain length l_m (which is the same for all geodesics).

The variation of a point on a geodesic sphere is differentiable with respect to the variation of the direction of the radius vector at the centre. The differential quotients here are continuous functions of the coordinates.

A geodesic sphere, therefore, has at every point a certain $(n-1)$-dimensional direction which is orthogonal to the radius. This $(n-1)$-dimensional direction, determined at every point by the geodesic spheres around P, is a continuous function of the coordinates.

If a certain geodesic q through P is chosen as '*axis*', it divides the bundle of initial directions into ∞^{n-2}meridian planes and also into ∞ parallel cones (each of which forms a certain angle with the axis); it also fixes a certain pole on an arbitrary geodesic sphere around P and divides this sphere into ∞^{n-2}meridional circles and ∞ parallel-$^{n-2}$spheres of which each cuts equal pieces off all meridional circles. Furthermore, these meridians and parallels have at every point a fixed direction: these two directions must therefore [18]) be orthogonal.

Let A be an arbitrary point on the axis a, and B an arbitrary point at a distance $< l_m$ from A; then the shortest geodesic γ_{AB} between A and B must lie in the meridional plane of B.

[18]) Because of the free mobility, every linear half-direction contained in the $(n-2)$-dimensional direction of the parallel must form the same angle with the direction of the meridian. (Moreover, we should notice that for $n = 3$, the anti-figures on a geodesic sphere have the same metric relations.)

[[190]]

To show this we consider every point of R_n as determined by the length r of its radius vector from P, the angle ϕ between this radius vector and a, and further $n-2$ coordinates which are constant for every meridional plane.

If γ_{AB} were not situated in the meridional plane of B we should be able to assign to anyone of its points a point of this meridional plane with the same coordinates r and ϕ. In this way a curve γ'_{AB} would be formed, which – like γ_{AB} – at every point would have a certain direction varying continuously from point to point. To every element ds of γ_{AB} there would correspond an element ds' of γ'_{AB}. If ds lies in a tangential plane then ds and ds' are equal; if not, ds can be split into a component in the meridional plane equal to ds', and a component *orthogonal to the first*, so that ds is larger than ds'.

If, therefore, γ_{AB} were not situated in the meridional plane of B, it would not be the shortest connection between A and B, which contradicts our supposition.

Therefore, if we call the sets of geodesic lines departing from P and determined by a pencil of initial directions *the geodesic planes* through P, we have the result that an *arbitrary geodesic line in* R_n *lies in a geodesic plane through* P; further:

Three arbitrary points of R_n, *which are not on a geodesic line, determine one geodesic plane with the property that every geodesic line which has any two points (which are not anti-points) in common with the geodesic plane, lies entirely in this plane.*

Similarly, we can show more generally that $p(\leqq n)$ arbitrary points of R_n, not contained in a geodesic R_{p-2}, determine a geodesic R_{p-1} with the property that every geodesic line which has two points in common with R_{p-1} (which are not anti-points) lies entirely in this R_{p-1}. This confirms the validity of projective geometry, and together with the congruence theorems it can be used to show the exclusive possibility of the four known forms of space.

To do so we notice that the transformation group of R_n has a subgroup which transforms a given geodesic plane according to one of the four known plane geometries [19]). If we denote by $f(r)d\phi$ the arc element between the r-endpoints of two geodesics parting from P with an angle $d\phi$, there are then three possibilities:

I. $f(r) = \dfrac{1}{k}\sin kr.$

II. $f(r) = r$ ⟦15⟧

III. $f(r) = \dfrac{1}{k}\sinh kr$

[19]) It is easier to derive these plane geometries synthetically as parts of R_n-geometries than to construct them in isolation (as above), since Desargues' Theorem is then guaranteed a priori.

One of these three formulae must in general hold for R_n. Formula I produces either the spherical geometry with distances π/k between every two anti-points, or the elliptic geometry with lengths π/k for all geodesics. Formulae II and III produce the Euclidean and the hyperbolic geometry, respectively.

NOTES

⟦1⟧ Translated from the original 'Karakterisering der Euclidische en niet-Euclidische bewegingsgroepen in R_n'.
⟦2⟧ B. Riemann 1868.
⟦3⟧ F. Klein 1873.
⟦4⟧ A. Cayley 1859.
⟦5⟧ D. Hilbert 1902.
⟦6⟧ E. B. Christoffel 1869 A, B.
⟦7⟧ R. Lipschitz 1869, 1870 A, 1870 B, F. Schur 1886, L. Bianchi 1898, F. Minding 1839.
⟦8⟧ S. Lie 1885, 1890.
⟦9⟧ W. Killing 1892.
⟦10⟧ W. Killing 1893.
⟦11⟧ F. Schur 1886.
⟦12⟧ This is not obvious. Probably some argument from group theory was meant.
⟦13⟧ G. Hessenberg 1905.
⟦14⟧ D. Hilbert 1902.
⟦15⟧ Corrections in Brouwer's hand have been made in the formulae.

CHAPTER 3

Toward the plane translation theorem

Mathematics. — "*Continuous one-one transformations of surfaces in themselves.*" By Mr. L. E. J. BROUWER. (Communicated Prof. D. J. KORTEWEG). 1909 F2 [[1]]

(Communicated in the meeting of February 27, 1909).

We shall treat in this paper an arbitrary surface, which in the sense of analysis situs is equivalent to a sphere, i. o. w. which is a continuous one-one image of a sphere. We shall submit that surface to an entirely arbitrary continuous one-one transformation in itself and we shall investigate whether this is possible without at least one point remaining in its place. [[2]]

To simplify we shall consider a continuous one-one image of the surface in a Cartesian plane; the infinity of it then of course forms an exception, because there the continuity of the correspondence is disturbed, and because it must answer as a whole to a single point; for that point we choose a point not remaining invariant in the transformation.

In the Cartesian plane we indicate the points of the untransformed figure by letters without a dash, the corresponding point of the transformed figure by equal letters with a dash. In particular we shall indicate infinity by H and K'.

We now construct in the untransformed figure the system of the circles around K as their centre. These closed curves k lie outside each other, expand from K to H and to each of those curves there is *uniform convergence* [1]) by the preceding ones as well as by the succeeding ones.

In the transformed figure to this answers a system of *single closed curves* [2]) k', which, lying inside each other, contract from K' to H', whilst there is again uniform convergence to each of the curves k' by the preceding ones as well as by the succeeding ones.

Let us now regard, starting from K and stopping in H, each curve k_α together with the corresponding curve k'_α; here α represents the radius of the circle k_α.

We shall assume that the transformation in the surface leaves the indicatrix invariant; then in k and k' opposite senses of circuit correspond to each other.

For values of α under a certain limit α_1 the curve k lies inside k'; beyond a certain value α_F however the curve k' lies continually inside k. Between α_1 and α_F there can be values for which the inner domains of k and k' lie entirely outside each other; these then have an upper limit α_2, situated between α_1 and α_F.

Between a certain nethermost limit α_0 (which with α_2 existing coincides with α_2, otherwise with α_1) and an upper limit α_F, the curves k and k' must continually penetrate each other.

If α_0 and α_F coincide, k_{α_0} and k'_{α_0} do so too, and, opposite senses of circuit corresponding for the transformation on this curve, we find *two points invariant for the transformation.*

We now assume that α_0 and α_F do not coincide. Then k_{α_0} and k'_{α_0} touch each other in one or more points or arcs, without penetrating each other, whilst for the rest they lie either outside each other, or k_{α_0} lies entirely inside k'_{α_0}. In the former case one domain is determined lying outside k_{α_0} and inside k'_{α_0}, in the latter case one or more domains.

We can now choose among the domains determined by k_α and k'_α in the following way for a certain segment of values α, starting at α_0 and stopping at α_F, every time a domain γ_α lying outside k_α and inside k'_α, in such a way, that to the boundary of each γ_α inside there is uniform convergence by the boundaries of the succeeding γ_α's, so that the domain γ_α continually contracts until at α_F it vanishes and its boundary is reduced to a point or arc of single curve.

To this end we choose in γ_{α_0} an arbitrary point, and fartheron we choose γ_α in such a way that this point lies inside it; this will be possible up to a certain value α'; we then stop at $^1/_2\,(\alpha_0 + \alpha') = \alpha''$. Next we choose inside the domain $\gamma_{\alpha''}$, determined in this way, an

[3] 1) SCHOENFLIES, Jahresber. d. D. M. V. XV, p. 560.
[4] 2) id., Mathem. Ann. 62, p. 305.

arbitrary point and for values of α, consecutive to α'', we determine γ_α in such a way, that the latter point lies inside it; if this is possible up to $\gamma^{(3)}$, we stop at $^1/_2\,(\alpha'' + \alpha^{(3)}) = \alpha^{(4)}$, and we continue this process in the same way until it leads after a denumerable number of steps to an end.

This series of γ_α's can in general be chosen in different ways; the value α_F, where it ends, lies then either *before* α_F, or it coincides with α_F.

We shall now investigate such a series of domains γ_α more closely and we shall suppose that from α_0 as far as and inclusive of α_F no invariant point is situated on the curve k_α resp. k'_α.

If we describe the boundary of a domain γ_α in *opposite direction* of the hands we find it consisting alternately of arcs and points belonging to both k_α and to k'_α and which we shall call *dividing arcs* (which are of course closed sets of points) resp. *dividing points*, of arcs belonging only to k_α (being not closed sets of points) which we shall call *inner arcs of* k_α, and of curves belonging only to k'_α, which we shall call *inner arcs of* k'_α. To the above mentioned circuit of γ_α corresponds an order of succession of the inner arcs of k_α, belonging to a circuit of k_α *with* the hands, and an order of succession of the inner arcs of k'_α, belonging to a circuit of k'_α in *opposite direction* of the hands.

The part of k_α resp. k'_α, not belonging to the dividing arcs, dividing

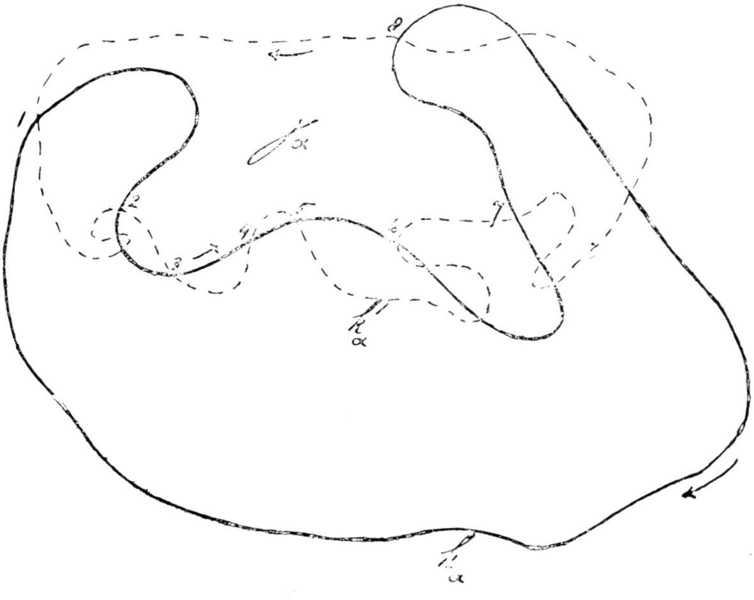

Fig. 1.

points, and inner arcs, consists of arcs (not closed sets of points), which we shall call *outer arcs of* k_α *resp.* k'_α. Between the end-points of an inner arc of k_α runs an outer arc of k'_α, and inversely.

Each inner arc of k_α encloses with the corresponding outer arc of k'_α a domain lying inside k_α and inside k'_α, thus outside γ_α.

Each inner arc of k'_α encloses with the corresponding outer arc of k_α a domain lying outside k_α and outside k'_α, thus likewise outside γ_α. To this however *one* inner arc of k'_α forms an exception: it encloses with the corresponding outer arc of k_α a domain containing the inner domain of k_α and also the domain γ_α.

We shall now run along the boundary of γ_α in opposite direction of the hands, starting somewhere on that *special* inner arc of k'_α: in this way the row of dividing points and dividing arcs, or, as we shall call it, the *row of elements* of γ_α gets a first and a last element.

Accordingly in fig. 1. (which is special in as far as the elements appear only as dividing points and only in finite number) the elements are numbered from 1 to 8.

We look upon all those dividing points and dividing arcs as elements of k_α and we determine their images on k'_α. After that we suppose each outer arc to be laid along the corresponding inner arc by a continuous one-one representation; in this way all the images of the elements of γ find their places on its boundary and we investigate for each element of γ_α, whether when describing the boundary of γ_α in the opposite direction of the hands we find it gaining or losing on its image on k'_α [1]). (That a dividing arc as a whole gains or as a whole loses, follows from the absence of an invariant point). To an element of the first species we assign the sign d; to one of the second species the sign p. These signs are unequivocally determined, except in the case, that all the elements should have to have the same sign, for which we can then take either p or d.

We divide this row of elements into a succession of groups as extensive as possible, containing each only equal signs; then before the first and after the last element of each group k_α and k'_α lie outside each other during arcs, whose extension does not fall below a certain minimum.

If such a group lies between two inner arcs of the same curve,

[1]) True, the difference in argument between an element and its image is determined only save an entire number of circuits; however a fixed choice for one of the elements includes a fixed choice for all. We can arrange that difference to be for all elements in absolute value smaller than a circuit; this is possible either in *one* or in *two* ways. In the first case is determined unequivocally, which elements gain on their images and which lose. In the second case they may be regarded arbitrarily as all gaining or as all losing on their images.

we call it an *even* group, and we represent it by two signs; if it lies between two inner arcs of different curves, we call it an *odd* group, which we represent by one sign.

The number of the groups must of necessity be finite; thus belongs to each α a *row of signs* containing an *even* number of signs p or d, of which not more than two equal ones follow on each other.

Let us now first consider a value α_1, for which the curve k'_{α_1} possesses in the immediate vicinity of each of the dividing points and dividing arcs points on both sides of k_{α_1}. On either side of such a value α_1 there is a segment of values of α, to each of which belongs the same row of signs as to α_1. For when α converges to α_1, the corresponding set of dividing points and dividing arcs converges uniformly to the *whole* set of elements belonging to α_1.

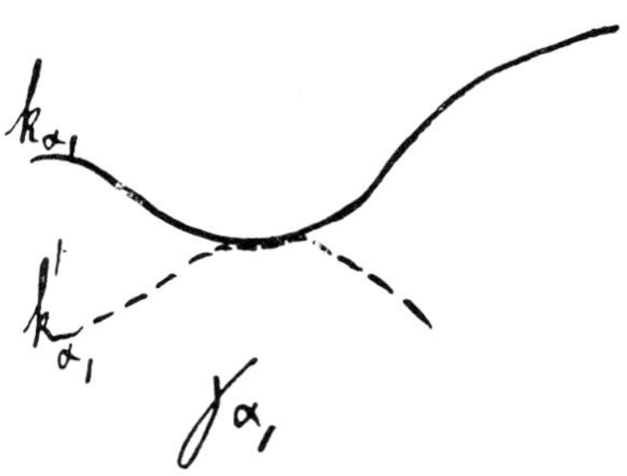

Fig. 2.

Let us next regard values α_1, where in the immediate vicinity of the elements which do not possess the just-mentioned property, the boundary of γ_{α_1} belongs exclusively to k_{α_1} or exclusively to k'_{α_1} (see fig. 2). For values α differing sufficiently little from that α_1 the corresponding set of elements uniformly differs indefinitely little from a *part* of the set of elements corresponding to α_1 (which *part* can be different for different α's). However, in such a part only even groups can be non-represented, and farther even groups belonging to that part are approximated by even groups and odd ones by odd ones.

Thus for the two regarded species of values α_1 within a certain vicinity each row of signs is obtained out of the row of α_1 by suppression of a certain number of pairs of equal successive signs.

We shall now understand by the *reduced row* of a given row of signs that one, which is obtained out of it by checking off a pair of equal successive signs, and repeating this process, until it is no longer possible. [1])

Thus e.g. of $\quad p\ d\ p\ d\ d\ p\ d\ p\ p$

the reduced row of signs is $\quad p$.

---

[1]) That this reduced row of signs is determined unequivocally is evident e.g. as follows: Let in a plane with rectangular system of coordinates p mean: *semi-revolution about the point* $(+1{,}0)$, and d: *semi-revolution about the point* $(-1{,}0)$, and let an arbitrary row of signs represent the product of the operations indicated by the signs. Different reduced rows of sign then furnish different results, and an arbitrary row of signs is equivalent to its reduced one.

We can now resume our results concerning the two regarded species of values α_1 as follows:

For each of these values α_1 *the reduced row of signs is invariant within a certain vicinity.*

We have now still to investigate a third species of values α_1, namely those for which dividing points appear, in whose immediate vicinity k_{α_1} and k'_{α_1} do not intersect each other, whilst the boundary of γ_{α_1} belongs on one side of that dividing point to k_{α_1}, on the other side to k'_{α_1}.

Such a dividing point may be endpoint of a *recurrent arc*, i. e. of an arc along which k_{α_1} and k'_{α_1} coincide, but in a direction, which corresponds for one of the two, but not for the other, to the sense of circuit of γ_{α_1}.

We shall now confine ourselves first to such values α_1 of the third species, as for all dividing points characterizing them as such possess that recurrent arc, and that in such a manner that in the immediate vicinity of it k_{α_1} and k'_{α_1} come on either side of each other (comp. fig. 3, where PQ is the recurrent arc and to the left of P the curves k_{α_1} and k'_{α_1} can be supposed to wind an infinite number of times round about each other, when approaching to P).

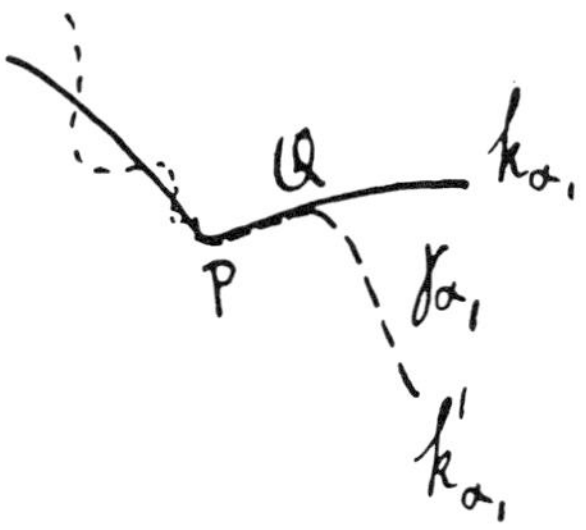

Fig. 3.

For values α differing sufficiently little from α_1 the corresponding set of elements uniformly differs indefinitely little from the sum of a *part* (with the same properties as for values α_1 of the second species) of the set of elements corresponding to α_1 and a *part* of the recurrent arc. On account of the absence of an invariant point on that arc here again the reduced row of signs must within a certain vicinity of α_1 be the same as that of α_1.

Fig. 4.

There remain still the values α_1 of the third species having dividing points characterizing them as such, which either possess *no* recurrent arc, or such a one that in its immediate vicinity the curves k_{α_1} and k'_{α_1} lie only on one side of each other (See fig. 4).

We here see clearly, that the proof for the invariability of the reduced row of signs *does* hold for a segment of values α *following* on α_1, but not for one preceding α_1.

We can however regard in this case of exception the domain γ_{α_1} as belonging to a *set of domains* G_{α_1}, whose character is that of a domain γ_α, not being in that case of exception, on whose boundary in one or more points or arcs an inner arc of k_α and an inner arc of k'_α touch each other without intersecting each other. An example is given in fig. 5, where γ_{α_1} with γ'_{α_1} and γ''_{α_1} composes the indicated set of domains G_{α_1}.

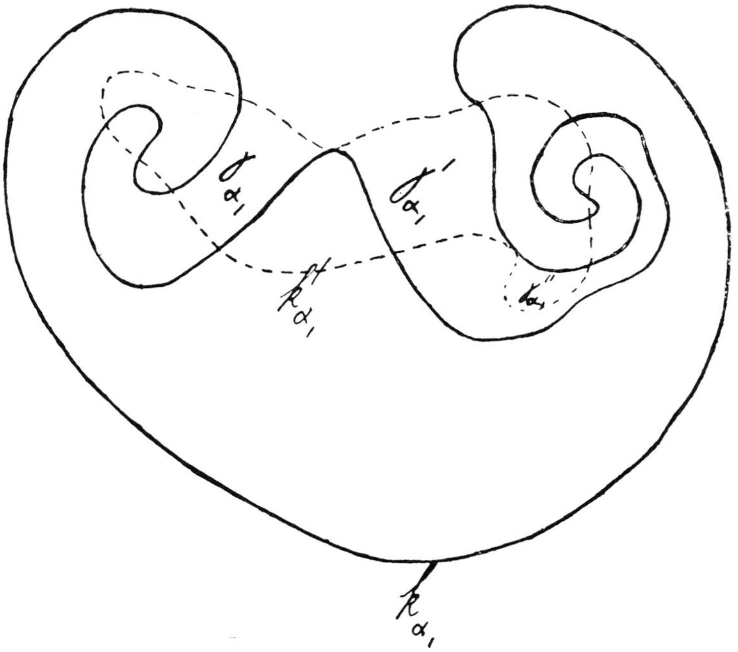

Fig. 5.

In the same way as for a domain γ_α we can define for a G_α the row of signs and the reduced row of signs.

For values of a preceding a_1 and converging to a_1 the boundary of γ_α converges uniformly to that of G_{α_1}, and *for a certain segment of values a preceding a_1 the reduced row of signs of γ_α is the same as that corresponding to G_{α_1}.*

The end we have in view is to prove that we can choose the successive domains γ_α in such a way, that for not one of those domains the reduced row of signs becomes *p d.*

If k_{α_0} and k'_{α_0} strike against each other on their outside, then we have for γ_{α_0} only one choice, for which the row of signs is *pp* (or *d d*) and the reduced row of signs *zero,* thus *not p d.*

If k_{α_0} lies entirely on or inside k'_{α_0}, then there is among the choices possible for γ_{α_0} certainly one whose row of signs is either *p p* or *dp*, and the reduced row of signs either *zero* or *d p*, so again *not p d*.

We can thus at all events take care that for γ_{α_0} the reduced row of signs is *not p d*.

We now continue the succession of γ_α's in an arbitrary way according to the method given above, until after a certain finite segment of values α a first α appears, for which the reduced row of signs changes. Then we have certainly there a value α which is in the case of exception, and which we call α_{u_1}. We there have a $G_{\alpha_{u_1}}$, which breaks up into a set of domains $\gamma_{\alpha_{u_1}}$, and with each of these we may continue the succession of the γ_α's. We know that the reduced row of signs of $G_{\alpha_{u_1}}$ is *not pd*, and we shall show how from this ensues that for at least *one* of the $\gamma_{\alpha_{u_1}}$'s the reduced row of signs is *not p d*.

As a breaking up of G_α into several γ_α's by contact points or contact arcs can always be reduced to divisions into *two* γ_α's by contact in a single point, we have but to show that if in the latter case the two composing domains γ_α possess the reduced row of signs *pd*, G_α must also have that same reduced row of signs.

Let to that end in fig. 6 R be the point of contact which makes the division; let I be the composing domain whose boundary with the assumed sense of circuit passes in R from k_α to k'_α; then in II it passes in R from k'_α to k_α.

On the boundary of I two elements P_1 and P_2 can now be indicated, whose images on k'_α lie between P_1 and P_2.

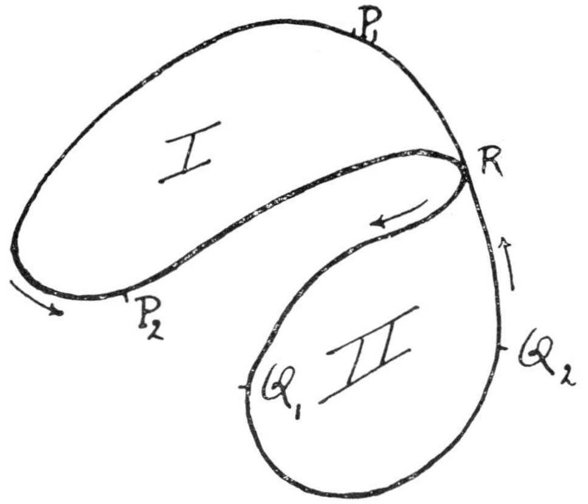

Fig. 6.

Likewise on the boundary of II we can choose two elements Q_1 and Q_2 whose images on k'_α lie between Q_1 and Q_2.

The element P_1 cannot coincide with R; for then it would be evident when describing the boundary of G_α as a whole, that all the elements of II had their images on k'_α on the same outer arc of II.

Neither can Q_2 coincide with R.

Now the image of R lies on k'_α between P_1 and Q_2, thus within the arc $P_1 R$ as well as within the arc $R Q_2$ of k'_α. So R gets for the domain I the sign d and for the domain II the sign p.

We must now distinguish three cases: *the special inner arc of* k'_α namely appears as such either in I or in II or (if it contains R) in neither of the two. We shall continue the proof only for the first case; the other two can be treated analogously.

By the point R the row of signs of G_α is broken up into three successive parts, which we shall call ϱ, σ and τ; then the row of signs of I becomes

$$\varrho \; d \; \tau,$$

and that of II

$$p \; \sigma.$$

The reduction of the row of signs of II, i.e. of $p\,\sigma$, having to give $p\,d$, the reduction of σ must give d.

Thus by substituting in an arbitrary row of signs a σ for a d, its reduced row of signs remains unchanged.

However, pd being the reduction of the row of signs of I, i.e. of $\varrho\, d\, \tau$, it is also the reduction of $\varrho\, \sigma\, \tau$, i.e. of the row of signs of G_α, the property which we had to demonstrate.

So we can choose $\gamma_{\alpha_{u_1}}$ in such a way, that its reduced row of signs is *not* pd, and, starting from that, we can continue the succession of γ_α's arbitrarily until in this way after a certain segment of values α the row of signs would change for the first time for α_{u_2}. However, we can then again choose $\gamma_{\alpha_{u_2}}$ in such a way that its row of signs is *not* pd, and we can keep going on in this way. If an α_{u_ω} is reached as the limit of the α_u's, then if the case of exception appears also for that, it can be treated similarly; we can choose $\gamma_{\alpha_{u_\omega}}$ thus, that its reduced row of signs is *not* pd, and we can continue the succession of γ_α's with this property till a value $\alpha_{u_{\omega+1}}$ is reached. The whole set of α_u's must, however, be denumerable, and finally α_F is reached with a γ_{α_F}, whose reduced row of signs is *not* $p\,d$, thus whose row of signs itself is *not* $p\,d$.

On the other hand: of a γ_{α_F} on whose boundary lies no invariant point, the row of signs must precisely be pd. Thus the assumption, that from α_0 up to and inclusive of α_F no invariant point appears, leads to an inconsistency, and we have proved:

THEOREM 1. *A continuous one-one transformation in itself with invariant indicatrix of a singly connected, twosided, closed surface*
[[2]] *possesses at least one invariant point.*

The above given proof fails for a transformation with inversion of the indicatrix; on the contrary, according to that proof we can show how such transformations may easily be constructed without invariant point.

For then the curves k_α and k'_α have the same sense of circuit, and with the aid of a method of representation of SCHOENFLIES [1]) we can arrange them to possess for all values α between α_0 and α_F only two points of intersection whilst originally k_{α_0} lies inside k'_{α_0}. Let us now regard the part of the boundary of γ_α belonging to k_α, then fig. 7 shows how the image of that arc can begin to

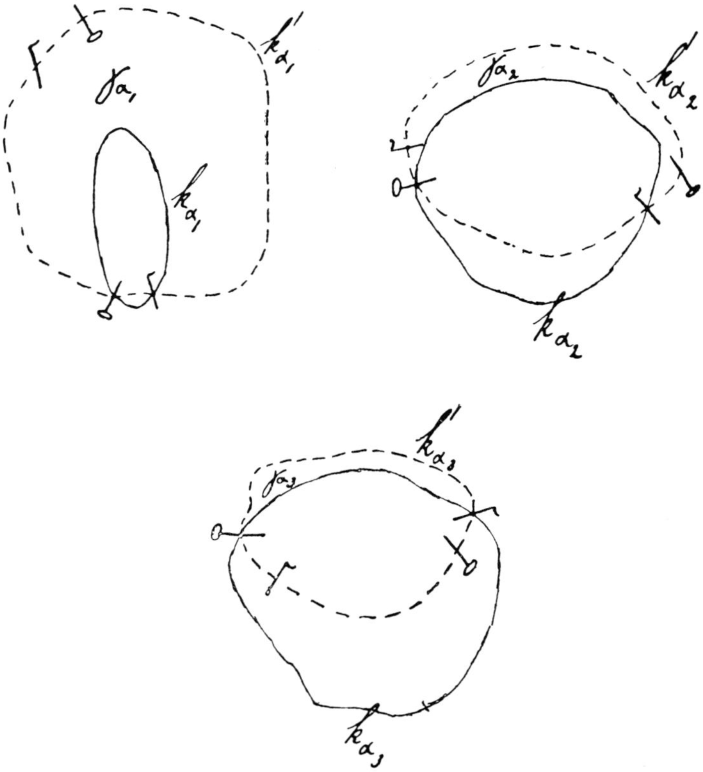

Fig. 7.

[[4]] 1) Mathem. Ann. 62, p. 319—324.

have its endpoints on the boundary of $\gamma_\varkappa$, but how it can lateron withdraw its endpoints from that boundary and fartheron remain entirely outside it, without an invariant point having to appear.

We thus formulate:

THEOREM 2. *A continuous one-one transformation in itself with inversion of the indicatrix of a singly connected, twosided, closed surface does not necessarily leave a point invariant*[1]).

An elementary special case of theorems 1 and 2 is furnished by a sphere in ordinary space, having always two invariant points for congruent transformations in itself, but not necessarily one for symmetric transformations in itself.

In the formulation of theorem 1 the restriction implied in the word *closed* is not superfluous; for, an ordinary Cartesian plane has in an arbitrary translation a continuous one-one transformation in itself with invariant indicatrix, *without an invariant point.*

Neither is superfluous the condition of *single connection*; for the ordinary tore of Euclidean space has in an arbitrary rotation about its axis a continuous one-one transformation in itself with invariant indicatrix *without an invariant point.*

The restriction of *twosidedness,* however, can be cancelled.

We can, namely, bring the points of a onesided, singly connected, closed surface into a continuous one-two correspondence with those of a twosided one; to an indicatrix on the onesided surface then correspond two opposite indicatrices on the twosided one. On the ground of such a correspondence answer to a continuous one-one transformation in itself of the onesided surface two such transformations in itself of the twosided one, so that for one of them the indicatrix remains invariant, whilst it is inverted for the other one. As for the former at least one point remains invariant, for both at least one pair of points and for the transformation of the onesided surface at least one point remains invariant.

As farthermore for that transformation of that onesided surface we cannot speak of remaining invariant or inversion of the indicatrix, we can formulate:

THEOREM 3. *A continuous one-one transformation in itself of a singly connected, onesided, closed surface leaves at least one point invariant.* [[5]]

An elementary special case of theorem 3 we find in an arbitrary plane, real-projective transformation having at least one invariant point.

[1]) For a closed *line* of theorems 1 and 2 exactly the contrary holds: here a transformation with invariant indicatrix can exist without an invariant point; but with inverted indicatrix such a transformation is impossible.

NOTES

⟦1⟧ The English version appeared in the Proceedings of the meeting of 27 March 1909. The present paper starts a series of eight with the same title published from 1909 to 1920. The first four *papers* 1909 F2, 1909 H2, 1911 H2, 1911 J2 constitute an abortive attempt at the plane translation theorem. A summary of these papers is 1910 F. The final paper on the plane translation theorem is 1912 B; an additional remark is found in 1919 M2.

In method these papers are point-set topology of the plane – Cantor–Schoenflies concepts and methods, although highly sophisticated. The second half of the above-mentioned series of eight are of quite another kind – set theory topology mixed with homological and homotopical topology such as developed by Brouwer himself. See chapter 7.

Comments on the translation theorem are made in 1909 H2 ⟦1⟧.

⟦2⟧ The particular aim of the present paper is the fixed point theorem for topological, orientation preserving mappings of the 2-sphere into itself. The approach is through point-set topology. Kerékjártó 1919 A, using the same kind of approach, simplified the proof. Brouwer 1919 J showed that this fixed point theorem does not hold for closed two-sided surfaces of genus > 1. J. Nielsen 1920 A, extended this to closed one-sided surfaces of genus > 1.

The 'true' Brouwer fixed point theorems, proved with more efficient methods, are found in 1911 D.

⟦3⟧ A. Schoenflies 1906 B.

⟦4⟧ A. Schoenflies 1906 A.

⟦5⟧ A proof of a more general theorem is to be found in Brouwer 1926 D2.

1909 H2

Mathematics. — "*Continuous one-one transformations of surfaces in themselves.*" (2nd communication).[1]) By Dr. L. E. J. BROUWER. (Communicated by Prof. D. J. KORTEWEG.) [[1]]

(Communicated in the meeting of June 26, 1909).

We shall now consider an arbitrary twosided[2]) surface and we shall submit it to an arbitrary continuous one-one transformation in itself with invariant indicatrix. [[2]]

Under a *limit region* of the transformation we shall understand a region of the surface lying entirely outside its image region, but losing that property by any extension.

However it may happen that a limit region allows of *enlargement*, i.e. can be united after an indefinitely small modification of its boundary with a finite adjacent region into a new limit region, whose surface, measured by a certain system of coordinates, is then of course greater than the old one's. This will be illustrated by the following developments.

Under a *transformation domain* we shall understand a limit region not capable of enlargement, and our intention is to construct such a transformation domain.

To this end we start from two arcs of simple curve[3]), which are each other's image, which have two and not more than two points in common, and which do not cross each other in those points. We shall suppose, that these arcs have no endpoint in common; their situation with respect to each other then still allows of various possibilities, indicated by fig. 1.

With the aid of these two arcs we now construct two regions G and G', bounded by simple closed curves, which regions are each other's image and lie entirely outside each other, whilst their boundaries have two arcs of simple curve in common. In fig. 2 this has been executed for the second possibility indicated by fig. 1.[4])

1) See these *Proceedings*, Vol. XI, page 788.

2) A onesided surface falls under our result only when brought into a continuous multivalent correspondence with a twosided one.

3) i.e. "einfache Kurvenbogen" after SCHÖNFLIES. In my preceding communications of this year (these *Proceedings* XI, p. 788, 850) I translated "einfach" by "single". Finding the term "simple closed curve" used by VEBLEN, I shall adopt in future this mode of expression. [[3]]

4) Only in the fifth and eighth case of fig. 1 this might give rise to some difficulty, namely if the two common points of the arcs are each other's image. By a slight modification of the figure this difficulty can then be cancelled.

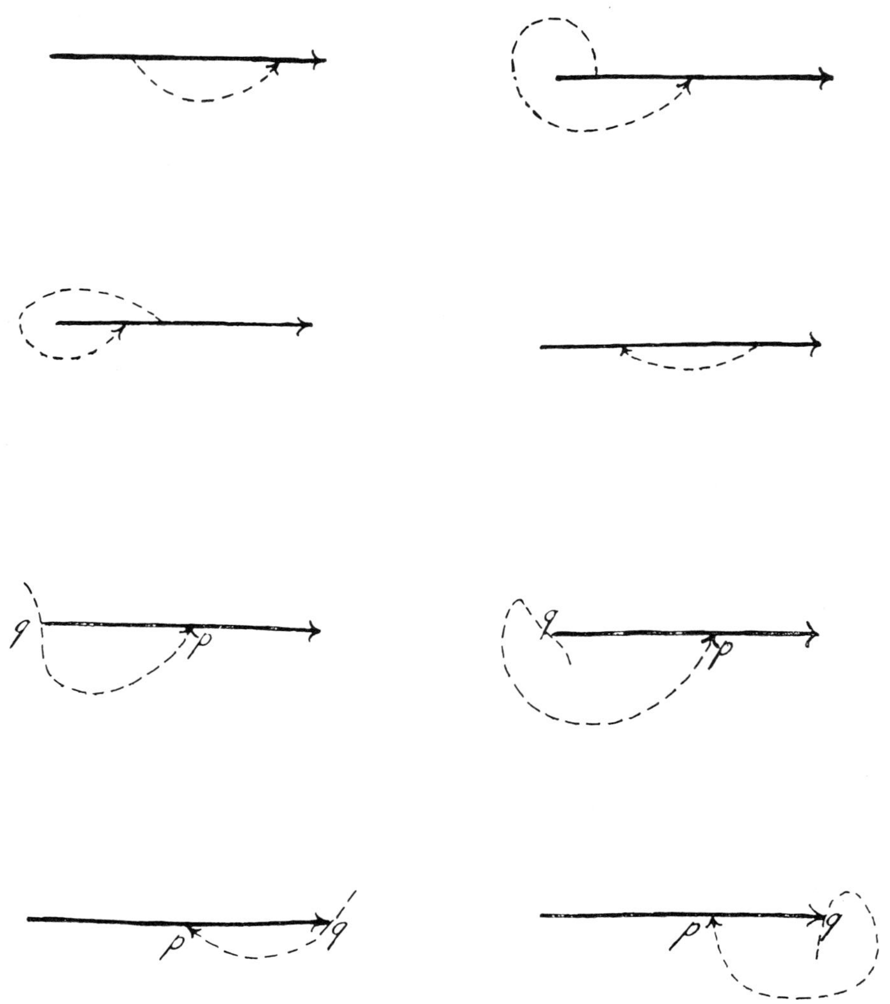

Fig. 1.

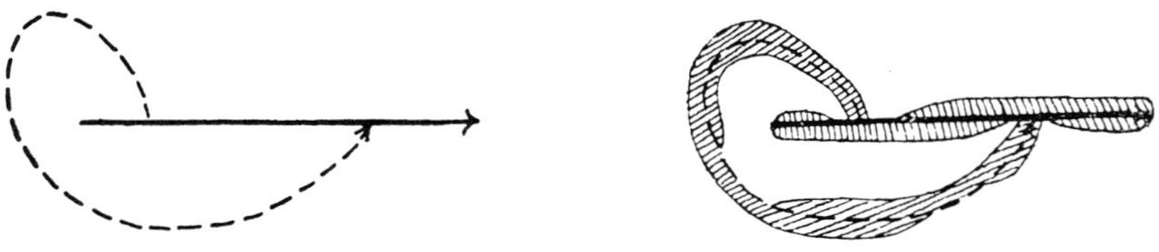

Fig. 2.

On the boundary of each of the two regions G and G' then directly lie some [1]) arcs of simple curve, which will remain *accessible* [2])

---

1) In the first five cases of fig. 1 this number is 4; in the sixth it is 4 or 3; in the seventh 3 or 2; in the eighth it is 3.

[[4]] 2) i. e. "erreichbar" after SCHÖNFLIES, Bericht über die Mengenlehre II, p. 176.

for any limit region, to which the region G may be extended, and likewise for its image region.

We now extend G in such a way, i. e. we replace it in such a way by a region in which it is contained, that, when G' has obtained the corresponding extension, the extended regions G and G' still lie outside each other. We repeat this extending process so often, until we get a limit region, and this will be the case after a denumerable number of extensions.

If possible, we then, after an indefinitely small modification of the boundary, execute an enlargement of this limit region; by this its property of being a limit region will in general be lost, but can be regained by a denumerable number of new extensions. This new limit region we again try to enlarge, and in this way go on, until by a denumerable number of operations a *transformation domain* O is obtained.

If in the surface no region exists, which at once with all points of its boundary is invariant for the transformation, the domain O can at most determine two rest regions, namely a rest region R_1, in which O' lies, and a rest region R_2, identical to R_1 or not, in whose image region R'_2 lies O.

If namely a third rest region G_3 existed, G_3 as well as G'_3 would be free of O, as well as of O'. Let P be an arbitrary point on the boundary of G_3 not coinciding with its image P'. Let us construct about P and P' simple closed curves, chosen as small as one likes, which are each other's image and bound regions π and π', then $\mathfrak{m}\ (O, G_3, \pi)$ and $\mathfrak{m}\ (O', G'_3, \pi')$ each contain one of a pair of regions, which are each other's image, and which one can make to contain of O and G_3 resp. of O' and G'_3 as closely as one likes *approximating* [1]) partial regions, to which O and O' might be *enlarged*, but this would clash with the property of a transformation domain. The same reasoning holds with a slight modification for the supposition that G_3 and G'_3 coincide. ⟦5⟧ ⟦6⟧

We shall say, that a region of a surface *does not break the connection of the surface*, if it determines only one rest region, possessing for analysis situs the character of a rest region of a trema. Then of course the region itself is singly connected. ⟦7⟧

We shall now assume, that O does not break the connection of the surface. Then it is bounded by a *closed curve* [2]) K, which on account of the commencement of the construction according to fig. 2 is a *non-singular closed curve*, and therefore does allow of

[1]) SCHÖNFLIES, Bericht über die Mengenlehre II, p. 104 sqq ⟦4⟧

[2]) Id., ibid., p. 118 sqq. ⟦4⟧

division into two proper arcs, and not into two improper ones [1]).

We shall now consider on each of the curves K and K' the cyclically ordered sets u and u' of their outwardly accessible points. These determine on each other certain segments σ_n and σ'_n, forming in pairs certain cyclically ordered sets $\varkappa_n$ of points accessible from a rest region of O and O', as is schematically illustrated by fig. 3.

If we make a circuit along u and u' together in such senses, that segments σ_n and σ'_n are reached simultaneously, this circuit is made *in opposite senses* for u and u'.

So if we make a circuit along u and u' together in such a way that points corresponding for the transformation are reached simultaneously, then with respect to the former order they must meet twice in that course, and at such a meeting either corresponding

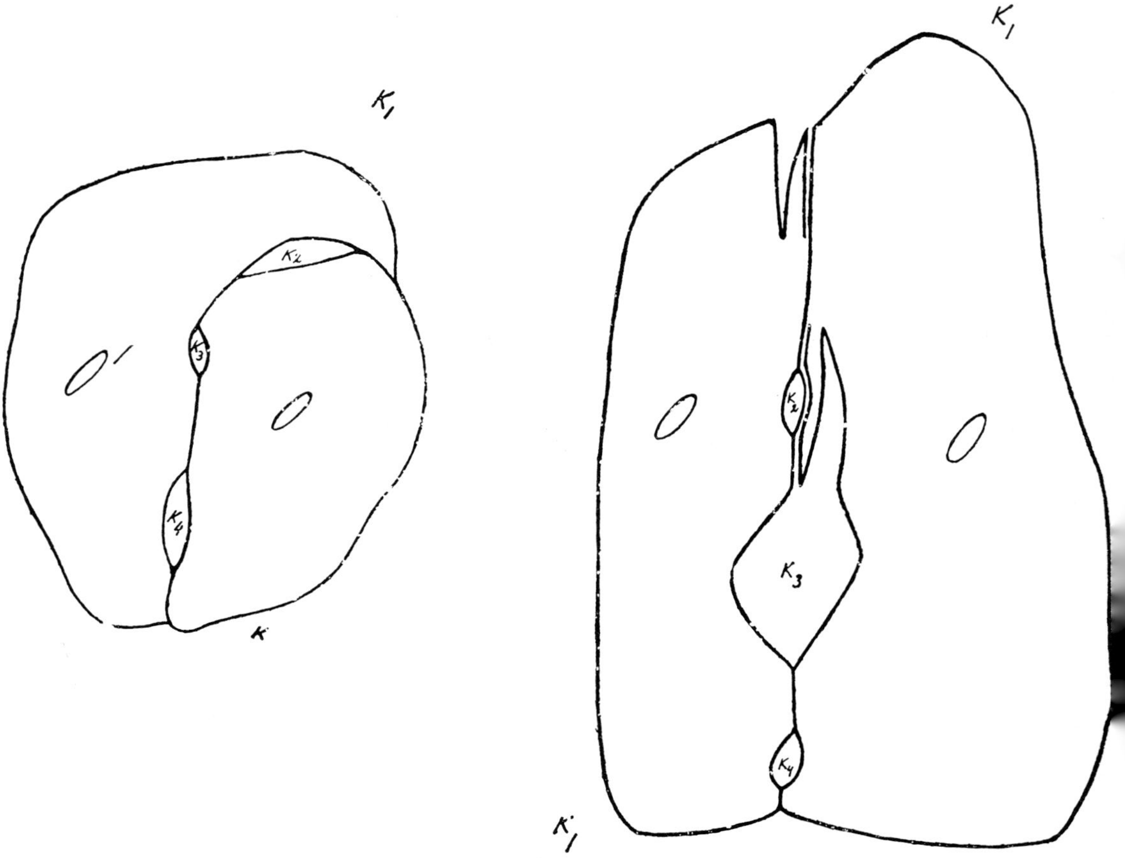

Fig. 3.

[1]) For these notions comp. L. E. J. BROUWER, "*Zur Analysis Situs*", and "*Ueber ein-eindeutige stetige Transformationen von Flächen in sich*", Mathem. Annalen,
[[8]] Bd. 68.

[[210]]

junctures [1]) Z (lying in u) and Z' (lying in u') form together a coherent set determining *one* rest region, or a point P of u, and its image P'' in u' will find themselves on the same $\varkappa_n$ in such a way, that P lies outside K', and P'' outside K. Then however in those points P and P'' an enlargement of O and O' would be possible, which would clash with the property of a transformation domain. So we are sure, that both meetings take place in the first-mentioned way.

Let us discern in a juncture Z, performing such a meeting, its *right end* Z_r and its *left end* Z_l. Let us represent by K_r resp. K_l the branch of u approaching to Z_r resp. Z_l; by ς_r resp. ς_l the part of the *circumference* [2]) of Z_r resp. Z_l, approximated by K_r resp. K_l; and by ς the complete circumference of Z. Let us further indicate (Z, Z') by T, the circumference of T by τ, (Z_r, Z'_l) by T_r, [[6]]
(Z_l, Z'_r) by T_l, and the part of the circumference of T_r resp. T_l, approximated by K_r and K'_l resp. K_l and K'_r, by τ_r resp. τ_l.

We at once see, that of the sets T_r and T_l at least one is coherent. For in the opposite case we could choose on K_r in the vicinity of Z_r a point of u outside K', whose image on u' would lie outside K, and an enlargement of O and O' would be possible there (see the schematic [[9]]
fig. 4).

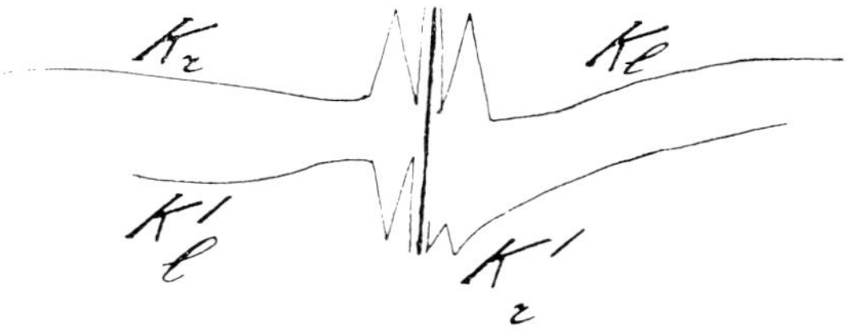

Fig. 4.

In the same way it is evident, that *either* ς_r and ς'_l or ς'_r and ς_l must be identical to each other, *or* Z must be a part of Z' or Z' a part of Z.

Otherwise namely *either* ς_r would have to lie partly outside ς'_l and at the same time ς'_r partly outside ς_l *or* ς_l partly outside ς'_r, and at the same time ς'_l partly outside ς_r, which free segments of circumference would partly correspond to each other, and would admit an enlargement of O and O' in their vicinity.

Of the two possibilities obtained we shall first discuss:

[1]) Two arcs of curve cohere to a new arc of curve by means of a "*juncture*", which contains an *end* of each of them. In the following reasoning we suppose the considered juncture to be composed of two non-singular ends. For singular ends it needs a slight modification.

[2]) i. e. the cyclically ordered set of its accessible points.

I. *ζ_r and ζ'_l are identical to τ_r.* We then further notice, that ζ_l and ζ'_r cannot leave a part s of the circumference of T_r free between K_l and K'_r (so that T_l also is coherent, though of τ_l this is not yet certain).

Otherwise namely s would be a part of the circumference of T_r, outside τ_r, ζ_l and ζ'_r; if we then regard T_r as Z_r, there will correspond to s a part s' of the circumference of Z'_r outside ζ'_r, ζ_l and τ_r. So about s and s' the domains O and O' could be enlarged.

We now distinguish the following two cases:

I*a*. *τ_r and ζ_l do not cohere on τ* [1]); then τ_r and ζ'_r do not cohere either. We can then represent Z schematically according to fig. 5, where each of the segments represents an arc of curve, to which further ramifications may be attached; however no ramifications cohere with the schema between ap and ah, and neither between bq and bk. Furthermore we notice that in a as well as in b ab unites itself with each of the two other arcs of curve ending there to a new arc of curve. In an analogous way the sets Z' and T are represented in this figure [2]).

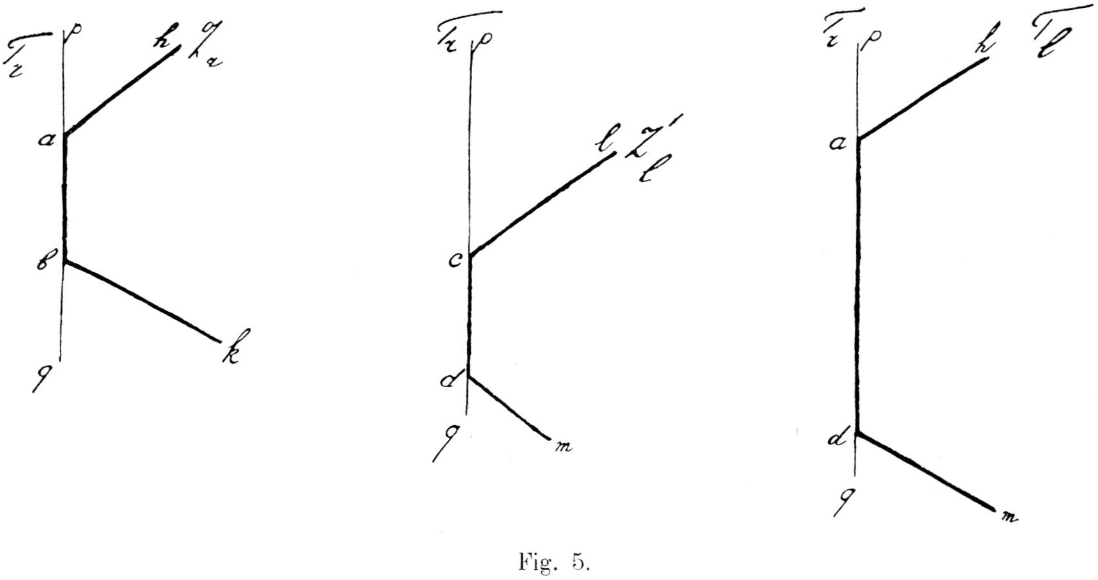

Fig. 5.

Now the situation of the three sets T_r, Z'_r, and Z_l with respect to each other can be in different ways in this case; the two essential possibilities are represented in fig. 6.

[1]) I*a* may also be treated in a less direct way, analogously to I*b*.

[2]) Some of the segments of these figures may shrink to zero; this however does not injure the correctness of our conclusions.

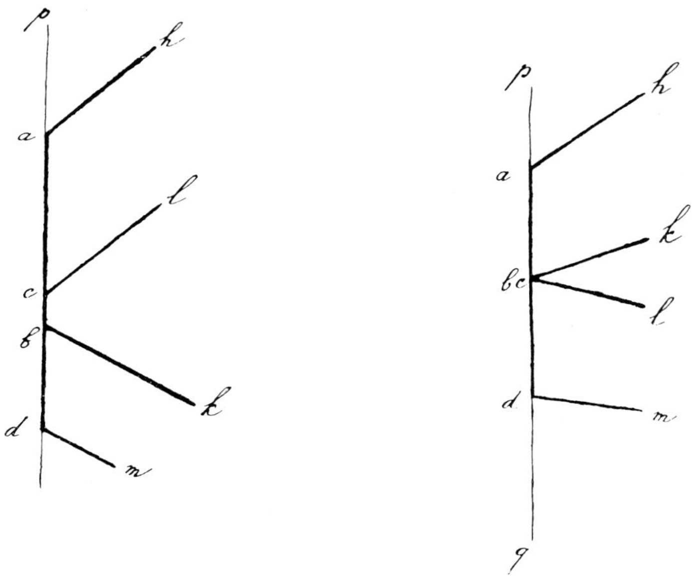

Fig. 6.

Iaα. When the first possibility of this figure appears, it follows from the correspondence of the arcs of curve *ab* and *dc* for the transformation, that on the arc of curve *bc* must lie somewhere a *juncture invariant for the transformation*.

Iaβ. When the second possibility appears, we notice that in *bc* the arcs of curve *dc* and *bk* certainly cohere. For otherwise *ab* and *cl* would cohere there neither, and from both would follow, that in *bc* between ς'_r and ς_l a segment *s* of the circumference of T_r would be free of both ς'_r and ς_l, the impossibility of which we have proved.

Let us furthermore in *bc* call L_{dc}, L_{lc}, and L_{ab} the ends of the arcs of curve *dc*, *lc*, and *ab*, then, *habq* having for the transformation as its image *qdcl*, and L_{lc} and L_{ab} cohering with each other, the situation of L_{dc}, L_{lc} and L_{ab} with respect to each other is quite the same, as that of τ_r, ς'_r and ς_l; the former thus allow of quite the same investigation as the latter, in which they are contained. If by this investigation we arrive again at the case *Iaβ*, we can investigate still further partial sets which are in the same case, and can go on in this way. After a denumerable number of repetitions of this process we must then either have reduced this case to an other, or have found *a point invariant for the transformation.*

Ib. τ_r and ς_l partly cover each other, or at least cohere on τ; then the same holds for τ_r and ς'_r.

Then, as τ_r is the image of ς_l, and ς'_r the image of τ_r, we can, starting at ζ_l, divide the part of τ, approximated by K and K', into successive "transformation domains", separated by Schnitte in this circumference, so that in a circuit each following domain is the image of the preceding one: of the last domain of the series then in general only a part appears, but three at least must be complete: fartheron the derived set of this last partial domain must cohere with that of the first domain; and we may suppose the number of the domains to be finite, as otherwise we should at once fall back on case II.

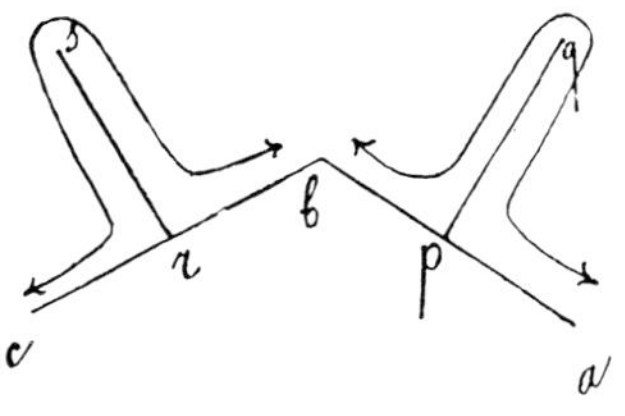

Fig. 7.

These domains of the circumference we can reduce by destroying in each of them the subsets of T, belonging to no other part of τ. If e.g. in fig. 7 the first domain extends from a to b, the second from b to c, we can deprive the first of the arc pq, the second of the arc rs.

On the circumference of the rest set R_1 we have then still a division into domains with the same properties, as the original division of the circumference of T. By a displacement of the separating Schnitte between the domains (after which in general the first as well as the last domain will be a partial one) it may occur, that the same process of reduction is once more applicable to R_1, and in this way it is repeated, until after a denumerable number of reductions a set R_0 is left, which by no displacement of the separating Schnitte can be made fit for further reduction.

A domain d then spreads over a part of the circumference h of an arc of curve k in such a manner, that the domain itself as well as the rest of the circumference possesses the whole k as its derived set.

Let the next domain d' follow of that rest of the circumference of k a part e', before it leaves k, and let f be the part of h belonging neither to d, nor to e'. Let e' be the image of e, and let us call g the segment of h approximated by k'. These notations are illustrated by two examples in fig. 8.

We now distinguish the following two cases:

Iba. f does not lie everywhere dense on k. Then e must lie everywhere dense on k.

For in the opposite case we might take the part of h enclosed between the end Schnitt of e and the end Schnitt of e' as a domain, which would be capable of reduction, and this is impossible.

But if e lies everywhere dense on k, the image of k is a part of k, so that we fall back on case II.

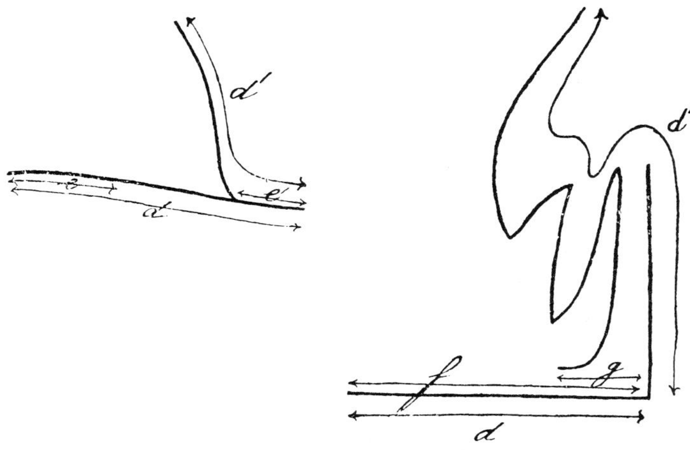

Fig. 8.

Ib β. f lies everywhere dense on k. Then g must also lie everywhere dense on k.

For in the opposite case the arcs of curve k, k', k'', etc. form an arc of curve B, in whose circumference the domains d, d', d'', etc. leave free a part everywhere dense in B, whilst each arc $k^{(n)}$ coheres with $k^{(n-1)}$ and $k^{(n+1)}$, but with none of the other. This however is impossible, as we have shown above, that the last and the first domain must cohere with each other.

But if g lies everywhere dense on k, k is a part of k', and again we are in case II.

II. *Z' is a part of Z.* We then, and also in the preceding cases reduced to this, have an *open system of curves*[1]) S, having as its image S' a part of itself. The image S'' of S' is again a part of S', etc. If $S^{(\omega)}$ is the set common to all sets $S^{(n)}$, it is *an open system of curves i, invariant for the transformation.*

From the commencement of the domain construction according to fig. 2 is evident, that the two sets T, in which u and u' meet each other, and therefore also the two invariant sets i, lie isolated from each other.

At the commencement of the domain construction we have presupposed, that the two arcs of curve of fig. 1 have no endpoint in common. If that *is* the case, the method needs but slight modifications, giving no difficulties. We then start not with common arcs of simple curve accessible for both O and O', but only with common points accessible for both O and O', which is sufficient too.

1) This means a perfect, coherent, nowhere dense set, determining in a region of the connection of the inner region of a circle or parabola only *one* rest region.

Furthermore we have supposed in our reasoning, that K_r etc. have nothing more in common with T, than Z_r etc. However it might be possible, that the last part of e. g. the branch of K_r approximating Z_r is contained in Z'. One can easily be convinced, that this pecularity does not harm the proof.

We can sum up our results as follows:

THEOREM 1. *An arbitrary continuous one-one transformation of a twosided surface in itself with invariant indicatrix possesses a transformation domain, which either breaks the connection of the surface,*
⟦10⟧ *or joins two isolated open systems of curves, invariant for the transformation.*

As furthermore such an invariant open system of curves possesses at least *one* invariant point, as I shall prove in another communication, the following holds likewise:

THEOREM 2. *An arbitrary continuous one-one transformation of a twosided surface in itself with invariant indicatrix possesses a transformation domain, which either breaks the connection of the surface, or joins two points invariant for the transformation.*

The fundamental importance of these theorems for the theory of transformations and transformation groups I shall show lateron. Here I only wish to indicate, how the theorem of the invariant point of the sphere, proved in my first communication upon this subject, is contained in them, and how this theorem can be extended to the Cartesian plane.

For, if the transformation domain of a sphere or Cartesian plane has more than two rest regions, then surely one of them must be invariant together with all the points of its boundary.

If we thus exclude this case, a transformation domain O on the sphere, which breaks its connection, is either annular, or singly connected; fig. 9 shows either of these possibilities.

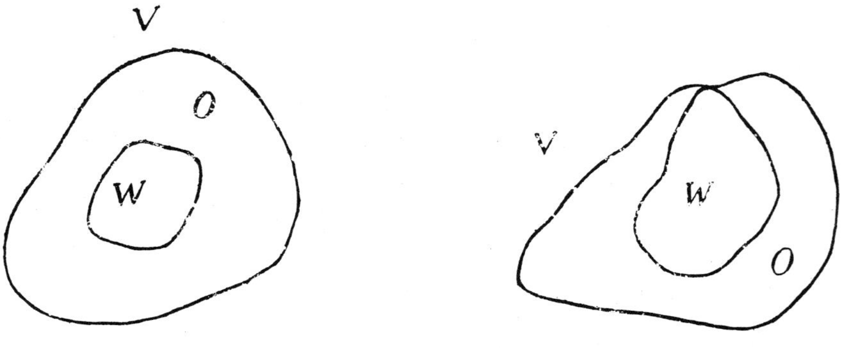

Fig. 9.

⟦216⟧

In both cases O determines on the sphere two regions, V and W, in one of which lies its image O'; then in the other lies O'_i, i.e. its image for the inverse transformation. If we now repeat the transformation itself as well as its inverse an indefinite number of times, we obtain on one hand in V a series of domains $O', O'', O''',\ldots$, and on the other hand in W a series of domains $O'_i, O''_i, O'''_i,\ldots$

In neither of these series exist intermediary regions between the successive terms, and each series converges to an invariant limit set, which causes the existence of at least one invariant point. These invariant limit sets may cohere with each other (in the second case of fig. 9): then we are sure of only *one* invariant point, otherwise always of *two*.

If in the Cartesian plane we have a transformation domain breaking its connection, either one of the cases of fig. 9 appears again, or the domain is bounded by an arc of curve running from infinite to infinite, or by two such arcs of curve.

If it is bounded by one arc, the existence of an invariant open system of curves can be deduced according to the proof of theorem 1.

If it is bounded by two such arcs B_1 and B_2, we can, if no

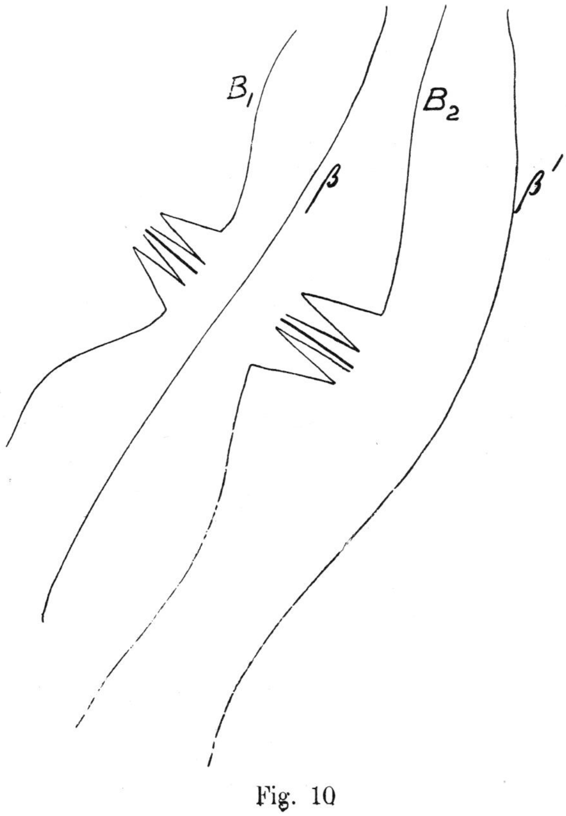

Fig. 10

invariant point exists, always arrange that these arcs do not cohere, and within O (see fig. 10) an arc of simple curve β can be constructed, running from infinite to infinite, lying entirely outside its image β', and enclosing with β' a new transformation domain ω. According to the SCHÖNFLIES process of representation [1]) the domains ω', ω'', ω''', ... , determined by a series of successive repetitions of the transformation, can then be represented on regions enclosed by straight lines $x = na$ and $x = (n+1)a$ in such a way that a series of corresponding points in $\omega, \omega', \omega'', \omega''',\ldots$ answers to a series of points (p,q), $(p+a,q)$,
[11] $(p+2a,q)$, $(p+3a,q),\ldots$; furthermore, if no invariant point exists, we can arrange, that the just-mentioned series of images of ω, continued indefinitely on both sides, covers the whole Cartesian plane, i. o. w. we have proved:

THEOREM 3. *An arbitrary continuous one-one transformation of the Cartesian plane in itself with invariant indicatrix either leaves at least one point invariant, or is a continuous one-one image of a translation.*

ERRATUM.

In my paper: *"The force field of the non-Euclidean spaces with*
[12] *positive curvature"*, these *Proceedings* IX 1, p. 261, l. 6 from top:
for: The unilateral elliptic Sp_n is enclosed by a plane Sp_{n-1},
read: The elliptic Sp_n is enclosed by a unilateral plane Sp_{n-1}.

1) Mathem. Ann. 62, p. 319—324

NOTES

[1] The English version appeared in the Proceedings of the meeting of 30 October 1909. Continuation of 1909 F2. Continued in 1911 H2, 1911 J2.
[2] The definitions and proofs of this paper are insufficient; the translation theorem as formulated at the end of the paper is even wrong. As Brouwer, 1912 K2, p. 360, explained, the paper rests on tacit applications of theorems of Schoenflies which were refuted at just the same time by Brouwer's sensational counter-examples (Brouwer 1910 C).

Brouwer made this statement after he had succeeded in proving the plane translation theorem along other lines (1912 B). In the twenties and thirties Brouwer stated orally his belief that the original proof could be salvaged. It would be worthwhile to try it because its idea, in particular the construction of a transformation field at one stroke, is more attractive than that of Brouwer's final proof.

At closer view the flaws indeed look less serious than Brouwer asserted in 1912 K2.

Brouwer's own copy of the Dutch version contains a large number of notes in ink and one in pencil notice. Some of the corrections were immediately carried out in the English version, others were published in 1911 H1, H2. Among them is a deletion which salvages the text of the translation theorem formally though even after this correction its formulation remains misleading.

The most revealing is the pencilled note, possibly of a later date, saying (in translation): 'Only those modifications which are provided with a ↑ are necessary; the others need not be carried out if the definition of "breaking the connection" is changed.' This rather than other, superficial, flaws, might be the key.

⟦3⟧ Brouwer, 1909 F2.

⟦4⟧ A. Schoenflies 1908 A.

⟦5⟧ The argument fails for instance if G_3 and G'_3 have the same boundary without coinciding. Even choosing P as an accessible point cannot salvage it.

⟦6⟧ In Cantor's notation set union is indicated by a gothic M, which has been rendered here by a script m. In some places this script m came out in print too faintly or not at all (thrice on p. 290, l. 13–14).

⟦7⟧ The pencil lines are probably to mark the key point.

⟦8⟧ According to Brouwer 1911 H2: for 68 read 68, 69. The papers are Brouwer 1910 C, 1910 F.

⟦9⟧ Restore the bracket to the left of 'see'.

⟦10⟧ According to a note in the Dutch version 1909 H1 an 'open system of curves' means a 'continuum, i.e. a perfect connected set which in a simply connected region determines only one complement region'. See also Brouwer 1911 H2, p. 767, footnote [2]).

⟦11⟧ According to Brouwer 1911 H2 the passage from 'furthermore' to 'proved' is to be replaced by 'i.o.w. we have proved'. This formally rescues the formulation of the translation theorem, though it does not prevent misleading interpretations. In general a topological orientation preserving mapping of the plane without a fixed point is not globally the topological image of a translation. The Dutch version contains in the margin a drawing of a counter example (see 1909 H1) which was published much later (Brouwer 1919 M2).

⟦12⟧ Brouwer 1906 C2.

1909 H1

[[1]]

van SCHOENFLIES[1]) kunnen we dan de door een reeks van opvolgende herhalingen der transformatie bepaalde gebieden $\omega', \omega'', \omega''', \ldots$ bij opvolging één-éénduidig continu afbeelden op het tusschen de rechte lijnen $x = na$ en $x = (n+1)a$ ingesloten gebied, zoodanig, dat een reeks van overeenkomstige punten in de domeinen $\omega, \omega', \omega'', \omega''', \ldots$ beantwoordt aan een puntrij (p,q), $(p+a,q)$, $(p+2a,q)$, $(p+3a,q), \ldots$; ~~verder kunnen we het, indien geen invariant punt optreedt, altijd zoo inrichten, dat de naar beide kanten onbepaald voortgezette domeinreeks het geheele Cartesiaansche vlak overdekt;~~ m. a. w. we hebben bewezen:

STELLING 3. *Een willekeurige één-éénduidige, continue transformatie van een Cartesiaansch vlak in zichzelf bezit òf minstens één invariant punt, òf is één-éénduidig, continu beeld eener translatie.*

1) Mathem. Ann. 62, p. 319—324.

Dat deze regels geschrapt moeten worden, blijkt uit dit voorbeeld:

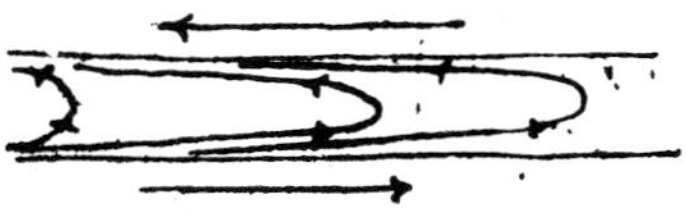

Hier vormen de *ω + ω-puntreeksen v niet eens een opp
vlak. Immers convergeert een puntrij tot een punt P, d
convergeeren hier de bijbehoorende puntreeksen v niet
tijd tot de puntreeks v van P.

NOTE

[[1]] This piece from 1909 H1 is copied as an illustration to 1909 H2 [[1]].

1911 H2

Mathematics. — "*Continuous one-one transformations of surfaces in themselves*". (3rd Communication [1])). By Dr. L. E. J. BROUWER. (Communicated by Prof. D. J. KORTEWEG). [[1]]

(Communicated in the meeting of December 24, 1910).

In the second communication several points of the argumentation were indicated but in short. We shall now treat some auxiliary theorems, of which the proof is necessary for a complete development of the theory. [[3]]

§ 1.

Definitions and lemmas.

On a surface we understand by a *finite continuum* a coherent set of points containing all its limiting points, and in which each fundamental series of points possesses a limiting point.

By a *continuum* we understand a coherent set of points containing all its limiting points and containing for every two of its points a finite continuum joining those two points.

A finite continuum determining only one rest region we shall call a *circular continuum*, if that rest region possesses for analysis situs the character of a rest region of a trema.

A continuum determining only one rest region, which is for analysis situs equivalent to the surface itself, we call a *parabolic continuum* [2]). [[4]]

A circular or a parabolic continuum together with a certain vicinity of it allows of a continuous one-one representation on a region of a Cartesian plane. There the circular continuum then lies entirel in a finite region, the parabolic continuum extends to the infinite Both determine in the Cartesian plane only one rest region and possess there a single circumference of *accessible* points, which lie in cyclic order for the circular continuum and in linear order for the parabolic continuum.

1) See these Proceedings Vol. XI, page 788, Vol. XII, p. 286. Compare also the abstract: "*Ueber eineindeutige, stetige Transformationen von Flächen in sich*" (Mathem. Annalen, Vol. 69, page 176), where the endresult of these researches is formulated. [[1]] [[2]]

2) I do not maintain the term "open system of curves", which I used in the preceding communication (p. 294 sq.) for a nowhere dense circular or parabolic continuum. [[5]]

A vector in the Cartesian plane being nowhere zero or infinite, of which the origin describes an arc of simple curve a and the endpoint as a continuous function of the origin another arc of simple curve
[[6]] b, starting in P and ending in Q, describes as integral of its infinitely small variations of direction a certain *total angular variation*.

If we substitute for b another arc of simple curve b' starting likewise in P and ending in Q, the two following theorems hold:

LEMMA 1. *If a has no point in common with $b + b'$, and if a is not separated from the infinite by $b + b'$, then the substitution of b' for b causes no change in the total angular variation of the vector.*

In that case we can namely construct a closed curve c containing a in its outer domain, $b + b'$ in its inner domain, and we can perform the modification of b into b' in a continuous way and entirely in the inner domain of c. The total angular variation of the vector can then on one hand undergo only continuous modifications, and on the other hand it can only vary by multiples of 2π; thus it remains unchanged.

LEMMA 2. *If b and b' form together a simple closed curve containing a in its inner domain, then by the substitution of b' for b the total angular variation of the vector increases or decreases by 2π, according to the positive sense of circuit of the closed curve corresponding to a movement of P to Q along b' or along b.*

If namely of a vector the endpoint describes a simple closed curve in a positive sense, whilst its origin describes as a continuous function of the endpoint a closed curve lying entirely inside that simple closed curve, we can by means of continuous modification, which does not change the total angular variation of the vector, transform the curve described by the endpoint into a circle, and reduce the curve described by the origin to the centre of that circle. Thus also before this modification the total angular variation is equal to 2π, from which lemma 2 immediately ensues.

§ 2.

The invariant point of the circular continuum.

We suppose a two-sided surface to be submitted with invariant indicatrix to a continuous one-one transformation in itself in such a way that a certain circular continuum φ' passes thereby into itself.

We represent φ' together with certain surroundings ψ' uni-univalently and continuously on a region of a Cartesian plane, whereby they become respectively the images φ and ψ, and we suppose that φ' possesses no point invariant for the transformation.

We can then surround φ by a polygon $\mathfrak{P}$ approximating φ at a distance ε[1]) and belonging entirely to ψ, in such a way that also the image of $\mathfrak{P}$ for the transformation lies entirely inside ψ, that in each point on or inside $\mathfrak{P}$ the length of the *transformation vector* (i.e. the vector joining the point with its image for the transformation) does not fall below a certain minimum b, and that each point of $\mathfrak{P}$ allows itself to be joined with φ by a path[2]) $< \frac{1}{32} b$.

On $\mathfrak{P}$ we then choose the points $P_1, P_2, \ldots P_n$, which have this order in the sense of a positive circuit, and possess the property that in each arc $P_k P_{k+1}$ (to these also belongs the arc $P_n P_1$) the distance of the endpoints lies between $\frac{1}{8} b$ and $\frac{3}{8} b$, and the distance of two arbitrary points does not exceed $\frac{3}{4} b$ [3]). Let us now draw from each point P_k to φ a path $P_k R_k < \frac{1}{32} b$ lying inside $\mathfrak{P}$, then the arc of simple curve $R_k P_k P_{k+1} R_{k+1}$ *cannot cut* its image for the transformation $\varrho_k \pi_k \pi_{k+1} \varrho_{k+1}$.

The arcs $R_k P_k P_{k+1} R_{k+1}$ we shall call *skeleton arcs;* the arcs $\varrho_k \pi_k \pi_{k+1} \varrho_{k+1}$ *image skeleton arcs.*

If we represent by ϑ the total angular variation described by the transformation vector for a positive circuit of $\mathfrak{P}$, and by α_k the total angle described by a vector of which the origin runs along the skeleton arc $P_k P_{k+1}$, and the endpoint as a continuous function of the origin along the image skeleton arc $\pi_k \pi_{k+1}$, then we have

$$\vartheta = \Sigma \alpha_k .$$

By τ_k we shall represent the point of the image skeleton arc $\pi_k \varrho_k$, which is, if this arc does not cut $\mathfrak{P}$, identical to π_k, and in the opposite case to its last point of intersection with $\mathfrak{P}$. If then β_k designs the total angular variation of a vector of which the origin runs along the skeleton arc $P_k P_{k+1}$ and the endpoint as a continuous function of the origin along the image skeleton arc $\tau_k \tau_{k+1}$, we have likewise

$$\vartheta = \Sigma \beta_k .$$

We now distinguish three cases:

1st. *On the circumference of* φ *the segments* $R_k R_{k+1}$ *and* $\varrho_k \varrho_{k+1}$

1) Schoenflies, Bericht über die Mengenlehre II, p. 104. ⟦7⟧

2) i.e. "Weg" in the sense of Schoenflies.

3) Schoenflies, Bericht über die Mengenlehre II, p. 183. ⟦7⟧

lie outside each other (see fig. 1). Then we can join τ_k and τ_{k+1} by an arc of simple curve (drawn splintered in the figure), which lies inside $\mathfrak{P}$ as well as inside the domain enclosed between φ and the image skeleton arc $\varrho_k \pi_k \pi_{k+1} \varrho_{k+1}$, and which we shall call the *path*

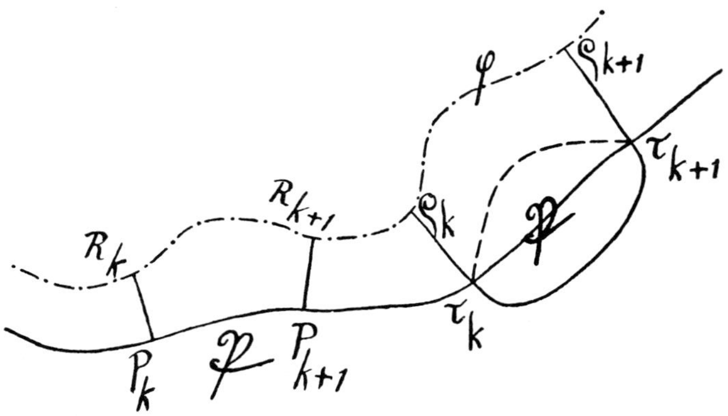

Fig. 1.

arc $\tau_k \tau_{k+1}$. This path arc together with the image skeleton arc $\tau_k \tau_{k+1}$ does not separate the skeleton arc $P_k P_{k+1}$ from the infinite.

So, if we represent by γ_k the total angle described by a vector of which the origin runs along the skeleton arc $P_k P_{k+1}$, and the end-point as a continuous function of the origin along the path arc $\tau_k \tau_{k+1}$, we have is this first case:

$$\beta_k = \gamma_k.$$

2nd. *On the circumference of* φ *the segment* $R_k R_{k+1}$ *is a part of the segment* $\varrho_k \varrho_{k+1}$. Then too we can construct a path arc $\tau_k \tau_{k+1}$ (drawn splintered in fig. 2), which lies inside $\mathfrak{P}$ as well as inside the domain enclosed between φ and the image skeleton arc

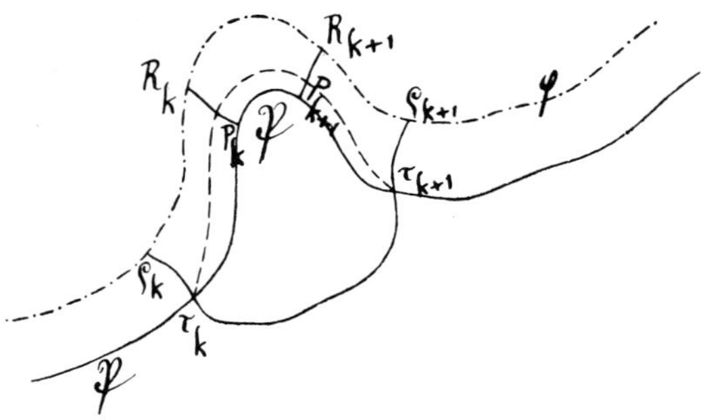

Fig. 2.

$\varrho_k \pi_k \pi_{k+1} \varrho_{k+1}$. But now this path arc forms with the image skeleton arc $\tau_k \tau_{k+1}$ a simple closed curve containing the skeleton arc $P_k P_{k+1}$ in its inner domain, whilst its positive sense of circuit corresponds to a movement from τ_k to τ_{k+1} along the image skeleton arc $\tau_k \tau_{k+1}$.

So we have in this second case, defining the angle γ_k in the same manner as in the first case:

$$\beta_k = \gamma_k + 2\pi.$$

3rd. *On the circumference of* φ *the segment* $\varrho_k \varrho_{k+1}$ *is a part of the segment* $R_k R_{k+1}$ (see fig. 3). Then the image skeleton arc $\varrho_k \pi_k \pi_{k+1} \varrho_{k+1}$ lies entirely inside $\mathfrak{P}$, and we choose as path arc

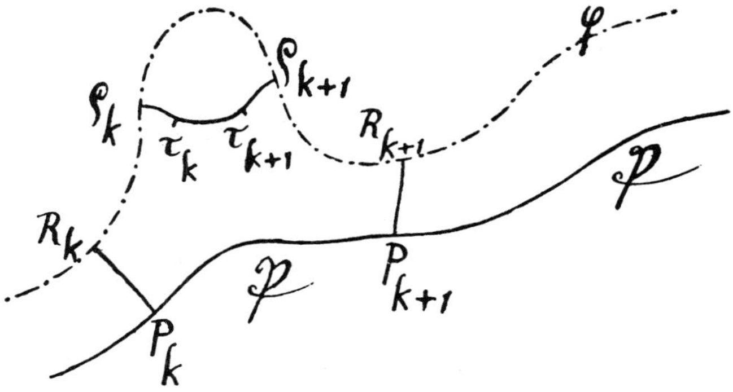

Fig. 3.

$\tau_k \tau_{k+1}$ the image skeleton arc $\tau_k \tau_{k+1}$ itself. Then we have in this third case, defining the angle γ_k in the same way as in the first and second cases:

$$\beta_k = \gamma_k.$$

Now we can take $\Sigma \gamma_k$ as the total angular variation of a vector nowhere becoming zero, of which the origin describes $\mathfrak{P}$ in a positive sense and the endpoint as a continuous function of the origin a closed curve passing nowhere outside $\mathfrak{P}$, so that we have

$$\Sigma \gamma_k = 2\pi.$$

From this ensues in connection with the preceding formulae:

$$\Sigma \beta_k = 2n\pi,$$

where n represents a positive integer ≥ 1.

Hence ϑ *cannot be equal to zero,* from which we conclude that the distribution of the transformation vectors must possess inside $\mathfrak{P}$ at least *one* singular point, i. o. w. that, contrary to the supposition at the commencement of this §, there must lie inside $\mathfrak{P}$ at least *one* point invariant for the transformation.

With this we have proved:

THEOREM 1. *For a continuous one-one transformation with invariant indicatrix of a two-sided surface in itself an invariant circular continuum contains at least one invariant point* [1]). [[8]]

---

1) Compare Mathem. Annalen, Vol. 69, p. 178; these Proceedings Vol. XII, p. 295. [[9]]

§ 3.

The invariant point of the parabolic continuum.

We suppose a two-sided surface to be submitted with invariant indicatrix to a continuous one-one transformation in itself in such a way that by it a certain nowhere dense parabolic continuum φ' is transformed into itself.

We suppose that φ' possesses no point invariant for the transformation. Then its circumference cannot contain an invariant Schnitt either; for, this would be a circular continuum and therefore according to theorem 1 would give rise to an invariant point.

We represent φ' together with certain surroundings ψ' uni-univalently and continuously on a region of a Cartesian plane whereby they become respectively the images φ and ψ. All the figures of the Cartesian plane to be constructed in the following and likewise their images for the transformation and their "counterimages" (i. e. their images for the inverse transformation) we suppose to lie in ψ. We suppose fartheron that for a positive circuit of the circumference of φ each accessible point precedes its image.

We surround φ by a fundamental series of polygonal lines $\mathfrak{P}_1$, $\mathfrak{P}_2$, $\mathfrak{P}_3$,... lying inside each other and approximating φ at indefinitely decreasing distances $\varepsilon_1, \varepsilon_2, \varepsilon_3, \ldots$, and we draw to an accessible point A lying on the circumference of φ a path w cutting each $\mathfrak{P}_k$ once and not more than once, and being $\leq 4\varepsilon_1$ [1]).

By a *circumference domain* of φ we understand such a segment of the linear type of order of its accessible points, as lies entirely outside its image segment, but for which each extension causes that property to be lost.

Let X and Y be two accessible points on the circumference of φ which are separated by A, whilst the order of succession XAY corresponds to a positive sense of circuit and between X and A as well as between A and Y there exist at least three circumference domains lying outside each other.

Let B be an accessible point of φ preceding X for a positive sense of circuit, possessing a finite distance p from the circumference segment XY, and belonging to the boundary of both regions determined by w between φ and $\mathfrak{P}_1$.

Let us understand by U an arbitrary accessible point of the

⟦10⟧ [1]) Compare SCHOENFLIES, Bericht über die Mengenlehre II, p. 127; L. E. J. BROUWER, "*Zur Analysis Situs*", Mathem. Annalen, Vol. 68, p. 428.

circumference of φ, whose distance from B does not exceed ε_1, then we suppose ε_1 to be so small that, independently of the choice of U, in all points whose distance from the circumference segment BU is $\leqq 32\,\varepsilon_1$, the length of the transformation vector is $\geqq 64\,\varepsilon_1$, whilst of an arc of simple curve $\leqq 32\,\varepsilon_1$, running between two accessible points of the circumference of φ, if one of its endpoints belongs to the circumference segment BU, the image and the counterimage do not intersect each other.

Let C be an accessible point of φ lying beyond Y for a positive sense of circuit, and possessing from B a distance $< p$, and $< \frac{1}{4}\,\varepsilon_1$.

Let S_1 be the first positively directed circumference Schnitt not preceding B, past which up to A all accessible points of φ can be reached from the infinite along paths not meeting the straight line segment BC.

We then choose (see fig. 4) on the polygonal line $\mathfrak{P}_1$ an arc D_1E_1, in which the distance of the endpoints is $8\,\varepsilon_1$ and the distance of two arbitrary points is $< 16\,\varepsilon_1$, in such a way that from D_1 and E_1

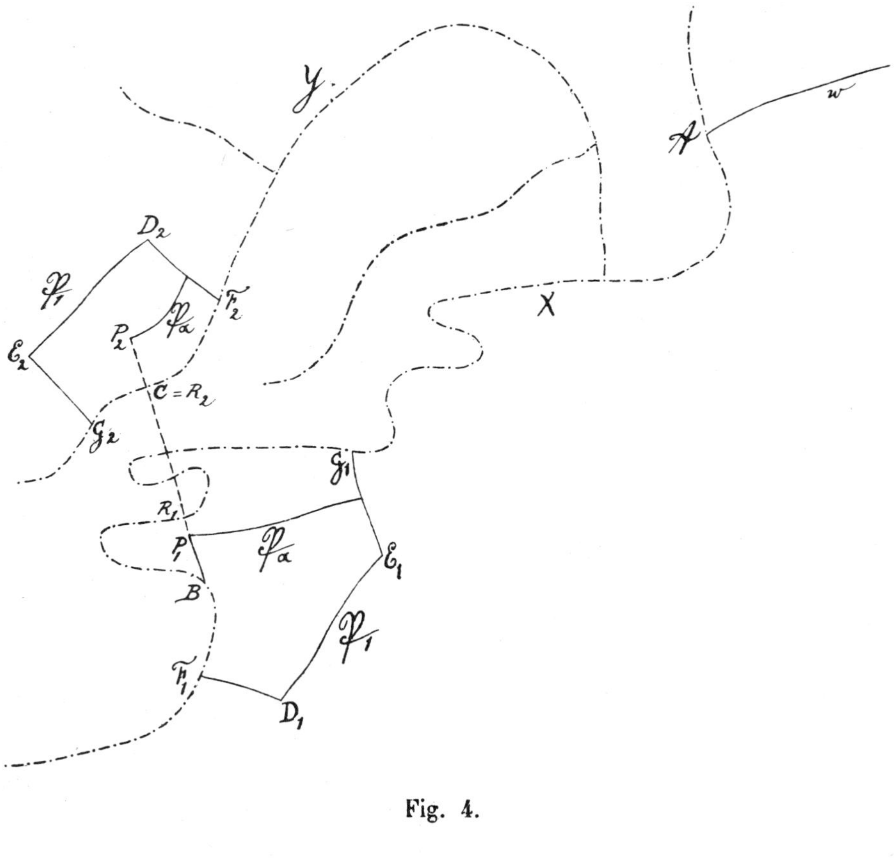

Fig. 4.

there can be laid to φ paths D_1F_1 and $E_1G_1 < 2\varepsilon_1$ not cutting the line segment BC and for $\alpha > m$ cutting each polygonal line $\mathfrak{P}_\alpha$ only once, whilst the Schnitt S_1 lies enclosed between these two paths. We may suppose ε_1 so small that not only the arc $F_1D_1E_1G_1$, but also its image *does not cut the path* w.

We now determine an approximating polygonal line $\mathfrak{P}_{\nu_1}$ $(\nu_1 > m)$, possessing the following property:

If a part of the circumference segment F_1G_1 is separated from the infinite by the line segment BC, $\mathfrak{P}_{\nu_1}$ cuts the line segment BC between D_1F_1 and E_1G_1; then the last point of intersection, which BC has there with $\mathfrak{P}_{\nu_1}$, we represent by P_1;

in the opposite case $\mathfrak{P}_{\nu_1}$ cuts a path $< \frac{1}{8}\varepsilon_1$ leading to B, and not cutting D_1F_1 and E_1G_1; then its last point of intersection with that path we represent by P_1.

Let S_2 be the last negatively directed circumference Schnitt not lying beyond C possessing the property that between A and S_2 all accessible points of φ can be reached from the infinite along paths not meeting the straight line segment P_1C.

We then choose on the polygonal line $\mathfrak{P}_1$ an arc D_2E_2 in which the distance of the endpoints is $8\varepsilon_1$ and the distance of two arbitrary points is $< 16\varepsilon_1$, in such a way that from D_2 and E_2 there can be laid to φ paths D_2F_2 and $E_2G_2 < 2\varepsilon_1$ not cutting the line segment P_1C and for $\alpha > m$ cutting each polygonal line $\mathfrak{P}_\alpha$ only once, whilst the Schnitt S_2 lies enclosed between those two paths. We may suppose ε_1 so small that not only the arc $F_2D_2E_2G_2$, but also its image and its counterimage *do not cut the path* w.

We now determine an approximating polygonal line $\mathfrak{P}_{\nu_2}$ $(\nu_2 > m)$ possessing the following property:

If a part of the circumference segment F_2G_2 is separated from the infinite by the line segment P_1C, $\mathfrak{P}_{\nu_2}$ cuts the line segment P_1C between D_2F_2 and E_2G_2; we then represent the first point of intersection, which P_1C has there with $\mathfrak{P}_{\nu_2}$, by P_2;

in the opposite case $\mathfrak{P}_{\nu_2}$ cuts a path $< \frac{1}{8}\varepsilon_1$ leading to C, and not cutting D_2F_2 and E_2G_2; we then represent its first point of intersection with that path by P_2.

Finally we impose on $\mathfrak{P}_{\nu_1}$ as well as on $\mathfrak{P}_{\nu_2}$ the condition that of their part contained between D_1F_1 and E_2G_2 the image as well as the counterimage lie inside $\mathfrak{P}_1$ and inside the image of $\mathfrak{P}_1$.

In the linesegment BC, eventually completed with the paths leading to B resp. to C that have been added to it, now lies an arc of simple curve $< \frac{1}{2} \varepsilon_1$ (drawn splintered fig. 4), joining P_1 and P_2, not cutting the arc of $\mathfrak{P}_{\nu_1}$ enclosed between P_1 and w neither the arc of $\mathfrak{P}_{\nu_2}$ enclosed between P_2 and w. On this arc of simple curve we represent the first resp. the last point of intersection with φ by R_1 resp. R_2.

By the *skeleton arc* $R_1 P_1 G_1$ we shall understand the arc of simple curve obtained by following from R_1 first the path $R_1 P_1$, then $\mathfrak{P}_{\nu_1}$ up to its point of intersection with $E_1 G_1$ and finally the path $E_1 G_1$ from that point of intersection to G_1.

This skeleton arc $R_1 P_1 G_1$ does not meet its image skeleton arc $\varrho_1 \pi_1 \gamma_1$, which image skeleton arc cuts neither the path w nor the path $E_1 G_1$, whilst the circumference segments $R_1 G_1$ and $\varrho_1 \gamma_1$ lie outside each other.

The arcs $P_1 G_1$ and $\varrho_1 \pi_1$ we join by an arc of simple curve $K_1 L_1$ (see fig. 5), belonging to an approximating polygonal line $\mathfrak{P}_{\tau_1}$ $(\tau_1 > \nu_1)$,

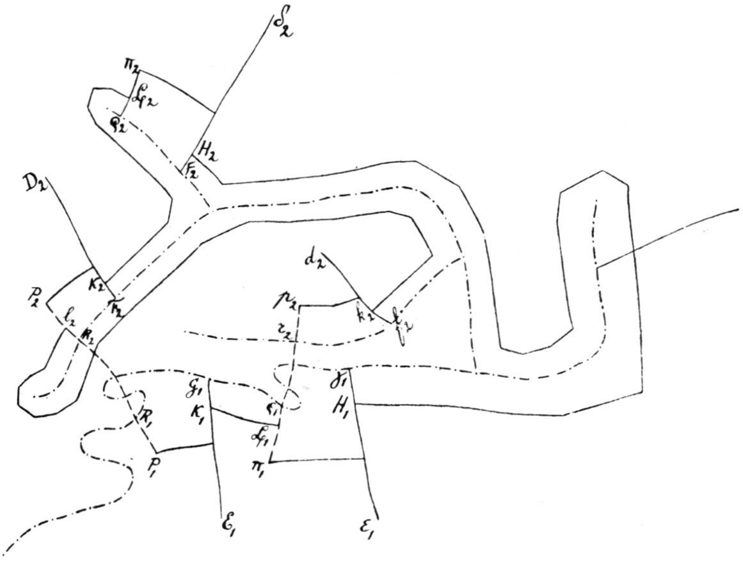

Fig. 5.

and abroad from its endpoints cutting neither the skeleton arcs $R\,P_1 G_1$ and $\varrho_1 \pi_1 \gamma_1$, nor the paths w and $E_1 G_1$.

The arc $K_1 L_1$ we divide into such partial arcs, that on each of them the distance of the endpoints lies between $4\varepsilon_1$ and $12\varepsilon_1$ and the distance of two arbitrary points does not exceed $24\varepsilon_1$. From the endpoints of these partial arcs we draw to φ straight paths $< \frac{3}{2} \varepsilon_{\tau_1}$, among which we regard each pair of two successive ones, together with the partial arc of $K_1 L_1$ connecting them, again as a skeleton arc.

By the *skeleton arc* $R_2P_2F_2$ we understand the arc of simple curve obtained by following from R_2 first the path R_2P_2, then recurring $\mathfrak{P}_{\nu_2}$ up to its point of intersection with D_2F_2 and finally following this path to F_2.

This skeleton arc $F_2P_2R_2$, its image $f_2p_2r_2$, and its counterimage $\varphi_2\pi_2\varrho_2$ meet neither each other, nor the path w. Farthermore $\varphi_2\pi_2\varrho_2$ cuts neither the path D_2F_2, nor its image d_2f_2, $f_2p_2r_2$ cuts neither the path D_2F_2, nor its counterimage $\delta_2\varphi_2$, and $F_2P_2R_2$ cuts neither the path $\delta_2\varphi_2$, nor the path d_2f_2.

The arcs P_2F_2 and $\varrho_2\pi_2$ we join by an arc of simple curve K_2L_2, belonging to an approximating polygonal line $\mathfrak{P}_{\tau_2}$ ($\tau_2 > \nu_2$), and abroad from its endpoints cutting neither the skeleton arcs $R_2P_2F_2$, $\varrho_2\pi_2\varphi_2$ and $r_2p_2f_2$, nor the paths w, D_2F_2, $\delta_2\varphi_2$ and d_2f_2, nor its own image k_2l_2.

The arcs $\pi_1\gamma_1$ and $\pi_2\varphi_2$ we join by an arc of simple curve H_1H_2, belonging to an approximating polygonal line $\mathfrak{P}_\tau$ ($\tau > \nu_1$ and $\tau > \nu_2$), and abroad from its endpoints cutting neither the paths E_1G_1 and D_2F_2, nor their images $\varepsilon_1\gamma_1$ and d_2f_2, nor the skeleton arcs $R_1P_1G_1$, $\varrho_1\pi_1\gamma_1$, $\varrho_2\pi_2\varphi_2$, $R_2P_2F_2$ and $r_2p_2f_2$, nor the joining arcs K_1L_1, K_2L_2 and k_2l_2.

Out of H_1H_2 and straight paths drawn from there to φ we finally construct skeleton arcs in the same way as above out of K_1L_1.

We have now built up a simple closed curve $P_1K_1L_1\pi_1H_1H_2\pi_2L_2K_2P_2R_2R_1P_1$, and, after addition of the image $\pi_1\varrho_1r_2p_2$ of the arc $P_1R_1R_2P_2$ drawn splintered in the figure, a second simple closed curve $\pi_1H_1H_2\pi_2L_2K_2P_2l_2k_2p_2r_2\varrho_1\pi_1$. These two closed curves have the arc $\pi_1H_1H_2\pi_2L_2K_2P_2$ in common, and this arc has for both closed curves the same inner side, so that it is *not separated from the infinite by the two completing arcs* $P_2R_2R_1P_1K_1L_1\pi_1$ and $P_2l_2k_2p_2r_2\varrho_1\pi_1$.

The first closed curve we represent by $\mathfrak{C}$, and it is our aim to find the total angular variation of the transformation vector for a positive circuit of $\mathfrak{C}$.

To this end we can during the description of the arc $P_1K_1L_1\pi_1H_1H_2\pi_2$ substitute in the curve described by the endpoint of the transformation vector for each image skeleton arc a path arc according to the method of § 2, with the restriction that here we are always in the case 1st of that §. After that substitution the curve described by the vector endpoint passes nowhere outside $\mathfrak{C}$, whilst its first and its last point have remained the same, and the total angular variation of the vector has not changed.

We now come to the total angular variation of the transformation vector during the description of the arc $\pi_2 L_2 K_2 P_2$; on the ground of § 1 it can also be obtained by carrying first the origin of the vector along the arc $\pi_2 L_2 K_2 P_2$ and the endpoint as a continuous function of the origin along the arc $P_2 R_2 R_1 P_1 K_1 L_1 \pi_1$; and then, whilst the origin remains in P_2, carrying the endpoint still along the arc $\pi_1 \varrho_1 r_2 p_2$.

Finally we can obtain the total angular variation of the transformation vector during the description of the arc $P_2 R_2 R_1 P_1$ by carrying first the endpoint of the vector along $p_2 r_2 \varrho_1 \pi_1$ whilst the origin remains in P_2, and then the origin along $P_2 R_2 R_1 P_1$, whilst the endpoint remains in π_1.

So the total angular variation of the transformation vector for a positive circuit of $\mathfrak{C}$ is obtained by carrying first the origin of a nowhere vanishing vector along the arc $P_1 K_1 L_1 \pi_1 H_1 H_2 \pi_2$ and the endpoint as a continuous function of the origin along a certain curve nowhere passing outside $\mathfrak{C}$; then, whilst the origin runs along the arc $\pi_2 L_2 K_2 P_2$, carrying the endpoint along the arc $P_2 R_2 R_1 P_1 K_1 L_1 \pi_1$; and finally, whilst the endpoint remains in π_1, carrying the origin along the arc $P_2 R_2 R_1 P_1$.

In none of the three parts of this movement the endpoint of the vector has passed outside $\mathfrak{C}$, so that the total angular variation amounts to $+ 2\pi$, from which we conclude that, contrary to the supposition, the distribution of the transformation vectors must possess inside $\mathfrak{C}$ at least one singular point.

With this we have proved:

THEOREM 2. *For a continuous one-one transformation with invariant indicatrix of a two-sided surface in itself an invariant nowhere dense parabolic continuum contains at least one invariant point.* [1])

ERRATA.

In the 2nd communication on this subject, these Proceedings Vol. XII [[12]]
p. 289, in the note

for: Mathem. Annalen, Bd. 68.

read: Mathem. Annalen, Bd. 68, 69.

p. 297, l. 10—13 from top

for: furthermore, if no invariant point exists, we can arrange, that the just-mentioned series of images of ω, continued indefinitely on both sides, covers the whole Cartesian plane, i. o. w. we have proved:

read: i. o. w. we have proved:

[1]) Compare Mathem. Annalen, Vol. 69, p. 178. [[11]]

NOTES

⟦1⟧ The English version appeared in the Proceedings of the meeting of 28 January 1911.

Continues 1909 F2, 1909 H2. Correction in 1911 J2. See the remark at the end of 1912 K2.

⟦2⟧ Brouwer 1910 F.

⟦3⟧ Delete with Brouwer the comma after 'indicated'. See 1911 J2.

⟦4⟧ Read: parabolic. See Brouwer 1911 J2.

⟦5⟧ Compare Brouwer 1909 H2, note ⟦10⟧.

⟦6⟧ Earlier topological uses of the circulation number occur in Brouwer 1910 A2, 1910 D2. See also Poincaré 1881.

⟦7⟧ A. Schoenflies 1908 A.

⟦8⟧ An extension of this theorem is in Brouwer 1911 J2.

⟦9⟧ Brouwer 1910 F, 1909 H2.

⟦10⟧ A. Schoenflies 1908 A, Brouwer 1910 C.

⟦11⟧ Brouwer 1910 F.

⟦12⟧ Brouwer 1909 H2.

1911 J2

Mathematics. — "*Continuous one-one transformations of surfaces in themselves.*" (4th communication [1])). By Dr. L. E. J. BROUWER. (Communicated by Prof. D. J. KORTEWEG). [[1]]

(Communicated in the meeting of May 27, 1911).

In this communication as in the preceding one we shall occupy ourselves with continuous one-one transformations with invariant indicatrix of a two-sided surface in itself.

If for such a transformation there is an invariant arc of simple curve, it contains at least *one* invariant point; more than one invariant point need not appear.

If, however, each of its two sides is invariant, then the arc contains at least *two* invariant points; more than two invariant points need not appear.

Of the former of these two evident theorems we have shown in § 2 of the third communication that it can be extended to the most general circular continuum (of which the arc of simple curve can be regarded as the simplest type); to the latter theorem we shall give the same extension in the following.

A segment of the circumference formed by the accessible points of a circular continuum will be called a *complete circumference segment,* if the set of its limiting points is identical to the circular continuum itself.

As the generalization of the arc of simple curve with two invariant sides we can consider a circular continuum q' whose circumference can be divided by two "Schnitte" into two complete circumference segments, both invariant for the transformation.

Of q' together with a certain vicinity ψ' we construct a continuous one-one representation on a finite region of a Cartesian plane, where they pass successively into q and ψ, and we draw in that Cartesian plane a simple closed curve $\varkappa$ lying together with its image and its counterimage in ψ, whilst its inner domain contains q.

All figures to be constructed in the following and likewise their images and their counterimages we suppose to lie in ψ.

According to the third communication q possesses a point I invariant for the transformation; we shall suppose that this point I is the only invariant point of q.

The two Schnitte determining on q the two invariant complete circumference segments o_1 and o_2, we shall represent by S_1 and S_2.

[1]) See these Proceedings Vol. XI, p. 788, Vol. XII, p. 286, Vol. XIII, p. 767. [[1]]

An arc of simple curve joining two points of the circumference of φ, and for the rest not meeting φ, will be called a *skeleton arc.*

We surround φ by a fundamental series of polygons $\mathfrak{P}_1, \mathfrak{P}_2, \mathfrak{P}_3, \ldots$ approximating φ at distances $\varepsilon_1, \varepsilon_2, \varepsilon_3, \ldots \ldots \left(\varepsilon_{k+1} < \frac{1}{8}\varepsilon_k\right)$. The side of the largest square whose inner domain lies between $\mathfrak{P}_n$ and φ, we represent by e_n; for indefinitely increasing n we find that e_n converges to zero.

Each polygon $\mathfrak{P}_k$ we divide into segments in which the distance of the endpoints lies between $4\varepsilon_k$ and $12\varepsilon_k$, and the distance of two arbitrary points does not exceed $24\varepsilon_k$, and we draw from the points which separate these segments, to φ paths $< 2\varepsilon_k$ not intersecting each other, and cutting each polygon $\mathfrak{P}_n$ $(n > k)$ only once. Each two of these paths which immediately succeed each other, form together with the segment of $\mathfrak{P}_k$ connecting them a skeleton arc.

We first suppose that the Schnitt S_1 is *not* determined by an accessible point, and we choose on a fundamental series of polygons $\mathfrak{P}_{\alpha_1}, \mathfrak{P}_{\alpha_2}, \ldots$ a fundamental series of skeleton arcs $s_{\alpha_1}, s_{\alpha_2}, \ldots$, not intersecting each other, converging to a single point P, and all containing between their endpoints the Schnitt S_1. The arc of $\mathfrak{P}_{\alpha_p}$ belonging to s_{α_p} we shall represent by q_{α_p}.

We then construct an arc of simple curve b ending in P, intersecting each element s_{τ_p} of a certain fundamental series $s_{\tau_1}, s_{\tau_2}, \ldots$ (contained in the series of the s_{α_p}) once and only once in a point P_{τ_p} of q_{τ_p}, and passing there from the outside of s_{τ_p} to its inner side. The part of b contained between $P_{\tau_{p-1}}$ and P_{τ_p} we represent by b_{τ_p}, the part of $\mathfrak{P}_{\tau_p}$ preceding resp. following q_{τ_p}, and lying inside $s_{\tau_{p-1}}$, by t_{τ_p} resp. v_{τ_p}. Then it is impossible that as well the part of t_{τ_p} lying to the right of b_{τ_p}, as the part of v_{τ_p} lying to the left of b_{τ_p}, converge to zero; for, in that case P would be an accessible point.

So out of the series of the τ_p we can select such a fundamental series $\beta_1, \beta_2, \ldots$ (preceded in the series of the τ_p successively by the elements $\gamma_1, \gamma_2, \ldots$), and determine to that series such a quantity c that for each β_p is attained on e. g. the part of t_{β_p} lying to the right of b_{β_p} a maximum distance $> 32c$ from P by a certain point Q_{β_p}, whilst neither s_{γ_p}, nor s_{β_p}, nor b_{β_p} reach a distance $> c$ from P, and ε_{γ_p} as well as e_{γ_p} are $< c$.

Then on v_{β_p} lies a point R_{β_p} which can be joined with Q_{β_p} inside

$\mathfrak{P}_{\beta_p}$ by a path $\leqq e_{\gamma_p} \sqrt{2}$, whilst farthermore Q_{β_p} and R_{β_p} may be connected with φ by paths $Q_{\beta_p} H_{\beta_p}$ and $R_{\beta_p} K_{\beta_p} < \frac{3}{2} \varepsilon_{\beta_p}$, lying outside $\mathfrak{P}_{\beta_p}$, and not cutting s_{β_p}, thus containing S_1 between them. These three paths form a skeleton arc $H_{\beta_p} Q_{\beta_p} R_{\beta_p} K_{\beta_p}$ whose size for indefinitely increasing p converges to zero, and which we shall represent by σ_{β_p}.

So out of the series of the β_p we can select a fundamental series $F_1, F_2, \ldots$, in such a way that *for indefinitely increasing p the skeleton arc σ_{F_p} converges to a single point V not identical to P.*

We shall now suppose that the Schnitt S_1 is determined by an accessible point P. Let in that case w be a path leading to P, and let $s_1, s_2, \ldots$ be a fundamental series of skeleton arcs separating S_1 from $\varkappa$, and whose size converges to zero. Then as soon as p has exceeded a certain value, all s_p must cut w, and that in points which for indefinitely increasing p uniformly converge to P, so that s_p converges *for indefinitely increasing p uniformly to P.*

So if S_1 resp. S_2 is not determined by an accessible point coinciding with I, we can construct a skeleton arc $U_1 V_1$ resp. $U_2 V_2$ as small as we like, separating S_1 resp. S_2 from $\varkappa$, and not cutting its image $U'_1 V'_1$ resp. $U'_2 V'_2$, so that either the circumference segment $U_1 V_1$ resp. $U_2 V_2$ is a part of the circumference segment $U'_1 V'_1$ resp. $U'_2 V'_2$, or the circumference segment $U'_1 V'_1$ resp. $U'_2 V'_2$ is a part of the circumference segment $U_1 V_1$ resp. $U_2 V_2$.

Farthermore it is impossible that S_1 and S_2 are determined by accessible points coinciding with each other, for, in that case the derived sets of o_1 and o_2 would have only that *one* point in common, so that o_1 and o_2 would not be complete circumference segments.

On o_1 we choose a point P not coinciding with I; the image of P we represent by P', the image of P' by P'', the counterimage of P by P_i. From $\varkappa$ we draw to P, P', P'', P_i paths w, z, u, v not meeting each other, and containing such endsegments e, e', e'', e_i that e' is the image of e, e'' the image of e', e_i the counterimage of e, and we construct an arc of simple curve k starting in P, not passing through I, cutting o_2, and not meeting w; the image of k we represent by k', the image of k' by k'', the counterimage of k by k_i, the size of k, k', k'', k_i successively by g, g', g'', g_i, the largest resp. smallest one of the latter four quantities by g_h resp. g_l. We describe circles a, a', a'', a_i containing in their inner domains j, j', j'', j_i at a distance g_h successively the arcs k, k', k'', k_i, and we take care to choose k

so small that two arbitrary ones of the sets of points $w+j$, $z+j'$, $u+j''$, $v+j_i$ possess a distance $>8\,g_h$ from each other, that the parts of w, z, u, v contained in j, j', j'', j_i belong entirely to e, e', e'', e_i, and that k cannot contain a skeleton arc separating a Schnitt S_1 or S_2 determined by an accessible point coinciding with I, from the infinite.

Either k or k' contains a point Q of o_2 accessible from $\varkappa$ along a path not cutting $\varphi+k+k'$. In the following we shall assume Q to belong to k; if it were to belong to k', we might consider instead of the given transformation its inverse, and then follow the reasoning of the text.

From $\varkappa$ to Q we lay a path m not cutting $\varphi+k+k'+w$.

The part of k contained between P and Q we represent by r, its image by r', the image of r' by r''. If we then approximate $\varphi+r$ at a sufficiently small distance by a polygon $\mathfrak{P}$, this polygon $\mathfrak{P}$ contains two arcs p_1 and p_2 both connecting w and m, and having no point in common. Together with certain parts of $w+r+m$ these arcs p_1 and p_2 form two polygons $\mathfrak{P}_1$ and $\mathfrak{P}_2$ whose inner domains have no point in common, so that the inner domain of e.g. $\mathfrak{P}_1$ does *not* contain the point I. We then determine the positive sense of circuit of the circumference of φ by a circuit from P to Q inside $\mathfrak{P}_1$.

The circumference segment PQ contains *one* and not more than *one* of the two Schnitte S_1 and S_2: we may assume the Schnitt S_1 to belong to the circumference segment PQ.

Then S_1 cannot be determined by an accessible point coinciding with I; for, in that case r could not contain a skeleton arc separating S_1 from the infinite, so that the point I would be accessible inside $\mathfrak{P}_1$, which is impossible, I lying outside $\mathfrak{P}_1$.

We represent the image of Q by Q', and according to the manner of succession of the points P, P', Q, Q' for a positive sense of circuit we distinguish four cases.

First case: P' precedes P, and Q' precedes Q.

In this case r contains a skeleton arc d separating Q' from the infinite, and accessible from the infinite without a crossing of $\varphi+r+r'$. Let M be the endpoint of d preceding Q' on the circumference of φ, t a segment of d containing M, c the part of r that remains after destroying in r all skeleton arcs separating Q' from the infinite.

Between the image w' of w and t we construct a polygonal line $\mathfrak{P}'_3$, and between t and the image m' of m a polygonal line $\mathfrak{P}'_4$ which both approximate $\varphi+c+r'+r''$ at a distance ε.

The segment cut off from w' resp. t by $\mathfrak{P}'_3$ we represent by

ρ' resp. τ'_3; the segment cut off from t resp. m' by $\mathfrak{P}'_4$ we represent by τ'_4 resp. μ'; the part of t contained between the endpoints of $\mathfrak{P}'_3$ and $\mathfrak{P}'_4$ we represent by τ'. The arcs r', ρ', $\mathfrak{P}'_3$, τ' $\mathfrak{P}'_4$, μ' form together a polygon $\mathfrak{P}'$; I lies outside this polygon. For the lengths of the transformation vector and of the inverse transformation vector inside $\mathfrak{P}'$ there exists a certain minimum i_ε. Let f be a quantity smaller than g_l and smaller than $\frac{1}{8} i_\varepsilon$; then we take care to choose ε so small that

$$\varepsilon < \frac{1}{32} f, \rho' < \frac{1}{32} f, \mu' < \frac{1}{32} f, \tau'_3 < \frac{1}{32} f, \tau'_4 < \frac{1}{32} f.$$

We divide $\mathfrak{P}'_3$ and $\mathfrak{P}'_4$ into segments in which the distance of the endpoints lies between $\frac{1}{8} f$ and $\frac{3}{8} f$, and the distance of two arbitrary points is smaller than $\frac{3}{4} f$. From the points separating these segments we draw to $q + c + r' + r''$ rectilinear paths whose lengths lie between $\frac{1}{2} \varepsilon$ and $\frac{3}{2} \varepsilon$, but among these paths we retain only those whose endpoints do not lie on r, r' or r''. These remaining paths determine together with w', m', τ'_3, and τ'_4 skeleton arcs lying against $\mathfrak{P}'_3$ and $\mathfrak{P}'_4$, and not meeting their counterimage skeleton arcs, whilst these counterimage skeleton arcs can meet neither r nor r'.

The last point of intersection with $\mathfrak{P}'_3$ of the counterimage skeleton arc s separating Q' from the infinite, we represent by L; the image of L we represent by L', the image of s by s', the first point of intersection of r with $\mathfrak{P}'$ by E, the image of E by E'.

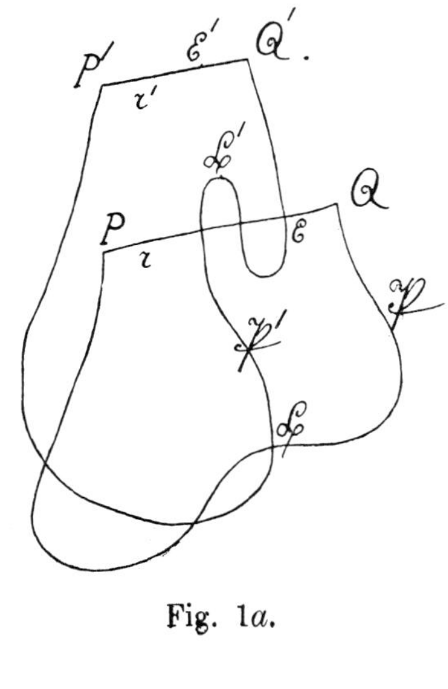

Fig. 1*a*.

A). s' *is separated by* s *from the infinite.* Our aim is to find the total angular variation ω_1 of the inverse transformation vector for a positive circuit of the polygon $\mathfrak{P}'$, and we represent by χ_1 the total angle described by the inverse transformation vector from P' to L' along $\mathfrak{P}'$; by χ_2 the total angular variation of a nowhere vanishing vector of which the origin runs from P' to L' along $\mathfrak{P}'$, and the endpoint as a continuous function of the origin from P to L along *path arcs* nowhere passing outside $\mathfrak{P}'$, constructed according to § 2 of the third communication[1]); by φ_1 the total angle described by the inverse transformation vector along the segment $L'\,E'$ of $\mathfrak{P}'$; by φ_2 the total angular variation of a nowhere vanishing vector of which the origin runs from L' to E' along $\mathfrak{P}'$, and the endpoint as a continuous function of the origin from L tot E along a curve p lying inside $\mathfrak{P}'$[2]); by ψ_1 the total angle described by the inverse transformation vector along the segment $E'\,P'$ of r'; by ψ_2 the total angular variation of a nowhere vanishing vector of which the origin runs from E' to P' along r', and the endpoint as a continuous function of the origin from E to P along a curve obtained by replacing in the segment EP of r each part lying outside $\mathfrak{P}'$ by the segment of $\mathfrak{P}'_4$ joining the same endpoints.

Then the following equations hold:

$$\chi_1 = \chi_2 + 2n\pi \; (n \geqq 0)$$
$$\varphi_1 = \varphi_2$$
$$\psi_1 = \psi_2$$
$$\omega_1 = \chi_1 + \varphi_1 + \psi_1 .$$

Now $\chi_2 + \varphi_2 + \psi_2$ represents the total angular variation of a nowhere vanishing vector of which the origin describes the polygon $\mathfrak{P}'$ in a positive sense, and the endpoint as a continuous function of the origin a closed curve nowhere passing outside $\mathfrak{P}'$, so that we have:

$$\chi_2 + \varphi_2 + \psi_2 = 2\pi .$$

Hence:

$$\omega_1 = 2n\pi \; (n \geqq 1),$$

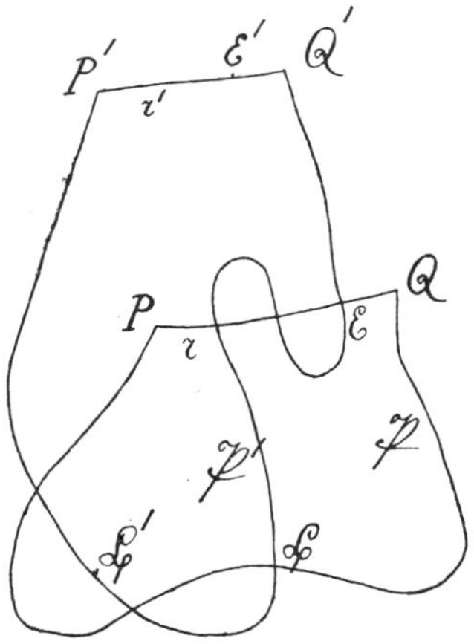

Fig. 1b.

⟦2⟧ 1) See these Proceedings Vol. XIII, p. 770.

2) If L' lies not on $\mathfrak{P}'$, but on one of the paths connecting $\mathfrak{P}'$ and φ, we must take care that p does not meet this path.

⟦238⟧

so that we arrive at the absurd result that inside $\mathfrak{P}'$ must lie an **invariant** point.

B). s' is not separated by s from the infinite. Then the two endpoints of s' as well as the two endpoints of s lie on o_2. Defining $\omega_1, \chi_1, \chi_2, \varphi_1, \varphi_2, \psi_1, \psi_2$ in the same way as just now, we arrive here at the following equations:

$$\chi_1 = \chi_2 + 2n\pi \; (n \geqq 1, \text{ because between } P' \text{ and } s' \text{ lies the Schnitt } S_1)$$
$$\varphi_1 = \varphi_2 - 2\pi$$
$$\psi_1 = \psi_2$$
$$\omega_1 = \chi_1 + \varphi_1 + \psi_1$$
$$\chi_2 + \varphi_2 + \psi_2 = 2\pi.$$

Thus again $\omega_1 = 2n\pi \; (n \geqq 1)$, so that inside $\mathfrak{P}'$ there would have to lie an invariant point.

Second case: P' follows P, and Q' precedes Q.

A). Q' is separated by r from the infinite. We construct the polygonal lines $\mathfrak{P}'_3$ and $\mathfrak{P}'_4$, and the polygon $\mathfrak{P}'$ with its skeleton arcs in the same way as in the first case. Then the counterimage of $\mathfrak{P}'$ is a simple closed curve $\mathfrak{P}$ bearing skeleton arcs which, like those of $\mathfrak{P}'$, cut neither r nor r'. We want to find the total angular variation ϑ_1 of the transformation vector for a positive circuit of $\mathfrak{P}$.

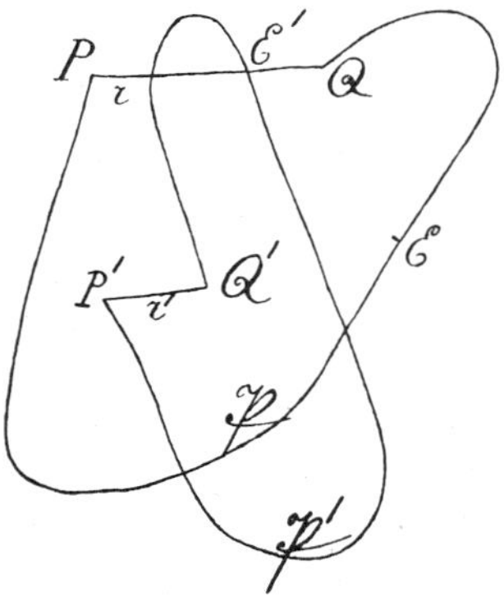

Fig. 2*a*.

We represent by E' the endpoint of $\mathfrak{P}'_3$ on t; by E the counterimage of E'; by χ_1 the total angle described by the transformation vector along the segment PE of $\mathfrak{P}$; by χ_2 the total angular variation of a nowhere vanishing vector of which the origin runs from P to E along $\mathfrak{P}$, and the endpoint as a continuous function of the origin from P' to E' along path arcs nowhere passing outside $\mathfrak{P}$; by ψ_1 the total angle described by the transformation vector along the segment EP of $\mathfrak{P}$; by ψ_2 the total angular variation of a nowhere vanishing vector of which the origin runs from E to P along $\mathfrak{P}$, and the endpoint as a continuous function of the origin along a curve obtained by replacing in the segment $E'P'$ of $\mathfrak{P}'$ each part lying outside $\mathfrak{P}$ by the segment of r joining the same endpoints.

From the equations

$$\chi_1 = \chi_2 + 2n\pi \; (n \geqq 0)$$
$$\psi_1 = \psi_2$$
$$\vartheta_1 = \chi_1 + \psi_1$$
$$\chi_2 + \psi_2 = 2\pi$$

then ensues $\vartheta_1 = 2n\pi\ (n \geqq 1)$, so that inside $\mathfrak{P}$ there would have to lie an invariant point.

B). *Q' is not separated by r from the infinite.* We construct between w' and m' a polygonal line approximating $\varphi + r + r' + r''$ at a distance ε, cutting off from w' resp. m' the segment $\digamma'$ resp. μ', and forming with $\digamma'$, r', and μ' a polygon $\mathfrak{P}'$. The determination of ε, and the construction of the skeleton arcs of $\mathfrak{P}'$ take place in the same way as in the first case. We want to find the total angular variation ϑ_1 of the transformation vector for a positive circuit of the counterimage $\mathfrak{P}$ of $\mathfrak{P}'$, and we understand by ϑ_2 the total angular variation of a nowhere vanishing vector of which the origin describes $\mathfrak{P}$, and the endpoint as a continuous function of the origin runs first from P' to Q' along path arcs nowhere passing outside $\mathfrak{P}$, and finally describes r'.

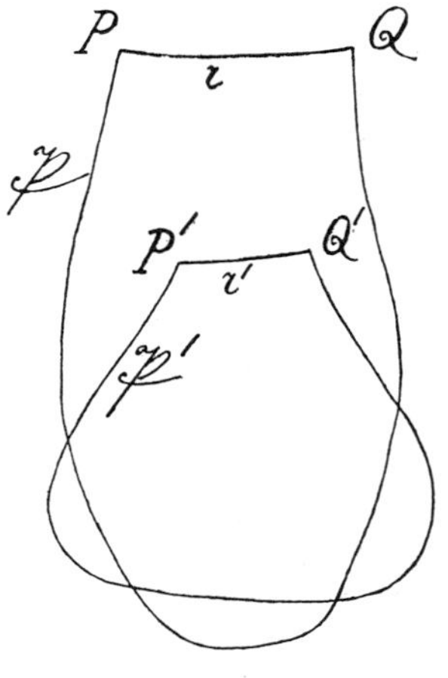

Fig· 2*b*.

Then we have:

$$\vartheta_1 = \vartheta_2 + 2n\pi\ (n \geqq 0)$$

$$\vartheta_2 = 2\pi.$$

Hence $\vartheta_1 = 2n\pi\ (n \geqq 1)$, so that inside $\mathfrak{P}$ there would have to lie an invariant point.

Third case: P' follows P, and Q' follows Q.

In this case r contains a skeleton arc d separating Q' from the infinite, and accessible from the infinite without a crossing of $\varphi + r + r'$. We determine c, t, and ε, and we construct $\mathfrak{P}'_3$, $\mathfrak{P}'_4$, $\mathfrak{P}'$, $\mathfrak{P}$, and the skeleton arcs of these polygons in the same way as in the second case under *A*).

The last point of intersection with $\mathfrak{P}$ of the skeleton arc s' of $\mathfrak{P}'_3$ separating Q from the infinite, we represent by L'; the counterimage of L' we represent by L, the counterimage of s' by s, the endpoint of $\mathfrak{P}'_3$ on t by E', the counterimage of E' by E.

A). *s is separated by s' from the infinite.* Our aim is to find the total angular variation ϑ_1 of the transformation vector for a positive circuit of $\mathfrak{P}$, and we represent by χ_1 the total angle described by the transformation vector from P to L along $\mathfrak{P}$; by χ_2 the total angular variation of a nowhere vanishing vector of which the origin runs from P to L along $\mathfrak{P}$, and the endpoint as a continuous function of the origin from P' to L' along path arcs nowhere passing outside $\mathfrak{P}$; by φ_1 the total angle described by the transformation vector from L to E along $\mathfrak{P}$; by φ_2 the total angular variation of a nowhere vanishing vector of which the origin runs from L to E along $\mathfrak{P}$, and the endpoint as a continuous function of the origin inside $\mathfrak{P}$ from L' to E' along an arc of simple curve p; by ψ_1 the total angle described by the transformation vector from E to P

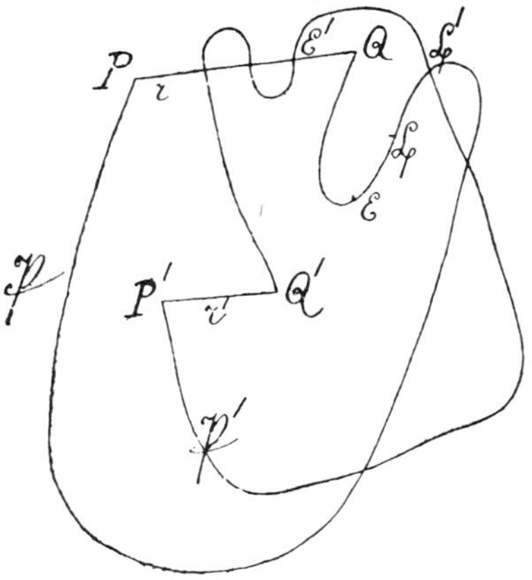

Fig. 3*a*.

along $\mathfrak{P}$; by ψ_2 the total angular variation of a nowhere vanishing vector of which the origin runs from E to P along $\mathfrak{P}$, and the endpoint as a continuous function of the origin along a curve obtained by replacing in the segment $E'P'$ of $\mathfrak{P}'$ each part lying outside $\mathfrak{P}$ by the segment of r joining the same endpoints.

Then the following equations hold:

$$\chi_1 = \chi_2 + 2n\pi \ (n \geqq 0)$$
$$\varphi_1 = \varphi_2 + 2\pi$$
$$\psi_1 = \psi_2$$
$$\vartheta_1 = \chi_1 + \varphi_1 + \psi_1$$
$$\chi_2 + \varphi_2 + \psi_2 = 2\pi.$$

Hence $\vartheta_1 = 2n\pi \ (n \geqq 2)$, so that inside $\mathfrak{P}$ there would have to lie an invariant point.

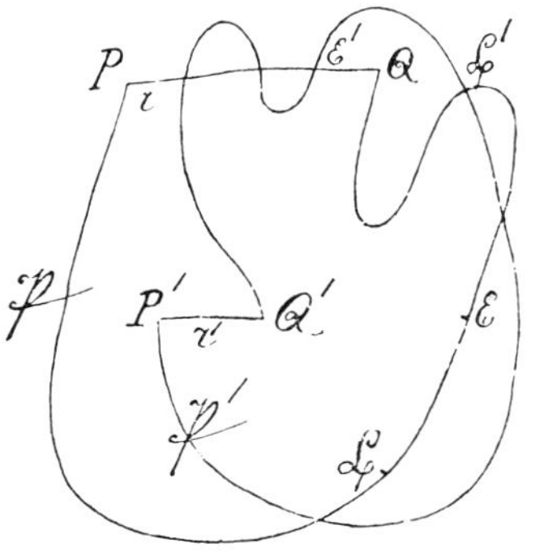

Fig. 3*b*.

B). *s is not separated by s' from the infinite*. Then the two endpoints of s as well as the two endpoints of s' lie on o_2. Defining $\vartheta_1, \chi_1, \chi_2, \varphi_1, \varphi_2, \psi_1, \psi_2$ in the same way as just now, we arrive here at the following equations:

$\chi_1 = \chi_2 + 2n\pi$ ($n \geqq 1$, because between P and s lies the Schnitt S_1)

$$\varphi_1 = \varphi_2$$
$$\psi_1 = \psi_2$$
$$\vartheta_1 = \chi_1 + \varphi_1 + \psi_1$$
$$\chi_2 + \varphi_2 + \psi_2 = 2\pi.$$

Thus again $\vartheta_1 = 2n\pi \ (n \geqq 2)$, so that inside $\mathfrak{P}$ there would have to lie an invariant point.

Fourth case: P′ precedes P, and Q′ follows Q.

A). *Q′ is separated by r from the infinite.* We construct the polygon $\mathfrak{P}'$ with its skeleton arcs in the same way as in the third case. We want to find the total angular variation ω_1 of the inverse transformation vector for a positive circuit of $\mathfrak{P}'$, and we represent by χ_1 the total angle described by the inverse transformation vector along the segment $P'Q'$ of $\mathfrak{P}'$; by χ_2 the total angular variation of a nowhere vanishing vector of which the origin runs from P' to Q' along $\mathfrak{P}'$, and the endpoint as a continuous function of the origin from P to Q along path arcs nowhere passing outside $\mathfrak{P}'$; by ψ_1 the total angle described by the inverse transformation vector from Q' to P' along r'; by ψ_2 the total angular variation of a nowhere vanishing vector of which the origin runs from Q' to P' along r', and the endpoint as a continuous function of the origin from Q to P along a curve obtained by replacing in r each part lying outside $\mathfrak{P}'$ by the segment of $\mathfrak{P}'_4$ joining the same endpoints.

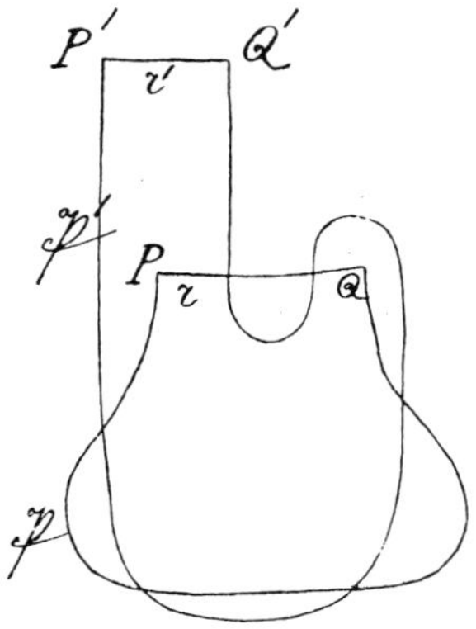

Fig. 4*a*.

From the equations

$$\chi_1 = \chi_2 + 2n\pi \; (n \geqq 0)$$
$$\psi_1 = \psi_2$$
$$\omega_1 = \chi_1 + \psi_1$$
$$\chi_2 + \psi_2 = 2\pi$$

then ensues $\omega_1 = 2n\pi \; (n \geqq 1)$, so that inside $\mathfrak{P}'$ there would have to lie an invariant point.

B). *Q′ is not separated by r from the infinite.* We construct the polygon $\mathfrak{P}'$ with its skeleton arcs in the same way as in the second case under *B*). We want to find the total angular variation ω_1 of the inverse transformation vector for a positive circuit of $\mathfrak{P}'$, and we understand by ω_2 the total angular variation of a nowhere vanishing vector of which the origin describes $\mathfrak{P}'$, and the endpoint as a continuous function of the origin runs first from P to Q along path arcs nowhere passing outside $\mathfrak{P}'$, and finally describes r.

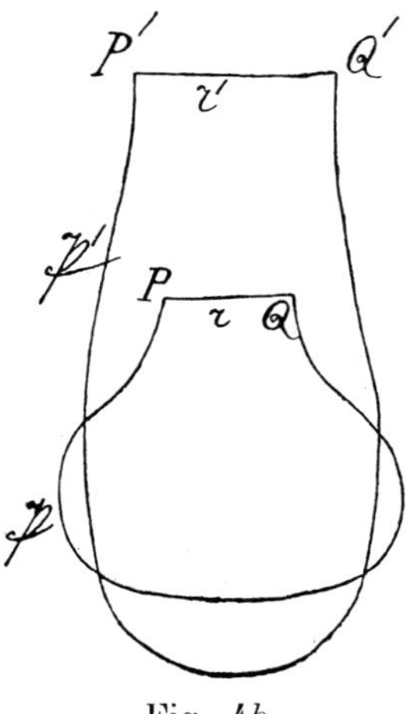

Fig. 4*b*.

Then we have:

$$\omega_1 = \omega_2 + 2n\pi \; (n \geqq 0)$$
$$\omega_2 = 2\pi.$$

Hence $\omega_1 = 2n\pi \; (n \geqq 1)$, so that inside $\mathfrak{P}'$ there would have to lie an invariant point.

With this we have completely proved the following

THEOREM. *For a continuous one-one transformation with invariant indicatrix of a two-sided surface in itself a circular continuum with two separated invariant complete circumference segments contains at least two invariant points.*

E R R A T A.

In the 3rd communication on this subject, these Proceedings Vol. XIII ⟦2⟧

p. 767, l. 6 from top	for: indicated, but	read: indicated but
l. 20 from top	for: *paraboli*	read: *parabolic*

NOTES

⟦1⟧ The English version appeared in the Proceedings of the meeting of 29 October 1911. Continues 1909 F2, 1909 H2, 1911 H2. See the remark at the end of 1912 K2.

⟦2⟧ Brouwer 1911 H2.

1910 F

Über eineindeutige, stetige Transformationen von Flächen in sich.

Von

L. E. J. Brouwer in Amsterdam.

Einer freundlichen Anregung von Herrn Hilbert folgend referiere ich hier kurz über Untersuchungen, welche ich in den Amsterdamer Berichten*) veröffentliche, und welche darauf abzielen, daß eine willkürliche eineindeutige, stetige, und ohne Verrückung der Fixpunkte stetig erreichbare Transformation einer zweiseitigen Fläche in sich in eine Translation umgesetzt wird, nachdem man die Fläche in Teilgebiete zerlegt und diese Teilgebiete nötigenfalls durch Abwickelung erweitert hat.

Die Untersuchungen stützen sich in ausgiebigster Weise auf die Schoenfliessche Theorie der Analysis Situs, erfordern aber eine eingehendere Analyse der Begriffe der *geschlossenen Kurve* und des *Kurvenbogens*, welche den eigentümlichen Gestalten, auf welche ich in meiner Note zur
⟦2⟧ Analysis Situs (Math. Ann., Bd. 68, S. 422) hingewiesen habe, Rechnung trägt.

Unter einer *geschlossenen Kurve* auf einer im Sinne der Analysis Situs willkürlichen Fläche verstehen wir eine geschränkte, perfekte, zusammen-
⟦3⟧ hängende Menge, welche die Punkte einer sie approximierenden Umgebung so in zwei Gebiete zerlegt, daß jeder Punkt der Menge gemeinsamer Grenzpunkt dieser beiden Gebiete ist.

M. a. W. die geschlossene Kurve besitzt zwei *Konturen* von erreichbaren Punkten; auf jeder von diesen kann man den zyklischen Ordnungstypus η der erreichbaren Punkte konstruieren. Durch zwei Schnitte wird dieser Ordnungstypus in zwei Teilmengen zerlegt; die Ableitung einer solchen Teilmenge nennen wir, wenn sie nicht mit der ganzen geschlossenen Kurve identisch ist, einen *eigentlichen Kurvenbogen*.

⟦1⟧ *) Amsterdamer Berichte, holländische Ausgabe XVII 2, S. 741; XVIII 1, S. 106; englische Ausgabe XI 2, S. 788; XII 1, S. 286; und weitere demnächst erscheinende Fortsetzungen.

Ein eigentlicher Kurvenbogen, oder kurzweg Kurvenbogen, bewirkt weder für die ganze Fläche, noch für die Punkte einer sie approximierenden Umgebung eine Gebietsteilung, m. a. W. besitzt nur *eine einzige Kontur*; es fragt sich nun, wann ein ~~geschränktes~~, abgeschlossenes, nirgends dichtes Kontinuum K mit einer einzigen Kontur als Kurvenbogen zu betrachten ist. Die Antwort lautet folgendermaßen: [[4]]

Man konstruiere auf der Kontur den zyklischen Ordnungstypus η_e der erreichbaren Punkte. Es kann nun vorkommen, daß ein zwischen einem nach rechts und einem nach links gerichteten Schnitte enthaltenes Teilsegment von η_e die ganze Menge K als Ableitung besitzt. Ein solches Teilsegment heiße eine *„vollständige Teilkontur“*. K ist nun dann und nur dann ein Kurvenbogen, wenn man in ihm zwei vollständige Teilkonturen bestimmen kann, welche entweder völlig getrennt sind, oder höchstens zwei sich unbeschränkt zusammenziehende Teilsegmente gemeinschaftlich haben.*)

Die Kurvenbogen lassen sich unterscheiden in *singuläre* und *nichtsinguläre*.

Letztere besitzen zwei *Enden* E_1 und E_2, welche voneinander getrennt sind und die Kontur außerhalb der Enden in zwei vollständige Teilkonturen zerlegen. Weiter kann man sie nach Tilgung der Enden in mannigfacher Weise zerlegen in einen Ordnungstypus $^*\omega + \omega$ von ebenfalls nichtsingulären Teilkurvenbogen, welche auf jeder Seite gegen ein Ende konvergieren, während jeder von ihnen wenigstens einen Punkt besitzt, welcher außerhalb aller der übrigen liegt.

Jede auf einer der beiden vollständigen Teilkonturen gegen ein Ende konvergierende Fundamentalreihe von erreichbaren Punkten bestimmt durch eine Teilreihe eine Fundamentalreihe von Teilkurvenbogen der eben genannten Art; jeder Teilkurvenbogen enthält sodann einen Punkt der Teilreihe, und jeder Punkt der Teilreihe liegt in einem und nur in einem der Teilkurvenbogen; weiter kann man mittels dieser Teilkurvenbogen leicht eine mit der ersten *zusammengehörige* Fundamentalreihe von erreichbaren Punkten auf der anderen vollständigen Teilkontur konstruieren.

Jede vollständige Teilkontur eines singulären Kurvenbogens enthält wenigstens einen nach rechts bzw. nach links gerichteten *singulären Schnitt*, d. h. einen Schnitt, der ein gewisses ihm beim Umlaufe rechts bzw. links herum unmittelbar vorangehendes Segment des Ordnungstypus η_e ganz enthält. Der singuläre Schnitt ist ein *spezieller singulärer Kurvenbogen* d. h. ein solcher, *der sich nicht in zwei Teilkurvenbogen zerlegen läßt*.

*) Es kommt auf dasselbe hinaus, einer gewissen vollständigen Teilkontur k die Bedingung aufzuerlegen, daß keine abgeschlossene zusammenhängende *Teil*menge von K existiert, welche von k sowohl ein Anfangs-, wie ein Endsegment besitzt.

Auf der Kontur eines *speziellen singulären Kurvenbogens* kann es mehr als zwei voneinander getrennte vollständige Teilkonturen geben. In meiner Note zur Analysis Situs finden sich Beispiele mit drei und mit vier vollständigen Teilkonturen. Eine solche vollständige Teilkontur enthält sodann immer einen Schnitt, welcher mit dem ganzen Kurvenbogen identisch ist.

Zu einer Fundamentalreihe von erreichbaren Punkten, welche auf einer vollständigen Teilkontur gegen einen singulären Schnitt konvergieren, gibt es niemals eine mit ihr zusammengehörige Fundamentalreihe von erreichbaren Punkten auf einer anderen vollständigen Teilkontur.

Diese vorangehenden Begriffe werden benutzt bei der Untersuchung der eineindeutigen, stetigen, und ohne Verrückung der Fixpunkte stetig erreichbaren Transformationen einer zweiseitigen Fläche in sich; mit ihrer Hilfe wird nämlich zunächst folgender Hilfssatz bewiesen:

„*Tritt bei einer Transformation der genannten Art ein invariantes Kontinuum mit einer einzigen Kontur auf, so enthält es dann wenigstens einen invarianten Punkt, wenn es entweder in einer Kreisfläche enthalten, oder nirgends dicht und in einer Parabelfläche enthalten ist.*“*)

Sodann konstruieren wir zu einem willkürlichen nicht invarianten Punkt der Fläche ein diesen Punkt enthaltendes *Transformationsfeld D*, d. h. ein Gebiet, das ganz außerhalb seines Bildgebietes liegt, aber diese Eigenschaft verliert durch jede Ausdehnung, welche entweder direkt oder nach unendlich kleiner Abänderung der Grenze mit ihm vorgenommen wird.

Wenn wir nun unter einer *J-Menge* eine nur invariante Punkte enthaltende Kontur verstehen (welche sich auch auf einen einzigen Punkt zusammenziehen kann), so ist unser wieder mittels der Theorie des Kurvenbogens erreichtes Endresultat, daß wir das Transformationsfeld immer in einer der folgenden Gestalten erhalten können:

Erster Fall. Es bestimmt ein Restgebiet R, welches den Zusammenhang des Restgebietes eines Loches besitzt.**)

Ist dann die Grenze von R die volle Grenze von D, so wird diese gebildet von zwei, zwischen denselben beiden J-Mengen als Enden oder zwischen einer und derselben J-Menge als Ende und einem und demselben Rande verlaufenden, außerhalb dieser Enden einfachen Kurvenbogen, deren einer das Bild des anderen ist.

Ist aber die Grenze von R nicht die volle Grenze von D, so wird

*) Die Forderung, daß die Transformation ohne Verrückung der Fixpunkte stetig erreichbar sein muß, ist übrigens für diesen Satz nicht notwendig.

**) Das Loch kann auch an einen Rand grenzen, und dementsprechend das Restgebiet den Zusammenhang der Fläche selbst besitzen.

diese gebildet entweder von zwei zusammenziehbaren, einfachen, geschlossenen Kurven, deren eine das Bild der anderen ist, oder von zwei sich schließenden einfachen Kurvenbogen, deren einer das Bild des anderen ist.

Diese beiden sich schließenden Kurvenbogen verlaufen entweder zwischen denselben beiden miteinander zusammenhängenden J-Mengen als Enden, oder zwischen demselben Rande und derselben mit diesem Rande zusammenhängenden J-Menge, oder schließlich rücken beide auf beiden Seiten in einen und denselben Rand. Ihre Konturen sind zusammenziehbar.

Zweiter Fall. Es bestimmt kein Restgebiet, welches den Zusammenhang des Restgebietes eines Loches besitzt. Es besitzt also wenigstens eine nicht zusammenziehbare Kontur.

Die Grenze von D wird dann gebildet entweder von zwei nicht zusammenziehbaren, einfachen, geschlossenen Kurven, deren eine das Bild der anderen ist, oder wieder von zwei ~~sich schließenden~~ einfachen Kurvenbogen, deren einer das Bild des anderen ist. ⟦5⟧

Diese beiden ~~sich schließenden~~ Kurvenbogen verlaufen hier entweder ⟦5a⟧ zwischen denselben beiden ~~miteinander zusammenhängenden~~ J-Mengen, ⟦6⟧ oder zwischen demselben Rande und derselben ~~mit einem Rande zusammenhängenden~~ ⟦7⟧ J-Menge, oder schließlich zwischen denselben beiden Rändern (welche auch identisch sein können). Ihre Konturen sind nicht zusammenziehbar.

Unterziehen wir nun das Feld D unbeschränkt immer wieder der Transformation selbst, sowie ihrer inversen, so konstruieren wir, indem wir eventuell die Fläche durch Abwickelung erweitern, einen Ordnungstypus $^*\omega + \omega$ von Bildern von D, welche außerhalb voneinander liegen, und zusammen ein Gebiet B ausfüllen, in dem außerhalb der Grenze die Transformation eineindeutiges und für jede abgeschlossene Teilmenge gleichmäßig stetiges Bild ist einer *Translation.* ~~der Cartesischen Ebene in sich oder des Kreiszylinders in sich.~~ ⟦8⟧

Und die ganze Fläche läßt sich ~~in~~ durch solche Gebiete ~~zerlegen~~ erschöpfen. ⟦9⟧

Als Nebenresultate kommen bei dieser Analyse folgende, in meiner zweiten Mitteilung über endliche kontinuierliche Gruppen*) zur Anwendung ⟦10⟧ gelangende Sätze heraus:

„Jede eineindeutige, stetige Transformation der Cartesischen Ebene in ⟦11⟧ *sich, welche den Umlaufssinn nicht ändert, ist entweder über die ganze*

*) Dieser Band, nächster Aufsatz.

Ebene eineindeutiges, stetiges Bild einer Translation, oder läßt wenigstens einen Punkt invariant.“

„Jede eineindeutige, stetige Transformation der Kugel in sich, welche den Umlaufssinn nicht ändert, läßt wenigstens einen Punkt invariant.“

„Jede eineindeutige, stetige Transformation der projektiven Ebene in sich läßt wenigstens einen Punkt invariant.“

Die Sätze gelten *nicht* für Transformationen, welche den Umlaufssinn ändern; Beispiele davon sind leicht beizubringen.

Der die Kugel betreffende Satz läßt sich in folgender Weise erweitern:

„Jede eineindeutige, stetige Transformation einer Kugel gerader Dimensionenzahl in sich, welche den Sinn der Indikatrix nicht ändert, oder einer Kugel ungerader Dimensionenzahl in sich, welche den Sinn der Indikatrix umkehrt, läßt wenigstens einen Punkt invariant.“

Amsterdam, August 1909.

NOTES

[[1]] Brouwer 1909 F2, 1909 H2, 1911 H2, 1911 J2.
[[2]] Brower 1910 C.
[[3]] Inserted: ‘zylinderförmigen’ after ‘einer’. All these corrections are in Brouwer’s hand. They were published in 1910 L and 1919 F.
[[4]] Changed: ‘geschränktes’ into ‘in einer Kreisfläche enthaltenes’.
[[5]] Changed: ‘sich schliessenden’ into ‘ausserhalb der Enden’.
[[5a]] Changed: ‘sich schliessenden’ into ‘einfachen’.
[[6]] Deleted: ‘mit einander zusammenhängenden’.
[[7]] Deleted: ‘mit einem Rande zusammenhängenden’.
[[8]] Deleted: ‘der Cartesischen Ebene in sich oder des Kreiszylinders in sich’. This deletion reduces the ‘general’ to the ‘plane’ translation theorem. According to Brouwer 1912 K2, p. 310 the proof of the general translation theorem (including all two-sided surfaces) in 1909 H2 was insufficient, whereas the plane translation theorem had been arrived at along different lines. See also 1909 H2 [[1]].
[[9]] Changed ‘in’ into ‘durch’, ‘zerlegen’ into ‘erschöpfen’.
[[10]] Brouwer 1910 H.
[[11]] Actual proof in 1912 B. See also 1909 H2 [[1]].

MA 69, 592 **1910 L**

Berichtigung

[[1]]

zu dem Aufsatz von L. E. J. Brouwer: „Über eineindeutige, stetige Transformationen von Flächen in sich", S. 176—180 dieses Bandes

S. 176, Z. 9 v. u.: statt „sie approximierenden" lies „zylinderförmigen sie approximierenden".

S. 177, Z. 4: statt „geschränktes" lies „in einer Kreisfläche enthaltenes".

S. 179, Z. 16: statt „sich schließenden einfachen" lies „außerhalb der Enden einfachen".

ibid., Z. 18: statt „sich schließenden" lies „einfachen".

NOTE

[[1]] Corrections to 1910 F.

MA 79, 403 **1919 F**

Berichtigung

[[1]]

zu dem Aufsatze von L. E. J. Brouwer: „Über eineindeutige, stetige Transformationen von Flächen in sich." Math. Ann. 69, S. 176—180.

S. 179, Z. 19 zu streichen: miteinander zusammenhängenden

Z. 20—21 zu streichen: mit einem Rande zusammenhängenden

Z. 30—31 zu streichen: der Cartesischen Ebene in sich oder des Kreiszylinders in sich

Z. 32 statt: in solche Gebiete zerlegen
zu lesen: durch solche Gebiete erschöpfen.

NOTE

[[1]] Corrections to 1910 F.

1912 B

Beweis des ebenen Translationssatzes.

Von

⟦1⟧ L. E. J. Brouwer in Amsterdam.

In der Cartesischen Ebene verstehen wir unter einer *einfachen offenen Linie* eine ihre Grenzpunkte enthaltende Punktmenge, welche eineindeutiges und stetiges Bild der geraden Linie ist.

In einer einfachen offenen Linie werden durch einen willkürlichen ihrer Punkte zwei zusammenhängende, ihre Grenzpunkte enthaltende Teilmengen bestimmt, welche nur den genannten Punkt gemeinschaftlich haben. Eine solche Teilmenge nennen wir eine *einfache offene Halblinie.*

Unter einem zu einer eineindeutigen, stetigen, den Umlaufssinn nicht ändernden Transformation der Cartesischen Ebene in sich gehörigen *Translationsfelde* verstehen wir ein Gebiet, welches außerhalb seines Bildgebietes liegt, und von zwei einander nicht treffenden einfachen offenen Linien, deren eine das Bild der anderen ist, begrenzt wird.

Eine eineindeutige, stetige, den Umlaufssinn nicht ändernde Trans-
⟦2⟧ formation der Cartesischen Ebene in sich soll *über die ganze Ebene eine Translation* heißen, wenn sich zu jedem Punkte der Ebene ein diesen Punkt enthaltendes Translationsfeld angeben läßt.

Es gilt nun folgender

Translationssatz. *Jede eineindeutige, stetige, den Umlaufssinn nicht*
⟦2⟧ *ändernde Transformation der Cartesischen Ebene in sich, welche keinen*
⟦4⟧ *Punkt invariant läßt, ist über die ganze Ebene eine Translation.**)

Von diesem Satze führe ich im folgenden den vollständigen Beweis.

⟦3⟧ *) Vgl. Math. Ann. 69, wo der Satz S. 179 formuliert und S. 199 angewandt wird.
⟦2⟧ Ich mache hier darauf aufmerksam, daß die Identifizierung jedes Punktes mit seinem Bildpunkte nicht immer aus der Ebene eine neue Fläche zu erzeugen braucht, wie dies bei der zitierten gruppentheoretischen Anwendung der Fall ist.

§ 1.

Theorie der Bahnkurven.

Wir denken also die Cartesische Ebene einer eineindeutigen, stetigen, den Umlaufssinn nicht ändernden Transformation, welche keinen Punkt invariant läßt, unterzogen.

Alsdann nennen wir einen einfachen Kurvenbogen AA', dessen Endpunkt A' Bildpunkt des Anfangspunktes A ist, und der außer dem Punkte A' mit seinem Bilde keinen Punkt gemeinschaftlich hat, einen *Translationsbogen.*

Indem wir einen Translationsbogen immer wieder der gegebenen Transformation sowie ihrer inversen unterziehen, erzeugen wir eine stetige Kurve, welche wir eine *Bahnkurve* nennen wollen.

Einen Teilbogen einer Bahnkurve, von dessen Endpunkten eines das Bild des anderen ist, nennen wir ein *Bahnsegment.*

Einen Teilbogen einer Bahnkurve, der ein Bahnsegment als Teil enthält, nennen wir einen *Bahnbogen.*

Ein einfacher Kurvenbogen, der sein Bild nicht trifft, soll ein *freier Bogen* heißen.

Unter dem *Transformationsvektor* eines Punktes der Ebene verstehen wir den Vektor, der diesen Punkt mit seinem Bildpunkte verbindet.

Satz 1. *Eine Bahnkurve kann keinen Doppelpunkt besitzen.*

Beweis. Falls ein Doppelpunkt existiert, so enthält gewiß auch der in A anfangende Teil der Bahnkurve einen Doppelpunkt. Sei also P' der Punkt, in dem dieser Teil *fürs erste Mal* sich selbst trifft.

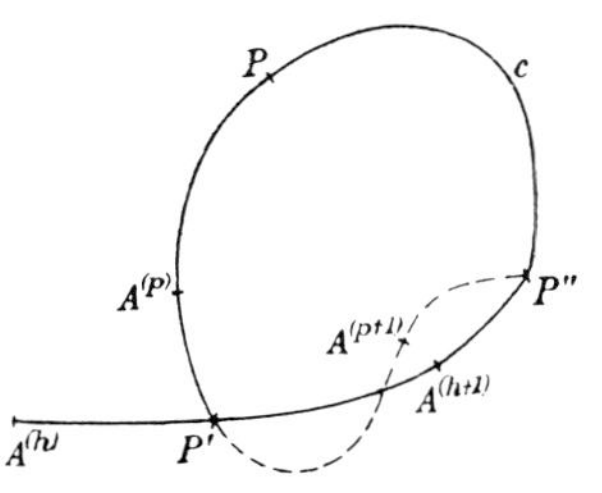

Fig. 1.

Auf der in dieser Weise gebildeten einfachen geschlossenen Kurve c müssen dann wenigstens zwei Punkte der durch beliebige Wiederholung der Transformation aus A erzeugten Punktreihe $A, A', A'', \cdots$ liegen. Der erste auf c liegende Punkt dieser Reihe sei $A^{(h+1)}$, der letzte $A^{(p)}$.

Sei P'' das Bild von P', P das inverse Bild von P'. Wenn wir die Bahnkurve noch über P' hinaus bis P'' verlängern, so kann dieses (in Fig. 1 gestrichelte) Bahnsegment $P'P''$ von der geschlossenen Kurve c nur den Bogen $P'A^{(h+1)}$ treffen. Der Kurvenbogen $P''A^{(p)}P'$, der sowohl mit dem Bahnsegmente $P'A^{(h+1)}P''$, wie mit dem Bahnsegmente $P'A^{(p+1)}P''$ eine einfache geschlossene Kurve bildet, hat für diese beiden geschlossenen Kurven dieselbe Innenseite, kann mithin durch die Bahnsegmente $P'A^{(h+1)}P''$ und $P'A^{(p+1)}P''$ zusammen nicht vom Unendlichen getrennt werden.

Wir führen nun von einem nirgends verschwindenden Vektor, dessen Anfangspunkt das Bahnsegment $PA^{(p)}P'$ durchläuft, den Endpunkt als stetige Funktion des Anfangspunktes von P' nach P'', ein erstes Mal auf dem Bahnsegmente $P'A^{(h+1)}P''$, ein zweites Mal auf dem Bahnsegmente $P'A^{(p+1)}P''$. Der dabei vom Vektor beschriebene Drehungswinkel ist, weil die Bogen $P'A^{(h+1)}P''$ und $P'A^{(p+1)}P''$ zusammen den Bogen $PA^{(p)}P'$ (oder, falls die Punktepaare $A^{(h+1)}$, P'' und P, $A^{(p)}$ einander auf c trennen, eine Seite des Bogens $PA^{(p)}P'$) *nicht* vom Unendlichen trennen, beide Male derselbe.

Der totale Drehungswinkel, den der Transformationsvektor bei einem vollen Umlauf von c beschreibt, wird somit auch erhalten, wenn wir von einem nirgends verschwindenden Vektor zunächst den Anfangspunkt den zu c gehörigen Bogen $P'A^{(h+1)}P$ und den Endpunkt als stetige Funktion des Anfangspunktes den Bogen $P''A^{(p)}P'$, und sodann den Anfangspunkt den Bogen $PA^{(p)}P'$ und den Endpunkt als stetige Funktion des Anfangspunktes den Bogen $P'A^{(h+1)}P''$ beschreiben lassen.

Der hierbei in den Punkten von c auftretende Vektor gehört aber als Transformationsvektor zu einer Rotation der geschlossenen Linie c in sich, sodaß sein totaler Drehungswinkel 2π beträgt.

Dies aber würde nach sich ziehen, daß die Transformation der Ebene innerhalb c einen Fixpunkt besäße, was ein Widerspruch ist.

Satz 2. *Eine Bahnkurve kann bei beliebig weiter Fortsetzung vorwärts und rückwärts nicht in beliebige Nähe eines ihrer Punkte zurückkehren.*

Beweis. Wir nehmen an, daß die Bahnkurve bei beliebig weiter Fortsetzung vorwärts in beliebige Nähe ihres Punktes P zurückkehrt. Dann existiert eine einfache geschlossene Kurve c, welche sich aus einem Bahnbogen PQ und einem freien Bogen PQ, der auch das P vorangehende Bahnsegment P_iP nicht trifft, zusammensetzt. Wir bezeichnen den Bahnbogen PQ mit β, den (in Fig. 2 gestrichelten) freien Bogen PQ mit α, die Bilder von c, P, Q, β, α der Reihe nach mit c', P', Q', β', α', die inversen Bilder von P resp. Q mit P_i resp. Q_i.

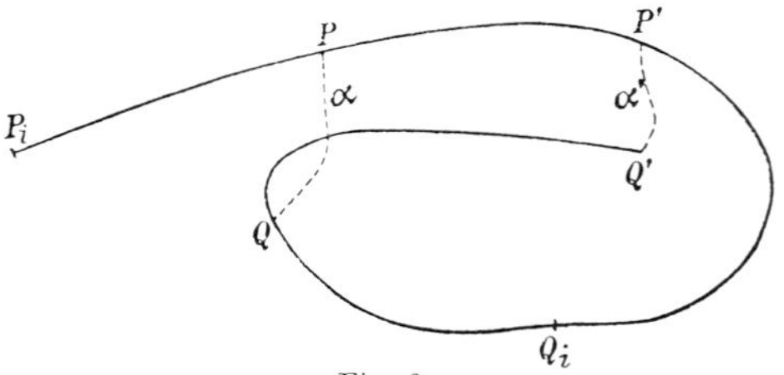

Fig. 2.

Der totale Drehungswinkel des Transformationsvektors für das Bahnsegment Q_iQ kann erzeugt werden, indem wir von einem nirgends verschwindenden Vektor zunächst den Anfangspunkt das Bahnsegment Q_iQ und den Endpunkt als stetige Funktion des Anfangspunktes den Bogen $QQ'P'$ beschreiben lassen, und sodann den Anfangspunkt in Q festhalten, und den Endpunkt auf α' von P' nach Q' zurückführen. In dieser Bahn

[[252]]

des Endpunktes dürfen wir aber den Bogen $QQ'P'$ durch QPP' ersetzen; die eine Seite des Bahnsegmentes $Q_i Q$ wird nämlich, auch falls die Punktepaare P, P' und Q_i, Q einander auf c trennen sollten, durch $QQ'P' + QPP'$ nicht vom Unendlichen getrennt.

Der totale Drehungswinkel des Transformationsvektors für den freien Bogen α kann erzeugt werden, indem wir von einem Vektor zunächst den Anfangspunkt in Q festhalten und den Endpunkt auf α' von Q' nach P' führen, und sodann den Endpunkt in P' festhalten und den Anfangspunkt auf α von Q nach P führen.

Der totale Drehungswinkel des Transformationsvektors für einen vollen Umlauf von c kann somit erzeugt werden, indem wir von einem nirgends verschwindenden Vektor den Anfangspunkt und den Endpunkt je einen vollen Umlauf von c vollbringen lassen Dieser totale Drehungswinkel beträgt somit 2π.

Dies aber würde wieder nach sich ziehen, daß die Transformation der Ebene innerhalb c einen Fixpunkt besäße, was ein Widerspruch ist.

Aus Satz 2 dürfen wir nicht folgern, daß eine Bahnkurve eine einfache offene Linie ist. Vielmehr können bei beliebig weiter Fortsetzung vorwärts sehr gut nicht zur Bahnkurve gehörige Grenzpunkte auftreten. Die von diesen Grenzpunkten gebildete Punktmenge nennen wir das *finale Ende* F der Bahnkurve. In analoger Weise wird bei beliebig weiter Fortsetzung rückwärts ein *initiales Ende* I der Bahnkurve bestimmt. Satz 2 besagt aber, daß *die Bahnkurve keines ihrer beiden Enden treffen kann.*

Wir wollen gleich ein Beispiel einer Bahnkurve, welche ein finales Ende besitzt, herstellen. Wir betrachten dazu die durch folgende Formeln ausgedrückte Transformation:

$$\text{für } x > \frac{1}{e}: \qquad x' = \frac{x}{e}, \qquad y' = 3y + 1,$$

$$\text{für } \frac{1}{e} > x > 0: \qquad x' = \frac{x}{e}, \qquad y' = y\left\{1 - \frac{2}{\lg(x)}\right\} + 1,$$

$$\text{für } x < 0: \qquad x' = x, \qquad y' = y + 1.$$

Sei A der Punkt mit den Koordinaten $\frac{1}{e}$ und 0, B der Punkt mit den Koordinaten $\frac{1}{e}$ und -2, C der Punkt mit den Koordinaten $\frac{1}{e^2}$ und 1. Aus den Strecken AB und BC können wir einen Translationsbogen zusammensetzen. Dieser Translationsbogen bestimmt eine Bahnkurve, welche die gerade Linie $x = 0$ zum finalen Ende besitzt.

Satz 3. *Kein Ende einer Bahnkurve kann ganz im Endlichen enthalten sein.*

Beweis. Ein im endlichen enthaltenes Ende wäre nämlich eine zusammenhängende, perfekte Punktmenge, welche ein die Bahnkurve enthaltendes Gebiet G_e bestimmen würde. Die Komplementärmenge dieses

Gebietes G_e wäre ein für die Transformation invariantes *zirkulares Kontinuum**), das wenigstens einen invarianten Punkt enthalten würde.**)

Ein Ende einer Bahnkurve erstreckt sich somit bis zum Unendlichen.

Wir werden nun für die weiteren Entwicklungen dieses Paragraphen die Ebene stereographisch auf eine Kugel projizieren, wobei das Unendliche in einen einzigen Punkt U übergeht; ein Ende einer Bahnkurve wird dann zu einer *zusammenhängenden perfekten* Punktmenge.

Hilfssatz 1. *Ein freier Bogen, welcher mit einem zwischen denselben Endpunkten verlaufenden Bahnbogen eine einfache geschlossene Kurve bildet, wird sowohl von dem vorangehenden, wie von dem folgenden Bahnsegment der Bahnkurve getroffen.*

Beweis. Wenn nämlich der freie Bogen z. B. von dem vorangehenden Bahnsegment der Bahnkurve nicht getroffen würde, so würde wieder der Fall der Fig. 2 eintreten, d. h. wir hätten eine einfache geschlossene Kurve c, welche sich aus einem Bahnbogen PQ, und einem freien Bogen PQ, der auch das P unmittelbar vorangehende Bahnsegment $P_i P$ nicht träfe, zusammensetzte. Hieraus gelangen wir genau wie im Beweise des Satzes 2 zu einem Widerspruch.

Wir zerlegen jetzt die Bahnkurve in ein wenigstens vier getrennte Segmente enthaltendes *Mittelstück*, ein *initiales Stück* und ein *finales Stück*, welche wir der Reihe nach mit m, i und f bezeichnen.

Sei G_f das von F bestimmte, die Bahnkurve enthaltende Gebiet. Unter der *Kontur von F* verstehen wir sodann den zyklischen Ordnungstypus seiner für G_f erreichbaren Punkte. Wir sagen, daß das finale Stück der Bahnkurve den Punkt P der Kontur von F *unbegrenzt approximiert*, wenn es jeden nach P führenden Weg in beliebiger Nähe von P trifft.

Hilfssatz 2. *Das finale Ende enthält, falls es sich nicht auf den Punkt U reduziert, wenigstens zwei erreichbare Punkte, welche vom finalen Stücke der Bahnkurve unbegrenzt approximiert werden.*

Beweis. Im entgegengesetzten Falle könnten wir nämlich einen freien Bogen β'' zeichnen, der einen Punkt H'' von f mit einem erreichbaren Punkte K'' von F verbände, der außerhalb seiner Endpunkte weder mit m, noch mit f, noch mit F einen gemeinschaftlichen Punkt besäße, und von dem auch das Bild und das inverse Bild einander nicht träfen.***)

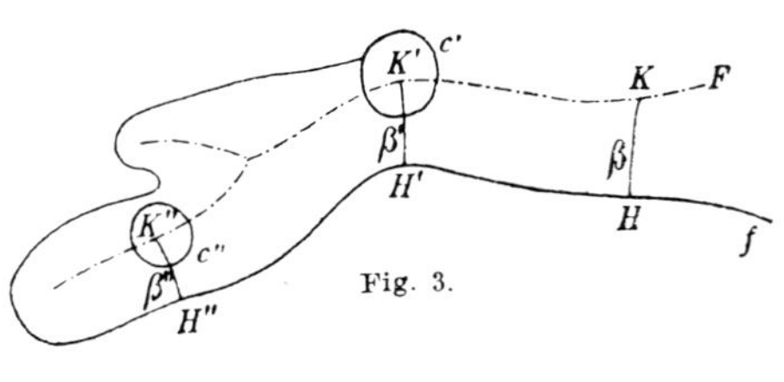

Fig. 3.

[[5]] *) Amsterd. Ber., holl. Ausg. 19, S. 737; engl. Ausg. 13, S. 767.

[[5]] **) ibid., holl. Ausg. 19, S. 741; engl. Ausg. 13, S. 771.

***) Letztere Bedingung ist übrigens für jeden freien Bogen von selbst erfüllt.

[[254]]

Die inversen Bilder von H'', K'', β'' bezeichnen wir der Reihe nach mit H', K', β', die inversen Bilder von H', K', β' der Reihe nach mit H, K, β, und wir beschreiben um K' einen Kreis c', der von seinem Bilde, von β, β'', und vom Bahnbogen HH'' nicht getroffen wird. Weil nun aber K' zu F gehört, müßte der auf H'' folgende Teil von f in c' eindringen, wobei eine einfache geschlossene Kurve von der durch Hilfssatz 1 ausgeschlossenen Art herauskäme.

Aus Hifssatz 2 folgern wir, daß die von f unbegrenzt approximierten erreichbaren Punkte von F ein gewisses *Segment* σ der Kontur von F einnehmen. Wir betrachten das von σ auf der Kontur von F bestimmte *Komplementärsegment* σ_c.

Hilfssatz 3. *Das Komplementärsegment σ_c reduziert sich auf höchstens einen einzigen erreichbaren Punkt.*

Beweis. Wir bemerken zunächst, daß die Schnitte, von denen das eventuelle Segment σ_c begrenzt wird, beide von einem mit U zusammenfallenden erreichbaren Punkte bestimmt werden müssen. In entgegengesetzten Falle könnten wir nämlich in endlicher Entfernung von U einen beliebig kleinen einfachen Kurvenbogen zeichnen, der einen Punkt von σ mit einem Punkte von σ_c verbände, und außerhalb seiner Endpunkte F nicht träfe.*) Dann aber könnten wir einen freien Bogen β'' von der im Beweise von Hilfssatz 2 auftretenden Art konstruieren, und dieser würde wie dort auf einen Widerspruch führen.

Wären nun diese beiden Schnitte nicht identisch, so hätten die Ableitungen von σ und σ_c den Punkt U als ihren einzigen gemeinschaftlichen Punkt, während f in beliebige Nähe eines nicht mit U identischen Punktes der Ableitung von σ_c gelangen müßte, sodaß wir wieder einen freien Bogen β'' von der durch den Beweis des Hilfssatzes 2 ausgeschlossenen Art konstruieren könnten.

Die Bahnenden I und F bilden zusammen eine zusammenhängende perfekte Punktmenge $I + F$. In einem der von ihr bestimmten Gebiete, das wir mit G bezeichnen, verläuft die Bahnkurve. Die Kontur von $I + F$ enthält ein Segment σ_i, das von i, und ein Segment σ_f, das von f unbegrenzt approximiert wird. Aus Hilfssatz 3 folgern wir, daß diese beiden Segmente auf der Kontur von $I + F$ kein Komplementärsegment übrig lassen.

Hilfssatz 4. *Die Segmente σ_i und σ_f besitzen kein gemeinschaftliches Teilsegment.*

Beweis. Wenn ein gemeinschaftliches Teilsegment existierte, könnten wir nach einem zu diesem Segmente gehörigen Punkte K einen Weg legen. Dieser Weg würde einen freien Bogen enthalten, der einen Punkt von i

*) Amsterd. Ber., holl. Ausg. 20, S. 26; engl. Ausg. 14, S. 302. [[6]]

mit einem Punkte von f verbände, und übrigens die Bahnkurve nicht träfe, was wider Hilfssatz 1 verstoßen würde.

Hilfssatz 5. *Die Schnitte, durch welche σ_i und σ_f getrennt werden, werden beide von einem mit U zusammenfallenden erreichbaren Punkte bestimmt.*

Beweis. Im entgegengesetzten Falle könnten wir nämlich in endlicher Entfernung von U einen beliebig kleinen, m nicht treffenden, freien Bogen zeichnen, der einen Punkt von σ_i mit einem Punkte von σ_f verbände, und außerhalb seiner Endpunkte F nicht träfe. Und dieser Kurvenbogen würde wieder einen freien Bogen enthalten, der einen Punkt von i mit einem Punkte von f verbände, und übrigens die Bahnkurve nicht träfe.

Aus Hilfssatz 5 folgern wir gleich, daß die Ableitung I_a von σ_i und die Ableitung F_a von σ_f *nur den Punkt U gemeinschaftlich haben.*

Satz 4. *Die Bahnenden I und F haben nur den Punkt U gemeinschaftlich.*

Beweis. Wenn nämlich i in beliebige Nähe eines nicht mit U identischen Punktes von F_a gelangte, so gäbe es in endlicher Entfernung von U einen beliebig kleinen nach F_a führenden Weg, der sowohl von i wie von f getroffen würde, und auf dem man einen i und f verbindenden und außerhalb seiner Endpunkte die Bahnkurve nicht treffenden freien Bogen bestimmen könnte, was durch Hilfssatz 1 ausgeschlossen ist.

Mithin ist I mit I_a und F mit F_a identisch, womit der Satz bewiesen ist.

Hilfssatz 6. *Es gibt eine einfache geschlossene Kurve, welche U enthält und die Bahnkurve in einem einzigen Punkte kreuzt, sonst aber weder die Bahnkurve noch ihre Enden trifft.*

Beweis. Wir bezeichnen $I - U$ mit I_r, $F - U$ mit F_r, und gehen aus von einer einfachen geschlossenen Kurve $\varkappa$, welche I und F nur in U

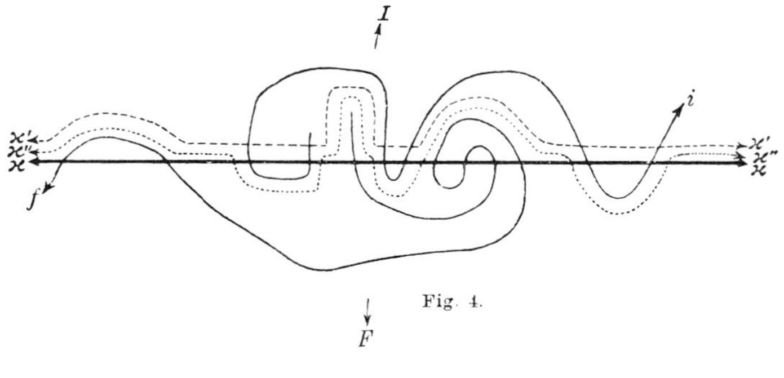

Fig. 4.

trifft, und die Kugel zerlegt in ein Gebiet β, welches I_r, oder, falls I_r nicht existiert, i enthält, und ein Gebiet η, welches F_r, oder, falls F_r nicht existiert, f enthält.

Sei β_m dasjenige von $\varkappa$ und f bestimmte Teilgebiet von β, welches I_r resp. i enthält; seine Grenze $\varkappa_m$ trifft $I+F$ nur in U. In beliebiger Nähe von $\varkappa_m$ läßt sich somit eine (in Fig. 4 gestrichelte) U enthaltende und übrigens in β_m verlaufende einfache geschlossene Kurve $\varkappa'$ zeichnen, welche die Kugel zerlegt in die Gebiete β' und η', von denen das erste I_r resp. i, das zweite sowohl F_r wie f enthält.

Sei η'_m dasjenige von $\varkappa'$ und i bestimmte Teilgebiet von η', welches F_r und f enthält; seine Grenze $\varkappa'_m$ trifft $I+F$ nur in U. In beliebiger Nähe von $\varkappa'_m$ läßt sich somit eine (in Fig. 4 punktierte) U enthaltende und übrigens in η'_m verlaufende einfache geschlossene Kurve $\varkappa''$ zeichnen, welche die Kugel zerlegt in die Gebiete β'' und η''; von denen das erste sowohl I_r wie i, das zweite sowohl F_r wie f enthält.

Seien H_1 und H_2 zwei solche Schnittpunkte von $\varkappa''$ und m, daß der U enthaltende Teilbogen H_1H_2 von $\varkappa''$ keinen weiteren Punkt von m enthält. Ersetzen wir jetzt in $\varkappa''$ den U *nicht* enthaltenden Teilbogen H_1H_2 durch den zu m gehörigen Bogen H_1H_2, so erhalten wir eine einfache geschlossene Kurve $\varkappa'''$, welche U enthält und die Bahnkurve in einem gemeinsamen Bogen H_1H_2 kreuzt, übrigens aber weder die Bahnkurve noch ihre Enden trifft. Nach einer beliebig kleinen Modifizierung genügt mithin $\varkappa'''$ den Anforderungen des zu beweisenden Satzes.

Satz 5. *U ist sowohl als Punkt von I wie als Punkt von F aus einem willkürlichen Punkte der Bahnkurve ohne weitere Durchsetzung der Bahnkurve erreichbar; übrigens enthält weder I noch F einen Punkt, der aus einem Punkte der Bahnkurve ohne weitere Durchsetzung der Bahnkurve erreichbar ist.*

Dieser Satz ist eine unmittelbare Folge der Hilfssätze 3 und 6.

Als Ergänzung zu Hilfssatz 1 können wir jetzt formulieren:

Satz 6. *Es gibt keine einfache geschlossene Kurve, welche sich aus einem freien Bogen und einem Bahnbogen mit denselben Endpunkten zusammensetzt.*

Beweis. Wir nehmen an, daß eine solche geschlossene Kurve c existiert, bezeichnen den freien Bogen und den Bahnbogen, aus denen c sich zusammensetzt, der Reihe nach mit α und β, den β vorangehenden Teil der Bahnkurve mit b, den auf β folgenden Teil der Bahnkurve mit e, die Komplementärmenge des Gebietes G_i mit I_z, die Komplementärmenge des Gebietes G_f mit F_z, die auf α von $b+I_z$ bestimmte abgeschlossene Punktmenge mit π_b, die auf α von $e+F_z$ bestimmte abgeschlossene Punktmenge mit π_e. Die auf α von $\pi_b+\pi_e$ bestimmte Intervallmenge enthält sicher ein Intervall i, von dessen Endpunkten der eine P zu π_b, der andere Q zu π_e gehört. Wenn nun P zu I_z gehörte, wäre er ein aus G ohne Durchsetzung der Bahnkurve erreichbarer Punkt von I, was unmöglich ist. Mithin gehört P zu b, ebenso Q zu e, und das Intervall i müßte zusammen mit dem Bogen PQ der Bahnkurve eine einfache geschlossene Kurve von der durch Hilfssatz 1 ausgeschlossenen Art bestimmen.

Satz 7. *Durch jeden Punkt der Ebene läßt sich eine Bahnkurve legen.*

Beweis. Sei P ein in der Ebene willkürlich angenommener Punkt, P' sein Bild, k ein P und P' verbindender einfacher Kurvenbogen, k' das Bild von k. Wir lassen einen beweglichen Punkt S von P aus den Kurvenbogen k beschreiben, bis seine Bahn und diejenige seines Bildpunktes S' einander treffen in einem Punkte Q'. Dort mündet dann entweder die Bahn von S auf eine Seite der Bahn von S', oder die Bahn von S' auf eine Seite der Bahn von S. Wir dürfen den ersten Fall voraussetzen; der zweite kann nämlich in den ersten übergeführt werden, indem wir die Transformation durch ihre inverse ersetzen, wobei die Rollen von P und P' verwechselt werden, unsere Aufgabe aber dieselbe bleibt.

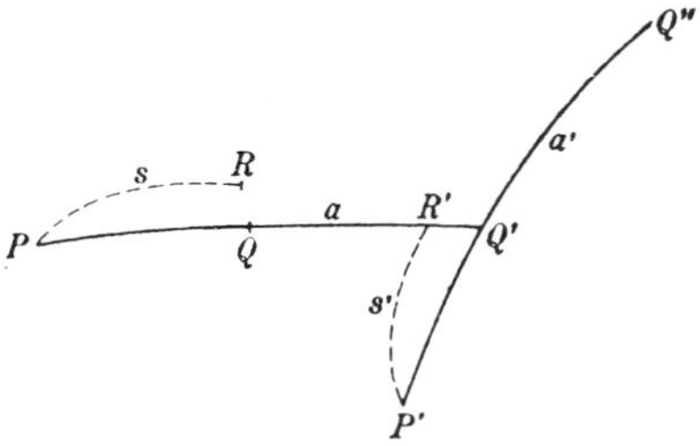

Fig. 5

Wir bezeichnen mit Q'' das Bild von Q', mit Q das inverse Bild von Q', mit a den Kurvenbogen PQ', mit a' das Bild von a.

Wir wählen auf a einen Punkt R' so nahe an Q', daß er sich mit P' verbinden läßt durch einen einfachen Kurvenbogen s', der außerhalb P' und R' weder sein inverses Bild s, noch a, noch a' trifft, während auch s außerhalb P weder a noch a' trifft. Alsdann ist aber $RPQR'$ ein Translationsbogen, der eine P enthaltende Bahnkurve erzeugt, womit unsere Aufgabe gelöst ist.

Satz 8. *Wenn wir einen in der Ebene willkürlich angenommenen Punkt immer wieder der gegebenen Transformation sowie ihrer inversen unterziehen, so erzeugen wir eine Punktmenge, welche keinen Grenzpunkt im Endlichen besitzt.*

Beweis. Wenn nämlich diese Punktmenge, welche wir mit t bezeichnen, eine Fundamentalreihe $P^{(n_1)}$, $P^{(n_2)}$, $\cdots$, welche gegen einen im Endlichen liegenden Grenzpunkt $P^{(\omega)}$ konvergierte, enthielte, so könnten wir nach der im Beweise von Satz 7 befolgten Methode durch $P^{(\omega)}$ und einen in genügender Nähe von $P^{(\omega)}$ liegenden Punkt $P^{(n_\alpha)}$ einen Translationsbogen legen; die von diesem Translationsbogen erzeugte Bahnkurve würde sodann die ganze Punktmenge t enthalten, müßte somit bei genügend weiter Fortsetzung vorwärts und rückwärts in beliebige Nähe von $P^{(\omega)}$ gelangen, was wider Satz 2 verstoßen würde.

[[258]]

§ 2.

Herstellung des Translationsfeldes.

Unser Ziel ist jetzt, zu einem in der Ebene willkürlich angenommenen Punkte P_0 ein diesen Punkt enthaltendes Translationsfeld zu konstruieren.

Wir ziehen durch P_0 eine willkürliche Bahnkurve b_0 und schlagen um P_0 als Mittelpunkt einen Kreis C_1.

Die Enden von b_0 bestimmen in der Ebene ein b_0 enthaltendes Gebiet, welches von b_0 in zwei Teilgebiete zerlegt wird. Diese Teilgebiete können wir als *das rechte* und *das linke Gebiet von* b_0 bezeichnen.

Einen zwischen zwei Punkten von b_0 im rechten bez. linken Gebiete von b_0 verlaufenden, im übrigen b_0 nicht treffenden und ohne Kreuzung von b_0 oder C_1 aus dem Unendlichen erreichbaren Teilbogen von C_1 nennen wir *einen rechten bez. linken Schließungsbogen von* C_1 *in bezug auf* b_0, *kurz einen* $Sb_r(C_1, b_0)$ *bez.* $Sb_l(C_1, b_0)$. Dieser Schließungsbogen soll *erster Art* heißen, wenn er sein Bild nicht trifft, *zweiter Art* im entgegengesetzten Falle.

Zu jedem Schließungsbogen gehört ein zwischen demselben Anfangspunkte und demselben Endpunkte verlaufender Teilbogen von b_0, den wir *einen rechten bez. linken Ausbiegungsbogen von* b_0 *in bezug auf* C_1, *kurz einen* $Ab_r(b_0, C_1)$ *bez.* $Ab_l(b_0, C_1)$ *nennen.* (In Fig. 6 sind die $Ab_r(b_0, C_1)$ gestrichelt gezeichnet.) Ein Ausbiegungsbogen soll *erster* oder *zweiter Art* heißen, je nachdem der entsprechende Schließungsbogen erster oder zweiter Art ist.

Sei A_0E_0 der kleinste die $Sb_r(C_1, b_0)$ enthaltende Teilbogen von C_1. Wir nehmen an, daß es $Sb_r(C_1, b_0)$ zweiter Art gibt; ihre Gesamtlänge sei l_0.

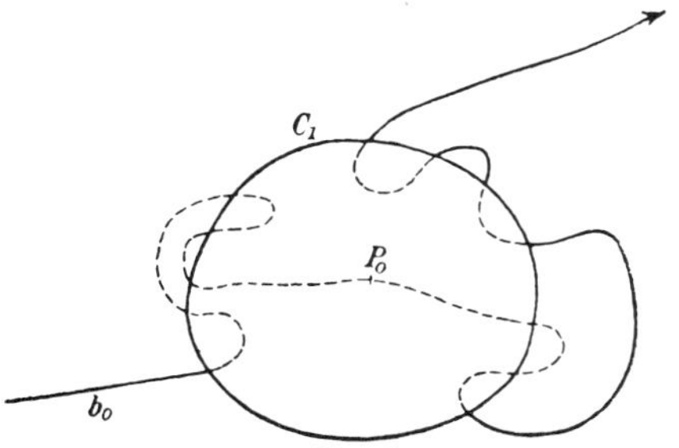

Fig. 6.

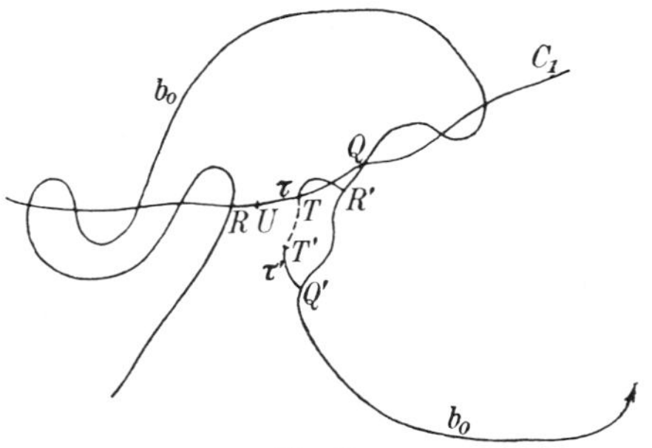

Fig. 7.

Wir bezeichnen (Fig. 7) einen willkürlichen dieser $Sb_r(C_1, b_0)$ zweiter Art mit τ, seinen Anfangspunkt mit R, seinen Endpunkt mit Q, sein Bild mit τ und lassen einen beweglichen Punkt S von Q aus den Bogen τ beschreiben, bis seine Bahn und diejenige seines Bildpunktes S' einander treffen in einem Punkte T. Dort mündet dann entweder die Bahn von S auf die rechte Seite der Bahn von S' oder die Bahn von S' auf die linke Seite der Bahn von S. Wenn wir uns auf die erste dieser beiden Möglichkeiten beschränken (die zweite läßt sich ganz analog behandeln), so ist der (in Fig. 7 gestrichelte) Teilbogen TT' von τ' ein Translationsbogen, der eine b_0 nicht treffende und rechts von b_0 verlaufende Bahnkurve ${}_1b_0$ erzeugt, während es einen in C_1 liegenden freien Bogen RU gibt, der b_0 und ${}_1b_0$ verbindet.

Wenn wir die Gesamtlänge der $Sb_r(C_1, {}_1b_0)$ zweiter Art mit ${}_1l_0$ und das Minimum der Länge des Transformationsvektors in den Punkten innerhalb C_1 mit m_1 bezeichnen, so ist $l_0 - {}_1l_0 > m_1$.

Ist ${}_1l_0 > 0$, so konstruieren wir in derselben Weise eine ${}_1b_0$ nicht treffende und rechts von ${}_1b_0$ verlaufende Bahnkurve ${}_2b_0$, zu welcher es wieder einen zu C_1 gehörigen freien Bogen gibt, der ${}_1b_0$ und ${}_2b_0$ verbindet.

In dieser Weise gelangen wir nach einer endlichen Zahl von Schritten zu einer Bahnkurve ${}_kb_0$, zu welcher es keine $Sb_r(C_1, {}_kb_0)$ zweiter Art mehr gibt.

Wir bezeichnen im weiteren ${}_kb_0$ mit b_{01} und wählen einen freien Teilbogen $P_{01}Q_{01}$ von $A_{01}E_{01}$, der mit $A_{01}E_{01}$ keinen Endpunkt gemeinschaftlich hat und der jeden rechten Schließungsbogen, von dem er einen Teilbogen enthält, ganz enthält.

Die in $P_{01}Q_{01}$ enthaltenen $Sb_r(C_1, b_{01})$ lassen sich dann in solcher Weise in eine *endliche* Zahl von Gruppen $s_1, s_2, \cdots$ ordnen, daß zu jeder Gruppe s_p eine in einem einzigen Bahnsegmente von b_{01} enthaltene Gruppe von Ausbiegungsbogen a_p gehört.

Ersetzen wir dann in b_{01} eine gewisse Gruppe a_p durch die entsprechende Gruppe s_p, so erzeugen wir eine Bahnkurve ${}_pb_{01}$, welche teilweise rechts von b_{01} verläuft und teilweise mit b_{01} zusammenfällt, während jeder nicht in b_{01} liegende Punkt von ${}_pb_{01}$ sich mit b_{01} durch einen freien Bogen, der keinen weiteren Schnittpunkt mit ${}_pb_{01}$ besitzt, verbinden läßt.

Die rechten Gebiete der verschiedenen ${}_pb_{01}$ besitzen ein größtes gemeinschaftliches Gebiet, welches seinerseits das rechte Gebiet einer Bahnkurve b_{02} bildet, in bezug auf welche im Kreisbogen $P_{01}Q_{01}$ keine rechten Schließungsbogen mehr enthalten sind, während jeder nicht zu b_{01} gehörige Punkt von b_{02} sich durch einen freien Bogen, der keinen weiteren Schnittpunkt mit b_{02} besitzt, mit b_{01} verbinden läßt.

Wir wählen jetzt einen freien Teilbogen $P_{02}Q_{02}$ von $A_{02}E_{02}$, der mit $P_{01}Q_{01}$ keinen Teilbogen gemeinschaftlich hat und nur volle rechte

[[260]]

Schließungsbogen in bezug auf b_{02} enthält. Mittels b_{02} und $P_{02}Q_{02}$ konstruieren wir dann eine Bahnkurve b_{03}, in bezug auf welche weder im Kreisbogen $P_{01}Q_{01}$ noch im Kreisbogen $P_{02}Q_{02}$ rechte Schließungsbogen enthalten sind, während jeder nicht zu b_{02} gehörige Punkt von b_{03} sich durch einen freien Bogen, der keinen weiteren Schnittpunkt mit b_{03} besitzt, mit b_{02} verbinden läßt.

In dieser Weise fortfahrend, gelangen wir bei passender Wahl der $P_{0m}Q_{0m}$ nach einer endlichen Zahl von Schritten zu einer Bahnkurve b_{0h}, in bezug auf welche die rechten Schließungsbogen von C_1 sich in zwei aufeinanderfolgende Gruppen zerlegen lassen, von denen die erste, die Gruppe der *„initialen rechten Schließungsbogen"*, mit ${}_iSb_r(C_1, b_{0h})$ zu bezeichnen, in einem freien kleinsten Kreisbogen $A_{0h}B_{0h}$, und die zweite, die Gruppe der *„finalen rechten Schließungsbogen"*, mit ${}_fSb_r(C_1, b_{0h})$ zu bezeichnen, in einem freien kleinsten Kreisbogen $D_{0h}E_{0h}$ enthalten ist. Jede dieser beiden Gruppen kann übrigens in Fortfall kommen.

Wir dürfen sogar annehmen, erstens daß zwischen den zu den ${}_iSb_r(C_1, b_{0h})$ gehörigen ${}_iAb_r(b_{0h}, C_1)$ und den zu den ${}_fSb_r(C_1, b_{0h})$ gehörigen ${}_fAb_r(b_{0h}, C_1)$ wenigstens ein volles Bahnsegment enthalten ist, und zweitens daß zu jedem der Bogen $A_{0h}B_{0h}$ und $D_{0h}E_{0h}$, der nicht in Fortfall kommt, unendlich viele gegen das entsprechende Ende von b_{0h} konvergierende $Ab_r(b_{0h}, C_1)$ gehören.

Um diese Behauptung zu beweisen, setzen wir zunächst voraus, daß der Kreisbogen $A_{0h}E_{0h}$ ein freier Bogen ist. Falls dann z. B. zu $A_{0h}B_{0h}$ nicht unendlich viele gegen das initiale Bahnende konvergierende $Ab_r(b_{0h}, C_1)$ gehören, können wir $A_{0h}B_{0h}$ die Rolle eines Bogens $P_{0h}Q_{0h}$ spielen lassen und in dieser Weise eine Bahnkurve $b_{0,h+1}$ konstruieren, für welche $A_{0,h+1}B_{0,h+1}$ in Fortfall kommt. Wir dürfen also voraussetzen, daß sowohl zu $A_{0h}B_{0h}$ wie zu $D_{0h}E_{0h}$ unendlich viele gegen das entsprechende Ende von b_{0h} konvergierende $Ab_r(b_{0h}, C_1)$ gehören, und bezeichnen mit $D_{0h}K_{0h}$ einen *kleinsten* Teilbogen von $D_{0h}E_{0h}$ mit der Eigenschaft, daß ihm ein in b_{0h} enthaltener Bahnbogen $D_{0h}K_{0h}$ entspricht. Indem wir jeden zu diesem Bahnbogen gehörigen $Ab_r(b_{0h}, C_1)$ durch den entsprechenden Schließungsbogen ersetzen, konstruieren wir eine Bahnkurve $b_{0,h+1}$, die einen nicht in das Innere von C_1 eindringenden, entweder den Punkt D_{0h} oder den Punkt K_{0h} enthaltenden Bahnbogen $m_{0,h+1}$ aufweist, mithin auf Grund von Satz 6 entweder nur initiale oder nur finale rechte Ausbiegungsbogen in bezug auf C_1 besitzt. Konvergieren diese Ausbiegungsbogen gegen das entsprechende Ende, so besitzt die Bahnkurve $b_{0,h+1}$ die geforderten Eigenschaften. Im entgegengesetzten Falle können wir aus ihr in der oben angegebenen Weise eine weitere Bahnkurve, welche die geforderten Eigenschaften besitzt, herleiten.

Wir setzen jetzt voraus, daß der Kreisbogen $A_{0h}E_{0h}$ nicht ein freier Bogen ist. Falls dann z. B. zu $A_{0h}B_{0h}$ nicht unendlich viele gegen das initiale Bahnende konvergierende $Ab_r(b_{0h}, C_1)$ gehören, können wir wieder $A_{0h}B_{0h}$ die Rolle eines Bogens $P_{0h}Q_{0h}$ spielen lassen, und in dieser Weise eine Bahnkurve $b_{0,h+1}$ konstruieren, für welche entweder $A_{0,h+1}B_{0,h+1}$ in Fortfall kommt, oder sowohl $A_{0,h+1}B_{0,h+1}$ wie $D_{0,h+1}E_{0,h+1}$ in $D_{0h}E_{0h}$ enthalten sind, mithin $A_{0,h+1}E_{0,h+1}$ ein freier Bogen ist, welchen Fall wir schon erledigt haben. Wir dürfen also voraussetzen, daß sowohl zu $A_{0h}B_{0h}$ wie zu $D_{0h}E_{0h}$ unendlich viele gegen das entsprechende Ende von b_{0h} konvergierende $Ab_r(b_{0h}, C_1)$ gehören, definieren in derselben Weise wie oben den Bogen $D_{0h}K_{0h}$ und die Bahnkurve $b_{0,h+1}$ mit dem Bahnbogen $m_{0,h+1}$ und bezeichnen mit $f_{0,h+1}$ bez. $i_{0,h+1}$ den auf $m_{0,h+1}$ folgenden bez. den $m_{0,h+1}$ vorangehenden Teil von $b_{0,h+1}$. Wenn es nun rechte Ausbiegungsbogen von $f_{0,h+1}$ in bezug auf $A_{0h}B_{0h}$ gäbe, so müßten auch rechte Ausbiegungsbogen von $i_{0,h+1}$ in bezug auf $A_{0h}B_{0h}$ existieren, und wir hätten eine einfache geschlossene Kurve, welche sich aus einem (in $A_{0h}B_{0h}$ liegenden) freien Bogen und einem (in $b_{0,h+1}$ liegenden) Bahnbogen zusammensetzte, was wider Satz 6 verstoßen würde. Ebensowenig können rechte Ausbiegungsbogen von $i_{0,h+1}$ in bezug auf $K_{0h}E_{0h}$ existieren. Besitzt nun sowohl $i_{0,h+1}$ wie $f_{0,h+1}$ gegen das entsprechende Ende konvergierende $Ab_r(b_{0,h+1}, C_1)$, so besitzt die Bahnkurve $b_{0,h+1}$ die geforderten Eigenschaften. Im entgegengesetzten Falle können wir aus ihr in der oben angegebenen Weise eine weitere Bahnkurve, welche die geforderten Eigenschaften besitzt, herleiten.

Wir dürfen mithin annehmen, daß zu jedem der Bogen $A_{0h}B_{0h}$ und $D_{0h}E_{0h}$, der nicht in Fortfall kommt, unendlich viele gegen das entsprechende Ende von b_{0h} konvergierende $Ab_r(b_{0h}, C_1)$ gehören, und daß zwischen den ${}_iAb_r(b_{0h}, C_1)$ und den ${}_fAb_r(b_{0h}, C_1)$ wenigstens ein volles Bahnsegment enthalten ist. Hieraus folgern wir mit Rücksicht auf Satz 6, daß die ${}_iAb_r(b_{0h}, C_1)$ bez. die ${}_fAb_r(b_{0h}, C_1)$ den Bogen $D_{0h}E_{0h}$ bez. $A_{0h}B_{0h}$ nicht treffen können, m. a. W., daß der Bogen $A_{0h}B_{0h}$ und der auf B_{0h} folgende Teil von b_{0h} bez. der Bogen $D_{0h}E_{0h}$ und der D_{0h} vorangehende Teil von b_{0h} einander nicht treffen können.

Wir bezeichnen mit $D'_{0h}E'_{0h}$ das Bild des Bogens $D_{0h}E_{0h}$, und mit ${}_f\gamma_{0h}$ das Gebiet, welches gebildet wird von denjenigen Punkten des rechten Gebietes von b_{0h}, die ohne Kreuzung von b_{0h} oder $D_{0h}E_{0h}$ aus dem Unendlichen erreicht werden können. Zur Grenze von ${}_f\gamma_{0h}$ gehört ein gewisser Teil ${}_f\tau_{0h}$ des finalen Endes von b_{0h}. Diese Punktmenge ${}_f\tau_{0h}$ bestimmt zusammen mit $D_{0h}E_{0h}$ und dem D_{0h} vorangehenden Teile von b_{0h} ein Gebiet ${}_fg_{0h}$, welches ${}_f\gamma_{0h}$ als Teil enthält.

Wir bezeichnen weiter mit $'A_{0h}'B_{0h}$ das inverse Bild des Bogens

$A_{0h} B_{0h}$, und mit ${}_i\gamma_{0h}$ das Gebiet, welches gebildet wird von denjenigen Punkten des rechten Gebietes von b_{0h}, die ohne Kreuzung von b_{0h} oder $A_{0h} B_{0h}$ aus dem Unendlichen erreicht werden können. Zur Grenze von ${}_i\gamma_{0h}$ gehört ein gewisser Teil ${}_i\tau_{0h}$ des initialen Endes von b_{0h}. Diese Punktmenge ${}_i\tau_{0h}$ bestimmt zusammen mit $A_{0h} B_{0h}$ und dem auf B_{0h} folgenden Teile von b_{0h} ein Gebiet ${}_i g_{0h}$, welches ${}_i\gamma_{0h}$ als Teil enthält.

Wir setzen nun einen Augenblick voraus, daß $D'_{0h} E'_{0h}$ innerhalb des Gebietes ${}_f g_{0h}$ liegt, bezeichnen das Bild von $D'_{0h} E'_{0h}$ mit $D''_{0h} E''_{0h}$, das Bild von $D''_{0h} E''_{0h}$ mit $D'''_{0h} E'''_{0h}$, usw., das Bild von ${}_f\gamma_{0h}$ mit ${}_f\gamma'_{0h}$, das Bild von ${}_f\gamma'_{0h}$ mit ${}_f\gamma''_{0h}$, usw., die Menge der allen ${}_f\gamma^{(m)}_{0h}$ gemeinsamen Punkte mit ${}_f\gamma^{(\omega)}_{0h}$, den kleinsten Teilbogen von $D_{0h} E_{0h}$, welcher alle zu ${}_f\gamma'_{0h}$ gehörigen ${}_f Sb_r(C_1, b_{0h})$ enthält, mit $D_{0h} H_{0h}$, das Bild von $D_{0h} H_{0h}$ mit $D'_{0h} H'_{0h}$, das Bild von $D'_{0h} H'_{0h}$ mit $D''_{0h} H''_{0h}$, usw. Alsdann ist das Gebiet derjenigen zum rechten Gebiete von b_{0h} gehörigen Punkte, welche sich innerhalb dieses Gebietes von dem D_{0h} vorangehenden Bahnsegmente aus ohne Kreuzung von $D_{0h} H_{0h}$, $D'_{0h} H'_{0h}$, $D''_{0h} H''_{0h}$, usw. erreichen lassen, identisch mit ${}_f\gamma^{(\omega)}_{0h}$. Wenn wir also $D_{0h} H_{0h}$ die Rolle eines Bogens $P_{0h} Q_{0h}$ spielen lassen, wird eine Bahnkurve $b_{0,h+1}$ erzeugt, deren rechtes Gebiet in ${}_f\gamma^{(\omega)}_{0h}$ enthalten ist, sodaß keine $Ab_r(b_{0,h+1}, D_{0h} E_{0h})$ existieren. Ebensowenig kann aber nach Satz 6 der auf H_{0h} folgende Teil von $b_{0,h+1}$ rechte Ausbiegungsbogen in bezug auf $A_{0h} B_{0h}$ besitzen, sodaß $D_{0,h+1} E_{0,h+1}$ in Fortfall kommt.

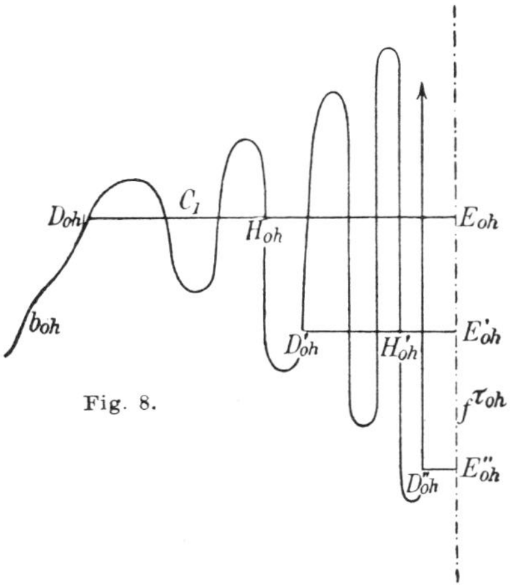

Fig. 8.

Wir dürfen mithin annehmen, daß zu jedem der Bogen $A_{0h} B_{0h}$ und $D_{0h} E_{0h}$, der nicht in Fortfall kommt, unendlich viele gegen das entsprechende Ende von b_{0h} konvergierende $Ab_r(b_{0h}, C_1)$ gehören, daß der Bogen $'A_{0h}{}'B_{0h}$ bez. $D'_{0h} E'_{0h}$, falls er existiert, außerhalb ${}_i g_{0h}$ bez. ${}_f g_{0h}$ liegt, und daß zwischen den ${}_i Ab_r(b_{0h}, C_1)$ und den ${}_f Ab_r(b_{0h}, C_1)$, falls beide existieren, wenigstens ein volles Bahnsegment von b_{0h} enthalten ist.

Wir bezeichnen im weiteren b_{0h} mit b_1, und schlagen um P_0 als Mittelpunkt einen Kreis C_2, der C_1 und einen gewissen Bahnbogen von b_1 in seinem Inneren enthält. In derselben Weise, wie aus b_0 und C_1 die Bahnkurve b_1 hergeleitet ist, gelangen wir von b_1 und C_2 aus zu einer Bahnkurve b_2 mit der Eigenschaft, daß zu jedem der die $Sb_r(C_2, b_2)$ enthaltenden freien Bogen $A_2 B_2$ und $D_2 E_2$, der nicht in Fortfall kommt, unendlich viele gegen das entsprechende Ende von b_{0h} konvergierende $Ab_r(b_2, C_2)$

gehören, daß der Bogen $'A_2'B_2$ bez. $D_2'E_2'$, falls er existiert, außerhalb ${}_i g_2$ bez. ${}_f g_2$ liegt, und daß zwischen den ${}_i Ab_r(b_2, C_2)$ und den ${}_f Ab_r(b_2, C_2)$, falls beide existieren, wenigstens ein volles Bahnsegment von b_2 enthalten ist. Die Kurven b_1 und b_2 können einander nicht treffen.

Wir werden nun zeigen, daß die ${}_i Ab_r(b_2, C_2)$ zwar die ${}_i Sb_r(C_1, b_1)$, nicht aber die ${}_f Sb_r(C_1, b_1)$ treffen können.

Im entgegengesetzten Falle würden nämlich die Bilder beliebiger Ordnung eines Schnittpunktes von b_2 mit einem ${}_f Sb_r(C_1, b_1)$ außerhalb des Gebietes ${}_f\gamma_1$ liegen, mithin würden auch die ${}_f Ab_r(b_2, C_2)$ die ${}_f Sb_r(C_1, b_1)$ treffen. Alsdann würde aber, weil die ${}_i Ab_r(b_2, C_2)$ und die ${}_f Ab_r(b_2, C_2)$ auf b_2 durch wenigstens ein volles Bahnsegment getrennt werden, eine einfache geschlossene Kurve existieren, welche sich aus einem (in $D_1 E_1$ liegenden) freien Bogen und einem (in b_2 liegenden) Bahnbogen zusammensetzte, was wider Satz 6 verstoßen würde.

Wir schlagen weiter um P_0 als Mittelpunkt einen Kreis C_3, der C_2 und einen gewissen Bahnbogen von b_2 in seinem Inneren enthält, gelangen in derselben Weise, wie aus b_0 und C_1 die Bahnkurve b_1 hergeleitet ist, von b_2 und C_3 aus zu einer Bahnkurve b_3 und fahren in dieser Weise fort. Den Radius des Kreises C_n lassen wir dabei mit n unbegrenzt wachsen. Für $q < p$ kann dann ein ${}_i Ab_r(b_p, C_p)$ zwar einen ${}_i Sb_r(C_q, b_q)$, nicht aber einen ${}_f Sb_r(C_q, b_q)$ treffen.

Im weiteren unterscheiden wir drei Fälle, von denen wir den kompliziertesten zunächst behandeln.

Erster Fall. *Es gibt einen Kreis um P_0, in den für jedes n die ${}_f Ab_r(b_n, C_n)$ eindringen; es gibt aber keinen Kreis um P_0, in den für jedes n die ${}_i Ab_r(b_n, C_n)$ eindringen.* Dann existiert eine gewisse positive ganze Zahl m, von welcher an zu jeder positiven ganzen Zahl n eine solche positive ganze Zahl α_n bestimmt werden kann, daß $\alpha_{n+1} \geqq \alpha_n$ und $\alpha_n < n$, daß α_n mit n unbegrenzt wächst, daß sowohl die ${}_f Ab_r(b_n, C_n)$ wie die ${}_f Ab_r(b_{n+1}, C_{n+1})$ in C_{α_n} eindringen, und daß weder die ${}_i Ab_r(b_n, C_n)$, noch die ${}_i Ab_r(b_{n+1}, C_{n+1})$ in C_{α_n} eindringen.

Sei V_n der erste Schnittpunkt der ${}_f Ab_r(b_n, C_n)$ mit C_{α_n}, und W_{n+1} der erste Schnittpunkt der ${}_f Ab_r(b_{n+1}, C_{n+1})$ mit C_{α_n}. Wir betrachten das Gebiet $\mathfrak{F}_n$, das gebildet wird von denjenigen zwischen b_n und b_{n+1} liegenden Punkten, welche aus dem dem Punkte V_n unmittelbar vorangehenden Bahnsegmente von b_n ohne Kreuzung von C_{α_n} erreichbar sind, und den zur Grenze von $\mathfrak{F}_n$ gehörigen, V_n und W_{n+1} verbindenden einfachen Kurvenbogen τ_n. Solche Teilbogen von τ_n, welche dem Kreise C_{α_n} angehören, nennen wir *Teilbogen erster Art*, solche, welche der Bahnkurve b_n angehören, *Teilbogen zweiter Art*, solche, welche der Bahnkurve b_{n+1} angehören, *Teilbogen dritter Art*. Bilder von Teilbogen erster Art

4*

[[264]]

haben weder mit dem Gebiete $\mathfrak{F}_n$, noch mit seiner Grenze einen Punkt gemeinschaftlich. Wenn wir den Kurvenbogen τ_n in der Richtung von V_n

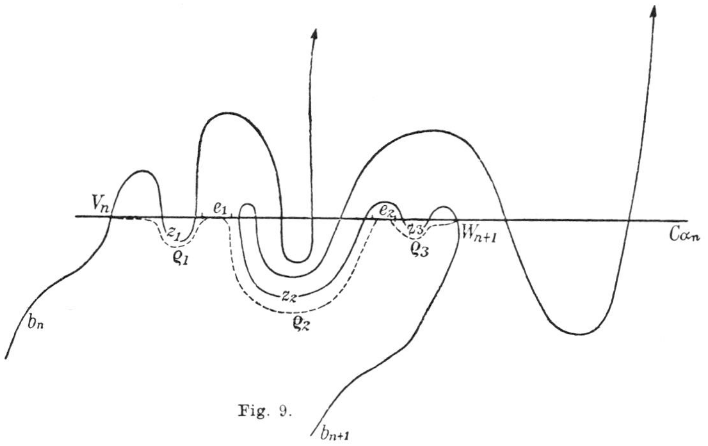

Fig. 9.

nach W_{n+1} beschreiben, so gehen die Teilbogen zweiter Art denjenigen dritter Art voran, besitzen die Teilbogen zweiter Art dieselbe Reihenfolge wie in der Bahnkurve b_n, die Teilbogen dritter Art dagegen die entgegengesetzte Reihenfolge wie in der Bahnkurve b_{n+1}.

Wir bestimmen in τ_n eine endliche Menge $e_1, e_2, \cdots, e_q, e_{q+1}, \cdots, e_{q+p}$ von in der angegebenen Reihenfolge aufeinanderfolgenden Teilbogen erster Art, welche in τ_n voneinander und von den Endpunkten getrennt werden durch die Zwischenbogen $z_1, z_2, \cdots, z_q, z_{q+1}, \cdots, z_{q+p+1}$, in solcher Weise, daß die Bogen $z_1, z_2, \cdots, z_q$ keinen Teilbogen dritter Art, und die Bogen $z_{q+1}, \cdots, z_{q+p+1}$ keinen Teilbogen zweiter Art enthalten, daß für $h \leqq q$ das Bild von z_h mit keinem der Bogen $z_1, \cdots, z_h$ einen gemeinschaftlichen Punkt besitzt, und daß für $h \geqq q+1$ das Bild von z_h mit keinem der Bogen $z_h, z_{h+1}, \cdots, z_{q+p+1}$ einen gemeinschaftlichen Punkt besitzt.

Sodann verbinden wir innerhalb $\mathfrak{F}_n$ den Punkt V_n und den Bogen e_1 durch einen einfachen Kurvenbogen ϱ_1, den Punkt W_{n+1} und den Bogen e_{q+p} durch einen einfachen Kurvenbogen ϱ_{q+p+1}, die Bogen e_1 und e_2 durch einen einfachen Kurvenbogen ϱ_2, die Bogen e_{q+p} und e_{q+p-1} durch einen einfachen Kurvenbogen $\varrho_{q+p}, \cdots$, die Bogen e_{q-1} und e_q durch einen einfachen Kurvenbogen ϱ_q, und die Bogen e_{q+1} und e_q durch einen einfachen Kurvenbogen ϱ_{q+1} in solcher Weise, daß die verschiedenen ϱ_h einander nicht treffen und daß von dem V_n unmittelbar vorangehenden Bahnsegmente von b_n aus die Bilder der ϱ_h nicht innerhalb $\mathfrak{F}_n$ ohne Kreuzung der ϱ_h erreichbar sind. (In Fig. 9 sind die drei daselbst auftretenden Bogen ϱ_h gestrichelt gezeichnet.)

Die ϱ_h bilden zusammen mit den sie der Reihe nach verbindenden Teilbogen der e_h einen sein Bild nicht treffenden, V_n und W_{n+1} verbindenden einfachen Kurvenbogen k_n, der zwischen b_n und b_{n+1} verläuft, und nicht in das Innere von C_{α_n} eindringt.

Nun dringt aber auch der W_{n+1} und V_{n+1} verbindende Teilbogen l_{n+1} von b_{n+1} (der sich übrigens auch auf einen einzigen Punkt reduzieren kann) nicht in das Innere von C_{α_n} ein. Wir können mithin einen sein Bild nicht treffenden, V_n und V_{n+1} verbindenden einfachen Kurvenbogen $\varkappa_n$ konstruieren, der zwischen b_n und b_{n+1} in beliebiger Nähe von k_n und l_{n+1} verläuft und nicht in das Innere von C_{α_n} eindringt.

Wenn wir n der Reihe nach alle ganzen positiven Werte $\geqq m$ annehmen lassen, so bildet die Gesamtheit der entsprechenden $\varkappa_n$ eine einfache offene Halblinie λ_r, welche ihr Bild nicht trifft, in V_m ihren Anfangspunkt hat und ganz im rechten Gebiete von b_m verläuft.

Zwischen b_0 und b_m liegt nach dem obigen Konstruktionsverfahren eine endliche Reihe von einander nicht kreuzenden Bahnkurven mit der Eigenschaft, daß ein auf einer dieser Bahnkurven b_s willkürlich angenommener Punkt, der nicht zugleich der folgenden Bahnkurve b_t angehört, sich mit b_t durch einen sein Bild nicht treffenden und ganz zwischen b_s und b_t verlaufenden einfachen Kurvenbogen verbinden läßt.

Mithin können wir, wenn wir mit M_0 den Anfangspunkt eines P_0 enthaltenden Bahnsegmentes von b_0 bezeichnen, einen sein Bild nicht treffenden einfachen Kurvenbogen k_{0m} konstruieren, der M_0 mit einem Punkte S_m von b_m verbindet und gänzlich zwischen b_0 und b_m verläuft. Mittels k_{0m} und des zwischen S_m und V_m enthaltenen Teilbogens l_m von b_m können wir dann weiter, weil die Kurven b_0 und b_m einander nicht treffen, einen sein Bild nicht treffenden, M_0 und V_m verbindenden einfachen Kurvenbogen $\varkappa_{0m}$ herstellen, der in beliebiger Nähe von k_{0m} und l_m und gänzlich zwischen b_0 und b_m verläuft. Dieser Kurvenbogen $\varkappa_{0m}$ bildet zusammen mit λ_r eine einfache offene Halblinie η_r, welche ihr Bild nicht trifft, in M_0 ihren Anfangspunkt hat und ganz im rechten Gebiete von b_0 verläuft.

Zweiter Fall. *Es gibt einen Kreis um P_0, in den für jedes n sowohl die ${}_fAb_r(b_n, C_n)$ wie die ${}_iAb_r(b_n, C_n)$ eindringen.* Dann existiert eine gewisse positive ganze Zahl m, von welcher an zu jeder positiven ganzen Zahl n eine solche positive ganze Zahl α_n bestimmt werden kann, daß $\alpha_{n+1} \geqq \alpha_n$ und $\alpha_n < n$, daß α_n mit n unbegrenzt wächst, und daß sowohl die ${}_fAb_r(b_n, C_n)$ und die ${}_fAb_r(b_{n+1}, C_{n+1})$, wie die ${}_iAb_r(b_n, C_n)$ und die ${}_iAb_r(b_{n+1}, C_{n+1})$ in C_{α_n} eindringen. Die ${}_fAb_r(b_{n+1}, C_{\alpha_n})$ können dann aber die ${}_iSb_r(C_{\alpha_n}, b_n)$ nicht treffen, und die ${}_iAb_r(b_{n+1}, C_{\alpha_n})$ können die ${}_fSb_r(C_{\alpha_n}, b_n)$ nicht treffen.

Mithin können die Punkte V_n und W_{n+1} genau wie im ersten Falle definiert und die Kurven λ_r, $\varkappa_{0m}$ und η_r genau wie im ersten Falle konstruiert werden.

Dritter Fall. *Es gibt keinen Kreis um P_0, in den für jedes n die Bahnkurve b_n eindringt,* d. h. für unbegrenzt wachsendes n konvergiert sowohl die Bahnkurve b_n, wie ihr rechtes Gebiet gleichmäßig gegen das Unendliche.

Nach dem obigen Konstruktionsverfahren liegt dann rechts von b_0 ein Ordnungstypus ω von einander nicht kreuzenden, gleichmäßig gegen das Unendliche konvergierenden Bahnkurven, mit der Eigenschaft, daß ein auf einer dieser Bahnkurven b_s willkürlich angenommener Punkt, der nicht zugleich der folgenden Bahnkurve b_t angehört, sich mit b_t durch einen sein Bild nicht treffenden und ganz zwischen b_s und b_t verlaufenden einfachen Kurvenbogen verbinden läßt.

Mithin können wir, wenn wir wieder mit M_0 den Anfangspunkt eines P_0 enthaltenden Bahnsegmentes von b_0 bezeichnen, eine ihr Bild nicht treffende einfache offene Halblinie η_r konstruieren, welche in M_0 ihren Anfangspunkt hat und ganz im rechten Gebiete von b_0 verläuft.

Diese Halblinie η_r existiert mithin für jeden der drei Fälle, welche wir unterschieden haben.

In derselben Weise können wir eine ihr Bild nicht treffende einfache offene Halblinie η_l herstellen, welche in M_0 ihren Anfangspunkt hat und ganz im linken Gebiete von b_0 verläuft.

Die Halblinien η_r und η_l bilden dann zusammen eine einfache offene Linie η, welche zusammen mit ihrem Bilde η' ein Translationsfeld begrenzt. Dieses Translationsfeld enthält das im M_0 anfangende Bahnsegment von b_0, also den Punkt P_0, womit der Translationssatz bewiesen ist.

NOTES

⟦1⟧ For an earlier attack on the plane translation theorem, see Brouwer 1909 H2.

⟦2⟧ The term 'a translation over the whole plane' suggests the meaning 'topologically equivalent to a translation'. The footnote is too ambiguous to prevent this interpretation. As late as 1919 M2 Brouwer gave an example to show that the two properties are not logically equivalent. This example together with a figure are already found in Brouwer's own copy of the Dutch version of 1909 H2. See the reproduction 1909 H1.

⟦3⟧ Brouwer 1910 F.

⟦4⟧ The proof is very complicated. Simpler proofs have been attempted; a few of them have been published. See an attempt to generalize Brouwer's theorem by W. Scherrer 1925, and the relatively simple proof of a weaker form of Brouwer's theorem by E. Sperner, 1934. It would still be worthwhile to reconsider Brouwer's original attempt, 1909 H2.

⟦5⟧ Brouwer 1911 H2.

⟦6⟧ Brouwer 1911 J2.

Mathematics. — "*Remark on the plane translation theorem*". By Prof. L. E. J. BROUWER. 1919M2

(Communicated in the meeting of December 28, 1918).

The *plane translation theorem* enunciated by me in Vol. XII of these Proceedings (p. 297) and completely proved for the first time [[1]]
in Vol. 72 of the Mathematische Annalen (p. 37—54), runs as follows: [[2]]

A continuous one-one transformation of the Cartesian plane Γ in itself with invariant indicatrix, but without an invariant point, is a translation all over the plane.

We mean by this that each point of Γ is situated in a *translation field*, i. e. in a region lying outside its image region and bounded by two simple open lines not meeting each other, one of which is the image of the other.

Let t be the given transformation, T a translation field belonging to t, ${}_nT$ for each positive or negative integer n including zero the image of T for the transformation t^n. The set of points $T' = \mathfrak{S}_n ({}_nT)$ can be represented biuniformly and continuously on a Cartesian plane C in such a way that the image of the transformation t of T' is a translation of C. Thus, *if by a convenient choice of T we can arrange T' to fill up the whole plane Γ, Γ can be represented biuniformly and continuously on a Cartesian plane C in such a way that the image of the transformation t of Γ is a translation of C.*

However the question, *whether for each transformation t a choice of T making T' to fill up the whole plane Γ, is possible*, must be answered in the negative, as was indicated by me in a footnote
on page 37 of the quoted paper of the Mathematische Annalen, [[2]]
and as appears from the following example: [[3]]

In Γ we define a Euclidean system of measurement, and a rectangular system of coordinates founded on it. The straight lines $y = 1$ and $y = -1$ divide Γ into three regions g_1 ($y > 1$), g_2 ($1 > y > -1$), and g_3 ($y < -1$). Each of the regions g_1 and g_3 we fill up with a
pencil of lines $y = c$, and the region g_2 we fill up with a pencil [[4]]
of lines $y^2 = \dfrac{x-c}{1+x-c}$. These three pencils together with the lines $y = 1$ and $y = -1$ form a pencil β of simple open lines not intersecting each other, and covering Γ entirely.

We shall now understand by t the transformation carrying each point P of Γ on the line of β passing through it along an arc of constant length l: *to the left*, if P lies in g_1 or on the boundary of g_1; *upward*, if P lies in g_2; *to the right*, if P lies in g_3 or on the boundary of g_3 This transformation t is duly biuniform and continuous, has no invariant point, and leaves the indicatrix invariant. But if for each positive or negative integer n including zero we represent the image of P for the transformation t^n by ${}_nP$ and $\underset{n}{\Xi}({}_nP)$ by P', *P' does not depend on P continuously* (for, if the sequence of points $P_1, P_2, P_3, \ldots$ lying in g_2 converges to the point P lying on the line $y = -1$, the sequence $P_1', P_2', P_3', \ldots$ does not converge exclusively to P', but also to other points of Γ).

Thus neither can Γ be represented biuniformly and continuously on a Cartesian plane C in such a way that the image of the transformation t of Γ is a translation of C. For, if such a representation were possible, P' would necessarily depend on P continuously.

NOTES

[[1]] Brouwer 1909 H2.
[[2]] Brouwer 1912 B.
[[3]] Brouwer 1912 B [[2]]. See the reproduction 1909 H1.
[[4]] Add '$x \geqq c$' to the definition of the lines within g_2 !

[[270]]

CHAPTER 4

Vector fields on surfaces

Mathematics. "*On continuous vector distributions on surfaces*". By Dr. L. E. J. Brouwer. (Communicated by Prof. D. J. Korteweg). 1909 G2 ⟦1⟧

(Communicated in the meeting of March 27, 1909).

The theorem, that a differential equation $\frac{dy}{dx}=f(x\,y)$, in which we suppose f to be univalent and continuous, possesses through each point (x_1, y_1) *one* integral curve was proved for the first time by Cauchy [1]) for a field, in which f possesses a continuous partial differential quotient with regard to one of the two variables, and then by Lipschitz [2]) for a field in which the difference quotients of f with regard to one of the two variables for increases of that variable below a certain maximum do not exceed in absolute value a certain maximum.

Peano [3]) finally has done away with all restrictions for f with the exception of its continuity, and has proved, that then still through any point *at least one* (but now in general more than *one*) integral curve exists. It is this result of Peano of which we shall make use to deduce a property of continuous vector distributions on a sphere (or on a surface equivalent to it in the sense of analysis situs, after it has been made measurable by a net of curves which is continuous one-one image of the net of principal circles of a sphere).

1) *Exerc. d'anal.* I, 1840, p. 327; comp. also Moigno, *Leç. sur le calc. diff. et int.*, Tm II (Paris 1844), leç. 26, 27, 28, 33. ⟦2⟧

2) Bull. des sc. math. 10, 1876, p. 149. ⟦3⟧

3) Mathem. Ann 37, 1890, p. 482; the proof has considerably been simplified by Arzelà, *Sull'esistenza degl'integrali nelle equazioni differenziali ordinarie*, Memorie della Acc. di Bologna (5) 6, 1896 p. 33. ⟦4⟧

We shall suppose that the vector becomes nowhere zero or infinite; in any point the direction is then univalently determined, and that direction varies continuously from point to point. By placing the sphere into a Euclidean space and by projecting there an arbitrary spherical shell upon its base plane, we then deduce from the theorem of PEANO that in an arbitrary point of the sphere we can at least start *one* single curve, which is *tangent curve* to the vector distribution. Let r be such a tangent curve; we shall then say, that we *pursue* r if we describe it in the direction of the vector and that we "*recur*" it, if we describe it in the direction opposite to the vector.

We now introduce a spherical distance β with the property, that within an arbitrary circle described on the sphere with radius β the *angle* [1]) of any two vectors is $< {}^1/_8\, \pi$. [2]).

Let us now start the curve r in A_0 and let us pursue it up to a point P in such a way, that all points of the described arc A_0P have a distance $< \beta$ from A_0, *then the radius vector drawn from A_0 to an arbitrary point of the arc A_0P will enclose with the direction of the vector in A_0 an angle $< {}^1/_8\, \pi$.* For, if one of the two arcs of principal circles, which in A_0 make an angle ${}^1/_8\, \pi$ with the vector, were transgressed by r between A_0 and P, then according to the supposition the vector is directed in that point of intersection to the inner side of the angle formed by those two circular arcs; so if we pursue r from A_0 to P, it can enter the just-mentioned angle, but it cannot leave it; then however it must always remain inside that angle.

It is likewise evident, that, if T is an arbitrary point on the arc A_0P, the radius vector drawn from T to an arbitrary point of the arc TP encloses with the vector direction in T an angle $< {}^1/_8\, \pi$.

Let Q now be an arbitrary point of r between A_0 and P, we then know that the vector direction has in Q a component in the direction of the radius vector A_0Q; thus, if Q moves along r from

---

1) For the definition of the angle between two vectors not starting from the
[[5]] sáme point in an arbitrary non-Euclidean space, comp. these *Proceedings* Vol. IX, 1906, page 121, 122. To determine that angle here on the sphere we transfer the two vectors to a point of the principal circle, which joins them, maintaining their angle with that circle.

2) That such a spherical distance β can always be indicated, is evident as follows: If a point D approaches indefinitely to a point C, in which the vector is not zero, then on account of the continuity of the vector distribution also the angle between the vectors in D and C converges to zero, and farthermore that convergence takes place for different points of convergence C *uniformly*, the vector distribution being uniformly *continuous* on account of the sphere being a closed set of points.

[[274]]

A_0 to P, the length of the radius vector A_0Q and likewise that of any radius vector TQ (if T is an arbitrary point already passed) increases continually.

From this we conclude in the first place that between A_0 and P the curve r cannot meet itself and then that, when pursuing r from A_0, *certainly a point* B_0 *is arrived at, possessing a distance* β *from* A_0.

For, if such a point were never reached, we could point out on r a series of points

$$G_1, G_2, \ldots G_\omega, G_{\omega+1}, \ldots G_\alpha, \ldots$$

which would not end at any number α of the second class of numbers, which points would possess from A_0 the distances

$$\varepsilon_1, \varepsilon_2, \ldots \varepsilon_\omega, \varepsilon_{\omega+1}, \ldots \varepsilon_\alpha, \ldots$$

which would continually increase in this order, but remain smaller than β. This however is impossible because the set of the differences

$$\varepsilon_{\alpha+1} - \varepsilon_\alpha$$

must be denumerable.

Any point A_0 is followed by a well-determined point B_0; such an arc A_0B_0 we shall call a β-*arc*; the distance between the end points of a β-arc is β; its length lies between β and $\beta\sqrt{2}$, as is easily seen.

Now in the first place it can occur that the pursuing branch and the recurrent one after a finite number of β-arcs have met either each other or one of the two itself.

In that case we possess a closed single tangent curve to the vector distribution.

If not, the pursuing branch and the recurrent one can be continued over an infinite number of β-arcs without a meeting taking place; this case we shall investigate more closely.

Let γ be $= {}^1/_2\beta$ and let $A_0, A_1, A_2, \ldots$ be a series of points on r in such a way, that each arc A_nA_{n+1} is a γ-arc. We are now sure that this series of points can reach each finite index. Let fartheron p be an integer greater than the quotient of the spherical surface by the surface of a circle of radius ${}^1/_2\gamma$ on the sphere.

Let us pursue r from A_0 and let us describe round each point A_n a circle with radius ${}^1/_2\gamma$, then two consecutive ones of those circles touch each other on their outside, and any four consecutive circles lie entirely outside each other; however, when A_p is reached, two circles intersecting each other must have appeared, and at the same time or already before, a *first* point F on r must have been reached possessing a distance γ from a point G lying more than a β-arc behind it. If G lies between A_{h-1} and A_h, then A_{h+1} and A_{h+2} have a distance from G which is $> \gamma$, whilst A_h is separated by less than a β-arc from G; so G and F are separated on r by at least three points A.

Let FE be a β-arc, then on FE lies a point H not coinciding with

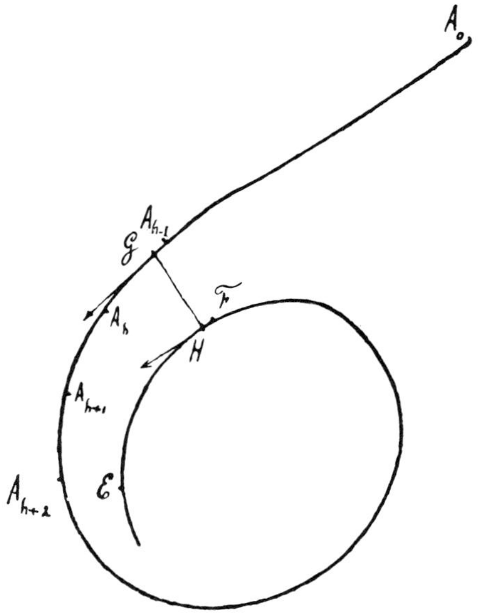

E in such a way, that no other point of that arc has a smaller distance from G. The arc of principal circle GH is then in H perpendicular to r, and as the vector directions in G and H form with each other an angle $< {}^1/_8\,\pi$, they are directed to the same side of that arc of circle. The arc of circle and the arc GH of r have farthermore only their endpoints in common, and they form together a *closed single curve k, of which the length is smaller than* $(p+3)\,\gamma\sqrt{2}$, and which divides the sphere into two domains.

If we pursue r from G, it first runs to H along the boundary of those two domains, and then at H enters one of those domains g_1, *and will leave it no more*, for it no more meets, its own arc GH according to the supposition, and if it were to meet the arc of principal circle GH, that would be at a distance $< \beta$ from H, thus with a pursuing tangent direction, which would *lead* it *into* g_1, but could not make it leave that domain; so this meeting will never be able to take place either [1]).

And analogously, if we recur r from H, it first runs to G along

[1]) On the same grounds it is clear that for r, independent of the choice of A_0, G and H, certainly a minimum distance δ can be pointed out, under which, after the β-arc beginning in H, r will never be able to approach, between the two β arcs beginning in G and in H, the arc of principal circle GH.

the boundary of the two domains, to enter at G the *other* domain g_2 and leave it no more.

Let now γ' be $= \frac{1}{2}\gamma$ and A'_0 (coinciding with H), A'_1, A'_2, ... a series of points on r in such a way, that each arc $A'_n A'_{n+1}$ is a γ'-arc. In the same way as we have constructed the curve k, we now construct a single closed curve k' consisting of an arc $G'H'$ of r, and an arc of principal circle $G'H'$, which is smaller than γ'.

This curve k' lies entirely inside g_1; for after the preceding the only way in which it might still leave it, is that the two arcs of principal circles GH and $G'H'$ should meet in two points, which is impossible, both arcs being $< \pi$.

So the curve k' divides g_1 into 1. an annular domain (which only in the special case that H and G' coincide becomes singly connected), in which lies the arc HG' of r, and 2. a singly connected domain g'_1, within which lies the pursuing branch of r past H'.

Repeating this process indefinitely and taking every time $\gamma^{(n+1)} = \frac{1}{2}\gamma^{(n)}$, we construct a type of order ω of single closed curves, of which each following one lies within the preceding ones, and it is easily proved, that as soon as $\gamma^{(n)}$ has fallen under a certain maximum [1]), the following curves $k^{(n)}$ have all a length smaller than $(p+3)\gamma\sqrt{2}$. Hence the lengths of *all* curves $k^{(n)}$ lie below a same finite limit.

We can now regard the place on the sphere of a variable point of $k^{(n)}$ as a function of the length of arc s between $G^{(n)}$ and that point. The different $k^{(n)}$'s are then represented by a system of *uniformly continuous* functions. Thus according to ARZELÀ [2]) a fundamental series

$$k^{(\tau_1)},\ k^{(\tau_2)},\ k^{(\tau_3)},\ \dots$$

can be indicated, *converging uniformly* to a *continuous* limit function $k^{(\omega)}$.

The differential quotients of the functions determining the curves $k^{(\tau)}$ are in any point indicated by the vector direction in that point; they are uniformly approximated by the functions of s determining the difference quotients with respect to s, and they are themselves uniformly continuous functions of s.

So they converge uniformly to a continuous limit function representing the differential quotient, i. e. the tangent direction of $k^{(\omega)}$.

---

[1]) Such a maximum is the quantity δ mentioned in the preceding note. For if we then take $G^{(n)}$ as A_0, and if we construct the corresponding curve k, then the length of the arc between $G^{(n)}$ and the point H belonging to that curve k is smaller than $(p+2)\gamma\sqrt{2}$. We know here however for sure, that, if before this point H no point on r has been reached possessing a distance smaller than δ from $G^{(n)}$, such a point will not appear farther on either.

[2]) „*Funzioni di linee*", Rendiconti Lincei (4) 5, 1 (1889), p. 342; "*Sulle funzioni di linee*", Memorie della Accademia di Bologna (5) 5 (1895), p. 225. [[6]]

So, as any point of $k^{(\omega)}$ is limit point of points of $k^{(n)}$ lying on r, *the limit curve $k^{(\omega)}$ is a tangent curve to the vector distribution.*

Farthermore s also represents the length of arc of $k^{(\omega)}$ and as the lengths of the $k^{(\nu)}$'s remain below a same finite limit, $k^{(\omega)}$ also returns into itself after having described a finite length of arc.

The curve $k^{(\omega)}$ cannot reduce to a single point, for then the whole of the directions of the tangents to a curve contracting to a single point would converge to a single direction, namely the vector direction in that limit point, which is impossible.

Neither can a point of $k^{(\omega)}$ belong to two different values of s, unless after a whole circuit; for otherwise $k^{(\omega)}$ would consist of a single closed curve plus points in its "inner domain" (if we call its "outer domain" that in which all curves $k^{(n)}$ lie); which is likewise impossible.

So it is evident that $k^{(\omega)}$ is a single closed curve to which the pursuing branch of r spirally converges uniformly.

In the same way it is evident, that also the recurrent branch of r spirally converges uniformly to a single closed curve $k'^{(\omega)}$ lying entirely outside $k^{(\omega)}$.

The tangent curve r has therefore for analysis situs the character of a double circular spiral circuit, whose two asymptotic closed curves are likewise tangent curves to the vector distribution.

So we possess for continuous vector distributions having in any point a definite direction, at any rate a single closed tangent curve.

Possessing now such a closed tangent curve r_1, we can start another tangent curve in one of the domains determined by the former, which domain we shall call its "inner domain". This curve can behave in different ways:

A. When sufficiently pursued on one side and recurred on the other side, it finally returns into itself without having met r_1.

In this case we possess a single closed tangent curve r_2 bounding a singly connected "inner domain" forming a part of the inner domain of r_1.

B. It does not return into itself inside r_1; here the following cases are possible:

α. When recurring we find it starting somewhere on r_1; when we pursue it, it does not meet r_1. Then according to what precedes it the pursuing branch converges spirally to a single closed tangent curve r_2, bounding a singly connected "inner domain", which is a part of the inner domain of r_1.

β. When recurring we do not find it starting on r_1; when we pursue it however it ends somewhere on r_1. Then the recurrent branch furnishes a single closed tangent curve r_2 with the same property as above.

γ. Neither when we pursue, neither when we recur it meets r_1. [[7]]
It then converges on both sides to a single closed tangent curve. One of these can coincide with r_1; the other however is a single closed tangent curve r_2 with the same property as above.

δ. When we pursue as well as when we recur it meets r_1. The [[7]]
arc lying between the first meeting-points on both sides forms then with an arc of r_1 joining those same points a single closed tangent curve r_2 with the same property as above.

In the same way we can now again enclose a part of the inner domain of r_2 by a single closed tangent curve r_3, and we can construct in this way a fundamental series of single closed tangent curves

$$r_1, r_2, r_3, r_4, r_5, \ldots$$

for which we show, in the same way as above for the curves $k^{(n)}$, that there exists an upper limit for their length of arc and fartheron, that they converge uniformly to a single closed tangent curve r_ω, whose inner domain is a part of that of any curve r_n.

But we can still again let r_ω lose a part of its inner domain by a single closed tangent curve $r_{\omega+1}$, and again $r_{\omega+1}$ by $r_{\omega+2}$, and this process can be continued after any index of the second class of numbers.

On the other hand, however, this is an absurdity, as the system of those losses of domain must remain denumerable.

The supposition, that the vector direction should be determined in any point, has thus proved to be impossible, so that we can formulate:

THEOREM 1. *A vector direction varying continuously on a singly connected, twosided, closed surface must be indeterminate in at least* [[8]]
one point.

And from this follows directly:

THEOREM 2. *A vector distribution anywhere univalent and continuous on a singly connected, twosided, closed surface must be zero or infinite in at least one point.*

If we represent the complex plane stereographically on the NEUMANN sphere, a complex function becomes a vector distribution on the sphere. So we can also interpret our result as follows:

THEOREM 3. *A univalent, continuous function of a complex variable being nowhere zero or infinite and without singular points cannot exist*[1]).

Between the above theorem 2 and the property deduced in a former communication [2]) that every continuous one-one transformation with invariant indicatrix of a sphere in itself shows at least *one* invariant point is a close connection. At first sight one might even suppose that they can be directly deduced out of each other. However this is not the case; on the contrary: they complete each other.

Let us namely suppose on one hand the theorem about the vector distribution to be proved. If then is given a continuous one-one transformation of the sphere in itself, we can join each point P with its image P' by an arc of principal circle PP', and consider that arc of circle in size and direction as a vector in P. But the univalence and the continuity of such a vector distribution is now assured only, if for no point P the image lies in the antipodic point; and as this may not be assumed for an arbitrary continuous one-one transformation, a direct appearance of the theorem of the invariant point is excluded.

Let us on the other hand regard as proved the theorem of the invariant point, and let a continuous vector distribution be constructed on the sphere. If we then make the points of the sphere undergo infinitesimal displacements proportionate to the vectors, and if we may suppose these displacements to generate at the limit a one-one transformation (of itself continuous and leaving the indicatrix invariant), we can conclude from it that the vector distribution must of necessity be somewhere zero or infinite. We are, however, sure of the one-one correspondence of that transformation only if by indefinite decrease of the vectors we can make the vector variation anywhere smaller than the corresponding point variation, thus if the infinitesimal difference quotients of the given vector distribution do not exceed a certain maximum. And as in general this condition is not satisfied, the theorem of the vector distribution in its general form does not appear directly as a consequence of the theorem of the invariant point.

A continuous vector distribution on the elliptic plane through a one-two correspondence determining a continuous vector distribution on the sphere, the two following theorems also hold:

1) For *monogenous* complex functions this is a well-known theorem; for, a constant has in the point of the NEUMANN sphere representing the infinite a singular point.

[9] 2) These Proceedings, page 797 of this volume.

THEOREM 4. *A vector direction varying continuously on a singly connected, onesided, closed surface must be indeterminate at least in one point.*

THEOREM 5. *A continuous vector distribution anywhere univalent on a singly connected, onesided, closed surface must vanish or become infinite at least in one point.*

By the following elementary example theorem 4 is illustrated: If we wish to adjoin in the projective plane by linear relations between the respective coordinates to any point P a straight line passing through that point, this is only possible by taking for that line the line which joins P with a fixed point Q. Theorem 4 informs us that if we wish to select in any point P one of the two half lines joining P and Q, this cannot be done continuously.

We really see that if P moves along a straight line and if at the same time the half line PQ varies continuously, after a circuit of P that half line has not remained the same.

Finally we notice that theorem 5 has as a direct consequence the theorem of the invariant point for the elliptic plane. [1])

For, for a continuous one-one transformation of the elliptic plane in itself the two straight line segments, which join P and its image P', determine, it is true, two oppositely directed vectors, but a selection out of them for *one* point determines a selection everywhere, varying continuously in the whole plane. This is immediately proved, if we let P move along a unilateral curve; P' describes then likewise a unilateral curve, and the selected segment PP', after having varied continuously during the circuit, has remained the same as before.

As farthermore the vector distribution, constructed in this way, becomes nowhere infinite, it must vanish at least in *one* point; this point is invariant for the transformation.

---

[1]) These Proceedings, page 798 of this vol. ⟦10⟧

NOTES

⟦1⟧ The English version appeared in the Proceedings of the meeting of 23 april 1909.

1909 G2, 1910 A2, 1910 D2 form a series of three papers. 1915 B is an application of the general theory to a special case.

In this study of singularities the emphasis is on solutions of differential equations rather than on vector fields. This field of research has been inaugurated a quarter of a century earlier by H. Poincaré, 1881, 1882, 1885 A. From the introduction to Poincaré 1881, one of the classics of mathematics, we quote: 'Finding out the properties of differential equations is a task of major importance. A first step has been made by studying the proposed function in the neighbourhood of one of the points of the plane. The time has come to go further and to study the functions in the whole plane. In this research one starting point will be our knowledge with respect to a certain region of the plane. The study of a function thus includes two parts: 1) The qualitative one, as it may be called, where the curve defined by the function is studied geometrically, 2) the quantitative one where values of the function are calculated numerically.'

For a long time Poincaré's papers – a seminal work in topology – failed to attract the attention they deserved. It is not clear whether Brouwer knew Poincaré's study when he started his own series of papers on vector fields. Not until the third paper (1910 D2, p. 184) did he mention Poincaré; perhaps Hadamard had drawn his attention to Poincaré's work. Brouwer's approach is genuinely topological and in this it differs from Poincaré's, where the algebraic element is strongly felt. It is not clear either whether Brouwer became acquainted with I. Bendixson 1901, which continues Poincaré's work.

In his first communication, however, Brouwer did refer to A.-L. Cauchy 1840, R. Lipschitz 1876, G. Peano 1890 B, Arzelà 1896. This circumstantial evidence may indicate how Brouwer came to analyze vector fields by topological means: it was Peano's discovery that continuity assumptions sufficed to ensure the existence of solutions of ordinary differential equations. It is, indeed, a remarkable fact that Brouwer's investigations are independent of uniqueness assumptions, which were implicit in Poincaré's approach. Poincaré even assumed algebraicity and simple singularities. Though I. Bendixson 1901 made more general assumptions, they still imply uniqueness. Brouwer on the contrary chose the utmost generality and even in specializing (see 1915 B) he admitted more general types of singularities than Poincaré had done. Thanks to its simpler design, Poincaré's work has been influential in the long run, but it is strange that Brouwer's contributions have been entirely overlooked. This is still felt in A. A. Andronov et al. 1973, who in their bibliography rather than Brouwer's work on vector distributions on surfaces mention Brouwer 1909 F2, 1909 H2, 1911 H1 – it is not clear why they do so and to which part of the text the citation is related.

⟦2⟧ A.-L. Cauchy 1840, F. N. M. Moigno 1844.

⟦3⟧ R. Lipschitz 1876.

⟦4⟧ G. Peano 1890 B, C. Arzelà 1896.

⟦5⟧ Brouwer 1906 B2, p. 120.

⟦6⟧ C. Arzelà 1889, 1896.

⟦7⟧ According to Brouwer 1910 D2 Erratum, change 'recure it, meets' into 'recur it, it meets'.

⟦8⟧ This theorem is usually ascribed to H. Poincaré, 1881, though in fact Poincaré asserted and proved it under quite restrictive conditions. Brouwer generalized it to higher dimensions and proved it by more adequate methods in 1911 D.

⟦9⟧ Brouwer 1909 F2, Theorem 1.

⟦10⟧ Brouwer 1909 F2, Theorem 3.

Mathematics. — *"On continuous vector distributions on surfaces".* (2nd communication)[1]). By Dr. L. E. J. BROUWER. (Communicated by Prof. D. J. KORTEWEG). 1910 A2

[[1]]

(Communicated in the meeting of February 26, 1910).

§ 1.

The tangent curves to a finite, uniformly continuous vector distribution with a finite[2]) *number of singular points in a singly connected inner domain of a closed curve.*

Let γ be the domain under consideration, then we can represent it on a sphere, so we can immediately formulate on account of the property deduced in the first communication (see there page 855): [[2]]

THEOREM 1. *A tangent curve, which does not indefinitely approach a point zero, is either a simple closed curve, or its pursuing as well as its recurring branch shows one of the following characters:* 1st. *stopping at a point of the boundary of* γ; 2nd. *spirally converging to a simple closed tangent curve;* 3rd. *entering into a simple closed tangent curve.*

We now shall farther investigate the form (in the sense of analysis situs) of a tangent curve r, of which we assume, that at least *one* of the two branches (e. g. the pursuing branch) approaches indefinitely one or more points zero, i. e singular points of the vector distribution.

We start the tangent curve in a point A_0 (not a point zero) and we pursue that curve in the following way: By β_ε we understand a distance with the property that in two points lying inside the same geodetic circle described with a radius β_ε, and possessing both a distance $> \varepsilon$ from the points zero, the vectors certainly make an angle $< \frac{1}{8}\pi$ with each other. We farther choose a fundamental series of decreasing quantities $\varepsilon_1, \varepsilon_2, \varepsilon_3, \ldots$ converging to 0, and of corresponding decreasing distances $\beta_{\varepsilon_1}, \beta_{\varepsilon_2}, \ldots$, which all we suppose, if a is the distance of A_0 from the points zero, to be smaller than $a - \varepsilon_1$.

We then prove in the manner indicated in the first communication p. 852, that, when pursuing r from A_0, a point B_0 is reached, possessing a distance β_{ε_1} from A_0; we call the arc A_0B_0 a β_{ε_1}-arc. According to our supposition there now exists a finite number n_1 in [[2]]

1) For the first communication see these Proceedings Vol. XI 2, p. 850. [[2]]

2) This restriction we shall drop in a following communication. [[3]]

such a way, that after having completed n_1 β_{ε_1}-arcs, but not yet n_1+1 β_{ε_1}-arcs, we reach a point A_1, where for the first time we have approached the points zero as far as a distance ε_1. Then again there is a finite number n_2 in such a way that, having completed from A_1 n_2, but not yet n_2+1 β_{ε_2}-arcs, we reach a point A_2, where for the first time we have approached the points zero as far as a distance ε_2. From there we pursue r with β_{ε_3}-arcs and continue this process indefinitely.

If we understand by $m(\varepsilon_n)$ the maximum distance from the points zero, which r reaches when being pursued *after* having for the first time approached the points zero as far as a distance ε_n, then a first possibility is, that $m(\varepsilon_n)$ converges with ε_n to zero.

In that case the pursuing branch converges to one single point zero and it is *an arc of simple curve, stopping at that point zero.*

We now suppose the second possibility, that $m(\varepsilon_n)$ surpasses for each ε_n a certain finite quantity e. Then we can effect (by eventually omitting a finite number of terms of the series of ε_n's), that each $\varepsilon < \frac{1}{2} e$ and each $\beta_{\varepsilon_n} < \frac{1}{2} e$.

On the pursuing branch then certainly two points P_1 and Q_1 can be indicated both at a distance e from the points zero, and separated on r by at least one point at a distance ε_1 from the points zero, whilst the distance between P_1 and Q_1 is $\leqq \frac{1}{4} \beta_{\varepsilon_1}$. Let P_1S and Q_1U be pursuing β_{ε_1}-arcs, and P_1R and Q_1T recurring β_{ε_1}-arcs.

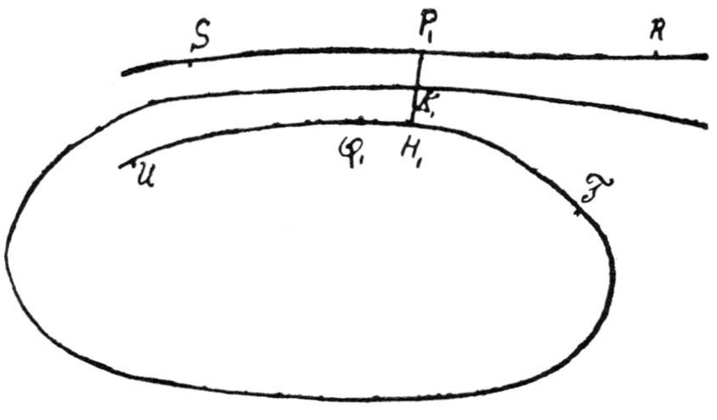

Fig. 1.

Let H_1 be a point of TU, having from P_1 the smallest possible distance, then H_1 cannot coincide with T or U, so that the geodetic arc P_1H_1 is in H_1 normal to the vector direction, and the vector directions in all points of that geodetic arc, forming with each other an angle $< \frac{1}{8}\pi$, are directed to the same side of the geodetic arc $P_1 H_1$.

Let K_1 be the last point of intersection of the arc $P_1 H_1$ of r with the geodetic arc $P_1 H_1$. Then the arc $K_1 H_1$ of r and the geodetic arc $K_1 H_1$ form a simple closed curve, and we prove in the manner indicated on page 853 of our first communication, that either the pursuing branch of r from H_1 lies in the inner domain, and the recurring branch from K_1 in the outer domain, or the pursuing branch from H_1 in the outer domain, and the recurring branch from K_1 in the inner domain.

Let us first assume that the pursuing branch lies in the *inner domain*, then certainly two points P_2 and Q_2 can be chosen on it, both at a distance e from the points zero and separated on r by at least one point at a distance ε_2 from the points zero, whilst the distance between P_2 and Q_2 is $\leqq \frac{1}{4} \beta_{\varepsilon_2}$. With the aid of those two points we construct in the same way as above now a simple closed curve, consisting of an arc $K_2 H_2$ of r and a geodetic arc $K_2 H_2$, in whose inner domain lies the pursuing branch of r from H_2.

Going on in this way we construct a fundamental series of closed curves $u_1, u_2, u_3, \ldots$ lying inside each other. If there is a domain or set of domains G, common to all the inner domains of these curves (which, as we shall presently show, is really the case) then the boundary of G can only be formed by points belonging to none of the curves $u_1, u_2, u_3, \ldots$ but being limit points of fundamental series of points lying on those curves

We assume $q > p$, and B to be a point of u_q having a distance $> 3\, \varepsilon_p$ and $> 3\, \beta_{\varepsilon_p}$ from the points zero. Let C be the first point when recurring from B, and D the first point when pursuing from B, which reaches a distance $\frac{1}{2} \beta_{\varepsilon_p}$ from B, then we shall assume for a moment that there exists on u_q, but not on the arc CD, a point S lying at a distance $\leqq \frac{1}{4} \beta_{\varepsilon_p}$ from B, and we shall show that this assumption leads to an absurdity.

Let SV be a recurring $\frac{1}{2} \beta_{\varepsilon_p}$-arc and SW a pursuing $\frac{1}{2} \beta_{\varepsilon_p}$-arc on u_q, then the arcs CD and VW can have no point in common, and the geodetic arc $K_q H_q$, belonging to u_q, has either no point in common with VW, or none with CD.

In the first case we determine on VW a point M, having from B a distance as small as possible. The geodetic arc BM is then in M normal to VW, and has a last point of intersection N with CD, so that the geodetic arc NM forms with one of the arcs NM of u_q, not containing e.g. the point C, a closed curve; u_q, taken with a certain sense of circuit, would at M enter one of the two domains determined by this closed curve, to leave it no more; further C would lie outside that domain; thus u_q would never be able to reach C, with which the absurdity of our assumption has been proved.

In the second case we determine on CD a point M having from S a distance as small as possible, and on the geodetic arc SM the last point of intersection N with VW. The further reasoning remains analogous to the one just followed: the parts of the arcs VW and CD are only interchanged.

Let now B_ω be the only limit point of a certain fundamental series of points $B_1, B_2, B_3, \ldots$, lying respectively on $u_1, u_2, u_3, \ldots$ We assume that B_ω is not a point zero; it has then for a suitably selected p a distance $> 4\,\varepsilon_p$ and $> 4\,\beta_{\varepsilon_p}$ from the points zero.

Let further each m_k be $> p$ and let $B_{m_1}, B_{m_2}, B_{m_3}, \ldots$ be a fundamental series contained in the series just mentioned, whose points have all from B_ω a distance $< \frac{1}{8}\,\varepsilon_p$ and $< \frac{1}{8}\,\beta_{\varepsilon_p}$.

If then further on the different u_{m_k} $B_{m_k} D_{m_k}$ are pursuing, $B_{m_k} C_{m_k}$ recurring $\frac{1}{2}\,\beta_{\varepsilon_p}$-arcs, we prove by the reasoning followed in the
⟦2⟧ first communication p. 854, that there exists a series $C_{n_1} D_{n_1}$, $C_{n_2} D_{n_2}$, $C_{n_3} D_{n_3}, \ldots$ converging uniformly to an arc $C_\omega D_\omega$ of a tangent curve u_ω in such a way, that all arcs $C_{n_k} D_{n_k}$ lie on the same side of $C_\omega D_\omega$.

If we describe round B_ω a geodetic circle with radius $\frac{1}{8}\,\beta_{\varepsilon_p}$, then it cuts from $C_\omega D_\omega$ an arc FI containing B_ω; this arc divides its inner domain into two regions, into one of which, to be called g, neither the arcs $C_{n_k} D_{n_k}$, nor any other parts of the curves u_{n_k} can penetrate, as they would get there a distance $\leq \frac{1}{4}\,\beta_{\varepsilon_p}$ from B_{n_k}.

As further the region g cannot lie outside all curves u_{n_k}, it must lie inside all curves u_{n_k}.

So there is certainly a domain or a set of domains G, common to all the inner domains of the curves u_k, and to the boundary of G belong all points of the limit set λ of the u_k's, which are not points zero, thus also all points of λ, which are points zero, as the latter are limit points of the former ones. So the boundary of G is identical to the limit set of the u_k's, is therefore coherent and identical to its *outer circumference,* whilst abroad from the points zero it consists of tangent curves to the vector distribution, which on account of the existence of the domain g can show nowhere in a non-singular point the character mentioned in theorem 1 sub 3.

We shall now show that a tangent curve r' belonging to the boundary of G cannot have the property of r, that its pursuing or recurring branch converges spirally to the boundary of a domain or set of domains G'.

We should then namely be able to form, in the same way as was done above and in the first communication for r, also for r' a closed curve u'_k consisting of a geodetic arc $\leq \frac{1}{4} \beta_{\varepsilon_k}$ and an arc φ' of r', joining the same two points K' and H'. And there would exist arcs of r which would converge uniformly to φ' from the same side, e.g. from the inner side of u'_k. But when pursuing such an arc ψ of r situated in sufficient vicinity of φ', we should never be able to return between ψ and φ'.

As farthermore in the case considered here, that the pursuing branch of r lies in the inner domain of u_1, it is also excluded, that r' reaches the boundary of γ, only *one* form remains possible for r', namely *that of an arc of simple curve, starting from a point zero, and stopping at a point zero.* (For the rest these two end points can very well be identical).

Of such tangent curves there can be in the boundary of G at most two, which possess the same end points, when these end points are different; but there can be an infinite number, which are closed in the same point zero. Of these however there are only a finite number, of which the extent surpasses an arbitrarily assumed finite limit. For, each of these contributes to G a domain with an area, which surpasses a certain finite value.

The curves r' whose extent surpasses a certain finite limit are run along by a u_k of sufficient high index in the same order, as they succeed each other on the outer circumference of G. From this ensues *that for all curves r' the pursuing sense belongs to the same sense of circuit of the outer circumference of G.*

If the pursuing branch of r lies in the *outer domain* of u_1, the preceding holds with slight modifications. A point of the limit set of the u_k's now necessarily bounds a region belonging to γ, and lying outside all u_k's, only then when it is not a point of the boundary of γ. The *inner circumference*, to which r *now* converges spirally *on the inner side*, consists here again of arcs of simple curve, which are tangent curves to the vector distribution, but these tangent curves can lie entirely or partially in the boundary of γ.

However they have all again a pursuing sense belonging to the same sense of circuit of the circumference.

We now agree about the following: When a pursuing branch of a tangent curve reaches a point zero, we continue it, if possible, along a pursuing branch, starting from that point zero, and not meeting the former within a certain finite distance; but if such a continuation is impossible, we stop the branch at that point zero, and so we do likewise when the branch has entered into a closed curve or has approximated spirally a circumference. Then we can resume the preceding reasonings as follows:

THEOREM 2. *A tangent curve is either a simple closed curve, or save its ends it is an arc of simple curve, of which the pursuing as well as the recurring branch shows one of the following characters*: 1st. *stopping at a point of the boundary of* γ; 2nd. *stopping at a point zero*; 3rd. *entering into a simple closed tangent curve*; 4th. *spirally converging to a circumference, consisting of one or more simple closed tangent curves.*

From this ensues in particular:

THEOREM 3. *A tangent curve cannot return into indefinite vicinity of one of its points, after having reached a finite distance from it, unless it be to close itself in that point.*

That the last theorem is not a matter of course, is evident from the fact that it does not hold for an annular surface. On this it is easy to construct tangent curves of the form pointed out by LORENTZ
[[4]] (Enz. der Math. Wiss. V 2, p. 120, 121).

We finally notice that the vector distribution considered in this §, does not possess of necessity a singular point (as is the case on the sphere). This is proved directly, by considering in the inner domain of a circle, situated in a Euclidean plane, a vector everywhere constant.

§ 2.

The structure of the field in the vicinity of a non-singular point.

To classify the singular points we shall surround each of them

[[288]]

with a domain which we shall cover entirely with tangent curves not crossing each other and we shall investigate the different ways in which that covering takes place in different cases. For the sake of more completeness and as an introduction we first do the same for a non-singular point.

Let P be the point under consideration, RS an arc of tangent curve r containing P, UV an arc containing P of an orthogonal curve of the vector distribution. We draw through U and V tangent curves α_0 and α_1, and through R and S orthogonal curves γ and δ, and we let the four points R, S, U, and V converge together to P. Before they have reached P, a moment comes when α_0, α_1, γ, and δ form a curvilinear rectangle, inside which lies P, and inside which lies no point zero of the vector distribution, thus inside which on account of the first communication no closed tangent curve can be drawn.

We shall cover this curvilinear rectangle with tangent curves not crossing each other.

We number α_0 with 0, r with $\frac{1}{2}$, α_1 with 1. Let $Q_{\frac{1}{4}}$ be a point inside or on the rectangle $A_0\,B_0\,S\,R$ (fig. 2) having from α_0 and r

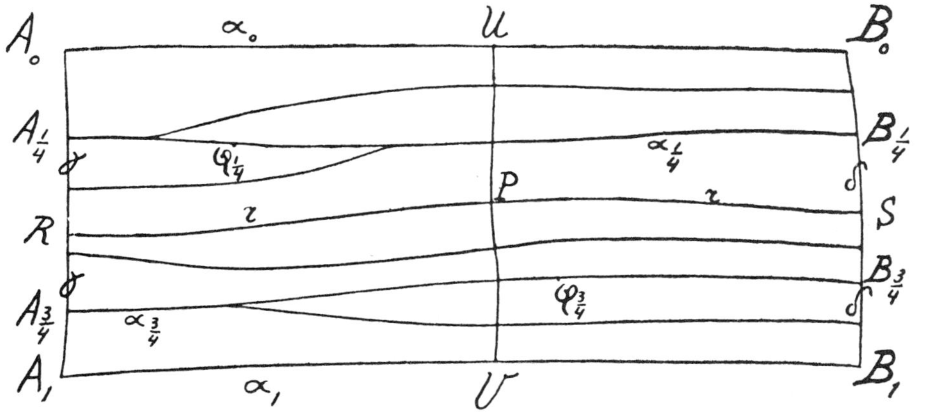

Fig. 2. Non-singular point.

a distance as large as possible. We draw through $Q_{\frac{1}{4}}$ a tangent curve $\alpha_{\frac{1}{4}}$, about which we agree, that, if it meets α_0 or r, we shall continue it, by pursuing or recurring α_0 or r, until we come upon γ or δ.

Then $\alpha_{\frac{1}{4}}$ is a tangent curve joining two points $A_{\frac{1}{4}}$ and $B_{\frac{1}{4}}$ of γ and δ between α_0 and r. In the same way we construct inside the rectangle $A_1 B_1 S R$ a tangent curve $\alpha_{\frac{3}{4}}$, joining two points $A_{\frac{3}{4}}$ and $B_{\frac{3}{4}}$ of γ and δ between r and α_1. The rectangle $A_0 B_0 B_1 A_1$ is then divided into four regions. In these we choose in the way described above successively the points $Q_{\frac{1}{8}}$, $Q_{\frac{3}{8}}$, $Q_{\frac{5}{8}}$, $Q_{\frac{7}{8}}$, draw through $Q_{\frac{1}{8}}$ a tangent curve $\alpha_{\frac{1}{8}}$ joining two points $A_{\frac{1}{8}}$ and $B_{\frac{1}{8}}$ of γ and δ, and we deal analogously with the other three points.

Going on in this manner we construct for each fraction $\frac{a}{2^n} < 1$ a tangent curve $\alpha_{\frac{a}{2^n}}$ joining two points of γ and δ; two of these curves chosen arbitrarily can coincide *partially*, but they cannot cross each other.

All these tangent curves must now cover everywhere densely the inner domain of the rectangle $A_0 B_0 B_1 A_1$. For, if they left there open a domain G, then a domain G'_n bounded by two tangent curves with indices $\frac{a}{2^n}$ and $\frac{a+1}{2^n}$ would converge to G. For n sufficiently great however the point $Q_{\frac{2a+1}{2^n+1}}$ would then lie inside G, thus in contradiction to the supposition also a tangent curve $\alpha_{\frac{2a+1}{2^n+1}}$ would pass through G.

From this ensues, that, if we add the limit elements of the tangent curves $\alpha_{\frac{a}{2^n}}$, which are likewise tangent curves, the inner domain of the rectangle $A_0 B_0 B_1 A_1$ is entirely covered, and further there is for each real number between 0 and 1 *one* and not more than *one* of these tangent curves having that number as its index.

§ 3.

The structure of the field in the vicinity of an isolated singular point. First principal case.

We surround the point zero P, supposed isolated, with a simple closed curve c, inside which lies no further point zero. And we assume as a first principal case that c can be chosen in such a way that inside c no simple closed tangent curve exists, inside which P lies. On account of the first communication there can exist inside c neither a simple closed tangent curve, outside which P lies. We now distinguish 2 cases:

a. There exists inside c a simple closed tangent curve ϱ through P. We can then choose c smaller, so that it meets ϱ, thus containing in its inner domain a tangent curve ϱ_1 which (in its pursuing direction) runs from P to c, and another ϱ_2 running from c to P, and we further look for such tangent curves inside c which cross neither ϱ_1 nor ϱ_2. Of the possible kinds of tangent curves mentioned at the conclusion of § 1 we shall agree about those, which enter into a closed tangent curve, to continue them along that tangent curve until they reach either P or c, and to stop there. Spirally converging to an inner circumference cannot appear, as the other end of such a tangent curve would be separated from P as well as from c, and so would determine a closed tangent curve, outside which P would be lying, which is impossible. Neither can appear spirally converging to an outer circumference, as P would have to lie in that outer circumference and the spiral would necessarily have to cross ϱ_1 and ϱ_2.

b. There exists inside c no simple closed tangent curve through P. Then inside c there exists no simple closed tangent curve at all, so that again spirally converging is excluded.

In any case, if we agree not to continue a tangent curve, when it reaches P or c, we can distinguish the tangent curves inside c, and not crossing ϱ_1 and ϱ_2 if the latter exist, into three categories:

1st. *Closed curves, containing P but not reaching c.*

2nd. *Arcs of curve, joining two points of c, but not containing P.*

3rd. *Arcs of curve which run from P to a point of c* (positive curves of the third kind) *or from a point of c to P* (negative curves of the third kind).

Of this third kind there must certainly exist tangent curves. For otherwise the closed sets determined by the curves of the first, and by those of the second kind would cover the whole inner domain of c, thus would certainly possess a point in common; through this point however a curve of the third kind would pass.

So we can commence by constructing one curve of the third kind and we choose eventually ϱ_1 for it. *If possible,* we then draw a second curve of the third kind not crossing the first and we choose eventually ϱ_2 for it. Into each of the two sectors, determined in this way inside c, we introduce if possible again a curve of the third kind, not crossing the already existing ones, and chosen in such a way that it reaches a distance as great as possible from the two curves of the third kind, which bound the sector, whilst, if the new curve terminates somewhere on one of the curves bounding the sector, we further follow the latter curve. In each of the sectors, determined after that in the inner domain of c, we repeat if possible this insertion, and we continue this process as often as possible, eventually to an indefinite number of insertions.

If in this manner we have obtained an infinite number of tangent curves of the third kind, they determine limit elements which each are either again a tangent curve of the third kind, or contain such a curve as a part. And in particular a fundamental series of positive respectively negative curves of the third kind determines in its limit elements again positive respectively negative curves of the third kind.

After addition of these limit curves of the third kind we are, however, quite sure that no new curves of the third kind not crossing the existing ones can be inserted. This is evident from a reasoning analogous to that followed in § 2. The whole of the curves of the third kind, obtained now, we shall call *a system of base curves of the vicinity of P*.

An arbitrary positive base curve and an arbitrary negative one enclose inside c a sector, of which the area cannot fall below a certain finite limit. For otherwise we should have a fundamental series of positive base curves, and a fundamental series of negative ones, possessing the same base curve as a limit element, which is impossible, as that limit base curve would have to be positive as well as negative.

So the inner domain of c is divided into a finite number of sectors which can be brought under the two following categories:

First category. Sectors bounded by a positive and a negative base curve, between which lie no further base curves. The areas of these sectors surpass a certain finite limit.

Second category. Sectors bounded by two positive (respectivaly two negative) base curves and containing only positive (respectively negative) base curves. A sector of this category can reduce itself in special cases to a single base curve.

We shall first treat a sector of the *first category* and to that end we first notice that *outside* a curve of the second kind lying in it (i. e. between that curve and c) lie only curves of the second kind, and *inside* a curve of the first kind lying in it only curves of the first kind.

If we draw in the sector a well-ordered series, continued as far as possible, of curves of the second kind enclosing each other, then it converges either to a curve of the second kind, or to two curves of the third kind and between them a finite or denumerable set of curves of the first kind, *not* enclosing each other, and *not* approaching c indefinitely.

If we can construct an infinite number of such series not enclosing each other, then there are among them which cut from the sector an area as small as one likes, and at the same time the maximum distance, which such a series reaches from c, decreases under each finite limit.

And analogously, if we draw in the sector a well-ordered series, continued as far as possible, of curves of the first kind enclosing each other, it converges either to a curve of the first kind, or to two curves of the third kind and between them a finite or denumerable set of curves of the second kind, *not* enclosing each other, and *not* approaching P indefinitely.

If we can construct an infinite number of such series not enclosing each other, then there are among them which enclose an area as small as one likes, and at the same time the maximum distance, which such a series reaches from P, decreases under each finite limit.

From this ensues that for the sectors of the first category we have to distinguish two cases:

First case. There are curves of the second kind in indefinite vicinity of P. Then the domain of the curves of the second kind is bounded by the two base curves which bound the sector, and a finite or denumerable number of curves of the first kind, *not* enclosing each other, and *not* approaching c indefinitely, in whose inner domains, which we call the *leaves* of the sector, can lie only curves of the first kind.

The region outside the leaves can be covered as follows with curves of the second kind not crossing each other: we first construct one which reaches a distance as great as possible from c and the boundary of the leaves; in this way two new regions are determined, in each of which we repeat this insertion. This process we continue indefinitely, and finally we add the limit curves. That then the region outside the leaves is entirely covered, is evident from the reasoning followed in § 2.

And in the same way we fill each of the leaves with curves of the first kind not crossing each other. The whole of the tangent curves filling the sector finally gets the form indicated in fig. 3. The sectors being in the discussed first case we shall call *hyperbolic sectors.*

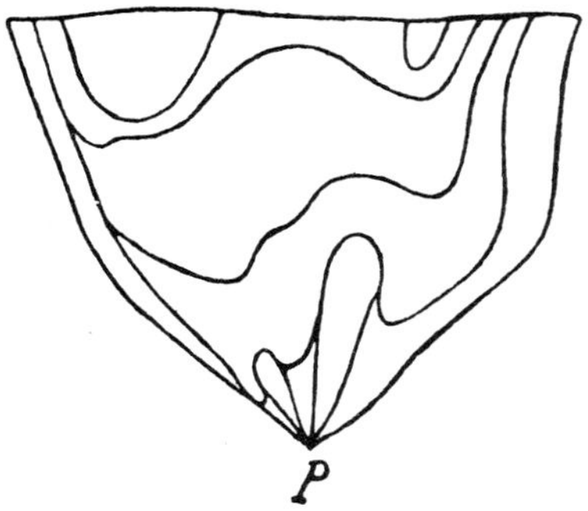

Fig. 3. Hyperbolic sector.

Second case. There are no curves of the second kind in indefinite vicinity of P. Then the domains covered by these curves are cut off from the sector by a finite or denumerable number of curves of the second kind, *not* enclosing each other, and *not* approaching P indefinitely. These domains we take from the sector (consequently modify an arc of c), and there remains a new sector, bounded by the same base curves as the old one, but consisting of one *leaf* inside which lie only curves of the first kind. This leaf we can fill with curves of the first kind not crossing each other (see fig. 4).

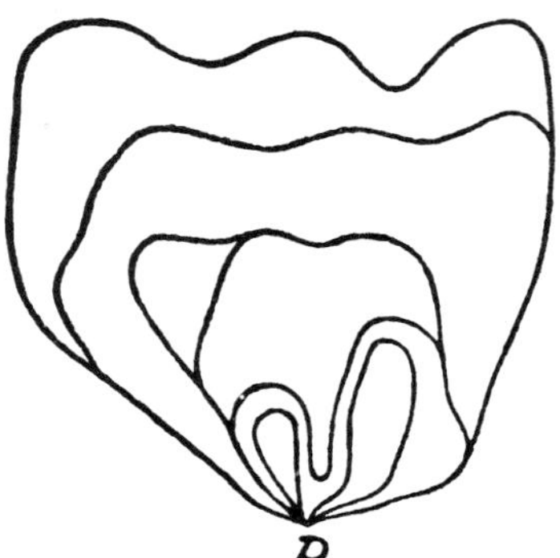

Fig. 4. Elliptic sector.

These sectors of the second case, which are reduced to a single leaf, we shall call *elliptic sectors*.

We now pass to the discussion of a sector of the *second category*, of which, to fix our ideas, we assume, that it is bounded by two positive base curves.

Let us consider the set of points lying in the sector or on its boundary, through which curves of the second kind not crossing the base curves can be drawn. This set of points cannot approach P indefinitely, as otherwise it would give rise to a negative curve of the third kind not crossing the base curves, which is excluded. In the same way as for the elliptic sectors we destroy the regions covered by this set of points, and there remains a sector of the second category bounded by a modified arc of c, inside which no curves of the second kind not crossing the base curves can be drawn.

In the modified sector we now consider the set of points, through which curves of the first kind not crossing the base curves can be drawn, and it is clear that this set of points cannot indefinitely approach the just now modified curve c. The regions covered by it are therefore bounded by a finite or denumerable number of curves of the first kind, *not* enclosing each other, *not* indefinitely approaching c, and each enclosing a domain which forms a *leaf*, not differing from those appearing in the hyperbolic sectors.

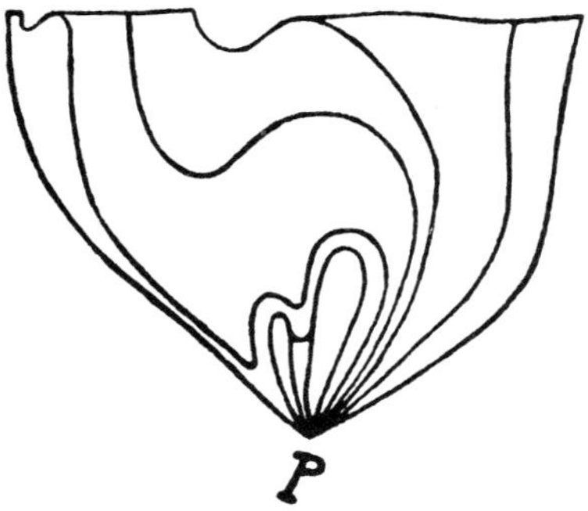

Fig. 5. Parabolic sector.

By the method applied above already several times the region outside the leaves can be filled with curves of the third kind (for instance we can choose for them the system of base curves present already in the sector), and finally each of the leaves with curves of the first kind (see fig. 5).

The sectors of the second category we shall call *positive* (resp. *negative*) *parabolic sectors.*

In special cases the whole inner domain of c can reduce itself to a single positive (resp. negative) parabolic sector. A point zero where this occurs we shall call a *source point* resp. *vanishing point.*

§ 4.

The structure of the field in the vicinity of an isolated singular point. Second principal case.

In this case any vicinity of P contains a simple closed tangent curve inside which P lies. We can then construct a fundamental series c, c', c'', ... of simple closed tangent curves converging to P, of which each following one lies inside each preceding one, and we can fill in the following way the inner domain of c with tangent curves not crossing each other.

In each annular domain between two curves $c^{(n)}$ and $c^{(n+1)}$ we choose a point having from the boundary of that domain a distance as great as possible and we lay through it a tangent curve situated in the annular domain. According to § 1 it is either closed or it gives rise to two closed curves, situated in the annular domain with its boundary, into which it terminates or to which it converges spirally, and which we draw likewise. (These closed tangent curves can entirely or partially coincide with $c^{(n)}$ or $c^{(n+1)}$). So the annular domain is either made singly connected or it is divided into two or three (annular or singly connected) new domains.

In each of these we again choose a point having from the boundary a distance as great as possible and we lay through it again a tangent curve. A singly connected domain is certainly divided by it into two singly connected domains; on an annular domain it has the effect just now mentioned.

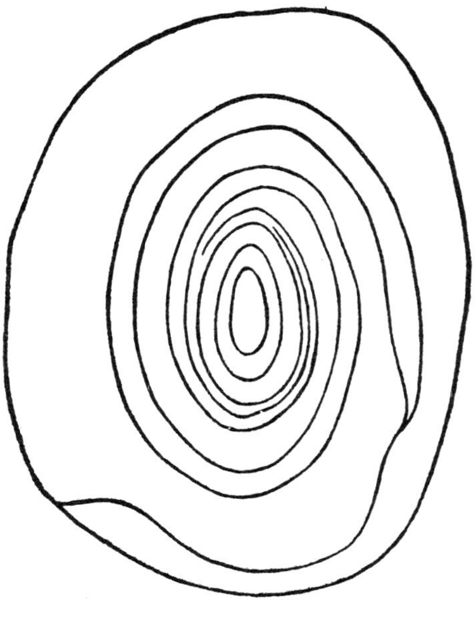

Fig. 6. Rotation point.

We repeat this process indefinitely. For each domain it can

happen only once that it undergoes no division; after that namely it becomes singly connected, so is divided at each new insertion of a tangent curve (see fig. 6).

We finally add the limit curves, and we prove in the same way as in § 2 that then through each point of the inner domain of c passes a tangent curve.

A point zero being in the second principal case we shall call a *rotation point.*

So we can say:

THEOREM 4. *An isolated singular point is either a rotation point, or a vicinity of it can be divided into a finite number of hyperbolic, elliptic, and parabolic sectors.*

The filling of a vicinity of a non-singular point in § 2 furnishes in this terminology two hyperbolic and two parabolic sectors.

We must add the observation that in the most general case, where neither in a singular, nor in a non-singular point the tangent curve is determined, sometimes by a modified method of construction, the structure of the first principal case can be given to a vicinity of a point zero being in the second principal case.

Even the form of the sector division of the first principal case is then not necessarily unequivocally determined. Out of the reasonings of the following § we can, however, deduce that, if modifications are possible in the form of the sector division, the difference of the number of elliptic sectors and the number of hyperbolic sectors always remains the same.

§ 5.

The reduction of an isolated singular point.

For what follows it is desirable to represent the domain γ on a Euclidean plane, and farther to substitute for the curve c a simple closed curve c' emerging nowhere from c, containing likewise P in its inner domain, and consisting of arcs of tangent curves and of orthogonal curves. In the second principal case this is already attained, and in the first principal case we have to modify in a suitable way only those arcs of c which bound the hyperbolic and the parabolic sectors.

In a hyperbolic sector we effect this by choosing a point on each of the two bounding base curves, and by drawing from those points H and K inside into the sector orthogonal arcs not intersecting one another. Then there is certainly an arc of a curve of the second kind joining a point B of one of these orthogonal arcs with a point C of the other, and we bound the modified sector by the orthogonal arcs HB and CK and the tangent arc BC.

If a parabolic sector is bounded by the base curves k and k', it is always possible to choose between them a finite number of base curves $k_1, k_2, \ldots k_n$ in such a way, that each k_p and k_{p+1} can be connected, inside the sector but outside the leaves lying in it, by an orthogonal arc. By those orthogonal arcs and the arcs of base curves joining their endpoints we bound the modified sector. The simple closed curve c' obtained in this way has a direction of tangents varying everywhere continuously, with the exception of a finite number of rectangular bends. To a definite sense of circuit of c', which we shall call the positive one, corresponds in each point of c' a definite tangent vector, and for a full circuit of c' that tangent vector describes a positive angle 2π. [[5]]

We shall now consider two successive parabolic sectors, π_1 and π_2, of which (for the positive sense of circuit) the first is positive, therefore the second negative, and we suppose them to be separated by a hyperbolic sector τ. On the orthogonal arcs belonging to the boundary of π_1 the given vector then forms with the tangent vector an angle $\left(2n - \frac{1}{2}\right)\pi$ (measured in the positive sense), on the orthogonal arcs belonging to the boundary of π_2 an angle $\left(2n + \frac{1}{2}\right)\pi$.

The transition takes place along the tangent arc belonging to the boundary of τ, by a negative rotation over an angle π of the given vector with respect to the tangent vector.

The same remains the case if we suppose π_1 to be negative, π_2 to be positive.

But if we suppose τ to be an elliptic sector, then the transition under discussion takes place along the tangent arc bounding τ, by a positive rotation over an angle π of the given vector with respect to the tangent vector.

As now the total angle, which the given vector describes for a full circuit of c', is equal to the total angle which the tangent vector describes plus the total angle which the given vector describes with respect to the tangent vector, the former angle is equal to $\pi\,(2 + n_1 - n_2)$, where n_1 represents the number of elliptic sectors, n_2 the number of hyperbolic ones.

Let further j be an arbitrary simple closed curve enveloping P, but enveloping no other singular point, then we can transform c' into j by continuous modification in such a way, that at every moment

P, but no other singular point, is enveloped by the modified curve. If we consider for each of the intermediary curves the total angle which the given vector describes by a positive circuit, then on one hand it can only have continuous modifications and on the other hand it must remain a multiple of 2π. Thus it remains unchanged, and we can formulate:

[[5]] THEOREM 5. *The total angle which, by a circuit of a simple closed curve enveloping only one point zero, the vector describes in the sense of that circuit, is equal to* $\pi(2+n_1-n_2)$, *where* n_1 *represents the number of elliptic sectors,* n_2 *the number of hyperbolic ones, which appear when a vicinity of the point zero is covered with tangent curves not crossing each other.*

In particular for source points, vanishing points and rotation points this angle is equal to $+2\pi$.

We now surround P with a simple closed curve $\varkappa$ which can be supposed as small as one likes, and we leave the vector distribution outside $\varkappa$ and on $\varkappa$ unchanged, but inside $\varkappa$ we construct a modified distribution in the following way:

Let us first suppose that for a positive circuit of $\varkappa$ the vector describes a positive angle $2n\pi$. From an arbitrary point Q inside $\varkappa$ we draw to $\varkappa$ n arcs of simple curve $\beta_1, \beta_2, \ldots \beta_n$, not cutting each other and determining in this order a positive sense of circuit. Let us call ${}_p\varkappa$ the arc of $\varkappa$ lying between β_p and β_{p+1}, and G_p the domain bounded by β_p, ${}_p\varkappa$ and β_{p+1}. Along β_1 we bring an arbitrary continuous vector distribution becoming nowhere zero and passing on $\varkappa$ into the original one. Then along β_2 such a one passing on $\varkappa$ and in Q into the already existing vectors, that along the boundary of G_1 positively described the vector turns a positive angle 2π. Then along β_3 such a one passing on $\varkappa$ and in Q into the existing vectors, that along the boundary of G_2 positively described the vector turns a positive angle 2π, etc.

As the angle described by the vector in a positive circuit of $\varkappa$ is equal to the sum of the angles described in positive circuits of the boundaries of the domains $G_1, G_2, \ldots . G_n$, it is finally evident, that also for a positive circuit of G_n the vector describes a positive angle 2π.

In each of the domains G_p with boundary $\varkappa_p$ we choose a simple closed curve c_p not meeting $\varkappa_p$, of which in a suitable system of coordinates the equation can be written in the form $x_p^2 + y_p^2 = r^2$.

Inside and on c_p we introduce a finite continuous vector distribution vanishing only in the point $(o,o)_p$, which is directed along the lines

$\frac{y_p}{x_p} = a$ and from the point $(o,o)_p$. This vector describes along c_p a positive angle 2π, just as the existing one along $\varkappa_p$. If then according to SCHOENFLIES we fill the annular domain between $\varkappa_p$ and c_p with simple closed curves enveloping each other and as functions of a cyclic parameter passing continuously into each other, then we can thereby at the same time make the vector distribution along $\varkappa_p$ pass continuously into that along c_p, and in this way give to the annular domain between $\varkappa_p$ and c_p a finite continuous vector distribution vanishing nowhere. Inside $\varkappa_p$ we have now obtained a finite continuous vector distribution, having but *one* point zero, namely the point $(o,o)_p$, and that a source point of very simple structure, which we shall call a *radiating point*.

And the inner domain of $\varkappa$ is covered with a finite continuous vector distribution passing on $\varkappa$ into the original one and possessing inside $\varkappa$, instead of the original point zero P, n radiating points.

Let us furthermore suppose that for a positive circuit of $\varkappa$ the vector describes a negative angle $2n\pi$. In an analogous way as above we then divide the inner domain of $\varkappa$ into n regions G_p with boundaries $\varkappa_p$, and we bring along each of these boundaries such a vector distribution, that for a positive circuit of $\varkappa_p$ the vector describes a negative angle 2π.

The curves c_p are introduced again as above, but inside and on c_p we introduce a finite continuous vector distribution vanishing only in the point $(0,0)_p$, which is directed along the lines $x_p y_p = a$. For a positive circuit this vector describes along c_p a negative angle 2π, just as the existing vector along $\varkappa_p$.

So the annular domain between $\varkappa_p$ and c_p can be filled up in an analogous way as just now with a finite continuous vector distribution vanishing nowhere, and the whole distribution inside $\varkappa_p$ possesses then only *one* point zero, namely the point $(0,0)_p$, having four hyperbolic sectors of very simple form (the four separating parabolic sectors are each reduced to a single line), which structure we characterize by the name of *reflexion point*.

After this the inner domain of $\varkappa$ is covered with a finite continuous vector distribution passing on $\varkappa$ into the original one and possessing inside $\varkappa$, instead of the original point zero P, n reflexion points.

Let us finally suppose that for a circuit of $\varkappa$ the total angle described by the vector is zero. We can then choose inside $\varkappa$ such a simple closed curve c, that in a suitable system of coordinates its equation can be written in the form $x^2 + y^2 = r^2$. Inside and on c we introduce a finite continuous vector distribution vanishing nowhere, which is directed along the lines $y = a$. The total angle described by this vector along c is zero, just as the one described by the existing vector along $\varkappa$. The annular domain between $\varkappa$ and c can thus be filled up as in the two preceding cases with such a finite continuous vector distribution, that the whole distribution inside $\varkappa$ is now free of points zero.

So we can formulate:

⟦6⟧ THEOREM 6. *A finite continuous vector distribution with a finite number of points zero can be transformed, by modifications as small as one likes inside vicinities of the points zero which can be chosen as small as one likes, into a new finite continuous vector distribution which has as points zero only a finite number of radiating points, and a finite number of reflexion points.*

In particular those points zero about which the angle, described by the vector for a positive circuit, is positive, are broken up into radiating points; those about which this angle is negative, are broken up into reflexion points; whilst those for which it is zero, vanish.

In a following communication we shall extend this theorem to distributions with an infinite (denumerable or continuous) number of points zero.

§ 6.

Remarks on the tangent curves and singular points on a sphere.

If we have on a sphere a finite continuous vector distribution with a finite number of singular points, then the reasonings of § 1 lead with small modifications to:

THEOREM 7. *A tangent curve to a finite continuous vector distribution with a finite number of singular points on a sphere is either a simple closed curve, or save its ends it is an arc of simple curve, of which the pursuing as well as the recurring branch either stops at a point zero, or enters into a simple closed tangent curve, or converges spirally to a circumference consisting of one or more simple closed tangent curves.*

From this ensues that also on a sphere a tangent curve cannot return into indefinite vicinity of one of its points, after having reached a finite distance from it, unless it be, to close itself in that point.

Out of the reasoning of § 1 we can deduce farthermore without difficulty that a fundamental series of closed tangent curves with the property that of the two domains determined by one of them, one contains no points of the preceding, the other no points of the following curves, converges either to a single singular point, or to the outer circumference, consisting of simple closed tangent curves, of a domain or set of domains.

Let now an arbitrary finite continuous vector distributionon a sphere be given. On account of § 5 we reduce it by means of indefinitely small modifications to a "reduced distribution", possessing as singular points only radiating points and reflexion points, and we investigate the tangent curves of that reduced distribution.

A closed tangent curve can possess no radiating points, but reflexion points it can possess (its tangent direction shows there a rectangular bend).

On the other hand a tangent curve can only stop at a radiating point.

We now consider an arbitrary tangent curve: according to theorem 7 it is either an arc of simple curve joining two radiating points, or it gives rise to a simple closed tangent curve j_0, which divides the sphere into two domains G and G'.

Then on j_0 no radiating point can lie, but we shall prove, that in G as well as in G' there must lie one.

If namely there were no radiating point in G, we could consider within G a new tangent curve, and as this would not be able to stop in G, it would on account of theorem 2 give rise to a new simple closed tangent curve j_1 enclosing a domain G_1 being a part of G. Within G_1 we could again consider an arbitrary tangent curve, and in this way we should arrive at a simple closed tangent curve j_2 enclosing a domain G_2 being a part of G_1.

Continuing this process indefinitely we construct a fundamental series of closed tangent curves $j_0, j_1, j_2, j_3, \ldots$, which cannot converge to a single singular point, as neither a radiating point nor a reflexion point contains closed tangent curves in an indefinitely small vicinity. On account of the remark made at the beginning of this § there must thus be at least *one* domain G_ω, bounded by a simple closed tangent curve j_ω, and contained in each of the domains $G_1, G_2, G_3, \ldots$

Within G_ω we could again construct a closed tangent curve $j_{\omega+1}$ bounding a domain $G_{\omega+1}$ being a part of G_ω, and we could continue this process to *any* index of the second class of numbers, which on the other hand is impossible, as the set of domains $G - G_1$, $G_1 - G_2, \ldots G_\omega - G_{\omega+1}, \ldots G_\alpha - G_{\alpha+1}, \ldots$ must remain denumerable.

So we finally formulate:

THEOREM 8. *A reduced distribution on a sphere possesses at least two radiating points.*

NOTES

⟦1⟧ Continuation of 1909 G2. Continued in 1910 D2. See also 1915 B.
⟦2⟧ Brouwer 1909 G2.
⟦3⟧ Brouwer 1910 D2.
⟦4⟧ H. A. Lorentz 1904. Brouwer means an example of a simple curve that is dense on the torus.
⟦5⟧ Though in a sense the circulation number goes back to Cauchy, this singularity index seems to be its first general occurrence in a topological context. See also Brouwer 1910 D2 ⟦5⟧, H. Poincaré 1881, Brouwer 1911 H2.
⟦6⟧ The first appearance of homotopy changes of vector fields.

Mathematics. — "*On continuous vector distributions on surfaces*" (3rd communication)[1]). By Dr. L. E. J. BROUWER. (Communicated by Prof. D. J. KORTEWEG). 1910 D2

[[1]]

(Communicated in the meeting of May 28, 1910).

§ 1.

The irrigating field on the sphere.

In order to get an insight into the structure of an arbitrary finite continuous vector field with a finite number of singular points on the sphere over its entire extent, we begin by investigating a particular case characterized by the *absence of simple closed tangent curves.*

In a field which possesses this property, and which we shall call an *irrigating field*, no spirals can appear as tangent curves and no rotation points as singular points. As farthermore a singular point can neither possess elliptic sectors or leaves, it is either a source point without leaves, or a vanishing point without leaves, or it possesses exclusively hyperbolic and parabolic sectors without leaves, in which case we shall speak of a *stroking point.*

The singular points of an irrigating field cannot all be stroking points. This follows from theorem 8 of the second communication [2]) in

1) For the first and second communication see these Proceedings Vol. XI 2, p. 850 and Vol. XII 2, p. 716. [[2]]

2) l. c. p. 734. [[3]]

connection with the observation, that the reduction of stroking points can lead only to reflexion points.

So there are certainly source points or vanishing points; to fix our thoughts we shall start from the existence of source points $B_1, B_2, \ldots\ldots B_m$.

In B_1 we start an arbitrary tangent curve which when pursued indefinitely can neither close itself, nor become a spiral. So it must stop at a singular point, which can be nothing but a vanishing point V_1.

If possible we then start in B_1 a second tangent curve, not crossing the first and stopping at an *other* vanishing point V_2.

If possible then in each of the two sectors generated in B_1 a tangent curve not crossing the two already existing ones and stopping either at a third vanishing point V_3, differing from V_1 and V_2, or, if that is excluded, stopping e.g. at V_1, but then in such a way that in B_1 a sector is determined limited by two tangent curves stopping at V_1, inside which we can draw a tangent curve not crossing the existing ones, starting from B_1 and stopping at V_2.

We continue this process of insertion as often as possible, whereby every time in each sector is inserted a tangent curve not crossing the existing ones which either stops at an other vanishing point as the two tangent curves limiting the sector, or,. if that is excluded, determines a new sector, in which such an insertion is possible.

In this way it is impossible that at some moment a sector should appear limited by two tangent curves stopping at the same vanishing point, and within which no other vanishing point should lie.

So the number of tangent curves stopping at one and the same vanishing point, and appearing in this process of insertion, must remain smaller than the total number of vanishing points and from this ensues that the process of insertion ends after a finite number of insertions.

Of the then constructed finite system of tangent curves starting from B_1, which we shall call a *system of skeleton curves of* B_1, no two consecutive ones have the same vanishing point as their end-point.

Let for a certain sense of circuit those skeleton curves be consecutively $r_1, r_2, \ldots\ldots r_n$, stopping respectively at the vanishing points $V_1, V_2, \ldots\ldots V_n$, which of course need not be all different.

We then if possible introduce between every r_ν and $r_{\nu+1}$ a tangent curve starting in B_1 and stopping at a certain vanishing point, not crossing the already existing ones and reaching a distance as great as possible from r_ν and $r_{\nu+1}$. In each of the sectors thereby generated at B_1 we repeat such an insertion, in each of the sectors thereby generated again and so on; finally after having repeated this process of insertion ω times, we add the limiting curves, which are likewise tangent curves starting from B_1 and stopping at certain vanishing points. After that, as ensues from the reasoning followed in § 2 of the second communication [1]), no new tangent curves starting from B_1 can be inserted, whilst the constructed tangent curves cover on the sphere a closed coherent set of points, to which belong all possible tangent curves starting from B_1, and which we shall call *the irrigation territory of* B_1.

The method according to which the skeleton curves have been constructed implies furthermore that between every r_p and r_{p+1} two tangent curves r''_p and r'_{p+1} appear, between which no further tangent curves starting from B_1 can be constructed, whilst all tangent curves, which have been constructed between r_p and r''_p, end in V_p, and all tangent curves, which have been constructed between r'_{p+1} and r_{p+1}, end in V_{p+1}.

From this ensues that these curves r''_p and r'_{p+1} coincide from B_1 up to a certain stroking point S_p, beyond which they diverge for good.

For, when diverging either in a non-singular point or immediately in B_1, insertion of new tangent curves starting from B_1 would be possible.

And also when rejoining after having previously diverged, an insertion of a new tangent curve starting from B_1 would be possible, namely of such a one that had with r''_p as well as with r'_{p+1} an arc in common.

So the irrigation territory of B_1, consisting of n sectors Σ_p, each limited by a tangent curve r'_p and a tangent curve r''_p, possesses an outer circumference $V_1 S_1 V_2 \ldots V_n S_n V_1$, consisting of $2n$ tangent arcs, which we shall call its "sides". It may happen here, that an even side $S_p V_{p+1}$, and an odd side $S_q V_q$ (p and q different) *touch each other outwardly* along an arc PV_{p+1} resp. PV_q (which can expand to an entire side $S_p V_{p+1}$ or $S_q V_q$, or reduce itself to a point V_{p+1} resp. V_q) but not in an other way.

For, when two such sides $S_p V_{p+1}$ and $S_q V_q$ have collided somewhere outwardly, they cannot leave each other any more before V_{p+1} resp. V_q has been reached. Otherwise a tangent curve coinciding partially with $S_p V_{p+1}$ and partially with $S_q V_q$ might be inserted, which would separate $S_p V_{p+1}$ and $S_q V_q$, so that these could not have collided with each other, but only with the newly inserted tangent curve. The sectors Σ_p connecting in this way B_1

[1]) l. c. p. 723.

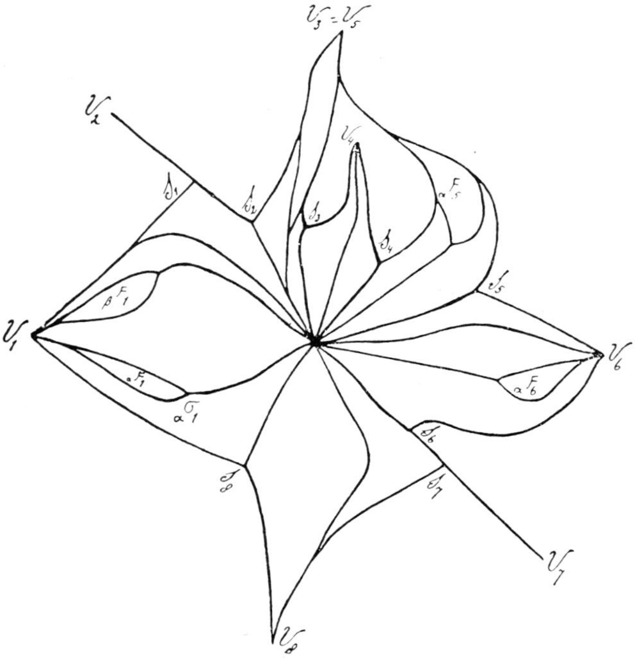

Fig. 1. Irrigation territory.

with one and the same vanishing point possess round about that vanishing point the same cyclic order as about B_1.

Let us consider a sector Σ_p. The limiting tangent curves r'_p and r''_p can collide inwardly in an arbitrary closed set of points (which in particular can entirely cover those curves). Farthermore it is not necessary that the entire inner domain determined by r'_p and r''_p belongs to Σ_p. However for each region ${}_\alpha F_p$ between r'_p and r''_p not belonging to Σ_p the property holds that it is limited by two tangent curves ${}_\alpha\varrho_p$ and ${}_\alpha\varrho'_p$ running from B_1 to V_p (between which no further tangent curves starting from B_1 can be constructed), which coincide from B_1 up to a certain stroking point ${}_\alpha\sigma_p$, then diverge, and finally after rejoining in a point ${}_\alpha H_p$ (which can also coincide with V_p) remain united to their end in V_p. If namely the latter property were lacking, then a new tangent curve starting from B_1 could be inserted. As finally the stroking point ${}_\alpha\sigma_p$ must give inside the region ${}_\alpha F_p$ two (and not more than two) hyperbolic sectors, only a finite number of points ${}_\alpha\sigma_p$ can coincide in one and the same stroking point, and from this ensues that there is only a finite number of regions ${}_\alpha F_p$.

The preceding shows that the residual regions determined on the sphere by the irrigation territory of B_1, are each bounded by a single inner circumference $V_{\alpha_1} S_{\alpha_1} V_{\alpha_2} S_{\alpha_2} \ldots V_{\alpha_n} S_{\alpha_n} V_{\alpha_1}$, whose sides each join a stroking point and a vanishing point, in such a way that two successive sides concurring in a vanishing point V_{α_p} can touch each other inwardly from a certain point P up to V_{α_p}, but other inner contacts are excluded, and farthermore that each stroking point S_{α_p} possesses in the considered residual region *two* hyperbolic sectors.

The irrigation territory s_1 of B_1 possesses a finite distance from all the remaining source points.

If we construct for B_2 the irrigation territory analogously as for B_1, these two irrigation territories can partially penetrate into each other. This can however, when constructing the irrigation territory of B_2, be prevented by enforcing on its tangent curves starting from B_2 the condition that they may neither cross each other nor any tangent curve starting from B_1, whilst for the rest we act in the same way as before.

In that manner we have the *irrigation territory s_2 of B_2, independent of B_1*, containing all those tangent curves starting from B_2 which do not cross any tangent curve starting from B_1. The structure of s_2 is entirely the same as of s_1. Between s_1 and s_2 outward contact may take place on account of the coincidence of an even (resp. odd) side $S_\alpha V_\gamma$ of s_1 and an odd (resp. even) side $S_\beta V_\gamma$ of s_2 along an arc PV_γ, which can expand to an entire side $S_\alpha V_\gamma$ or $S_\beta V_\gamma$ or can reduce itself to the point V_γ. Farthermore s_2 lies entirely in *one* of the residual regions determined by s_1, however in such a way, that between two successive sides of this region which are inwardly pressed together, s_2 can very well penetrate to the vanishing point in which those sides concur. Together s_1 and s_2 contain all tangent curves starting from B_1 or B_2. For the residual regions which are determined on the sphere by s_1 and s_2 together the same properties hold as for the residual regions of s_1 alone.

In one of those residual regions lies B_3 at a finite distance from s_1 and s_2, and in that region we construct the *irrigation territory s_3 of B_3, independent of B_1 and B_2*, containing all those tangent curves starting from B_3 which do not cross any tangent curve starting from B_1 or B_2. Together s_1, s_2, and s_3 contain then all the tangent curves starting from B_1, B_2 or B_3. Outward contact between s_3 and s_1 or s_2 can take place in the same way as between s_1 and s_2.

In a quite analogous way we construct s_4 in one of the residual regions determined by s_1, s_2, and s_3. And in this way we go on. When we have constructed $s_1, s_2, \ldots s_{m-1}$, then the sphere is not yet quite covered. For, the system of the tangent curves starting from

B_1, B_2, ... B_{m-1} cannot approach B_m within a certain finite distance. But after insertion of s_m the sphere is completely covered, for the set of the tangent curves starting from B_1, B_2, ... B_{m-1}, B_m is identical with the set of *all* tangent curves, so must cover the sphere entirely, and we have proved:

THEOREM 1. *An irrigating field divides the sphere into a finite number of irrigation territories each of which contains in its interior one of the source points.*

A clear example of an irrigating field is the force field of a finite number of positive and negative divergency points [1]).

The notion of irrigating field can be extended in the following manner: Let be given on the sphere a multiply connected region γ, bounded by a finite number of coherent boundaries, and in γ a finite, continuous vector distribution, which continuity is uniform with the exception of a finite number of points. We then can construct of the region γ exclusive of its boundaries a continuous one-one representation on a sphere β in such a way that to the boundaries of γ correspond on β single points. The tangent curves of γ are thereby represented on a set of simple curves ϱ described in a certain sense. If among these curves ϱ no simple closed curves appear, they determine on β the structure of an irrigating field. In that case we shall call the given vector field in γ likewise an irrigating field.

This more general irrigating field differs thereby from the particular kind first considered that a boundary can play the part of a singular point. We accordingly distinguish *source boundaries*, *vanishing boundaries*, and *stroking boundaries*. From this ensues that in the more general irrigating field also spirals can appear as tangent curves, namely such whose windings converge uniformly to a source boundary or to a vanishing boundary.

§ 2.

The most general field with a finite number of singular points.

Let there be given an arbitrary finite continuous vector field on the sphere with a finite number of singular points. Let N be one of the singular points, then we shall say that a closed tangent curve *flows round about* N, if it does not contain a singular point, and encloses a region in which lies N but no other singular point. Fartheron we shall say that a closed tangent curve *flows round*

[4] [1]) Compare my paper: "*The force field of the non-Euclidean spaces with positive curvature*", these Proceedings Vol. IX 1, p. 250.

[308]

against N, if it contains N but no other singular point, and encloses a region in which no singular point lies.

If there is neither a tangent curve flowing round about N nor a tangent curve flowing round against N, then we shall call N a *naked singular point*, otherwise a *wrapped singular point*.

We shall assume that N is a wrapped singular point and we shall distinguish two cases:

First case. There is no tangent curve flowing round about N. Let then ϱ be a tangent curve flowing round against N and let us agree about an arbitrary tangent curve r inside ϱ, that, when it reaches ϱ, we shall pursue resp. recur it along ϱ, until it reaches N; in this way r also becomes a tangent curve flowing round against N. We can thus fill the inner domain of ϱ with tangent curves flowing round against N and not crossing each other in the same way as in the second communication p. 727 was executed for an elliptic sector.

If we now construct a well-ordered series continued as far as possible of tangent curves flowing round against N, enclosing ϱ and bounding outside ϱ an ever increasing area, then it converges either to a tangent curve flowing round against N, or to a *circumference* consisting of simple closed tangent curves which can contain besides N still other singular points and which possesses all the properties deduced in the second communication p. 720 and 721 for the limiting

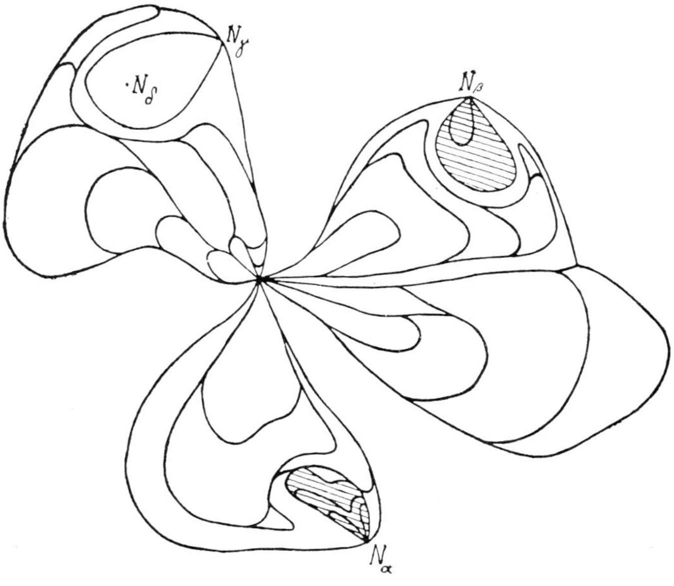

Fig 2. Circumfluence territory with (shaded) additional territories.
First case.

circumference of a spiral tangent curve. The inner region of that circumference, which can be entirely filled with tangent curves flowing round against N and not crossing each other, we shall call a *circumfluence sector of* N.

The singular point N can possess an infinite number of circumfluence sectors lying outside each other, but amongst these there are only a finite number, which reach an arbitrarily assumed finite distance from N.

The set of regions covered by the different circumfluence sectors of N we shall call the *circumfluence territory* of N.

We shall now regard of this circumfluence territory those residual regions which are bounded by a tangent curve flowing round against an other singular point $N_\varkappa$, and we shall fill them with tangent curves flowing round against $N_\varkappa$ and not crossing each other. The set of regions filled in this way with tangent curves possesses at each of the points $N_\varkappa$ entirely the structure of a circumfluence territory, and we shall call it an *additional circumfluence territory of* N. The point N possesses then only a finite number of additional circumfluence territories.

The circumfluence territory of N determines with its additional territories together a finite number of residual regions on the sphere.

Second case. There exists a tangent curve flowing round about N. Let ϱ be that curve, we then construct from ϱ outwards a well-ordered series continued as far as possible of tangent curves flowing round about N, enclosing ϱ and bounding outside ϱ an ever increasing area. The limit τ_1 to which this series converges is either a tangent curve flowing round about N, or a *circumference* containing singular points, consisting of simple closed tangent curves, and possessing all the properties deduced in the second communication p. 720, 721 for the limiting circumference of a spiral tangent curve.

Let us construct likewise from ϱ inwards a well-ordered series continued as far as possible of tangent curves flowing round about N, enclosed by ϱ, and limiting around N an ever decreasing area, then the limit τ_0 to which this series converges is either the point N, or a tangent curve flowing round about N, or a circumference consisting of a finite or countable set of tangent curves flowing round against N.

If τ_0 is a circumference containing N, we can fill up its inner regions with tangent curves flowing round against N and not crossing each other.

If τ_0 is a tangent curve flowing round about N, there can exist no tangent curve flowing round against N and having with a τ_0

point in common. For then in the terminology of § 3 of the second communication we should possess between N and τ_0 a positive as well as a negative curve of the third kind, from which we could start to fill the inner region of τ_0 with tangent curves not crossing each other. We should then have to find there the number of elliptic sectors equal to the number of hyperbolic sectors; so there would have to be at least *one* hyperbolic sector inside τ_0; this would however give rise to tangent curves flowing round about N and lying inside τ_0, which is excluded.

So if τ_0 is a tangent curve flowing round about N, then there exists inside τ_0 at a finite distance from τ_0 a circumference τ'_0 containing N, consisting of a finite or countable set of tangent curves flowing round against N, and inside which lie all existing tangent curves flowing round against N. If τ'_0 does not reduce itself to the single point N, its inner regions can be filled with tangent curves flowing round against N and not crossing each other.

The tangent curves not crossing each other with which the annular region between τ_0 and τ'_0 can be filled, must on one side either all enter into τ_0 or all converge spirally to τ_0, and on the other side either all enter into τ'_0 or all converge spirally to τ'_0.

In order to fill up the annular region between τ_0 and τ_1 with tangent curves not crossing each other, we construct in it a tangent curve $r_{\frac{1}{2}}$ flowing round about N and reaching from τ_0 and τ_1 a distance as great as possible. Between τ_0 and $r_{\frac{1}{2}}$ we then if possible insert a tangent curve $r_{\frac{1}{4}}$ flowing round about N and reaching from τ_0 and $r_{\frac{1}{2}}$ a distance as great as possible; likewise between $r_{\frac{1}{2}}$ and τ_1 if possible a tangent curve $r_{\frac{3}{4}}$ flowing round about N and reaching from $r_{\frac{1}{2}}$ and τ_1 a distance as great as possible. This inserting process we repeat as often as possible, eventually ω times, and finally we add the limiting curves. We are then sure that no more tangent curves flowing round about N can be inserted, so that eventually the regions between τ_0 and τ_1 remained empty of tangent curves must be annular regions.

Let a be such an annular region bounded by the tangent curves r_p and r_q flowing round about N, then a can be filled with tangent curves not crossing each other, which on one side either all enter into r_p or all converge spirally to r_p, and on the other side either all enter into r_q or all converge spirally to r_q.

The inner region of τ_1, in this manner entirely filled with tangent curves not crossing each other, we shall call the *circumfluence territory* of N.

We shall farther of this circumfluence territory fill each residual

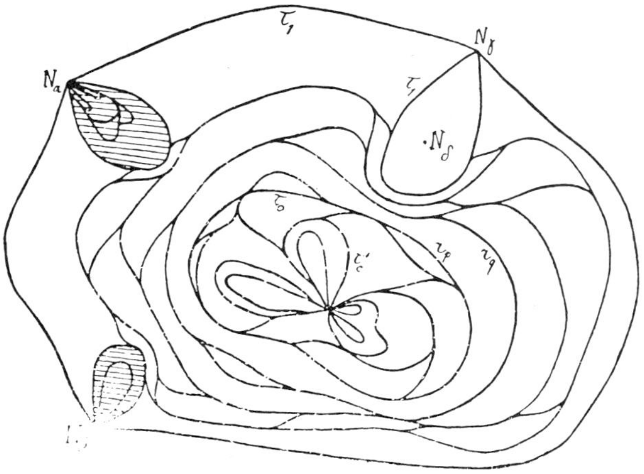

Fig. 3. Circumfluence territory with (shaded) additional territories. Second case.

region, bounded by a tangent curve flowing round against a singular point $N_\varkappa$, with tangent curves flowing round against $N_\varkappa$ and not crossing each other. In this manner we add to the circumfluence territory of N a finite number of *additional circumfluence territories*, after which there remain on the sphere only a finite number of residual regions.

Let us now consider on the sphere a finite and, with the exception of a finite number of points, uniformly continuous vector distribution in a multiply connected region γ with a finite number of coherent boundaries. By a *closed tangent curve* we shall understand here, besides each tangent curve to which we have formerly given this name, each system of n simple tangent arcs not meeting each other and n cyclically ordered boundaries or singular points not contained in a boundary $N_1, N_2, N_3 \ldots N_n$, between which those tangent arcs run consecutively from N_1 to N_2, from N_2 to $N_3, \ldots$ and from N_n to N_1. In particular thus a simple tangent arc whose endpoints lie on one and the same boundary forms together with that boundary a closed tangent curve. Fartheron we shall understand by the *boundaries* of such a field for shortness' sake also the singular points which are not contained in a boundary. Finally we shall call a closed tangent curve not containing a boundary, and enclosing in γ a region in which lies N but no other boundary, *a tangent curve flowing round about N*, and we shall call a closed tangent curve containing N but no other boundary, and enclosing in γ a region in which lies no boundary, *a tangent curve flowing round against N*. *Naked* and *wrapped boundaries* we then define analogously as before naked and wrapped singular points.

For a wrapped boundary the circumfluence territory can be constructed in the same way as was done above for a wrapped singular point; the whole of its structure undergoes in this more general case no change, we only have to replace closed tangent curves in the narrower sense by closed tangent curves in the wider sense. The filling with tangent curves not crossing each other and the completion of the territories by means of its additional territories needs no modification either.

We shall understand by the *order* of the field twice the number of its naked boundaries plus three times the number of its wrapped boundaries.

We shall now start from a finite and, with the exception of only a finite number of points, uniformly continuous vector field in a region of the sphere with a finite number of coherent boundaries, each of which either reduces itself to a single point, or consists of tangent arcs turning one of their sides to the field, whilst in the latter case we assume that each fundamental series of consecutive points in a segment of a boundary determines only *one* limiting point, which property we express by calling the boundary *simple*. So the appearance of spirals in the boundaries is excluded.

We shall indicate two operations, both of which reduce this field to a finite number of fields of the same kind but of a lower order:

First reducing operation: We construct in the given field such a closed tangent curve which together with each of the two partial fields determined by it contains at least two of the boundaries of the given field.

Then namely each of the two partial fields is of a lower order than the original field.

Second reducing operation: we construct to a wrapped boundary the circumfluence territory with its eventual additional territories.

Then namely each of the residual fields is of a lower order than the original field.

It is clear that after a finite number of applications of these reducing operations either nothing of the original field is left or there remain only such fields to which neither of the two operations can any more be applied.

Then however in these residual fields there exists no closed tangent curve, so that they are irrigating fields.

If these last remaining residual fields are lacking, then the original field can be divided by *simple* boundaries consisting of tangent arcs into a finite number of circumfluence territories with additional territories which property we shall express by calling it a *circumfluent field.*

The circumfluent field can be regarded as the counterpiece to the irrigating field analysed in § 1.

A clear example of a circumfluent field is the force field without divergences of a finite number of positive and negative rotation points. [1])

We now have proved:

THEOREM 2. *A finite continuous vector field on the sphere with a finite number of singular points can be divided by simple boundaries consisting of tangent arcs into a finite number of irrigating fields and a finite number of circumfluence territories.*

At the same time we notice that among the tangent curves not crossing each other, with which in the preceding pages we have filled the field, spirals *cannot* appear in the *boundaries* of the irrigating fields or circumfluence territories meant in theorem 2, and in their *interior* exclusively in the following two ways:

1st. A circumfluence territory of the second kind can contain annular regions filled with spirals.

2nd. An irrigating field can possess source boundaries or vanishing boundaries round about which all tangent curves arrive resp. depart spirally.

§ 3.

The theorem of the invariant point on the sphere.

⟦2⟧ In the first communication on this subject (these Proceedings Vol. XI 2) we have on page 857 brought an arbitrary continuous one-one transformation of the sphere in itself into relation with the vector distribution for which in each point the vector direction is determined by the shortest arc of principal circle joining that point with its image point, for which distribution appear as singular points: 1st. the points invariant for the transformation. 2nd. the points having their antipodic points as their image points. The singular points of the latter kind form for transformations with inversion of the indicatrix as well as for transformations with invariant indicatrix a closed set of points of the most general kind which makes it pretty well

⟦4⟧ 1) Compare my paper quoted above: "*The force field of the non-Euclidean spaces with positive curvature*".

impossible to deduce out of the properties of the vector distribution, either by means of theorem 2 of the first communication, or by means of theorem 8 of the second communication, the existence of at least *one* invariant point for transformations with invariant indicatrix.

The difficulty caused by this inconvenient set of points disappears however for an other vector distribution deduced from the transformation.

To construct this distribution we bring through each point P a circle containing its image point P' and a fixed point O, and we determine the vector direction in P by the arc of circle PP' *not* containing O. Let Q be the point having O as its image point, then as singular points of this vector distribution appear 1st. the point O. 2nd. the point Q. 3rd. the points invariant for the transformation.

If this vector distribution has an infinite number of singular points, then there are certainly points invariant for the transformation; so we assume in the following that the number of singular points is finite, and we investigate first the nature of the singularity in O.

For a point P in sufficient proximity of O the vector direction differs indefinitely little from the direction of the geodetic arc of circle OP. So by a circuit of a small circle about O the total angle which the vector turns with respect to the tangent to the small circle is zero, *so that when reduced the singularity gives rise to a radiating point.*

To investigate the nature of the singularity in Q, we represent the sphere stereographically on a Euclidean plane in such a way that O represents the infinite of the plane. Then in this plane the vector distribution is determined in each point by the straight line segment joining the point with its image point.

In the Euclidean plane the image of an infinitesimal circle about Q is an infinitely large circle; the infinitesimal circle and the infinitely large circle possess for transformations with invariant indicatrix *opposite senses of circuit*; for transformations with inversion of the indicatrix *equal senses of circuit.*

In the former case the vector describes in a circuit of the infinitesimal circle an angle 2π *in a sense opposite to the circuit*; in the latter case an angle 2π *in the same sense as that of the circuit.*

So when reduced the singularity in Q gives rise for transformations with invariant indicatrix to a reflexion point, for transformations with inversion of the indicatrix to a radiating point.

Thus the two radiating points, which according to theorem 8 of the second communication (p. 734) must be present in the reduced [[3]]

distribution, appear for a transformation with inversion of the indicatrix in the points O and Q; for a transformation with invariant indicatrix however the second radiating point can be furnished only by a point invariant for the transformation, *which therefore must necessarily exist.*

§ 4.

The index relation on the sphere for a finite number of singular points.

We shall now discuss the questions whether the number of singular points of a finite continuous vector distribution on the sphere, which according to theorem 2 of the first communication cannot be zero, is arbitrary for the rest, and farther whether the structure of the singuiar points, which according to theorem 8 of the second communication is not entirely free, is liable to still other restrictions than those expressed in that theorem.

These questions can be fully answered by means of the following reasoning, which is analogous to the proof of EULER's law, and which was indicated to me by Prof. HADAMARD.

The total angle which for a finite stereographic representation of the inner region of a simple closed curve enveloping only *one* singular point on a Euclidean plane the vector describes by a circuit in the sense of that circuit, and which according to theorem 5 of
[[3]] the second communication (page 731) is equal to $\pi\ (2 + n_1 - n_2)$, where n_1 represents the number of elliptic sectors, n_2 the number of hyperbolic sectors of the singular point, can be written in the form $2k\pi$, where k is an integer, which we call the *index* [1]) of the singular point.

For a simple closed curve, enveloping n singular points with indices $k_1, k_2, k_3, \ldots\ldots k_r$, the total angle which, for a finite stereographic representation of the inner domain of that curve on a Euclidean plane, the vector describes by a circuit in the sense of that circuit, is equal to $2\pi\,(k_1 + k_2 + \ldots\ldots k_n)$, as is immediately evident when we divide the inner domain under observation by means of arcs of simple curve into n inner domains of such simple closed curves, which each envelop only *one* of the singular points.

[1]) This expression is used (not for the singular point itself but for a curve by
[[5]] which it is enclosed) by POINCARÉ: "*Sur les courbes définies par une équation différentielle*", 1er mémoire, Journ. de Math. (3) 7, p. 400. The univalent continuous vector distributions treated there are of a particular algebraic kind, so that only indices $+1$ and -1 appear for the singular points.

[[316]]

We now make on the sphere a circuit along a certain principal circle on which lies no singular point; the total angle, which in the sense of a certain indicatrix on the sphere the vector direction describes by that circuit with respect to the tangent direction, is equal to $2h\pi$, where h is an integer.

The sense of that circuit is with respect to one of the hemispheres, into which the sphere is divided by that principal circle, the same as the sense of the indicatrix, with respect to the other opposite to the sense of the indicatrix; so for a circuit of the first hemisphere the vector describes with respect to the tangent direction an angle $2h\pi$ in the sense of the circuit, for a circuit of the second hemisphere an angle $2h\pi$ opposite to the sense of the circuit.

The total angle which, for finite stereographic representation of the first resp. the second hemisphere on a Euclidean plane, the vector describes by a circuit in the sense of that circuit, is thus equal to $2(1+h)\pi$ resp. $2(1-h)\pi$.

If in the first hemisphere lie m singular points with indices $k_1, k_2, \ldots k_m$, in the second hemisphere $n-m$ singular points with indices $k_{m+1}, k_{m+2}, \ldots k_n$, we have

$$\begin{aligned} k_1 + k_2 + \ldots k_m &= 1 + h, \\ k_{m+1} + k_{m+2} + \ldots k_n &= 1 - h, \\ \hline k_1 + k_2 + \ldots + k_{n-1} + k_n &= 2, \end{aligned}$$

so that the sum of the indices of the singular points is equal to 2, a generalisation of the relation deduced by Poincaré for the particular case treated by him [1]), whilst the structure of the singular points is submitted to the following restrictive property:

Theorem 3. *Twice the number of singular points plus the number of elliptic sectors is equal to the number of hyperbolic sectors plus four.*

The necessary existence of at least one singular point before reduction as well as of at least two radiating points after reduction lies included in this theorem and finds there its simplest proof.

We shall finally show that the set of singular points (supposed finite) is submitted to no other restriction than the one expressed in theorem 3.

Let us namely assume an arbitrary finite set of points as singular points, let us enclose them each by a suchlike simple closed curve that these curves do not intersect each other, and let us give inside and on these curves to the vector field a structure satisfying theorem 3 but for the rest arbitrary. We must then show that the outer domain

[1]) l.c. p. 405.

of these curves can be filled up with a finite continuous vector distribution *without* singular points and passing into the already existing ones.

To that end we take for the closed curves a certain cyclic order and join each of them with the succeeding one by such an arc of simple curve that these arcs do not intersect each other, so that on the sphere two free regions γ_1 and γ_2, bounded by simple closed curves, are determined. We then construct along the inserted arcs of curve suchlike finite continuous vector distributions without singular points and passing into the existing ones that the total angle, which for finite stereographic representation of γ_1 on a Euclidean plane the vector describes in a circuit, is zero. Then γ_1 can be filled, in the
⟦3⟧ manner indicated in the second communication p. 732, 733, with a finite continuous vector distribution *without* singular points and passing into the existing ones.

As now however the singularities have been chosen in such a way that they satisfy theorem 3, the vector describes in a circuit of the complementary domain of γ_2, stereographically represented on a finite region, a total angle 4π in the sense of the circuit; thus by a circuit of the region γ_2 itself, when stereographically represented on a finite region, a total angle zero. Therefore γ_2 also can be filled with a finite continuous vector distribution *without* singular points and passing on its boundary into the existing ones, with which the lack of other restrictions than those expressed in theorem 3, has been proved.

As for the singular points (supposed to form a finite set) of a finite continuous vector distribution in the Euclidean plane, neither their number, nor their structure is submitted to any restriction.

ERRATUM.

In the first communication on this subject, these Proceedings Vol. XI 2,
⟦6⟧ p. 856, l. 3 and 7 from top

for: recure it, meets read: recur it, it meets

NOTES

⟦1⟧ The English version appeared in the Proceedings of the meeting of 25 June 1910. Continuation of 1909 G2, 1910 A2. See also 1915 B.

⟦2⟧ Brouwer 1909 G2, 1910 A2.

⟦3⟧ Brouwer 1910 A2.

⟦4⟧ Brouwer 1906 C2.

⟦5⟧ H. Poincaré 1881. About the singularity index or circulation number see also Brouwer 1910 A2 ⟦5⟧, Brouwer 1911 H2 ⟦6⟧.

⟦6⟧ Brouwer 1909 G2.

On the orthogonal trajectories of the orbits of a one parameter plane projective group 1915 B ⟦1⟧

As the orbits of a one parameter plane projective group are generated by an infinitesimal transformation, which is a linear combination of the infinitesimal transformations

$$p;\ q;\ xp;\ yp;\ xq;\ yq;\ x^2p+xyq;\ xyp+y^2q,$$

their orthogonal trajectories on a rectangular system of axes can be generated by an infinitesimal transformation which is a linear combination of the infinitesimal transformations

$$p;\ q;\ xp;\ yp;\ xq;\ yq;\ xyp-x^2q;\ y^2p-xyq.$$

In other words, these trajectories are the integral curves in x and y of a system of differential equations of the form

$$\begin{aligned} \dot{x} &= -hxy-ky^2+ax+by+c \\ \dot{y} &= hx^2+kxy+ex+fy+g \end{aligned} \tag{1}$$

thus are identical, as Kasner[1]) has noticed, with the lines of force of a plane field of forces where every straight line is a tautochrone.

In the following we will make a few remarks on the topological structure[2]) of this field of trajectories. We will not consider the degenerate case

$$\dot{x} = ax+by+c$$

$$\dot{y} = ex+fy+g$$

which has already completely been treated by Poincaré[3]).

Moreover, for greater perspicuity we will add to the Cartesian plane one 'point at infinity' Ω, by which it becomes a variety of the connection of the sphere.

We start by introducing a few terms for the different ways in which a singular point S of a simple and directed cover of the plane by curves can be characterized by the topological structure of that system of curves in a neighbourhood of S:

α) S is called a *source point*, if for example all halflines starting from S belong to the curves system. ⟦4⟧

[1]) Amer. Bull. 15 (1909), p. 479, 480. ⟦2⟧

[2]) that is, about the complex of properties invariant under one-to-one continuous transformations of the plane.

[3]) Journ. de Mathématiques (3) 7 (1881), 385–391. ⟦3⟧

β) S is called a *vanishing point*, if for example all halflines arriving at S belong to the system of curves.

γ) S is called a *rotation point*, if for example all circles of centre S belong to the curves system.

δ) S is called a *collision point*, if for example in a cartesian coordinate system with origin S the positive and negative X-axis departing from S and the positive and negative Y-axis arriving at S belong to the system of curves and otherwise no meeting of curves takes place in the neighbourhood of S.

ε) S is called a *magnet point*, if for example all circles touching a given line through S at S in the same sense, belong to the system of curves.

ζ) S is called a *brush point*, if for example on a cartesian coordinate system with origin S the negative X-axis arriving at S (the 'shaft' of the brush) and all half-lines departing from S for which x is positive or zero belong to the system of curves, whereas otherwise no meeting of curves of the system takes place in the neighbourhood of S.

First Case

The three invariant points of the one parameter projective group coincide into one single point P.

If we assume as *case A* that P is a *proper point* of the cartesian plane, then there is only one line λ through P invariant under the group and no other line through P can be tangent to an orbit of the group. Thus if l is an arbitrary line through P, then at the intersection with l the distance of a trajectory from P either increases for all trajectories, or decreases for all trajectories, or is stationary for all trajectories, the third case *being possible only if* $l = \lambda$.

From this it follows that by inverting the sense of all trajectories, if need be, we can manage that along all trajectories the distance from P is permanently decreasing. *So there are no closed trajectories, P is a vanishing point and Ω is a source* – three properties which, topologically speaking, characterize the field of trajectories completely.

If now we assume as *case B* that P is an improper point of the cartesian plane, *then the cartesian plane contains no singular point, hence no closed trajectory*, from which finally it follows that Ω *is a magnet point.* Again this characterizes the trajectories field completely.

Second Case

The one parameter group has a double invariant point Q and a simple invariant point P.

Let us take the *case A* where the singular point Q is a *proper point* of the cartesian plane and choose Q as the origin of a rectangular coordinate system, the *X-axis* of which is passing through P; the group has two invariant lines, the line λ connecting P and Q, and a line κ through Q, not coinciding with λ. If κ and λ are not orthogonal to each other, then the differential equations of the trajectories can be written as follows:

$$\begin{aligned} \dot{x} &= -y(hx+ky) \\ \dot{y} &= x(hx+ky)+cx+y \end{aligned} \tag{2}$$

where $c \neq 0$ and $c \neq -h/k$ may be assumed because otherwise we would get degeneracy. Putting $-cx+y=\xi$ and $x+cy=\eta$ we get the equations in the form

$$\begin{aligned} \dot{\xi} &= \eta(p\eta-q\xi)+\xi \\ \dot{\eta} &= -\xi(p\eta-q\xi)+\xi \end{aligned} \tag{3}$$

where

$$p = \frac{h+kc}{1+c^2}, \qquad q = \frac{hc-k}{1+c^2}.$$

We may assume c to be negative and p positive; since, if we replace x and y by $-x$ and $-y$, respectively, the value of c remains unchanged while h and k change their sign as does p. It further follows immediately from the equations (2) that in Q there cannot be any other tangents to trajectories, except $\xi=-cx+y=0$ and $x=0$.

For trajectories which touch $\xi=0$ it follows from the equations (3) that $\lim(p\eta^2/\xi)=-1$, so that no trajectories can touch the η-axis from above (that is on the side where ξ is positive).

If we choose η_1 very small, then in the intersection of the line $\eta=\eta_1$ with the parabola $\xi=\alpha\eta^2$ we get for the trajectory through that point in first approximation

$$\begin{aligned} \dot{\xi} &= (\alpha+p)\eta_1^2 \\ \dot{\eta} &= c\alpha\eta_1^2. \end{aligned}$$

Therefore if we let α decrease from 0 to a very large negative value, then the direction of the trajectory in the corresponding intersection turns in the sense of a rotation of the positive ξ-axis towards the positive η-axis, therefore also of the positive Y-axis into the positive X-axis, from the direction $\dot{\xi}=1$, $\dot{\eta}=0$ towards a direction which does not differ much from $\dot{\xi}=-1$, $\dot{\eta}=-c$.

From this it follows that from the side of negative η one single trajectory arrives touching the η-axis from below whereas from the side of positive η a continuous sequence of trajectories departs touching at the lower side of the η-axis.

From

$$\frac{\mathrm{d}x}{\mathrm{d}y} = -\frac{y(hx+ky)}{x(hx+ky)-cx+y}$$

it follows

$$\frac{\mathrm{d}\delta x/\mathrm{d}y}{\delta x} = \frac{\partial}{\partial x}\left(\frac{\mathrm{d}x}{\mathrm{d}y}\right) = \frac{y(hx+ky)(2hx+ky-c)-\{x(hx+ky)-cx+y\}hy}{\{x(hx+ky)-cx+y\}^2}.$$

Therefore by first approximation:

$$\frac{\mathrm{d}\delta x/\mathrm{d}y}{\delta x} = \frac{-cky^2-hy^2}{y^2} = -(h+ck).$$

Therefore if $|y| < \varepsilon_1$ and $\left|\frac{x}{y}\right| < \varepsilon_2$, then $\left|\frac{\mathrm{d}\log\delta x}{\mathrm{d}y}\right| < j$ where j is a certain positive magnitude. Then, however, also

$$|(\log\delta x)_{y=0} - (\log\delta x)_{y=y_1}| < j|y_1|.$$

Neither at the side of positive y nor at that of negative y can two *different* trajectories touch $x = 0$ at Q; indeed between them $|(\log\delta x)_{y=0} - (\log\delta x)_{y=y_1}|$ would necessarily become infinite.

If now we consider on the one hand the segments of the trajectory starting in points of the line $x = y/2c$ for which $x^2 \leqq y^2/4c$ and y is positive and $\leqq \varepsilon$, and on the other hand the segments of the trajectory starting in points of the line $x = -2y/c$ for which $x^2 \leqq y^2/4c$ and y is positive and $\leqq \varepsilon$, then the transition between them must be formed by *at least one* trajectory parting from Q towards positive y and touching at $x = 0$.

Similarly, it is clear that there exists at least one trajectory which departs from Q towards the negative direction of y and touches the line $x = 0$.

Since further between two curves parting from an isolated singular point of the system, there must be necessarily a third curve of the system which either arrives at that singular point or departs from it, it follows that Q is a *brush point*, the shaft of which touches the negative η-axis.

As *case* $A\alpha$ we take the case where h is positive. Then the singular point P is on the negative X-axis. As in the first case it can be shown that either along all trajectories the distance from P is permanently increasing, or along all trajectories it is permanently decreasing; therefore in our case along all trajectories the distance from P is permanently increasing, so that *there are no closed trajectories, P is a source and Ω a vanishing point and the shaft of the brush point Q passes through P*, which, topologically speaking, characterizes the field of trajectories completely.

As *case* $A\beta$ we take the case where h is negative. Then P is lying on the positive X-axis, and along all trajectories the distance from P must permanently decrease, so that *P is a vanishing point and Ω is a source, while there are no closed trajectories and the shaft of the brush point Q passes through Ω*, which finishes the topological characterization of the field of trajectories.

As *case Aγ* we take $h = 0$. Then P is an improper point of the cartesian plane, so that *apart from the brush point Q the cartesian plane has no singular point.* Since $\dot{x}$ is nowhere negative, there can be again *no closed trajectories.* Therefore the line of force arriving in Q comes from Ω, and the lines of force departing from Q go to Ω, whereas all other lines of force start from Ω and finish in Ω, such that Ω is *a magnet point*, which settles the topological structure of the field of trajectories.

Let us now consider the assumption that κ *and* λ are orthogonal to each other. Then the differential equations of the lines of force can be written as

$$\begin{aligned} \dot{x} &= y(hx+ky) \\ \dot{y} &= -x(hx+ky)+x \end{aligned} \tag{4}$$

where we may take k to be positive.

The only possible tangent of trajectories at Q is now the Y-axis. Moreover, from (4) it follows that in the neighbourhood of Q for a trajectory arriving at or departing from Q we have $\mathrm{d}y/\mathrm{d}x$ approximately equal to x/ky^2; therefore for this trajectory approximately the relation $x^2 = \frac{2}{3}ky^3$ holds, which is the equation of a cubic parabola with Q as a vertex and the positive Y-axis as the tangent at the vertex, so that *in Q no trajectory can touch the negative Y-axis*, whereas a trajectory that touches the positive Y-axis at Q, if existent, must be enclosed between the two cubic parabolas $x^2 = \frac{2}{3}(k+\varepsilon)y^3$ and $x^2 = \frac{2}{3}(k-\varepsilon)y^3$ for y smaller than a very small positive y_1.

If we consider the points of the line $y = y_2$ ($y_2 < y_1$) lying between the two cubic parabolas, then the direction of the trajectory at these points is by first approximation given by $\mathrm{d}y/\mathrm{d}x = x/ky_2^2$, so that $\mathrm{d}y/\mathrm{d}x$ increases with x; therefore both to the right and to the left of the Y-axis the trajectories *diverge towards Q*, and both to the right and to the left of the Y-axis *at most one* trajectory can touch the Y-axis.

If however, we consider the trajectories departing upwards from the segment of the line $x = \varepsilon_1 y$ (ε_1 positive), for which $0 < y < \varepsilon_1^3$, then they cannot intersect either the above mentioned segment of the line $x = \varepsilon_1 y$ or the cubic parabola $x^2 = \frac{1}{3}ky^3$ as long as they keep $0 < y < \varepsilon_1^3$; therefore all of them have a segment between $x = \varepsilon_1 y$ and $y = \varepsilon_1^3$ which except for its endpoints, has no point in common with $x = \varepsilon_1 y$ nor with $y = \varepsilon_1^3$ nor with $x^2 = \frac{1}{3}ky^3$. Therefore as limit of these segments of trajectories there exists *at least one* segment that connects the point Q between $x = \varepsilon_1 y$ and $x^2 = \frac{1}{3}ky^3$ with the line $y = \varepsilon_1^3$.

In the same way one shows the existence of *at least one* trajectory segment connecting the line $y = \varepsilon_1^3$ with the point Q between $x^2 = \frac{1}{3}ky^3$ and $x = -\varepsilon_1 y$. Therefore since there exists one and only one trajectory arriving at Q and one and only one trajectory departing from Q, it follows that *Q is topologically a non-singular point for the field of trajectories.*

As *case Aδ* we now take $h \neq 0$, so that P is a *proper point* of the cartesian plane.

Then P is the *only singular point of the cartesian plane, and the topological structure of the field of trajectories is the same as in the first case A.*

As *case Aε* we take $h = 0$, so that P is an *improper point* of the cartesian plane. *Then the cartesian plane contains no singular point and the topological structure of the field of trajectories is the same as in the first case B.*

If finally we take as *case B* that where the singular point Q is an *improper point* of the cartesian plane. We may then assume P to be a proper point of the cartesian plane (because otherwise it would be a degenerate case), from which *the same topological structure of the field as in the first case A* may be derived.

Third Case

The one parameter group has two conjugate imaginary invariant points and one real invariant point.

This case can be dealt with in the same way as the first even with some simplifications.

If we first take *case A* where P is a *proper point* of the cartesian plane, then we find for the field *the same topological structure as in the first case A.*

If then we take *case B* where P is an *improper point* of the cartesian plane, then we find for the field *the same topological structure as in the first case B.*

Fourth Case

The one parameter group has three different real invariant points.

We may assume that at most one of these three invariant points is an improper point of the cartesian plane; otherwise we would meet with the degenerate case. We may then suppose that the orbital system of the projective group in the cartesian plane possesses a singular point A_1 with index[1]) $+1$ which is always a proper point, further a singular point A_2 with index $+1$ and a singular point A_3 with index -1; of the last two *at most one* is an improper point.

As *case A* let us assume that *both A_2 and A_3 are proper points* of the cartesian plane, such that the three singular points form a *proper triangle.* By reversing, if need be, the direction of all trajectories, we can make sure that the trajectories on the triangle side A_1A_2 are directed to the exterior, on the triangle sides A_2A_3 and A_3A_1 are directed to the interior, while on the extension of any side of the triangle the directions of the trajectories are opposite to those on the side itself. Further we may assume that the order $A_1A_2A_3$ indicates the anti-clockwise circulation of the triangle.

[[5]] [1]) Cp. Poincaré l.c. p. 400. Brouwer, Versl. K. Akad. v. Wet. 19 (1910) 49.

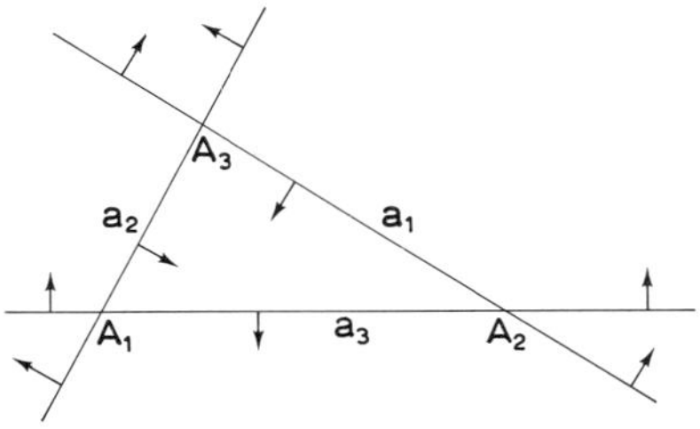

Fig. 1.

From the formulae (1) it follows that along every radius vector from A_3 which does not have the direction $hx+ky = 0$, at very large distances the direction of the trajectory is approximately orthogonal to the radius vector, whereas the sign of the turning sense around A_3 corresponding to the direction of the trajectory cannot change except on the radius vector of direction $hx+ky = 0$. Therefore we read from fig. 1 immediately that the direction $hx+ky = 0$ cannot lie outside the angle A_3 of the triangle. Neither can this direction be parallel with the sides A_3A_1 or A_3A_2 of the triangle. For if we take A_1 to be the origin of a rectangular coordinate system and place the X-axis along A_1A_3, then we can write the differential equations of the trajectories as

$$\dot{x} = y(hx+ky-b)$$

$$\dot{y} = -x(hx+ky-a)+gy,$$

where on the one hand h should vanish if the direction in which at large distances the turning sense reverses (that is the direction $hx+ky = 0$) were parallel to A_3A_1; on the other hand $h \neq 0$, if A_3 is a proper point of the cartesian plane.

Among the two pairs of opposite angles at A_3 there must be one in which the distance from A_3 along the trajectories decreases and one where it increases; from fig. 1 it follows that its increase takes place in the angle A_3 of the triangle and its opposite angle.

Once a trajectory has reached a large distance from A_3, it can neither stay in the top domains of A_1 and A_2, nor in the bottom domains of a_1 and a_2, as is shown in fig. 1; on the contrary, it must finally stay either in the bottom domain of a_3 or in the top domain of A_3, where its distance from A_3 can only increase, so that the trajectory must necessarily move to infinity. This shows *that the point at infinity Ω is a vanishing point.*

We will now investigate *the singular point A_3*. First of all, it is a simple singular point (in the sense of Poincaré), and since it has index -1 for the orbits of the projective group, it has index -1 also for their orthogonal trajectories, so that *by first approximation it is a collision point for the field of trajectories*, and in a suitably oblique system of axes the trajectories in the neighborhood of A_3 satisfy by first approximation the differential equations

$$\dot{x} = x$$
$$\dot{y} = -cy$$

where c is positive. Thus for the first approximation of the trajectories in the neighbourhood of A_3 the relation

$$\frac{\mathrm{d}(\mathrm{d}y/\mathrm{d}x)}{\mathrm{d}(y/x)} = -c$$

holds; therefore

$$\left\{\frac{\partial(\mathrm{d}y/\mathrm{d}x)}{\partial(y/x)}\right\}_{r\,\mathrm{const}}$$

is finite and negative for the exact trajectories in the neighbourhood of A_3, so that in the exact field of trajectories at A_3 along one of the tangents at A_3 from both sides only one trajectory arrives, and along the other tangent only one trajectory leaves towards both sides.

Therefore A_3 is a collision point for the exact field of trajectories, too, and as follows from fig. 1, the departing trajectories lie within the angle A_3 of the triangle and its opposite angle, and the arriving trajectories lie within the two exterior angles.

For the *topological investigation of the singular point A_1* we *first* assume that the angle $A_3A_1A_2$ *is acute.* We choose a rectangular coordinate system with origin A_1 and the X-axis along A_1A_2; on this system we can write the differential equations of the trajectories as

$$\begin{aligned} \dot{x} &= -y(hx+ky-b) \\ \dot{y} &= x(hx+ky-a)+gy. \end{aligned} \tag{5}$$

From the directions of the trajectories in points of the X-axis at large distances it follows that h is positive. Since a/h is the abscissa of A_2, a is positive. Since the line through A_3 in the direction $hx+ky=0$ has the equation $hx+ky-b=0$ and must intersect the side A_1A_2 of the triangle, b is positive and $a-b$ is positive. Since finally A_1A_3 has the equation $gy-(a-b)x=0$ and the angle $A_3A_1A_2$ is acute, g is positive.

If we put $\frac{1}{2}(ax^2+by^2)=\rho$, then

$$\dot{\rho} = -(a-b)xy(hx+ky)+bgy^2,$$

such that $\dot{\rho}$ is permanently positive along an arbitrary trajectory in the neighbourhood of A_1 if we disregard certain elements of time near the moments of intersection with the X-axis. For each of these intersections with the X-axis, however, we can indicate two lines $y=\varepsilon x$ and $y=-\varepsilon x$ between which $\int\frac{\mathrm{d}\rho}{\mathrm{d}y}\,\mathrm{d}y$ – approximately equal to $\int \frac{bgy^2 - h(a-b)x^2 y}{-ax} \mathrm{d}y$ – along the trajectory takes a positive value. *Thus* A_1 *must be a source.*

Secondly, we assume *angle* $A_3 A_1 A_2$ *to be obtuse.* We choose our coordinate system in the same way, so that the differential equations of the trajectories have the same form as above. Then h, a, b and $a-b$ again appear to be positive, whereas g is negative, so that from the equation $\dot{\rho} = -(a-b)xy(hx+ky)+bgy^2$ it follows that A_1 *is a vanishing point.*

Thirdly, we assume the *angle* $A_3 A_1 A_2$ *to be a right angle.* We choose our coordinate system as before and obtain the differential equations of the trajectories in the form

$$\begin{aligned} \dot{x} &= -y(hx+ky-b) \\ \dot{y} &= x(hx+ky-a) \end{aligned} \tag{6}$$

where h, k, a, b, $a-b$ are all positive.

By a suitable coordinate transformation of the form

$$\begin{aligned} x &= \alpha x' \\ y &= y' \end{aligned}$$

where α is positive, these differential equations become

$$\begin{aligned} \dot{x} &= -y\{q(hx+ky)-1\} \\ \dot{y} &= x\{p(hx+ky)-1\} \end{aligned} \tag{7}$$

where h, k, p, q, and $q-p$ are positive. For the new coordinate system the trajectories in the neighbourhood of A_1 are by first approximation circles with centre A_1, so that *in first approximation of the field of trajectories* A_1 *is a rotation point.*

To investigate the *character of* A_1 *in the exact field of trajectories,* we derive from (7)

$$\frac{\mathrm{d}(r^2)}{\mathrm{d}t} = \frac{\mathrm{d}(x^2+y^2)}{\mathrm{d}t} = -2(q-p)xy(hx+ky).$$

Or, putting $\operatorname{arc\,tg}(y/x) = \phi$,

$$\frac{\mathrm{d}r}{\mathrm{d}t} = -(q-p)r^2 \sin\phi \cos\phi(h\cos\phi + k\sin\phi)$$

Further we have

$$\frac{\mathrm{d}\phi}{\mathrm{d}t} = \frac{x(\mathrm{d}y/\mathrm{d}t)-y(\mathrm{d}x/\mathrm{d}t)}{x^2+y^2} = \frac{(px^2+qy^2)(hx+ky)}{x^2+y^2} - 1 =$$
$$= r(p\cos^2\phi + q\sin^2\phi)(h\cos\phi + k\sin\phi) - 1.$$

Therefore

$$\frac{\mathrm{d}r}{\mathrm{d}\phi}=\frac{\dot{r}}{\dot{\phi}}=\frac{(q-p)r^2\sin\phi\cos\phi(h\cos\phi+k\sin\phi)}{1-r(p\cos^2\phi+q\sin^2\phi)(h\cos\phi+k\sin\phi)}\,.$$

In the neighbourhood of A_1 we may put $r=r_1+\Delta r$ along a trajectory, where r_1 is a constant and $\Delta r/r_1$ becomes infinitely small with r_1. Then we get *by first approximation*

$$\frac{\mathrm{d}r}{\mathrm{d}\phi}=r_1^2(q-p)\{h\cos^2\phi\sin\phi+k\sin^2\phi\cos\phi\}$$

$$\Delta r=\tfrac{1}{3}r_1^2(q-p)\cdot\Delta\{-h\cos^3\phi+k\sin^3\phi\},$$

or, choosing for r_1 the value of r for $\phi=\operatorname{arc\,tg}\sqrt[3]{h/k}$,

$$\Delta r=\tfrac{1}{3}r_1^2(q-p)(-h\cos^3\phi+k\sin^3\phi).$$

We now have *in second approximation:*

$$\begin{aligned}\frac{\mathrm{d}r}{\mathrm{d}\phi}&=(r_1^2+2r_1\Delta r)(q-p)(h\cos^2\phi\,\sin\phi+k\sin^2\phi\,\cos\phi)\\&\qquad\times\{1+r_1(h\cos\phi+k\sin\phi)(p\cos^2\phi+q\sin^2\phi)\}\\&=r_1^2(q-p)(h\cos^2\phi\sin\phi+k\sin^2\phi\cos\phi)\\&\qquad\times\{1+\tfrac{2}{3}t_1(q-p)(-h\cos^3\phi+k\sin^3\phi)\}\{1+r_1(h\cos\phi+k\sin\phi)\\&\qquad\qquad\times(p\cos^2\phi+q\sin^2\phi)\}\\&=r_1^2(q-p)(h\cos^2\phi\sin\phi+k\sin^2\phi\cos\phi)\\&\qquad\times\{1+\tfrac{2}{3}r_1(q-p)(-h\cos^3\phi+k\sin^3\phi)+r_1(h\cos\phi+k\sin\phi)\\&\qquad\qquad\times(p\cos^2\phi+q\sin^2\phi)\}.\end{aligned}$$

Only the terms of the second part of the last equation contribute to $\int_0^{2\pi}\frac{\mathrm{d}r}{\mathrm{d}\phi}\mathrm{d}\phi$ in such a way that *the exponents of the powers of both* $\sin\phi$ *and* $\cos\phi$ *are even.* These terms are the following:

$$2r_1^3(q-p)hk\sin^2\phi\cos^2\phi(p\cos^2\phi+q\sin^2\phi).$$

They therefore make a *positive* contribution to $\int_0^{2\pi}\frac{\mathrm{d}r}{\mathrm{d}\phi}\mathrm{d}\phi$. On the other hand $\frac{\mathrm{d}t}{\mathrm{d}\phi}$ and $\int_0^{2\pi}\frac{\mathrm{d}t}{\mathrm{d}\phi}\mathrm{d}\phi$ are *negative*, such that after every turn of the trajectory under consideration the length of the radius vector has decreased, and thus A_1 *is a vanishing point.*

A topological investigation of *the singular point A_2* in the same way reveals *that A_2 is a source for an acute angle $A_3A_2A_1$, and a vanishing point for an obtuse or right angle $A_3A_2A_1$.*

In order to get some insight into the structure of the field of trajectories globally, we notice that closed trajectories surrounding all the three points A_1, A_2 and A_3 are impossible because the top domain of A_3 cannot be crossed by trajectories, and that closed trajectories which of the points A_1, A_2 and A_3 surround only two points or no points or only the point A_3, are equally impossible by Poincaré's index theory.

Let us first assume that *there is no closed trajectory through A_3*. Then every closed trajectory surrounds of the three points A_1, A_2, A_3 either only A_1, or only A_2. The outermost trajectory of the first kind, or if it does not exist, the point A_1, is denoted by γ_1, the outermost trajectory of the second kind, or if it does not exist, the point A_2, is denoted by γ_2. The field of trajectories is divided by γ_1 and γ_2 into an *inner field* b_1 containing A_1 (which may vanish into the point A_1) and an *inner field* b_2 containing A_2 (which may vanish into the point A_2), and an *outer field* β.

With respect to the *inner fields* it can be remarked that all closed trajectories of b_1 are travelled in a clockwise sense and surround each other and the point A_1, whereas all closed trajectories of b_2 are travelled in an anti-clockwise sense and surround each other and the point A_2. The domains left open by the closed trajectories in b_1 and b_2, as far as they contain A_1 or A_2 respectively, must be simply connected and filled with spiraling trajectories that connect A_1 or A_2, respectively, with the asymptotically approached boundary, and as far as they contain neither A_1 nor A_2, are ringshaped and filled with spiraling trajectories connecting the two asymptotically approached boundaries with each other.

The *outer field* has as singularities the vanishing point Ω, the collision point A_3 and the source- or vanishing borders γ_1 and γ_2; in order to investigate the topological structure of this field we distinguish two cases:

In *case $A\alpha$* we assume that *the trajectory departing from A_3 inside the triangle bends around γ_1 or γ_2*, and we may restrict the further investigation by the supposition that it bends around γ_1. Therefore since γ_1 is a vanishing border, both of the two trajectories arriving at A_3 come necessarily from γ_2 and must together surround γ_1. About the four trajectories uniting at A_3 we now know that two of them connect A_3 with γ_2, a third with γ_1, and the fourth with Ω; from this and the topological character of the singularities follows the topological structure of the outer field completely.

In order to show the actual existence of the case Aα, we take the angle $A_3A_1A_2$ very near 180° by making the values of h, a, b in the equations (5) very small and all of the same order, for a fixed negative value of g, so that we may replace them by $h\varepsilon$, $a\varepsilon$ and $b\varepsilon$ where ε is very small. Along the trajectory departing from A_3 within the triangle $A_1A_2A_3$ we see the radius vector from A_1 permanently decrease up to its intersection with A_1A_2 in a point P which may, or may not, coincide with A_1;

and since the perpendicular line from A_1 on the side $A_2 A_3$ is of the order ε, also $A_1 P$ cannot be larger than a magnitude of order ε. If then the trajectory enters the angle between $A_1 A_2$ and the extension of $A_1 A_3$, the radius vector from A_1 will again increase but, on the other hand, inside this angle, it holds approximately that

$$\frac{dx}{dy} = \frac{by}{-ax+gy}$$

and

$$\left|\frac{y}{x}\right| < \left|\frac{a-b}{g}\right| < \left|\frac{a}{g}\right|,$$

thus dx/dy is of the same order as y/x, that is of the order ε, so that in each of its points the trajectory is approximately orthogonal to the radius vector from A_1, thus intersects the extension of $A_1 A_3$ in such a point Q that by first approximation $A_1 Q = A_1 P$. The trajectory subsequently enters the angle between the extensions of $A_1 A_3$ and $A_1 A_2$, where along its course the radius vector permanently decreases, so that it necessarily intersects the extension of $A_1 A_2$ in such a point R different from A, that $A_1 R$ is not larger than a magnitude of order ε. If then it enters the angle between $A_1 A_3$ and the extension of $A_1 A_2$, then along its course the radius is again going to increase, but $A_1 A_3$ must be intersected in such a point that by first approximation $A_1 S = A_1 R$. The further course of the trajectory can only cause it to finish around γ_1; in fact, from the equations (5) it can be derived for this case that γ_1 is necessarily identical with A_1.

As *case* $A\beta$ we assume that *the trajectory departing from* A_3 *inside the triangle moves to infinity*. Then the two trajectories departing from A_3 to infinity form together a curve that separates γ_1 and γ_2 from each other, as well as the two trajectories arriving at A_3. Then one of the last two trajectories comes from γ_1 and the other from γ_2, so that both γ_1 and γ_2 are source borders. This establishes completely the topological structure of the outer field.

In order to show that the case Aβ really exists, we assume the triangle $A_1 A_2 A_3$ to be isoceles with A_3 as its apex, and further we choose the one parameter group such that up to the directions its orbits are symmetric with respect to the perpendicular from A_3 on $A_1 A_2$. Then the system of trajectories must possess the same symmetry, and this can only happen if one of the trajectories arriving at A_3 comes from γ_1 and the other from γ_2.

As *case* $A\gamma$ we investigate the possibility we have disregarded so far that *there is a closed trajectory through* A_3. Since the top domain at A_3 cannot be crossed by trajectories, it first is clear that the trajectory departing in that domain from A_3 cannot be part of a closed trajectory through A_3 so that therę cannot possibly exist two such trajectories; moreover that such a trajectory arrives in A_3 either between $A_1 A_3$ and the extension of $A_2 A_3$ and then surrounds the point A_1 but not the point A_2, or between $A_2 A_3$ and the extension of $A_1 A_3$ and then surrounds the point A_2 but not the point A_1. If we restrict ourselves to the first case and denote the closed trajectory through A_3 by γ_1 and by γ_2 the outermost closed trajectory surrounding A_2 which may vanish into A_2, then the field of trajectories is again divided by γ_1 and γ_2 into an *inner field* b_1 that contains A_1, an *inner field* b_2 that contains A_2 (it may vanish into A_2), and an *outer field* β.

The topological structure of the outer field is again completely determined: All trajectories depart from the 'source border' γ_2 to the vanishing point Ω; one single trajectory κ passes through the point A_3; on one side of κ the trajectories converge to $\kappa+\gamma_1$, on the other side to κ.

With respect to the topological structure of the *inner fields* the same holds as in the cases Aα and Aβ.

To show that the case Aγ is indeed possible, we choose fixed points A_1 and A_3 and have A_2 moving along a circle around A_1 the radius of which is smaller than twice the length of A_1A_3 and choose the line $hx+ky=0$ always in the direction of the bisector of the angle $A_1A_3A_2$. Then A_2 will have both positions belonging to case Aα and to case Aβ; certainly, A_2 gets a position A_2^0 which either creates the case Aβ or is limit of positions $'A_2, ''A_2, \cdots$ which lead to the case Aα as well as of positions $A_2', A_2'', \cdots$ which lead to the case Aβ. If under the last assumption we consider the field of trajectories belonging to A_2^0 with the outermost closed trajectories γ_1^0 and γ_2^0, then the outermost closed trajectories ${}^{(v)}\gamma_1$ and ${}^{(v)}\gamma_2$, and $\gamma_1^{(v)}$ and $\gamma_2^{(v)}$, respectively, belonging to ${}^{(v)}A_2$ and $A_2^{(v)}$, respectively, converge to limit curves ${}^{(\omega)}\gamma_1$ and ${}^{(\omega)}\gamma_2$, and $\gamma_1^{(\omega)}$ and $\gamma_2^{(\omega)}$, respectively, which are *not lying outside* γ_1^0 and γ_2^0. Since Ω is a vanishing point, the infinite series of trajectories ${}^{(v)}\kappa$ belonging to ${}^{(v)}A_2$ and leading from A_3 to ${}^{(v)}\gamma_1$ cannot possibly converge to a combination of a trajectory passing from A_3 to infinity and a trajectory going from infinity to ${}^{(\omega)}\gamma_1$. Therefore the field of trajectories belonging to A_2^0 cannot be in the case Aβ. Since the infinite series of trajectories $\kappa^{(v)}$ belonging to $A_2^{(v)}$ and departing from A_3, crossing $A_1A_2^{(v)}$ and going to infinity cannot possibly converge to the combination of a trajectory from A_3 to γ_1^0 and a trajectory from γ_1^0 to infinity, the field of trajectories cannot be in the case Aα. The supposition that the field of trajectories belonging to A_2^0 is *not* in the case Aγ, consequently leads to a contradiction.

Finally we will show that *closed trajectories not passing through* A_3 are not only possible but that *their existence is not a special case but as general as their possible absence*. To prove this we consider the same continuous variation of the data as we did in the existence proof for case Aγ. There is then a position A_2^{00} of A_2 such that the angle $A_2^{00}A_1A_3$ is right. The corresponding field of trajectories has a vanishing point at A_1 and in the neighbourhood of A_1 on A_1A_3 a point P with the property that, after a circulation around A_1, the trajectory departing from P intersects the side A_1A_3 at a point Q such that $A_1Q < A_1P$.

By a small variation of the data one can obtain an *acute angle* $A_2A_1A_3$ which differs little from $\frac{1}{2}\pi$, and a corresponding field of trajectories such that after a circulation around A_1 the trajectory departing from P still intersects the side $A_1 A_3$ in such a point R that $A_1 R < A_1 P$. Since A_1 is now a source, this trajectory cannot finish at A_1; it must therefore necessarily converge in an asymptotic spiral to *a curve surrounding A_1 and not passing through A_3*.

Let us now consider the case B where A_2 is *an improper point* of the cartesian plane. We may then assume that the topological situation of the invariant lines of the projective group and of the directions of the trajectories on these lines agrees with fig. 2.

If we choose a rectangular coordinate system with origin A_1 and the X-axis along $A_1 P$ and write the differential equations of the trajectories in the form (5), then with A_2 at infinity, we have $h = 0$, so that the equations become

$$\begin{aligned} \dot{x} &= -y(ky-b) \\ \dot{y} &= x(ky-a)+gy \end{aligned} \tag{8}$$

From the sign of $\dot{x}$ at large distances in the direction of a_2 it follows that k is positive. From the direction of the trajectories on the segment $A_1 P$ it follows that a is positive. Since $ky-b = 0$ must be fulfilled at A_3, it results that b is positive and $ky-b = 0$ is the equation of a_1. Since $\dot{y}$ for $ky-b = 0$ and indefinitely increasing x finally will have the sign of $ky-a$, the sign of $ky-a$ must be negative for $ky-b = 0$ according to fig. 2 so that $a-b$ is positive. Since $\dot{y}$ has the sign of g for $x = 0$ and $ky-b = 0$, according to fig. 2, g will be positive, zero, or negative, depending on whether the angle $A_3 A_1 P$ is acute, right, and obtuse.

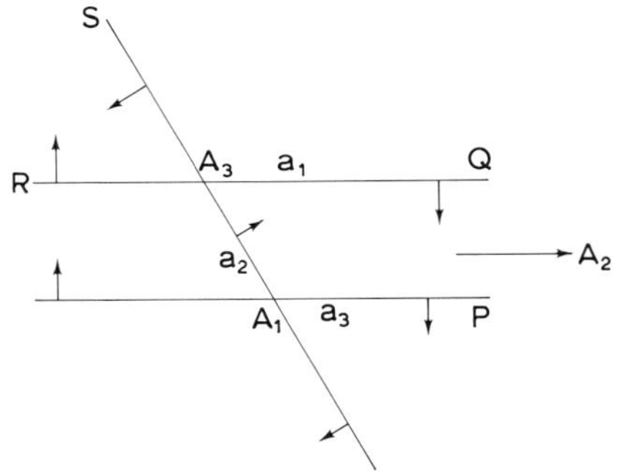

Fig. 2.

From fig. 2 it follows that a trajectory which is inside the angle SA_3Q, cannot have been outside this angle in its earlier course, therefore must have come from infinity inside this angle. Likewise a trajectory which is inside the angle RA_3S cannot in its further course leave this angle, and must therefore reach infinity inside this angle.

If we let x^2+y^2 grow indefinitely, then from

$$\frac{dy}{dx}=\frac{x(ky-a)+gy}{-y(ky-a)-(a-b)y}$$

follows that the direction of the trajectory approaches the perpendicular upon the radius vector from A_1 provided we see to it that $ky-a$ does not become indefinitely small, that is, the line $y=a/k$ (*lying above* a_1) is not approached indefinitely. Therefore with A_1 as a centre one can draw a circle with a radius sufficiently large that every trajectory with a point outside this circle, originates from the infinity of the angle SA_3Q, and moves to angle RA_3S and there to infinity. *Thus the point at infinity* Ω *is a magnet point* [1]).

In the same way as in case A it results that *the singular point* A_3 *is a collision point*; it follows from fig. 2 that the trajectories departing from A_3 lie inside the angles RA_3S and A_1A_3Q, and that the trajectories arriving at A_3 lie inside the angles RA_3A_1 and SA_3Q.

Likewise it follows in the same way as in case A that *the singular point* A_1 *is a source for an acute angle* A_3A_1P *and a vanishing point for an obtuse angle* A_3A_1P.

If the angle A_3A_1P *is a right angle*, then in the same way as in case A it follows that *the singular point* A_1 *is a rotation point for the first approximation of the field of trajectories*. Since, however, the absence of g results in a_2 becoming a line of symmetry of the system of the non-directed trajectories, A_1 *must be a rotation point for the exact field of trajectories, too.*

To investigate the global structure of the field of trajectories, we notice that every closed trajectory must either pass through the point A_3, or separate A_1 from A_3 and surround A_1.

In *case Bα* we assume that *there is no closed trajectory through* A_3. If then we denote the outermost trajectory or if there is none, the point A_1 by γ_1, then the field of trajectories is divided by γ_1 into an inner field b_1 containing A_1 (possibly vanishing into A_1), and an outer field β containing A_3.

As to the topological character of the *inner field* the same general remarks apply as were made regarding the inner fields of case A.

In order to describe the topological character of the *outer field*, we may assume that *the trajectory departing from* A_3 *inside the angle* A_1A_3Q *bends around* γ_1, *so that the trajectory arriving at* A_3 *inside the angle* A_1A_3R *cannot come from* γ_1, *and must therefore come from infinity*. Indeed, in the opposite case the trajectory departing from A_3 inside the angle A_1A_3Q would go to infinity, so that the trajectory arriving at A_3 inside the angle A_1A_3R could not have come from infinity but must have come from γ_1. By reflection at the *Y*-axis and reversion of all trajectories our suppositions would again be fulfilled.

[1]) In this magnet point – as, in fact, in the vanishing point Q of case A – all trajectories also possess a common asymptote, but this is not a topological property.

Among the four trajectories meeting at A_3 there are now two coming from infinity, one moving to the vanishing border γ_1 and the other to infinity. From this and from the character of Ω as a magnet point the topological character of the outer field follows completely.

In *case Bβ* we assume *the existence of a closed trajectory through A_3*. We denote it by γ_1; since the trajectories cannot cross the part of the plane above a_1, the trajectory γ_1 must depart from A_3 inside the angle A_1A_3Q, arrive at A_3 inside A_1A_3R and surround the point A_1. It divides the field of trajectories into an *inner field b_1* and an *outer field β*.

As to the topological character of the inner field the same general remarks apply as in the cases A and Bα.

The topological structure of the *outer field* is completely determined: All trajectories start and finish at the magnet point Ω; one of them, κ, goes through point A_3; on one side of κ the trajectories converge to $\kappa+\gamma_1$, on the other side to κ.

To find out when each of the cases Bα and Bβ apply, we first assume *A_3A_1P to be a right angle*; because of the symmetry of the field of trajectories with respect to a_2, the case Bα is excluded; we therefore *have case Bβ* and, because of the symmetry, *the inner field contains closed trajectories only.*[1])

Let us then consider *the angle A_3A_1P to be acute* and the field of trajectories represented by the equations (8) where, consequently, g is positive. Let us compare this field with another field of trajectories represented on the same coordinate system with the same values of a, b, and k by the equations

$$\begin{aligned}\dot{x} &= -y(ky-b)\\ \dot{y} &= x(ky-a).\end{aligned} \tag{9}$$

This field possesses an inner field around A_1, filled with closed trajectories surrounding A_1. If we consider an arbitrary point H, not on the X-axis, of an arbitrary closed trajectory τ of the second field, then comparing (8) and (9) we see immediately that at H the directions of the trajectory of the first field *point towards the outside of τ*. If now K is a point of τ on the X-axis and l_1 and l_2 are two lines parallel to the X-axis and on either side at a small distance from the X-axis, then through all points of τ in the neighbourhood of K there are segments of trajectories of the first field joining l_1 and l_2 and intersecting τ only once, so that there is also a segment of trajectory of the first field that passes through K, joins l_1 and l_2 and intersects τ only once. Therefore τ can be intersected by an arbitrary trajectory of the first field only in such a way that the trajectory passes from the interior to the exterior of τ. From this it follows immediately that the first field contains no closed trajectories (which if existent would surround A_1 without coming above the

[1]) F. Schuh drew my attention to the fact that the differential equations of the trajectories can be very simply integrated in this case and so the same results can be reached straightforwardly.

[[334]]

line $ky-b=0$), *hence it is necessarily in the case $B\alpha$, with an inner field b_1 vanishing into A_1.*
For an obtuse angle A_3A_1P the same result is obtained in a similar way.

In *case C* we assume that *A_3 is an improper point* of the cartesian plane. We may then assume that the topological situation of the invariant lines of the projective group and the directions of the trajectories on these lines agrees with fig. 3.

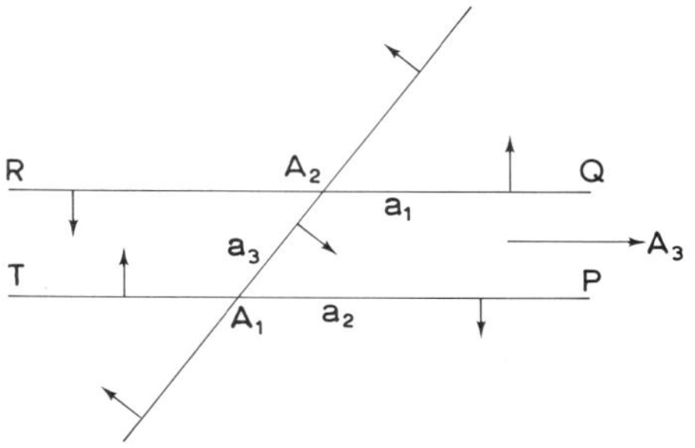

Fig. 3.

If we choose a rectangular coordinate system with the origin at A_1 and the X-axis along A_1P, then we can again write the differential equations of the trajectories in the form

$$\begin{aligned} \dot{x} &= -y(ky-b) \\ \dot{y} &= x(ky-a)+gy. \end{aligned} \tag{8}$$

From the sign of $\dot{x}$ at large distances in the direction of a_3 it follows that k is positive. From the directions of the trajectories on the segment A_1P it follows that a is positive. Since $ky-b=0$ must hold at A_2, we find b positive and $ky-b=0$ as equation of a_1. Since for $ky-b=0$ and indefinitely increasing x the sign of $\dot{y}$ will ultimately be the same as the sign of $ky-a$, the sign of $ky-a$ must be positive for $ky-b=0$ according to fig. 3, such that $b-a$ is positive. Since $\dot{y}$ for $x=0$ and $ky-b=0$ has the sign of g, it follows from fig. 3 that g is positive, zero, and negative according to whether the angle A_2A_1P is obtuse, a right angle, or acute.

If we let x^2+y^2 increase indefinitely, it follows from

$$\frac{dy}{dx} = \frac{x(ky-a)+gy}{-y(ky-a)+(b-a)y},$$

that the trajectory direction indefinitely approaches the perpendicular to the radius vector from A_1 provided we make sure that $ky-a$ does not become indefinitely small, that is the line $y=a/k$ (*situated between a_1 and a_2*) is not approached indefinitely. Therefore, a trajectory can come from, or go to, infinity in no other way but with the line $y = a/k$ as its asymptote; further, since in the neighbourhood of this asymptote $\dot{x}$ is positive, the trajectories can descend from infinity only on the left and go to infinity only on the right; finally, since it follows immediately from the shape of the orbits of the projective group that between a_1 and a_2 the trajectories *diverge to infinity* both right and left of a_3, there is only one trajectory that originates from infinity on the left of the asymptote and also only one trajectory that goes to infinity on the right of the asymptote. *Therefore the point at infinity Ω is topologically a non-singular point.*

By putting temporarily the X-axis of the coordinate system along a_3 we can prove with the method of case A that *for an acute angle TA_1A_2 the singular point A_1 is a source and the singular point A_2 a vanishing point, whereas for an obtuse angle TA_1A_2 the point A_1 is a vanishing point and the point A_2 a source.*

Likewise it follows by the method of case A that *for a right angle TA_1A_2 both points A_1 and A_2 are rotation points by first approximation.* Since, however, the absence of g implies that a_3 is a line of symmetry of the field of undirected trajectories, *A_1 and A_2 must be rotation points also for the exact field of trajectories.*

To investigate the structure of the field of trajectories globally, we distinguish two cases.

As *case $C\alpha$* we assume *that the trajectories originating from, and going to, infinity are different.* Then the trajectory originating from infinity must either finish around a closed trajectory surrounding A_1 (which may vanish into the point A_1), or finish around a closed trajectory surrounding the point A_2 (which may vanish into the point A_2); we may assume that the first of these two completely equivalent cases is actualized. The closed trajectory surrounding A_1 round which the trajectory originating from infinity finishes, is denoted by γ_1. For the trajectory going to infinity the only possible behaviour is now that it either unwinds in a spiral from a closed trajectory surrounding A_2, or that it originates from the point A_2 itself. If in the first case we denote the closed trajectory surrounding A_2 and in the second case the point A_2 itself by γ_2, then γ_1 and γ_2 divide the plane into an *inner field* b_1 containing A_1 (which may vanish into the point A_1), and an *inner field* b_2 containing A_2 (which may vanish into the point A_2), and an *outer field* β.

As to the topological character of the *inner fields* the general remarks of cases A and B hold also here.

In the *outer field* each trajectory must originate from γ_2 except the one that comes from infinity, and each trajectory must bend around γ_1 except the one that goes to infinity. If we close the cartesian plane by the point Ω, which for the field of trajectories is a non-singular point, we can therefore say that *all trajectories bend around γ_1 and come from γ_2*, which defines the topological structure of the outer field completely.

As *case $C\beta$* we assume that *the trajectories coming from and going to infinity are identical*, and therefore form a single trajectory γ. Then γ divides the plane into an *inner field* b_1 containing A_1 and an *inner field* b_2 containing A_2 both of which extend to infinity; the general remarks of cases A and B on the topological structure of the inner field hold for both of them.

To investigate when each of the cases Cα and Cβ occurs, we first suppose *the angle* TA_1A_2 *to be a right angle*; because of the symmetry of the field of trajectories with respect to a_3 the case Cα is then impossible; *we have therefore the case* Cβ, and in particular, because of the symmetry, with *inner fields containing only closed trajectories.*[1])

We now assume *the angle* TA_1A_2 *to be acute* and the field of trajectories to be represented by the equations (8), where g is, therefore, negative. Since $\dot{y}$ is negative for $ky-a \leqq 0$, $x \geqq 0$ and $y > 0$, the trajectories going to infinity must lie entirely above the asymptote $ky-a = 0$ for $x \geqq 0$. Since $\dot{y}$ is negative for $ky-a \geqq 0$, $x \leqq 0$ and $y > 0$, the trajectory originating from infinity must lie entirely below the asymptote $ky-a$ for $x \leqq 0$. Since therefore the trajectory going to infinity and the trajectory coming from infinity cannot possibly be identical, we can be sure we meet here with the case Cα.

Let us now compare this field with another field of trajectories, which is represented on the same coordinate system by the equations (9) with the same values of a, b, and k so that the part of the plane below $ky-a = 0$ is filled with closed trajectories surrounding A_1. If we consider an arbitrary point H, not on the X-axis, of a closed trajectory of the second field τ surrounding the point A_1, we will see immediately by comparing (8) and (9) that the trajectories of the first field at H are directed towards the inside of τ. From this we can derive that τ can be intersected by an arbitrary trajectory of the first field only in such a way that the trajectory passes from the exterior to the interior of τ, so that the first field cannot possibly contain a closed trajectory surrounding A_1. Since, however, for this field A_1 and A_2 play completely equivalent roles, a closed trajectory surrounding A_2 cannot exist either, so that the field is in *the case* Cα *in such a way that the inner fields* b_1 *and* b_2 *have disappeared into* A_1 *and* A_2.

For an *obtuse angle* TA_1A_2 the same result is obtained in a similar way.

[1]) The remark of F. Schuh mentioned on p. ⟦334⟧ applies also here.

NOTES

⟦1⟧ This is an application of the general theory developed in 1909 G2, 1910 A2, 1910 D2. Translated from the Dutch.
⟦2⟧ Ed. Kasner 1909.
⟦3⟧ H. Poincaré 1881.
⟦4⟧ The 'examples' should be interpreted in a topological sense. Thus a singular point in which there is only one arriving and one departing trajectory is considered as non-singular.
⟦5⟧ H. Poincaré 1881. Brouwer 1910 D2, p. 184.

CHAPTER 5

Cantor–Schoenflies style topology

Mathematics. — "*On the structure of perfect sets of points*". By Dr. L. E. J. BROUWER. (Communicated by Prof. KORTEWEG). 1910 B2 [[1]]

(Communicated in the meeting of March 26, 1909). [[2]]

§ 1.

Sets of points and sets of pieces.

The sets of points discussed in the following lines are supposed to be lying within a finite domain of a Sp_n.

By a *piece* of a closed set of points μ we understand a *single point or closed coherent set of points*, belonging to μ, and not contained in an other closed coherent set of points belonging to μ. [[3]] [[4]]

We can regard as elements of μ its pieces as well as its points, in other words we can consider μ on one hand as a *set of points*, on the other hand as a *set of pieces*.

Let us choose among the pieces of μ a fundamental series S_1, S_2, $S_3, \ldots$, then to μ belong one or more pieces ${}_1S_\omega$, ${}_2S_\omega, \ldots$ with the property that S_n lies entirely within a for indefinitely increasing n indefinitely decreasing distance ε_n from one of the pieces ${}_\alpha S_\omega$. These parts ${}_\alpha S_\omega$ we shall call the *limiting pieces* of the fundamental series $S_1, S_2, S_3, \ldots$

As thus the set μ possesses to each of its fundamental series of pieces at least one limiting piece, *a closed set of points is likewise closed as set of pieces.*

By an *isolated piece* of μ we understand a piece having from its rest set in μ a finite distance, in other words a piece, the rest set of which is closed.

THEOREM 1. *Each piece of μ is either a limiting piece, or an isolated piece.*

Let namely S be a non-isolated piece, then there exists in μ a fundamental series of points $t_1, t_2, t_3, \ldots$ not belonging to S, converging to a single point t of S. If t_1 lies on S_1, then S_1 has a certain distance ε_1 from S. There is then certainly a point t'_2 of the fundamental series possessing a distance $< \varepsilon_1$ from S, lying therefore not on S_1 but on an other piece S_2. Let ε_2 be the distance of S_2 from S, then there is certainly a point t'_3 of the fundamental series possessing a distance $< \varepsilon_2$ from S, lying thus neither on S_1, nor on S_2, but on a third piece S_3. Continuing in this manner we determine a fundamental series of pieces $S_1, S_2, S_3, \ldots$, containing consecutively the points $t_1, t'_2, t'_3, \ldots$ converging to t. So the pieces $S_1, S_2, S_3, \ldots$ converge to a single limiting piece which can be no other than S.

By a *perfect set of pieces* we understand a closed set, of which each piece is a limiting piece.

A perfect set of pieces is also perfect as set of points; but the inverse does not hold. For, a perfect set of points can very well contain isolated pieces.

We shall say that two sets of pieces *possess the same geometric type of order,* when they can be brought piece by piece into such a one-one correspondence, that to a limiting piece of a fundamental series in one set corresponds a limiting piece of the corresponding fundamental series in the other set. So in general a closed set considered as a set of pieces possesses not the same geometric type of order as when considered as a set of points.

A closed set we shall call *punctual,* when it does not contain a coherent part, in other words when all its pieces are points.

§ 2.

Cantor's fundamental theorem and its extensions.

The fundamental theorem of the theory of sets of points runs as follows:

If we destroy in a closed set an isolated point, in the rest set again an isolated point, and so on transfinitely, this process leads after a denumerable number of steps to an end.

The discoverers of this theorem, CANTOR [1]) and BENDIXSON [2]) proved

[[5]] 1) Mathem. Annalen 23, p. 459—471.

[[6]] 2) Acta Mathematica 2, p. 419—427.

it with the aid of the notion of the *second transfinite cardinal* Ω, which is however not recognised by all mathematicians. LINDELÖF [1]) gave a proof independent of this notion, where, however, the process of destruction itself remaining non-considered, the result is more or less obtained by surprise.

Only for linear sets there have been given proofs of the fundamental theorem, which at the same time follow the process of destruction and are independent of Ω [2]).

The rest set which remains after completion of the process of destruction and which we may call the *Cantor residue*, is after CANTOR [3]) *a perfect set of points*, however of the most general kind, thus *in general not a perfect set of pieces.*

An extension of the fundamental theorem, enunciated by SCHOENFLIES[4]) and proved by me [5]), can be formulated as follows:

If we destroy in a closed set an isolated piece, in the rest set again an isolated piece, and so on transfinitely, this process leads after a denumerable number of steps to an end.

My proof given formerly for this theorem was a generalisation of LINDELÖF'S method, but at the same time I announced a proof which follows the process of destruction, and which I give now here; in it is contained a proof of the fundamental theorem, which in simplicity surpasses by far the existing ones, is independent of Ω, and follows the process of destruction:

By means of Sp_{n-1}'s belonging to an orthogonal system of directions we divide the Sp_n into n-dimensional cubes with edge a, each of these cubes into 2^n cubes with edge $\frac{1}{2}\,a$, each of the latter into 2^n cubes with edge $\frac{1}{4}\,a$, etc.

All cubes constructed in this way form together a denumerable set of cubes K.

Let now μ be the given closed set, then K possesses as a part a likewise denumerable set K_1 consisting of those cubes which contain in their interior or on their boundary points of μ.

---

1) Acta Mathematica 29, p. 183—190. [[7]]

2) SCHOENFLIES, Bericht über die Mengenlehre I, p. 80, 81; Gött. Nachr. 1903, p. 21—31; HARDY, Mess. of Mathematics 33, p. 67—69; YOUNG, Proceedings of the London Math. Soc. (2) 1, p. 230—246. [[8]]

3) l. c. p. 465. [[5]]

4) Mathem. Annalen 59; the proof given there p. 141—145, and Bericht über die Mengenlehre II, p. 131—135 does not hold. [[9]]

5) Mathem. Annalen 68, p. 429. [[10]]

To each destruction of an isolated point or isolated piece in μ now answers a destruction of at least *one* [4]) cube in K_1; but of the latter destructions only a denumerable number is possible, thus also of the former, with which CANTOR's theorem and SCHOENFLIES's theorem are proved both together.

Let us call the rest set, which remains after destruction of all isolable pieces, the *Schoenflies residue*, then on the ground of theorem 1 we can formulate:

THEOREM 2. *A Schoenflies residue is a perfect set of pieces.*

§ 3.

The structure of perfect sets of pieces.

Let S_1 and S_2 be two pieces of a perfect set of pieces μ. Let it be possible to place a finite number of pieces of μ into a row having S_1 as its first element and S_2 as its last element in such a way, that the distance between two consecutive pieces of that row *is smaller than a*. Then we say, *that S_2 belongs to the a-group of S_1*.

If S_2 and S_3 both belong to the a-group of S_1, then S_3 belongs also to the a-group of S_2, so that μ breaks up into a certain number of "a-groups". This number is finite, because the distance of two different a-groups cannot be smaller than a.

If $a_1 < a_2$, and if an a_1-group and an a_2-group of μ are given, then these are either entirely separated or the a_1-group is contained in the a_2-group.

If two pieces S_1 and S_2 of μ are given, then there is a certain maximum value of a, for which S_1 and S_2 lie in different a-groups of μ. That value we shall call the *separating bound of S_1 and S_2 in μ*, and we shall represent it by $\sigma_\mu(S_1, S_2)$.

If fartheron we represent the *distance of S_1 and S_2* by $a(S_1, S_2)$, then $\sigma_\mu(S_1, S_2)$ converges with $a(S_1, S_2)$ to zero, but also inversely $a(S_1, S_2)$ with $\sigma_\mu(S_1, S_2)$. For otherwise convergency of $\sigma_\mu(S_1, S_2)$ to zero would involve the existence of a coherent part of μ, in which two different pieces of μ were contained, which is impossible.

The maximum value of a for which μ breaks up into different a-groups we shall call the *width of dispersion* of μ, and shall represent it by $\delta(\mu)$. This width of dispersion of μ is at the same time the greatest value which $\sigma_\mu(S_1, S_2)$ can reach for two pieces S_1 and S_2 of μ.

---

[4]) Even of an infinite number.

The maximum value of a, for which μ breaks up into *at least n* different a-groups we shall call the *n-partite width of dispersion* of μ, and shall represent it by $\delta_n(\mu)$. Clearly $\delta_n(\mu)$ is $\leq \delta(\mu)$.

For μ exists furthermore a series of increasing positive integers $n_1(\mu)$, $n_2(\mu)$, $n_3(\mu)$, ... in such a way that $\delta_n(\mu)$ for n between $n_{k-1}(\mu)$ and $n_k(\mu)$ is equal to $\delta_{n_k(\mu)}(\mu)$. This quantity $\delta_{n_k(\mu)}(\mu)$ we call the kth *width of dispersion* of μ and as such we represent it by $\delta^{(k)}(\mu)$.

We now assert that it is always possible to break up μ into m_1 perfect sets of pieces $\mu_1, \ldots \mu_{m_1}$ so as to have $\delta(\mu_h) \leq \delta_{m_1}(\mu)$ and $a(\mu_{h_1}, \mu_{h_2}) \geq \delta_{m_1}(\mu)$.

Let namely be $\delta_{m_1}(\mu) = \delta^{(k)}(\mu)$; we can then obtain the required number m_1 by composing each μ_h of a certain number of $\delta^{(k)}(\mu)$-groups belonging to a same $\delta^{(k-1)}(\mu)$-group. We are then also sure of having satisfied the condition $a(\mu_{h_1}, \mu_{h_2}) \geq \delta_{m_1}(\mu)$.

Fartheron we can place the $\delta^{(k)}(\mu)$-groups of a same $\delta^{(k-1)}(\mu)$-group into such a row that the distance between two consecutive ones is equal to $\delta^{(k)}(\mu)$. If we take care that each μ_h consists of a non-interrupted segment of such a row, then the condition $\delta(\mu_h) \leq \delta_{m_1}(\mu)$ is also satisfied.

Let us now break up in the same way each μ_h into m_2 perfect sets of pieces $\mu_{h1}, \ldots \mu_{hm_2}$ in such a way that $\delta(\mu_{hi}) \leq \delta_{m_2}(\mu_h)$ and $a(\mu_{hi_1}, \mu_{hi_2}) \geq \delta_{m_2}(\mu_h)$, and let us continue this process indefinitely.

If then we represent by F_ν an arbitrary row of ν indices, then we shall always find

$$\delta_{m_\nu}(\mu_{F_{\nu-1}}) \leq \delta_{m_1+m_2+\ldots+m_\nu+1-\nu}(\mu) \quad . \quad . \quad . \quad . \quad (A)$$

As μ is a perfect set of pieces, the width of dispersion $\delta(\mu_{F_\nu})$ can converge to zero only for indefinite increase of ν; out of the formula (A) follows, however, that for indefinite increase of ν that convergency to zero always takes place and, indeed, uniformly for all ν^{th} elements of decomposition together.

At the same time the separating bound of every two pieces lying in one and the same ν^{th} element of decomposition converges uniformly to zero; so these elements of decomposition converge themselves uniformly each to a single piece.

If finally a variable pair of pieces of μ is given, then their distance can converge to zero only when the order of the smallest element of decomposition, in which both are contained, increases indefinitely.

The simplest mode in which this process of decomposition can be executed is by taking all m_k's equal to 2. If then we represent the two elements of decomposition of the first order by μ_0 and μ_2, those of the second order by $\mu_{00}, \mu_{02}, \mu_{20}, \mu_{22}$, and so on, then in this way the different pieces of μ are brought into a one-one correspondence with the different fundamental series consisting of figures 0 and 2. And two pieces converge to each other then, and only then, when the commencing segment which is common to their fundamental series, increases indefinitely.

Let us consider on the other hand, in the linear continuum of real numbers between 0 and 1, the perfect punctual set π of those numbers which can be represented in the triadic system by an infinite number of figures 0 and 2. The geometric type of order of π we shall represent by ζ.

Two numbers of π converge to each other then and only then, when the commencing segment which is common to their series of figures, increases indefinitely.

So, if we realize such a *one-one* correspondence between the pieces of μ and the numbers of π, that for each piece of μ the series of indices is equal to the series of figures of the corresponding number of π, then to a limiting piece of a fundamental series of pieces of μ corresponds a limiting number of the corresponding series of numbers in π, so that we can formulate:

[[11]] THEOREM 3. *Each perfect set of pieces possesses the geometric type of order* ζ.

For the case that the set under discussion *is punctual and lies in a plane,* this theorem ensues immediately from the following well-known property:

Through each plane closed punctual set we can bring an arc of simple curve.

Combining SCHOENFLIES's theorem mentioned in § 2 with theorem 3 we can say:

[[12]] THEOREM 4. *Each closed set consists of two sets of pieces; one of them possesses, if it does not vanish, the geometric type of order* ζ, *and the other is denumerable.*

§ 4.

[[13]] *The groups which transform the geometric type of order* ζ *in itself.*

Just as spaces admit of groups of continuous one-one transformations, whose geometric types of order [1]) are again spaces, namely

---

[1]) In this special case formerly called by me "Parametermannigfaltigkeiten"
[[14]] Comp. Mathem. Annalen 67, p. 247.

the finite continuous groups of LIE, the geometric type of order ζ admits of groups of continuous one-one transformations, which possess likewise the geometric type of order ζ.

In order to construct such groups we start from a decomposition according to § 3 of the set μ into m_1 "parts of the first order" $\mu_1, \mu_2, \ldots \mu_{m_1}$, of each of these parts of the first order into m_2 "parts of the second order" $\mu_{h1}, \mu_{h2}, \mu_{h3}, \ldots \mu_{hm_2}$, etc.

The parts of the first order we submit to an arbitrary transitive substitution group of m_1 elements, of which we represent the order by p_1, and which we represent itself by g_1.

After this we submit the parts of the second order to a transitive substitution group g_2 of $m_1 m_2$ elements which possesses the parts of the first order as systems of imprimitivity and g_1 as substitution group of those systems into each other. We can then represent the order of g_2 by $p_1 p_2$.

The simplest way to construct such a group g_2, is to choose it as the direct product of g_1 and a substitution group γ_2, which of the parts of the second order leaves the first index unchanged and transforms the second index according to a single transitive substitution group of m_2 elements.

We then submit the parts of the third order to a transitive substitution group g_3 of $m_1 m_2 m_3$ elements which possesses the parts of the second order as systems of imprimitivity and g_2 as substitution group of those systems into each other. We can represent the order of g_3 by $p_1 p_2 p_3$.

In this way we construct a fundamental series of substitution groups $g_1, g_2, g_3, \ldots$

Let τ_1 be an arbitrary substitution of g_1; τ_2 a substitution of g_2 having on the first index of the parts of the second order the same influence as τ_1; τ_3 a substitution of g_3 having on the first two indices of the parts of the third order the same influence as τ_2; and so on.

The whole of the substitutions τ_n then determines a substitution of the different fundamental series of indices into each other, in other words *a transformation* τ *of the pieces of* μ *into each other.*

This transformation is in the first place a *one-one transformation*; for, two different pieces of μ lie in two different parts of a certain, e.g. of the rth order, and these are transformed by τ into again two different parts of the rth order.

If fartheron $S_1, S_2, S_3, \ldots$ is a fundamental series of pieces, possessing S_ω as its only limiting piece, then, if $\lambda(n)$ is the lowest possible order with the property that S_n and S_ω lie in different parts of that order, $\lambda(n)$ must increase indefinitely with n.

So by the transformation τ the fundamental series passes into a new fundamental series having as its only limiting piece the piece into which S_ω passes by τ.

As a set of pieces μ is thus *continuously* transformed by τ.

Let $\tau'_1, \tau'_2, \tau'_3, \ldots$ be a series of substitutions satisfying the same conditions as the series $\tau_1, \tau_2, \tau_3, \ldots$ If then $\tau_1 \tau'_1 = \tau''_1$; $\tau_2 \tau'_2 = \tau''_2$; etc., then the series $\tau''_1, \tau''_2, \tau''_3, \ldots$ likewise satisfies the same conditions.

If farthermore τ' and τ'' are defined analogously to τ, then $\tau \tau'$ is equal to τ''.

So the transformations satisfying the conditions put for τ form a group, which we shall represent by g.

To investigate the geometric type of order of this group, we decompose in the way indicated in § 3 a perfect set of pieces ϱ into p_1 parts of the first order $\varrho_1, \varrho_2, \ldots, \varrho_{p_1}$; each of these into p_2 parts of the second order $\varrho_{h1}, \varrho_{h2}, \ldots, \varrho_{hp_2}$; and so on.

The p_1 substitutions of g_1 we bring into a one-one correspondence to the parts of the first order of ϱ. Then the $p_1 p_2$ substitutions of g_2 into such a one-one correspondence to the parts of the second order of ϱ, that, if a substitution of g_2 and a substitution of g_1 have the same influence on the first index of the parts of the second order of μ, the part of the second order of ϱ corresponding to the former lies in the part of the first order of ϱ corresponding to the latter. In like manner we bring the $p_1\ p_2\ p_3$ substitutions of g_3 into such a one-one correspondence to the parts of the third order of ϱ, that, if a substitution of g_3 and a substitution of g_2 have the same influence on the first two indices of the parts of the third order of μ, the part of the third order of ϱ corresponding to the former lies in the part of the second order of ϱ corresponding to the latter; and so on.

The parts of ϱ corresponding to a series $\tau_1, \tau_2, \tau_3, \ldots$ then converge to a single piece of ϱ, which we let answer to the transformation τ deduced from the series. Then also inversely to each piece of ϱ answers a transformation τ, and the correspondenceattained in this manner is *a one-one correspondence.*

Farthermore two transformations τ and τ' converge to each other then and only then, when their generating series $\tau_1, \tau_2, \tau_3, \ldots$ and $\tau'_1, \tau'_2, \tau'_3, \ldots$ have an indefinitely increasing commencing segment in common, in other words when the corresponding pieces of ϱ converge to each other. So the correspondence between the transformations τ and the pieces of ϱ is *continuous.*

The transformations τ, in other words the transformations of the group g, have thus been brought into a continuous one-one correspondence to the pieces of ϱ, *so that g possesses the geometric type of order ζ.*

If now we adjoin to each substitution group g_n a finite group g'_n of continuous one-one transformations of μ as a set of pieces in itself, tranforming of the pieces of μ the first n indices according to g_n, but leaving unchanged all their other indices, *then the fundamental series of the groups g'_1, g'_2, g'_3, . . . converges uniformly to the group g.*

The set whose elements are the groups g of the geometric type of order ζ constructable in the indicated manner possesses the cardinal number of the continuum. For, already the set of those series m_1, m_2, m_3, . . . , which consist of prime numbers, possesses this cardinal number, and any two different series of this set give rise to different groups g.

We can sum up the preceding as follows:

THEOREM 5. *The geometric type of order ζ allows of an infinite number of groups consisting of a geometric type of order ζ of continuous one-one transformations and being uniformly approximated by a fundamental series of groups consisting each of a finite number of continuous one-one transformations.*

If in particular we consider those groups g for which each g_n is chosen in the way described at the commencement of this § as the direct product of g_{n-1} and a group γ_n, we can formulate in particular:

THEOREM 6. *The geometric of order ζ allows of an infinite number of groups consisting of a geometric type of order ζ of continuous one-one transformations and being uniformly convergent direct products each of a fundamental series of finite groups of continuous one-one transformations*

§ 5.

The sham-addition in the geometric type of order ζ.

Let us choose the factor groups indicated in theorem 6 as simply as possible, namely g_1 as the group of cyclic displacements corresponding to a certain cyclic arrangement of the first indices, and likewise each γ_n as the group of cyclic displacements corresponding to a certain cyclic arrangement of the nth indices; g is then commutative, and transitive in such a way that a transformation of g is determined uniformly by the position which it gives to one of the elements of μ.

Let us further choose an arbitrary piece of μ as *piece zero*. Let us represent this piece by S_0, and the transformation, which transfers S_0 into S_x and is thereby determined, by "$\mp S_x$". That the

53*

piece S_β is transferred by this transformation into S_γ, we shall express by the formula

$$S_\beta \overset{\frown}{+} S_\alpha = S_\gamma,$$

which operation is associative and commutative.

Let us finally choose, in order to make the resemblance to ordinary ciphering as complete as possible, all m_n's equal to 10, let us take for each system of nth indices the digits 0, 1, 2, 3, 4, 5, 6, 7, 8, 9 in this order, and let us give to the piece zero only indices 0.

The different pieces of μ we can then represent biuniformly by the different infinite decimal fractions lying between 0 and 1, in such a way, however, that *finite* decimal fractions do not appear and that $\cdot 3\dot{0}$ is not equal to $\cdot 2\dot{9}$, whilst each group γ_n consists of the different ways in which one can add the same number to all n^{th} decimals, modulo 10.

Now according to the above we understand by $\cdot 5473\ldots \overset{\frown}{+} \cdot 9566\ldots$ the decimal fraction, into which $\cdot 5473\ldots$ is transferred by the transformation which transfers $\cdot\dot{0}$ into $\cdot 9566\ldots$, or, what comes to the same, the decimal fraction, into which $\cdot 9566\ldots$ is transferred by the transformation which transfers $\cdot\dot{0}$ into $\cdot 5473\ldots$

We shall call the operation furnishing this result, on the ground of its associativity and commutativity, the "*sham-addition*" of $\cdot 9566\ldots$ to $\cdot 5473\ldots$; it takes place just as ordinary addition, with this difference that in each decimal position the surplus beyond 10 is neglected, thus that different decimal positions do not influence each other. So we have:

$$\cdot 5473\ldots \overset{\frown}{+} \cdot 9566\ldots = \cdot 4939\ldots$$

Let us understand analogously by $\cdot 5473\ldots \overset{\frown}{-} \cdot 9566\ldots$ the decimal fraction, into which $\cdot 5473\ldots$ is transferred by the transformation which transfers $\cdot 9566\ldots$ into $\cdot\dot{0}$, and let us call the operation furnishing this decimal fraction the "*sham-subtraction*" of $\cdot 9566\ldots$ from $\cdot 5473\ldots$; then this sham-subtraction is performed in the same way as ordinary subtraction with this difference, that "borrowing" does not take place at the cost of the preceding decimal positions, so that here again different decimal positions do not influence each other. So we have:

$$\cdot 5473\ldots \overset{\frown}{-} \cdot 9566\ldots = \cdot 6917\ldots$$

By operating only with a finite number, great enough, of consecutive figures directly behind the decimal sign, sham-addition and sham-subtraction furnish in the type of order ζ a result agreeing with the exact one up to any desired degree of accuracy. In this too they behave like ordinary addition and subtraction of real numbers.

NOTES

⟦1⟧ This is the first of a series of Cantor–Schoenflies style investigations in topology, published in various journals. See 1910 C, 1910 J, 1910 E, 1911 B2, 1913 B2, 1915 A2, 1917 B2.

⟦2⟧ '1909' is a misprint. It should be: 1910.

⟦3⟧ 'piece': the present term is component.

⟦4⟧ 'Closed' seems to include here 'bounded' as is often the case in that period (see also 1911 E ⟦11⟧ and 1913 A ⟦8⟧.) Sometimes it is not clear which use of 'closed' is meant. 'Coherent': a term due to O. Veblen 1905. In the present context it means the same as the modern 'connected'. (See also 1911 E ⟦12⟧.) In the Dutch version Brouwer uses the term 'samenhangend'.

⟦5⟧ G. Cantor 1884.

⟦6⟧ I. Bendixson 1883.

⟦7⟧ E. Lindelöf 1905.

⟦8⟧ A. Schoenflies 1900, 1903 A; G. H. Hardy 1903; W. H. Young 1904 A.

⟦9⟧ A. Schoenflies 1904, 1908 A.

⟦10⟧ Brouwer 1910 C.

⟦11⟧ See S. Mazurkiewicz 1920, p. 62.

⟦12⟧ See C. Kuratowski 1952 II, p. 122.

⟦13⟧ The first example of non-Lie topological groups. See D. van Dantzig, 1936. Cp. the *p*-adic numbers of K. Hensel 1908.

⟦14⟧ Brouwer 1909 C.

1910C

Zur Analysis Situs.

Von

[1] L. E. J. Brouwer in Amsterdam.

Mit 2 Tafeln

I. Einleitung.

[2] In diesen Annalen, Bd. 58, 59 und 62 hat Herr Schoenflies wichtige Beiträge zur Klärung der Analysis Situs der Ebene geliefert, welche er später im vierten und fünften Kapitel des zweiten Teiles seines mengentheoretischen Berichtes zusammengefaßt hat. Unter seinen Resultaten sind besonders die Präzisierung und Umkehrung des Jordanschen Kurvensatzes als äußerst wertvoll zu bezeichnen. Sie eröffnen die Möglichkeit zu allgemeinerer und strengerer Begründung von verschiedenen Gebieten der Mathematik.

Wenn nun aber die Schoenfliessche Theorie der einfachen geschlossenen Kurven, sowie der von solchen begrenzten Gebiete einwandfrei ist, von seinen Ergebnissen, insoweit sie *geschlossene Kurven und Gebiete allgemeiner Art* betreffen, läßt sich dasselbe nicht behaupten. Hier sind mehrere seiner Resultate falsch, andere zwar richtig, aber ungenügend begründet. Ich glaube dies am deutlichsten zu beleuchten, indem ich zunächst einige Punktmengen konstruiere, welche Eigenschaften besitzen, die nach der Schoenfliesschen Theorie unmöglich sein würden, und nachher diese Theorie systematisch kritisiere; dabei schließe ich mich hauptsächlich dem Berichte an, wo der Stoff natürlicher geordnet und vollständiger dargestellt erscheint, und verweise auf die bezüglichen Stellen der Annalen nur gelegentlich.*)

*) Ich hebe ausdrücklich hervor, daß diese Arbeit den hohen Wert der Schoenfliesschen Untersuchungen keineswegs zu beeinträchtigen sucht. Nur ihre große Tragweite hat mich zu dieser Kritik veranlaßt, welche übrigens den umfangreichsten Teil, nämlich die Theorie der einfachen Kurven, nicht wesentlich trifft.

II. Einfach zusammenhängende Gebiete ohne geschlossene äußere Randkurve.

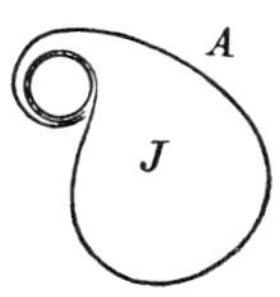

Fig. 1.

Ein Beispiel eines solchen Gebietes liefert das Gebiet J in Fig. 1. Es wird erhalten, indem man einen einfachen Kurvenbogen nach beiden Seiten fortsetzt und beide Fortsetzungen, ohne daß sie einander treffen, spiralartig um einen und denselben Kreis herumwindet.

III. Geschlossene Kurven, welche sich in zwei uneigentliche Kurvenbogen*) zerlegen lassen.

Für die in Fig. 2 (Tafel I) schematisch gezeichnete geschlossene Kurve wird das äußere Gebiet gebildet von demjenigen Teil der Ebene, welcher außerhalb des Hauptrechteckes liegt, sowie von den schwarz schraffierten Teilgebieten des Hauptrechteckes; das innere Gebiet ist das rot schraffierte Teilgebiet des Hauptrechteckes. Diese schraffierten Gebiete werden in folgender Weise erhalten.

Das erste schwarze Gebiet ist ein in der Mitte der Grundlinie des Hauptrechteckes in der Weise aufgerichtetes Teilrechteck, daß der zweimal umgebogene weiße Streifen, welcher übrig bleibt, in seinen drei Teilen dieselbe Breite besitzt.

Das Verhältnis seiner Grundlinie zu der des Hauptrechteckes sei $\frac{1}{2\alpha+1}$; dann hat der weiße Streifen eine Breite $\frac{\alpha}{2\alpha+1}a$, wenn a die Länge der Grundlinie des Hauptrechteckes ist.

Jetzt zeichnen wir den zwischen den Querschnitten $P_1 P_1'$ und $Q_1 Q_1'$ enthaltenen Teil des roten Gebietes. Er besteht aus einem Streifen von der Breite $\frac{1}{2\alpha+1} \cdot \frac{\alpha}{2\alpha+1}a$, und seine Grenze verläuft überall im Ab-

---

*) Für „Kurvenbogen" möchte ich einen weiteren Begriff, als den von Schoenflies im Bericht II S. 128 eingeführten, vorschlagen: ich möchte nämlich einen Kurvenbogen definieren durch zwei *Schnitte* in einem zyklischen, abzählbaren, überall dichten Ordnungstypus von für das innere (bezw. äußere) Gebiet erreichbaren Punkten. Der Schoenfliessche Begriff läßt sodann unter diesen Schnitten nur durch erreichbare Punkte gebildete zu. Eine völlig verschiedene Definition gibt Schoenflies in den Gött. Nachr. 1904 S. 516 und in den Math. Annalen Bd. 59, S. 147. Diese scheint mir aber durchaus zu verwerfen, weil sich nach ihr jede nirgends dichte perfekte zusammenhängende Menge mit einem einzigen Komplementärgebiete als Kurvenbogen auffassen läßt. ⟦3⟧ ⟦4⟧

Unter einem *uneigentlichen Kurvenbogen* möchte ich einen solchen durch zwei Schnitte definierten Kurvenbogen verstehen, welche die ganze geschlossene Kurve enthält, und im entgegengesetzten Falle sprechen von einem *eigentlichen Kurvenbogen.* ⟦5⟧

Q'1
Q1
P1
P'1
Q2
Q'2
P2
P'2

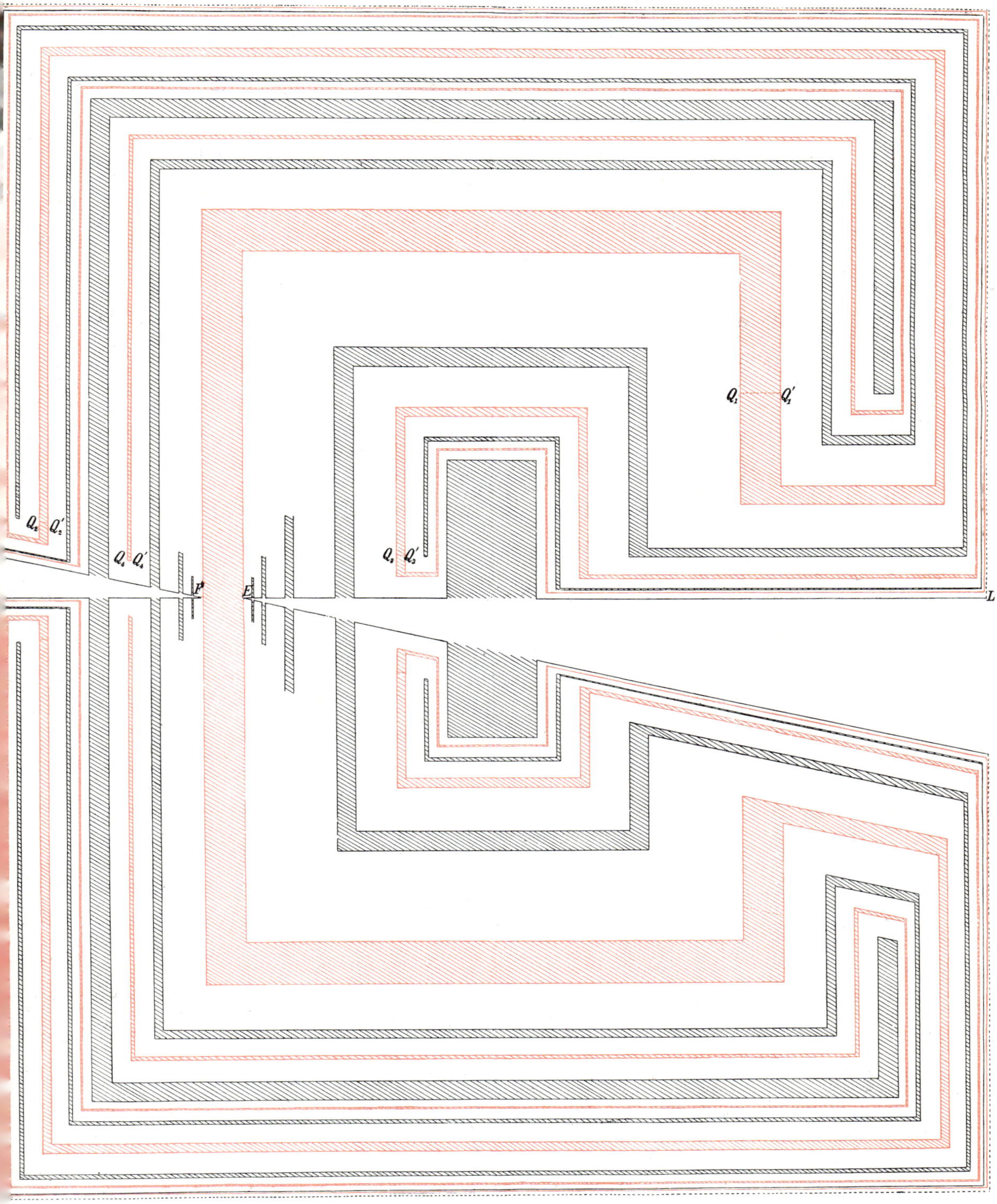

Fig. 3.

stande $\left(\frac{\alpha}{2\alpha+1}\right)^2 a$ parallel der Grenze des ebengenannten weißen Streifens, in dem er enthalten ist.

Das zweite schwarze Gebiet wird jetzt in der Weise um den bisher gezeichneten Teil des roten Gebietes außen herumgelegt, daß links auf der Grundlinie des Hauptrechteckes angefangen, und rechts auf der Höhe von $Q_1\, Q_1'$ aufgehört wird. Dieser Streifen hat wieder eine Breite gleich $\frac{1}{2\alpha+1}$ von der des weißen Streifens, in die Mitte dessen er hineingelegt wird.

Die beiden jetzt existierenden schwarzen Gebiete lassen im Hauptrechteck einen sechs Mal umgebogenen Streifen frei, und das rote Gebiet wird jetzt in der Weise fortgesetzt, daß es auch die bisher von ihm freien Teile durchzieht, und zwar diese in ihrer Mitte und unter Hinwegnahme des $\frac{1}{2\alpha+1}$ten Teiles ihrer Breite. Das rote Gebiet wird also fortgesetzt vom Querschnitt $Q_1\, Q_1'$ bis zum Querschnitt $Q_2\, Q_2'$, welcher denselben Abstand von der Grundlinie des Hauptrechteckes besitzt, wie die vertikale Grenze in Q_2 vom zweiten gestreift schraffierten Gebiet.

In dieser Weise wird fortgegangen. Abwechselnd wird ein neues schwarzes Gebiet um den schon erhaltenen Teil des roten Gebietes außen herumgelegt, wobei ganz links auf der Grundlinie des Hauptrechteckes angefangen, und auf derselben Höhe mit dem bisher fertigen Teile des roten Gebietes aufgehört wird, und sodann das rote Gebiet in der Weise fortgesetzt, daß es in alle Teilkanäle des von den schwarzen Flächen freigelassenen Streifens eindringt. Diese Verlängerungen des roten Gebietes geschehen abwechselnd auf beiden Seiten: nach einer Fortsetzung von $Q_n\, Q_n'$ bis $Q_{n+1}\, Q_{n+1}'$ folgt eine Fortsetzung von $P_n\, P_n'$ bis $P_{n+1}\, P_{n+1}'$. Von jedem weißen Streifen, welcher neu durchzogen wird, es sei von „rot" oder von „schwarz", wird der $\frac{1}{2\alpha+1}$te Teil der Breite in Anspruch genommen.

Jedem Punkte der Grenze eines schwarzen Gebietes kommt nach genügender Fortsetzung des Verfahrens das rote Gebiet beliebig nahe, und jedem Punkte der Grenze des roten Gebietes kommen schließlich die schwarzen Gebiete beliebig nahe, während durch die volle Grenze K des roten Gebietes „rot" und „schwarz" völlig voneinander getrennt werden.

Die Komplementärmenge von K enthält nur zwei Gebiete, nämlich das rote, und das aus den schwarzen und dem äußeren Gebiete des Hauptrechteckes zusammengesetzte Gebiet; K ist gemeinsame Grenze dieser beiden Gebiete, also eine geschlossene Kurve.

Wählt man nun aber Q_1 und Q_1' als Schnitte der für das innere Gebiet erreichbaren Punkte, so werden dadurch zwei uneigentliche Kurvenbogen bestimmt.

IV. Eigentliche Kurvenbogen, welche sich in zwei mit dem ganzen identische Kurvenbogen zerlegen lassen.

Das im vorigen Paragraphen geschilderte Verhalten kann auch bei eigentlichen Kurvenbogen auftreten. Um dies zu erläutern, betrachten wir die obere Hälfte von Fig. 3 (Tafel II), deren Erklärung wir, da in mancher Hinsicht volle Analogie mit Fig. 2 vorliegt, kurz fassen können.

Wir gehen wieder aus von einem Hauptrechteck*) und konstruieren ein erstes schwarzes Gebiet, wie in Fig. 2.

Der erste Teil des roten Gebietes unterscheidet sich nun aber von dem der Fig. 2 dadurch, daß er links auf der Grundlinie des Hauptrechteckes anfängt; rechts endet er wie in Fig. 2 bei $Q_1\, Q_1'$.

Das zweite schwarze Gebiet ist wie in Fig. 2; ebenso der zweite Teil des roten Gebietes, welcher zwischen $Q_1\, Q_1'$ und $Q_2\, Q_2'$ enthalten ist.

Das dritte schwarze Gebiet wird zwar wieder an dem bisher erhaltenen Teil des roten Gebietes entlang gelegt, aber es fängt jetzt auf der Grundlinie zwischen dem roten Gebiete und dem ersten schwarzen Gebiete an, was ermöglicht, daß nachher die Fortsetzungen des roten Gebietes in einer einzigen Reihe aneinander schließen.

Im allgemeinen wird für das n^{te} schwarze Gebiet auf der Grundlinie zwischen dem roten und dem $(n-2)^{\text{ten}}$ schwarzen Gebiete angefangen, und wird es in derselben Weise wie in Fig. 2 an dem schon fertigen Teile des roten Gebietes entlang gelegt, während das rote Gebiet auch hier stets in der Weise fortgesetzt wird, daß es jedesmal in alle Teilkanäle des vorliegenden von „schwarz" freien Streifens eindringt.

Die von den Grenzen aller roten und schwarzen Gebiete**) nebst ihren Grenzpunkten gebildete Menge K_1 läßt sich durch den Teil der Grundlinie, über dem das rote Gebiet anfängt, zu einer geschlossenen Kurve ergänzen; sie läßt sich also als ein eigentlicher Kurvenbogen jener geschlossenen Kurve betrachten. Als solcher kann sie aber sowohl aus dem inneren wie aus dem äußeren Gebiete in zwei Kurvenbogen zerlegt werden, welche ihr selbst gleich sind.

*) Die Verzerrung des Rechteckes auf der linken Seite ist mit Rücksicht auf die spätere Zusammensetzung mit der unteren Hälfte vorgenommen; sie soll hier außer Acht bleiben.

**) In der Grundlinie begrenzen wir diese Gebiete nicht; dadurch hängen sie mit dem äußeren Gebiete des Hauptrechteckes zusammen.

V. Geschlossene Kurven, welche sich nicht in zwei eigentliche Kurvenbogen zerlegen lassen.*)

Als Beispiel einer solchen kann wieder Fig. 2 dienen. Ich hebe indessen hervor, daß die Eigenschaft dieser Figur, welche in III beleuchtet wurde, nicht notwendig die jetzt in Betracht gezogene nach sich zieht. Es gibt nämlich auch geschlossene Kurven, welche sich *sowohl in zwei eigentliche, wie in zwei uneigentliche Kurvenbogen zerlegen lassen.*

Dies zeigt die volle Fig. 3, welche keiner näheren Erklärung bedarf, da die untere Hälfte die gleiche Struktur hat, wie die obere. Die volle Grenze des roten Gebietes ist eine geschlossene Kurve. Trennen wir die aus „rot" erreichbaren Punkte in diejenigen, welche auf der rechten, und diejenigen, welche auf der linken Grenzlinie des roten Streifens liegen, so zerlegen wir die Kurve in zwei uneigentliche Kurvenbogen. Trennen wir ihre erreichbaren Punkte aber in die in der oberen und die in der unteren Hälfte der Figur enthaltenen, so zerlegen wir sie in zwei eigentliche Kurvenbogen.

Andererseits gibt es auch geschlossene Kurven, welche sich *weder in zwei eigentliche, noch in zwei uneigentliche Kurvenbogen zerlegen lassen.***)

Um eine solche zu konstruieren, reduziert man in der oberen Hälfte der Fig. 3 die Breite des roten Streifens auf Null (so daß beim Aufbaue der Figur nur ein in einem Punkte H der Grundlinie anfangender roter Streckenzug immer wieder verlängert wird), während die schwarzen Gebiete in derselben Weise, wie zuvor, hineingelegt werden. Sodann wird das Rechteck in der Weise gebogen, daß der rechte Endpunkt der Grundlinie mit H zusammenfällt, sonst aber die Figur eineindeutiges stetiges Bild ihrer ursprünglichen Lage bleibt. Dabei kommt eine Teilung der Ebene in zwei Gebiete zustande, welche getrennt werden durch eine geschlossene Kurve, welche die geforderte Eigenschaft besitzt.

*) Auch gibt es Kurvenbogen, welche sich nicht in zwei Teilkurvenbogen zerlegen lassen. Dies zeigt wieder die obere Hälfte der Fig. 3. Diese Kurven bezw. Kurvenbogen geben zugleich Beispiele ab von solchen zusammenhängenden Mengen, welche nach Hinwegnahme einer willkürlichen zusammenhängenden Teilmenge stets eine Restmenge übrig lassen, welche dieselbe Ableitung besitzt, wie die ursprüngliche Menge; diese läßt sich also nicht durch eine endliche Zahl von zusammenhängenden perfekten Teilmengen erschöpfen.

**) Eine kurze Überlegung zeigt weiter, daß in dieser Hinsicht gerade fünf Arten von geschlossenen Kurven existieren. Für die erste Art ist von den durch zwei Schnitte bestimmten Kurvenbogen entweder der eine eigentlich und der andere uneigentlich, oder beide eigentlich, oder beide uneigentlich; für die zweite entweder der eine eigentlich und der andere uneigentlich, oder beide uneigentlich; für die dritte stets der eine eigentlich und der andere uneigentlich; für die vierte stets beide

VI. Nirgends dichte zusammenhängende Mengen, welche die Ebene in mehr als zwei Gebiete zerlegen, deren gemeinschaftliche Grenze sie sind.

Seien in der oberen Hälfte von Fig. 3 K und L die Endpunkte der Grundlinie des Hauptrechteckes; F und E die Endpunkte jenes Segmentes der Grundlinie, über dem das rote Gebiet anfängt, so kann man die Figur so biegen, daß F mit K und E mit L zusammenfällt, und daß sie übrigens ein-eindeutiges Bild ihrer ursprünglichen Lage bleibt Die Ebene ist dann in drei Teilgebiete zerlegt, welche ihre Grenze gemeinsam haben.

Durch passende Abänderungen des Verfahrens kann man analoge Zerlegungen in beliebig viele, und sogar in unendlich viele Teilgebiete mit gemeinsamer Grenze herstellen.

VII. Bemerkungen über das vierte Kapitel des Schoenfliesschen Berichtes. [[3]]

Zu § 11. Im Beweise des Satzes XIII findet sich S. 123 die unrichtige Behauptung, daß $\mathfrak{J} = \mathfrak{G}' + \mathfrak{H} + \mathfrak{T}_g + \mathfrak{T}_{gh}$ ein zusammenhängendes Gebiet ist. Daß diese Behauptung sich ebensowenig wie der Satz XIII selbst aufrecht erhalten läßt, zeigt die oben in II angeführte Fig. 1. Mit Satz XIII fallen zugleich die Sätze XIV und XV, wie auch die Betrachtungen über Gebietsteilung in Ann. Bd. 59, besonders S. 154, 155 [[6]] u. 159. Der Beweis des Satzes XIV versagt sogar auch dann noch, wenn die äußere Randkurve existiert. Denn es ist kein Grund, weshalb $\mathfrak{G}'$ nicht in dem zwischen $\mathfrak{P}$ und $\mathfrak{Q}$ enthaltenen Ringgebiet liegen könnte. Dies läßt sich z. B. verwirklichen für ein durch den Querschnitt $Q_1 Q_1'$ bestimmtes Teilgebiet des roten Gebietes der Fig. 2. Dieses Gebiet ist denn auch, *obwohl eine äußere Randkurve existiert, durch seine Grenze nicht eindeutig bestimmt.*

Zu § 12. S. 127 wird behauptet, daß man die in den erreichbaren Punkten mündenden Wege immer so wählen kann, daß sie jedes approximierende Polygon nur einmal kreuzen. Der Beweis wählt zunächst einen beliebigen Weg und beabsichtigt, daraus durch sukzessive Abänderungen den geforderten Weg herzustellen. Jede Abänderung für sich ist zwar einwandfrei, aber sowohl hier, wie in Math. Ann. Bd. 62, S. 294 kommt es [[7]] darauf an, daß die Größe dieser Abänderung für hinreichend großes ν beliebig

eigentlich; für die fünfte entweder der eine eigentlich und der andere uneigentlich, oder beide eigentlich. Nach der Schoenfliesschen Theorie würden nur die beiden letzten Arten möglich sein.

klein werde; dies würde jedenfalls erhebliche weitere Ausführungen erfordern, wird aber weder erwähnt, noch bewiesen. Ich erlaube mir nun hier eine Abänderungsmethode anzugeben, welche diese notwendige Eigenschaft besitzt.

Sei l der beliebig gezeichnete aus m zu t führende Weg. Sei $\mathfrak{P}_\nu$ ein approximierendes Polygon, seien ${}_1L_\nu$ und ${}_2L_\nu$ zwei seiner Kreuzungspunkte mit l, $\mathfrak{p}_\nu$ ein zwischen ${}_1L_\nu$ und ${}_2L_\nu$ enthaltener Streckenzug von $\mathfrak{P}_\nu$, und l_ν der zwischen ${}_1L_\nu$ und ${}_2L_\nu$ enthaltene Streckenzug von l. Dann besteht $\mathfrak{p}_\nu$, insoweit er nicht in l_ν fällt, aus zwischen je zwei Punkten A_n und A_{n+1} von l_ν verlaufenden und l_ν sonst nicht treffenden Streckenzügen ${}_n\mathfrak{p}_\nu$. Jeder Streckenzug ${}_n\mathfrak{p}_\nu$ weist sodann dem zwischen A_n und A_{n+1} enthaltenen Teilstreckenzug ${}_nl_\nu$ von l_ν entweder die Innenseite oder die Außenseite von $\mathfrak{P}_\nu$ zu. Im zweiten Falle kann er sich nach § 3 nicht weiter als $\frac{3}{2}\varepsilon_\nu$ von ${}_nl_\nu$ entfernen. Im ersten Falle aber grenzt an seine Innenseite ein von ${}_nl_\nu$ und $\mathfrak{p}_\nu$ bestimmtes Gebiet, dessen Grenze, insoweit sie nicht zu ${}_n\mathfrak{p}_\nu$ gehört, sich nicht weiter als $\frac{3}{2}\varepsilon_\nu$ von ${}_nl_\nu$ entfernen kann. Wir können also zwischen ${}_1L_\nu$ und ${}_2L_\nu$, dem Verlaufe von $\mathfrak{p}_\nu$ folgend, einen Streckenzug λ_ν zeichnen, welcher sich nicht weiter als $2\varepsilon_\nu$ von l_ν entfernt, überall in beliebiger Nähe entweder von l_ν oder von $\mathfrak{p}_\nu$ ist, und ganz innerhalb $\mathfrak{P}_\nu$ verläuft. Wir wählen nun auf l stets für ${}_1L_\nu$ den ersten auf ${}_2L_{\nu-1}$ folgenden Kreuzungspunkt mit $\mathfrak{P}_\nu$, und für ${}_2L_\nu$ den letzten Kreuzungspunkt mit $\mathfrak{P}_\nu$. Sodann ersetzen wir jedes l_ν durch das bezügliche λ_ν, und erhalten einen wirklichen *Weg*, welcher jedes $\mathfrak{P}_\nu$ nur in dem bezüglichen Punkte ${}_2L_\nu$ kreuzt. Außerdem berührt er $\mathfrak{P}_\nu$ noch in dem Punkte ${}_1L_\nu$, aber mit ν zu Null konvergierende Abänderungen heben auch diese Berührungspunkte auf, womit der geforderte Weg fertig ist.

Durch diese Methode ist im § 13 des Kap. V und im § 3 des Kap. VI (vgl.
[[7]] dazu die §§ 11 u. 13 in Annalen 62) auch sogleich ersichtlich, daß man aus m nach allen Punkten c_i Wege legen kann, welche alle approximierenden Polygone nur einmal, und einander nicht kreuzen, während der nach c_i führende Weg alle Punkte p_i, p_i', p_i'' usw. enthält, und sich zwischen $\mathfrak{P}_{\nu-1}$ und $\mathfrak{P}_\nu$ nicht weiter als $2\varepsilon_\nu$ von der Strecke $p_i^{(\nu-1)}\,p_i^{(\nu)}$ zu entfernen braucht.

Seite 128 wird aus dem falschen Satze XIV das unrichtige Resultat erhalten, daß die Kurvenbogen, in welche eine geschlossene Kurve durch zwei erreichbare Punkte zerlegt wird, nicht beide mit der Kurve selbst identisch sein können. Daß dies sehr gut möglich ist, haben wir oben in III gezeigt.

Zu § 13. Der Beweis des Satzes XVII wird wieder auf § 11 gestützt. Seine Unrichtigkeit läßt sich also erwarten, und leuchtet aus fast allen oben in II—VI angeführten Beispielen ein.

Zu § 14. Der Satz, daß das Abspaltungsverfahren nach einer abzählbaren Reihe von Schritten zu Ende führt, ist richtig, der Beweis aber läßt sich nicht aufrecht erhalten. Die S. 134 bewiesene Möglichkeit, nach einer Fundamentalreihe von vorangehenden Bestandteilen S_ν ein S_ω passend zu wählen, läßt sich nämlich schon nicht erweitern zu einer Möglichkeit, nach einem Ordnungstypus ω^2 von vorangehenden Bestandteilen S ein S_{ω^2} passend zu wählen. Es könnte ja hier der Fall auftreten, daß jeder gewählte Teilordnungstypus ω^2 *alle* isolierbaren Bestandteile von $\mathfrak{S}_{\omega^2}$ enthielte. Dann würde mit der wohlgeordneten Menge der S_α keine wohlgeordnete Menge von außerhalb voneinander liegenden Gebieten $\mathfrak{G}_\alpha$ parallel laufen, und der Beweis also versagen.

Ein zwingender Beweis kann wie folgt geführt werden: Die isolierbaren Bestandteile bilden in der Rangordnung, in der sie abgespalten werden, eine wohlgeordnete Menge, und jeder isolierbare Bestandteil S_α, dessen Index α zur ersten oder zweiten Zahlenklasse gehört, hat, wenn er abgespalten wird, einen endlichen Abstand δ_α von der dann noch übrigen Restmenge; er läßt sich also mit einem endlichen Gebiete $\mathfrak{G}_\alpha$ umgeben, außerhalb dessen die Restmenge liegt, in welches also nur eine abzählbare Menge von Bestandteilen S_α eindringt. Wir fassen nun diejenigen Bestandteile ${}_1S_\alpha$ ins Auge, für welche ${}_1\delta_\alpha > \varepsilon_1$. Es läßt sich sodann unter den Gebieten ${}_1\mathfrak{G}_\alpha$ eine *endliche* Teilmenge derart auswählen, daß jeder ${}_1S_\alpha$ in wenigstens *eines* dieser Gebiete eindringt. Da nun aber in jedes Gebiet dieser endlichen Menge nur eine abzählbare Menge von Bestandteilen ${}_1S_\alpha$ durchdringen kann, so ist die Menge der ${}_1S_\alpha$ abzählbar. Ebenso zeigt man, daß die Menge der ${}_2S_\alpha$, für welche $\varepsilon_1 > {}_2\delta_\alpha > \varepsilon_2$, abzählbar ist. Läßt man nun eine Reihe abnehmender Zahlen $\varepsilon_1, \varepsilon_2, \varepsilon_3, \cdots$ gegen Null konvergieren, so findet sich die Menge der S_α in eine Fundamentalreihe von abzählbaren Mengen zerlegt.*)

Zu § 15. Die hier gegebenen Andeutungen unterliegen natürlich einer analogen Kritik, wie die Paragraphen, deren Ausdehnung auf den Raum sie beabsichtigen.

VIII. Bemerkungen über das fünfte Kapitel des Schoenfliesschen Berichtes. [[3]]

Zu § 3. Der Gedankengang dieses Paragraphen enthält mehrere Fehlschlüsse; er läßt sich aber durch passende Modifizierung und Ergänzung umformen zu einem Beweise des folgenden Satzes:

*) Diese Methode entspricht dem Lindelöfschen Beweisverfahren für das Cantorsche Haupttheorem. Der Beweis kann aber auch so geführt werden, daß er dem Abspaltungsprozesse selber folgt. Darauf gehe ich jedoch an dieser Stelle nicht weiter ein.

„Wenn eine geschlossene Kurve C sich in zwei eigentliche Kurvenbogen C_1 und C_2 zerlegen läßt, welche miteinander zusammenhängen in zwei getrennten Mengen c_1 und c_2*), so bestimmt ihr eineindeutiges stetiges Bild C' mehr als ein Gebiet."

Approximieren wir nämlich C_1' und C_2' durch äußere Randpolygone ${}_{\varepsilon_1}P_1$ und ${}_{\varepsilon_2}P_2$. Diese können sich nur in mit den ε gegen Null konvergierender Nähe von c_1' und c_2' durchkreuzen. Wählen wir einen solchen Kreuzungspunkt in der Nähe von c_1' und einen zweiten in der Nähe von c_2', so werden diese beiden Punkte auf ${}_{\varepsilon_1}P_1$ verbunden durch zwei Streckenzüge, auf welchen beiden je ein Punkt gewählt werden kann, welcher außerhalb von ${}_{\varepsilon_2}P_2$ liegt. Im Polygone ${}_{\varepsilon_1}P_1$ sind also *wenigstens zwei* Streckenzüge enthalten, welche auf ${}_{\varepsilon_2}P_2$ in mit den ε gegen Null konvergierender Nähe von c_1' anfangen, ganz außerhalb ${}_{\varepsilon_2}P_2$ verlaufen, und wieder auf ${}_{\varepsilon_2}P_2$ in mit den ε gegen Null konvergierender Nähe von c_2' enden. Unter diesen gibt es wieder *wenigstens zwei*, deren jeder ein für ${}_{\varepsilon_1}P_1$ und ${}_{\varepsilon_2}P_2$ gemeinsames äußeres Gebiet bestimmt, welches begrenzt wird vom genannten Streckenzuge von ${}_{\varepsilon_1}P_1$, einem Streckenzuge von ${}_{\varepsilon_2}P_2$ mit der reziproken Eigenschaft (diese beiden Streckenzüge nennen wir Hauptstreckenzüge), und weiter nur solchen Streckenzügen, welche ihren Anfangspunkt und Endpunkt beide in der Nähe von c_1' oder beide in der Nähe von c_2' haben. Diese so bestimmten äußeren Gebiete nennen wir die zu ${}_{\varepsilon_1}P_1$ und ${}_{\varepsilon_2}P_2$ gehörigen *Hauptgebiete*. Lassen wir nun zunächst ${}_{\varepsilon_1}P_1$ ungeändert, ersetzen aber ${}_{\varepsilon_2}P_2$ durch ein nächstfolgendes approximierendes Polygon ${}_{\varepsilon_2'}P_2$, und betrachten ein zu ${}_{\varepsilon_1}P_1$ und ${}_{\varepsilon_2'}P_2$ gehöriges Hauptgebiet H. Sein Hauptstreckenzug auf ${}_{\varepsilon_1}P_1$ kann nur für gewisse Teile in der Nähe von c_1' und c_2' innerhalb ${}_{\varepsilon_2}P_2$ fallen. Es führt also innerhalb H ein Streckenzug von ${}_{\varepsilon_2}P_2$ aus der Nähe von c_1' in die von c_2'; H enthält also als Teilgebiet ein zu ${}_{\varepsilon_1}P_1$ und ${}_{\varepsilon_2}P_2$ gehöriges Hauptgebiet. Wir fahren nun fort, indem wir jetzt ${}_{\varepsilon_2'}P_2$ ungeändert lassen, ${}_{\varepsilon_1}P_1$ ersetzen durch ein nächstfolgendes approximierendes Polygon ${}_{\varepsilon_1'}P_1$ und die zu ${}_{\varepsilon_1'}P_1$ und ${}_{\varepsilon_2'}P_2$ gehörigen Hauptgebiete betrachten. Jedes von ihnen enthält als Teilgebiet ein zu ${}_{\varepsilon_1}P_1$ und ${}_{\varepsilon_2'}P_2$ gehöriges, also auch ein zu ${}_{\varepsilon_1}P_1$ und ${}_{\varepsilon_2}P_2$ gehöriges Hauptgebiet. Wenn wir nun weiter immer wieder abwechselnd ${}_{\varepsilon_2^{(\nu)}}P_2$ durch ${}_{\varepsilon_2^{(\nu+1)}}P_2$ und ${}_{\varepsilon_1^{(\nu)}}P_1$ durch ${}_{\varepsilon_1^{(\nu+1)}}P_1$ ersetzen, so müssen immer wieder wenigstens zwei Hauptgebiete existieren. Unter den zu ${}_{\varepsilon_1}P_1$ und ${}_{\varepsilon_2}P_2$ gehörigen Hauptgebieten gibt es also wenigstens zwei, welche im weiteren zwar fortwährend wachsen, aber stets voneinander getrennt bleiben.

*) Auf diese Voraussetzungen, von denen man übrigens zeigen kann, daß die zweite eine Folge der ersten ist, stützt sich im Grunde auch Herr Schoenflies, obgleich er nur die Voraussetzung, daß C_1 und C_2 nicht identisch sind, ausdrücklich erwähnt.

Die Grenzgebilde der approximierenden Polygone, welche in C_1' und C_2', also in C' enthalten sind, bestimmen also mindestens zwei getrennte Gebiete, womit der Satz bewiesen ist.

Dieser Satz sagt nur eine Eigenschaft geschlossener Kurven besonderer Art aus; daß seine Voraussetzungen für die allgemeine geschlossene Kurve nicht notwendig erfüllt sind, haben wir in V erkannt. Aber sogar für diese besondere Kurvenart läßt sich aus ihm nicht auf die Invarianz schließen. Der Schluß benutzt nämlich die äußere Randkurve der Bildmenge, welche nach II nicht zu existieren braucht.

Zu § 4. Das Resultat dieses Paragraphen kann der Hauptsache nach an dieser Stelle gefolgert werden, die daselbst gegebene Herleitung ist aber nicht einwandfrei, zunächst weil § 3 benutzt wird, was wir als unerlaubt erkannt haben, und sodann, weil S. 161 oben ohne Grund stillschweigend vorausgesetzt wird, daß, wenn P' sich unbeschränkt der Kurve $\mathfrak{C}'$ nähert, ein willkürlicher innerer Punkt von $\mathfrak{C}'$ schließlich durch P' von $\mathfrak{C}'$ getrennt wird.

Dem ersten Mangel kann abgeholfen werden, indem wir statt geschlossener Kurven Polygone P betrachten, und uns auf den Jordanschen Kurvensatz, welcher unabhängig bewiesen werden kann, stützen.

Der zweite Mangel läßt sich beseitigen, indem wir den Beweis wie folgt führen:

Wir konstruieren ein Polygon Q, welches P in zwei Punkten kreuzt. So werden vier Streckenzüge P_1, P_2, Q_1 und Q_2 bestimmt, und ebenso vier Gebiete G_{11}, G_{12}, G_{21}, G_{22}. Nehmen wir nun an, daß in dem Bilde Q_1' und Q_2' beide innerhalb P' liegen. $\mathfrak{J}(P')$ wird sodann durch Q_1' in zwei Gebiete zerlegt*); und eines dieser beiden Gebiete wieder durch Q_2'. So werden drei Teilgebiete von $\mathfrak{J}(P')$ bestimmt, deren eines nur Punkte von Q_1' und Q_2' auf seiner Grenze besitzt, also von keinem der Gebiete G das Bild enthalten kann. Von den beiden übrigen Teilgebieten enthält aber jedes in seiner Grenze das Bild der Grenze nur eines der vier Gebiete G, kann also auch das Bild nur eines der vier Gebiete in sich enthalten. Es blieben also wenigstens zwei der Gebiete G übrig, deren Bilder wir nicht unterzubringen wüßten, und die Annahme hat sich als unmöglich herausgestellt.

Der Satz VI läßt sich also hier beweisen, wenn wir das Wort „Kurven“ durch „Polygone“ ersetzen.

---

*) Nämlich das von P_1' und Q_1' bestimmte Innengebiet, welches für jeden Punkt von P_1' alle in einer gewissen Umgebung dieses Punktes enthaltenen Punkte von $\mathfrak{J}(P')$ enthalten muß, und das von P_2' und Q_1' bestimmte Innengebiet. Ein drittes Teilgebiet würde nämlich auf seiner Grenze nur Punkte von Q_1' besitzen können, und dies ist unmöglich.

Die S. 162 oben genannte Möglichkeit könnte gleich ausgeschlossen werden; wenn sie nämlich für zwei beliebige Polygone eintreten würde, so gäbe es auch ein Paar von Polygonen mit derselben Eigenschaft in beliebiger Nähe voneinander; sodann würden aber von einer der Bildkurven zwei Gebiete bestimmt werden, deren eines sich nur beliebig wenig von ihr entfernen könnte, was unmöglich ist.

Zu § 5. Der Satz, daß eine geschlossene Kurve, welche innerhalb eines Gebietes, das eineindeutig und stetig abgebildet wird, liegt, dabei wieder in eine geschlossene Kurve übergeht, und eventuell ihre Fläche in die der Bildkurve, ist richtig; der Beweis aber bedarf der Ergänzung.

Zunächst sollten die Polygone P_λ (z. B. nach Kap. VI, S. 210) genauer festgelegt werden.

Unrichtig ist dann aber, aus den Relationen (S. 162):

$$\varrho(p'_\nu,\ P'_{\nu+1}) < \varepsilon, \qquad \varrho(p'_{\nu+1},\ P'_\nu) < \varepsilon$$

zu folgern, daß die geschlossenen Kurven P'_ν schließlich überall dicht liegen. Als Beispiel des Gegenteiles betrachte man in Fig. 1 in J und in A je ein in hinreichender Nähe approximierendes Polygon. Sie genügen den obenstehenden Relationen; dennoch lassen sie zwischen sich ein endliches Gebiet frei.

In unserem Falle aber kann man bemerken, daß, wenn in $\mathfrak{J}(\mathfrak{C}')$ ein freies Gebiet übrig bliebe, man auf seiner Grenze einen erreichbaren Punkt Q' wählen könnte; sei Q' das Bild von Q, so könnte man um Q einen genügend kleinen Kreis legen derart, daß die ihm entsprechende geschlossene Kurve um Q' teilweise außerhalb von allen Kurven P' verliefe, während alle Punkte des Kreises den Polygonen P angehörten, was unmöglich ist.

Der fehlende Nachweis, daß $\mathfrak{C}'$ wieder eine geschlossene Kurve ist, kann nun in analoger Weise geliefert werden, indem man zeigt, daß zwischen den Bildkurven der Polygone, welche $\mathfrak{C}$ von außen und von innen approximieren, kein freies Gebiet enthalten sein kann, sodaß $\mathfrak{C}'$ gerade zwei Gebiete bestimmt, welche beide alle Punkte von $\mathfrak{C}'$ als Grenzpunkte besitzen.

Satz VIII ist richtig, aber für Satz VII fehlen genügende Gründe. Aus dem Vorhergehenden läßt sich nämlich nur dieses folgern: „Das umkehrbar eindeutige und stetige Abbild einer Kurvenfläche ist ein einfach zusammenhängendes Gebiet mit seiner Grenze, und alle Punkte dieser Grenze sind zugleich Grenzpunkte der Restgebiete.“

Zu § 8. Auch hier in dem Beweise des Jordanschen Kurvensatzes wird die geschlossene Kurve gefolgert aus der Eigenschaft, daß jeder Punkt der von außen und von innen approximierenden Polygone schließlich von der Kurve, und jeder Punkt der Kurve schließlich von den beiden Polygonen einen beliebig kleinen Abstand erhält. Es ist wieder eine (übrigens leicht zu gebende) Ergänzung erforderlich, um zu zeigen, daß zwischen den von außen und den von innen approximierenden Polygonen kein endliches Gebiet eingeschlossen bleiben kann.*)

Zu § 15. Die Behauptung, daß die Invarianz der Erreichbarkeit auch in dem Sinne stattfindet, daß sie nicht bloß für stetige Transformationen der ganzen Ebene, sondern auch für stetige Abbildungen von bloßen Kurven oder sonstigen linienhaften Mengen aufeinander statt hat, wird von Herrn Schoenflies, wie der Satz XVII zeigt, in dem Sinne gemeint, daß unter „Erreichbarkeit" die Erreichbarkeit aller Punkte der Kurve zusammen verstanden werden soll. Ich erlaube mir hier die Bemerkung, daß, wenn man die Erreichbarkeit eines einzelnen Punktes ins Auge faßt, die Invarianz nicht bestehen bleibt, wie folgendes Beispiel zeigt:

In Fig. 4a konvergiert eine Näherungskurve nur gegen *eine* Seite eines einfachen Kurvenbogens; in Fig. 4b im eineindeutigen und stetigen

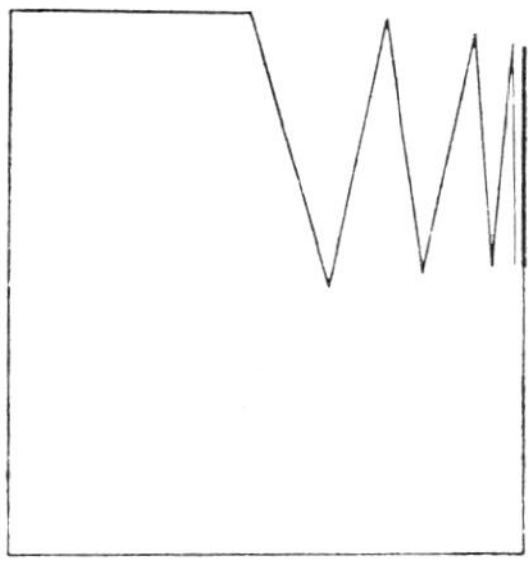

Fig. 4a.

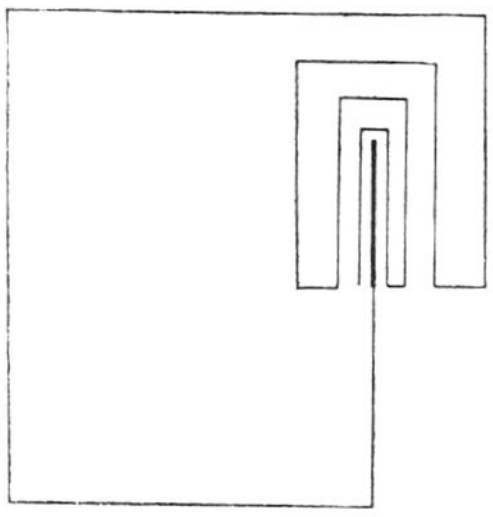

Fig. 4b.

Abbilde gegen *beide* Seiten. In der Weise kommen Punkte heraus, welche *vor* der Abbildung erreichbar, *nachher* unerreichbar sind.

Der Beweis des Satzes XVII bedarf einer wesentlichen Ergänzung**); die Mengen T_ν', deren unbegrenzte Abnahme hier bewiesen wird, sind nämlich nicht die Mengen T_ν, aus deren unbegrenzter Abnahme im § 10 die Erreichbarkeit des Grenzpunktes gefolgert wird. Man stellt die Sache richtig***), indem man bemerkt, daß T_ν zwar außer T_ν' noch mit T_ν' zusammenhängende Bestandteile ${}_\alpha S_\nu$ enthalten kann, daß aber je zwei

*) Herrn Schoenflies' Einwendungen gegen den Jordanschen Beweis scheinen mir unbegründet.

**) Der in Annalen Bd. 62 § 12 enthaltene Beweis des engeren Satzes ist in anderen Punkten unzutreffend. [[7]]

***) Wie Herr Schoenflies mir brieflich bemerkt, kann die notwendige Ergänzung auch durch die Überlegungen des § 5 von Kap. VI geliefert werden.

solchen Bestandteilen zwei verschiedene Bogenmengen des Kreises entsprechen. Mit den entsprechenden Bogenmengen konvergieren nun auch die zu einem selben ν gehörigen Bestandteile ${}_\alpha S_\nu$ zu Null, also auch die Mengen T_ν.

Zu § 16. Zum Beweise des Satzes XVIII ist zu bemerken, daß zwar von § 3 Gebrauch gemacht wird, jedoch nur von demjenigen Bestandteil seines Resultates, den wir oben als richtig herausgehoben haben.

Wesentlich unerlaubt ist aber wieder die Voraussetzung der Existenz der äußeren Randkurve eines endlichen Gebietes.

Man kann indes den Satz XVIII unmittelbar folgern aus dem unabhängig bewiesenen Jordanschen Satze, in Zusammenhang mit § 12 und § 15.

Zu § 17. Hier wird als ein nach dem Vorhergehenden evidenter Satz ausgesprochen, daß jede perfekte nirgends dichte Menge eineindeutiges stetiges Bild einer aus Strecken aufgebauten Menge ist. Dazu möchte ich bemerken, daß der Satz zwar in einem gewissen Sinne richtig ist, aber eines ausführlichen Beweises bedarf, wie ich an anderer Stelle ausführen werde.

Zusammenfassung. Von den Schoenfliesschen (entweder hergeleiteten oder als selbstverständlich angewendeten) Sätzen sind folgende unrichtig:

daß jede geschlossene Kurve sich in zwei eigentliche Kurvenbogen zerlegen läßt,

daß bei der Zerlegung einer geschlossenen Kurve in zwei Kurvenbogen wenigstens einer von ihnen eigentlich ist,

daß jedes geschränkte, einfach zusammenhängende Gebiet eine geschlossene äußere Randkurve aufweist,

daß jedes geschränkte, einfach zusammenhängende Gebiet durch seine Grenze eindeutig bestimmt ist,

daß jede geschränkte, abgeschlossene Menge, welche gemeinsame Grenze von zwei Gebieten ohne gemeinschaftliche Punkte ist, eine geschlossene Kurve ist;

während folgende zunächst unsicher bleiben:

daß das eineindeutige stetige Bild einer geschlossenen Kurve wieder eine solche ist,

⟦8⟧ *daß das eineindeutige stetige Bild einer Kurvenfläche wieder eine Kurvenfläche ist.*

Amsterdam, 14. Mai 1909.

⟦9⟧

NOTES

⟦1⟧ The origin of this paper is explained in Y10. See also 1910 C ⟦9⟧.

The paper contains Brouwer's sensational discovery of a curve which splits the plane into three domains of which it is the common boundary. (See also, for more than three domains, Brouwer 1911 B2.) It also is the first example of an indecomposable continuum, that is, a continuum that cannot be obtained as a union of two true subcontinua. There is an enormous literature on indecomposable continua; most of these investigations were published in Fundamenta Mathematicae. Bibliographies are to be found in C. Kuratowski 1922, pp. 230–231, 1927, pp. 274–275, 1928, pp. 41–42, 1952 II, pp. 143 footnote [1]). A general method of constructing indecomposable continua was developed by A. van Heemert 1943. As solenoids indecomposable continua have become important in group theory – see L. Vietoris 1927, D. van Dantzig 1930, A. van Heemert 1937.

Another early example of a pathological splitting of the plane into three domains was given by A. Denjoy 1910. A particularly attractive version by Wada of Brouwer's construction is reported by K. Yoneyama 1917:

Suppose that there is a land surrounded by sea, and that in this land there is a fresh lake. Also suppose that, from these lake and sea, canals are built to introduce the waters of them into the land according to the following scheme.

Let $$E_1, E_2, \cdots\cdots, E_n, E_{n+1}, \cdots\cdots$$

be a sequence of positive numbers, monotonously converging to zero, namely let

$$E_1 > E_2 > \cdots\cdots > E_n > E_{n+1} > \cdots\cdots$$

and $$\lim_{n\to\infty} E_n = 0.$$

On the first day, a canal is built from the lake, such that it does not meet the sea-water, and the shortest distance from any point on the shore of the sea to that of the lake and canal does not exceed E_1. The end point of this canal is denoted by L_1.

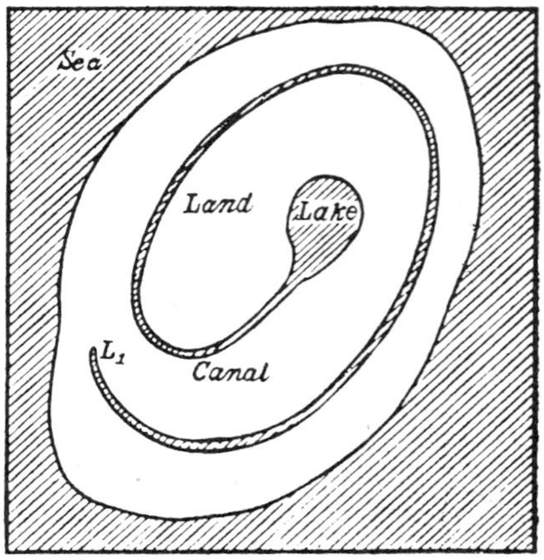

On the second day, a canal is built from the sea, never meeting the fresh water of the lake and canal constructed the day before, and the work is continued until the shortest distance from any point on the shore of the lake and canal filled with fresh water to that of the sea and canal filled with salt water does not exceed E_2. The end point of this canal is denoted by S_2.

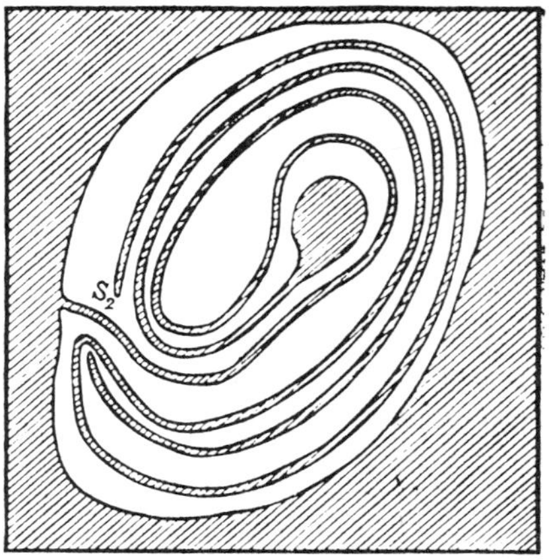

On the third day, the work is begun from L_1, never cuting the canals already built, and the work is continued until the shortest distance from any point on the shore of the sea and canal filled with salt water to that of the lake and canal filled with fresh water does not exceed E_3. The end point of this canal is denoted by L_2.

Now it is clear that we can continue the work day by day in the above way, by adequately narrowing the breadth of the canals, since the land is always semi-continuous at the end of the work of every day. If we proceed in this way indefinitely, we get at last an everywhere dense set of waters, fresh and salt, which never mingle together at any place. Now denote by M_L the shore of the lake and canal filled with fresh water, and by M_S that of the sea and canal filled with salt water, and by M_P the set of limiting points of M_L and M_S, not contained in them. Then the sum of M_L, M_S, and M_P forms a continuous set, and any three points, each taken from the above different sets, form a system of three points, every two of which form a pair of principal points of the set. Thus the set is the required continuous one.

If we suppose that there are many lakes in the land, we may obtain by the similar method a continuous set consisting of many aggregates having the properties stated in Theorem 12.

Thus having proved the existence of a continuous set, which has a system of three points, every two of which form a pair of principal points, we give here the following definition.

⟦2⟧ A. Schoenflies 1903 B, 1904, 1906 A.
⟦3⟧ A. Schoenflies 1908 A.
⟦4⟧ A. Schoenflies 1903 A, 1904.
⟦5⟧ For 'welche' read 'welcher'.
⟦6⟧ A. Schoenflies 1904.
⟦7⟧ A. Schoenflies 1906 A.
⟦8⟧ Invariance of the closed curve proved in 1912 L.
⟦9⟧ This paper is followed in Mathematische Annalen by A. Schoenflies 1910. The two authors ordered and mailed combined reprints. Correspondence of Brouwer with Hilbert (14 May 1909, 24 June 1909, 26 July 1909) shows that some resistance by Schoenflies to Brouwer's criticism had first to be overcome. Schoenflies was a brilliant lecturer (in a report on the Nauheim meeting of the Deutsche Mathematiker-Verenigiung to F. Klein, on 14 October 1920 – Univ. Bibl. Göttingen) Blumenthal called him the 'lion' of the conference) but he had difficulties in understanding Brouwer's subtlety and in competing with his awe inspiring rigour.

At that time Schoenflies prepared the second edition, Schoenflies 1913, of his *Bericht*, Schoenflies 1900. Thanks to Fricke's mediation Schoenflies had accepted Brouwer's unofficial supervisorship. This turned out to be a hard job for both of them. In his correspondence with Hilbert (16 April 1913, 23 April 1913, 5 September 1913, 11 September 1913 – Univ. Bibl. Göttingen) Brouwer complained about Schoenflies, and asked Hilbert to intervene; a picture card of 11 September 1913 bears the funny postscript of Mrs. Brouwer: 'Wenn mein Mann von Schoenflies verrückt wird, verdankt er es Ihnen.' A copy of a letter of Brouwer to Schoenflies of 4 September 1913 exists among Hilbert's papers (Univ. Bibl. Göttingen); Brouwer's papers contain two letters of Schoenflies to Brouwer (12 November 1909, and 14 August with no year indication, probably 1913), and printer's proofs of A. Schoenflies 1931, with numerous corrections by Brouwer. Often Schoenflies misinterpreted Brouwer's remarks, which in fact were not easy to understand; sometimes he inserted them in a wrong place, or in a corrupted form, or as a second proof along with his own wrong proof. However, the personal relations between the Brouwer and Schoenflies families were not disturbed by the scientific quarrels. (See also Schoenflies 1913, VII, Vorwort.)

An isolated jotting, in Brouwer's hand, says in Dutch about A. Schoenflies 1913 'As a matter of fact all original non-trivial ideas in the *Bericht* have come from me', which is probably true. In the preface of A. Schoenflies 1913 Brouwer's 'unselfish and efficient' cooperation is acknowledged. It appears from fragments that Brouwer would have liked Schoenflies to mention in the preface that his support was given although '*Herrn B.s persönliche Anschauungen über die Grundlagen der Mengenlehre zu den leitenden Prinzipien des Berichts vielfach in schroffem Gengensatz stehen*'. Brouwer's review, Brouwer 1914, of A. Schoenflies 1913 is mainly an exposition of the intuitionist view on set theory and topology.

Letter to Hilbert Y 10

⟦Univ. Bibl. Göttingen⟧

Amsterdam, 14 Mai 1909.

Overtoom 565.

Sehr geehrter Herr Geheimrat

Als ich im vergangenen Winter meine zweite Mitteilung über endliche kontinuier-
liche Gruppen zur Zusendung an die Redaktion der Mathematischen Annalen ⟦1⟧
schon fertig hatte, entdeckte ich auf einmal, dass die Schoenfliesschen Unter- ⟦2⟧
suchungen über Analysis Situs der Ebene, auf welche ich mich in ausgiebigster
Weise gestützt hatte, sich nicht in allen ihren Teilen aufrecht erhalten lassen, so
dass auch meine gruppentheoretischen Ergebnisse in Frage kamen.

Um wieder Klarheit zu erlangen, war zunächst notwendig, die bezügliche
Schoenfliessche Theorie gründlich durchzuarbeiten, und genau festzustellen, auf
welche ihrer Resultate sich in vollem Vertrauen weiter bauen lässt. So entstand die
hier eingeschlossene Arbeit, welche in mehrere Teile der Schoenfliesschen Theorie ⟦3⟧
eingrifft ⟦sic⟧, und einige gewissermassen neu gestaltet.

Ihre Publikation womöglich an derselben Stelle, wo Herr Schoenfliess seine Abhandlungen ursprünglich veröffentlicht hat, schien mir angebracht, weshalb ich sie der Redaktion der Mathematischen Annalen zur gefälligen Aufnahme zusende, während ich zugleich Herrn Schoenflies eine Kopie zugehen lasse.

Was nun weiter meine Gruppentheorie angeht, ich habe sie wieder aufrichten
können, aber nur indem ich zuvor mehrere Sätze über Transformationen von
Flächen in sich ableitete; welche ausserhalb der Gruppentheorie liegen, und welche
ich deshalb, um die Annalenartikel, Ihrem Wunsche gemäss, nicht weiter aus- ⟦4⟧
zudehnen als notwendig ist, in den Verslagen der Amsterdamer Akademie ver- ⟦5⟧
öffentliche. Eine erste solche Notiz wird Sie in englischem Abdruck hoffentlich ⟦6⟧
vor Kurzem erreicht haben. Dieser werde ich in den nächsten Monten noch einige
weitere folgen lassen, und sodann den sie voraussetzenden zweiten Artikel über
endliche, kontinuierliche Gruppen definitiv abfassen. ⟦1⟧

Mit ausgezeichneter Hochachtung

L E J Brouwer.

NOTES

⟦1⟧ Brouwer 1910 H.
⟦2⟧ A. Schoenflies 1903 B, 1904, 1906 A, 1908 A.
⟦3⟧ Brouwer 1910 C.
⟦4⟧ Brouwer 1910 F.
⟦5⟧ Brouwer 1909 F2, 1909 H2, 1911 H2, 1911 J2.
⟦6⟧ Brouwer 1909 F2.

SUR LES

CONTINUS IRRÉDUCTIBLES DE M. ZORETTI, 1910 J

PAR M. L.-E.-J. BROUWER.

Dans son étude sur la *Notion de la ligne*, insérée dans le Tome XXV (3e série) de ce Journal (p. 485-497), M. Zoretti définit comme *ensemble continu* un ensemble parfait bien enchaîné et appelle *continu irréductible entre a et b* un ensemble continu contenant a et b, mais ne possédant pas de portion continue qui contienne a et b. Se proposant alors de prouver qu'il y a à peu près identité entre les notions de ligne simple de M. Jordan et de continu irréductible, M. Zoretti s'appuie sur le lemme suivant : [[1]]

(p. 489). *Soient* C *un continu irréductible entre a et b, c un point quelconque de* C; *il est possible et d'une seule manière de décomposer* C *en deux continus ayant c pour seul point commun.*

Pour se convaincre que ce lemme est faux, il suffit de considérer, de la courbe $y = \sin \frac{1}{x}$, la portion comprise entre $x = 0$ et $x = 1$ complétée par ses points limites, et de choisir pour a, b et c respectivement les points $(0,1)$, $(1,0)$ et $(0,0)$.

Aussi la structure générale du continu irréductible est loin d'être aussi simple que ne le ferait croire l'étude de M. Zoretti. Pour bien mettre cela en évidence, examinons encore deux autres théorèmes qui y sont énoncés :

1° (p. 495). *La frontière extérieure d'un domaine peut se décomposer en deux continus ayant seulement deux points communs.*

L'inexactitude de ce théorème est démontré par l'existence des *geschlossene Kurven, welche sich nicht in zwei eigentliche Kurvenbogen zerlegen lassen*, que j'ai étudiées dans ma Note *Zur Analysis Situs* (*Ma-*
[[2]] *them. Annalen*, Bd. LXVIII, p. 422-434).

2° (p. 496). *L'ensemble des points qui n'appartiennent pas à un continu irréductible donné est un continuum.*

Pour montrer la fausseté de ce théorème, considérons deux spirales ayant un même cercle limite et ne se rencontrant pas. Soient a un point de l'une des deux, b un point de l'autre, et laissons de côté les parties des spirales qui sont situées au delà de a et de b respectivement. Alors les parties restantes complétées par le cercle limite forment un continu irréductible entre a et b ne jouissant pas de la propriété énoncée par M. Zoretti.

[[3]] Amsterdam, 5 juin 1910.

NOTES

[[1]] L. Zoretti 1909.
[[2]] Brouwer 1910C.
[[3]] Brouwer's paper is followed by a note, L. Zoretti 1910B, referring to the correction of the mistakes in L. Zoretti 1910A.

Letter to Hilbert Y 15

⟦Univ. Bibl. Göttingen⟧

Amsterdam, 15.10.09.

Sehr geehrter Herr Geheimrat ⟦1⟧

Beiliegend lasse ich Ihnen nochmal meinen Beweis des Kurvensatzes zugehen, ⟦2⟧
und hoffe sehr, dass Sie die Darstellung befriedigend finden werden.

Darf ich Ihnen bemerken, dass wir über die Ausdehnung des Satzes *dass das ein-eindeutige stetige Bild einer geschlossenen Kurve wieder eine geschlossene* ⟦3⟧
Kurve ist

noch gar keine Gewissheit haben, und hier ein recht schwieriges Problem vorzuliegen scheint. (Unter einer „geschlossenen Kurve" wird verstanden eine Punktmenge, welche in der Ebene zwei Gebiete bestimmt, und mit der Grenze jedes dieser Gebiete identisch ist.)

'Ebensowenig liegt noch ein Beweis vor, dass im Raume das ein-eindeutige stetige Bild der Kugel zwei und nur zwei Gebiete bestimmt. Allerdings bezweifelt ⟦4⟧
wohl niemand die Richtigkeit dieser Behauptung.

Neulich fand ich, dass, wenn man fragt nach denjenigen stetigen Transformationen der Kugel in sich, welche einerseits eindeutig, andererseits *höchstens* zweideutig sind, eine solche Transformation stets ein-eindeutiges stetiges Bild ⟦5⟧
der Funktion $x' = x^2$ ist. Dies gibt mir Hoffnung, dass sich die ganze Riemannsche Theorie der algebraischen Funktionen auf die Analysis Situs wird gründen lassen.

Vielleicht hat man sogar für die Theorie der analytischen Funktionen einen Ansatz, wenn man fragt nach solchen Korrespondenzen, welche aus *gebietsweise* ⟦6⟧
ein-eindeutigen und stetigen Korrespondenzen aufgebaut sind. Singuläre Punkte sind dann solche, um welche keine Umgebung existiert, in der die Korrespondenz ein-eindeutig und stetig ist. Jedenfalls kommt hier gleich eine Fülle von Problemen auf, welche innerhalb der Analysis Situs liegen: welcher Art die Menge der singulären Punkte sein kann, und welcher Art die Singularität in jedem Punkt; welche Punktmenge, in der nur die Stetigkeit bekannt ist, man zulassen kann, damit aus der Stetigkeit die Nichtsingularität dieser Punkte zu folgern ist; und an erster Stelle, ob man jeder dieser Korrespondenzen eine analytische Funktion entsprechen lassen kann.

So komme ich noch einmal auf die Frage, welche Sie in unseren Unterhaltungen berührten, nämlich nach dem Verhalten einer analytischen Funktion in der

Umgebung einer nirgends dichten perfekten Punktmenge μ von Ausnahmepunk-
[[7]] ten. Ich finde darüber etwas behandelt bei *Pompeiu*: „Sur la continuité des fonctions de variables complexes", Annales de Toulouse 1905. Er beweist u.m.[[sic]], dass, wenn die Funktion gleichmässig stetig ist, und die Menge μ den Inhalt Null hat, die Stetigkeit genügt, um auch das Analytisch-sein sicherzustellen, dass aber, wenn der Inhalt von μ nicht Null ist, die Funktion sehr gut in diesen Punkten stetig, nicht aber analytisch, sein kann.

Er führt auch noch einen neuen Begriff, den der „longueur" einer Menge ein, und leitet auf Grund dieses Begriffes ein zweites Kriterium her.

Erlauben Sie, an Ihr Versprechen zu erinneren, mir Separate von Ihnen (ü[[ber]] das Dirichletsche Prinzip, Integralgleichungen, und Nachruf auf Minkowski), und von den Herren Zermelo (Grundlagen der Mengenlehre) und Koebe (Uniformisierung) schicken zu lassen? Im voraus spreche ich Ihnen dafür vielen Dank aus.

Meine Frau lässt herzlich grüssen. Hoffend auf ein Wiedersehen, mit vielen Grüssen für die liebe Frau Geheimrat, deren Heilung zu vernehmen uns sehr freuen wird

Ihr herzlich ergebener

L E J Brouwer

NOTES

[[1]] Brouwer's topological activity on the eve of his great discoveries is fairly well characterized by this letter.
[[2]] Brouwer 1910 E.
[[3]] Proved in Brouwer 1912 F, 1912 L.
[[4]] Proved in Brouwer 1911 F.
[[5]] Closely related to more general work that was published much later. See Brouwer 1919 S, 1919 N2, 1919 G.
[[6]] See Brouwer 1912 K2, 1912 M. The general problem was attacked and solved by S. Stoïlow 1924.
[[7]] D. Pompeiu 1905. See also the remarkable result of P. Urysohn 1923.

Beweis des Jordanschen Kurvensatzes. 1910 E

Von

L. E. J. Brouwer in Amsterdam. [[1]]

Der Beweis, den Herr Jordan von seinem grundlegenden Kurvensatze gibt,*) operiert mit gegen die Kurve konvergierenden Polygonfolgen, und in allen mir bekannten seitdem veröffentlichten Beweisen ist, insoweit sie sich nicht auf besondere Fälle beschränken oder unrichtig sind, diese Methode beibehalten.

Erst von Veblen**) wurde ein Beweis gegeben, der die Kurve, statt sie als Limes von Polygonen aufzufassen, gleich selbst in Betracht zieht, und mittels ihrer inneren Eigenschaften zum Resultate gelangt.

In dieser direkteren Weise werden auch wir im folgenden verfahren, aber auf einem völlig anderen und, wie ich glaube, erheblich kürzeren Wege das Ziel erreichen.

Der Satz läßt sich in folgender Form aussprechen: *Das eineindeutige stetige Bild eines Kreises bestimmt in der Ebene zwei Gebiete, und ist mit der Grenze jedes dieser Gebiete identisch; es ist m. a. W. in der Schoenfliesschen Terminologie eine geschlossene Kurve.*

Um die Aussage des Satzes und seinen Beweis völlig elementar verständlich zu machen, erinnere ich in den §§ 1 und 2 an einige einfache, längst bekannte mengentheoretische Definitionen und Theoreme; § 3 bringt einen Hilfssatz, und die §§ 4, 5 und 6 je einen der drei Bestandteile, in welche ich den Jordanschen Satz zergliedere.

§ 1.

Unter einem *Weg* wird verstanden ein aus einer endlichen Zahl von geradlinigen Strecken gebildeter, sich selbst nicht kreuzender Streckenzug.

Unter einem *Gebiete* wird verstanden eine ebene Menge mit den beiden folgenden Eigenschaften: 1. um jeden seiner Punkte läßt sich ein

*) Cours d'Analyse, I, 2me éd., S. 91—99. [[2]]

**) Transactions of the Amer. Math. Soc., vol. 6, S. 83. [[3]]

Kreis schlagen, dessen Inneres ebenfalls zum Gebiete gehört; 2. je zwei seiner Punkte sind durch einen ganz zum Gebiete gehörigen Weg verbindbar.

Diejenigen Grenzpunkte eines Gebietes, welche nicht zum Gebiete gehören, bilden eine abgeschlossene, nirgends dichte Menge, welche die *Grenze* des Gebietes heißt.

[[4]] Die Restmenge einer abgeschlossenen ebenen Menge T zerfällt in eine endliche oder abzählbare Menge von Gebieten, welche wir als *von der Menge T bestimmt* bezeichnen.

Eine Menge heißt nach Schoenflies *geschränkt*, wenn sie ganz im Endlichen liegt.

Eine abgeschlossene Menge heißt *zusammenhängend*, wenn sie sich nicht in zwei abgeschlossene Teilmengen zerlegen läßt. Der Kürze halber nennen wir im folgenden eine zusammenhängende abgeschlossene Menge ein *Kontinuum*. Von einem Kontinuum gilt dann folgende Eigenschaft:

Theorem 1. *Wenn ein Gebiet und seine Restmenge beide Punkte eines Kontinuums Z besitzen, so besitzt auch die Grenze des Gebietes wenigstens einen Punkt von Z; und insbesondere: wenn sowohl das Innere wie das Äußere eines Polygons Punkte eines Kontinuums Z besitzen, so besitzt auch das Polygon selbst wenigstens einen Punkt von Z.*

Im entgegengesetzten Falle würden nämlich die im Gebiete und die in der Restmenge enthaltenen Punkte von Z zwei abgeschlossene Teilmengen von Z ausmachen.

Theorem 2. *Zu einer willkürlichen geschränkten, abgeschlossenen, nicht zusammenhängenden Punktmenge K kann man ein K nicht treffendes Polygon bestimmen, das einen Teil von K in seinem Inneren, einen anderen Teil von K in seinem Äußeren enthält.*

Beweis. Seien K_1 und K_2 zwei abgeschlossene Teilmengen von K. Sie haben voneinander einen endlichen Abstand a. Sei P_1 ein Punkt von K_1, P_2 ein Punkt von K_2. Wir wählen nun $e < \frac{1}{2}a$, konstruieren eine quadratische Teilung der Ebene mit der Quadratseite e, heben diejenigen Quadrate q heraus, welche nebst ihrem Umfange von K_1 frei sind, und nehmen die Quadratseiten, welche zwei Quadrate q trennen, hinzu. Wir erhalten dann eine Menge von Gebieten γ, in denen K_2 enthalten ist, während K_1 weder in ihnen selbst noch auf ihrer Grenze einen Punkt besitzen kann. Sei nun G dasjenige unter den Gebieten γ, das P_2 enthält, so gibt es ein Grenzpolygon von G, das P_1 von P_2 trennt; dieses Polygon besitzt also die verlangte Eigenschaft.

Theorem 3. *Die Grenze Z eines von einem geschränkten Kontinuum Z' bestimmten Gebietes $\mathfrak{G}$ ist wieder ein Kontinuum.*

[[378]]

Beweis. Falls sie kein Kontinuum wäre, könnten wir nach Theorem 2 ein Z nicht treffendes Polygon π bestimmen, das sowohl in seinem Inneren, wie in seinem Äußeren Punkte von Z, also auch von $\mathfrak{G}$ enthielte. Verbinden wir sodann einen innerhalb π und einen außerhalb π liegenden Punkt von $\mathfrak{G}$ durch einen zu $\mathfrak{G}$ gehörigen Weg w, so sehen wir, daß π wenigstens einen Punkt von $\mathfrak{G}$ besitzt, also, da π die Grenze von $\mathfrak{G}$ nicht trifft, daß π ganz zu $\mathfrak{G}$ gehören müßte, also auch Z' nicht treffen könnte. Dies ist aber nach Theorem 1 ein Widerspruch.

§ 2.

Es seien K_1 und K_2 zwei geschränkte Kontinua ohne gemeinschaftliche Punkte. Nach Theorem 2 kann man sie trennen durch ein Polygon π, das weder K_1 noch K_2 trifft, und eines von beiden in seinem Inneren, das andere in seinem Äußeren enthält. Dasjenige von K_1 und K_2 bestimmte Gebiet g, das π enthält und, wie man unmittelbar einsieht, von der Wahl von π unabhängig ist, nennen wir das *Zwischengebiet* von K_1 und K_2.

Sei w ein Weg, der, während von seinen Endpunkten der eine auf K_1, der andere auf K_2 liegt, ganz in g verläuft, kurz, der *K_1 und K_2 innerhalb g verbindet.* Dieser Weg besitzt zwei Seiten, welche wir seine *rechte* und *linke Seite* nennen. Wir wollen untersuchen, wieviel Teilgebiete w in g bestimmt.

Aus jedem Punkte P von g läßt sich ein Weg nach w ziehen, welcher entweder auf die rechte oder auf die linke Seite von w mündet. Jedenfalls gehören nun alle diejenigen Punkte P, für welche diese Mündung auf derselben Seite von w stattfindet, zum selben von K_1, K_2 und w bestimmten Gebiete $_r\gamma$ resp. $_l\gamma$.

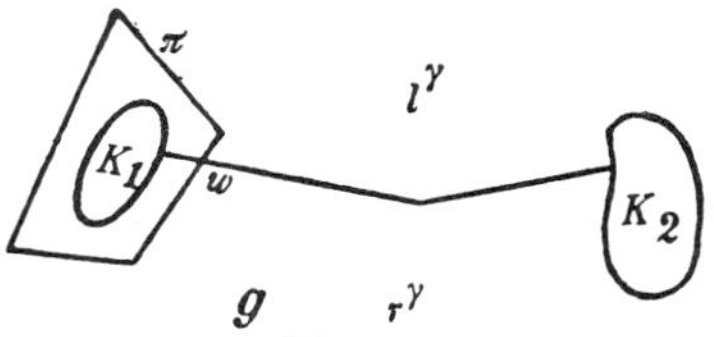

Fig. 1.

Weil nun aber von den Endpunkten von w der eine innerhalb π, der andere außerhalb π liegt, muß π wenigstens einen w nicht treffenden Teilstreckenzug besitzen, der die beiden Seiten von w verbindet. Da nun die Punkte dieses Streckenzugs sowohl zu $_r\gamma$ wie zu $_l\gamma$ gehören müssen, sind $_r\gamma$ und $_l\gamma$ identisch. Mithin gilt:

Theorem 4. *Im Zwischengebiete zweier Kontinua wird von einem sie verbindenden Wege nur ein Gebiet bestimmt.*

Wir setzen weiter voraus, daß K_1 und K_2 innerhalb g durch *zwei* einander nicht treffende Wege w_1 und w_2 verbunden sind.

Aus jedem Punkte P von g läßt sich nach Theorem 4 ein w_1 nicht treffender Weg nach w_2 ziehen, welcher entweder auf die rechte oder auf die linke Seite von w_2 mündet, und alle Punkte P, für welche die Mündung auf derselben Seite von w_2 stattfindet, gehören zum selben von K_1, K_2, w_1 und w_2 bestimmten Gebiete ${}_r\gamma_2$ resp. ${}_l\gamma_2$.

Wären nun diese beiden Gebiete identisch, so könnten wir die beiden Seiten von w_2 durch einen Weg F verbinden, der in seinem Verlaufe weder w_1 noch w_2 träfe und mit einem Teilstreckenzuge τ von w_2 ein Polygon p bilden würde, welches von den Endpunkten von w_2 den einen in seinem Inneren, den anderen in seinem Äußeren enthielte. Von dem aus K_1, K_2 und w_1 gebildeten Kontinuum μ würden also sowohl in dem Inneren, wie in dem Äußeren von p, aber nicht auf p selbst Punkte liegen, was nach Theorem 1 ein Widerspruch ist. Wir haben also bewiesen:

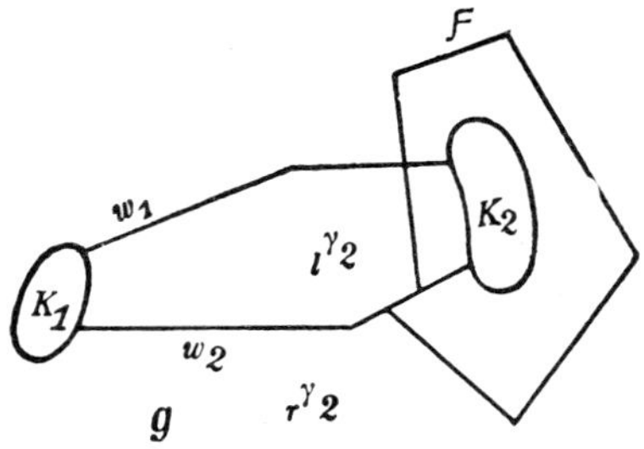

Fig. 2.

Theorem 5. *Im Zwischengebiete zweier Kontinua werden von zwei sie verbindenden und einander nicht treffenden Wegen zwei Teilgebiete bestimmt.*

Denken wir nunmehr K_1 und K_2 noch durch einen *dritten*, w_1 und w_2 nicht treffenden Weg w_3 verbunden, so liegt dieser in einem der beiden von w_1 und w_2 in g bestimmten Teilgebiete. Aus jedem Punkte P dieses Teilgebietes läßt sich ein weder w_1 noch w_2 treffender Weg nach w_3 ziehen, welcher entweder auf seine rechte oder auf seine linke Seite mündet, und alle Punkte P, für welche die Mündung auf derselben Seite von w_3 stattfindet, gehören zum selben von K_1, K_2, w_1, w_2 und w_3 bestimmten Gebiete ${}_r\gamma_3$ resp. ${}_l\gamma_3$; daß diese beiden Gebiete nicht identisch sein können, zeigt sich genau so wie beim Theorem 5, sodaß in g von w_1, w_2 und w_3 drei Gebiete bestimmt werden. Und da man dieselben Überlegungen für jeden neu hinzugefügten Weg w_n anstellen kann, gilt allgemein:

Theorem 6. *Im Zwischengebiete zweier Kontinua werden von n sie verbindenden und einander nicht treffenden Wegen n Teilgebiete bestimmt.*

§ 3.

Es sei nun C ein Kreis, C' sein eineindeutiges stetiges Bild c ein Teilbogen von C mit den Endpunkten A und B; c', A' und B' die entsprechenden Bilder. Wir wollen auch c' kurz als Teilbogen von C' bezeichnen.

Sei G ein Gebiet, dessen Grenze eine c' enthaltende Teilmenge von C' ist, und H ein nicht mit A' oder B' zusammenfallender Punkt von c'.

Dieser Punkt H hat sodann einen endlichen Abstand a von der Menge $C' - c'$, und man kann in G einen Punkt L bestimmen, dessen Abstand von $H < \frac{1}{2}a$, dessen Abstand von $C' - c'$ also $> \frac{1}{2}a$ ist. Die Strecke LH kann dann nicht von $C' - c'$ getroffen werden. Sei N der erste Punkt, in dem diese Strecke c' trifft, so wird die Strecke LN weder von c' noch von $C' - c'$, also von C' überhaupt nicht getroffen. Mithin ist bewiesen:

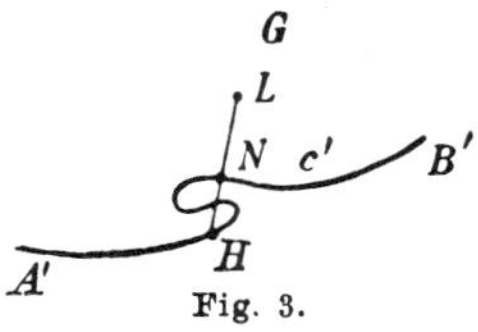

Fig. 3.

Hilfssatz. *Wird die Grenze eines Gebietes von einer Jordanschen Kurve oder einer Teilmenge einer solchen gebildet, so kann man aus ihm nach jedem Teilbogen seiner Grenze einen Weg legen.*

§ 4.

Bestimmt C' nur ein einziges, in der ganzen Ebene überall dicht liegendes Gebiet G, so ist die Grenze von G sicher mit C' identisch.

Sei aber G_1 ein von C' bestimmtes, in der Ebene nicht überall dicht liegendes Gebiet, und G_2 ein weiteres von der Grenze von G_1 bestimmtes Gebiet.

Nach Theorem 3 ist dann die Grenze von G_2 ein Kontinuum C'_a, von dem wir beweisen wollen, daß es mit C' identisch ist, indem wir zeigen, daß die entgegengesetzte Annahme auf einen Widerspruch führt.

Entspricht nämlich C'_a auf dem Kreise ein Teilbogen C_a, so können wir diesen mittels zweier Punkte D und E wieder in drei Teilbogen c_1, c_2 und c_3 zerlegen. Die entsprechenden Bilder seien D', E', c_1', c_2' und c_3'. Auf Grund des Hilfssatzes von § 3 können wir c_1' und c_3' sowohl innerhalb G_1 wie innerhalb G_2 durch einen Weg verbinden. Diese beiden Wege v_1 und v_2 verlaufen notwendig im Zwischengebiete von c_1' und c_3', und bestimmen in ihm nach Theorem 5 zwei Teilgebiete, γ_1 und γ_2. Innerhalb jedes dieser Gebiete können wir v_1 und v_2 durch Wege u_1 resp. u_2 verbinden, welche sowohl zu G_2, wie zur Restmenge von G_2 gehörige Punkte besitzen, also nach Theorem 1 beide notwendig die Grenze von G_2, d. h. C'_a treffen müssen; c_1' und c_3' treffen sie aber sicher nicht, sie treffen also c_2'. *Mithin liegen sowohl in γ_1 wie in γ_2 Punkte von c_2'.*

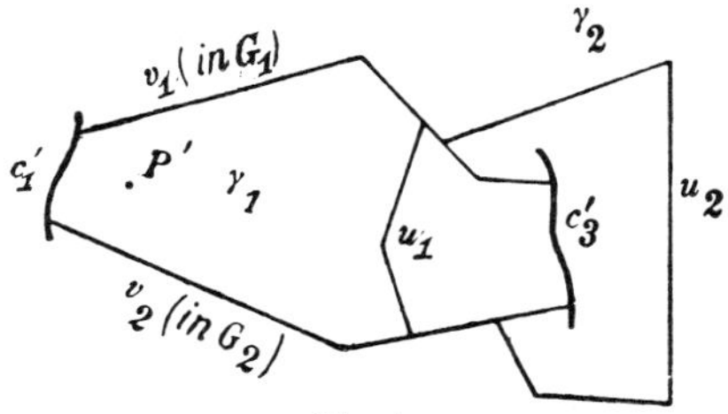

Fig. 4.

Sei nun P' ein in γ_1 liegender Punkt von c_2', dem im Kreise der Punkt P entspricht. Sei c_4 ein willkürlicher Teilbogen von c_2, der P

enthält, aber D und E nicht enthält. Weil sodann c_4' keinen Punkt der Grenze von γ_1 enthält, muß nach Theorem 1 c_4' gänzlich zu γ_1 gehören. Dies gilt unabhängig von der Wahl des c_4, sodaß mit Ausnahme der Punkte D' und E' das ganze Kontinuum c_2' in γ_1 liegt, *und* γ_2 *keinen Punkt von* c_2' *enthalten kann*, was ein Widerspruch ist.

Es gilt also auf jeden Fall:

Satz 1. *Die Grenze eines von einer Jordanschen Kurve bestimmten Gebietes ist mit der ganzen Kurve identisch.*

In der Aussage dieses Satzes ist enthalten:

Satz 1a. *Eine Jordansche Kurve ist eine nirgends dichte Punktmenge.*

§ 5.

Wir nehmen nun an, daß C' $n(>2)$ Gebiete bestimmt. Dann zerlegen wir C in vier Teilbogen c_1, c_2, c_3 und c_4, welche sich in der angegebenen Reihenfolge aneinander schließen. Seien c_1', c_2', c_3' und c_4' ihre Bilder. Innerhalb jedes der n durch C' bestimmten Gebiete verbinden wir sodann c_1' und c_3' durch einen Weg, was nach dem Hilfssatze und Satz 1 möglich ist. Diese Wege bestimmen im Zwischengebiete von c_1' und c_3' n Teilgebiete. Genau nach der Methode des § 4 zeigt sich, daß einerseits jedes dieser Teilgebiete Punkte von c_2' oder c_4' enthalten muß, und andererseits sowohl c_2' wie c_4' nur in je eines der Teilgebiete eindringen kann. Aus diesem Widerspruche folgern wir:

Satz 2. *Eine Jordansche Kurve kann nicht mehr als zwei Gebiete bestimmen.*

§ 6.

Schließlich nehmen wir an, daß C' nur *ein* Gebiet G bestimmt.

Dann legen wir aus einem Punkte M von G zwei einander nicht treffende Wege u_1 und u_2 nach C', welche in den Punkten P_1 und P_2 von C' münden. Nach Theorem 5 bestimmt die von C', u_1 und u_2 gebildete Menge τ in G zwei Gebiete g_1 und g_2.

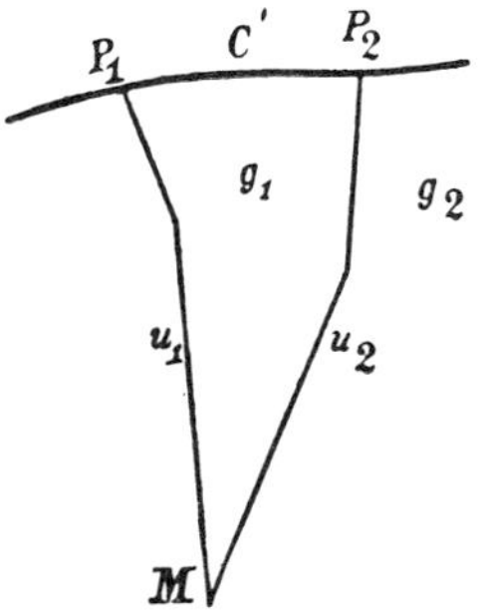

Fig. 5.

Seien z' und z'' die beiden P_1 und P_2 verbindenden Teilbogen von C', und τ' und τ'' die von z', u_1 und u_2 resp. z'', u_1 und u_2 gebildeten Jordanschen Kurven. Nach Theorem 5 bestimmt τ' zwei Gebiete g_1' und g_2'. Ein Weg, der einen Punkt von g_1' mit einem Punkte von g_2' verbindet, trifft notwendig τ', also τ; die beiden Punkte können also weder beide in g_1, noch beide in g_2 liegen. Mithin ist beispielsweise g_1 in g_1' und g_2 in g_2' enthalten.

[[382]]

Nach Satz 1a liegt dann g_1 in g_1' und g_2 in g_2' überall dicht. Nach Satz 1 ist weiter jeder Punkt von τ' gemeinsamer Grenzpunkt von g_1' und g_2', also auch von g_1 und g_2.

Sei Q ein Punkt von τ'', so liegt er entweder in g_1' oder in g_2'; wir nehmen an, daß er in g_1' liegt. Dann gehört er aber weder zu g_2' noch zu seiner Grenze, also auch weder zu g_2 noch zu dessen Grenze. Andererseits muß aber jeder Punkt von τ'', ebenso wie jeder Punkt von τ', zur Grenze von g_2 gehören. Wir werden somit auf einen Widerspruch geführt und sprechen aus:

Satz 3. *Eine Jordansche Kurve muß wenigstens zwei Gebiete bestimmen.*

NOTES

⟦1⟧ A most elegant proof of Jordan's theorem (for the plane), which found its way into F. Hausdorff 1914, pp. 354–358. Meanwhile a great many proofs of Jordan's theorem were produced, in particular in the twenties and early thirties, by people who believed Jordan's theorem was indispensable in complex function theory. The only proof by point-set topology worthy of comparison with Brouwer's is E. Schmidt's, 1923.

The method of the present proof is usually attributed to 'Phragmén–Brouwer'. E. Phragmén 1885, Phragmén's only paper in topology, contains nothing to justify this terminology; in fact, it does not contain any of the so-called Phragmén theorems. In that paper Phragmén proved only that a closed point set in the plane with no connected subset does not split the plane.

Though the method of the present paper is point-set topology, it is on the verge of what we have called 'The new methods in topology'. A modern reader will very probably translate the proof into homological language by means of addition theorems and generalize it to n dimensions. This was in fact done by J. W. Alexander, 1922.

See also Alexandroff–Hopf 1935, pp. 450–457.

⟦2⟧ C. Jordan 1893, pp. 91–99.

⟦3⟧ O. Veblen 1905.

⟦4⟧ Strokes in Brouwer's hand.

1911 B2 **Mathematics.** — "*On the structure of perfect sets of points*" (second communication [1])). By Dr. L. E. J. Brouwer. (Communicated by Prof. Korteweg).

[[1]] (Communicated in the meeting of April 28, 1911).

§ 1.

A further extension of Cantor's fundamental theorem.

The proof of Cantor's fundamental theorem and of its Schoenflies extension, given in § 2 of the first communication, holds also for the following property :

[[1]] [1]) For the first communication see these Proceedings, Vol. XII, p. 785.

Theorem 1. *A well-ordered set of points in Sp_n each point of which possesses a finite distance from the set formed by all the following points, is denumerable.*

Out of this is obtained in the following form a generalization of Cantor's fundamental theorem, probably the widest one of which it is capable:

When a closed set of points is replaced by a closed set contained in it, we shall say that the first set is *lopped.*

A fundamental series of closed sets of points will be called a *lopping series,* if each following set is contained in the preceding one. The greatest common part of the terms of such a series is a closed set, which we shall call the *limiting set* of the lopping series.

By an *inductible property of closed sets of points* we shall understand a property which, when possessed by each term of a lopping series, holds also for the limiting set of that series.

From theorem 1 now follows:

Theorem 2. *Let μ be a closed set of points of Sp_n possessing the inductible property α; we can reduce it by a denumerable number of* ⟦2⟧
loppings of a definite kind β to a closed set of points μ_i possessing still the property α, but losing it by any new lopping of kind β.

This theorem can be specialized in many directions.

If we choose as property α the simple property of being closed, and as lopping of kind β the destruction of an isolated point resp. of an isolated piece, then Cantor's fundamental theorem resp. its Schoenflies extension appears.

An other special case is obtained in the following way:

After Zoretti[1]) a continuum C is called *irreductible between P and Q,* if the pair of points (P,Q) belongs to C, but to no other continuum contained in C, and Janiszewski[2]) and Mazurkiewicz[3]) have proved the following theorem:

Let C be an arbitrary continuum and P and Q two of its points, then in C is contained a continuum irreductible between P and Q.

This property appears likewise as a special case of theorem 2, namely by choosing as property α the property of containing P and Q and being continuous, and as lopping of kind β the most general lopping.

§ 2.

The structure of closed sets of pieces.

In § 3 of the first communication it has been proved that all perfect

---

1) Annales de l'École Normale, 1909, p. 485. ⟦3⟧
2) Comptes Rendus, t. 151, p. 198. ⟦4⟧
3) ibid., p. 296. ⟦5⟧

sets of pieces possess the same geometric type of order, namely the common type of order of linear, perfect, punctual sets of points. An analogous theorem exists for closed sets of pieces.

In § 3 of the first communication the set of pieces μ, taken there as perfect, was broken up into two such closed sets μ_0 and μ_2, that $\delta(\mu_h) \leqq \delta(\mu)$ and $a(\mu_0, \mu_2) \geqq \delta(\mu)$; then each μ_h into two such closed sets μ_{h0} and μ_{h2}, that $\delta(\mu_{hk}) \leqq \delta(\mu_h)$ and $a(\mu_{h0}, \mu_{h2}) \geqq \delta(\mu_h)$; and so on. In this way the μ_F's converged for indefinite accrescence of the rows of indices F uniformly to the pieces of μ, and we could construct a continuous one-one correspondence between the pieces of μ and a nowhere dense perfect set of real numbers between 0 and 1, where for each of those numbers the row of figures in the numerical system of base 3 was identical to the row of indices of the corresponding piece of μ.

If, however, μ is a closed, not perfect set of pieces, then the breaking up of an arbitrary μ_F into μ_{F0} and μ_{F2} can take place in the same way with the only exception that a μ_F consisting of a single piece also appears as μ_{F0}, whilst μ_{F2} falls out. Then too the μ_F's converge for indefinite accrescence of the rows of indices F uniformly to the pieces of μ, and we can construct a continuous one-one correspondence between the pieces of μ and a nowhere dense closed set of real numbers between 0 and 1, where for each of those numbers the row of figures in the numerical system of base 3 is identical to the row of indices of the corresponding piece of μ.

So we have proved:

THEOREM 3. *Each closed set of pieces in Sp_n possesses the geometric type of order of a linear, closed, punctual set of points.*

§ 3.

The division of the plane into more than two regions with a common boundary.

On a former occasion [1]) I constructed a division of the plane into three regions with a common boundary, and I communicated at the same time that by a suitable modification of the method followed there a division into an arbitrary finite number, and even into an infinite number of regions with a common boundary can be obtained. That modified method I shall now explain.

[6] [1]) Compare "*Zur Analysis Situs*", Mathem. Annalen, Vol. 68, p. 422—434.

Fig. 1.

[[6]] Our starting-point is figure 2 (Plate I) explained l. c. p. 423 and 424 of Vol. 68 of the Mathem. Annalen, which figure in the following will be called the *primitive figure*.

We first simplify the primitive figure by leaving out the red band, and by reducing the breadth of the black bands to zero. These contracted black bands we shall call "*supporting threads*", and we draw each of them through the middle of the white band determined by all the preceding supporting threads, as is executed here in figure 1 for the first four supporting threads (this figure is to be looked at in the position indicated by the subscription: Fig. 1).

The rectangular circumference of figure 1 we shall indicate by k, the circumference together with its inner domain by F. The circumference together with the supporting threads we shall call the *skeleton* of the figure. Two arbitrary points of the skeleton possess the property of being contained in a perfect coherent part of the skeleton.

We now consider a horizontal section l of figure 1 cutting all the vertical line segments of the supporting threads, and we determine the points of l by their *abscis*, i. e. their distance from the left endpoint of l. The length of l we choose as unity of length.

Then the abscis of the point of intersection of l with the first supporting thread is $\frac{1}{2}$; the abscissae of the points of intersection of l with the second supporting thread are $\frac{1}{4}$ and $\frac{3}{4}$; those with the third supporting thread are $\frac{1}{8}$, $\frac{3}{8}$, $\frac{5}{8}$ and $\frac{7}{8}$; and so on.

So the set of points determined on l by the system of supporting threads possesses as their abscissae the set of dual fractions between 0 and 1.

Two points of F will be called *directly coherent*, if they are contained in a perfect coherent part of F having no point in common with the skeleton. Two points directly coherent with a third point are also directly coherent with each other. The points directly coherent with a given point form a set which will be called a *coherence thread*.

The abscissae of the points of intersection of l with a coherence thread form a set of numbers to be called a *directly coherent set of numbers*. Two abscissae then and only then belong to the same directly coherent set of numbers, if either their sum or their difference is a dual fraction.

The set of coherence threads possesses the power of infinity of the continuum.

[[388]]

Of this figure 1 we shall now construct a generalization ; in the white band determined by the first n supporting threads we shall namely draw the $(n+1)^{\text{th}}$ supporting thread *not through the middle*; thereby each supporting thread segment gets an arbitrary distance from the corresponding white band edge segments; however we take care firstly that the new supporting thread penetrates into each segment of the corresponding white band, and secondly that each vertical supporting thread segment cuts the line l.

In the more general figure it may happen that some segments of coherence threads have expanded to bands, so that for this figure we shall replace the name of coherence threads by *coherence strips*.

And if each point of F which can be joined to the skeleton by a line segment meeting the skeleton only in its endpoint, is added to the skeleton, then also in the skeleton, just as in the coherence threads, certain segments may expand to bands, so that for the new skeleton we shall replace the name of supporting threads by *supporting strips*.

In the more general figure we assign to the points and intervals in which l is cut by the nth supporting strip as their *coordinates* the same numbers $\frac{2k+1}{2^n}$ which in figure 1 appeared as the abscissae of the corresponding points of intersection of l with the nth supporting thread, and each point or interval determined on l by a coherence strip gets as its coordinate the number corresponding to the Schnitt determined in the coordinates belonging to the supporting strips. Then along l the coordinate is a nowhere decreasing continuous function of the abscis, and like the abscis it has the initial value 0 and the endvalue 1.

Now for a moment we abstract from the figure, and set apart a finite or a denumerable infinite system of directly coherent sets of numbers. The numbers belonging to these sets we shall call *special numbers* and we determine a coordinate function of the just now described kind possessing each special numerical value over a certain interval of abscissae, but each other numerical value (0 and 1 included) only for a single abscis.

Our aim is to construct the generalized figure 1 in such a way that the coherence strips corresponding to the special directly coherent sets of numbers get everywhere a finite breadth, whilst all segments of the other coherence strips and of the skeleton get a breadth zero. Starting from the coordinate function just now constructed on l we succeed in this in the following manner:

The first supporting thread we construct through the point of l with coordinate $\frac{1}{2}$, and each pair of points resp. intervals of l with mean coordinate $\frac{1}{2}$ we join within k and round about the first supporting thread by threads resp. bands twice rectangularly bent and not meeting each other, thereby taking care that the horizontal segments of these threads and bands determine an everywhere dense set of points and intervals on the perpendicular let down from the endpoint of the first supporting thread on the horizontal upper limit of F.

The second supporting thread we construct in such a way through the points of l with coordinates $\frac{1}{4}$ and $\frac{3}{4}$ that it does not cross the threads and bands already constructed, and each pair of points resp. intervals of l with coordinates $> \frac{1}{2}$ and with mean coordinate $\frac{3}{4}$ we join within k and round about the second supporting thread by threads resp. bands twice rectangularly bent and not meeting each other, thereby taking care that the horizontal segments of these threads and bands determine an everywhere dense set of points and intervals on the perpendicular let down from the endpoint of the second supporting thread on the baseline of F.

The third supporting thread we construct in such a way through the points of l with coordinates $\frac{1}{8}$, $\frac{3}{8}$, $\frac{5}{8}$ and $\frac{7}{8}$, that it does not cross the threads and bands already constructed, and each pair of points resp. intervals of l with coordinates between $\frac{1}{4}$ and $\frac{1}{2}$ and with mean coordinate $\frac{3}{8}$ we join within k and round about the third supporting thread by threads resp. bands twice rectangularly bent and not meeting each other, thereby taking care that the horizontal segments of these threads and bands determine an everywhere dense set of points and intervals on the perpendicular let down from the endpoint of the third supporting thread on the baseline of F.

Continuing in this way we secure that the "special" coherence strips get everywhere a finite breadth and that the other coherence strips and the skeleton get everywhere a breadth zero. The inner domains of the special coherence strips form together a set of points everywhere dense in F, and *all these inner domains have the same boundary*.

If we choose the coordinate function on l in such a way that it possesses not only each special numerical value but also the value of each dual fraction not identical to 0 or 1 over a certain interval of abscissae, then the just now described construction of the generalized figure 1 can be repeated without modification with the only difference that the supporting threads are replaced by supporting bands. These supporting bands determine together with the rest region of F a region G possessing the same boundary g as the inner domains of the coherence bands.

So this continuum g divides the plane into regions with a common boundary; whether the number of these regions is finite or infinite, depends on the choice of the special directly coherent sets of numbers.

Let us call two points contained in a perfect coherent part of g not identical to g, *directly coherent in* g, and let us call the set formed by the points directly coherent in g with a given point, **a** *nerve of* g, then the skeleton of the figure and likewise each coherence thread furnishes *one* nerve of g, and each coherence band furnishes *two* nerves of g.

If we choose only *one* special directly coherent set of numbers, then our construction furnishes a closed curve (in the sense of SCHOENFLIES) which can be divided into two improper arcs of curve but not into two proper ones, in which category is included the primitive figure from which we started.

§ 4.

The impossibility of a linear arrangement of the points of an irreductible continuum. [[7]]

By ZORETTI lately a method has been explained of arranging the points of an irreductible continuum linearly, analogously to those of a line segment [1]).

His method is however inapplicable to several continua constructed in the article "*Zur Analysis Situs*" cited above. [[6]]

This having been pointed out to him, ZORETTI has based a method of more restricted aim on the following theorem [2]):

"*Given an irreductible continuum C and a point c of C, then C can be divided in one definite manner into three sets of points C_1, C_2 and Γ, possessing the following properties: C_1 and C_2 are coherent and have c as their only common point; Γ consists of the common limiting points of C_1 and C_2. Both sets of points $C_1 + \Gamma$ and $C_2 + \Gamma$ are irreductible continua.*"

1) Annales de l'École Normale, 1909, p. 485—497. [[3]]

2) Comptes Rendus, t. 151, p. 202. [[8]]

From this theorem would follow that, if not all the points of C, yet a considerable part of them would be capable of linear arrangement, and that it would be possible to crumble C in the same way as a line segment into an indefinitely large number of linearly arranged "*partial arcs*", two arbitrary ones of which then and only then cohere, if in that linear order they succeed each other immediately.

But neither this theorem can be maintained, if we try to apply it to our primitive figure.

If namely we choose this closed curve as the irreductible continuum C, then either $C_1 + \Gamma$ or $C_2 + \Gamma$ must be identical to C, and either C_2 or C_1 reduces to the single point c, so that the division of C becomes illusory.

[[7]] It is a priori certain that all attempts to arrange the points of such a continuum linearly by repeated crumblings must fail, the crumbling being practicable only for a single system of directly coherent points, and therefore the linear arrangement being restricted in any case to points of a single *nerve*.

And even of this we are not sure for the most general irreductible continuum. For, in a system of points directly coherent in C again may be contained an irreductible continuum C'' breaking up into a set of the power of infinity of the continuum of systems of points directly coherent in C''. And so on.

§ 5.

[[9]] *A generalization of* JORDAN'S *theorem.*

JORDAN'S theorem runs that a continuous one-one image of a circle is a closed curve, i.e. divides the plane into two regions of which it is the common boundary.

[[10]] The extension lying at hand that a continuous one-one image of a closed curve is again a closed curve, has not yet been proved. However, for a special kind of closed curves a partial result can be arrived at, as we shall explain in the following.

Let C be an arbitrary closed curve, and let us represent by $c\eta_i$ the cyclic type of order of its points accessible from its inner region. A Schnitt s arbitrarily given in $c\eta_i$ determines two "*Schnitt-continua*" σ_l and σ_r, to which $c\eta_i$ converges on the left resp. on the right of s. The points common to σ_l and σ_r form a closed set of points σ, to be called "*the juncture belonging to the Schnitt s*".

LEMMA. *In the inner region of C we can construct an arc of curve which abroad from its ends is simple, and of which one end reduces to a single point of the inner region of C and the other is contained in the juncture σ.*

[[392]]

PROOF. Let M be a point taken arbitrarily in the inner region of C, and let $a_1, a_2, a_3, \ldots$ be a fundamental series of indefinitely decreasing arcs of simple curve lying abroad from their endpoints in the inner region of C, whilst the endpoints belong to ${}_C\eta_i$, and are separated by the Schnitt s. In this series is contained a series $b_1, b_2, b_3, \ldots$ converging to a single point P of σ, and in which each b_n is separated from M by $C + b_{n-1}$. We can then join M and P by an arc of simple curve z cutting an infinite number of the b_α in the order of their indices each in one point and passing there from their side turned to M to their side turned to s. The b_α intersected in this way form a series $d_1, d_2, d_3, \ldots$ Let z_n be the part of z enclosed between d_n and d_{n+1}, D_n the point of intersection of z and d_n, A_n resp. B_n the left resp. right endpoint of d_n, φ_n resp. ψ_n the arc of curve determined on C by the Schnitte corresponding to A_n and A_{n+1} resp. to B_n and B_{n+1}, ϱ_n resp. τ_n the part of the inner region of C cut off by the arc of simple curve $A_n D_n D_{n+1} A_{n+1}$ resp. $B_n D_n D_{n+1} B_{n+1}$, u_n resp. v_n the part of φ_n resp. ψ_n lying in τ_n resp. ϱ_n. We then can join D_n and D_{n+1} by an arc of simple curve t_n lying entirely in the part of the inner region of C enclosed between d_n and d_{n+1} and moving away from $z_n + u_n + v_n$ no farther than a certain maximum distance ε_n indefinitely decreasing for indefinitely increasing n. These arcs t_n form together an arc of curve possessing the properties required.

THEOREM 4. *If the closed curve C is divided by the Schnitte s_1 and s_2 of ${}_C\eta_i$ into two proper (i.e. not identical to C) arcs of curve C_1 and C_2, then the points common to C_1 and C_2 form a non-coherent set of points C_{12}.*

PROOF. Let σ_1 resp. σ_2 be the juncture belonging to s_1 resp. s_2, then according to the lemma just now proved we can draw from a point M taken arbitrarily in the inner region of C to ends e_1 and e_2 contained in σ_1 resp. σ_2 two arcs of curve which abroad from their ends are simple, do not meet each other, and lie entirely in the inner region of C. These arcs of curve we represent by F_1 and F_2 (see the schema in fig. 2), and the largest perfect coherent part of C_{12}, containing e_1 resp. e_2, by p_1 resp. p_2. Let Q_1 resp. Q_2 be a point of ${}_C\eta_i$ belonging to C_1 but not to C_2, resp. to C_2 but not to C_1. Then from M to Q_1 and Q_2 we can draw paths w_1 and w_2 which abroad from their ends lie entirely in the inner region of C, and meet neither each other nor F_1 or F_2. In the inner region of C these paths w_1 and w_2 are separated by F_1 and F_2.

About Q_1 as centre we describe a small circle $\varkappa_1$ which together with its inner region has no point in common with $C_2 + F_1 + F_2$, and we draw a path w'_1 joining the infinite with a point Q'_1 of $\varkappa_1$, and abroad from Q'_1 lying entirely in the outer region of C as well

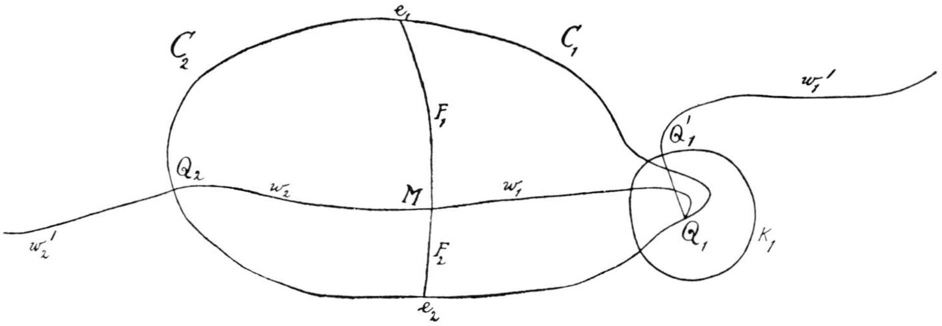

Fig. 2.

as in the outer region of $\varkappa_1$. Then w'_1 forms together with some parts of w_1 and of the radius $Q_1 Q'_1$ of $\varkappa_1$ a path l_1 joining M with the infinite, and abroad from M not meeting $C_2 + F_1 + F_2$.

In the same way we can construct a path l_2 joining M with the infinite, coinciding for a certain initial part with a part of w_2, and abroad from M not meeting $C_1 + F_1 + F_2$.

As in the vicinity of M the arcs F_1 and F_2 are separated by w_1 and w_2, in the complete plane e_1 and e_2 are separated by $l_1 + l_2$ (whether l_1 and l_2 meet each other abroad from M or not).

So, since $l_1 + l_2$ contains no point of C_{12}, also p_1 and p_2 are separated in the complete plane by $l_1 + l_2$. Hence p_1 and p_2 cannot be identical, and C_{12} cannot be a continuum.

As furthermore two finite continua whose common points form a non-coherent set determine more than one region in the plane [1]), from theorem 4 ensues immediately:

THEOREM 5. *A continuous one-one image of a closed curve divisible into two proper arcs of curve determines in the plane more than one region.*

---

[1]) This may be proved by breaking up the boundary of a region determined by the common points of these continua into two closed sets c_1 and c_2 possessing a finite distance from each other, and then applying the reasoning of Mathem. Annalen, Vol. 68, p. 430.

[[6]]

[[394]]

NOTES

⟦1⟧ The English version appeared in the Proceedings of the meeting of 24 June 1911. See Brouwer 1910 B2.

⟦2⟧ Brouwer's irreducibility theorem. See C. Kuratowski 1952, p. 27, Hurewicz–Wallmann 1948, p. 161.

⟦3⟧ L. Zoretti 1909. See also Brouwer 1910 J.

⟦4⟧ S. Janiszewski 1910.

⟦5⟧ E. Mazurkiewicz 1910.

⟦6⟧ Brouwer 1910 C.

⟦7⟧ Quoted by C. Kuratowski 1927, p. 225 footnote [3]), p. 227 footnote [3]).

⟦8⟧ L. Zoretti 1910.

⟦9⟧ This is related in method to Brouwer 1910 E.

⟦10⟧ The invariance of the closed curve, proved in Brouwer 1912 L.

1913 B2 **Mathematics.** — "*Some remarks on the coherence type* η." By
[[1]] Prof. L. E. J. BROUWER.

[[2]] In order to introduce the notion of a "coherence type" we shall say that a set M is *normally connected*, if to some sequences f of elements of M are adjoined certain elements of M as their "limiting elements", the following conditions being satisfied:

1st. each limiting element of f is at the same time a limiting element of each end segment of f.

[[3]] **2nd**. for each limiting element of f a partial sequence of f can be found of which it is the *only* limiting element.

3rd. each limiting element of a partial sequence of f is at the same time a limiting element of f.

4th. if m is the only limiting element of the sequence $\{m_k\}$ and

m_μ for μ constant the only limiting element of the sequence $\{m_{\mu\nu}\}$, then each of the latter sequences contains such an end segment $\{m_{\mu\lambda}\}$, that an arbitrary sequence of elements $m_{\mu\lambda}$ for which μ continually increases, possesses m as its only limiting element.

The sets of points of an n-dimensional space form a special case of normally connected sets.

Another special case we get in the following way: In an *n-ply ordered* set [1]) we understand by an *interval* the partial set formed by the elements u satisfying for $\varrho \leq n$ different values of i a relation of the form

$$b_i \overset{i}{<} u \overset{i}{<} c_i \quad \text{or} \quad b_i \overset{i}{<} u \quad \text{or} \quad u \overset{i}{<} c_i \;; \qquad [\![3a]\!]$$

we further define an element m to be a *limiting element* of a sequence f, if each interval containing m, contains elements of f not identical to m, and the given set to be *everywhere dense*, if none of its intervals reduces to zero. Then the *everywhere dense, countable, n-ply ordered* sets which will be considered more closely in this paper, likewise belong to the class of normally connected sets.

A representation of a normally connected set preserving the limiting element relations, will be called a *continuous representation.*

If of a normally connected set there exists a continuous one-one representation on an other normally connected set, the two sets will be said to possess *the same coherence type.*

One of the simplest coherence types is the type η already introduced by CANTOR [2]). From a proof of CANTOR follows namely:

THEOREM 1. *All countable sets of points lying everywhere dense on the open straight line, possess the same coherence type* η.

The proof is founded on the following construction of a one-one correspondence *preserving the relations of order*, between two sets of points $M = \{m_1, m_2, \ldots\}$ and $R = \{r_1, r_2, \ldots\}$ of the class considered: To r_1 CANTOR makes to correspond the point m_1; to r_2 the point m_{i_2} with the smallest index, having with respect to m_1 the same situation (determined by a relation of order), as r_2 has with respect to r_1; to r_3 the point m_{i_3} with the smallest index, having with respect to m_1 and m_{i_2} the same situation (determined by two relations of order), as r_3 has with respect to r_1 and r_2; and so on. That in this way not only all points of R, but also all points of M have their turn, i.o.w. that if among $m_1, m_{i_2}, \ldots m_{i_\lambda}$ appear $m_1, m_2, \ldots m_\nu$, *but not* $m_{\nu+1}$, there exists a number σ with the property that $m_{\nu+1} = m_{i_{\lambda+\sigma}}$,

[1]) Comp. F. RIESZ, Mathem. Annalen 61, p. 406. [\![4]\!]
[2]) Mathem. Annalen 46, p. 504. [\![5]\!]

is evident by choosing for $r_{\nu+1}$ the point of R with the smallest index, having with respect to $r_1, r_2, \ldots r_\nu$ the same situation, as $m_{\nu+1}$ has with respect to $m_1, m_{i_2}, \ldots m_{i_\nu}$. The correspondence constructed in this way, is at the same time *continuous*; for, the limiting point relations depend exclusively on the relations of order, as a point m is then and only then a limiting point of a sequence f, if each interval containing m contains an infinite number of points of f.

The above proof shows at the same time the independence of the coherence type η of the linear continuum. For, after CANTOR it leads also to the following more general result:

THEOREM 2. *All everywhere dense, countable, simply ordered sets possess the coherence type* η.[1])

Theorem 1 may be extended as follows:

THEOREM 3. *If on the open straight line be given two countable, everywhere dense sets of points M and R, a continuous one-one transformation of the open straight line in itself can be constructed, by which M passes into R.*

In order to define such a transformation, we first by CANTOR's method construct a continuous one-one representation of M on R. Then the order of succession of the points of M is the same as the order of succession of the corresponding points of R. We further make to correspond to each point gm of the straight line *not* belonging to M, the point gr having to the points of R the same relations of order, as gm has to the corresponding points of M. In this way we get a one-one transformation of the straight line in itself, preserving the relations of order. On the grounds indicated in the proof of theorem 1 this transformation must also be a continuous one.

Analogously to theorem 3 is proved:

THEOREM 4. *If within a finite line segment be given two countable, everywhere dense sets of points M and R, a continuous one-one transformation of the line segment, t h e e n d p o i n t s i n c l u d e d, in itself can be constructed, by which M passes into R.*

We shall now treat the question, to what extent the theorems 1, 2, 3, and 4 may be generalized to polydimensional sets of points

1) The possibility of a definition founded exclusively on relations of order, shewn by CANTOR not only for the coherence type ν, but likewise for the coherence type ϑ of the complete linear continuum, holds also for the coherence type ζ of the
[[6]] perfect, punctual sets of points in R_n (comp these Proceedings XII, p. 790). As is easily proved, this coherence type belongs to *all perfect, nowhere dense, simply ordered sets of which the set of intervals is countable* (an "interval" is formed here by each pair of elements between which no further elements lie).

on one hand, and to multiply ordered sets on the other hand. In the first place the following theorem holds here:

THEOREM 5. *All countable sets of points lying everywhere dense in a cartesian R_n, possess the same coherence type η^n.* [1])

For, to an arbitrary countable set of points, lying everywhere dense in R_n, we can construct a cartesian system of coordinates C_m with the property that no R_{n-1} parallel to a coordinate space contains more than *one* point of the set. If now two such sets, M and R, are given, then in the special case that C_m and C_r are identical, a one-one representation of M on R preserving the n-fold relations of order as determined by $C_m = C_r$, can be constructed by CANTOR's method cited above, only modified in as far as the "situation" of the points with respect to each other is determined here not by simple, but by n-fold relations of order. As on the grounds indicated in the proof of theorem 1 this representation must also be a continuous one, theorem 5 has been established in the special case that C_m and C_r are identical. From this the general case of the theorem ensues immediately.

If on the other hand we have an arbitrary *everywhere dense, countable, n-ply ordered* set Z, then its n simple *projections* [2]), being everywhere dense, countable, and simply ordered, admit of one-one representations preserving the relations of order, on n countable sets of points lying everywhere dense on the n axes of a cartesian system of coordinates successively; these n representations determine together a one-one representation preserving the relations of order, thus a continuous one-one representation of Z on a countable set of points, everywhere dense in R_n. From this we conclude on account of theorem 5:

THEOREM 6. *All everywhere dense, countable, n-ply ordered sets possess the coherence type η^n.*

As the n-dimensional analogon of theorem 3 the following extension of theorem 5 holds:

THEOREM 7. *If in a cartesian R_n be given two countable, everywhere dense sets of points M and R, a continuous one-one transformation of R_n in itself can be constructed, by which M passes into R.* ⟦8⟧

In the special case that C_m and C_r are identical, we can namely first construct a continuous one-one correspondence between M and R in the manner indicated in the proof of theorem 5, and then make to correspond to each point gm *not* belonging to M, the point gr having to the points of R the same (n-fold) relations of order, as gm has

[1]) This theorem and its proof have been communicated to me by Prof. BOREL. ⟦7⟧
[2]) Comp. F. RIESZ, l.c. p. 409. ⟦3⟧

to the corresponding points of M. In this way we get a one-one transformation of R_n in itself *preserving the relations of order as determined by* $C_m = C_r$. As on the grounds indicated in the proof of theorem 1 this transformation is also a continuous one, theorem 7 has been established in the special case that C_m and C_r are identical. From this the general case of the theorem ensues immediately.

The n-dimensional extension of theorem 4 runs as follows:

[[9]] THEOREM 8. *If within an n-dimensional cube be given two countable, everywhere dense sets of points M and R, a continuous one-one transformation of the cube, the boundary included, in itself can be constructed, by which M passes into R.*

The proof of this theorem is somewhat more complicated than those of the preceding ones. We choose in R_n such a rectangular system of coordinates that the coordinates $x_1, x_2, \ldots x_n$ of the cube vertices are all either $+1$ or -1, and for $p = 1, 2, \ldots n$ successively we try to form a continuous transition between the $(n-1)$-dimensional spaces $x_p = -1$ and $x_p = +1$ by means of a onedimensional continuum s_{mp} of plane $(n-1)$-dimensional spaces meeting each other neither in the interior nor on the boundary of the cube, and containing each at most one point of M. In this we succeed as follows: Let $S = \sum_{p=1}^{n} a_p x_p = c$ be a plane $(n-1)$-dimensional space containing *no* straight line parallel to a line r_m joining two points of M, and through each point $(x_1 = x_2 = \ldots = x_{p-1} = 0,$ $x_p = a, x_{p+1} = x_{p+2} = \ldots = x_n = 0)$ let us lay an $(n-1)$-dimensional space: $x_p + e(1 - a^2) S = a + e a_p a (1 - a^2)$; in this way we get a continuous series σ_e of plane $(n-1)$-dimensional spaces, and we can choose a magnitude e_1 with the property that for $e < e_1$ two arbitrary spaces of σ_e meet each other neither in the interior nor on the boundary of the cube. As further an $(n-1)$-dimensional space belongs to *at most one* σ_e, thus a line r_m is contained in an $(n-1)$-dimensional space belonging to σ_e for *at most one* value of e, and the lines r_m exist in countable number only, it is possible to choose a suitable value for $e < e_1$ with the property that no space of σ_e contains a line r_m, i.o.w. that σ_e satisfies the conditions imposed to s_{mp}.

If for each value of p we choose out of s_{mp} an arbitrary space, then these n spaces possess one single point, lying in the interior of the cube, in common. For, by projecting an arbitrary space of s_{m1} together with the sections determined in it by $s_{m2}, s_{m3}, \ldots s_{mn}$, into the space $x_1 = 0$, we reduce this property of the n-dimensional cube to the analogous property of the $(n-1)$-dimensional cube. So if we introduce as the coordinate x_{mp} of an arbitrary point H lying in the

[[400]]

interior or on the boundary of the cube, the value of x_p in that point of the X_p-axis which lies with H in one and the same space of s_{mp}, then to each system of values > -1 and ≤ 1 for $x_{m1}, x_{m2}, \ldots x_{mn}$ corresponds one and only one point of the interior or of the boundary of the cube, which point is a biuniform, continuous function of x_{m1}, $x_{m2}, \ldots x_{mn}$. I.o.w. the transformation $\{x'_p = x_{mp}\}$, to be represented by T_m, is a continuous one-one transformation of the cube with its boundary in itself, by which M passes into a countable, everywhere dense set of points M_1 *of which no* $(n-1)$*-dimensional space parallel to a coordinate space contains more than one point.*

In the same way we can define a continuous one-one transformation T_r of the cube with its boundary in itself, by which R passes into a countable, everywhere dense set of points R_1 *of which no* $(n-1)$*-dimensional space parallel to a coordinate space contains more than one point.*

Further after the proof of theorem 7 a continuous one-one transformation T of the cube with its boundary in itself exists, by which M_1 passes into R_1, so that the transformation

$$T_r^{-1} \, . \, T \, . \, T_m$$

possesses the properties required by theorem 8.

We now come to a property which at first sight seems to clash with the conception of dimension:

THEOREM 9. *The coherence types* η^n *and* η *are identical.*

To prove this property, in an n-dimensional cube for which the rectangular coordinates of the vertices are all either 0 or 1, we consider the set M_n of coherence type η^n consisting of those points whose coordinates when developed into a series of negative powers of 3, from a certain moment produce exclusively the number 1, and together with this we consider the set M of coherence type η consisting of those real numbers between 0 and 1 which when developed into a series of negative powers of 3^n, from a certain moment produce exclusively the number $\frac{3^n-1}{2}$. The continuous PEANO representation [1]) of the real numbers between 0 and 1 on the n-dimensional cube with edge 1, then determines a *continuous one-one representation of* M *on* M_n establishing the exactness of theorem 9.

That in reality theorem 9 *does not* clash with the conception of dimension, is elucidated by the remark that *not every continuous one-one correspondence between two countable sets of points M and R*,

[1]) Comp. Math. Annalen 36, p. 59, and SCHOENFLIES, Bericht über die Mengenlehre I, p. 125. [10]

lying everywhere dense in R_n, admits of an extension to a continuous one-one transformation of R_n in itself. If e.g. the set of the rational points of the open straight line is submitted to the continuous one-one transformation $x' = \frac{1}{\pi - x}$, this transformation *does not* admit of an extension to a continuous one-one transformation of the open straight line in itself.

A more characteristic example, presenting the property moreover that in no partial region an extension is possible, we get as follows: Let t_1 denote the set of those real numbers between 0 and 1 of which the development in the nonal system from a certain moment produces exclusively the digit 4, t_2 the set of the finite ternal fractions between 0 and 1. Let T denote a continuous one-one transformation of the set of the real numbers between 0 and 1 in itself, by which t_1 passes into $t_1 + t_2$, thus a part t_3 of t_1 into t_1, and a part t_4 of t_1 into t_2. By a PEANO representation T_1 the sets t_1, t_2, t_3, t_4 successively pass into countable sets of points s_1, s_2, s_3, s_4, lying everywhere dense within a square with side unity, and, so far as are concerned, s_1, s_3, and s_4, containing no points of the boundary of this square. The continuous one-one representation T of t_3 on t_1 now determines a *continuous one-one representation $T_2 = T_1 T T_1^{-1}$ of s_3 on s_1, not capable of an extension to a continuous one-one representation of the interior of the square in itself.* For, if such an extension would exist, it would be, for each set of points in the interior of the square, the only possible continuous extension of T_2. For s_1, however, $T_1 T T_1^{-1}$ furnishes itself such a continuous extension, which we know to be *not* a one-one representation.

The conception of dimension can now be saved, at least for the everywhere dense, countable *sets of points*, by replacing the notion of *coherence type* by the notion of *geometric type* [1]). Two sets of points will namely be said to possess the same *geometric type*, if a *uniformly continuous* one-one correpondence exists between them. And it is for uniformly continuous representations that the following property holds:

THEOREM 10. *Every uniformly continuous one-one correspondence between two countable sets of points M and R, lying everywhere dense in an n-dimensional cube, admits of an extension to a continuous one-one transformation of the cube with its boundary in itself.*

[[6]] 1) For closed sets the two notions are equivalent. For these they were introduced formerly under the name of *geometric type of order*, these Proceedings XII, p. 786.

[[402]]

For, on account of the uniform continuity of the correspondence between M and R, to a sequence of points of M possessing only one limiting point, a sequence of points of R likewise possessing only one limiting point, must correspond, and reciprocally. On this ground the given correspondence already admits of an extension to a one-one transformation of the cube with its boundary in itself of which we have still to prove the continuity in the property that a sequence $\{g_{m\nu}\}$ of limiting points of M converging to a single limiting point $g_{m\omega}$, the sequence $\{g_{r\nu}\}$ of the corresponding limiting points of R converges likewise to a single limiting point. For this purpose we adjoin to each point $g_{m\nu}$ a point m_ν of M possessing a distance $< \varepsilon_\nu$ from $g_{m\nu}$, the distance between $g_{r\nu}$ and the point r_ν corresponding to m_ν likewise being $< \varepsilon_\nu$, and for ν indefinitely increasing we make ε_ν to converge to zero. Thus $\{m_\nu\}$ converging exclusively to $g_{m\omega}$, $\{r_\nu\}$ likewise possesses a single limiting point $g_{r\omega}$, and also $\{g_{r\nu}\}$ must converge exclusively to $g_{r\omega}$.

On account of the invariance of the number of dimensions [1]) we can enunciate as a corollary of theorem 10:

THEOREM 11. *For $m < n$ the geometric types η^m and η^n are different.*

As, however, for normally connected sets in general the notion of uniform continuity is senseless, the *indeterminateness of the number of dimensions of everywhere dense, countable, multiply ordered sets*, as expressed in theorem 9, must be considered as irreparable.

[1]) Comp. Math. Annalen 70, p. 161.

[[11]]

NOTES

⟦1⟧ The Dutch version was communicated at the meeting of 25 April 1913. The English version appeared in the Proceedings of the meeting of 25 April 1913.

⟦2⟧ The strange looking term 'coherence' is probably derived from G. Cantor's 'Cohärenz' (= set of accumulation points).

⟦3⟧ Brouwer's copy of the Dutch text indicates the changes (translated):

2nd. for each limiting element of f a partial sequence can be found every partial sequence of which has f as its *only* limiting element

4th. if m is the only limiting element of all partial sequences of the sequence $\{m_\mu\}$ and $m_{\mu\nu}$ for μ constant the only limiting element of all partial sequences of the sequence $\{m_{\mu\nu}\}$ then $\cdots$

⟦3a⟧ Added in Brouwer's copy of the Dutch text: $(b_i \overset{i}{<} c_i)$ before the first 'or'.

⟦4⟧ F. Riesz 1905.

⟦5⟧ G. Cantor 1895.

⟦6⟧ Brouwer 1910 B2.

⟦7⟧ This was probably an oral communication, at the International Congress of Mathematicians in Cambridge in August 1912 (see Y58). The theorem itself is superseded by Brouwer's theorem 9. Brouwer's formulations seem to indicate that Borel did not remark that his method allowed him to prove the much stronger theorem 7. Brouwer's papers contain a correspondence with E. Borel, of a later date and related to theorem 8, which he had discussed with Borel at Cambridge, and for which Borel gave an incorrect proof. See 1916 A, 1916 B, Y57, Y58.

⟦8⟧ M. Fréchet 1910, p. 159 had enunciated a theorem which essentially coincides with theorem 7, but the proof was wrong as noticed by P. Urysohn 1925, p. 83–89. Clearly neither Borel nor Brouwer knew about Fréchet's statement nor Urysohn about Brouwer's paper.

M. Fréchet 1928, p. 49 reacted to Urysohn's remark; he had noticed his mistake earlier and found a way of correcting it. M. Fréchet 1928, p. 49 said: 'Dans l'intervalle, d'ailleurs, M. Brouwer avait publié en 1913 une communication de M. Borel, énonçant et démontrant le même théorème.' This is not correct; in any case it does not follow from Brouwer's text I interpret to mean that Borel gave Brouwer a proof of theorem 5 and that Brouwer noticed that the method even allows one to prove theorem 7. In this connection M. Fréchet also mentioned Borel 1922, but not Borel 1913 nor Brouwer's criticism (see ⟦9⟧ and 1916 B ⟦2⟧).

⟦9⟧ This theorem is related to Borel 1913 (see Brouwer 1916 A).

⟦10⟧ G. Peano 1890 A, A. Schoenflies 1900.

⟦11⟧ Brouwer 1911 C.

L. E. J. BROUWER. Eenige opmerkingen over het samenhangstype η. Amst. Ak. Versl. **21**, 1412-1419. [[1]]

In der Terminologie des Verf. besitzen zwei Mengen denselben „Zusammenhangstypus", wenn sie sich eineindeutig und stetig aufeinander abbilden lassen, und denselben „geometrischen Typus", wenn diese Abbildung überdies gleichmäßig stetig ist. Es wird gezeigt, daß alle mehrfach geordneten, abzählbaren und überall dichten Mengen denselben Zusammenhangstypus besitzen, nämlich den Typus η der rationalen Zahlen. Weil bei diesen Mengen vom Begriff des geometrischen Typus keine Rede sein kann, so kommt für sie die Invarianz der Dimensionenzahl in Fortfall. Sodann beweist der Verf. folgendes geometrische Theorem: „Wenn im Innern eines n-dimensionalen Kubus zwei abzählbare, überall dichte Punktmengen gegeben sind, so lassen sie sich ineinander überführen mittels einer eineindeutigen und stetigen Transformation des Kubus einschließlich der Grenze in sich." Brw.

NOTE

[[1]] This is an author's review of 1913 B1. See also 1913 B2, 1916 B, Y57. Brouwer's papers contain a manuscript of this review. Deviations are probably due to the editor of *Jahrbuch*.

[[1]] É. Borel. Les ensembles de mesure nulle. S. M. F. Bull. **41**, 1-19.

Eine Punktmenge M wird „regulär" genannt, wenn sie sich in folgender Weise definieren läßt: Es sei $A_1, A_2, \ldots, A_n, \ldots$ eine abzählbare Punktmenge, die als „Menge der Fundamentalpunkte" bezeichnet werden möge, und $C_1^{(h)}$, $C_2^{(h)}, \ldots, C_n^{(h)}$ für jedes ganze positive h eine abzählbare Menge von Quadraten, deren Flächeninhalte eine konvergierende Reihe bilden, während $C_n^{(h)}$ in seinem Inneren $C_n^{(h+1)}$ enthält und sich für festes n und unbeschränkt wachsendes h auf A_n zusammenzieht. Wenn noch die Vereinigungsmenge von $C_1^{(h)}, C_2^{(h)}, \ldots$ mit E_h bezeichnet wird, so wird M als der gemeinsame Durchschnitt aller E_h definiert. Der Verf. zeigt zunächst, daß jede Punktmenge vom Maße Null in einer regulären Punktmenge enthalten ist, wendet sich darauf zum Studium derjenigen regulären Punktmengen, welche in einem Bereiche überall dicht liegen, und stellt folgendes Lemma auf: „Es seien C und C' zwei kongruente Kreise, A und B zwei abzählbare Punktmengen, von denen die erste auf C und im Inneren von C, die zweite auf C' und im Inneren von C' überall dicht liegt, und ε eine beliebig klein vorgegebene positive Zahl; alsdann lassen die Punkte von A und B sich derart numerieren, daß für jedes p und jedes q :
[[2]] $1-\varepsilon<\frac{A_p A_q}{B_p B_q}<1+\varepsilon$." Der von diesem Lemma gegebene Beweis ist nicht stichhaltig: damit die S. 9 des Aufsatzes entwickelte Konstruktion sich unbeschränkt fortsetzen lasse, genügt es nicht, bei jedem Schritt die Bedingung zu erfüllen, daß die Verhältnisse der Längen der sich entsprechenden Seiten des bis dahin konstruierten Bereichs zwischen $(1-\varepsilon_1)\ldots(1-\varepsilon_n)$ und $(1+\varepsilon_1)\ldots(1+\varepsilon_n)$ liegen; denn wenn für das Verhältnis der Abszissen- und Ordinatendifferenzen

von zwei willkürlichen sich entsprechenden Eckpunktepaaren der vollständigen
Bereichsysteme nicht dasselbe der Fall ist, kann es sehr gut vorkommen, daß
bei Fortsetzung der Konstruktion auch den vom Verf. vorgeschriebenen Bedin- ⟦3⟧
gungen nicht mehr genügt werden kann. Anders ausgedrückt, der Verf. beachtet
bei der Wahl eines neuen Punktes A_n nur die Abszisse und die Ordinate von A_n,
während in Wirklichkeit die Abszissen der Punkte des Kreisumfanges, welche
dieselbe Ordinate wie A_n, und die Ordinaten der Punkte des Kreisumfanges,
welche dieselbe Abszisse wie A_n besitzen, auf die Fortsetzbarkeit der Konstruktion
ebensogut von Einfluß sind. Das Lemma ist indessen richtig und kann leicht
bewiesen werden mittels der Abbildung, welche Ref. zur Begründung des am ⟦4⟧
Schlusse des vorstehenden Referates formulierten Theorems benutzt hat. Aus
dem Lemma zieht der Verf. die Folgerung, daß zu einer willkürlichen regulären ⟦5⟧
Punktmenge R, deren Menge der Fundamentalpunkte B im Inneren von C ⟦6⟧
überall dicht liegt, eine reguläre Punktmenge S existiert, deren Menge der
Fundamentalpunkte A im Inneren von C überall dicht liegt, und welche der
Menge R eineindeutig und stetig mit einem zwischen $1 - \varepsilon$ und $1 + \varepsilon$ schwanken-
den Ähnlichkeitsverhältnis entspricht. Im Schlußparagraphen schlägt der Verf.
eine auf das asymptotische Dekrement der einschließenden Intervalle gegründete
Klassifizierung der regulären Punktmengen vor und zieht die physikalische
Folgerung, daß die Hypothese, die Gleichgewichtslagen der Schwerpunkte der
Moleküle eines festen Körpers bilden eine überall dichte Punktlage, gleich-
wertig ist mit der Annahme, daß dieselben mit der Menge der rationalen Punkte
zusammenfallen.

Brw.

NOTES

⟦1⟧ This is a review of E. Borel 1913 by Brouwer. A manuscript of this review is among Brouwer's papers. There are a few deviations, probably due to the editor of *Jahrbuch*.

⟦2⟧ The error had also been indicated to Borel, in a letter, Y57. Strange enough, the whole proof was literally reproduced in the book Borel 1922, pp. 25–28.

⟦3⟧ The word 'auch' in line 3, which is meaningless, does not occur in the earlier mentioned manuscript.

⟦4⟧ Brouwer 1916 A, 1913 B2.

⟦5⟧ The formulation is not clear. Of course S has to be constructed as a regular set with respect to a *given A*.

⟦6⟧ C should be C' in line 13 of p. 557 which, indeed, is shown by the manuscript.

Y 57 Letter to E. Borel, copy

Blaricum, 7.11.13.

⟦1⟧ Cher Monsieur Borel

En regardant de plus près à votre démonstration, il me semble qu'elle aurait besoin d'une seconde modification, garantissant qu'à chaque étappe de la construction de la page 9 non seulement le rapport *des dimensions homologues des régions construites* soit compris entre $(1-\varepsilon_1)\cdots(1-\varepsilon_n)$ et $(1+\varepsilon_1)\cdots(1+\varepsilon_n)$, mais qu'il en soit de même pour le rapport *des différences homologues des abscisses et des ordonnées de tous les sommets des régions construites.*

Sans cela, dans le cours de la construction, il peut arriver que tout d'un coup on se heurte à des régions homologues de la forme suivante:

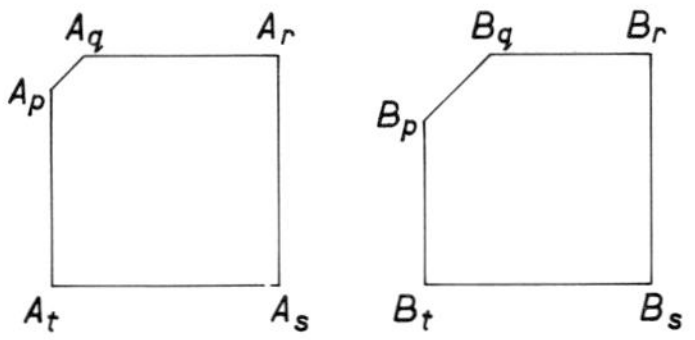

où A_qA_r et B_qB_r, A_rA_s et B_rB_s, A_sA_t et B_sB_t, A_tA_p et B_tB_p ont à peu près la même longueur, mais où B_pB_q est beaucoup des ⟦sic⟧ fois plus grand que A_pA_q. Et en ce cas-là, dans la suite de la construction, il devient impossible de satisfaire aux conditions imposées.

Autrement dit, il faudrait, au choix de chaque point nouveau A_n, faire attention non seulement à l'abscisse et à l'ordonnée de A_n, mais aussi bien aux abscisses des points de la circonférence ayant même ordonnée que A_n et aux ordonnées des points de la circonférence ayant même abscisse que A_n.

Autant que je vois, il y a là une difficulté plus grave que celles que je vous signalais dans ma lettre précédente; c'est pour cette raison que la construction
⟦2⟧ donnée dans ma note de l'Académie d'Amsterdam, me semble présenter quelques avantages, car c'est d'une manière très simple qu'on peut la ⟦sic⟧ faire satisfaire aux conditions

$$1-\varepsilon < \frac{A_pA_q}{B_pB_q} < 1+\varepsilon \quad \text{(p. 7 de votre article)}$$

et $p^\alpha < n < p^\beta$ (p. 13 de votre article)

En espérant que vous me pardonnez mes remarques et que, si je me trompe, vous voudrez bien me désabuser en deux mots, je vous prie d'agréer mes salutations respectueuses et cordiales.

L. E. J. Brouwer.

NOTES

⟦1⟧ With respect to the correspondence with E. Borel, Brouwer's papers contain the present copy of a letter of Brouwer to Borel, and two letters of Borel to Brouwer dated 7 November 1913 and 8 November 1913, related to Brouwer's letter but with no relevant information. See, however, Y58. The subject of the present letter was published in 1916 B, but the exposition in the present letter is more to the point. As remarked in 1916 B⟦2⟧ the proof was literally reproduced in Borel 1922, pp. 25–28.

⟦2⟧ Brouwer 1913 B2.

Y 58

Letter to H. Blaschke, draft

Blaricum, 19.11.15

Lieber Herr Blaschke

Vor einem paar Jahren bat mich schon Blumenthal, für seine bei Teubner erscheinende Sammlung: "Fortschritte der mathematischen Wissenschaften in Monografieën" ein Buch zu redigieren. Ich meinte damals es noch nicht versprechen zu können; jetzt haben sich meine Umstände etwas ⟨steht es schon ein wenig⟩ geändert, und glaube ich schon zusagen zu können. Sie können das in
⟦1⟧ meinem Namen Herrn Ackermann mitteilen; weisen Sie ihn aber bitte darauf hin, das ⟦sic⟧ Blumenthal ein altes Recht hat, das Buch für seine Sammlung zu bekommen. Der Titel würde etwa lauten: "Neuere Untersuchungen über Analysis Situs"̣. Es ist meine Absicht, auch Arbeiten anderer (Tietze, Carathéodory, Lebesgue, Sierpinski) mit hineinzuverarbeiten.

Ihr Brief hat mich plötzlich daran erinnert, dass Sie mich schon gegen Weih-
⟦2⟧ nachten des vorigen Jahres ein Referat für die Fortschritte über meine Arbeit
⟦3⟧ "Eenige Opmerkingen over het samenhangstype η" gefragt haben ⟦sic⟧. Ich bin nun ganz trostlos darüber, dass ich damals so sehr von Prüfungen in Anspruch genommen war, das ich einfach vergessen habe, Ihnen zu antworten. Ich bitte Ihnen ⟦sic⟧ von ganzem Herzen um Entschuldigung dafür. Es wäre übrigens in meinem Interesse gewesen, Ihnen sofort sachlich wie folgt zu antworten:

⟦4⟧ Der Inhalt der Arbeit steht angegeben in der Revue Semestrielle des Publications Mathématiques XXI, 2, p. 99. Der wesentliche Punkt wird gebildet von dem
⟦5⟧ daselbst in den letzten drei bis vier Zeilen ausgesprochenen Satze. Der Satz kam
⟦6⟧ zur Sprache in einer mündlichen Unterhaltung in Cambridge zwischen Borel und mir über Mengen vom Maass Null; keiner von uns hatte damals einen Beweis. Wir publizierten darauf beide unabhängig von einander gleichzeitig je einen Be-
⟦3⟧ weis, ich in der Arbeit "Eenige opmerkingen enz.", Borel in dem Aufsatz: "Les en-
⟦7⟧ sembles de mesure nulle", erschienen im Bulletin de la Soc. Math. de France, 41
⟦8⟧ (1913), S. 6–14. Der Borelsche Beweis ist aber falsch, weil er S. 9 bei jedem neuen Schritt nur sicherstelt, dass das Verhältnis der sich entsprechenden Dimensionen der jedesmal konstruiert vorliegenden Bereiche, nicht aber dass das Verhältnis der sich entsprechenden Abszissen- und Ordinaten-Differenzen der Eckpunkte dieser Bereiche zwischen $(1-\varepsilon_1)\cdots(1-\varepsilon_n)$ und $(1+\varepsilon_1)\cdots(1+\varepsilon_n)$ liegt; dies bewirkt, dass es sehr gut für ein gewisses n passieren kann, dass den Borelschen Bedingungen unmöglich genügt werden kann, sodass die Konstruktion versagt. Dies ist nicht etwa eine ergänzbare Beweislücke, sondern ein wirklicher Fehler, der den ganzen Beweis stürzt.

⟦410⟧

Es wäre mir nun sehr angenehm, wenn in den Referaten der Fortschritte ⟨n⟩ ⟦8⟧
über die beiden Arbeiten der obige Sachverhalt klargestellt werden könnte. Wäre das noch möglich? Borel selbst habe ich den Fehler schon kurz nach Erscheinen ⟦8⟧
der Arbeit auseinandergesetzt.

Haben Sie Nachrichten von Weitzenböck? Im ersten Kriegsjahre erhielt ich ein paar Karten von ihm; jetzt seit mehreren Monaten nichts mehr.

Zu Pfingsten habe ich Study besucht. Herzliche Grüsse

von Ihrem

Brouwer.

NOTES

⟦1⟧ Ackermann, of the Publishing House Teubner.
⟦2⟧ Jahrbuch über die Fortschritte der Mathematik. See Brouwer 1916 A.
⟦3⟧ Brouwer 1913 B1. See Brouwer 1913 B2.
⟦4⟧ A reviewing periodical.
⟦5⟧ Theorem 8 of 1913 B2.
⟦6⟧ Probably at the International Congress of Mathematicians in August 1912.
⟦7⟧ E. Borel 1913.
⟦8⟧ See Brouwer 1916 B, Y57.

1915 A2 **Mathematics**. — "*Remark on inner limiting sets*". By Prof. L. E. J.
⟦1⟧ BROUWER.

(Communicated in the meeting of April 23, 1915).

The notion of *inner limiting set* i. e. the set of all the points common to a series of sets of regions, was prepared by BOREL [1]), and fully developed by YOUNG [2]). The two principal theorems about this class of sets are the following:

1. *An inner limiting set containing a component dense in itself, has the continuous potency.*

2. *A countable set containing no component dense in itself, is an inner limiting set.*

The former theorem has been proved by YOUNG, first for the linear domain, then for the space of n dimensions [3]). The latter theorem has been proved for the first time by HOBSON [4]). It is true that this theorem can be considered as a corollary of the following theorem enunciated somewhat before by YOUNG [5]):

3. *If Q be an arbitrary set of points, an inner limiting set exists containing besides Q only limiting points of the ultimate coherence* [6]) *of Q*;

but this theorem was deduced by YOUNG [5]) from the property: "*Each of the successive adherences* [7]) *of a set of points consists entirely of points which are limiting points of every preceding adherence*", and the proof given by YOUNG for this property is erroneous [8]), so that undoubtedly the priority for the proof of theorem 2 belongs to HOBSON.

We can, however, arrive at theorem 2 in a much simpler way

⟦2⟧ 1) Leçons sur la théorie des fonctions, p. 44.

⟦3⟧ 2) Leipziger Ber. 1903, p. 288; Proc. London M. S. (2) 3, p. 372.

⟦3⟧ 3) Leipziger Ber. 1903, p. 289—292; Proc. London M. S. (2) 3, p. 372—374.
⟦4⟧ These proofs are referred to not quite exactly by SCHOENFLIES, Bericht über die Mengenlehre II, p. 81 and Entwickelung der Mengenlehre I, p. 356.

⟦5⟧ 4) Proc. London M. S. (2) 2, p. 316—323.

⟦6⟧ 5) Proc. London M. S. (2) 1, p. 262—266.

⟦7⟧ 6) YOUNG, Quarterly Journ of Math., vol. 35, p. 113.

⟦8⟧ 7) CANTOR, Acta Mathematica 7, p. 110.

⟦7⟧ 8) Quarterly Journ. of Math., vol. 35, p. 115. The error is contained in the sentence (line 8—6 from the bottom): "Thus P, being a limiting point of every one of the derived coherences, is a limiting point of F". A correct proof of the property in question was communicated to me about two years ago by G. CHISHOLM YOUNG.

⟦412⟧

than HOBSON and YOUNG did, by means of the following [1]) proof of theorem 3, which is valid for the space of n dimensions:

For each positive integer ν we describe round each point q of Q as centre with a radius smaller than ε_ν ($lim\ \varepsilon_\nu = 0$) a sphere which, if q is a point of the adherence $Qc^\beta a$, excludes all points of the derived set of Qc^β. In this way for each positive integer ν a set of regions J_ν containing Q is determined.

The inner limiting set $\mathfrak{D}(J_\nu)$ then possesses the property required. For, if p be a limiting point of Q not belonging to Q and not being a limiting point of the ultimate coherence of Q, a transfinite number τ_p exists with the property that p is not a limiting point of Qc^{τ_p}, but for any $\alpha < \tau_p$ is a limiting point of Qc^α. Then on one hand p is excluded by every sphere described round a point of $\sum_{\alpha<\tau_p} Qc^\alpha a$, on the other hand a positive integer σ_p exists so that p is excluded by every sphere described for a $\nu > \sigma_p$ round a point of Qc^{τ_p}. Hence p lies outside every J_ν for which $\nu > \sigma_p$, so that p cannot belong to $\mathfrak{D}(J_\nu)$. Thus the theorem has been established.

---

1) This proof was communicated about two years ago to SCHOENFLIES, who on p. 356 of his Entwickelung der Mengenlehre I, applies it to prove the following special case of theorem 2: "*Every component of a countable closed set is an inner limiting set*". Comp. HOBSON, l. c. p. 320: "*Every reducible set is an inner limiting set*". ⟦9⟧ ⟦5⟧ ⟦10⟧

NOTES

⟦1⟧ The English version appeared in the Proceedings of the meeting of 29 May 1915. 'Inner limiting set' means G_δ in more modern terminology.
⟦2⟧ E. Borel 1898.
⟦3⟧ W. H. Young 1903 A, 1905.
⟦4⟧ A. Schoenflies 1908 A, 1900.
⟦5⟧ E. W. Hobson 1904.
⟦6⟧ W. H. Young 1904 B.
⟦7⟧ W. H. Young 1903 B.
⟦8⟧ G. Cantor 1885.
⟦9⟧ A. Schoenflies 1900.
⟦10⟧ An addition to this footnote in Brouwer 1917 B2, p. 1192, footnote.

1917 B2

⟦1⟧ **Mathematics.** — *"On linear inner limiting sets"* [1]). By Prof. L. E. J. BROUWER.

(Communicated in the meeting of April 27, 1917).

We consider an inner limiting set I, determined inside the unit interval as the intersection (greatest common divisor) of the sets of (non-overlapping) intervals $i_1, i_2, \ldots$, each point of $i_{\nu+1}$ being also a point of i_ν. Then the complementary set C of I with regard to the closed unit interval is the union (common measure) of the closed sets $a_1, a_2, \ldots$, each $a_{\nu+1}$ containing a_ν. *We shall suppose that I as well as C is uncountable in each sub-interval of the unit interval*; then we may assume that each a_ν contains as its nucleus a perfect set p_ν. The difference of a_ν and p_ν will be indicated by v_ν, the complementary set of p_ν, considered as a set of intervals, by u_ν, and the inner limiting set determined as the intersection of $u_1, u_2, \ldots$ by U. Then the points of each u_ν lie everywhere dense, and each u_ν is a set of order-type η of intervals, whose length does not exceed a certain value ε_ν having the limit zero for indefinitely increasing ν.

Let us assume that we dispose of such a set j_ν of order-type η of intervals each being an element of one of the sets $u_\nu, u_{\nu+1}, u_{\nu+2}, \ldots$, that j_ν contains no point of v_ν, but does contain all points of U not belonging to v_ν. We shall indicate a method leading from j_ν to such a set $j_{\nu+1}$ of order-type η of intervals each lying inside an interval of j_ν, and being an element of one of the sets $u_{\nu+1}, u_{\nu+2}, u_{\nu+3}, \ldots$, that $j_{\nu+1}$ contains no point of $v_{\nu+1}$, but does contain all points of U not belonging to $v_{\nu+1}$, each interval of j_ν containing a subset of $j_{\nu+1}$ of order-type η.

Let AB be an arbitrary element of j_ν being at the same time an element of u_μ $(\mu \geq \nu)$, let F be the subset of $u_{\mu+1}$ lying inside AB,

---

1) To the last footnote of my former communication on inner limiting sets (these
⟦2⟧ *Proceedings* XVIII, p. 49) must be added that the changed form in which SCHOENFLIES has referred to my reasoning (applying it to a special case only, and deducing the general theorem from this special case) is irrelevant. The error is contained
⟦3⟧ in the sentence (Entwickelung der Mengenlehre I, p. 359, line 5—8 from the top): "Ist nämlich P irgend eine abzählbare Menge, die nicht dicht in bezug auf eine perfekte Menge ist, und geht man durch Hinzufügung sämtlicher Grenzpunkte zu einer abgeschlossenen Menge Q über, so kann diese keinen perfekten Bestandteil enthalten, ist also ebenfalls abzählbar".

and w the set of intervals which is left from F after destroying all points of $v_{\nu+1}$ contained in F. Let PQ be an arbitrary element of w, s_ρ the set of intervals determined as the intersection of u_ρ and PQ, t_ρ the set of intervals which is left from s_ρ after destroying its first and its last element, *in so far those elements exist*, γ the set of intervals determined as the union of $t_1, t_2, t_3, \ldots$, and φ the set of intervals which is generated by constructing in each element of w a set of intervals in the same way as γ has been constructed in PQ. Then the required set of intervals $j_{\nu+1}$ is generated by constructing in each element of j_ν a set of intervals in the same way as φ has been constructed in AB.

If we understand by u_0 as well as by j_0 the unit interval itself, then we arrive from j_0 at j_1 by the same process which has led us from j_ν to $j_{\nu+1}$.

The inner limiting set determined as the intersection of $j_1, j_2, \ldots$ contains all points of U belonging to none of the sets v_ν, so a fortiori all points of U belonging to none of the sets a_ν, so also all points of the unit interval belonging to none of the sets a_ν. As, on the other hand, this inner limiting set can neither contain a point of a v_ν, nor (as a subset of U) a point of a p_ν, it finally cannot contain a point of a a_ν either. *So it is identical to the complementary set of C, i.e. to I.*

If we construct a ternal scale on the unit interval, and if (designing by ϱ_ν an arbitrary finite series of digits 0, 1 or 2, among which ν digits 1 occur) we understand by $d_{\nu+1}$ the set of the intervals ι_{ρ_ν} whose end-points have the coordinates $\varrho_\nu 1$ and $\varrho_\nu 2$, then we can first represent the set of intervals j_1, biuniformly and with invariant relations of order, on the set of intervals d_1; thereupon we can in each interval of j_1 represent the subset of j_2 contained in it, biuniformly and with invariant relations of order, on the subset of d_2 contained in the corresponding interval of d_1; and so on. In this way we determine a continuous one-one transformation of the unit interval in itself by which I passes into the set τ_2 of the points expressible in the ternal scale by means of a sequence of digits containing an infinite number of digits 1, whilst C passes into the set τ_1 of the points expressible in the ternal scale by means of a sequence of digits containing only a finite number of digits 1. Thus, indicating the *geometric types*[1]) of τ_1 and τ_2 by $\bar{\mu}$ and $\bar{\nu}$ respectively, we have proved the following

Theorem 1. *Each inner limiting set contained in a linear interval,* ⟦5⟧

[1]) Comp. these *Proceedings* XV, p. 1262. ⟦4⟧

and, as well as its complementary set, uncountable in each sub-interval, possesses the geometric type $\bar{\nu}$, *and its complementary set possesses the geometric type* $\bar{\mu}$.

Let H and K be two arbitrary points of τ_1, we can choose ν in such a way that neither H nor K is an endpoint of an interval of d_ν. Let us indicate the set of points which is the complementary set of d_ν, by e_ν, and the set whose elements are the intervals of d and the points of e_ν, by r_ν. Then we can construct a one-one trans formation of d_ν and e_ν each in itself, by which the relations of order between the elements of r_ν remain invariant, and H passes into K. This transformation can be extended to a continuous one-one transformation of the unit interval in itself, for which the subsets of τ_1 and τ_2 contained in corresponding intervals of d_ν, correspond to each other. We thus have generated a continuous one-one trans ormation of the unit interval in itself, by which τ_1 passes into itself, and the point H chosen arbitrarily in τ_1, passes into the point K chosen
[[6]] likewise arbitrarily in τ_1, *so that* τ_1 *is a homogeneous set of points.*

Let H and K be two arbitrary points of τ_2 contained in the intervals h_ν and k_ν of d_ν respectively, we can construct a one-one transformation of the set of intervals d_1 in itself leaving invariant the relations of order, by which h_1 passes into k_1; this transformation can be extended to a one-one transformation of the set of intervals d_2 in itself leaving invariant the relations of order, by which h_2 passes into k_2; continuing indefinitely in this way, we generate a continuous one-one transformation of the unit interval in itself, by which each d_ν, so also τ_2 passes into itself, and the point H chosen arbitrarily in τ_2, passes into the point K chosen likewise arbitrarily in τ_2, *so that* τ_2 *too is a homogeneous set of points,* and we have proved the following

[[7]] THEOREM 2. *Each inner limiting set contained in a linear interval, and, as well as its complementary set, uncountable in each sub-interval, is homogeneous, and its complementary set is likewise homogeneous.*

NOTES

[[1]] 'Inner limiting set' means G_δ in more modern terminology.
[[2]] Brouwer 1915 A2.
[[3]] A. Schoenflies 1900.
[[4]] Brouwer 1913 B2.
[[5]] See Alexandroff–Urysohn 1928, footnote 16a.
C. Kuratowski 1952, I, p. 349.
[[6]] This concept of homogeneity differs from that of Hausdorff 1914, p. 173. Brouwer posed the problem of determining all linear homogeneous sets. See B. P. Haalmeijer 1919, 1923.
[[7]] See Alexandroff–Urysohn 1928, footnote 20.

CHAPTER 6

The new methods in topology

Paris, Neujahrsmorgen 1910

Mein lieber Herr Geheimrat

Ihnen und Ihrer lieben Gattin die herzlichsten Wünsche für das neue Jahr, für Ihre Gesundheit und für Ihre wissenschaftliche Tätigkeit. Ich logiere hier während der Winterferien bei meinem Bruder, der Geologe ist; meine Frau konnte leider mich nicht begleiten. Mitte Januar fangen meine Vorlesungen wieder an, und kehre ich zurück.

Mit Schoenflies ist, wohl hauptsächlich durch Ihre Zwischenkunft, das Einverständnis wiederhergestellt. Ich füge hier seine beiden letzten Briefe bei, welche ich mit einer letzten Fassung zufrieden beantwortet habe, dass ich die Sache als erledigt betrachte.

Darf ich noch einige Bemerkungen über die eindeutigen (umkehrbar eindeutigen) stetigen Abbildungen einer Kugel auf eine Kugel hinzufügen. Legt man der die Bedingung auf, dass ... stetig sein soll, dass ... jeder geschlossenen Jordanschen Kurve ... einen Punkt K ... welche gegen K konvergiert, für jeden Punkt K ... eine geschlossene Jordankurve um K, welche gegen K konvergiert, entsprechen soll.

Haben wir nun zwei eindeutige Abbildungen einer Kugel (oder allgemeiner geschlossenen Fläche) K auf eine Kugel L, und zwischen auf eine Kugel M, so erhebt sich die Frage, welche weiteren Bedingungen genügt ... damit die Beziehung zwischen L und M ... im Sinne der Analysis ... eine algebraische ... ist. [illegible] Kehre ich wieder zur allgemeinen eindeutigen und stetigen Beziehung zwischen zwei Kugeln zurück, so lässt sich für jede solche eine endliche Zahl n als ihr Grad angeben, ... alle Beziehungen gleichen Grades, und nur diese, lassen sich stetig ineinander überführen lassen. Insbesondere lassen sich alle Beziehungen n^ten Grades stetig in rationale Funktionen n^ten Grades der komplexen Variablen überführen.

Letter, very likely to Hilbert, draft or copy Y 16

Paris, Neujahrsmorgen 1910

Mein lieber Herr Geheimrat ⟦1⟧
Ihnen und Ihrer lieben Gattin die herzlichsten Wünsche für das neue Jahr, für Ihre Gesundheit und für Ihre wissenschaftliche Tätigkeit.

Ich logiere hier während der Winterferien bei meinem Bruder, dem Geologen; meine Frau konnte leider mich nicht begleiten. Mitte Januar fangen meine Vorlesungen wieder an, und kehre ich zurück.

Mit Herrn Schoenflies ist, wohl hauptsächlich durch Ihre Zwischenkunft, das Einverständnis wiederhergestellt. Ich füge hier seine beiden letzten Briefe bei, auf welche ich geantwortet habe, dass ich mit seiner letzten Fassung zufrieden bin, und die Sache als erledigt betrachte.

Darf ich noch einige Bemerkungen über die eindeutige (nicht notwendig eineindeutige) stetige Abbildung einer Kugel κ auf eine Kugel λ hinzufügen? Legt man ihr die Bedingung auf, dass sie *beiderseits* stetig sein soll, so ist sie ein-eindeutiges stetiges Bild einer rationalen Funktion der komplexen Variabelen⟦.⟧ ⟦2⟧
Unter dieser Bedingung der beiderseitigen Stetigkeit verstehe ich, dass einer geschlossenen Jordanschen Kurve um einen Punkt L von λ, welche gegen L konvergiert, für jeden Punkt K von κ, von dem L das Bild ist, eine geschlossene Jordansche Kurve um K, welche gegen K konvergiert, entsprechen soll.

Haben wir nun *zwei* dieser Bedingung genügende eindeutige Abbildungen einer Kugel (oder allgemeineren geschlossenen Fläche) K auf eine Kugel L und auf eine Kugel M, so erhebt sich die Frage welcher weiteren Bedingung genügt werden muss, damit die Beziehung zwischen L und M im Sinne der Analysis Situs eine komplexe algebraische ist. Kehre ich wieder zur allgemeinen einerseits eindeutigen und stetigen Beziehung zwischen zwei Kugeln zurück, so lässt sich für jede solche eine endliche Zahl n als ihr *Grad* angeben, derart dass alle Beziehungen gleichen Grades, und nur diese, sich stetig ineinander überführen lassen. ⟦3⟧
Insbesondere lassen sich alle Beziehungen n^{ten} Grades stetig in rationale Funktionen n^{ten} Grades der komplexen Variabeln überführen.

Um diesen Grad zu definieren, führen wir auf κ homogene Koördinaten ⟦sic⟧ x, y, z ein, und auf λ homogene Koördinaten ⟦sic⟧ ξ, η, ζ, und betrachten zunächst die eindeutige Abbildung, welche gebietsweise bestimmt ist durch eine Beziehung

$$\xi : \eta : \zeta = f_1(x, y, z) : f_2(x, y, z) : f_3(x, y, z), \qquad ⟦4⟧$$

wo f_1, f_2, f_3 Polynome sind.

Nehmen wir sodann auf beiden Kugeln einen positiven Umlaufssinn an, und wählen wir in jedem Punkte von κ diesen positiven Umlaufssinn, so tritt jeder Punkt allgemeiner Lage von λ p Male auf mit positivem Umlaufssinn und q Male mit negativem Umlaufssinn. Man kann dann zeigen, dass $p-q$ für alle Punkte allgemeiner Lage von λ eine Konstante ist, welche wir den *Grad* der Abbildung nennen.

Ist die Beziehung zwischen x, y, z und ξ, η, ζ nicht durch Polynome bestimmt, so kann man sie durch solche polynomiale Beziehungen approximieren, und man zeigt leicht, dass diese approximierenden Beziehungen einen konstanten Grad besitzen, den wir somit auch der Limesbeziehung zuweisen. Dieser Grad ist immer eine endliche, positive oder negative Zahl. Insbesondere hat eine *ein-eindeutige*
⟦5⟧ – ⟦in margine⟧ strenger formuleeren – stetige Transformation der Kugel in sich, wenn sie den Umlaufssinn nicht ändert, den Grad $+1$, sonst -1.

⟦6⟧ Sie kennen nun meinen Satz, das jede *ein-eindeutige* stetige Transformation der Kugel in sich, welche den Umlaufssinn *nicht* ändert, immer wenigstens einen Fixpunkt besitzt. Dieser Satz lässt sich in folgender Weise erweitern, dass jede *eindeutige* stetige Transformation der Kugel in sich, deren Grad *nicht* -1 ist, immer wenigstens einen Fixpunkt besitzt.

⟦7⟧ Und es ist mir gelungen, den Satz in dieser Form auf die n-dimensionale Kugel zu erweitern. Er lautet dann: Jede eindeutige stetige Transformation der n-dimensionalen Kugel in sich besitzt wenigstens einen Fixpunkt. Eine Ausnahme machen für n ungerade die Transformationen $+1^{\text{ten}}$ Grades, und für n gerade die Transformationen -1ten Grades.

Ein-eindeutige Transformationen besitzen mithin notwendig den Fixpunkt, einerseits für n ungerade und umgekehrten Umlaufssinn, andererseits für n gerade
⟦8⟧ und ungeänderten Umlaufssinn. Noch allgemeiner lautet das Resultat für eindeutige stetige Transformationen des n-dimensionalen Kugelinneren in sich, denn solche besitzen auf jeden Fall einen Fixpunkt.

Nochmals an Sie beide die besten Wünsche und Grüsse

Ihres immer verehrenden

L E J Brouwer.

NOTES

⟦1⟧ Originally the present chapter started with Brouwer 1911 C, the famous proof of invariance of dimension. When putting the finishing touches to the edition of Brouwer's topological work, I noticed an exercise-book bearing the label 'Potentiaaltheorie en Vectoranalyse' which I did not remember I had ever seen before, and in this book I discovered two pieces of paper which at first sight

I identified as drafts or copies of letters to Hilbert and to Hadamard, written in Paris on 1 January 1910 and 4 January 1910 respectively. These two letters shed unexpected new light on the origin and development of Brouwer's great discoveries. They are edited here as Y16 and Y17 to which I added a paragraph of a letter to Hilbert of 18 March 1910 from Univ. Bibl. Göttingen as Y18. The first six pages of the exercise book in which I found the drafts of the two letters discuss the subject announced on the label; then on page 7 the text changes to topology written with the same pen and ink; this continues up to p. 19, where the text ends. All must have been written within a short period, except for some marginal notes on p. 10 and 12, which may be dated a few months later. The exercise book included three more pieces of paper; they are related to the first subject though one of them is to a certain extent linked to the letter to Hadamard. Brouwer's estate includes about 10 more fragments which thanks to this discovery can be nicely dated and fairly well understood. These fragments are all related to what I have called the New Methods in Topology. Both exercise book and fragments are in Dutch. This is the reason why they are not edited along with the German and French inedita, though they deserve to be. Translating a carefully edited and printed text is quite different from translating jottings which can easily be interpreted too formally by the translator. Y3 and Y4, which have been translated from Dutch, were relatively easy cases. I did not dare repeat this experiment with the material I have just mentioned. In the sequel, however, I will mention and use some data from these unedited pieces.

After this discovery chapter 6 now starts with what might be called the prelude of the New Methods in Topology. I should really have rewritten parts of what follows but I preferred to confine myself to small changes.

As far as the *printed* evidence indicates, The New Methods in Topology started with 1911 C, which is the invariance of dimension. The mapping degree became explicit in the next paper, 1911 D, on mapping of manifolds, though it is implicit in 1911 C as are all the fundamental concepts Brouwer would later display.

The new discovery shows that the historical order was just the reverse. The invariance of dimension was not the germ. Its proof is a streamlined byproduct, a mere derivative of richer material.

Actually Brouwer started with vector fields on manifolds. Vector fields were a classical subject. The index of a vector field at a point, defined by an integral had been known to Kronecker 1869 if not already to Gauss 1840. H. Poincaré 1881, 1882, 1885 A, had taken up these ideas, and so had Bohl 1904. What they studied, were vector fields in Cartesian space, possibly restricted to hypersurfaces, though not necessarily tangential ones.

Brouwer 1906 A2, 1906 B2, 1906 C2 had studied vector fields in an analytic setting. When he turned to study vector fields on surfaces topologically in 1909 G2, 1910 A2, he probably was not acquainted with Poincaré's related work. He discovered the existence of singularities on spheres. In his parallel work on topological mappings of surfaces 1909 F2, 1909 H2, he found the existence of an invariant point under orientation preserving mappings. His letter to Hadamard shows that he was then puzzled by this analogy – perhaps he had already noticed it in higher dimensions. According to Y17 he learned the deeper reason for it in the correspondence with Hadamard, which does not mean that Hadamard revealed it to him. On the contrary, according to Hadamard 1910 the connection between vector fields on the sphere and one-to-one orientation preserving mappings is due to Brouwer himself. His first great discovery in this domain according to the new evidence is the total index of tangential vector fields to hyperspheres: 2 in even and 0 in odd dimensions. Since he communicated it to Hilbert and Hadamard in the first days of 1910, one may conclude that the proof had been found in the first weeks of his Christmas holiday stay in Paris. Since it is also the first topological subject of the exercise book, I am prone to conclude that its greater part had been written during this holiday stay in Paris. There is another possibility: perhaps the Paris part begins on p. 14, which is written in a different ink. The proof of the theorem on the total index of tangential vector fields on spheres is essentially the same as that in Brouwer 1911 D, pp. 107–112, a marvelously profound and highly sophisticated proof which indicates a long incubation period. In the proof Brouwer uses algebraic instead of simplicial approximations to vector fields; a marginal note of a later date introduces the simplicial method. In the last lines of p. 12 of the exercise book Brouwer turns to continuous mappings of spheres. Here the terminology shows French influence (*univoque* rather than *éénduidig*, derived from German *eindeutig*), and a private Hadamard expression (*inhoud* or *volume* of a sphere, meaning its interior with boundary). Pages 13–14 of the exercise book deal with the same subject as the letters to Hilbert and Hadamard, though they are less formal. The mapping degree is more clearly exhibited in the two letters. Page 16 of the exercise book reveals an attempt at proving the invariance of dimension; p. 18, which probably is of a later date, again deals with the invariance of dimension, and the Jordan theorem in space as well. Page 19 is a new approach to invariance of dimension. Moreover it contains the same remark as that of the letter to Hilbert of 18 March 1911, Y18: it follows from the theorem on vector fields that even and odd dimensions can be distinguished topologically. Fortunately Brouwer did not notice that the validity of the theorem of invariance of dimensions for an *infinite number* of dimensions implies the same thing for *all* of them, so he continued looking for a general proof. There are many approaches to the invariance theorem in Brouwer's above mentioned fragments: one using separations, ('coupures'), as suggested by ideas of H. Poincaré set forth in Poincaré 1903, Poincaré 1912 A, which might have reached Brouwer orally and were elaborated later on by Brouwer 1913 A, the other using the mapping degree directly, leading to Brouwer 1911 C. It is a pity that the final switch to the simple version of 1911 C does not appear in Brower's fragments and that no handwritten manuscript of 1911 C exists. Part of the typewritten manuscript is extant.

The idea of the mapping degree is fully present in this period, which extends from Christmas 1909 to March 1910. What is still lacking is the idea of simplicial approximation, although simplicial subdivisions are used. Algebraic approximations, strangely enough, play the role which is assumed by simplicial ones later on. Nevertheless I venture to date the birth of the New Methods in Topology by 1 January 1910 and identify as its birthplace: 6 rue de l'Abbé de l'Epée in the Latin Quarter of Paris.

What has now also become clearer is Hadamard's rôle as midwife. Before I had never understood how Hadamard figured in this story and why Brouwer regarded him so highly in this context. Hadamard's work in this part of topology, Hadamard 1910, looks rather old-fashioned. Its strong dependence on analytic tools would have hampered rather than stimulated true topology. It was Brouwer's achievement to have shaken off the yoke of analysis from topology. In the genesis and maturation of his ideas, however, his intercourse with Hadamard must have meant more to Brouwer than can be expressed by mere citations and quotations.

The only possible addressee of the present letter is D. Hilbert. However, before this date and still in a letter of 18 March 1911 (Univ. Bibl. Göttingen), Brouwer addressed Hilbert as 'Sehr geehrter Herr Geheimrat' and only from 31 March 1911 (Univ. Bibl. Göttingen) we find the address 'Lieber Herr Geheimrat' in Brouwer's authentic letters to Hilbert. (It changes into 'Lieber Herr Hilbert' on 4 July 1913 after Brouwer's appointment as a professor.) The letter of 1 January 1910 is not extant in Univ. Bibl. Göttingen. If Brouwer had never dispatched it, he would have put a notice upon the draft. Flushed with his new discoveries, perhaps Brouwer addressed Hilbert in a more familiar way only to return to the former mode of address when he calmed down, but it is also possible that Brouwer's mode of address was determined by Hilbert's, which owing to the lack of the relevant material is unknown.

⟦2⟧ This kind of mapping is studied in Brouwer 1912 K2, 1912 M.

⟦3⟧ The first appearance of the mapping degree. In the exercise book 'Potentiaaltheorie en Vectoranalyse' Brouwer also uses the vector field term 'index'.

⟦4⟧ The use of homogeneous coordinates is not correct. Y17 gives a correct formulation. See Y17 ⟦3⟧.

⟦5⟧ The notice 'strenger formuleeren' was very likely to indicate a gap in Brouwer's arguments. A crucial point was to prove there are not more than two homotopy classes of topological mappings of the sphere.

⟦6⟧ Brouwer 1909 F2.

⟦7⟧ Brouwer 1911 D, Satz 3. This shows that Brouwer is in possession of a general definition of mapping degree. It is confirmed by the exercise book 'Potentiaaltheorie en Vectoranalyse'.

⟦8⟧ Brouwer 1911 D, Satz 4.

Y 17 Letter certainly to J. Hadamard, draft or copy

Paris, 4.1.10.

6 Rue de l'Abbé de l'Epée

⟦1⟧ Cher Monsieur

Je puis à présent vous communiquer quelques extensions du théorème du point invariant dans les transformations biunivoques et continues de la sphère. Elles se rapportent aux transformations univoques et continues de la sphère. A une telle
⟦2⟧ transformation on peut attribuer un nombre fini n comme son *degré*. En partant d'une transformation de degré n, on peut atteindre moyennant des variations continues toute autre transformation de degré n, mais pas d'autres. En particulier on peut toujours atteindre de cette manière une transformation rationnelle de degré n de la sphère complexe.

⟦3⟧ Pour déterminer ce *degré*, introduisons des coordonnées homogènes (à double sens), écrivons les $x\ y\ z$ pour la sphère primitive et ξ, η, ζ pour l'image, divisons la sphère en un nombre fini de régions, et considérons d'abord les transformations définies par les relations:

$$\xi : \eta : \zeta = f_1(x, y, z) : f_2(x, y, z) : f_3(x, y, z),$$

où f_1, f_2, f_3 sont des polynômes, qui d'ailleurs peuvent différer pour les différentes régions de la sphère. Appelons cette transformation une transformation polynomiale. Définissons une indicatrice sur la sphère: alors chaque point P de l'image *de position générale* apparaîtra un nombre r_P de fois avec indicatrice positive, et un nombre s_P de fois avec indicatrice négative. On démontre alors que $r_P - s_P$ est une constante: c'est le degré de la transformation polynomiale.

Revenons à une transformation univoque et continue générale. Elle peut être approximée par une série de transformations polynomiales; on démontre que ces dernières ont toutes le même degré: c'est encore le degré de la transformation-limite.

Le degré est toujours un nombre entier fini, positif ou négatif. Le degré d'une transformation *bi*univoque est de $+1$, si l'indicatrice reste la même, et de -1, si elle est renversée.

Maintenant le théorème du point invariant généralisé devient le suivant: Toute transformation univoque et continue de la sphère en elle-même, dont le degré n'est pas -1, possède au moins un point invariant.

J'ai encore étendu ce théorème aux sphères à m dimensions. Il s'énonce alors de la manière suivante: Toute transformation univoque et continue de la sphère m-dimensionale ⟦sic⟧ en elle-même possède au moins un point invariant, excepté a) quand m est impair et si le degré n est $+1$, b) quand m est pair et si le degré n est -1. ⟦4⟧

En particulier si la transformation est *bi*univoque – ⟦in margine⟧ strenger formuleeren – il existe au moins un point invariant a) si m est impair et si l'indicatrice est renversée, b) si m est pair et si l'indicatrice est invariante. ⟦5⟧

Pour le volume d'une sphère m-dimensionale ⟦sic⟧ dans l'espace à $m+1$ dimensions (si nous y comprenons la sphère elle-même) j'ai réussi dernièrement à établir un résultat plus général encore, à savoir: Toute transformation univoque (pas nécessairement biunivoque) et continue du volume d'une sphère m-dimensionale ⟦sic⟧ en lui-même possède au moins un point invariant. ⟦6⟧

Sur les distributions vectorielles continues générales de la sphère paraîtront ⟦7⟧
bientôt encore deux articles de ma main, où j'étudie quelques questions qui se rattachent au principe de Dirichlet et à la décomposition du champ en une partie
qui est 'quellenfrei' et une partie qui est 'wirbelfrei'. Pour cela j'établis d'abord la ⟦8⟧
forme la plus générale que peuvent affecter les courbes tangentes (ou caractéristi-
ques d'après Poincaré). Comme résultat principal du premier article il faut ⟦9⟧
considérer la propriété que toute caractéristique qui ne se rapproche pas indéfiniment d'un point singulier est une spirale, dont les deux cycles limites sont euxmêmes des caractéristiques. La propriété que l'existence d'un point singulier au moins est nécessaire, n'y est au fond qu'un corollaire accessoire, sur lequel je n'ai insisté que parce que c'était le premier résultat qui se laissait formuler simplement, et parce qu'il me semblait exister une relation intime entre ce théorème et celui du
point invariant de la sphère, relation qui ne s'est éclaircie que dans votre cor- ⟦10⟧
respondance. Dans le second article j'ai inséré votre belle démonstration directe ⟦11⟧
et plus complète de l'existence d'un point singulier au moins.

Mon adresse sera à Paris jusqu'au 15 janvier. Peut-être l'occasion se présenterait de vous voir? Agréez, monsieur, mes salutations distinguées.

L E J Brouwer.

NOTES

⟦1⟧ The name of the addressee is determined by Brouwer's mentioning Hadamard's 'belle démonstration' published in Brouwer 1910 D2 (See ⟦11⟧.) Some letter like the present one was certainly sent to Hadamard, since in his fragments Brouwer often mentions 'the theorems communicated to Hadamard'.

⟦2⟧ See Y16 ⟦3⟧.

⟦3⟧ This is the correct formulation. (See Y16 ⟦4⟧.)

⟦4⟧ Brouwer 1911 D.

⟦5⟧ See Y16 ⟦5⟧.

⟦6⟧ The expression 'volume', meaning closed ball, betrays Hadamard's influence. See Hadamard 1910.

⟦7⟧ Brouwer 1910 A2, 1910 D2.

⟦8⟧ This problem seems to have been one of Brouwer's starting points on his way to topology. See also 1910 D2, Theorem 2.

⟦9⟧ This formulation may indicate that Brouwer became acquainted with this work of Poincaré after he had published Brouwer 1909 G2, thanks, perhaps, to Hadamard's intervention.

⟦10⟧ This need not mean that Brouwer learned it from Hadamard. In Hadamard 1910 this discovery was attributed to Brouwer.

⟦11⟧ Brouwer 1910 D2.

Letter to Hilbert

Y 18

⟦Univ. Bibl. Göttingen⟧

18.3.10

Amsterdam

Sehr geehrter Herr Geheimrat

⟦· · ·⟧

Ausser einer neuen gruppentheoretischen Mitteilung präpariere ich zur Einsendung an die ⟦An⟧nalenredaktion einen Artikel, in dem ich das Invarianzproblem der Dimension insoweit löse, dass ich zeige dass jedenfalls Räume gerader und ungerader Dimensionenzahl sich nicht stetig und ein-eindeutig auf einander abbilden lassen. ⟦1⟧

⟦· · ·⟧

L E J Brouwer.

NOTE

⟦1⟧ Also found in the exercise book mentioned in Y16 ⟦1⟧.

1911 C

Beweis der Invarianz der Dimensionenzahl.

Von

[[1]] L. E. J. Brouwer in Amsterdam.

Das Problem der Invarianz der Dimensionenzahl, d. h. der Unmöglichkeit, zwischen einer m-dimensionalen Mannigfaltigkeit und einer $(m+h)$-dimensionalen Mannigfaltigkeit $(h > 0)$ eine eineindeutige und stetige Beziehung herzustellen, ist für den Fall $m \leqq 3$ von Lüroth gelöst worden.*) Im Folgenden soll der allgemeine Fall erledigt werden.

§ 1.

Wir denken uns in einer q-dimensionalen Mannigfaltigkeit einen q-dimensionalen Kubus K mit der Kantenlänge $2a$, welcher durch Verrückungen seiner Punkte, deren Größe ein gewisses Maximum $b < a$ nicht übersteigt, einer eindeutigen und stetigen Abbildung α unterliegt, und bezeichnen mit K_1 einen bestimmten mit K konzentrischen und homothetischen Kubus, dessen Kantenlänge $2a_1 < 2(a-b)$ ist.

Wir unterziehen K einer *simplizialen Zerlegung*, d. h. wir wählen eine willkürliche ganze Zahl n, zerlegen K zunächst in n^q mit ihm homothetische Teilkuben mit der Kantenlänge $\frac{a}{n}$, und sodann jeden dieser Teilkuben in $q!$ Simplexe, deren Eckpunkte zugleich Eckpunkte der Teilkuben sind. Diese Simplexe nennen wir die *Grundsimplexe* der Zerlegung,

---

[[2]] *) Vgl. „Über Abbildung von Mannigfaltigkeiten", Math. Annalen 63 S. 222—238. Hinsichtlich der Fehlschlüsse in den von Thomae, Netto und Cantor für den allgemeinen Fall versuchten Beweisen vgl. E. Jürgens, „Der Begriff der n-fachen stetigen Mannigfaltigkeit", Jahresber. der Deutschen Math.-Ver. 7_1, S 50—55, und A. Schoenflies, Bericht über die Lehre von den Punktmannigfaltigkeiten, Bd. II, S. 167. Über die Orientierung des Problems vgl. auch R. Baire, „Sur la nonapplicabilité de deux continus à n et $n+p$ dimensions", Comptes Rendus 144, S. 318—321; Bull. des Sc. Math. (2), **31**, S. 94—99, und M. Fréchet, „Les dimensions d'un ensemble abstrait", Math. Ann. 68, S. 145—168.

ihre Eckpunkte die *Grundpunkte* der Zerlegung. Unter *Grundseiten* der Zerlegung verstehen wir die in den Grenzen der Grundsimplexe auftretenden $(q-1)$-dimensionalen Simplexe, unter *Grundkanten* der Zerlegung die in den Grenzen der Grundsimplexe auftretenden $(q-1-p)$-dimensionalen Simplexe $(p>0)$. Diejenigen Grundseiten, welche in der Grenze von K liegen, sollen *Randseiten* heißen, die übrigen *innere Seiten.*

Wir setzen für K einen positiven Sinn der Indikatrix fest, und in jedem Grundsimplex bringen wir eine positive Indikatrix an.

Sei π eines der Grundsimplexe mit den Eckpunkten $A_1, A_2, \cdots, A_{q+1}$, welche durch die Abbildung α übergehen in $B_1, B_2, \cdots, B_{q+1}$. Ein im Inneren oder auf der Grenze von π liegender Punkt P läßt sich als Schwerpunkt von gewissen in $A_1, A_2, \cdots, A_{q+1}$ konzentrierten positiven Massen auffassen. Bringen wir dieselben Massen der Reihe nach in $B_1, B_2, \cdots, B_{q+1}$ an, so bestimmen sie einen Schwerpunkt Q, den wir P entsprechen lassen. Alsdann wird, falls die Punkte $B_1, B_2, \cdots, B_{q+1}$ nicht in einem ebenen $(q-1)$-dimensionalen Raume liegen, das Grundsimplex π eineindeutig und stetig abgebildet auf ein Bildsimplex ϱ, dessen Inhalt wir, je nachdem die Bildindikatrix positiv oder negativ ausfällt, positiv bzw. negativ rechnen werden. Wenn aber die Punkte $B_1, B_2, \cdots, B_{q+1}$ in einem $(q-1)$-dimensionalen Raume liegen, so wird das Bildsimplex singulär, und bekommt den Inhalt Null.

Indem wir mit jedem der Grundsimplexe in analoger Weise verfahren, erhalten wir eine neue eindeutige und stetige Abbildung β von K, deren spezielle Art wir mit dem Namen „*simpliziale Abbildung*" bezeichnen wollen.

Falls es für β singuläre Bildsimplexe gibt, können wir mittels eine beliebig klein gewählte Größe nicht übersteigender Herabsetzungen der Verrückungen der Grundpunkte eine „*modifizierte*" *simpliziale Abbildung* γ, für welche es keine singulären Bildsimplexe mehr gibt, herstellen.

Wir beschäftigen uns zunächst mit der Abbildung γ und bezeichnen mit μ die Menge derjenigen im Inneren von K_1 enthaltenen Punkte, welche *nicht* dem Bilde einer Grundkante angehören. Diese Menge μ besitzt die Eigenschaft, daß je zwei ihrer Punkte sich durch einen aus einer endlichen Zahl von Strecken bestehenden, ganz in μ verlaufenden Streckenzug verbinden lassen.

Weiter bemerken wir, daß die Bilder der Randseiten nicht in das Innere von K_1 eindringen können, und nennen diejenigen Punkte von μ, welche *nicht* dem Bilde einer inneren Seite angehören, *die gewöhnlichen Punkte von* K_1.

Seien P_1 und P_2 zwei willkürliche gewöhnliche Punkte von K_1, welche wir durch einen ganz in μ verlaufenden Streckenzug s verbinden, und sei P ein variabler gewöhnlicher Punkt von K_1. Die Zahl der positiven bzw. negativen Bildsimplexe, welche P_1 bedecken, bezeichnen wir mit p_1 bzw. p_1', die analogen Zahlen für P_2 mit p_2 und p_2', die analogen Zahlen für P mit p und p'.

Diese Zahlen p und p' können sich bei Bewegung von P an s entlang nur dann ändern, wenn das Bild σ einer inneren Seite passiert wird. Wenn die beiden Bildsimplexe, welche in σ zusammenstoßen, auf verschiedenen Seiten von σ liegen, so besitzen sie dasselbe Vorzeichen, und die Kreuzung von σ hat auf die Zahlen p und p', also auch auf die Zahl $p - p'$, keinen Einfluß. Wenn aber die beiden Bildsimplexe auf derselben Seite von σ liegen, so besitzen sie entgegengesetzte Vorzeichen, und die Kreuzung von σ führt entweder eine Zunahme von p und p' je um eins, oder eine Abnahme von p und p' je um eins herbei, läßt also wieder die Zahl $p - p'$ ungeändert. Mithin ist $p_1 - p_1' = p_2 - p_2'$, und *für die Abbildung γ ist die Zahl $p - p'$ in den gewöhnlichen Punkten von K_1 eine Konstante.*

Wenn nun für die Abbildung β, in bezug auf welche wir die gewöhnlichen Punkte von K_1 und die Zahlen p und p' in analoger Weise wie in bezug auf γ definieren, zwei gewöhnliche Punkte P_1 und P_2 existierten, für welche $p_1 - p_1'$ und $p_2 - p_2'$ verschieden wären, so könnten wir β mit solcher Genauigkeit durch eine „modifizierte" simpliziale Abbildung γ_ν approximieren, daß die Punkte P_1 und P_2 für γ_ν wieder gewöhnliche Punkte von K_1 wären und die Zahlen p_1, p_1', p_2, p_2' für γ_ν dieselben Werte wie für β besäßen, daß mithin für γ_ν die Zahl $p - p'$ keine Konstante sein könnte. Aus diesem Widerspruche folgern wir, daß *auch für die Abbildung β die Zahl $p - p'$ in den gewöhnlichen Punkten*
[[3]] *von K_1 eine Konstante c ist.*

Um den Wert dieser Konstante c zu ermitteln, bezeichnen wir den Inhalt von K_1 mit I_1 und bemerken, daß der Gesamtinhalt derjenigen Teile der Bildsimplexe, welche in K_1 enthalten sind, $c \cdot I_1$ beträgt. Wenn wir dann die der simplizialen Abbildung β zugrunde liegenden Verrückungen der Grundpunkte stetig verkleinern, bis sie alle den Wert Null erreichen, und die Abbildung β entsprechend stetig in die identische Abbildung übergehen lassen, so kann dabei der genannte Gesamtinhalt keine Sprünge erleiden, und somit die ganze Zahl c sich nicht ändern. Hieraus folgern wir, daß die Zahl c für β denselben Wert, wie für die identische Abbildung, d. h. den Wert 1, besitzt.

Dann aber können für β in keinem gewöhnlichen Punkte von K_1 die Zahlen p und p' beide Null sein, sodaß sich herausstellt, daß *die durch β erzeugte Menge der Bildsimplexe den ganzen Kubus K_1 enthält.*

Indem wir nun weiter bemerken, daß die Abbildung α sich durch eine aus hinreichend groß gewähltem n hervorgegangene simpliziale Ab-

11*

bildung mit jedem beliebigem Grade der Genauigkeit approximieren läßt, zeigt sich, daß auch *die durch α bestimmte Bildmenge von K den ganzen Kubus K_1 enthält.*

Mithin ist bewiesen folgender

Hilfssatz. *Wenn in einer q-dimensionalen Mannigfaltigkeit bei einer eindeutigen und stetigen Abbildung eines q-dimensionalen Kubus das Maximum der Verrückungen kleiner ist als die halbe Kantenlänge, so existiert ein konzentrischer und homothetischer Kubus, der ganz in der Bildmenge enthalten ist.*

§ 2.

Wir nehmen nun an, daß in einer m-dimensionalen Mannigfaltigkeit das eineindeutige und stetige Bild eines m-dimensionalen Bereiches in einer gewissen Umgebung eines seiner Punkte nirgends dicht liegt. Alsdann existiert eine eineindeutige und stetige Beziehung ϑ zwischen einem m-dimensionalen Kubus K und einer in einer m-dimensionalen Mannigfaltigkeit liegenden, nirgends dichten, zusammenhängenden perfekten Menge C.

Wir schließen C in einen m-dimensionalen Kubus k ein, den wir einer simplizialen Zerlegung ζ unterziehen, und heben aus den in dieser Weise bestimmten Grundsimplexen diejenigen heraus, welche jedes wenigstens einen Punkt von C in ihrem Inneren oder auf ihrer Grenze enthalten. Die in dieser Weise bestimmte Simplexmenge bezeichnen wir mit $\mathcal{F}$ und bestimmen in folgender Weise eine eindeutige und stetige Abbildung η von C auf eine nirgends dichte Teilmenge von K:

Jedem zu $\mathcal{F}$ gehörigen Grundpunkte E der Zerlegung von k lassen wir einen solchen Punkt von K entsprechen, dessen Bildpunkt für die Beziehung ϑ in einem der Grundsimplexe, welche E zum Eckpunkte besitzen, enthalten ist. Wenn dann ein Punkt P von C dem Grundsimplexe $A_1 A_2 \cdots A_{m+1}$ angehört, dessen Eckpunkten die Punkte $B_1, B_2, \cdots, B_{m+1}$ von K entsprechen, so läßt er sich als Schwerpunkt von gewissen in $A_1, A_2, \cdots, A_{m+1}$ konzentrierten positiven Massen auffassen. Dieselben Massen bringen wir der Reihe nach in $B_1, B_2, \cdots, B_{m+1}$ an, und lassen ihren Schwerpunkt Q für die Abbildung η dem Punkte P entsprechen.

Sei R derjenige Punkt von K, der für die Beziehung ϑ dem Punkte P entspricht, so können wir, indem wir die Zahl der Grundsimplexe der simplizialen Zerlegung ζ hinreichend groß wählen, das Maximum der Abstände QR unter jede Grenze, a fortiori also unter die halbe Kantenlänge von K herabsinken lassen. Die Beziehung zwischen den Punkten R und den Punkten Q wird dann aber *eine solche eindeutige und stetige Ab-*

bildung des Kubus K auf eine nirgends dichte Teilmenge von sich, bei der das Maximum der Verrückungen kleiner ist als die halbe Kantenlänge, was nach dem Resultate des § 1 ein Widerspruch ist.

Die versuchte Annahme hat sich mithin als unstatthaft erwiesen, und wir haben gezeigt:

[[4]] Satz 1. *In einer m-dimensionalen Mannigfaltigkeit enthält das eineindeutige und stetige Abbild eines m-dimensionalen Bereiches in beliebiger Nähe eines beliebigen seiner Punkte einen Bereich.*

Wir nehmen weiter an, daß eine m-dimensionale Mannigfaltigkeit R_m von einem $(m+h)$-dimensionalen Bereiche B_{m+h} $(h>0)$ ein eineindeutiges und stetiges Bild B'_{m+h} enthält. Dann würde von einem in B_{m+h} nirgends dicht liegenden m-dimensionalen Bereiche B_m die Bildmenge B'_m in B'_{m+h}, also auch in R_m nirgends dicht liegen, was wider Satz 1 verstoßen würde. Mithin gilt:

Satz 2. *Eine m-dimensionale Mannigfaltigkeit kann nicht das eineindeutige und stetige Abbild eines Bereiches höherer Dimensionenzahl enthalten.*

Schließlich nehmen wir an, daß in einer m-dimensionalen Mannigfaltigkeit das eineindeutige und stetige Bild eines $(m-h)$-dimensionalen Bereiches B_{m-h} $(h>0)$ einen gewissen Bereich B_m enthält. Dann aber würde B_{m-h} ein eineindeutiges und stetiges Bild von B_m enthalten, was nach Satz 2 ausgeschlossen ist, sodaß wir bewiesen haben:

Satz 3. *In einer m-dimensionalen Mannigfaltigkeit ist das eineindeutige und stetige Abbild eines Bereiches geringerer Dimensionenzahl eine nirgends dichte Punktmenge.*

Die Sätze 2 und 3 enthalten beide die Invarianz der Dimensionenzahl als unmittelbare Folgerung.

[[434]]

NOTES

⟦1⟧ When B. Riemann 1868 defined what is now called an n-dimensional manifold (he called it *n-fach ausgedehnte Grösse*; others spoke about *Zahlenmannigfaltigkeit*), he did not care about the topological invariance of this concept. In fact, he had only differentiable manifolds in mind; but even in the sense of differential topology the invariance question was not raised.

G. Cantor's unprecedented discovery in 1877 of the equivalence of cartesian spaces of all dimensions under *one-to-one* mappings struck like lightning. At the *Naturforscherversammlung* at Cassel 1878 it must have been one of the major topics of the day. (See Tageblatt 1878.) Mathematicians rushed to shore up a conceptual structure they felt to be tottering. Yet there was hope left: one could reasonably guess that one-to-one *continuous* mappings would discriminate between the dimensions. Abortive attempts to prove this by J. Thomae 1878, E. Netto 1878, G. Cantor 1879 show the interest aroused by this problem. The common mistake was too much faith in what analytic mappings suggested: deriving invariance of dimension from a tacitly assumed invariance of domain. J. Lüroth was more cautious and more successful. He proved the invariance of n-dimensionality (that is, the impossibility of a topological mapping of an n-dimensional upon a *higher* dimensional figure) for $n \leqq 2$, and $n \leqq 3$ in J. Lüroth 1878 and J. Lüroth 1899, respectively (see also J. Lüroth 1906) – the proof of the latter case is extremely complicated. In the meantime the mathematical world had been shocked once again, namely, by G. Peano's 1890 discovery of *continuous* mappings of domains of m-dimensional onto domains of n-dimensional cartesian space for $m < n$. In 1899, E. Jürgens 1899 gave a critical account of the state of the problem. (He claimed a correct proof of invariance of 2-dimensionality for himself: E. Jürgens 1879.) See also A. Schoenflies 1908, pp. 165–168 and R. Milesi 1892, mentioned by Schoenflies. A. Schoenflies 1908 A, pp. 162–164, also formulated the so-called Schoenflies theorem, the *invariance of the n-dimensional domain* and proved it for $n = 2$.

In the mean time the problem had become even more urgent because the invariance of n-dimensionality or even of n-dimensional domain had come to be used and considered as an indispensable tool in automorphic functions and uniformisation. R. Baire 1907 A, 1907 B underpinned invariance of dimension by invariance of domain, and invariance of domain by a weak form of Jordan's theorem in n-space, but Baire's proof of the latter theorem is restricted to an insufficient sketch for the case of two dimensions. Baire's ideas were sound but he grossly underestimated the technical difficulties of the problem. The continuing interest in related problems is demonstrated by J. Hadamard 1910; he showed how to use classical integral invariants such as introduced by C. F. Gauss 1840 and L. Kronecker 1869 in topology.

Brouwer 1911 C (submitted as early as June 1910) marks the onset of a new period in topology. It is ostensibly a short and simple proof of the invariance of dimension but it is in fact much more than this – the paradigm of an entirely new and highly promising method, now known as algebraic topology. It exhibits the ideas of simplicial mapping, barycentric extension, simplicial approximation, small modification, and, implicitly, the mapping degree and its invariance under homotopic change, and the concept of homotopy class.

This is so brilliant a record that I do not fear to say in the same breath what is lacking not only in this first paper but in all the papers of this series. It will aid in understanding the historical course of topology.

What is in fact lacking are all the tools of homology, not only cycles and homology classes, but even algebraic chains and combinatorial boundaries. Of course Brouwer knew the ideas of H. Poincaré 1895, but there is no indication that he ever contemplated using them. Even what looks like one-dimensional homology in Brouwer 1912 L, and the use of the term *Zyklosis*, betrays the influence of Listing 1847 rather than of Poincaré. In his proof of Lebesgue's paving principle Brouwer 1913 was so close to the combinatorial boundary concept that nothing but the explicit formulation was lacking, but this last step was not taken by him. The lack of homological concepts led to general clumsiness. The order of a point in an n-sphere with respect to a closed manifold or pseudomanifold of codimension 1 or with respect to what we would call a $(n-1)$-cycle, is defined by the degree of the mapping of the cycle on the vector field joining the point with the cycle, and analogous inefficient devices serve later on to define the linking coefficient of two cycles. It is strange that at the same time H. Lebesgue 1911 B used equally inefficient tools though he, too, should have known Poincaré's work. Or was there someting in Poincaré's work that prevented others from using it?

Anyhow the first to successfully link Brouwer's method to Poincaré's concepts was J. W. Alexander 1922. Still later – L. Vietoris 1927 – was Brouwer's *Zyklosis* method transferred to homology and higher dimensions to yield Betti numbers.

The language of 1911 C is less intricate than Brouwer's usual style. To modern readers it offers few problems of understanding but to Brouwer's contemporaries it seems to have been an unsurmountable barrier. Even F. Hausdorff 1914, p. 468, complained: 'Die den Leser zu vielen eigenen Ergänzungen zwingende Kürze der Brouwerschen Publikationen ist bei dem Mangel einer sonstigen einwandfreien und ausführlichen Darstellung sehr zu bedauern.' Today we would say that most of Brouwer's work suffers from prolixity rather than shortness.

1911 C, submitted June 1910, issued 14 February 1911, was the first of a series of papers which appeared in rapid succession:

1911 D, also dating from 1910, issued 25 July 1911: mappings of manifolds and the mapping degree, vector fields on spheres, and singularities in the case of even

dimensions, fixed points of mappings of spheres and balls – belatedly extended in:
1926 D (communicated 29 May 1926) to projective spaces.

1911 E, dating from 1910 or early 1911, mailed to Hilbert 14 July 1911, printer's proof 20 October 1911, ready for print 21 October 1911, issued 16 November 1911: a piece of a Jordan manifold does not separate n-space, with the invariance of n-dimensional domain as a consequence.

1911 F, summer 1911, promised to Hilbert on 14 July 1911, issued 16 November 1911: the full Jordan theorem for n-space, (A brief note on the same:
1911 L, read on 11 September 1911.)

1911 G, issued 16 November 1911: *Erreichbarkeit*, *Unbewalltheit*, *Zweiseitigkeit* of Jordan manifolds, new approach to the mapping degree and its topological invariance. See also:
1931: a correction of 1911 G.

1912 C, issued 15 March 1912: New, simple proof of the invariance of domain, by direct use of the mapping degree.

1912 E2, read on 24 February 1912: definition of the looping coefficient.

1912 F, read on 1 April 1912, 1912 L, issued 19 July 1912: the invariance of the closed curve – forerunning the homology theory of compact spaces.

1912 K2, read on 28 September 1912, 1912 M: the mapping degree is shown to characterise the homotopy class for mappings of a 2-sphere into itself.

1913 A: The general dimension concept – the paper goes back to 1911. See also:
1923 D2, 1924 J2, 1924 M: corrections and remarks regarding 1913 A; 1924 L: partial reprint of 1924 J2.
1928 C: historical exposition.

1912 D (read 13 January 1912), 1912 G (read 22 June 1912), 1912 H (read September 1911, issued August 1912) are applications of the invariance of domain to automorphic functions and uniformisation.

Methodologically related papers:
1918 C, 1919 E: Extending a continuous function.
1918 D: the absence of topological properties of Lebesgue measure.
1924 K2: simplex stars.

Brouwer's tremendous activity in the years 1910–1912 is also attested to by a large number of fragments among his papers, in particular numerous attempts at proving invariance of domain. A closer study would certainly shed more light on the origin and development of his ideas.

As far as we know from the literature the first mathematicians who profoundly understood and self-confidently employed Brouwer's method, were Erhard Schmidt (see H. Hopf 1966) and J. W. Alexander. Even in the twenties when point-set topology progressed rapidly, Brouwer's method, though highly praised, was little known and less appreciated. According to a slogan of about 1930 set theory lavished beautiful methods on ugly results while combinatorial topology boasted beautiful results obtained by ugly methods. Not until the late forties did algebraic topology, as it is called now, become a broad stream.

Brouwer 1911 C is followed in Mathematische Annalen by H. Lebesgue 1911 A. O. Blumenthal, coeditor of Mathematische Annalen had met with Lebesgue and told him about a forthcoming proof of the invariance of dimension by Brouwer, and on his part Lebesgue had indicated to Blumenthal *le principe de quelques preuves du même théorème*. According to Blumenthal himself in a letter to Hilbert of 27 October 1910 (Univ. Bibl. Göttingen):

> 'Wir haben in den Ferien eine sehr schöne Reise nach Paris gemacht. Mathematiker freilich habe ich leider nicht gesehen, die waren noch alle in Ferien. D.h. ich habe doch die Bekanntschaft von Lebesgue gemacht, der zufällig in Paris war. Er ist sehr interessant und sagte mir, dass er schon seit langer Zeit nicht nur einen, sondern mehrere Beweise des Satzes von der Invarianz der Dimensionenzahl besitzt, den Brouwer jetzt in den Annalen bewiesen hat. Einen dieser Beweise, der sehr witzig aussieht, hat er mir für die Annalen eingeschickt. Ich habe ihn nicht genau auf Richtigkeit der Durchführung, sondern nur auf Richtigkeit der Idee angesehen: in Einzelheiten kann man sich doch auf einen so scharfsinnigen Mann verlassen. Wenn Sie aber noch genau nachprüfen wollen, steht die Arbeit Ihnen zur Verfügung.'

Lebesgue's publication 1911 A has the form of a letter to Blumenthal in which he explained *la plus simple de ces démonstrations*. Its core is the celebrated paving principle 'if an n-dimensional domain is covered by a set of sufficiently small closed sets, there are $n+1$ among them with non empty intersection', which indeed suffices to characterise dimension topologically. Lebesgue's sketch of a proof, however, is so wrong that one hardly understands how it could have ever been submitted and printed, but it shows clearly how Lebesgue arrived at his paving principle: The lower and upper face of an n-dimensional cube are separated from each other by a $(n-1)$-dimensional surface which looks like a $(n-1)$-dimensional cube, to which the same procedure applies; continuing in this way, the number of steps until single points are reached, means the dimension. Now Lebesgue shifted the emphasis from the unmanageable surfaces to the closed domains bounded by them and finally from this special covering of the cube to arbitrary ones. It was a splendid idea, but when Lebesgue tried to elaborate it, he could not get rid of those heuristic separating surfaces. The paving principle was a flash of genius, and when Lebesgue enunciated it, he certainly believed that its proof was a mere technicality. He had the courage to enunciate this idea a bit too early – fortunately, I would say, because if he had waited, he would not have published it at all. His letter to Blumenthal has the sympathetic ring of boyish naïveté; there is a whisper of 'what a fuss about dimension, I have got several proofs'. At the end of Lebesgue 1911 A, Baire 1907 A, 1907 B were mentioned: 'Sans doute M. Baire n'a pas développé sa démonstration; mais il me semble que, si l'on tient compte des indications données par M. Baire, il ne reste plus à trancher que des difficultés de détail peu sérieuses'. Lebesgue was not on speaking terms with Baire; it is hard to say whether citing Baire meant teasing Baire or belittling Brouwer. Anyhow Brouwer fought back – a lifelong struggle. Yet Lebesgue could not abide Brouwer's reaction.

Brouwer 1911 C and Lebesgue 1911 A appeared in February 1911. At the 27 March 1911 meeting of the Académie des Sciences Picard read the note Lebesgue 1911 B – again chock full of marvelous ideas. The first part is a sketch of a direct proof of the invariance of dimension, not using the paving idea, in which all honour is paid to Brouwer. The second is a *proof by induction* of what Lebesgue called a generalisation of Jordan's theorem – in fact it only states the existence of *at least* two complement domains. Lebesgue here introduced the concept of linking of two varieties (an odd number of crossings if one of them is shrunk into a point), and given a variety T_n in E_{n+p+1} he claimed the existence of a variety T_p linked with it, which, indeed for $p = 0$ means the earlier mentioned part of Jordan's theorem (actually the theorem is incorrect in this form).

Lebesgue 1911 B is much less naïve than Lebesgue 1911 A though the difficulties are still grossly underestimated. The sketches are insufficient or even wrong though not in the same way as 1911 A was, and were never elaborated by him. The idea of the second part of Lebesgue 1911 B is close to Brouwer's 1910 E proof of the plane Jordan theorem. It was elaborated much later by J. W. Alexander 1922.

Let us return to Brouwer. Why dwell on these quarrels? The reason is that there are numerous allusions to Lebesgue in Brouwer's publications which remain unintelligible without some background knowledge. Lebesgue 1911 A led to correspondences of Brouwer with Lebesgue, with Mathematische Annalen, with R. Baire, and possibly with others. Nothing is left of the Lebesgue correspondence although indirect evidence shows that Lebesgue sent Brouwer new proofs that were equally wrong, and promised to publish proofs that never appeared. Nevertheless, it is a pity that all this is lost.

Brouwer's first reaction was a note for Mathematische Annalen, which, however, was not published. No manuscript of the note is preserved. On 14 March 1911 Blumenthal, according to a letter to Hilbert (Univ. Bibl. Göttingen) still believed that Lebesgue's proof was virtually correct, and he was not even sure whether there were no gaps in Brouwer's proof. He hesitated to accept Brouwer's note because of its '*unfreundlichen und unangenehmen Ton*'. On 31 March 1911 Brouwer wrote to Hilbert (Univ. Bibl. Göttingen):

'Beiliegend sende ich Ihnen zur gefl. Kenntnisnahme die Fortsetzung des Briefwechsels mit Herrn Lebesgue, während welcher ich meine eingesandte Bemerkung zurückgezogen habe.

Diese Zurücknahme war mir sehr angenehm, weil die Briefe von Lebesgue (ebenso wie später der von Herrn Blumenthal) mir zeigten, dass meine Bemerkung vom Leser als eine Prioritätsklage aufgefasst wurde, als welche sie durchaus nicht gemeint war.

Warum Lebesgue die in seinem letzten Briefe enthaltene Ausarbeitung (deren Inhalt mir übrigens nach einer ersten Durchsicht noch dunkel geblieben ist) nicht zur Kenntnis der Annalenleser bringen will, bleibt mir unerklärlich.

⟦· · ·⟧'

On 14 July 1911 Brouwer wrote to Hilbert (Univ. Bibl. Göttingen)

'Beiliegend finden Sie den tragischen Schluss des Briefwechsels mit Lebesgue.'

Brouwer's papers include a draft letter (probably the final version) to one of the editors of Mathematische Annalen (probably Blumenthal[1]) dated 19 June 1911 (see Y1), and a complementary draft letter (undated, but probably 20 June 1911 (see Y2). The first part of the first letter (not printed) is a *plan de campagne* to press Lebesgue hard either to send Mathematische Annalen a complete proof of his paving theorem, or to plead guilty – it appears that Brouwer drafted the French letters of Mathematische Annalen to Lebesgue. Brouwer says he could prove Lebesgue's theorem but would abstain from publishing a proof in order to give Lebesgue a fair chance. In the second part of this letter Brouwer refuted Lebesgue's proof idea by a counterexample. Brouwer never published that proof as such. On 26 August 1911 Blumenthal mentioned in a letter to Brouwer the *Lebesgue-beweis*, which, however, can also mean Brouwer's refutation of Lebesgue's proof. On 16 January 1912 Blumenthal told Brouwer he would like the 'absolut kürzeste und beste Beweis der Invarianz der Dimension' to appear in the Mathematische Annalen, which can only mean the proof by means of the paving principle. We have printed from Brouwer's papers piece Y3 which he probably considered as an abortive attempt at Lebesgue's theorem, and Y4 which shows Brouwer struggling toward, and finally reaching, the goal. Another version with the same basic idea has been incorporated into Brouwer 1913 A (the inductive approach to dimension), which also includes a version of Brouwer's counterexample – I feel the original version of Y1 was more lucid.

(Moreover Brouwer's papers contain the last page of another letter, Y5, prob-

[1]) The addressee is *Lieber Herr Professor*. This cannot mean *Geheimrat* Klein. It looks too formal to mean Blumenthal. However, since Brouwer was only a *Privatdozent*, Blumenthal rightly wrote 'Lieber Herr Brouwer', while Brouwer called him Professor. Moreover the mention of Aachen and some topic of automorphic functions indicates Blumenthal as addressee.

ably addressed to the same editor of Mathematische Annalen as Y1. It is a refutation of the second part of Lebesgue 1911 B, which we will deal with in due course.)

Brouwer's correspondence with R. Baire is represented by three letters of R. Baire of October 28, November 2, and December 5, 1911, and a Dutch summary of a letter of Brouwer to Baire, November 5, 1911, along with an allusion to a 'next letter'.

The Brouwer–Baire correspondence of 1911 can be reconstructed as follows:

(1) Brouwer sent Baire reprints including his proof of the invariance of dimension, perhaps at the same time asking some question about Baire's own proof (Baire 1907 A, 1907 B).

(2) Baire (October 28) thanked Brouwer while complaining that up to then illness had prevented him from elaborating his proof.

(3) In a second letter (not available) Brouwer seems to have insisted on the fact that some people had tried to violate his priority rights.

(4) Baire (November 2) wondered which people were involved, and ventured to guess that Lebesgue were one of them. Baire had not kept up personal relations with Lebesgue for years. He recognised the need for more details in his own proof, though he believed they would be mere technicalities. He again mentioned his illness and complained about his ignorance of foreign languages.

(5) The following is the summary of Brouwer's response, translated from the Dutch original.

Summary of a letter to R. Baire, translated from the Dutch:

Letter to Baire, 5 November 1911.

Explication of the 'tiefer liegen' than the invariance. Poincaré's continuity method. Last summer I found the proof of Baire's lemma, but rediscovered the proof in the 2nd part of Lebesgue's C⟦omptes⟧. R⟦endus⟧.-note. *Earlier* I had found my other 'Beweis der Invarianz des n-dim. Gebiets'. *Later* the proof of the complete Jordan theorem. – *On Lebesgue's Annalen paper*. I indicate to Leb⟦esgue⟧. the inadequacy of the characterization of I_p. Leb⟦esgue⟧. tries in his first letter to repair the charact⟦erisations⟧. For me they are still unsatisfactory. Then Leb⟦esgue⟧. tries to succeed without the I_p. This last proof is unintelligible, both to me and to Blumenthal. Leb⟦esgue⟧., however, refuses 1° further informations, 2° to correct his error in the Annalen. As for me, I had already proved the Leb⟦esgue⟧. Annalen theorem a few days after its appearence, but I do not publish the proof, because I wish to give Leb⟦esgue⟧. the opportunity to do his duty.

The priority of my 'Invarianz des Gebiets' is not disturbed.

a) *publicly*, because not yet all is published with Baire–Lebesgue but necessary ⟦or: connected?⟧ reasonings of Lebesgue are to be expected, which he expressly *promised*, so one may assume that he does not regard them as trivial.

b) *privately*, because, when Lebesgue wrote me that with the 2nd half of his C⟦omptes⟧. R⟦endus⟧.-note he could fix up Baire, I already was in possession of my 'Invarianz des Gebietes.'

c) *publicly and privately*, because neither in C⟦omptes⟧. R⟦endus⟧. nor in his letters did Lebesgue enunciate the 'Invarianz des Gebiets' and even by Baire only the 'invariance of the domain sets ⟦sets possessing an interior point?⟧' is enunciated.

In the next letter to Baire indicate that Leb⟦esgue⟧. had written his Annalen rem⟦arks⟧. 'à cause de Baire, qu'il savait très neurasthénique' [1]), and that when correcting his Annalen paper in the C⟦omptes⟧. R⟦endus⟧. Leb⟦esgue⟧ should have mentioned my name.

The reason why I do not mention the Baire–Lebesgue proof in the first of my 3 Annalen papers is that I did not understand that *formally wrong* proof until I had found it myself anew; but then my first Annalen paper was already submitted.

(6) Brouwer seems to have sent Baire more reprints and used this opportunity to inform Baire about Lebesgue's confidential remark about Baire's illness.

(7) Baire (December 5) is more explicit about his bad relations with Lebesgue. He still believed that modulo technicalities his method was sound though he agrees that Brouwer's version (1911 F) of Jordan's theorem was more complete.

* * *

Some explanations regarding the summary of Brouwer's letter to Baire of 5 November 1911:

It is not clear what 'Baire's lemma' (hulpstelling) means. In any case it is not the proposition called *lemme* in Baire 1907 B. The most likely assumption is that it means the part of Jordan's theorem (at least two complement domains) used by Baire as well as by Lebesgue. It is the only relevant thing Brouwer could rediscover in the second part of Lebesgue 1911 B.

Brouwer's other proof of invariance of domain may be the one he published in 1911 E, though it is not impossible, indeed quite probable that he meant one that was not yet published at that moment, that is 1912 C. There is still a third possibility, that is, a proof in between 1911 E and 1912 C, using, as Baire and Lebesgue

[1]) Baire was indeed seriously ill in the sense indicated by the euphemism 'neurasthénique'.

did, the weak Jordan theorem of 'at least two complement domains'. This would explain the allusion to the complete Jordan theorem in the next sentence. Moreover, Brouwer's fragments of 1911 (undated but indirectly dated by references to Lebesgue 1911 A) reveal a large number of approaches to the invariance of domain.

It is indeed a strange thing that Lebesgue never mentioned the invariance of domain. The uncertainty of my interpretation of 'domain sets' (which in Brouwer's work usually means 'open set') is caused by Brouwer's assertion that Baire did not enunciate the invariance of domain. Baire did, but perhaps Brouwer meant that his method did not extend past 'domain sets'.

Brouwer's remark on Lebesgue's not mentioning his name, looks strange. Lebesgue 1911 B gave Brouwer full credit as regards Brouwer 1911 C. Or did Brouwer mean that Lebesgue should have mentioned Brouwer 1910 E too? It would not have been unreasonable.

The 'three Annalen papers' are obviously those that appeared in a connected setting in the third issue of Mathematische Annalen 1911, that is 1911 E, 1911 F, 1911 G. Indeed, in 1911 E Brouwer cited the reference of Lebesgue 1911 A to Baire 1907 A, 1907 B, though not Lebesgue 1911 B. A copy of the last proofreadings of Brouwer 1911 E among Brouwer's papers bears the note 'Nach Korrektur druckfertig 21.10.11 LEJB.'. So Brouwer could have added a reference to Lebesgue 1911 B in the proofsheets but in fact he did no such thing.

The whole piece reveals Brouwer's pronounced anxiety about his priority rights with regard to the invariance of domain. Much depended on whether Lebesgue was able to elaborate the sketches of Lebesgue 1911 B. This explains the way the struggle continued.

Brouwer's public reaction to Lebesgue 1911 A and 1911 B is found in Brouwer 1911 E, footnote *): 'the first proof is insufficient, the second is factually identical with mine, but the deviation complicates the course of thought.' There is little doubt that Brouwer means the first part of Lebesgue 1911 B. But then his remark looks strange. In fact Lebesgue's proof is *not correct*, and his method *differs* quite a bit from Brouwer's. But Brouwer read Lebesgue's proof through the spectacles of his own knowledge, sympathetically filled in the gaps with ideas of his own and then stated that it was his own proof, though needlessly complicated.

The second part of Lebesgue 1911 B, though privately attacked in Brouwer Y5, was ignored by Brouwer 1911 E, but publicly recognised by Brouwer 1911 F, which is about Jordan's theorem in n dimensions. There Brouwer omitted the proof of 'at least two domains' because 'it could be proved by a method sketched by Lebesgue'. In a letter of 26 August 1911 Blumenthal had asked Brouwer to provide the missing part, but the actual publication shows that Brouwer did not comply.

Brouwer's behaviour looks irrational. Clearly he had meanwhile succeeded in saving the method of the second part of Lebesgue 1911 B, but why, rather than providing the full proof, did he attempt to appease the reader with the reference to a vague and incomplete sketch? In fact, as long as this gap was not solidly filled, the proof of Jordan's theorem was not complete.

Brouwer's strange behaviour cannot be understood without the background knowledge presented above. Citing Lebesgue's sketch meant challenging his adversary. It meant urging him either to produce a watertigtht proof, or if (as Brouwer anticipated) he could not produce one, to admit he had sinned.

Brouwer 1913 A, which contains Brouwer's inductive approach to dimension along with a proof of Lebesgue's paving theorem, provided a new opportunity of mentioning the insufficiency of Lebesgue 1911 A. At the end of his own proof Brouwer announced a forthcoming proof of Lebesgue's theorem by Lebesgue himself; this proof, however, has never been published.

It is strange that Brouwer 1913 A was published in 'Crelle' rather than in 'Mathematische Annalen' though it had been promised this journal in the letter Y1, and probably also on other occasions, as indicated in the earlier cited pieces (Blumenthal 26 August 1911, 16 January 1912). There must have been subtle reasons why Brouwer ceded this paper to 'Crelle' rather than to 'Mathematische Annalen'. There is no sign of estrangement between Brouwer and Mathematische Annalen at that time. On the contrary, after having served as a referee for several years, he became a co-editor in 1914 or 1915. Anyhow the fact that Brouwer 1913 A was not published in Mathematische Annalen had momentous consequences, which we will examine later on.

The public reaction to the quarrel between Brouwer and Lebesgue was not free from nationalist sentiments. In a review of A. Schoenflies 1908 A by L. Z⟦oretti⟧ 1911, the reviewer said:

'Le Chapitre V contient une intéressante étude de la correspondance entre deux domaines à n et $n+p$ dimensions et de l'invariance de la notion de dimension. Des travaux tout récents de MM. Baire, Lebesgue et Brouwer ont fait faire à la question un pas décisif, en sorte que, sur cette question, l'ouvrage n'est plus au courant.'

The reviewer seems to bestow priority on Lebesgue above Brouwer. One year earlier Brouwer 1910 J had indicated mistakes in a paper of Zoretti 1909.

In the 'Rapport sur les travaux de M. R. Baire' read by E. Picard on 12 June 1911 in the Secret Committee of the Académie des Sciences, it is stated that,

'En 1907, dans deux notes intitulées "Sur la non-applicabilité de deux continus à n et $n+p$ dimensions" M. Baire a donné une méthode pour étudier la question visée dans le titre. Quelques lacunes restaient dans les démonstrations que l'auteur se proposait de combler. Il a été devant par MM. Brouwer et Lebesgue qui, par d'autres voies, sont arrivés au théorème clairement pressenti par M. Baire, à savoir que l'on ne peut établir une correspondance *biunivoque* entre un continuum à n dimensions et un continuum à $n+p$ dimensions.'

On the other hand the Jahrbuch über die Fortschritte der Mathematik 42, Jahrbuch 1914, was less friendly. In an Appendix Lesbesgue 1911 A was re-examined: 'In dem Bericht auf S. 419 über die Arbeit von Lebesgue ist zu bemerken, dass der *Lebesguesche* Beweis nicht richtig ist. Er enthält eine Lücke, die der Verf. bisher noch nicht ausgefüllt hat. Näheres findet man in einer Arbeit von L. E. J. Brouwer in J. für Math. *142*, 151.

Finally Lebesgue 1921 provided a proof of Lebesgue's theorem, which according to Brouwer 1923 D2, 1924 J2, 1924 L was correct but essentially identical with that of Brouwer 1913 A though needlessly complicated. In any case, it was not a direct proof but one by methods similar to those meanwhile developed by Brouwer. Lebesgue 1921 also contains a proof of the invariance of domain, which is patently incorrect. A year later Lebesgue published a review of his own total work, Lebesgue 1922. In the chapter on Analysis situs he gave Brouwer full credit for the invariance of dimension but claimed for himself, without citing Brouwer or even, for that matter, any paper of his own, the invariance of domain, which he had unsuccessfully tackled as late as ten years after Brouwer. Lebesgue 1924 is an (unconvincing) correction of this part of Lebesgue 1921.

The earlier mentioned improved version Brouwer 1923 D2 of Brouwer 1913 A gave Brouwer another opportunity to renew the old quarrel, and in Brouwer 1924 J2, 1924 L he made Lebesgue responsible for the fact that the badly needed correction had been delayed for so long. It is a far fetched accusation, but the story itself is not as improbable as it looks.

This is the saga of Lebesgue's paving principle which happily ends with a glorious outlook on general dimension theory, in which context it was fully exploited and profoundly understood, first by P. Urysohn and K. Menger and then by P. Alexandroff, to whom it became the path to algebraic topology in topological spaces.

Since P. Alexandroff 1928 A, Lebesgue's principle is most lucidly expressed as a fact of homology theory. However, for a homology-free proof of Lebesgue's theorem, see E. Sperner 1928. A similar but less well-known proof was published by W. Hurewicz 1929.

⟦2⟧ J. Lüroth 1906, J. Thomae 1878, E. Netto 1878, G. Cantor 1879, E. Jürgens 1899, A. Schoenflies 1908 A, R. Baire 1907 A, 1907 B, M. Fréchet 1910.

⟦3⟧ This is essentially the mapping degree.

⟦4⟧ A weak form of the invariance of domain.

Y 1 Draft letter of Brouwer, very likely to O. Blumenthal

Amsterdam, 19.6.1911.

⟦1⟧ Lieber Herr Professor,

⟦· · ·⟧

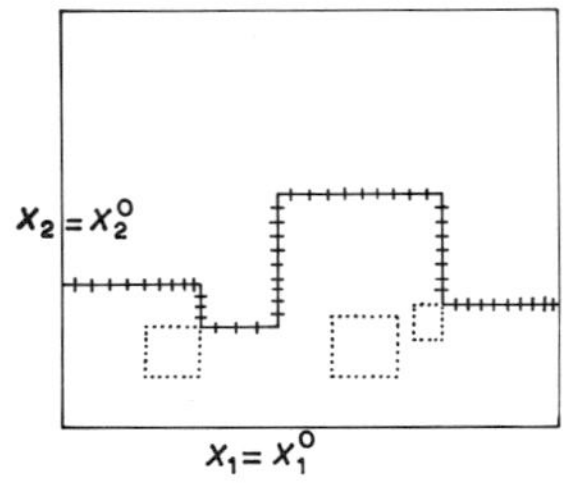

Jetzt zur Sache, und zwar zunächst Ihre ebene Figur. Im Falle zweier Dimensionen gibt es in der Grenze von e_1 nur *ein* ⟦sic⟧ *einzigen* (in der Figur strichpunktiert gezeichneten) eindimensionalen Raum, welcher sich von $x_2 = x_2^0$ bis $x_2 = x_2^0 + 2l$ erstreckt; dieser wird als I_1 gewählt; die übrigen in der Grenze von e_1 enthaltenen, in der Figur punktiert gezeichneten eindimensionalen Räume können sich nicht von $x_2 = x_2^0$ bis $x_2 = x_2^0 + 2l$ erstrecken.

Weil nun aber I_1 zusammenhängend ist, und sich von $x_2 = x_2^0$ bis $x_2 = x_2^0 + 2l$ erstreckt, müssen zwei Teilmengen von I_1, von denen die erste das Teilintervall, welches an $x_2 = x_2^0$ grenzt, und die zweite das Teilintervall, welches an $x_2 = x_2^0 + 2l$ grenzt, enthält, wenigstens einen Punkt gemeinschaftlich haben. Es existiert also sicher ein Punkt I_2, *sodass für zwei Dimensionen keine Schwierigkeiten auftreten.*

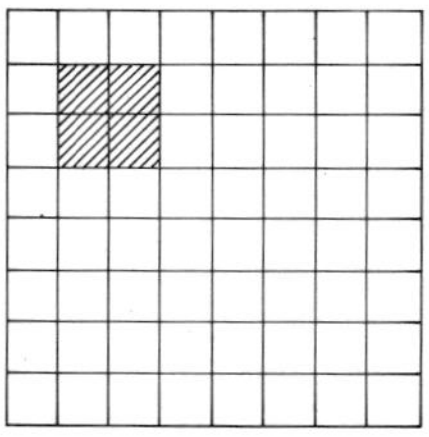

Für drei Dimensionen wird es schon schlimmer. Zerlegen wir z.B. die Kanten des Hauptkubus je in 8 gleiche Teile, der Kubus wird dann in 8^3 Teilkuben zerlegt und lässt sich einer willkürlicher Seitenfläche parallel in 8 Teilschichten je von ⟦8^2⟧ Teilkuben zerlegen. Ich nehme nun an, dass (wie ja möglich ist) e_1 sich zusammensetzt aus *erstens* der ganzen ersten der Fläche $x_1 = x_1^0$ parallelen Schicht, *zweitens* von der zweiten der Fläche $x_1 = x_1^0$ parallelen Schicht die in nebenstehender Figur gestrichelten Teilkuben; *drittens* der ganzen dritten der Fläche $x_1 = x_1^0$ parallelen Schicht, während e_2 resp. e_3 sich zusammensetzt aus den beiden ersten der Fläche $x_2 = x_2^0$ resp. $x_3 = x_3^0$ parallelen Schichten.

Die Grenze von e_1 innerhalb des Hauptkubus setzt sich dann zusammen aus zwei zusammenh⟦ängenden⟧. 2-dim. Räumen (einem einfach zusammenhängenden Raum ${}_\alpha I_1$, und einem Raum ${}_\beta I_1$ vom Zus⟦ammen⟧h⟦ang⟧. der Zylinderfläche, welche⟦r⟧ sich je sowohl von $x_2 = x_2^0$ bis $x_2 = x_2^0 + 2l$ wie von $x_3 = x_3^0$ bis $x_3 = x_3^0 + 2l$ erstreckt. *Einen* und *nur einen* dieser Räume müssen wir als I_1 wählen; beide zusammen behalten geht deshalb nicht an, weil alle weiteren Schlüsse auf dem *Zusammenhang* von I_1 beruhen; *welchen* wir zu wählen haben, darüber gibt Lebesgue keinen Aufschluss. Von den beiden Räumen, welche hier in Betracht kommen, führt aber nur einer, nämlich ${}_\alpha I_1$ schliesslich zu einem I_3 ⟦;⟧ ${}_\beta I_1$ führt, wie man gleich sieht, zu einem nicht zusammenhängenden I_2, während I_3 gar nicht existiert. Dennoch besitzt ${}_\beta I_1$ die von Lebesgue zu Grunde gelegte Eigenschaft, sich von $x_\alpha = x_\alpha^0$ bis $x_\alpha = x_\alpha^0 + 2l$ $(\alpha = 2, \cdots, n)$ zu erstrecken. Diese Eigenschaft ist also für die Argumentierung völlig wertlos, und gibt für den n-dimensionalen Raum *keine* Sicherheit, dass unter den verschiedenen $_{\nu}I_1$ immer einer existiert, welcher bei passender Wahl der sukzessiven I_α schliesslich zu einem I_n führt. Die Auswahl von I_1 unter den $_{\nu}I_1$ würde sich noch dadurch bestimmen lassen, dass I_1 den *Aussenrand* von e_1 darstellt, aber für I_2, I_3 usw. versagt auch dieses Kriterium, sodass ich nicht glaube, dass sich auf diesem Wege etwas erreichen lässt.

Wie ich schon oben sagte, enthalten diese Auseinandersetzungen für Herrn Lebesgue nichts neues. Ich hoffe sehr mich Ihnen jetzt vollkommen verständlich ausgedrückt zu haben und möchte schliesslich gerne nochmal von Ihnen vernehmen, welchen Satz aus der Analysis Situs Sie mir in Aachen nannten als unbedingt notwendig für den Continuitätsbeweis der Existenz polym⟦orpher⟧. ⟦2⟧
Funktionen auf Riem⟦annschen⟧. Flächen? Aus Ihrem letzten Briefe scheine ich schliessen zu müssen, dass es nicht der Jordansche Satz ist.

Mit den besten Grüssen

NOTES

⟦1⟧ A draft letter found among Brouwer's papers. The addressee must be O. Blumenthal, acting editor of Mathematische Annalen. The first part of the letter is omitted; its content is summarized in 1911 C ⟦1⟧. The second part, as printed here, is a counterexample to the proof in Lebesgue 1911 A. The notations are mainly those of Lebesgue 1911 A. It is essentially the same counterexample as in Brouwer 1913 A, p. 151, footnote *.
⟦2⟧ This foreshadows Brouwer 1912 H, 1912 D.

Y 2 Draft letter of Brouwer, very likely to O. Blumenthal

⟦1⟧ ⟦20 June 1911⟧

Zur weiteren Erläuterung meines gestrigen Briefes füge ich noch hinzu, dass, wenn in meinem dreidimensionalen Beispiele ${}_{\beta}I_1$ als I_1 gewählt wird, wir als "Grenze desjenigen Teiles von I_1, das ⟦sic⟧ in den nicht ganz zu e_1 gehörigen Elementen von e_2 enthalten ist", folgende in der Ebene $x_2 = x_2^0 + \frac{1}{4}l$ liegende Punktmenge erhalten:

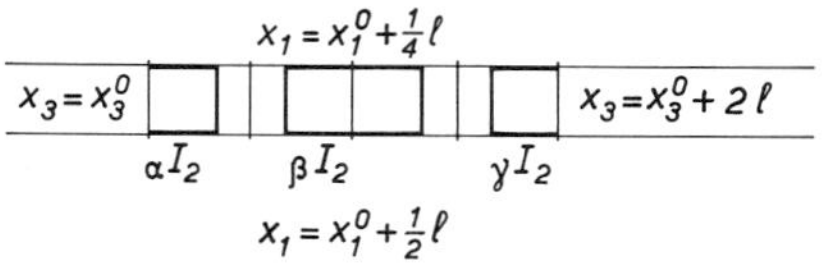

Von den drei zusammenhängenden Teilmengen dieser Punktmenge ist ${}_{\alpha}I_2$ ganz enthalten in "*den nicht ganz zu* e_1 *oder* e_2 *gehörigen Elementen von* e_3"; ${}_{\beta}I_1$ und ${}_{\gamma}I_2$ aber liegen gänzlich ausserhalb "*der nicht ganz zu* e_1 *oder* e_2 *gehörigen Elemente von* e_3".

Als I_3 hätte man somit zu wählen die Grenze zwischen ${}_{\alpha}I_2$ und ${}_{\beta}I_2 + {}_{\gamma}I_2$. Diese Grenze existiert aber nicht, und *es gibt keinen* I_3.

NOTE

⟦1⟧ Continuation of Y1.

⟦*English translation of the Dutch text*⟧ ⟦1⟧

It is to be proved that if in the cube K_n a separation between the x_1-planes is made which separates a single positive domain from the remainder, and on that positive domain we make a separation between the x_2-planes which separates a single positive domain from the remainder, and so on, then there is a point common to all those boundaries.

To this end we extend the separations from the second one onwards to obtain complete separations like the first one.

Then we attach an x_1-vector, positive above the x_1-separation, negative below, and zero on it, namely ⟦a vector whose length is⟧ always equal to the distance from the separation.

Then likewise ⟦we attach⟧ an x_2-vector which, however, we allow actually to approach zero when approaching the boundary from the x_1-x_2-positive domain, but when approaching from an x_1-negative x_2-positive domain, we turn it just before reaching the boundary through the negative x_1-direction in such a way that at the boundary itself it is negative and directed according to x_2.

We continue in this way. We choose the x_p-vector so that it really approaches zero when approaching the boundary from an x_1-x_2-$\cdots$-x_p-positive domain, but when approaching from an x_1-$\cdots$-x_q-negative and x_{q+1}-$\cdots$-x_p positive domain we turn it into the negative x_p-direction through directions obtained from combining $x_1, \cdots, x_q$. On the boundaries, where positive domains pass into negative ones, we can modify these combined turn directions. ⟦2⟧

The vector distribution has a finite integral over the surface of the cube so there is a singular point in the interior, and this is the point we required. ⟦3⟧

NOTES

⟦1⟧ The text, in Dutch, is written on a 10 by 13 cm piece of paper. According to the summary of a letter to Baire of 5 November 1911 (see Brouwer 1911 C ⟦1⟧) it must be assigned to early 1911, more precisely shortly after Brouwer saw Lebesgue 1911 A, which bears the date of issue 14 February 1914. The translation tries to imitate Brouwer's fragmentary style.

⟦2⟧ In the terminology then usual 'finite' means 'non-zero'.

⟦3⟧ Clearly this approach was not regarded as satisfactory.

Y 4

⟦1⟧ ⟦*English translation of the Dutch text*⟧

⟦· · ·⟧

⟦2⟧ The nucleus of the proof of Lebesgue must be as follows: Let A be the bottom face, B the top face of the n-dimensional cube K. By definition an r-set of small cubes in K covers the entire top face, is connected, may cover itself, forms intrinsically an ^{n}space, and neither penetrates as far as A nor returns to B. Then the boundary of this r-set contains a connected part S^{n-1}, which is intrinsically an ordinary $^{n-1}$ space, and if projected upon A covers this bottom face with degree 1. The other connected pieces cover A with degree 0. This connected part is intrinsically connected, *even* if we omit all its points in the vertical faces of K. But if we add again the points in the vertical faces (the lateral boundary spaces) of K, then in each of those $^{n-1}$cubes, there is now a connected part S^{n-2} with the same properties as S^{n-1}.

However, in the above method of proving Lebesgue's theorem we have to require two-sidedness of the first r-set. Then S^{n-1} is also two-sided and so is the $(n-2)$-dimensional coupure to be made in that S^{n-1}, etc.

In S^{n-1} two incident edges are considered as identical *only* if they are so for the r-set of n-dimensional cubes (i.e. the r-set considered *as ordinary* space with singularities neither in the interior nor on the boundary).

The question may also be formulated as follows: Suppose we are given an R_n that, without singularities, covers a K_n (n-dimensional cube) with degree 1 and is connected outside the boundary of K_n. Let A_1, A_2 be two opposite faces of K_n and let us single out a part D of R_n which near A_1 contains *all* of R_n and near A_2 *nothing* of R_n. Then the boundary of D consists of one or more spaces R_{n-1}, which are connected within K_n and without singularities. After projection each of them covers A_1 with a certain degree, which is unity for at least one of the R_{n-1}.

Perhaps we can also argue in the Lebesgue case as follows:

If near each $^{n-1}$face of the original cube K_n we single out only the first layer of partial cubes, then the theorem is obvious. Now we must only show that it remains
⟦3⟧ valid if we repeat the operation of adding a cube to one of the layers.

Questions for Intermédiaire: 1) Can each R_n occur as the *unique* boundary of an R_{n+1}? 2) How many geometrical order types of everywhere dense countable point sets are there in R_2? ⟦4⟧

Better for the proof of Lebesgue. Choose for I_1 the entire set of boundaries of e_1 (namely as far as not belonging to the bottom face and the vertical faces); the projection of I_1 upon $x_1 = x_1^0 + 2l$ covers this with degree 1. Thus its boundaries in $x_2 = x_2^0$ cover by projection upon $(x_1 = x_1^0 + 2l,\ x_2 = x_2^0)$ also this face with degree 1. Now repeatedly take away a little element from I_1, then its boundaries (as far as not belonging to the vertical faces) still cover that face with degree 1. Finally this system of boundaries is I_2. Thus I_2, too, covers by projection the face in question with degree 1. Finally I_{n-1} covers a one-dimensional edge with degree 1. Thus I_{n-1} *exists* and *so does its coupure* I_n. ⟦5⟧

Here, too, divide all within into simplexes. ⟦6⟧
⟦· · ·⟧

NOTES

⟦1⟧ These are four connected pages of about 19 by 13 cm. The fragment must be dated shortly after Y3. It starts with a few lines on some easy question of point-set topology. Then comes the attempt to prove Lebesgue's theorem. It develops in a natural way and ends in complete success. From the start the essential tool is the mapping degree. The piece printed here covers $2\frac{1}{3}$ pages but a text of $\frac{1}{3}$ of a page has been deleted. Then on the back page follows an attempt to prove the Jordan theorem in 3-space, which continues on a fifth separate page. The translation from Dutch tries to imitate the fragmentary style of the Dutch original.
⟦2⟧ 'Lebesgue' of course means 'Lebesgue's theorem'.
⟦3⟧ Here a third of a page of text has been deleted. The parts still readable show that 'the bell has rung'.
⟦4⟧ It is a characteristic feature that Brouwer interrupts his activity, and formulates two problems, the first related to what later will be known as Antoine's example and J. W. Alexander's 'horned sphere', the second related to Brouwer 1913 B2. 'Intermédiaire des mathématiciens' refers to a periodical devoted to queries and problems.
⟦5⟧ The next paragraph is written in quite small letters. Probably it was squeezed in after the next and a fifth page had been filled. It is the final proof in a few lines of Brouwer 1913 A, though formulated in terms of cubes rather than simplexes. The proof cries out for homological tools.
⟦6⟧ This is written in the vertical direction in the left margin of the last paragraph in letters which are less than 1 mm high. It foreshadows the final version of Brouwer 1913 A.

Y 5 Last page of a draft letter, very likely to O. Blumenthal

⟦1⟧ Denken wir uns nun aber T_n als eine Ringfläche im drei-dimensionalen Raume, ϑ als eine zusammenziehbare geschlossene Jordansche Kurve auf T_n, A als das von ϑ auf T_n bestimmte einfach zusammenhängende Gebiet, B als das Restgebiet von T_n. Γ trifft dann auf Gr⟦un⟧d der Definition der variétés enlacées sicher A (resp. A'), weil sich ϑ innerhalb A auf einen einzigen Punkt zusammenziehen lässt. Innerhalb B lässt sich ϑ aber *nicht* auf einen einzigen Punkt zusammenziehen. Die Existenz von Schnittpunkten von Γ mit B ist also *nicht sicher*. Wenn dann weiter die Invarianz der Dimension *nicht* feststeht, könnte Γ ganz in A enthalten sein, und würde dann nur *ein einziges Gebiet* α bilden. Dieses Gebiet α besässe *alsdann keine Grenze, welche man durch mit* T_n *verschlungene "Punktepaare"* T_p approximieren könnte.

Für der genannten einfachen Fall kann man übrigens leicht die Existenz von Schnittpunkten von Γ auch mit B (resp. B') nachweisen. Um für beliebiges n und p und für beliebige T_n dasselbe zu erreichen, hätte man zu zeigen, dass die geschlossenen polyedrischen Mannigfaltigkeiten Γ mit $T_n = A' + B'$ immer eine gerade Zahl von Schnittpunkten haben. Die strenge Durchführung des Beweises dieser "evidenten" Tatsache scheint mir aber äusserst mühsam. Ubrigens findet sich eine analoge Schwierigkeit schon in der Begründung der *Definition* der variétés enlacées für beliebiges p und n.

Meine Stellung dem zweiten Teile der Lebesgueschen Note gegenüber ist also diese, dass er für den dreidimensionalen Raum einen sehr schönen Satz ganz richtig bewiesen hat, dass er für höhere Räume aber nur "evidente" Erweiterungen ausgesprochen hat, ohne etwas zu beweisen. Den Invarianzbeweis braucht man aber gerade für die höheren Räume, sodass der genannte Teil der Leb⟦esgueschen⟧. Note meines Erachtens für die Invarianz garnichts enthält.

Besten Gruss

Ihrem ⟦sic⟧ L E J Brouwer

NOTE

⟦1⟧ The last page of a draft letter found among Brouwer's papers; the whole page is covered by several cross-outs. The addressee must again be O. Blumenthal (see Y1). It is a refutation of the second part of Lebesgue 1911 B though later on Brouwer 1911 F recognised the method itself as correct (see Brouwer 1911 C ⟦1⟧). The notation is that of Lebesgue 1911 B. The piece should be dated shortly after the appearance of Lebesgue 1911 B, which was read 27 March 1911. It may precede Y1.

From left to right: P. Alexandroff, L. E. J. Brouwer and P. Urysohn.
(Courtesy of P. Alexandroff.)

1911 D

Über Abbildung von Mannigfaltigkeiten.*)

Von

⟦1⟧ L. E. J. Brouwer in Amsterdam.

§ 1.

Der Grad einer stetigen Abbildung einer geschlossenen zweiseitigen Mannigfaltigkeit.

Unter einem *Simplexsterne* des n-dimensionalen Zahlenraumes verstehen wir eine in einer Umgebung eines Punktes O überall dicht liegende, endliche Menge von nicht in das Innere voneinander eindringenden und den Punkt O als Eckpunkt besitzenden Simplexen, deren je zwei eine p-dimensionale $(0 \leqq p \leqq n-1)$ Seite gemeinsam haben, sonst aber keinen gemeinschaftlichen Punkt besitzen.

Unter einem *n-dimensionalen Elemente E* verstehen wir das eineindeutige und stetige Bild eines Simplexes S des n-dimensionalen Zahlenraumes.

Unter den *Eckpunkten* bez. *p-dimensionalen Seiten* von E verstehen wir alsdann die Bilder der Eckpunkte bez. der p-dimensionalen Seiten von S.

Wir bilden nun aus n-dimensionalen Elementen eine solche zusammenhängende Punktmenge Z, daß je zwei dieser Elemente entweder keinen gemeinschaftlichen Punkt besitzen, oder eine p-dimensionale $(0 \leqq p \leqq n-1)$ Seite (und dann zugleich alle in ihr liegenden Seiten geringerer Dimensionenzahl) gemeinsam haben, übrigens aber keinen gemeinschaftlichen Punkt besitzen, während in jedem Eckpunkte die daselbst zusammenstoßenden Elemente in derselben Weise, wie die Simplexe eines gewissen Simplexsternes des n-dimensionalen Zahlenraumes, aneinander schließen.

Die in dieser Weise konstruierte Punktmenge Z soll eine *n-dimen-*
⟦3⟧ *sionale Mannigfaltigkeit* heißen, und zwar, wenn die Zahl ihrer Elemente

*) Während der Drucklegung dieser Abhandlung ist im zweiten Bande der „Introduction à la théorie des fonctions d'une variable" von J. Tannery erschienen:
⟦2⟧ J. Hadamard, „Sur quelques applications de l'indice de Kronecker". Die daselbst ausgeführte Theorie berührt sich mannigfach mit den vorliegenden Entwicklungen.

endlich ist, eine *geschlossene n-dimensionale Mannigfaltigkeit;* wenn die Zahl ihrer Elemente unendlich ist, eine *offene n-dimensionale Mannigfaltigkeit.*

Diese Definition läßt sowohl endliche, wie unendliche Zusammenhangszahlen zu*).

Eine *geschlossene* n-dimensionale Mannigfaltigkeit ist eine *abgeschlossene* Punktmenge**); eine *offene* n-dimensionale Mannigfaltigkeit aber eine *nicht abgeschlossene* Punktmenge. Sei nämlich $P_1, P_2, P_3, \cdots$ eine solche Fundamentalreihe von Punkten einer offenen Mannigfaltigkeit, daß keine zwei dieser Punkte demselben Elemente angehören, so existiert, weil es kein Element gibt, in dessen beliebiger Nähe Punkte unendlich vieler verschiedener Elemente liegen, für diese Fundamentalreihe kein Grenzpunkt. [[5]]

Eine n-dimensionale Mannigfaltigkeit soll eine *gemessene n-dimensionale Mannigfaltigkeit* heißen, wenn in jedem Elemente die Punkte in solcher Weise durch $n + 1$ nicht negative, homogene Koordinaten (als „*Normalkoordinaten*" zu bezeichnen) eineindeutig und stetig repräsentiert sind, daß in jeder p-dimensionalen Seite die Punkte für jedes Element, dem diese Seite angehört, dieselben Koordinaten besitzen.

Wir behaupten nun, daß jede n-dimensionale Mannigfaltigkeit zu einer gemessenen n-dimensionalen Mannigfaltigkeit gemacht werden kann.

Zum Beweise dieser Behauptung weisen wir zunächst den Punkten der eindimensionalen Seiten ihre Normalkoordinaten zu, und zwar soll,

*) Dasselbe leistet die Konstruktion, welche ich meinen gruppentheoretischen Untersuchungen zugrunde gelegt habe (vgl. Math. Ann. 67, S. 247, und die Berichtigung in 69, S. 180). Dort wird von einem willkürlichen endlichen Gebiete G des n-dimensionalen Zahlenraumes mit der Grenze $\mathfrak{G}$ ausgegangen; eine zusammenhängende Teilmenge von $\mathfrak{G}$ mit der Eigenschaft, daß, wenn ein Punkt P zu ihr gehört, auch der innerhalb einer gewissen im n-dimensionalen Zahlenraume um P beschriebenen Kugel liegende Teil von $\mathfrak{G}$ zu ihr gehört, wird ein *Teilgebiet* γ von $\mathfrak{G}$, und das innerhalb einer solchen Menge von Kugeln liegende Teilgebiet von G eine *Innenseite* von γ genannt. Sodann werden gewisse Gebiete γ zusammen mit ihren Grenzen g derart paarweise eineindeutig und stetig aufeinander abgebildet und identifiziert, daß je zwei in dieser Weise zusammentreffende Innenseiten zusammen mit dem entsprechenden Gebiete γ eine auf ein Gebiet des n-dimensionalen Zahlenraumes eineindeutig und stetig abbildbare Punktmenge darstellen, und es werden erstens die paarweise identifizierten Gebiete γ, und zweitens diejenigen Punkte der Grenzen g, welche durch die Identifizierungen auf Gebiete des n-dimensionalen Zahlenraumes eineindeutig und stetig abbildbare Umgebungen bekommen haben, zum Gebiete G hinzugerechnet. Mit einer solchen Definition läßt sich aber, so lange die Theorie der gestaltlichen Invarianten des n-dimensionalen Zahlenraumes nicht genügend weit entwickelt ist, in vielen Fällen nicht weiter operieren; so sind wir z. B. für den Gegenstand der vorliegenden Arbeit auf den oben im Texte gewählten Ausgangspunkt angewiesen. [[4]]

**) d. h. eine Punktmenge, in welcher jede Fundamentalreihe einen ebenfalls zur Punktmenge gehörigen Grenzpunkt besitzt.

wenn wir die nicht Null werdenden Normalkoordinaten der Punkte der eindimensionalen Seite $A_p A_q$ mit u_p und u_q bezeichnen, das Verhältnis $\frac{u_p}{u_q}$ in A_p unendlich sein, und von dort an stetig abnehmen, bis in A_q der Wert Null erreicht wird.

Um sodann die Normalkoordinaten der Punkte der zweidimensionalen Seiten zu bestimmen, bilden wir eine solche zweidimensionale Seite $A_p A_q A_r$ derart eineindeutig und stetig auf ein Euklidisches ebenes Dreieck FGH ab, daß den Seiten des Dreiecks die eindimensionalen Elementseiten $A_p A_q$, $A_q A_r$ und $A_r A_p$ entsprechen, repräsentieren die Punkte dieses Euklidischen Dreiecks durch homogene Schwerpunktskoordinaten in bezug auf die Eckpunkte F, G, H, und wählen innerhalb des Dreiecks einen willkürlichen Punkt O.

Sei B ein Punkt der zweidimensionalen Seite $A_p A_q A_r$, B' sein Bild im Dreiecke FGH. Das jenseits B' fortgesetzte geradlinige Segment OB' trifft eine der Seiten des Dreiecks, beispielsweise die Seite GH, in einem Punkte C'. Sei C derjenige Punkt der eindimensionalen Seite $A_q A_r$, der C' entspricht, ${}_C u_q$ und ${}_C u_r$ seine Normalkoordinaten, C'' der Punkt der Dreiecksseite GH mit den Schwerpunktskoordinaten 0, ${}_C u_q$ und ${}_C u_r$, und B'' der Schnittpunkt der geraden Linie OC'' mit der B' enthaltenden Parallellinie zu GH. Dann wählen wir als Normalkoordinaten von B die Schwerpunktskoordinaten von B''.

Nachdem in dieser Weise von den Punkten aller zweidimensionalen Seiten die Normalkoordinaten bestimmt sind, betrachten wir eine dreidimensionale Seite $A_p A_q A_r A_s$, und bilden sie derart auf ein Euklidisches Tetraeder $FGHK$ ab, daß den Seiten und Kanten des Tetraeders der Reihe nach die zu $A_p A_q A_r A_s$ gehörigen zweidimensionalen und eindimensionalen Elementseiten entsprechen. Die Punkte dieses Euklidischen Tetraeders repräsentieren wir durch homogene Schwerpunktskoordinaten in bezug auf die Eckpunkte F, G, H, K, und wählen innerhalb des Tetraeders einen willkürlichen Punkt M.

Sei D ein Punkt der dreidimensionalen Seite $A_p A_q A_r A_s$, D' sein Bild im Tetraeder $FGHK$. Das jenseits D' fortgesetzte geradlinige Segment MD' trifft eine der Seiten des Tetraeders, beispielsweise die Seite GHK in einem Punkte E'. Sei E derjenige Punkt der zweidimensionalen Seite $A_q A_r A_s$, der E' entspricht, ${}_E u_q$, ${}_E u_r$, ${}_E u_s$ seine Normalkoordinaten, E'' der Punkt der Tetraederseite GHK mit den Schwerpunktskoordinaten 0, ${}_E u_q$, ${}_E u_r$, ${}_E u_s$, und D'' der Schnittpunkt der geraden Linie ME'' mit der D' enthaltenden Parallelebene zur Ebene GHK. Dann wählen wir als Normalkoordinaten von D die Schwerpunktskoordinaten von D''.

In dieser Weise fahren wir fort, bis wir schließlich mittels der Normalkoordinaten der Punkte der $(n-1)$-dimensionalen Elementseiten die Normalkoordinaten *aller* Punkte der Mannigfaltigkeit derart bestimmt haben, daß diese in der Tat als eine gemessene n-dimensionale Mannigfaltigkeit vorliegt.

Wir wählen nun zu jedem Elemente einer gemessenen n-dimensionalen Mannigfaltigkeit ein Euklidisches reguläres Simplex fester Kantenlänge als sein *„repräsentierendes Simplex"*, und bilden es darauf in solcher Weise eineindeutig und stetig ab, daß die Schwerpunktskoordinaten in Bezug auf die Eckpunkte eines Punktes des repräsentierenden Simplexes den Normalkoordinaten des entsprechenden Punktes des n-dimensionalen Elementes gleich sind.

Unter einem *Teilsimplex* bez. einem *ebenen p-dimensionalen Raumstück* des Elementes verstehen wir alsdann das Bild eines Teilsimplexes bez. eines ebenen p-dimensionalen Raumstückes des repräsentierenden Simplexes.

Unter der *Länge* eines in einem Elemente liegenden geradlinigen Segmentes verstehen wir die Länge, welche das entsprechende Segment im repräsentierenden Simplexe besitzt. Unter der *Entfernung* zweier Punkte der Mannigfaltigkeit verstehen wir das Minimum der Länge eines sie verbindenden (eventuell mehrere Elemente durchziehenden) Streckenzuges.

Unter dem *Schwerpunkte* gewisser in Punkten desselben Elementes angebrachter Massen verstehen wir den Bildpunkt des Schwerpunktes von in den entsprechenden Punkten des repräsentierenden Simplexes angebrachten, jenen gleichen Massen.

Unter dem *Inhalte* eines Teilgebietes eines Elementes verstehen wir den Inhalt des entsprechenden Teilgebietes des repräsentierenden Simplexes.

Als *Indikatrix eines Elementes* bezeichnen wir eine gewisse Reihenfolge seiner Eckpunkte, wobei solche Reihenfolgen, welche durch eine gerade Zahl von Vertauschungen zweier Eckpunkte auseinander hervorgehen, als äquivalent betrachtet werden. Somit sind nur zwei Indikatrizen des Elementes möglich, von denen wir eine beliebige als die *positive* auswählen können; dann bezeichnen wir die andere als die *negative*. Durch die Wahl der positiven Indikatrix des Elementes ist zugleich die positive Indikatrix des repräsentierenden Simplexes, durch diese die positive Indikatrix jedes Teilsimplexes des repräsentierenden Simplexes, und durch letztere schließlich die positive Indikatrix jedes Teilsimplexes des Elementes bestimmt.

[[6]] Wir betrachten nur eine solche endliche geschlossene Kette von untereinander verschiedenen Elementen der gemessenen n-dimensionalen Mannigfaltigkeit, bei der je zwei aufeinanderfolgende Elemente n Eckpunkte gemeinsam haben, und nehmen für eines dieser Elemente eine positive Indikatrix an. Mittels dieser bestimmen wir die *negative* Indikatrix des folgenden Elementes, indem wir den für diesen neu auftretenden Eckpunkt an die Stelle des für ihn in Fortfall kommenden setzen. In dieser Weise bestimmen wir der Reihe nach für jedes Element der Kette eine positive Indikatrix, und finden schließlich, wenn wir zum Ausgangselemente zurückkommen, für dieses eine neue positive Indikatrix. Wenn für *jede* geschlossene Kette von Elementen diese neue positive Indikatrix mit derjenigen, von der wir ausgegangen sind, identisch ist, so soll die Mannigfaltigkeit *zweiseitig*, im entgegengesetzten Falle *einseitig* heißen.

In einer zweiseitigen gemessenen n-dimensionalen Mannigfaltigkeit ist somit durch die Wahl der positiven Indikatrix eines Elementes zugleich für jedes Element und für jedes Teilsimplex eines Elementes die positive Indikatrix festgelegt, sodaß wir von einer *positiven Indikatrix der Mannigfaltigkeit* sprechen können.

Wir denken uns mit einer zweiseitigen, geschlossenen, gemessenen, n-dimensionalen Mannigfaltigkeit μ eine eindeutige und stetige Abbildung α auf eine gemessene n-dimensionale Mannigfaltigkeit μ' vorgenommen, und setzen zunächst auch μ' als zweiseitig und geschlossen voraus.

Wir unterziehen μ einer *simplizialen Zerlegung* ζ *von der Dichte* ε, d. h. wir zerlegen jedes Element von μ in solcher Weise in eine endliche Zahl von Teilsimplexen, daß je zwei davon entweder keinen gemeinschaftlichen Punkt besitzen, oder eine p-dimensionale ($0 \leqq p \leqq n-1$) Seite (und dann zugleich alle in ihr liegenden Seiten geringerer Dimensionenzahl) gemeinsam haben, im übrigen aber keinen gemeinschaftlichen Punkt besitzen, und daß die Breite der Teilsimplexe die Größe ε als Maximum besitzt. Diese Teilsimplexe sollen die *Grundsimplexe*, ihre Eckpunkte die *Grundpunkte*, ihre Seiten die *Grundseiten* der Zerlegung heißen. Für die Dichte ε wählen wir eine obere Grenze ε_1, welche sie nicht übersteigen soll.

In der Mannigfaltigkeit μ wird eine positive Indikatrix angenommen, und dementsprechend jedes Grundsimplex mit einer positiven Indikatrix versehen.

Ein solches Grundsimplex, dessen Eckpunktsbilder alle demselben Elemente von μ' angehören, wollen wir ein *„gewöhnliches Grundsimplex“* nennen.

Jetzt definieren wir, was unter einer *zu* ζ *gehörigen simplizialen Ab-*

bildung β, welche die Abbildung α approximiert, verstanden werden soll. Es ist eine Abbildung, welche sich nur auf die gewöhnlichen Grundsimplexe erstreckt, und zwar in folgender Weise:

Sei π ein gewöhnliches Grundsimplex mit den Eckpunkten $A_1, A_2, \cdots, A_{n+1}$, denen für die Abbildung α die demselben Elemente von μ' angehörigen Punkte $B_1, B_2, \cdots, B_{n+1}$ entsprechen. Ein im Inneren oder auf der Grenze von π liegender Punkt P läßt sich als Schwerpunkt von gewissen in $A_1, A_2, \cdots, A_{n+1}$ konzentrierten positiven Massen auffassen. Wenn wir dieselben Massen der Reihe nach in $B_1, B_2, \cdots, B_{n+1}$ anbringen, so bestimmen sie einen Schwerpunkt Q, den wir für die Abbildung β dem Punkte P entsprechen lassen. Alsdann wird, falls die Punkte $B_1, B_2, \cdots, B_{n+1}$ nicht in einem ebenen $(n-1)$-dimensionalen Raumstücke liegen, das Grundsimplex π auf ein Bildsimplex ϱ, dessen Inhalt wir, je nachdem die Bildindikatrix positiv oder negativ ausfällt, positiv oder negativ rechnen werden, eineindeutig und stetig abgebildet. Wenn aber die Punkte $B_1, B_2, \cdots, B_{n+1}$ in einem $(n-1)$-dimensionalen Raumstücke enthalten sind, so wird das Bildsimplex singulär und bekommt den Inhalt Null.

Die Entfernung zwischen den Bildpunkten für α und für β desselben Punktes von μ besitzt eine gewisse obere Grenze e_1, welche mit ε_1 unter jede Grenze herabsinkt.

Wir definieren weiter eine *zu ζ gehörige modifizierte simpliziale Abbildung γ, welche die Abbildung α approximiert*. Sie wird bestimmt durch die Größe ε_1 nicht übersteigende Verrückungen der Grundpunktsbilder, und wird sodann aus ihren Grundpunktsbildern in derselben Weise wie β konstruiert, wobei für γ im allgemeinen nicht dieselben gewöhnlichen Grundsimplexe auftreten, wie für β.

Wir bestimmen nun beliebig in jedem Elemente von μ' als sein *„Innensimplex"* ein solches Teilsimplex, dessen Grenze die Grenze des Elementes nicht trifft, und wählen ε_1 in solcher Weise, daß jeder Punkt von μ, der für *eine* Abbildung β oder γ in einem Innensimplexe abgebildet wird, für *jede* Abbildung β oder γ nur gewöhnlichen Grundsimplexen angehört, welche dann natürlich stets in demselben Elemente von μ' abgebildet werden.

Wir beschäftigen uns zunächst mit einer solchen Abbildung γ, welche keine singulären Bildsimplexe aufweist, und zwar in bezug auf das Innensimplex J eines gewissen Elementes E. Die Menge derjenigen im Inneren von J enthaltenen Punkte, welche *nicht* dem Bilde einer $(n-2)$-dimensionalen Grundseite angehören, bezeichnen wir mit σ, und bemerken, daß je zwei Punkte von σ sich durch einen aus einer endlichen Zahl von Strecken bestehenden, ganz in σ verlaufenden Streckenzug verbinden lassen.

Einen solchen zu σ gehörigen Punkt, welcher *nicht* dem Bilde einer $(n-1)$-dimensionalen Grundseite angehört, nennen wir einen *gewöhnlichen Punkt von J.*

Seien P_1 und P_2 zwei beliebige gewöhnliche Punkte von J, welche durch einen ganz in σ verlaufenden Streckenzug s verbunden sind, und sei P ein variabler gewöhnlicher Punkt von J. Die Zahl der positiven bez. negativen Bildsimplexe, welche P_1 bedecken, bezeichnen wir mit p_1 bzw. p_1', die analogen Zahlen für P_2 mit p_2 und p_2', die analogen Zahlen für P mit p und p'.

Diese Zahlen p und p' können sich bei Bewegung von P an s entlang nur dann ändern, wenn das Bild τ einer $(n-1)$-dimensionalen Grundseite passiert wird. Wenn dabei die beiden Bildsimplexe, welche in τ zusammenstoßen, auf verschiedenen Seiten von τ liegen, so besitzen sie dasselbe Vorzeichen, und die Kreuzung von τ hat auf die Zahlen p und p', also auch auf die Zahl $p-p'$, keinen Einfluß. Wenn aber die beiden Bildsimplexe, welche in τ zusammenstoßen, auf derselben Seite von τ liegen, so besitzen sie entgegengesetzte Vorzeichen, und die Kreuzung von τ führt entweder eine Zunahme von p und p' je um eins, oder eine Abnahme von p und p' je um eins herbei, läßt also wieder die Zahl $p-p'$ ungeändert. Mithin ist $p_1-p_1'=p_2-p_2'$, und *für die Abbildung γ ist in den gewöhnlichen Punkten von J die Zahl $p-p'$ eine Konstante.*

Wir wenden uns nunmehr zu einer Abbildung β, für welche wir die gewöhnlichen Punkte von J, und die Zahlen p und p' in analoger Weise, wie in Bezug auf γ, definieren. Wenn dann für β in J zwei gewöhnliche Punkte P_1 und P_2 existierten, für welche p_1-p_1' und p_2-p_2' verschieden wären, so könnten wir β mit solcher Genauigkeit durch eine zu ζ gehörige und keine singulären Bildsimplexe aufweisende modifizierte simpliziale Abbildung γ_ν approximieren, daß die Punkte P_1 und P_2 auch in bezug auf γ_ν gewöhnliche Punkte von J wären, und die Zahlen p_1, p_1', p_2, p_2' in bezug auf γ_ν dieselben Werte, wie in bezug auf β, besäßen, sodaß für γ_ν die Zahl $p-p'$ in den gewöhnlichen Punkten von J keine Konstante sein könnte. Aus diesem Widerspruche folgern wir, *daß auch für die Abbildung β in den gewöhnlichen Punkten von J die Zahl $p-p'$ eine Konstante ist.*

Wir behaupten weiter, daß *der Wert dieser Konstante, den wir mit c bezeichnen, für jede simpliziale Abbildung, welche die Abbildung α approximiert, derselbe ist,* und zwar zeigen wir zunächst für zwei derartige simpliziale Abbildungen β_1' und β_2', daß die β_1' zugrunde liegende simpliziale Zerlegung ζ_1' aus der β_2' zugrunde liegenden simplizialen Zerlegung ζ_2' durch simpliziale Zerlegung der Grundsimplexe von ζ_2' hervorgeht, wir sagen kurz, daß ζ_1' eine *Unterteilung* von ζ_2' ist.

In diesem Falle können wir nämlich die Bilder der Grundpunkte von ζ_1' für β_1', insofern die entsprechenden Bilder für β_2' existieren, in diese auf den kürzesten verbindenden Streckenzügen stetig überführen, und für die Folge festhalten, wobei, was J betrifft, die Abbildung β_2' stetig in β_1' übergeht. Der Gesamtinhalt der in J enthaltenen Teile der Bildsimplexe, welcher c mal so groß ist als J, kann bei diesem stetigen Übergange keine Sprünge erleiden, sodaß die ganze Zahl c sich nicht ändern kann.

Es seien jetzt β_1 und β_2 zwei willkürliche simpliziale Abbildungen, welche die Abbildung α approximieren. Seien ζ_1 und ζ_2 die ihnen zugrunde liegenden simplizialen Zerlegungen von μ. Wir konstruieren eine so dichte Unterteilung ζ_3 von ζ_1, daß es für die zugehörige simpliziale Abbildung β_3, welche die Abbildung β_2 approximiert, möglich ist, in J einen solchen sowohl für β_2, wie für β_3 gewöhnlichen Punkt P zu wählen, daß die Eckpunkte eines willkürlichen, P bedeckenden Bildsimplexes für β_3 in jedem P bedeckenden Bildsimplexe für β_2 enthalten sind. Dann aber gehört zu jedem P bedeckenden Bildsimplex für β_2 *ein und nur ein* P bedeckendes Bildsimplex für β_3, und zwar besitzen je zwei einander in dieser Weise entsprechende Bildsimplexe denselben Sinn der Indikatrix. Hieraus folgern wir, daß die Zahl c, welche schon für β_1 und β_3 dieselbe ist, auch für β_2 und β_3, und deshalb schließlich auch für β_1 und β_2 dieselbe ist.

Wir werden jetzt zeigen, daß die Werte c_1 und c_2, welche c in den Innensimplexen J_1 und J_2 zweier solcher Elemente E_1 und E_2, welche eine $(n-1)$-dimensionale Seite S gemeinsam haben, besitzt, einander gleich sind.

Dazu wählen wir zwei solche Euklidische reguläre Simplexe T_1 und T_2, welche in bezug auf eine gemeinsame $(n-1)$-dimensionale Seite Σ die Spiegelbilder voneinander sind, der Reihe nach als repräsentierende Simplexe von E_1 und E_2, sodaß die $(n-1)$-dimensionalen Seiten S und Σ einander entsprechen.

Sodann werden die Begriffe „*Teilsimplex*", „*p-dimensionales Raumstück*", „*Schwerpunkt*" und „*Inhalt*" von $T_1 + T_2$ (d. h. von der von T_1 und T_2 zusammen gebildeten Punktmenge) in genau derselben Weise auf $E_1 + E_2$ übertragen, wie sie früher von einem einzigen repräsentierenden Simplex auf das von ihm repräsentierte Element übertragen worden sind.

Wir bestimmen in E_1 und E_2 solche der Reihe nach J_1 und J_2 enthaltende Teilsimplexe U_1 und U_2, welche eine in S liegende $(n-1)$-dimensionale Seite gemeinsam haben, deren Grenzen aber die übrigen Seiten von E_1 und E_2 nicht treffen.

Diejenigen Grundsimplexe, deren Eckpunktsbilder für die Abbildung α

alle zum Inneren der Punktmenge $E_1 + E_2$ gehören, rechnen wir nunmehr zu den gewöhnlichen Grundsimplexen, und bestimmen eine simpliziale Abbildung β', welche α approximiert, und welche sich auch auf die neu hinzugekommenen gewöhnlichen Grundsimplexe erstreckt; zu ihr gehören wieder modifizierte simpliziale Abbildungen γ'. Die obere Grenze für ε wählen wir jetzt derart, daß in bezug auf $U_1 + U_2$ und auf die gewöhnlichen Grundsimplexe in der neuen Bedeutung dieselbe Eigenschaft, welche früher den Innensimplexen und den gewöhnlichen Grundsimplexen im alten Sinne auferlegt ist, gültig bleibt. Dann können wir nach der schon oben angewandten Methode folgern, daß *für die Abbildung β' in den gewöhnlichen Punkten von $U_1 + U_2$ die Zahl $p - p'$ eine Konstante ist.*

Diese zu β' und $U_1 + U_2$ gehörige Konstante ist aber einerseits mit der zu β und J_1 gehörigen Konstante c_1, andererseits mit der zu β und J_2 gehörigen Konstante c_2 identisch, sodaß in der Tat die Konstante c in J_1 und J_2, *dann aber auch in allen Elementen von μ'*, denselben Wert hat.

Die Konstante c, welche einerseits in allen Elementen von μ', und andererseits für alle simplizialen Abbildungen, welche α approximieren, denselben Wert besitzt, *stellt mithin eine Eigenschaft von α dar.* Wir
[[7]] nennen sie *den Grad der eindeutigen und stetigen Abbildung α.*

Zwei eindeutige und stetige Abbildungen von μ auf μ', welche sich stetig
[[8]] *ineinander überführen lassen, besitzen denselben Grad.*

Approximieren wir sie nämlich beide durch solche simpliziale Abbildungen, denen paarweise dieselbe simpliziale Zerlegung von μ zugrunde liegt, so können wir zwischen jedes Paar dieser Abbildungen eine solche endliche Reihe von auf dieselbe Zerlegung gegründeten simplizialen Abbildungen einschalten, daß jede von ihnen den oben der simplizialen Abbildung β auferlegten Bedingungen genügt, und daß je zwei aufeinanderfolgende von ihnen sich nur im Bildpunkte *eines* der Grundpunkte, und zwar beliebig wenig, unterscheiden, sodaß es sicher ein Innensimplex von μ' gibt, in dem sie beide genau dieselbe Bildmenge, also denselben Wert der Konstante c bestimmen. Mithin kann sich dieser Wert an der genannten Reihe von Abbildungen entlang nicht ändern, was nur möglich ist, wenn die beiden eindeutigen und stetigen Abbildungen, von denen wir ausgegangen sind, von gleichem Grade sind.

Daß jede endliche, positive oder negative, ganze Zahl als Grad einer Abbildung auftreten kann, zeigen am einfachsten diejenigen eindeutigen und stetigen Abbildungen einer Kugel auf eine andere Kugel, welche rationale Funktionen der komplexen Variabeln darstellen. Der absolute Wert des Grades der Abbildung ist hier mit dem Grade der entsprechenden Funktion identisch.

[[462]]

Schließlich bemerken wir, daß, *falls für die Abbildung α die Bildmenge nicht überall dicht in μ' ist, der Grad von α sicher Null ist.*

Dann existiert nämlich sicher auch eine simpliziale Abbildung β, welche α approximiert, zu der wir in μ' ein solches Innensimplex J wählen können, in dem die von ihr bestimmte Bildmenge nicht überall dicht liegt. In einem Teilgebiete eines Innensimplexes, wo die Bildmenge von μ für β nicht eindringt, sind aber sowohl p wie p', also auch c gleich Null.

Wir haben bis jetzt μ' als zweiseitig und geschlossen angenommen. Wenn aber μ' einseitig und geschlossen ist, so bleiben die vorstehenden Überlegungen sowohl in bezug auf ein einziges Innensimplex, wie in bezug auf die Innensimplexe zweier in einer $(n-1)$-dimensionalen Seite aneinander grenzenden Elemente, vollständig in Kraft. Betrachten wir dann aber weiter eine solche geschlossene Kette von Elementen, in welcher der Sinn der Indikatrix umkehrt, so muß in dieser Kette die Zahl $c = p - p'$ einerseits konstant sein, und andererseits bei einem vollen Umlaufe ihr Zeichen wechseln, was nur möglich ist, wenn sie *gleich Null ist.*

Wenn schließlich μ' offen ist, so kann man, weil die Bildmenge von μ in μ' abgeschlossen ist, in μ' eine solche *endliche* Menge μ'' von Elementen angeben, daß sowohl für α, wie für die verschiedenen Abbildungen β und γ die Bildmenge von μ ganz in μ'' enthalten ist, ohne jedoch in μ'' überall dicht zu liegen. Erstere Eigenschaft bringt mit sich, daß sowohl hinsichtlich eines einzigen Innensimplexes, wie hinsichtlich der Innensimplexe zweier in einer $(n-1)$-dimensionalen Seite aneinander grenzenden Elemente die obigen Überlegungen wieder völlig gültig bleiben. Aus der Eigenschaft, daß die Bildmenge von μ in μ'' *nicht* überall dicht liegt, folgern wir dann weiter, daß auch hier der Grad von α *nur Null sein kann.*

Wir fassen das Ergebnis dieses Paragraphen wie folgt zusammen:

Satz 1. *Wenn eine zweiseitige, geschlossene, gemessene n-dimensionale Mannigfaltigkeit μ auf eine gemessene n-dimensionale Mannigfaltigkeit μ' eindeutig und stetig abgebildet wird, so existiert eine bei stetiger Modifizierung der Abbildung sich nicht ändernde endliche ganze Zahl c mit der Eigenschaft, daß die Bildmenge von μ jedes Teilgebiet von μ' im ganzen c Male positiv überdeckt. Ist μ' einseitig oder offen, so ist c stets gleich Null.**)

*) Man bemerkt gleich, daß dieser Abbildungsgrad in hohem Maße von der speziellen Zerlegung von μ und μ' in Elemente, sowie von der speziellen Herstellung ihrer Messungsskalen unabhängig ist. Für $n = 2$ kann sogar leicht bewiesen werden, [[9]]

§ 2.

[[10]] **Die stetigen Vektorfelder auf n-dimensionalen Kugeln.**

Wir betrachten eine n-dimensionale Kugel K, welche wir in einem $(n+1)$-dimensionalen Euklidischen Raume R_{n+1} durch die Gleichung $\Sigma x_h^2 = 1$ in rechtwinkligen Cartesischen Koordinaten darstellen. Daß diese Kugel unter den im vorigen Paragraphen entwickelten Begriff der gemessenen n-dimensionalen Mannigfaltigkeit fällt, zeigt sich, wenn wir die 2^{n+1} sphärischen Simplexe, in welche sie durch die n-dimensionalen ebenen Räume $x_h = 0$ des R_{n+1} zerlegt wird, als ihre Elemente, und die Teilgebiete der in ihr liegenden p-dimensionalen großen Kugeln als ihre ebenen p-dimensionalen Raumstücke auffassen, während wir in jedem Elemente denjenigen Punkt, dessen Cartesische Koordinaten im R_{n+1} alle gleich $\pm\sqrt{\frac{1}{n}}$ sind, als Einheitspunkt der Normalkoordinaten wählen. Nachdem weiter für eines der Elemente eine positive Indikatrix gewählt ist, ist zugleich für jedes sphärische Simplex (ein solches wird von $n+1$ verschiedenen $(n-1)$-dimensionalen Großkugeln gebildet, und ist gänzlich in einer gewissen Halbkugel von K enthalten) eine positive Indikatrix festgelegt.

In jedem Punkte P von K existiert nun eine $(n-1)$-dimensionale Richtungskugel λ_P, in der wir mittels Großkugeln analog wie für K selbst, die Elemente, die Simplexe, die ebenen p-dimensionalen Raumstücke und Normalkoordinaten definieren können. Weiter soll die positive Indikatrix von λ_P in folgender Weise aus der positiven Indikatrix von K hergeleitet werden: Sei s ein zu λ_P gehöriges $(n-1)$-dimensionales sphärisches Simplex; wir bestimmen in K ein solches n-dimensionales Simplex S, das P als Eckpunkt besitzt, während seine P enthaltenden $(n-1)$-dimensionalen Seiten durch die $(n-2)$-dimensionalen Seiten von s bestimmt werden, und die letzte $(n-1)$-dimensionale Seite beliebig ist. Sodann schreiben wir die positive Indikatrix von S in solcher Weise, daß in der Reihe der Eckpunkte P die letzte Stelle erhält. Die übrigen Eckpunkte dieser Reihe bestimmen dann eine gewisse Reihenfolge der in P zusammenkommenden Kanten von S, mithin ebenso eine gewisse Reihenfolge der Eckpunkte von s. Letztere Reihenfolge wählen wir als positive Indikatrix von s.

---

daß diese Unabhängigkeit vollkommen ist. Der entsprechende Nachweis für höhere Dimensionenzahlen dürfte ziemlich tief liegen, doch läßt sich auch ohne ihn Satz 1 verwenden, wie die folgenden Paragraphen zeigen.

[[464]]

Wir denken uns in K ein solches stetiges Vektorfeld, daß nur die Richtung, nicht die Größe der Vektoren in Betracht kommt, während nur eine endliche Zahl von singulären Punkten (d. h. Punkten, in denen die Stetigkeit der Vektorrichtung gestört wird) existiert, und zerlegen K mittels einer keinen singulären Punkt des Vektorfeldes enthaltenden, $(n-1)$-dimensionalen Großkugel $\varkappa$, mit den Polen π_1 und π_2, in eine π_1 enthaltende Hälfte H_1 und eine π_2 enthaltende Hälfte H_2 (die Großkugel $\varkappa$ betrachten wir als zu beiden Hälften gehörig). Sowohl H_1 wie H_2 zerlegen wir mittels $(n-1)$-dimensionaler Großkugeln keinen singulären Punkt des Vektorfeldes enthalten, die in eine endliche Zahl von n-dimensionalen sphärischen Simplexen $S_{11}, S_{12}, S_{13}, \cdots$ bez. $S_{21}, S_{22}, S_{23}, \cdots$, welche alle ebenso, wie die Elemente einer n-dimensionalen Mannigfaltigkeit, aneinander schließen, während die Teilsimplexe von H_2 die Spiegelbilder in bezug auf $\varkappa$ der Teilsimplexe von H_1 sind.

Sei σ eine $(n-1)$-dimensionale Seite eines Teilsimplexes $S_{\alpha\beta}$. Aus der positiven Indikatrix von $S_{\alpha\beta}$ leiten wir in folgender Weise eine „positive Indikatrix von σ, betrachtet als Seite von $S_{\alpha\beta}$" her: Wir schreiben die positive Indikatrix von $S_{\alpha\beta}$ in solcher Weise, daß in der Reihe der Eckpunkte der *nicht* in σ liegende Eckpunkt die letzte Stelle erhält. Die Reihenfolge der übrigen Eckpunkte bestimmt die positive Indikatrix von σ, betrachtet als Seite von $S_{\alpha\beta}$. In dieser Weise wird zugleich für den *Umfang* von $S_{\alpha\beta}$, welcher sich derart als zweiseitige, geschlossene, gemessene $(n-1)$-dimensionale Mannigfaltigkeit auffassen läßt, daß die $(n-1)$-dimensionalen Seiten von $S_{\alpha\beta}$ seine Elemente, und die Teilgebiete der in ihm liegenden p-dimensionalen Großkugeln seine ebenen p-dimensionalen Raumstücke repräsentieren, eine positive Indikatrix festgelegt.

Wir projizieren im R_{n+1} den Umfang $U_{\alpha\beta}$ eines Simplexes $S_{\alpha\beta}$ zusammen mit den in seinen Punkten angebrachten Vektoren aus einem außerhalb $S_{\alpha\beta}$ liegenden Punkte Q von K auf den im Q diametral entgegengesetzten Punkte O angebrachten n-dimensionalen ebenen Berührungsraum ϑ. Dadurch wird eine eindeutige und stetige Abbildung von $U_{\alpha\beta}$ auf die Kugel der Richtungen von ϑ, deren positive Indikatrix durch die positive Indikatrix der Kugel λ_O festgelegt ist, bestimmt; wir behaupten von dieser Abbildung, daß ihr Grad unabhängig ist von der speziellen Wahl des Projektionszentrums Q.

Zum Beweise dieser Behauptung bemerken wir, daß durch die stereographische Projektion die $(n-1)$-dimensionale Richtungskugel λ_P eines willkürlichen zu $U_{\alpha\beta}$ gehörigen Punktes P in eine kongruente Beziehung zur $(n-1)$-dimensionalen Richtungskugel von ϑ gesetzt wird, wodurch zugleich zwischen den Richtungskugeln λ_P und λ_R von zwei willkürlichen zu $U_{\alpha\beta}$ gehörigen Punkten P und R eine kongruente Beziehung b_{PR} hergestellt wird, und zwar in folgender Weise:

Sei V eine willkürliche zu λ_P gehörige Richtung. Sie bestimmt mit Q und R eine in K liegende zweidimensionale Kugel l, in welcher die Punkte P, Q und R einen Kreis k bestimmen. In R gibt es dann eine solche zu l gehörige Richtung, welche in l mit k denselben Winkel bestimmt wie V. Diese Richtung korrespondiert mit V für die Beziehung b_{PR}.

Stetige Bewegung von Q in endlicher Entfernung von P und R kann diese kongruente Beziehung b_{PR} nur stetig ändern; mithin kann stetige Bewegung von Q in endlicher Entfernung von $S_{\alpha\beta}$ das ganze System der kongruenten Beziehungen zwischen den $(n-1)$-dimensionalen Richtungskugeln der Punkte von $U_{\alpha\beta}$ nur stetig ändern, sodaß der Grad der durch das Vektorfeld bestimmten Abbildung von $U_{\alpha\beta}$ auf die $(n-1)$-dimensionale Richtungskugel des Projektionsraumes keine Sprünge erleiden kann,
[[11]] was nur möglich ist, wenn er eine Konstante $c_{\alpha\beta}$ ist, welche wir den *Grad des Simplexes* $S_{\alpha\beta}$ nennen. Wir stellen uns als Ziel, die Summe dieser Grade der verschiedenen $S_{\alpha\beta}$ zu ermitteln.

Wir projizieren die Hälfte H_1 stereographisch aus π_2, und H_2 aus π_1. Die entsprechenden n-dimensionalen ebenen Projektionsräume bezeichnen wir mit ϑ_1 und ϑ_2, betrachten die durch diese stereographische Projektion bestimmten Abbildungen der $U_{1\beta}$ auf die Richtungskugel von ϑ_1, und suchen die Summe der Grade dieser Abbildungen.

In dieser Summe liefert jede solche $(n-1)$-dimensionale Seite eines $S_{1\beta}$, welche *nicht* in $\varkappa$ enthalten ist, weil die in ihr enthaltenen Grundsimplexe zweimal mit entgegengesetzter Indikatrix abgebildet werden, zwei einander zerstörende Beiträge, sodaß nur die Beiträge der in $\varkappa$ enthaltenen $(n-1)$-dimensionalen Seiten $\sigma_\varkappa$ von Einfluß sind.

Fassen wir nun aber $\varkappa$ in solcher Weise als eine zweiseitige, geschlossene, gemessene $(n-1)$-dimensionale Mannigfaltigkeit auf, daß die $\sigma_\varkappa$ ihre Elemente, und die Teilgebiete der in ihr liegenden p-dimensionalen Großkugeln ihre ebenen p-dimensionalen Raumstücke repräsentieren, während ihre positive Indikatrix von der positiven Indikatrix der $\sigma_\varkappa$, betrachtet als Seiten der $S_{1\beta}$, bestimmt wird, so ruft die stereographische Projektion von $\varkappa$ auf ϑ_1 eine eindeutige und stetige Abbildung von $\varkappa$ auf die Richtungskugel von ϑ_1 hervor, zu deren Grad c_1 die $\sigma_\varkappa$ genau dieselben Beiträge liefern, wie zur Summe der $c_{1\beta}$, sodaß c_1 der Summe der $c_{1\beta}$ gleich ist.

Versehen wir aber $\varkappa$ mit derjenigen positiven Indikatrix, welche

[[466]]

der positiven Indikatrix der $\sigma_\varkappa$, betrachtet als Seiten der $S_{2\beta}$, entspricht, und projizieren wir sodann $\varkappa$ stereographisch auf ϑ_2, so zeigt sich in derselben Weise, daß der Grad c_2 der in dieser Weise hervorgerufenen eindeutigen und stetigen Abbildung von $\varkappa$ auf die Richtungskugel von ϑ_2 der Summe der $c_{2\beta}$ gleich ist.

Sei ϱ der $\varkappa$ enthaltende n-dimensionale ebene Raum des R_{n+1}, so entspricht bei einer Spiegelung des R_{n+1} in bezug auf ϱ die Projektionskugel $\varkappa_2$ von $\varkappa$ in ϑ_2 der Projektionskugel $\varkappa_1$ von $\varkappa$ in ϑ_1 mit entgegengesetzter Indikatrix, die Richtungskugel von ϑ_2 der Richtungskugel von ϑ_1, ebenfalls mit entgegengesetzter Indikatrix, und die vorliegende Vektorverteilung in $\varkappa_2$ der *Reflexionsverteilung* in $\varkappa_1$, d. h. derjenigen Verteilung, welche aus der vorliegenden in jedem Punkte durch Spiegelung am ebenen in ϑ_1 liegenden Berührungsraum erhalten wird.

Um die Summe der $c_{\alpha\beta}$, d. h. die Summe von c_1 und c_2 zu ermitteln, haben wir mithin die Frage zu erörtern: *wieviel in einem ebenen Euklidischen n-dimensionalen Raume ϑ die Summe der Grade derjenigen Abbildungen δ und ϱ einer $(n-1)$-dimensionalen Kugel $\mathfrak{f}$ auf die Richtungskugel λ von ϑ beträgt, welche durch eine stetige Vektorverteilung in den Punkten von $\mathfrak{f}$ und ihre Reflexionsverteilung bestimmt wird.* [[12]]

Zu diesem Zwecke bestimmen wir in $\mathfrak{f}$ mittels eines $(n-1)$-dimensionalen ebenen Raumes f eine $(n-2)$-dimensionale Großkugel mit den Polen q_1 und q_2, und nehmen mit $\mathfrak{f}$ eine solche Fundamentalreihe von simplizialen Zerlegungen ζ_1', ζ_2', ζ_3', $\cdots$, ζ_m', $\cdots$ vor, daß für keine von ihnen p_1 in einer Grundseite liegt, und daß ihre Dichte mit steigendem m unter jede Grenze herabsinkt. Wir bezeichnen mit z_m dasjenige Grundsimplex, welches q_1 für ζ_m' enthält, und mit u_m den Umfang von z_m.

Für jedes m leiten wir in folgender Weise aus δ eine neue eindeutige und stetige Abbildung δ_m von $\mathfrak{f}$ auf λ her: Sei j ein willkürlicher halber Großkreis, welcher in $\mathfrak{f}$ die Punkte q_1 und q_2 verbindet, und h sein Schnittpunkt mit u_m. Den Bogen q_1h von j bilden wir auf den Halbkreis j ähnlich ab. Wird dabei ein willkürlicher Punkt F in F' abgebildet, so soll dem Punkte F für die Abbildung δ_m derjenige Punkt von λ entsprechen, welcher für die Abbildung δ dem Punkte F' entspricht, während die nicht zum Bogen q_1h gehörigen Punkte von j für δ_m alle in den Bildpunkt g von q_2 für δ abgebildet werden.

Die Reflexionsabbildung (d. h. die zur Reflexionsverteilung gehörige Abbildung) von δ_m bezeichnen wir mit ϱ_m.

Weil δ sich stetig in δ_m überführen läßt, wobei zugleich ϱ stetig in ϱ_m übergeht, so ist die Gradsumme von δ und ϱ derjenigen von δ_m und ϱ_m gleich.

Durch weitere simpliziale Zerlegungen der Grundsimplexe von ζ_m' gelangen wir zu einer solchen simplizialen Zerlegung ζ_m von $\mathfrak{F}$, daß die entsprechenden simplizialen Abbildungen δ_m' und ϱ_m', welche δ_m und ϱ_m approximieren, den Bedingungen, welche im § 1 der approximierenden Abbildung β auferlegt sind, genügen.

Wir suchen zunächst die Beiträge, welche die von δ_m' und ϱ_m' erzeugten Bildmengen von z_m zur Gradsumme von δ_m und ϱ_m liefern. Wenn wir mit g' den Reflexionsvektor der Richtung g im Punkte q_1 von $\mathfrak{F}$ bezeichnen, so ist außerhalb einer für unbegrenzt wachsendes m unter jede Grenze herabsinkenden Umgebung von g' sowohl für die von δ_m' erzeugte Bildmenge von z_m, wie für die von ϱ_m' erzeugte Bildmenge von z_m, die Zahl $p - p'$ eine Konstante. Von diesen beiden Konstanten ändert die letztere sich nicht, wenn wir die von ϱ_m' erzeugte Bildmenge derart modifizieren, daß als Bildpunkt eines jeden in z_m liegenden Grundpunktes P von ζ_m, dem für δ_m die Richtung e_P entspricht, an Stelle des Reflexionsvektor von e_P in P der Reflexionsvektor von e_P in q_1 tritt. Nach dieser Modifizierung ist aber die von ϱ_m' erzeugte Bildmenge von z_m in das Spiegelbild in bezug auf f der von δ_m' erzeugten Bildmenge von z_m übergegangen, und zwei solche Spiegelbilder besitzen entgegengesetzte Indikatrizen, sodaß die beiden Bildmengen der Reihe nach in zwei Punkten von λ, welche die Spiegelbilder voneinander in Bezug auf f sind, entgegengesetzte Zahlen $p - p'$ bestimmen; dann aber bestimmen außerhalb der unter jede Grenze herabsinkenden Umgebung von g' die von δ_m' und ϱ_m' erzeugten Bildmengen von z_m *überall* entgegengesetzte Zahlen $p - p'$, und zerstören mithin die beiderseitigen Beiträge zur Gradsumme von δ_m und ϱ_m.

Wir haben jetzt die Beiträge zu ermitteln, welche die von δ_m' und ϱ_m' erzeugten Bildmengen des von z_m in $\mathfrak{F}$ bestimmten Restgebietes t_m zur Gradsumme von δ_m und ϱ_m liefern. Zunächst reduziert sich die von δ_m' erzeugte Bildmenge auf den einzigen Punkt g, liefert also keinen Beitrag. Sodann ist für die von ϱ_m' erzeugte Bildmenge außerhalb einer für unbegrenzt wachsendes m unter jede Grenze herabsinkenden Umgebung von g' *die Zahl $p - p'$ dem Grade derjenigen Abbildung von $\mathfrak{F}$ auf λ gleich, welche durch die Reflexionsverteilung eines konstanten Vektors in den Punkten von $\mathfrak{F}$ bestimmt wird.*

Um diesen Grad zu ermitteln, bezeichnen wir mit χ_1 denjenigen Punkt von $\mathfrak{F}$, dessen Radius dem konstanten Vektor entgegengesetzt ist, mit χ_2 den χ_1 diametral entgegengesetzten Punkt von $\mathfrak{F}$, mit w die zu $\mathfrak{F}$ gehörige $(n-2)$-dimensionale Großkugel, welche χ_1 und χ_2 zu Polen hat, mit a_1 bez. a_2 die χ_1 bez. χ_2 enthaltende Hälfte, welche w in $\mathfrak{F}$ bestimmt. Wir können dann mit der Hälfte a_1 leicht eine solche simpliziale Zerlegung vornehmen, daß die Bilder der mit positiver Indikatrix versehenen Grundsimplexe die Kugel λ genau einmal, und zwar mit positiver Indikatrix überdecken. Zerlegen wir weiter die Hälfte a_2 in solche Grundsimplexe, welche denjenigen von a_1 diametral entgegengesetzt sind, und versehen wir diese je mit einer solchen Indikatrix, welche derjenigen des entsprechenden Grundsimplexes von a_1 ebenfalls diametral entgegengesetzt ist, was bewirkt, daß diese Indikatrix für gerades n positiv, für ungerades n negativ ausfällt, so bestimmen zwei entsprechende Grundsimplexe von a_1 und a_2 in λ dasselbe mit positiver Indikatrix versehene Bildsimplex.

Wenn wir also auch die Grundsimplexe von a_2 mit einer positiven Indikatrix versehen, so überdecken ihre Bildsimplexe die Kugel λ wieder genau einmal, und zwar für gerades n mit positiver, für ungerades n mit negativer Indikatrix, *sodaß der gesuchte Abbildungsgrad für n gerade gleich zwei, für n ungerade gleich Null ist.*

Also sind auch die Gradsumme von δ_m und ϱ_m, die Gradsumme von δ und ϱ, und schließlich die Summe der $c_{\alpha\beta}$ für n gerade gleich zwei, für n ungerade gleich Null.

Wenn das Vektorfeld in K *keine* singulären Punkte aufweist, so ist seine Stetigkeit gleichmäßig. Wir können dann die $S_{\alpha\beta}$ so klein wählen, daß die Richtungsvariation der auf ϑ_1 resp. ϑ_2 stereographisch projizierten Vektoren desselben $U_{\alpha\beta}$ eine beliebig klein gewählte Größe nicht übersteigt, daß mithin jedes $c_{\alpha\beta}$ gleich Null ist. Dies ist aber nach dem obigen Ergebnisse für gerades n unmöglich, sodaß wir bewiesen haben:

Satz 2. *Ein stetiges Vektorfeld auf einer Kugel gerader Dimensionenzahl besitzt wenigstens einen singulären Punkt.* [[13]]

Daß dieser Satz für Kugeln ungerader Dimensionenzahl *nicht* zutrifft, erhellt am einfachsten aus der Betrachtung eines solchen Vektorfeldes auf der dreidimensionalen Kugel, für welche die Berührungskurven von einer [[14]] (zweifach unendlichen) Schar von untereinander im Cliffordschen Sinne parallelen großen Kreisen gebildet werden.

§ 3.

Die eindeutigen und stetigen Transformationen n-dimensionaler Kugeln und n-dimensionaler Elemente in sich.

Wir denken uns jetzt die n-dimensionale Kugel K einer eindeutigen und stetigen Transformation r unterworfen, welche keinen Punkt fest läßt. Wir nehmen mit K solche beliebig dichte simpliziale Zerlegungen vor, für welche die Grundsimplexe in derselben Weise, wie im § 2 die Simplexe $S_{\alpha\beta}$, gebildet werden, und konstruieren die entsprechenden simplizialen Transformationen von K in sich, welche r approximieren. Dann existiert sicher eine simpliziale r approximierende Transformation t, welche ebenfalls keinen Punkt fest läßt. Sonst nämlich gäbe es eine Fundamentalreihe von solchen gegen r konvergierenden simplizialen Transformationen, welche der Reihe nach die Punkte $F_1, F_2, F_3, \cdots$ fest ließen; dann aber wäre jeder Grenzpunkt dieser Fundamentalreihe $F_1, F_2, F_3, \cdots$ Fixpunkt für die Transformation r.

Wir wählen einen willkürlichen, für t gewöhnlichen und nicht in einer Seite eines Grundsimplexes liegenden Punkt O von K, verbinden jeden Punkt P von K mit seinem Bildpunkte P' für die Transformation t durch einen O enthaltenden Kreis, und bringen in P denjenigen Vektor an, welcher von dem O *nicht* enthaltenden Bogen dieses Kreises bestimmt wird. Dann entsteht auf K ein stetiges Vektorfeld, das nur eine endliche Zahl von singulären Punkten aufweist, nämlich erstens die Punkte, welche O zum Bildpunkt haben, zweitens den Punkt O selbst.

Wir wählen in K eine positive Indikatrix, bezeichnen mit $S_1, S_2, S_3, \cdots, S_p$ diejenigen Grundsimplexe, deren Bildsimplexe ${}_bS_1, {}_bS_2, {}_bS_3, \cdots, {}_bS_p$ den Punkt O mit positiver Indikatrix bedecken, mit $S_1', S_2', S_3', \cdots, S_{p'}'$ diejenigen Grundsimplexe, deren Bildsimplexe ${}_bS_1', {}_bS_2', {}_bS_3', \cdots, {}_bS_{p'}'$ den Punkt O mit negativer Indikatrix bedecken, mit S'' das O enthaltende Grundsimplex. Weiter dürfen wir die t zugrunde liegende simpliziale Zerlegung von K so dicht voraussetzen, daß kein Grundsimplex mit seinem Bildsimplex einen gemeinschaftlichen Punkt besitzt; dann ist auch sicher S'' von allen S_α und von allen S_α' verschieden.

Um die Grade der Simplexe S_α und S_α' zu ermitteln, projizieren wir im R_{n+1} die Kugel K samt ihren Vektorfelde stereographisch aus O auf den in dem O diametral entgegengesetzten Punkte O' angebrachten n-dimensionalen ebenen Berührungsraum ϑ.

Die Projektion des Simplexes S_α bezeichnen wir mit $\mathfrak{S}_\alpha$, seine Grenze mit $\mathfrak{U}_\alpha$, die Projektion der Grenze des Simplexes ${}_bS_\alpha$ mit ${}_b\mathfrak{U}_\alpha$. Die Bildindikatrix in ${}_b\mathfrak{U}_\alpha$ gehört nun aber in ϑ zu einer *negativen* Indikatrix des von ${}_b\mathfrak{U}_\alpha$ begrenzten sphärischen Simplexes ${}_{b'}\mathfrak{S}_\alpha$.

Die Vektoren in den Punkten $\mathfrak{P}$ von $\mathfrak{U}_\alpha$, denen die Punkte $\mathfrak{P}'$ von ${}_b\mathfrak{U}_\alpha$ entsprechen, werden durch die geradlinigen Verbindungsstrecken $\mathfrak{P}\mathfrak{P}'$ bestimmt; durch gleichmäßig stetige Abänderung lassen diese sich in diejenigen Vektoren überführen, welche in jedem Punkte $\mathfrak{P}$ der geradlinigen Verbindungsstrecke $O'\mathfrak{P}'$ parallel sind. Letztere Vektoren überdecken aber die Richtungskugel $\lambda_{O'}$ genau einmal mit negativer Indikatrix, sodaß der Grad von S_α gleich -1 ist.

In derselben Weise wird für den Grad jedes Simplexes S_α' der Wert $+1$ gefunden.

Um den Grad von S'' zu ermitteln, führen wir die Vektorverteilung in der Grenze U'' von S'' durch stetige Abänderung in eine solche über, für welche in jedem Punkte P von U'' der Vektor durch den O nicht enthaltenden Bogen eines O und P verbindenden Großkreises bestimmt wird. Projizieren wir sodann U'' samt seinem Vektorfelde im R_{n+1} stereographisch aus O' auf den ebenen n-dimensionalen Raum ϑ', welcher K in O berührt, in $\mathfrak{U}''$, so wird in jedem Punkte $\mathfrak{P}$ von $\mathfrak{U}''$ der Vektor bestimmt durch die Richtung der geradlinigen Verbindungsstrecke $O\mathfrak{P}$; somit überdeckt die Vektorverteilung in $\mathfrak{U}''$ die Richtungskugel λ_0 genau einmal mit positiver Indikatrix, und der Grad von S'' ist gleich $+1$.

Was die übrigen Grundsimplexe angeht, so lassen sie sich in solche Teilsimplexe zerlegen, sodaß innerhalb jedes einzelnen bei stereographischer Projektion die Variation der Vektorrichtung eine beliebig klein gewählte Größe nicht übersteigt. Jedes dieser Teilsimplexe besitzt also den Grad Null, und dasselbe gilt für die sich aus ihnen zusammensetzenden Grundsimplexe.

Die Summe der Grade aller Grundsimplexe ist mithin gleich $-p+p'+1$, was für gerades n gleich zwei, für ungerades n gleich Null sein muß. Hieraus folgern wir, daß $p-p'$, d. h. der Grad der Transformationen r und t, für gerades n gleich -1, für ungerades n gleich $+1$ ist, womit wir zu folgendem Ergebnis gelangt sind:

Satz 3. *Eine eindeutige und stetige Transformation einer n-dimensionalen Kugel in sich, welche keinen Fixpunkt aufweist, besitzt für gerades n den Grad -1, für ungerades n den Grad $+1$.* [[15]]

Von diesem Satze formulieren wir folgende besonderen Fälle:

Folgerung 1. *Wenn bei einer eindeutigen und stetigen Transformation einer n-dimensionalen Kugel in sich die Bildmenge nicht in der ganzen Kugel überall dicht liegt, so existiert sicher ein Fixpunkt.*

Folgerung 2. *Jede eineindeutige und stetige Transformation einer Kugel gerader Dimensionenzahl in sich, welche sich stetig in die Identität überführen läßt, besitzt sicher einen Fixpunkt.**)

Folgerung 3. *Jede eineindeutige und stetige Transformation einer Kugel ungerader Dimensionenzahl in sich, welche sich stetig in eine Spiegelung überführen läßt, besitzt sicher einen Fixpunkt.*

---

*) Von diesem Satze habe ich früher den speziellen Fall bewiesen, daß jede eineindeutige und stetige Transformation der zweidimensionalen Kugel in sich, welche den Umlaufssinn nicht ändert, sicher einen Fixpunkt aufweist. Vgl. Amsterd. Ber., holl. Ausg. XVII 2, S. 750, XIX 1, S. 48; engl. Ausg. XI 2, S. 797, XIII 1, S. 184. [[16]] [[17]]

Daß Transformationen -1ten Grades für gerades n, und Transformationen $+1^{\text{ten}}$ Grades für ungerades n *nicht* notwendig einen Fixpunkt aufweisen, erhellt am einfachsten aus der Betrachtung der Rotationen und Spiegelungen eines Euklidischen R_{n+1} um einen festen Punkt.

Wir betrachten nunmehr eine eindeutige und stetige Transformation $\mathfrak{T}$ eines n-dimensionalen Elementes E in sich. Wir bringen E in eine ein-eindeutige und stetige Beziehung zu einer Kugelhälfte H_1, welche in einer n-dimensionalen Kugel K von einer $(n-1)$-dimensionalen Großkugel $\varkappa$ bestimmt wird. Dabei entspricht der eindeutigen und stetigen Transformation von E in sich eine eindeutige und stetige Transformation $\mathfrak{A}$ von H_1 in sich. Erweitern wir nun die Transformation $\mathfrak{A}$ in solcher Weise auf die andere Hälfte H_2 von K, daß je zwei Punkte von K, welche die Spiegelbilder voneinander in Bezug auf $\varkappa$ sind, für $\mathfrak{A}$ in denselben Punkt von H_1 transformiert werden, so liegt eine solche eindeutige und stetige Transformation von K in sich vor, bei welcher die Bildmenge nicht überall dicht in K ist, für welche also sicher ein, notwendig in H_1 liegender, Fixpunkt existiert. Diesem Fixpunkte muß aber ein Fixpunkt der Transformation $\mathfrak{T}$ von E in sich entsprechen, sodaß wir bewiesen haben:

Satz 4: *Eine eindeutige und stetige Transformation eines n-dimensionalen Elementes in sich besitzt sicher einen Fixpunkt.*

Amsterdam, Juli 1910.

[[18]] Satz 3 lässt sich auch unmittelbar aus § 1 herleiten durch die Bemerkung, dass eine Transformation ohne Fixpunkt sich in die antipodische Transformation der Identität stetig überführen lässt, nämlich durch Bewegung jedes Punktes auf dem kürzesten seine beiden Lagen verbindenden Grosskreisbogen.

NOTES

⟦1⟧ It would be infeasible to illustrate the impact of this paper by mere quotations of literature. A good survey is provided by Alexandroff–Hopf 1935, G. Feigl 1928. See also, as to the most straightforward continuations, H. Hopf 1926 A, 1926 B, 1928 A, B, 1929, 1930. The paper goes back at least to January 1910. Its essentials, in particular the proof of *Satz 2*, are already found in letters to Hilbert (Y16) and Hadamard (Y17), and in an exercise book (see Y16 ⟦1⟧).

An intermediate manuscript of this paper is extant. Corrections to the present paper: 1911 M, 1921 E.

⟦2⟧ J. Hadamard 1910. According to a remark of H. Hopf 1926 B, p. 225–226 there was some mutual influence. This is also shown by Y16, see Y16, ⟦1⟧.

⟦3⟧ The concept of simplicial manifold – compare Dehn–Hegaard 1907.

⟦4⟧ Brouwer 1909 C, 1910 G.

⟦5⟧ This use of *abgeschlossen* is found at various places in work of A. Schoenflies, e.g. 1908 A.

⟦6⟧ p. 101, l. 1, *nun* instead of *nur*.

⟦7⟧ The first explicit mention of the mapping degree. See also 1911 C ⟦3⟧.

⟦8⟧ The homotopy concept for mappings. Terms such as transformation class appear first in Brouwer 1912 K2, 1912 M. A converse of this theorem in Brouwer 1912 K2.

⟦9⟧ See Brouwer 1911 G, Satz 6. See also H. Hopf 1926 B. W. Wilson 1928, p. 578.

⟦10⟧ Earlier topological research on vector fields by H. Poincaré 1881, 1882, 1885. Brouwer 1909 G2.

⟦11⟧ The singularity index defined as a mapping degree – owing to the lack of homological concepts, involved constructions are needed to relate the mappings and mapping degrees to one sphere. These methods were simplified by H. Hopf 1926 A, 1926 B, 1928 B, 1929.

⟦12⟧ p. 110, l. 19 *werden* instead of *wird*.

⟦13⟧ The case $n = 2$: Brouwer 1909 G2.

⟦14⟧ p. 112, l. 27 *welches* instead of *welche*.

⟦15⟧ Extended to projective spaces in 1926 D.

⟦16⟧ In the footnote replace the first line by 'Aus dem zweidimensionalen Spezialfall dieses Satzes ergibt sich mittels einer einfachen Überlegung ein neuer Beweis der Eigenschaft, dass jede' (according to 1921 E).

⟦17⟧ Brouwer 1909 F2, 1910 D2.

⟦18⟧ This strip, affixed to Brouwer's own copy (and probably to all, or the greater part, of the reprints) seems to be a private print of a version of a footnote of Brouwer 1911 G, p. 325. It is, indeed, a most natural simplification.

MA 71, p. 598

1911 M

Berichtigung

zu dem Aufsatze von L. E. J. Brouwer: „Über Abbildung von Mannigfaltigkeiten". Math. Ann. 71, S. 97—115.

S. 101, Z. 1	statt: nur	lies: nun
S. 110, Z. 19	statt: *wird*	lies: *werden*
S. 112, Z. 27	statt: welche	lies: welches

⟦See 1911 D⟧

MA 82, p. 286

1921 E

Berichtigung

zu dem Aufsatze von L. E. J. Brouwer: „Über Abbildung von Mannigfaltigkeiten". Math. Ann. **71**, S. 97—115.

S. 114, Z. 1 der Fußnote statt: Von diesem Satze habe ich früher den speziellen Fall bewiesen, daß . . .

lies: Aus dem zweidimensionalen Spezialfall dieses Satzes ergibt sich mittels einer einfachen Überlegung ein neuer Beweis der Eigenschaft, daß . . .

⟦See 1911 D⟧

Mathematics. — *"On transformations of projective spaces"*. By Prof. L. E. J. BROUWER. 1926 D2

(Communicated at the meeting of May 29, 1926).

In my article: *"Ueber Abbildung von Mannigfaltigkeiten"* (Math. Ann. 71, p. 97—115) I have proved the following theorem: [[1]]

A uniform continuous transformation without fixed point of an n-dimensional sphere into itself, has degree -1 *for even n and* $+1$ *for odd n.*

Since it appears that the specialisation of this theorem which gives an analogous property of uniform continuous transformations of n-dimensional projective spaces into themselves, and of which a particular case had already been published in these Proceedings XI, p. 798, is not to [[2]] that degree, as I then thought permissible to assume, self evident to all readers, it shall be set forth here in full detail.

In the first place let there be given a uniform continuous transformation τ of a $(2n-1)$-dimensional projective space E into itself. We provide E with a positive indicatrix and with an elliptic metric. Let S be the $(2n-1)$-dimensional sphere provided with a positive indicatrix and a metric, obtained by duplication of E. Let P be a point of E, P_1 and P_2 the corresponding points of S, P' the image of P in E under τ, and P_1' and P_2' the points of S corresponding to P'. Let τ_1 be the uniform continuous transformation of S into itself which brings P_1 into P_1', and which becomes τ by the folding of S into E [1]). Then the volume of the image of S under τ_1 (measured by the volume of the image of S under the simplicial approximations of τ_1) is twice as much as the volume of the image of E under τ. Since, however, the volume of S is also twice that of E, the degree of τ_1 appears to be equal to that of τ. Further, since the absence of a fixed point for τ implies the absence of one for τ_1, *the transformation* τ, *if exhibiting no fixed point must be necessarily of degree* $+1$.

On the other hand there exist arbitrarily small *congruent* transformations of S into itself without fixed point. To these correspond arbitrarily

1) The existence of τ_1 is due to the fact that to a circuit of S passing through P_1 corresponds a contractible circuit of E passing through P the image of which under τ is a contractible circuit of E passing through P' which corresponds to a circuit of S passing through P_1'.

The antipodal point-pairs of S become under τ_1 again antipodal point-pairs or simple points, according as uncontractible circuits of E become under τ again uncontractible or contractible.

small congruent transformations of E into itself without fixed point. *Thus there exist uniform continuous transformations of E into itself of degree $+1$ and without fixed point.*

In the second place let there be given a uniform continuous transformation τ of a $2n$-dimensional projective space E into itself. We provide E with an elliptic metric. Let S be the $2n$-dimensional sphere provided with a metric, obtained by duplication of E. We provide S with a positive indicatrix. Let P be a point of E, P_1 and P_2 the corresponding points of S, P' the image of P in E under τ, and P_1' and P_2' the points of S corresponding to P'. Let τ_1 and τ_2 be the uniform continuous transformations of S into itself which carry P_1 into P_1' and P_2' respectively, and which become τ by the folding of S into E. Then corresponding image simplexes of corresponding simplicial approximations of τ_1 and τ_2 have equal volumes of opposite signs; thus τ_1 and τ_2 have equal degrees of opposite signs, and thus either τ_1 or τ_2 has a fixed point. *Then, however, τ also must have a fixed point* [2]).

2) Dr. HOPF points out to me that a uniform continuous transformation of a $(2n-1)$-dimensional projective space E into itself possesses at least two invariant points, if its degree is odd and $\neq +1$. We can add that on the other hand a uniform continuous transformation of a $2n$-dimensional projective space E into itself has at least two invariant points, if its *absolute degree* (i.e. the absolute value of the degrees of the correspondent transformations of the sphere S duplicating E) is $\geqq 2$. [[3]]

NOTES

[[1]] Brouwer 1911 D.

[[2]] Brouwer 1909 F2.

[[3]] The proof is easy by methods developed by H. Hopf 1926 A, B: it suffices to prove that for n odd a mapping of an n-sphere with degree $\neq 1$ has both a fixed point and a point with an antipodal image. Obviously a mapping with an antipodal pair is homotopic to the identical mapping; by composition with the antipodal mapping (which for n odd is of degree 1) the existence of a fixed point results. Brouwer's assertion is proved in an analogous way.

Beweis der Invarianz des n-dimensionalen Gebiets. 1911 E

Von

L. E. J. Brouwer in Amsterdam. [[1]]

Zur Begründung der Invarianz der Dimensionenzahl*) habe ich folgenden Satz bewiesen:

In einer n-dimensionalen Mannigfaltigkeit enthält das eineindeutige und stetige Abbild eines n-dimensionalen Gebiets in beliebiger Nähe eines beliebigen seiner Punkte ein Gebiet.

Zweck der vorliegenden Abhandlung ist, diesen Satz folgendermaßen zu verschärfen:

Satz von der Invarianz des n-dimensionalen Gebiets: *In einer n-dimensionalen Mannigfaltigkeit ist das eineindeutige und stetige Abbild eines n-dimensionalen Gebiets wiederum ein Gebiet.*

Diese für $n = 1$ evidente Invarianz ist für $n = 2$ von Schoenflies**) bewiesen worden, während Baire***) und Hadamard†) gezeigt haben, daß sie sich für beliebiges n aus dem auf n Dimensionen erweiterten Jordanschen Satze fast unmittelbar folgern lassen würde.

§ 1.

Unter einer *n-dimensionalen Pseudomannigfaltigkeit* verstehen wir eine [[6]]
von derart aneinander schließenden n-dimensionalen *Elementen* (vgl. Math.

---

*) Vgl. Math. Ann. 70, S. 161—165. Dieser (Juni 1910 eingereichten, Februar 1911 erschienenen) Abhandlung hat Herr Lebesgue zwei weitere Beweise folgen lassen. Der erste (Math. Ann. 70, S. 166—168, datiert 14. Oktober 1910) ist ungenügend. Der zweite (C. R., 27 mars 1911) ist sachlich mit dem meinigen identisch: die Abweichungen machen den Gedankengang nur verwickelter. [[2]]

Was die von Herrn Lebesgue angeführten Entwicklungen des Herrn Baire betrifft, so liegen die Sätze, auf welche dort das Problem zurückgeführt wird, tiefer als die Invarianz der Dimensionenzahl.

**) Gött. Nachr. 1899, S. 282—290. Einfachere Beweise gaben Osgood (ibid. 1900, S. 94—97) und F. Bernstein (ibid. 1900, S. 98—102). [[3]]

***) Bull. des Sc. Math. (2) 31, S. 97—99. [[4]]

†) Vgl. das S. 97 dieses Annalenbandes zitierte Buch. [[5]]

[Schliesslich sollen je zwei Punkte der Pseudomannigfaltigkeit sich durch einen ganz der Pseudomannigfaltigkeit angehörigen und keine $(n-p)$-dimensionale Seite $(p>1)$ treffenden einfachen Kurvenbogen verbinden lassen.

[[7]] Ann. 71, S. 97) gebildete Punktmenge, daß jede $(n-1)$-dimensionale Seite zwei und nur zwei Elementen gemeinsam ist, während jeder nicht in einer $(n-1)$-dimensionalen Seite enthaltene Punkt nur zu einem einzigen Elemente und jeder andere Punkt nur zu einer endlichen Zahl von Ele-
[[8]] menten gehört/

Dieser Begriff umfaßt offenbar mehr, als der an der erwähnten Stelle definierte Begriff der *n-dimensionalen Mannigfaltigkeit*; diejenigen Punkte einer n-dimensionalen Pseudomannigfaltigkeit, welche einer p-dimensionalen $(p \leqq n-3)$ Elementseite angehören, brauchen nämlich nicht auf Gebiete des n-dimensionalen Zahlenraumes eineindeutig und stetig abbildbare Umgebungen zu besitzen. In Hinsicht auf das *Meßbarmachen*, die Bestimmung der *Indikatrizen* und die Definition der Begriffe *offen* und *geschlossen*, *einseitig* und *zweiseitig* verhalten sich jedoch die n-dimensionalen Mannigfaltigkeiten und die n-dimensionalen Pseudomannigfaltigkeiten genau gleich.

Unter einem *n-dimensionalen Fragmente* verstehen wir eine von einer endlichen Zahl von einer oder mehreren n-dimensionalen Pseudomannigfaltigkeiten angehörigen Elementen gebildete Punktmenge.

Diejenigen $(n-1)$-dimensionalen Elementseiten, welche nur *einem* Elemente des Fragmentes angehören, sollen die *Grenze* des Fragmentes bilden.

Die Grenze eines *zweiseitigen n-dimensionalen Fragmentes* setzt sich zusammen aus einer endlichen Zahl von *geschlossenen zweiseitigen $(n-1)$-dimensionalen Pseudomannigfaltigkeiten.* In dieser Grenze können übrigens mehrere p-dimensionale Seiten $(p \leq n-2)$, welche als verschieden zu betrachten sind, und welche derselben oder verschiedenen Pseudomannigfaltigkeiten angehören, zusammenfallen.

§ 2.

Sei $\varkappa$ eine im n-dimensionalen Cartesischen Raume μ liegende $(n-1)$-dimensionale Kugel, K_2 und K_3 zwei ihrer Punkte, I ihr inneres Gebiet. Wir konstruieren eine simpliziale Zerlegung von I, deren Grundseiten von K_3 enthaltenden Kugeln gebildet werden. Seien g und g' zwei Gruppen von Grundsimplexen, welche sowohl voneinander, wie von K_2 und K_3 einen endlichen Abstand besitzen. Wir nehmen an, daß *K_2 und K_3 innerhalb $\varkappa$ weder durch g noch durch g' getrennt werden.*

Sei γ resp. γ' dasjenige von g resp. g' innerhalb $\varkappa$ bestimmte Gebiet, zu dessen Grenze K_3 gehört. Der nicht in $\varkappa$ enthaltene Teil seiner Grenze bildet ein zweiseitiges $(n-1)$-dimensionales Fragment $\mathfrak{F}$ resp. $\mathfrak{F}'$; die Grenze von $\mathfrak{F}$ resp. $\mathfrak{F}'$ ist ganz in $\varkappa$ enthalten. Ein willkürlicher, innerhalb $\varkappa$ liegender und nicht zu $\mathfrak{F}$ resp. $\mathfrak{F}'$ gehöriger Punkt K gehört dann und nur dann zu γ resp. γ', wenn ein willkürlicher, K und K_3 innerhalb $\varkappa$ ver-

[[478]]

bindender Weg, der keine $(n-2)$-dimensionale Seite von $\mathfrak{F}$ resp. $\mathfrak{F}'$ trifft, eine gerade Zahl von Kreuzungspunkten mit $\mathfrak{F}$ resp. $\mathfrak{F}'$ aufweist. Sei f resp. f' ein solches $(n-1)$-dimensionales Teilfragment von $\mathfrak{F}$ resp. $\mathfrak{F}'$, dessen Grenze ebenfalls ganz in $\varkappa$ enthalten ist. Alsdann besitzt, falls der Punkt K zu γ resp. γ' gehört, ein willkürlicher, K und K_3 innerhalb $\varkappa$ verbindender Weg, der keine $(n-2)$-dimensionale Seite von f resp. f' trifft, eine gerade Zahl von Kreuzungspunkten mit f resp. f'. Indem wir für f resp. f' dasjenige Teilfragment von $\mathfrak{F}$ resp. $\mathfrak{F}'$ wählen, welches mit K_3 zur Grenze desselben von $g+g'$ innerhalb $\varkappa$ bestimmten Gebietes γ'' gehört, ersehen wir, daß ein willkürlicher, K_2 und K_3 innerhalb $\varkappa$ verbindender Weg, der weder von f noch von f' eine $(n-2)$-dimensionale Seite trifft, mit $f+f'$ eine gerade Zahl von Kreuzungspunkten aufweist, sodaß K_2 zur Grenze von γ'' gehört *und* K_2 *und* K_3 *innerhalb* $\varkappa$ *durch* $g+g'$ *nicht getrennt werden.* [[9]]

Unter einer *innerhalb* $\varkappa$ *abgeschlossenen Punktmenge* werden wir eine solche innerhalb $\varkappa$ liegende Punktmenge verstehen, welche jeden innerhalb $\varkappa$ liegenden ihrer Grenzpunkte enthält. [[10]]

Seien α und α' zwei innerhalb $\varkappa$ abgeschlossene Punktmengen ohne gemeinsame Punkte, in endlicher Entfernung sowohl von K_2 wie von K_3. Wir nehmen an, daß K_2 *und* K_3 *innerhalb* $\varkappa$ *weder durch* α *noch durch* α' *getrennt werden.*

Sei $\varkappa_1$ eine willkürliche, die Kugel $\varkappa$ im Punkte K_3 berührende und übrigens ganz im Innern von $\varkappa$ enthaltene $(n-1)$-dimensionale Kugel, so können wir die simpliziale Zerlegung von I immer in solcher Weise konstruieren, daß das Innere eines willkürlichen Grundsimplexes entweder ganz innerhalb oder ganz außerhalb $\varkappa_1$ liegt.

Wir bezeichnen mit K_4 denjenigen Punkt von $\varkappa_1$, welcher K_2 am nächsten liegt, und wählen $\varkappa_1$ so nahe an $\varkappa$, daß erstens die geradlinige Strecke K_2K_4 weder von α noch von α' getroffen wird und zweitens K_3 und K_4 innerhalb $\varkappa_1$ weder durch α noch durch α' getrennt werden. Weiter konstruieren wir eine der genannten Wahl von $\varkappa_1$ entsprechende simpliziale Zerlegung von I, bezeichnen mit β resp. β' die von den entweder in ihrem Innern oder auf ihrer Grenze wenigstens einen Punkt von α resp. α' besitzenden Grundsimplexen bestimmte Punktmenge, mit δ resp. δ' die innerhalb $\varkappa_1$ liegende Teilmenge von β resp. β', und wählen die genannte simpliziale Zerlegung so dicht, daß erstens die geradlinige Strecke K_2K_4 weder von β noch von β' getroffen wird, zweitens K_3 und K_4 innerhalb $\varkappa_1$ weder durch δ noch durch δ' getrennt werden, und drittens δ und δ' einen endlichen Abstand voneinander besitzen.

Alsdann werden nach dem Obigen K_4 und K_3 innerhalb $\varkappa_1$ durch $\delta+\delta'$

nicht getrennt. Mithin werden K_2 *und* K_3 *innerhalb* $\varkappa$ *durch* $\beta + \beta'$ *nicht getrennt, also auch durch* $\alpha + \alpha'$ *nicht.*

Eine innerhalb $\varkappa$ abgeschlossene Punktmenge soll *zusammenhängend*
[[11]] heißen, wenn sie sich nicht in zwei innerhalb $\varkappa$ abgeschlossene Teilmengen
[[12]] zerlegen läßt.

Eine zusammenhängende, innerhalb $\varkappa$ abgeschlossene Teilmenge einer abgeschlossenen Punktmenge $\mathfrak{C}$, welche nicht in einer anderen zusammenhängenden, innerhalb $\varkappa$ abgeschlossenen Teilmenge von $\mathfrak{C}$ enthalten ist, soll ein *Teilkontinuum von* α *in bezug auf* $\varkappa$ heißen.

Den obigen Entwickelungen gemäß läßt sich aus einer innerhalb $\varkappa$ abgeschlossenen, *nicht* zusammenhängenden, die Punkte K_2 und K_3 innerhalb $\varkappa$ trennenden Punktmenge α eine innerhalb $\varkappa$ abgeschlossene, ebenfalls die Punkte K_2 und K_3 innerhalb $\varkappa$ trennende Teilmenge herausheben. Diese Teilmenge enthält jedes Teilkontinuum von α, von welchem sie einen Punkt enthält.

Durch unbeschränkte Wiederholung desselben Verfahrens gelangen wir nach einer abzählbaren Zahl von Schritten zu einer Teilmenge von α, auf welche das Verfahren nicht weiter angewendet werden kann, welche somit ein einziges Teilkontinuum von α darstellt und die Punkte K_2 und K_3 noch immer innerhalb $\varkappa$ trennt.

§ 3.

Im n-dimensionalen Cartesischen Raume μ soll eine abgeschlossene Punktmenge r ein *Jordansches Raumstück* heißen, wenn sie sich auf eine Teilmenge $\mathfrak{R}$ einer geschlossenen $(n-1)$-dimensionalen Mannigfaltigkeit R eineindeutig und stetig abbilden läßt (die Teilmenge $\mathfrak{R}$ darf *nicht* mit R identisch sein).

Wir nehmen nun an, daß *das Jordansche Raumstück* r *in* μ *mehr als ein Gebiet bestimmt.*

Sei G_1 eines dieser Gebiete, r' die Grenze von G_1, G_2 ein weiteres von r' bestimmtes Gebiet, r'' die Grenze von G_2. Alsdann bestimmt das Jordansche Raumstück r'' in μ zwei Gebiete G_2 und G_3, von denen es die gemeinsame Grenze ist, und ist mithin *zusammenhängend.*

Sei $\mathfrak{R}''$ das Bild von r'' in R, $\mathfrak{R}'''$ ein von $\mathfrak{R}''$ in R bestimmtes Restgebiet, $\mathfrak{P}$ ein zur Grenze von $\mathfrak{R}'''$, also zugleich zu $\mathfrak{R}''$ gehöriger Punkt, P das Bild von $\mathfrak{P}$ in r.

Wir schlagen in μ um P eine solche $(n-1)$-dimensionale Kugel $\varkappa$, welche weder G_2 noch G_3 ganz in ihrem Innern I enthält, wählen in G_2 und G_3 je einen außerhalb $\varkappa$ liegenden Punkt, ziehen von dort nach r'' Wege, deren Endpunkte S_2 und S_3 je einen Abstand $< \varepsilon$ von P besitzen,

[[480]]

und bezeichnen mit K_2 und K_3 die letzten Schnittpunkte dieser Wege mit $\varkappa$. Nach dem Resultat des § 2 existiert ein mit τ zu bezeichnendes *Teilkontinuum von r'' in bezug auf $\varkappa$*, welches K_2 und K_3 innerhalb $\varkappa$ trennt. Die in $\varkappa$ liegende Grenze von τ bezeichnen wir mit v, die Bilder von τ und v in R mit T und U. Alsdann ist T in einem einzigen durch U in R bestimmten Gebiete $\mathfrak{G}$ enthalten. Wir wollen einen Augenblick annehmen, daß T das ganze Gebiet $\mathfrak{G}$ erfüllt, und gleich zeigen, daß diese Annahme auf einen Widerspruch führt.

Weil es Punkte in beliebiger Nähe von S_2 und S_3 gibt, welche durch τ innerhalb $\varkappa$ getrennt werden, muß die Entfernung, welche P von τ besitzt, also auch die Entfernung, welche $\mathfrak{P}$ von T besitzt, mit ε unter jede Grenze herabsinken. Durch genügend kleine Wahl von ε kann man mithin erreichen, daß $\mathfrak{P}$ sich in R mit einem Punkte von T durch einen U nicht treffenden einfachen Kurvenbogen verbinden läßt, mithin im Gebiete $T \equiv \mathfrak{G}$ enthalten sein muß. Dann wäre aber $\mathfrak{P}$ in einem zu $\mathfrak{R}''$ gehörigen Gebiete von R enthalten, was ein Widerspruch ist.

Wenn wir also in R eine hinreichend dichte simpliziale Zerlegung ζ_1 konstruieren, *so existiert in $\mathfrak{G}$ ein Grundsimplex σ, dem kein Punkt von T angehört.*

§ 4. [[13]]

Wir unterziehen μ einer Transformation mittels reziproker Radien mit dem Zentrum K_3, durch welche $\varkappa$, I, K_2, τ und v der Reihe nach in k, ω, O, t und u übergehen. In dem vom ebenen $(n-1)$-dimensionalen Raume k in μ bestimmten Gebiete ω wird dann O durch t vom Unendlichen getrennt.

Sei $\mathfrak{K}$ eine in k um O beschriebene $(n-2)$-dimensionale Kugel. Indem wir u auf $\mathfrak{K}$ projizieren, wird u, und damit zugleich sein eineindeutiges und stetiges Bild U, auf $\mathfrak{K}$ *eindeutig und stetig* abgebildet. Unser Ziel ist, diese Abbildung von U auf $\mathfrak{K}$ zu einer eindeutigen und stetigen Abbildung von $U + \mathfrak{G} - \sigma$ auf $\mathfrak{K}$ zu erweitern.

Wir nehmen mit R eine solche Fundamentalreihe $\zeta_1, \zeta_2, \zeta_3, \cdots$ von simplizialen Zerlegungen vor, daß $\zeta_{\nu+1}$ eine Unterteilung von ζ_ν ist, und daß die Größe der zu ζ_ν gehörigen Grundsimplexe für unbeschränkt wachsendes ν unter jede Grenze herabsinkt. Jedes ζ_ν bestimmt ein größtes von Grundsimplexen gebildetes, σ enthaltendes, in $\mathfrak{G}$ enthaltenes Teilgebiet $\mathfrak{G}_\nu$ von R; die Grenze von $\mathfrak{G}_\nu$ konvergiert für unbeschränkt wachsendes ν gleichmäßig gegen U.

Jedem Grundpunkte der verschiedenen ζ_ν weisen wir einen solchen Punkt von U zu, welcher von ihm die kleinstmögliche Entfernung besitzt, und bestimmen eine ganze Zahl p mit der Eigenschaft, daß die den Eck-

[[481]]

punkten eines außerhalb $\mathfrak{G}_p$ willkürlich gewählten $\mathfrak{G}_\nu$-Elementes zugewiesenen Punkte von U innerhalb einer Halbkugel von $\mathfrak{K}$ abgebildet werden.

Wenn wir die Menge derjenigen Elemente von $\mathfrak{G}_\nu$, welche außerhalb $\mathfrak{G}_{\nu-1}$ liegen, mit $\mathfrak{g}_\nu$ bezeichnen, so setzt $\mathfrak{G}$ sich zusammen aus $\mathfrak{G}_p$ und der von den verschiedenen $\mathfrak{g}_\nu$ für $\nu > p$ gebildeten Punktmenge.

Jedem Elementeckpunkte eines $\mathfrak{g}_\nu$ $(\nu > p)$ lassen wir in $\mathfrak{K}$ den Bildpunkt des ihm zugewiesenen Punktes von U entsprechen, bilden in Anschluß daran und der Euklidischen Metrik von k gemäß die von den $\mathfrak{g}_\nu$ $(\nu > p)$ gebildete Punktmenge simplizial auf k ab, und projizieren schließlich jeden in dieser Weise in k bestimmten Bildpunkt aus O auf $\mathfrak{K}$. Hierdurch wird die von den $\mathfrak{g}_\nu$ $(\nu > p)$ gebildete Punktmenge, mithin auch die Grenze von $\mathfrak{G}_p$, eindeutig und stetig auf $\mathfrak{K}$ abgebildet. Diese Abbildung geht in die früher bestimmte Abbildung von U auf $\mathfrak{K}$ stetig über.

Nunmehr lassen wir jedem Elementeckpunkte von $\mathfrak{G}_p$, der nicht in der Grenze von $\mathfrak{G}_p$ enthalten ist, einen willkürlichen Punkt von $\mathfrak{K}$ entsprechen, mit der einzigen Beschränkung, daß die Eckpunktsbilder einer willkürlichen $(n-3)$-dimensionalen Elementseite keinen O enthaltenden ebenen $(n-3)$-dimensionalen Raum bestimmen dürfen. In Anschluß daran bilden wir die $(n-3)$-dimensionalen Elementseiten von $\mathfrak{G}_p$ simplizial auf k ab und projizieren schließlich jeden in dieser Weise in k bestimmten Bildpunkt aus O auf $\mathfrak{K}$.

Wir bauen nun $\mathfrak{G}_p$ in solcher Weise aus seinen durch ζ_p bestimmten Elementen auf, daß wir ausgehen von σ, diesem ein Element $\mathfrak{s}_1$, von dem eine $(n-2)$-dimensionale Seite in der Grenze von σ liegt, hinzufügen, sodann der Punktmenge $\sigma' = \sigma + \mathfrak{s}_1$, ein weiteres Element $\mathfrak{s}_2$, von dem eine $(n-2)$-dimensionale Seite in der Grenze von σ' liegt, hinzufügen, und so fortfahren, bis mit einem gewissen Elemente $\mathfrak{s}_d$ das Gebiet $\mathfrak{G}_p$ wiederhergestellt ist.

Weil von der herzustellenden Abbildung von $\mathfrak{G}_p$ auf $\mathfrak{K}$ bisher nur die Bilder der in der *Grenze* von $\mathfrak{G}_p$ enthaltenen $(n-2)$-dimensionalen Elementseiten festgelegt sind, gibt es wenigstens *eine* $(n-2)$-dimensionale Seite $\mathfrak{y}_d$ von $\mathfrak{s}_d$, über deren Bildpunkte noch nicht verfügt ist. Auf die übrigen $(n-2)$-dimensionalen Seiten von $\mathfrak{s}_d$ läßt sich aber die Abbildung leicht in Anschluß an die schon existierende erweitern. Die Bildpunkte des Innern von $\mathfrak{s}_d$ und der Seite $\mathfrak{y}_d$ werden sodann bestimmt durch die Forderung, daß jeder in einem repräsentierenden Simplexe (vgl. Math. Ann. 71, S. 100) von $\mathfrak{s}_d$ auf $\mathfrak{y}_d$ errichteten Lote nur ein einziger Punkt von $\mathfrak{K}$ entsprechen soll.

In $\sigma^{(d-1)}$ sind jetzt nur von denjenigen $(n-2)$-dimensionalen Elementseiten, welche in der Grenze von $\sigma^{(d-1)}$ enthalten sind, die Bilder festgelegt, sodaß es wieder wenigstens *eine* $(n-2)$-dimensionale Seite $\mathfrak{y}_{d-1}$

von $\mathfrak{s}_{d-1}$ gibt, über deren Bild noch nicht verfügt ist, und wir die Abbildung über $\mathfrak{s}_{d-1}$ in derselben Weise wie über $\mathfrak{s}_d$ erweitern können. So fahren wir fort, bis jedes $\mathfrak{s}_\nu$, und damit zugleich die ganze Punktmenge $U + \mathfrak{G} - \sigma$, *also auch die Punktmenge* $u + t$ *eindeutig und stetig auf* $\mathfrak{K}$ *abgebildet ist.*

§ 5.

Wir bezeichnen mit G dasjenige von t in ω bestimmte Gebiet, zu dessen Grenze O gehört, und konstruieren in ω eine solche Fundamentalreihe $z_1, z_2, z_3, \cdots$ von Euklidischen simplizialen Zerlegungen, daß $z_{\nu+1}$ eine Unterteilung von z_ν ist, daß die Größe der zu z_ν gehörigen Grundsimplexe für unbeschränkt wachsendes ν unter jede Grenze herabsinkt, und daß O niemals einer $(n-2)$-dimensionalen Grundseite angehört. Jedes z_ν bestimmt ein größtes, von Grundsimplexen gebildetes, an O grenzendes, in G enthaltenes Teilgebiet G_ν von ω; der nicht in k liegende Teil l_ν der Grenze von G_ν ist ein zweiseitiges $(n-1)$-dimensionales Fragment, welches für unbeschränkt wachsendes ν gleichmäßig gegen t konvergiert; die Grenze o_ν von l_ν ist in k enthalten, besteht aus einer endlichen Zahl von geschlossenen zweiseitigen $(n-2)$-dimensionalen Pseudomannigfaltigkeiten, und konvergiert für unbeschränkt wachsendes ν gleichmäßig gegen u.

Eine für G_ν angenommene positive Indikatrix bestimmt in bekannter Weise (vgl. Math. Ann. 71, S. 108) den positiven Sinn der Indikatrix für ein willkürliches Element e_ν von l_ν, für eine willkürliche $(n-2)$-dimensionale Seite von e_ν, *betrachtet als Seite von* e_ν, und schließlich für ein willkürliches Element von o_ν.

Sei A ein willkürlicher Punkt von o_ν, A' ein solcher Punkt von t, welcher von A die kleinstmögliche Entfernung besitzt, B die Projektion von A auf $\mathfrak{K}$ aus O, B' der Bildpunkt von A' auf $\mathfrak{K}$ für die im § 4 hergestellte Abbildung, ϱ_A die Entfernung der Punkte B und B', ε_ν das Maximum von ϱ_A für einen in o_ν variierenden Punkt A. Alsdann sinkt ε_ν für unbeschränkt wachsendes ν unter jede Grenze herab.

Jeden Elementeckpunkt C von l_ν bilden wir in einen Punkt von $\mathfrak{K}$ ab, der vom Bildpunkte eines von C die kleinstmögliche Entfernung aufweisenden Punktes von t einen Abstand $\leqq \varepsilon_\nu$ besitzt. Liegt C in o_ν, so wird insbesondere die Projektion von C auf $\mathfrak{K}$ aus O als Bildpunkt gewählt.

Wir bestimmen eine ganze Zahl q von der Eigenschaft, daß die Eckpunktsbilder jedes Elementes von l_q innerhalb einer Halbkugel von $\mathfrak{K}$ liegen, bilden in Anschluß an diese Eckpunktsbilder und der Euklidischen Metrik von ω und k gemäß l_q simplizial auf k ab, und projizieren schließlich jeden in dieser Weise in k bestimmten Bildpunkt aus O auf $\mathfrak{K}$. Hierdurch wird von jedem Elemente von l_q der Umfang *mit dem Grade Null* (vgl. Math. Ann. 71, S. 105) auf $\mathfrak{K}$ abgebildet; *die entsprechende Abbildung von o_q auf $\mathfrak{K}$ besitzt mithin ebenfalls den Grad Null.*

§ 6.

Sei $\mathfrak{S}_0$ dasjenige Element von G_q, welches O auf seinem Umfange enthält. Wir bauen G_q in solcher Weise aus seinen durch z_q bestimmten Elementen auf, daß wir ausgehen von $\mathfrak{S}_0$, diesem ein Element Σ_1, von dem eine $(n-1)$-dimensionale Seite in der Grenze von $\mathfrak{S}_0$ liegt, hinzufügen, sodann der Punktmenge $\mathfrak{S}_1 = \mathfrak{S}_0 + \Sigma_1$ ein weiteres Element Σ_2, von dem eine $(n-1)$-dimensionale Seite in der Grenze von $\mathfrak{S}_1$ liegt, hinzufügen und so fortfahren, bis mit einem gewissen Elemente Σ_h das Gebiet G_q wiederhergestellt ist.

Der nicht in k enthaltene Teil der Grenze von $\mathfrak{S}_m$ bildet ein zweiseitiges $(n-1)$-dimensionales Fragment, dessen Grenze ψ_m sich aus einer endlichen Zahl von in k liegenden geschlossenen zweiseitigen $(n-2)$-dimensionalen Pseudomannigfaltigkeiten zusammensetzt.

Eine für G_q angenommene positive Indikatrix bestimmt in bekannter Weise den positiven Sinn der Indikatrix für ein willkürliches ψ_m.

Aus O projizieren wir von jedem ψ_m die Elementeckpunkte auf $\mathfrak{K}$, bilden in Anschluß an diese Eckpunktsbilder und gemäß der Euklidischen Metrik von ω und k die ψ_m simplizial auf k ab, und projizieren schließlich jeden in dieser Weise in k bestimmten Bildpunkt aus O auf $\mathfrak{K}$.

Weil es für jedes m Teilgebiete von $\mathfrak{K}$ gibt, in welchen die in dieser Weise bestimmten Bilder von ψ_m und ψ_{m+1} identisch sind, ist der entsprechende, auf Grund der Euklidischen Metrik von k und der sphärischen Metrik von $\mathfrak{K}$ bestimmte Abbildungsgrad für ψ_{m+1} derselbe wie für ψ_m, mithin für ψ_h derselbe wie für ψ_0, d. h. ± 1.

Nun ist aber nicht nur ψ_h identisch mit o_q, sondern auch die in diesem Paragraphen konstruierte Abbildung von ψ_h auf $\mathfrak{K}$ identisch mit der im vorigen Paragraphen konstruierten Abbildung von o_q auf $\mathfrak{K}$, müßte mithin wie letztere den Grad Null besitzen.

Unsere Annahme, daß r in μ mehr als ein Gebiet bestimmt, hat somit auf einen Widerspruch geführt, und wir können folgenden Satz aussprechen:

Im n-dimensionalen Cartesischen Raume bestimmt ein Jordansches Raumstück nur ein einziges Gebiet.

§ 7.

Wir betrachten nunmehr das in einer n-dimensionalen Mannigfaltigkeit η liegende eineindeutige und stetige Bild χ' eines n-dimensionalen Gebiets χ, und nehmen an, daß in χ' ein Punkt P' existiert, in dessen be-

[[484]]

liebiger Nähe nicht zu χ' gehörige Punkte von η liegen. Wir wählen in χ' einen willkürlichen, nicht mit P' identischen Punkt Q', und bezeichnen die in χ liegenden Bilder von P' und Q' mit P resp. Q.

Wir schlagen in η um P' eine nicht ganz in χ' enthaltene und Q' ausschließende $(n-1)$-dimensionale Kugel $\varkappa$ mit beliebig kleinem Radius ε und bezeichnen mit $\mathfrak{R}$ die in χ' enthaltene abgeschlossene Teilmenge von $\varkappa$, mit r das in χ liegende Bild von $\mathfrak{R}$. Nun ist es unmöglich, P' und Q' innerhalb η durch einen $\varkappa$ nicht treffenden einfachen Kurvenbogen zu verbinden. Mithin ist es auch unmöglich, P' und Q' innerhalb χ' durch einen $\mathfrak{R}$ nicht treffenden einfachen Kurvenbogen zu verbinden.

Für genügend kleines ε ist das Jordansche Raumstück r ganz in einem zu χ gehörenden Kugelinnern enthalten. Mithin lassen sich P und Q nach dem Resultate des § 6 innerhalb χ durch einen r nicht treffenden einfachen Kurvenbogen verbinden. Als Bild dieses Kurvenbogens würde dann aber in χ' ein $\mathfrak{R}$ nicht treffender, P' und Q' verbindender einfacher Kurvenbogen existieren, was ein Widerspruch ist, sodaß die versuchte Annahme sich als unstatthaft erweist.

Mithin existiert in η zu jedem Punkte von χ' eine um diesen Punkt beschriebene Kugel, deren Inneres ganz in χ' enthalten ist. Weil weiter je zwei Punkte von χ' sich innerhalb χ' durch einen einfachen Kurvenbogen verbinden lassen, *ist χ' in η ein Gebiet.*

W. z. b. w.

NOTES

⟦1⟧ One of the last manuscripts of the paper is extant. Corrections: 1921 F. A simpler proof in Brouwer 1912 C.

⟦2⟧ 1911 C. Lebesgue 1911 A, 1911 B.

⟦3⟧ A. Schoenflies 1899, Osgood 1900, F. Bernstein 1900.

⟦4⟧ R. Baire 1907 B.

⟦5⟧ J. Hadamard 1910.

⟦6⟧ The concept of pseudo-manifold is crucial in this context. It has been adjusted in 1921 F.

⟦7⟧ Brouwer 1911 D.

⟦8⟧ The appended correction agrees with 1921 F. According to the handwriting it must be dated much nearer to 1911 than to 1921.

⟦9⟧ A copy of the last proofreadings (21 October 1911) of the present paper (with more comments on a separate sheet) contains a lengthy proof in Dutch of this assertion. It shows how Brouwer proved such elementary facts. Because of the lack of homological concepts such notions as the order of a point in an n-sphere with respect to what is now called a $(n-1)$-cycle is defined by the 'angle' under which the cycle is seen from the point.

⟦10⟧ Probably the first occurrence of *relative* closedness.
⟦11⟧ p. 308, l. 3 'π' added after 'Punktmenge'; l. 4 'oder: relativ π abgeschlossene)' added after 'abgeschlossene'. According to the handwriting, dating from the twenties.
⟦12⟧ The history of the concept of connectedness suffers from a certain confusion. The first definition of connectedness (by -chains) was given by G. Cantor 1883. C. Jordan 1893, p. 25, defined connectedness ('d'un seul tenant') for closed sets (of the plane or space): not divisible into two non-empty closed subsets. The concept of domain goes back to K. Weierstrass 1880; connection here means joining by polygons. Attempts at a unifying definition before 1911 were sporadic and less successful: E. Study 1903: 'das abgeschlossene Continuum, als perfecte Menge, die nicht in zwei ebenfalls perfecte Mengen zerlegt werden kann ⟦· · ·⟧ den allgemeinen Begriff des Continuums, als Menge, in der irgend zwei Stellen durch (mindestens) ein abgeschlossenes Continuum verbunden werden können'. A. Schoenflies several times dealt with connectedness: A. Schoenflies 1903 B, pp. 209–210, 1904, p. 139 (which contains grave errors, even after its correction in Schoenflies 1908 B), Schoenflies 1908 A, pp. 108, 109, 110, 116 (again with mistakes [1]). His general definition agrees with Study's. However, this general definition was never operational. In the history of mathematical concepts there are more examples of 'wrong' definitions which disappear as soon as the desired concepts are really needed. Even C. Caratheodory 1918, p. 208, 222, still defined connectedness separately for closed and open sets (not divisible into two open sets).

At the present place Jordan's concept of connectedness for closed sets is interpreted on the basis of *relative* closedness (see ⟦10⟧). This essentially yields the now accepted unified concept of connectedness. It is so natural a consequence of relative closedness that it makes the controversial explanation in Brouwer 1924 J2 of the use of another concept of connectedness in Brouwer 1913 A as a slip of the pen, look quite reasonable. Another proof that Brouwer considered this notion of connectedness as the true one at that time, is found in a letter to F. Engel, Y44.

An explicit formulation of the general concept of connectedness is found at about the same time in J. Lennes 1911. Brouwer could hardly have known Lennes'

[1]) The confusion is nicely illustrated by the following occurrences:

p. 108: 'eine zusammenhängende abgeschlossene Menge ⟦ist⟧ perfekt' – forgetting about the one point sets.

p. 108: 'Eine abgeschlossene zusammenhängende Menge soll als *abgeschlossenes* Kontinuum bezeichnet werden. Sie wird immer als geschränkt angenommen'. – Why the double terminology?

p. 117: 'Eine Punktmenge P soll ein weiteres Kontinuum heissen. wenn es für je zwei ihrer Punkte m_1 und m_2 ein abgeschlossenes Kontinuum gibt, das Teilmenge von P ist und m_1 und m_2 enthält'.

p. 117: 'Jedes nicht abgeschlossene Kontinuum geht durch Hinzufügung seiner Grenzpunkte in ein abgeschlossenes Kontinuum über' – which is wrong unless the non-closed continua are also supposed bounded.

definition when he wrote the present paper, but he did become acquainted with it at least as early as 1912. [1]).

Lennes 1911 was reviewed in Jahrbuch über die Fortschritte der Mathematik by H. Hahn. F. Hausdorff 1914 defined connectedness in the general sense but did not cite Lennes 1911. To those who in the twenties tied Hausdorff's name to the concept of connectedness, Brouwer used to indicate Lennes as its first author.

[1]) Among Brouwer's papers I found a registered envelope from Blumenthal – mailed 3 February 1912, arriving Amsterdam 4 February 1912 – which very probably contained a manuscript of Lennes' continuing J. Lennes 1911, and submitted to Mathematische Annalen; this envelope is covered with an analysis in Dutch of what might have been Lennes' manuscript, which was eventually turned down by Mathematische Annalen. There is also an analysis of parts of Lennes 1911 among Brouwer's papers, which, according to the handwriting, appears to date from about 1920.

Two letters of Blumenthal to Brouwer relating to this affair are witnessed by typewritten copies, appended to a letter of Brouwer to H. Hahn of 4 August 1929. Unfortunately the letters seem to have disappeared after they were removed from Brouwer's Blumenthal file in order to be copied (which is not unusual). This is the text of these letters:

3.2.1912.

⟦*Blumenthal to Brouwer*⟧

Lieber Herr Brouwer!
Ich bitte Sie um einen Dienst fuer die Annalen. Ich schicke Ihnen gleichzeitig eine Arbeit von Lennes "Curves and surfaces in non-metrical spaces", die mir in enger Beruehrung mit Ihren Arbeiten zu stehen scheint. Insbesondere sehe ich darin einen Teil des Jordanschen Satzes ausgesprochen und zwar gerade denjenigen, den Sie in den Annalen ausfuehrlich bewiesen haben. Ich bitte Sie daher um Ihr Urteil ueber die Arbeit: ob sie richtig ist, in welchem Verhaeltnis sie zu den Ihrigen steht und ob Sie sie fuer wert halten, in die Annalen aufgenommen zu werden.

Besten Dank im voraus und viele Gruesse!

Ihr
(gez.) O. Blumenthal

12.2.1912.

⟦*Blumenthal to Brouwer*⟧

Lieber Herr Brouwer!
Ich habe die Lennessche Arbeit und Ihre Kritik mit bestem Dank erhalten. Ich danke Ihnen besonders, dass Sie Ihr Urteil so eingehend begruendet und scharf formuliert haben. Ich habe Herrn Lennes daraufhin seine Arbeit natuerlich zurueckgeschickt, Ihre Bemerkungen beigelegt und ihn ausserdem auf Ihre Arbeit ueber den Jordanschen Satz verwiesen, die er noch nicht zu kennen scheint.

Nochmals besten Dank und viele Gruesse.

Ihr

(gez.) O. Blumenthal

Obviously he did not insist that he had originated the concept because Lennes' definition was certainly independent and more explicit. [1]).

This matter gained some, exaggerated, attention in a controversy we will deal with sub 1928 C2 ⟦1⟧. The confusion we mentioned at the beginning consists in believing that there was some older concept of connectedness in real use before 1911 or 1914 that was superseded by a modern one, invented by J. Lennes or F. Hausdorff. This distorted view, partly deriving from Brouwer himself, created serious misunderstandings.

⟦13⟧ This construction is used in 1918 C.

1921 F

MA 82, p. 286

Berichtigung

zu dem Aufsatze von L. E. J. Brouwer: „Beweis der Invarianz des n-dimensionalen Gebiets". Math. Ann. **71**, S. 305–313.

S. 306, Z. 5 ist hinzuzufügen: Schließlich sollen je zwei Punkte der Punktmenge sich durch einen ganz der Punktmenge angehörigen und keine $(n-p)$-dimensionale Seite $(p > 1)$ treffenden einfachen Kurvenbogen verbinden lassen.

⟦See 1911 E⟧

[1]) The present concept of connectedness is even older. It is already found in F. Riesz 1906/07 (submitted to the Hungarian Academy of Sciences, 3d class, as early as 22 January 1906). This is a most remarkable paper, related to Fréchet's work in General Analysis (see for example M. Fréchet 1906) and to Poincaré's discussion of dimension (H. Poincaré 1903), and anticipating N. J. Lennes 1911 and F. Hausdorff 1914 with respect to connectedness and other general concepts. The paper, published at a somewhat unusual place, seems to have remained unnoticed for a long time. As far as I see its first quotation is found in Tietze–Vietoris 1931 who consequently speak of 'Riesz–Lennes–Hausdorff connectedness'.

Beweis des Jordanschen Satzes für den n-dimensionalen Raum. 1911 F

Von

L. E. J. Brouwer in Amsterdam.

Der zu beweisende Satz lautet folgendermaßen: ⟦1⟧

Im n-dimensionalen Raume R_n bestimmt eine Jordansche Mannigfaltigkeit, d. h. das eineindeutige und stetige Bild einer geschlossenen $(n-1)$-dimensionalen Mannigfaltigkeit zwei Gebiete, und ist mit der Grenze jedes dieser Gebiete identisch.

Wir bezeichnen die Jordansche Mannigfaltigkeit mit J, und zerlegen (analog wie früher in der Ebene*)) den Satz in folgende drei Bestandteile: ⟦2⟧

1. *Die Grenze eines von J bestimmten Gebietes ist mit J identisch.*
2. *J bestimmt höchstens zwei Gebiete.*
3. *J bestimmt mindestens zwei Gebiete.*

Alsdann ist der erste Teil im Resultate des § 6 meiner vorstehenden Arbeit enthalten, während der dritte Teil sich nach einer von Lebesgue skizzierten Methode**) herleiten läßt. Der noch restierende zweite Teil soll im Folgenden erledigt werden. ⟦3⟧ ⟦4⟧

---

*) Math. Ann. 69, S. 169—175.

**) C. R., 27 mars 1911. Das daselbst erhaltene Resultat liefert zusammen mit den Entwicklungen von Baire in Bull. des Sc. Math. (2), 31 einen zweiten Beweis der Invarianz des n-dimensionalen Gebiets. ⟦5⟧

Für diejenigen Leser, deren Interesse mit dem dreidimensionalen Raume aufhört, lasse ich hier einen sehr einfachen, aber nur für $n = 3$ gültigen Beweis dieses dritten Teiles des Jordanschen Satzes folgen. ⟦6⟧

Seien j_1 und j_2 zwei geschlossene stetige Kurven (im Sinne von Schoenflies) des R_3, so können wir den vollen Umlauf von j_1 resp. j_2 in unendlich vielen Weisen als eindeutiges und stetiges Bild β_1 resp. β_2 eines Kreises auffassen. Für eine bestimmte Wahl von β_1 und β_2 besitzt die Entfernung zweier entsprechender Punkte von j_1 und j_2 ein Maximum M; die bei Variierung von β_1 und β_2 auftretende untere Grenze von M soll die *Parameterdistanz von j_1 und j_2* heißen.

Wenn wir eine endliche Reihe von geschlossenen stetigen Kurven, in welcher j das erste und ein einziger Punkt das letzte Element ist, in solcher Weise konstruieren, daß das Maximum der Parameterdistanzen zweier aufeinanderfolgender Elemente den

§ 1.

Sei E das von J bestimmte unendliche Gebiet, I ein von J bestimmtes endliches Gebiet, P ein Punkt von I. Wir wählen in J ein willkürliches Element J' heraus, bezeichnen die von den übrigen Elementen gebildete Punktmenge mit J'', den Umfang von J' mit j, das repräsentierende
[[7]] Simplex (vgl. Math. Ann. 71, S. 100) von J' mit S, den Umfang von S mit s, wählen in J' und dementsprechend (vgl. l. c. S. 108) in j einen positiven Sinn der Indikatrix, verbinden P und $J''-j$ innerhalb I durch einen Weg w', und verbinden $J'-j$ und $J''-j$ innerhalb E durch einen Weg w_e.

Die Menge derjenigen Punkte von J', welche von J'' eine Entfernung $\geqq \frac{\sqrt{n}}{2^{\tau-1}}$ besitzen, bezeichnen wir mit J'_τ.

Wir zerlegen den R_n in homothetische n-dimensionale Kuben $\mathfrak{q}_0$ mit der Kantenlänge 1, jeden der $\mathfrak{q}_0$ in 2^n homothetische Teilkuben $\mathfrak{q}_1$ mit

---

Wert ε besitzt, so werden wir sagen, *daß j mit dem Unstetigkeitsgrade ε zusammengezogen wird.*

Eine endliche Menge von geschlossenen stetigen Kurven nennen wir kurz ein *Kurvensystem*.

Sei nun J' eine Jordansche Fläche im R_3, $\varkappa$ eine solche um einen ihrer Punkte beschriebene Kugel, die einen Teil von J' in ihrem Äußeren und nur zweiseitige Teilgebiete von J' in ihrem Inneren enthält. Durch eine geeignete Inversion des R_3 geht $\varkappa$ in eine Ebene k, J' in eine Jordansche Fläche J über. Sei $\mathfrak{G}$ ein durch k in J ausgeschnittenes zweiseitiges Gebiet, γ die Grenze von $\mathfrak{G}$, $\mathfrak{S}$ ein γ im Abstande ε_1 approximierendes Polygonsystem in $\mathfrak{G}$, S ein von $\mathfrak{S}$ die Parameterdistanz ε_2 besitzendes, in k liegendes Kurvensystem. Wenn wir ε_2 mit ε_1 gegen Null konvergieren lassen, so existiert in k ein Gebiet G, welches für hinreichend kleines ε_1 eine nicht verschwindende Zahl c Male von S umlaufen wird.

Im entgegengesetzten Falle könnten wir nämlich das Kurvensystem S in k in einer Entfernung $< \varepsilon_3$ von γ mit dem Unstetigkeitsgrade ε_4, und auf Grund davon das Kurvensystem $\mathfrak{S}$ in J in einer Entfernung $\leqq \varepsilon_5$ von γ mit dem Unstetigkeitsgrade ε_6 zusammenziehen, wo ε_3, ε_4, ε_5 und ε_6 mit ε_1 gegen Null konvergieren. Dies ist aber nach der Definition von $\mathfrak{S}$ ein Widerspruch.

Ein in einem Punkte von G auf k errichtetes Lot l wird sowohl von S wie von $\mathfrak{S}$ c Male umlaufen, sodaß die Differenz der Anzahlen der positiven und der negativen Kreuzungen von l mit einer willkürlichen hinreichend genauen simplizialen Approximierung von $\mathfrak{G}$ gleich $\pm c$ ist. Mithin können wir eine solche nicht von $J-\mathfrak{G}$ getroffene und von nicht in J liegenden Endpunkten Q_1 und Q_2 begrenzte Teilstrecke von l bestimmen, für welche die Differenz der Zahlen der positiven und der negativen Kreuzungen mit einer willkürlichen hinreichend genauen simplizialen Approximierung von $\mathfrak{G}$ *nicht gleich Null* ist. Dann aber müssen ein willkürlicher, Q_1 und Q_2 verbindender Streckenzug und eine willkürliche hinreichend genaue simpliziale Approximierung von J notwendig einander treffen, *sodaß Q_1 und Q_2 durch J getrennt werden.*

[[490]]

der Kantenlänge $\frac{1}{2}$, jeden der $\mathfrak{q}_1$ in 2^n homothetische Teilkuben $\mathfrak{q}_2$ mit der Kantenlänge $\frac{1}{4}$, usw.

Die von denjenigen $\mathfrak{q}_\tau$, welche in ihrem Inneren oder auf ihrem Umfange wenigstens einen Punkt von J_τ' enthalten, gebildete Punktmenge bezeichnen wir mit μ_τ, die Menge der zu $\mu_\tau, \mu_{\tau+1}, \mu_{\tau+2}, \cdots$ gehörigen Punkte mit π_τ; wir wählen τ so groß, daß w' von π_τ *nicht* getroffen wird. Das P enthaltende von $J + \pi_\tau$ bestimmte Gebiet bezeichnen wir mit I_τ, den nicht in J'' enthaltenen Teil der Grenze von I_τ mit g.

Dasjenige von $J'' + g$ bestimmte Gebiet, in dem E enthalten ist, bezeichnen wir mit E_τ, und ziehen in E_τ einen Weg w''' aus E nach g. Der Endpunkt R dieses Weges liegt in einer gewissen zur Grenze von E_τ gehörigen zweiseitigen $(n-1)$-dimensionalen Pseudomannigfaltigkeit (vgl. meine vorstehende Arbeit S. 305) γ mit ebenen Elementen und einem ⟦3⟧
in j enthaltenen Rande*). Hierbei werden zwei $(n-2)$-dimensionale Elementseiten von γ, welche in R_n zusammenfallen, dann und nur dann auch für γ als identisch betrachtet, wenn die beiden entsprechenden Elemente daselbst einen zu E_τ gehörigen Winkel einschließen.

Wir dürfen annehmen, daß der Punkt R *nicht* einer $(n-2)$-dimensionalen Elementseite von γ angehört.

Wir wählen in γ einen positiven Sinn der Indikatrix, ziehen in I_τ aus P einen w' nicht treffenden Weg w'' nach R, und bezeichnen den von w' und w'' gebildeten Streckenzug mit w_{ir}.

Die zu E_τ gehörige Seite von γ soll ihre *linke*, die andere ihre *rechte* Seite heißen. Alsdann verbindet der Weg w_{ir} innerhalb I die rechte Seite von γ mit $J'' - j$, während der Weg w''' ein Endsegment w_{il} besitzt, welches innerhalb I die linke Seite von γ mit $J' - j$ verbindet.

Mittels zweier in beliebiger Nähe von J' resp. von J'' verlaufender, j nicht treffender Streckenzüge v' und v'' können wir die Wege w_{ir}, w_{il} und w_e ergänzen zu einem γ nur in einem einzigen Kreuzungspunkte, nämlich im Punkte R treffenden Polygon w.

§ 2.

Im R_n verstehen wir unter einem p-dimensionalen *Netze* resp. *Netzfragmente* das simpliziale Bild einer p-dimensionalen Pseudomannigfaltigkeit resp. eines p-dimensionalen Fragmentes (vgl. meine vorstehende Arbeit S. 306). ⟦3⟧

---

*) Unter dem Rande von γ verstehen wir die Menge der nicht in γ enthaltenen Grenzpunkte von γ. Ob solche Grenzpunkte existieren, bleibt in diesem Paragraphen noch dahingestellt.

Unter den *Grundsimplexen*, *Grundpunkten* und *Grundseiten* eines Netzes resp. Netzfragmentes verstehen wir die Bilder der Grundsimplexe, Grundpunkte und Grundseiten der entsprechenden Pseudomannigfaltigkeit resp. des entsprechenden Fragmentes.

Wenn im R_n ein mit einem positiven Umlaufssinne versehenes Polygon $\mathfrak{P}$ und ein mit einer positiven Indikatrix versehenes geschlossenes zweiseitiges $(n-1)$-dimensionales Netz $\mathfrak{N}$ in solcher Weise gelegen sind, daß kein Eckpunkt und keine Teilstrecke von $\mathfrak{P}$ in $\mathfrak{N}$ liegt und keine Seite von $\mathfrak{P}$ eine $(n-2)$-dimensionale Grundseite von $\mathfrak{N}$ trifft, so sind die Anzahlen der positiven und der negativen Kreuzungen von $\mathfrak{P}$ mit $\mathfrak{N}$ einander gleich.*)

Für den Fall, daß $\mathfrak{N}$ einseitig ist, läßt sich nur aussagen, daß die absolute Anzahl der Kreuzungen von $\mathfrak{P}$ und $\mathfrak{N}$ gerade ist.

Wir denken uns nun im R_n ein mit einer positiven Indikatrix versehenes geschlossenes zweiseitiges $(n-2)$-dimensionales Netz $\mathfrak{N}$ und ein mit einem positiven Umlaufssinne versehenes, $\mathfrak{N}$ nicht treffendes Polygon $\mathfrak{P}$.

Sei $\mathfrak{C}$ ein solches mit einer positiven Indikatrix versehenes zweiseitiges $(n-1)$-dimensionales Netzfragment, welches $\mathfrak{N}$ als seinen einzigen Rand und die positive Indikatrix von $\mathfrak{N}$ als positive Randindikatrix**) besitzt, während kein Eckpunkt und keine Teilstrecke von $\mathfrak{P}$ in $\mathfrak{C}$ liegt und keine Seite von $\mathfrak{P}$ eine $(n-2)$-dimensionale Grundseite von $\mathfrak{C}$ trifft. Wenn wir dann mit p die Anzahl der positiven, mit p' die Anzahl der negativen Kreuzungen von $\mathfrak{P}$ und $\mathfrak{C}$ bezeichnen, so ist für gegebenes $\mathfrak{P}$ und $\mathfrak{N}$ die Zahl $c = p - p'$ unabhängig von der Wahl von $\mathfrak{C}$.

Die Zahl c stellt mithin eine Beziehung zwischen $\mathfrak{P}$ und $\mathfrak{N}$ dar. Wir nennen sie *den Grad von $\mathfrak{N}$ in bezug auf $\mathfrak{P}$*.

§ 3.

Diejenigen Elemente von γ, welche von j eine Entfernung $\geqq \frac{1}{2^\nu}$ besitzen, bilden ein mit γ_ν zu bezeichnendes $(n-1)$-dimensionales Fragment, dessen Grenze η_ν für unbeschränkt wachsendes ν gleichmäßig gegen j konvergiert.

---

*) Sei f der Endpunkt der kreuzenden Polygonseite für einen positiven Umlaufssinn von $\mathfrak{P}$, i eine positive Indikatrix des gekreuzten Grundsimplexes von $\mathfrak{N}$. Falls if für den R_n eine positive Indikatrix darstellt, heißt die Kreuzung positiv. Um die im Texte ausgesprochene Eigenschaft einzusehen, braucht man $\mathfrak{P}$ nur in solcher Weise aus dem Unendlichen heranrücken zu lassen, daß die Bahnen der Punkte von $\mathfrak{P}$ keine $(n-3)$-dimensionale Grundseite, und speziell die Bahnen der Eckpunkte von $\mathfrak{P}$ keine $(n-2)$-dimensionale Grundseite von $\mathfrak{N}$ treffen.

**) Hinsichtlich der Beziehung zwischen „positiver Indikatrix" und „positiver
[[7]] Randindikatrix" vgl. Math. Ann. 71, S. 108.

[[492]]

Wir nehmen mit J eine solche Fundamentalreihe $z_1, z_2, \cdots$ von simplizialen Zerlegungen vor daß für unbeschränkt wachsendes ν die Breite der zu z_ν gehörigen Grundsimplexe unter jede Grenze herabsinkt, während jedes $z_{\nu+1}$ eine Unterteilung von z_ν ist. Jedes z_ν bestimmt im R_n ein $(n-1)$-dimensionales Netzfragment J_ν' resp. J_ν'' als simpliziales Bild von J' resp. J'', und ein geschlossenes zweiseitiges $(n-2)$-dimensionales Netz j_ν als simpliziales Bild von j. Jedem Elementeckpunkte von η_ν weisen wir einen solchen in j liegenden Grundpunkt von z_ν zu, welcher von ihm die kleinstmögliche Entfernung besitzt, und konstruieren dementsprechend im repräsentierenden Simplexe S von J' eine simpliziale Abbildung von η_ν, welche wir aus dem Mittelpunkte von S auf s projizieren.

Sodann nehmen wir mit η_ν eine solche simpliziale Zerlegung ζ_ν vor, daß durch die genannte Projektion jedes Grundsimplex von ζ_ν innerhalb eines einzigen zu z_n gehörigen Grundsimplexes von s abgebildet wird. Den Eckpunkten eines willkürlichen Grundsimplexes von ζ_ν sind dann solche Punkte von j_ν zugewiesen, welche in einem einzigen Grundsimplexe von j_ν enthalten sind.

Durch eine passende simpliziale Zerlegung der Grenzelemente von γ_ν erweitern wir ζ_ν zu einer simplizialen Zerlegung von γ_ν, und indem wir von dieser Zerlegung alle nicht zu η_ν gehörigen Grundpunkte festlassen, jeden zu η_ν gehörigen Grundpunkt aber durch den ihm entsprechenden Punkt von j_ν ersetzen, wird als simpliziales Bild von γ_ν ein zweiseitiges $(n-1)$-dimensionales Netzfragment $\mathfrak{F}_\nu$ bestimmt, dessen Grenze λ_ν sich als simpliziales Bild von η_ν aus einer endlichen Zahl von geschlossenen zweiseitigen $(n-2)$-dimensionalen Netzen zusammensetzt und in j_ν enthalten ist.

§ 4.

Für hinreichend großes ν hat das Polygon w außer R keinen Punkt mit $\mathfrak{F}_\nu$ gemeinsam, ist mithin der totale Grad von λ_ν in bezug auf w gleich ± 1.

Den Grad der Abbildung von λ_ν auf j_ν (vgl. Math. Ann. 71, S. 105) ⟦7⟧
bezeichnen wir mit c, den Grad von j_ν in bezug auf w mit c'.

Sei $\mathfrak{C}_1$ ein solches mit einer positiven Indikatrix versehenes zweiseitiges $(n-1)$-dimensionales Netzfragment, welches j_ν als Grenze und die negative Indikatrix von j_ν als positive Randindikatrix besitzt, während kein Eckpunkt und keine Teilstrecke von w in $\mathfrak{C}_1$ liegt und keine Seite von w eine $(n-2)$-dimensionale Grundseite von $\mathfrak{C}_1$ trifft.

Seien $\mathfrak{C}_2, \mathfrak{C}_3, \cdots, \mathfrak{C}_c$ weitere Netzfragmente derselben Art, so sind die Anzahlen der positiven und der negativen Kreuzungen von w mit $\mathfrak{F}_\nu + \mathfrak{C}_1 + \mathfrak{C}_2 + \mathfrak{C}_3 + \cdots + \mathfrak{C}_c$ einander gleich.

Der totale Grad von λ_ν in bezug auf w wird somit erhalten, indem wir den Grad von j_ν in bezug auf w mit c multiplizieren, d. h. es gilt folgende Formel:

$$c\,c' = \pm 1.$$

Dieser Formel kann aber nur dadurch genügt werden, daß sowohl c wie c' gleich ± 1 ist.

§ 5.

Wir nehmen nun dem zu beweisenden Satze entgegen an, daß außer I noch ein zweites von J bestimmtes endliches Gebiet I' existiert. Alsdann konstruieren wir γ' und $\mathcal{F}_\nu'$ in I' analog wie γ und $\mathcal{F}_\nu$ in I, wobei die Grenze λ_ν' von $\mathcal{F}_\nu'$ ebenso wie die Grenze λ_ν von $\mathcal{F}_\nu$ das Netz j_ν mit dem Grade ± 1 überdeckt. Weiter dürfen wir die Streckenzüge v' und v'' in solcher Weise konstruiert denken, daß sie γ' ebensowenig wie γ treffen, sodaß γ' mit w keinen Punkt gemeinsam hat.

Nun muß einerseits jedes Polygon, von dem kein Eckpunkt und keine Teilstrecke in $\mathcal{F}_\nu + \mathcal{F}_\nu'$ liegt und keine Seite eine $(n-2)$-dimensionale Grundseite von $\mathcal{F}_\nu + \mathcal{F}_\nu'$ trifft, mit $\mathcal{F}_\nu + \mathcal{F}_\nu'$ eine gerade Anzahl von Kreuzungspunkten aufweisen, andererseits aber können wir ν so groß wählen, daß das Polygon w mit $\mathcal{F}_\nu'$ keinen Punkt gemeinsam hat, mithin $\mathcal{F}_\nu + \mathcal{F}_\nu'$ nur in einem einzigen Punkte, nämlich im Punkte R kreuzt.

Aus diesem Widerspruche folgern wir, *daß J nur ein einziges endliches Gebiet I bestimmen kann.*

W. z. b. w.

NOTES

⟦1⟧ C. Jordan 1893 proved the theorem for the plane. A. Schoenflies (see 1908 A) considerably sharpened the statement. Brouwer 1910 E gave an elegant proof for the 2-dimensional case. Jordan's theorem in space occupied Brouwer's attention from at least 1909 onwards (see Y15) and attempts at proving it date from spring 1910 at the latest, as indicated by an excerise book (see Y16 ⟦1⟧). Brouwer's ultimate attack on the general Jordan theorem was stimulated by simultaneous work of Lebesgue (see 1911 C ⟦1⟧). There is an intermediate manuscript of the present paper.

⟦2⟧ Brouwer 1910 E.

⟦3⟧ Brouwer 1911 E.

⟦4⟧ The strange fact that Brouwer did not elaborate Lebesgue's sketch was explained in 1911 C ⟦1⟧.

⟦5⟧ Lebesgue 1911 B, R. Baire 1907 B. The 'first proof' of the invariance of domain is in 1911 E.

⟦6⟧ The method is related to that of Lebesgue 1911 B. It is restricted to $n = 3$ for reasons which look no more than technical today but were not so in 1911. Owing to the lack of homological concepts (or rather the lack of attention to Poincaré's work on homology) such concepts as singularity index (in vector fields) and linking coefficient were defined in an inefficient way (see 1911 C ⟦1⟧). Rather than being regarded as a combinatorial cycle, the closed curve was endowed with a circulation sense – a device seemingly lacking in higher dimensions.
⟦7⟧ Brouwer 1911 D.

⟦See 1911 G⟧

1911 L GÉOMÉTRIE. — *Sur le théorème de M. Jordan dans l'espace à n dimensions.* Note de M. **L.-E.-J. Brouwer.**

⟦1⟧

Le théorème de M. Jordan, complété par M. Schœnflies, s'énonce ainsi : *Une variété fermée à une dimension sans points multiples détermine dans le plan deux régions; chaque point de la variété est accessible (erreichbar) pour chacune de ces régions.*

On généralise ce théorème en disant : *Une variété fermée* F_{n-1} *à* $n-1$ *dimensions sans points multiples détermine, dans l'espace* E_n *à n dimensions, deux régions ; chaque point de* F_{n-1} *est accessible pour chacune de ces régions.*

Une partie de ce théorème, savoir que F_{n-1} détermine dans E_n *au moins*
⟦2⟧ deux régions, a été prouvée par M. Lebesgue (voir *Comptes rendus* du 27 mars 1911), j'ai divisé la partie restante en trois énoncés dont la démonstration complète se trouve dans une série de Mémoires qui paraîtront dans un autre recueil.

Voici ces énoncés :

1° La frontière d'une région déterminée par F_{n-1} est identique à F_{n-1}.
2° F_{n-1} ne détermine que deux régions I et S.
3° Chaque point de F_{n-1} est accessible pour I et pour S.

Je résume brièvement mes raisonnements.

Soient E_{n-1} un espace plan à $n-1$ dimensions situé dans E_n, D l'une des deux régions déterminées dans E_n par E_{n-1}, P un point de E_{n-1} non appartenant à F_{n-1}. Soit R une région de F_{n-1} située dans D; la frontière de R se compose d'une partie F située dans E_{n-1} et une partie F′ située dans D. Je dis que si R + F′ sépare dans D le point P de l'infini, F′ doit se réduire à zéro.

Pour le démontrer, décrivons dans E_{n-1} une hypersphère S_{n-2} à centre P et construisons moyennant une projection centrale à centre P une représentation univoque et continue de F sur S_{n-2}. Essayons d'étendre cette représentation à une representation univoque et continue de R + F′ + F sur S_{n-2} ; nous trouverons d'une part que cette extension est impossible à cause de la séparation de P de l'infini, d'autre part qu'elle serait possible s'il existait une frontière F'.

On déduit de là qu'une partie fermée de F_{n-1} ne peut pas diviser E_n, ce qui entraîne la propriété 1°.

Soient S la région infinie, I une région finie déterminée dans E_n par F_{n-1}. Divisons F_{n-1} en deux régions, F'_{n-1} et F''_{n-1}, moyennant l'image biunivoque et continue j d'une hypersphère à $n-2$ dimensions. Au moyen d'un ensemble d'intervalles adhérents à F'_{n-1} on réussit à construire une variété bilatérale $(n-1)$-dimensionale g_{n-1} située dans I, ayant j pour frontière, se composant d'éléments plans et divisant I en deux régions dont l'une a pour frontière $F'_{n-1}+g_{n-1}+j$, l'autre $F''_{n-1}+g_{n-1}+j$. Si la propriété 2° n'était pas satisfaite, F_{n-1} déterminerait une seconde région finie I', on construirait dans I' une variété g'_{n-1} analogue à g_{n-1}, et l'ensemble $g_{n-1}+g'_{n-1}+j$ séparerait F'_{n-1} de F''_{n-1}, résultat absurde, puisque dans S on peut mener un arc simple joignant F'_{n-1} et F''_{n-1}.

Soit Q un point arbitraire de F_{n-1}. Construisons dans I une suite de variétés $g'_{n-1}, g''_{n-1}, g'''_{n-1}, \ldots$, ne se rencontrant pas et convergeant vers Q; joignons chaque $g^{(p)}_{n-1}$ à $g^{(p+1)}_{n-1}$ par un arc simple $\beta^{(p)}$, et chaque $\beta^{(p-1)}$ à $\beta^{(p)}$ par un arc simple $\vartheta^{(p)}$ situé dans $g^{(p)}_{n-1}$. L'ensemble des arcs $\beta^{(p)}$ et $\vartheta^{(p)}$ forme un arc simple situé dans I et aboutissant en Q; la propriété 3° est démontrée.

Parmi les conséquences de notre théorème je signale la bilatéralité intrinsèque de F_{n-1}, l'invariance de la région n-dimensionale, l'invariance du degré de transformation d'une variété bilatérale fermée, la généralisation de la notion d'indicatrice.

Quant à la proposition inverse, démontrée dans le plan par M. Schœnflies, pour $n > 2$ elle n'est plus valable.

(*Comptes rendus*, t. 153, p. 542, séance du 11 septembre 1911.)

NOTES

[[1]] Sketch of the proof of 1911 F.

[[2]] Lebesgue 1911 B.

1911 G

Über Jordansche Mannigfaltigkeiten.

Von

L. E. J. Brouwer in Amsterdam.

---

§ 1.

[[1]] **Die Erreichbarkeit.**

[[2]] Die im § 1 des vorstehenden Aufsatzes im Gebiete I konstruierte Pseudomannigfaltigkeit γ besitzt die Eigenschaft, daß je zwei Punkte ihrer linken Seite sich durch einen in beliebiger Nähe von γ verlaufenden, $J+\gamma$ außerhalb seiner Endpunkte nicht treffenden Weg verbinden lassen, und daß jeder Punkt ihrer rechten Seite sich durch einen außerhalb seiner Endpunkte $J+\gamma$ nicht treffenden Weg mit dem Punkte P verbinden läßt. Mithin wird I von γ in höchstens zwei Gebiete zerlegt.

Nach § 4 des vorstehenden Aufsatzes muß jedes Polygon, von dem kein Eckpunkt und keine Teilstrecke in $\mathcal{F}_\nu + J_\nu''$ liegt und keine Seite eine $(n-2)$-dimensionale Grundseite von $\mathcal{F}_\nu + J_\nu''$ trifft, mit $\mathcal{F}_\nu + J_\nu''$ eine gerade Zahl von Kreuzungspunkten aufweisen, sodaß die linke und die rechte Seite von $\mathcal{F}_\nu$ durch $\mathcal{F}_\nu + J_\nu''$ voneinander getrennt werden. Dann aber werden auch die linke und die rechte Seite von γ durch $\gamma + J''$ voneinander getrennt, sodaß von $\gamma + J''$ ein endliches, die rechte Seite von γ enthaltendes Gebiet I_r bestimmt wird. Dieses Gebiet I_r ist ein Teilgebiet von I, und nach der im vorstehenden Aufsatze „Beweis der Invarianz des n-dimensionalen Gebiets“ angewandten Methode läßt sich zeigen, daß die Grenze von I_r mit $J'' + \gamma$ identisch ist.

Ebenso wird von $\gamma + J'$ ein die linke Seite von γ enthaltendes Teilgebiet I_l von I bestimmt, dessen Grenze mit $J' + \gamma$ identisch ist.

In diese zwei Gebiete I_r und I_l wird I von γ zerlegt.

Sei Q ein willkürlicher Punkt von J; ${}_1J'$, ${}_2J'$, ${}_3J'$, $\cdots$ eine solche Folge von Q enthaltenden, gegen Q konvergierenden Grundsimplexen von J, deren jedes im Innern des vorangehenden enthalten ist. Zu jedem ${}_pJ'$ konstruieren wir eine Mannigfaltigkeit ${}_p\gamma$, welche I in ${}_pI_r$ und ${}_pI_l$ zerlegt, und

[[498]]

wir sorgen dafür, daß einerseits die verschiedenen ${}_p\gamma$ einander nicht treffen, sodaß ${}_{p+1}I_l$ in ${}_pI_l$ enthalten ist, und andererseits ${}_pI_l$ für unbeschränkt wachsendes p gleichmäßig gegen Q konvergiert.

In jedem Gebiete ${}_pI_l$ verbinden wir ${}_p\gamma$ und ${}_{p+1}\gamma$ durch einen Weg ${}_p\sigma$, und in jedem ${}_p\gamma$ verbinden wir den Endpunkt von ${}_{p-1}\sigma$ und den Anfangspunkt von ${}_p\sigma$ durch einen Weg ${}_p\varrho$. Alsdann bilden die ${}_p\sigma$ und ${}_p\varrho$ zusammen einen in I nach Q führenden einfachen Kurvenbogen, sodaß wir bewiesen haben:

Satz 1. *Jeder Punkt einer Jordanschen Mannigfaltigkeit ist sowohl für das innere wie für das äußere Gebiet erreichbar.*

Die Schoenfliessche Erweiterung des Jordanschen Satzes bleibt mithin im n-dimensionalen Raume in Kraft. Sie verliert hier aber ihre in der Ebene bestehende Umkehrbarkeit, denn diese Umkehrung würde folgendermaßen lauten:

Eine geschränkte abgeschlossene Punktmenge, welche im R_n zwei Gebiete bestimmt, und von der jeder Punkt für jedes dieser Gebiete erreichbar ist, ist eine Jordansche Mannigfaltigkeit.

Um einzusehen, daß dieser Satz schon im dreidimensionalen Raume nicht mehr zutrifft, betrachte man die Gesamtheit F der in zylindrischen Koordinaten durch folgende Formeln bestimmten Punktmengen: [[3]]

(1) $\quad 1 > \varrho > 0 \qquad z = 2 + \cos\varphi + \left(1 - \sqrt{\cos^2\varphi}\right)\sin\frac{\pi}{\varrho}\,.$

(2) $\quad 1 > \varrho > 0 \qquad z = -2 - \cos\varphi - \left(1 - \sqrt{\cos^2\varphi}\right)\sin\frac{\pi}{\varrho}\,.$

(3) $\quad \varrho = 1 \qquad 2 + \cos\varphi \geqq z \geqq -2 - \cos\varphi\,.$

(4) $\quad \varrho = 0 \qquad 3 \geqq z \geqq 1\,.$

(5) $\quad \varrho = 0 \qquad -1 \geqq z \geqq -3\,.$

Diese abgeschlossene Punktmenge F *zerlegt den Raum in zwei Gebiete, und jeder Punkt von F ist für jedes dieser Gebiete erreichbar.* Aber ein gegen den Punkt H ($\varrho = 0$, $z = 2$) konvergierender Punkt von F läßt sich in F nicht durch einen gegen H konvergierenden einfachen Kurvenbogen mit H verbinden, *sodaß F keine Jordansche Fläche ist.*

§ 2.

Die Unbewalltheit.

Sei G eine abgeschlossene Punktmenge des R_n, $\mathfrak{G}$ ein von G bestimmtes Gebiet, Q und Q' zwei erreichbare Punkte der Grenze von $\mathfrak{G}$, k ein Q und Q' innerhalb $\mathfrak{G}$ verbindender einfacher Kurvenbogen, M das Maximum der Entfernungen, welche die zu k gehörigen Punkte von Q

und von Q' besitzen, m das Minimum von M für die verschiedenen bei gegebenen Q und Q' möglichen Wahlen von k. Diese Größe m bezeichnen wir mit Schoenflies als die *Wegdistanz* von Q und Q' für das Gebiet $\mathfrak{G}$. Wenn diese Wegdistanz mit der Entfernung von Q und Q' gegen Null konvergiert, so werden wir G für das Gebiet $\mathfrak{G}$ *unbewallt* nennen.

Sei nun (Q_1, Q_1'), (Q_2, Q_2'), $\cdots$ eine solche Folge von in J liegenden Punktepaaren, daß der Abstand von Q_p und Q_p' für unbeschränkt wachsendes p gegen Null konvergiert. Zu jedem Paare (Q_p, Q_p') können wir ein dieses Paar in seinem Innern enthaltendes Grundsimplex J_p', und zu jedem J_p' ein γ_p in solcher Weise konstruieren, daß die Breiten sowohl von J_p' wie von γ_p für unbeschränkt wachsendes p unter jede Grenze herabsinken. Im Gebiete I_{lp} ziehen wir einfache Kurvenbogen χ_p resp. χ_p' von γ_p nach Q_p resp. Q_p', bezeichnen ihre Anfangspunkte mit φ_p resp. φ_p', und verbinden in γ_p die Punkte φ_p und φ_p' durch einen Weg ω_p.

Alsdann bilden χ_p, χ_p' und ω_p zusammen einen Q_p und Q_p' innerhalb I verbindenden einfachen Kurvenbogen, dessen Breite für unbeschränkt wachsendes p unter jede Grenze herabsinkt, d. h. es gilt:

Satz 2. *Eine Jordansche Mannigfaltigkeit ist sowohl für das innere wie für das äußere Gebiet unbewallt.*

Auch dieser Satz ist für die Ebene von Schoenflies bewiesen worden.

§ 3.

Die Zweiseitigkeit.

Sei J_1', J_2', $\cdots$, J_1' eine solche geschlossene Kette von Elementen von J, in welcher je zwei aufeinanderfolgende eine $(n-2)$-dimensionale Seite gemeinsam haben. Wir bezeichnen die von J_p' und J_{p+1}' gebildete Punktmenge mit ${}_{p+1}J_p'$, den Umfang von J_p' resp. ${}_{p+1}J_p'$ mit j_p resp. ${}_{p+1}j_p$, wählen wir J_1', und dementsprechend nach Math. Ann. 71, S. 108 für j_1 und nach ibid. S. 101 für J_2' und ${}_2J_1'$ einen positiven Sinn der Indikatrix, und verbinden einen im Innengebiete I liegenden Punkt P_i und einen im Außengebiete E liegenden Punkt P_e durch solche Wege $\mathfrak{t}_1$, $\mathfrak{t}_2$, $\cdots$, $\mathfrak{t}_1$, daß jedes $\mathfrak{t}_p$ die Mannigfaltigkeit J nur in Punkten von $J_p' - j_p$ trifft. Für diese $\mathfrak{t}_p$ nehmen wir die Richtung von P_i nach P_e als die positive an, und wir bezeichnen das vom im positiven Sinne durchlaufenen $\mathfrak{t}_p$ mit dem im negativen Sinne durchlaufenen $\mathfrak{t}_{p+2}$ gebildete Polygon mit w_p.

Zu jedem J_p' resp. ${}_{p+1}J_p'$ konstruieren wir nach der Methode des vorstehenden Aufsatzes ein γ_p, ein $j_{\nu p}$ und ein $\mathcal{F}_{\nu p}$ resp. ein ${}_{p+1}\gamma_p$, ein ${}_{p+1}j_{\nu p}$ und ein ${}_{p+1}\mathcal{F}_{\nu p}$, wobei wir annehmen dürfen, daß $\mathfrak{t}_p$ von den γ nur γ_p, ${}_{p+1}\gamma_p$ und ${}_p\gamma_{p-1}$, und von den $\mathcal{F}_\nu$ nur $\mathcal{F}_{\nu p}$, ${}_{p+1}\mathcal{F}_{\nu p}$ und ${}_p\mathcal{F}_{\nu,p-1}$ trifft, daß kein Eckpunkt und keine Teilstrecke eines $\mathfrak{t}_p$ in einem der γ oder $\mathcal{F}_\nu$ liegt,

[[500]]

und daß keine Seite eines t_p eine $(n-2)$-dimensionale Grundseite von einem der γ oder $\mathfrak{F}_\nu$ trifft.

Alsdann ist die Differenz der Anzahlen der positiven und der negativen Kreuzungen mit ${}_2\gamma_1$ entweder sowohl für t_1 wie für t_2 gleich $+1$, oder sowohl für t_1 wie für t_2 gleich -1, sodaß für hinreichend großes ν der Grad von ${}_2j_{\nu 1}$ in bezug auf w_2 dasselbe Vorzeichen hat wie der Grad von ${}_2j_{\nu 1}$ in bezug auf w_1. Mithin hat für hinreichend großes ν auch der Grad von $j_{\nu 2}$ in bezug auf w_2 dasselbe Vorzeichen wie der Grad von $j_{\nu 1}$ in bezug auf w_1.

Wenn wir nun weiter die positive Indikatrix von J_3' aus der positiven Indikatrix von J_2' herleiten, so finden wir in derselben Weise, daß der Grad von $j_{\nu 3}$ in bezug auf w_3 dasselbe Vorzeichen hat wie der Grad von $j_{\nu 2}$ in bezug auf w_2. Indem wir so fortfahrend schließlich zu J_1' zurückkehren, wird einerseits für J_1' eine neue positive Indikatrix bestimmt, andererseits aber finden wir, daß der Grad von $j_{\nu 1}$ in bezug auf w_1 sein ursprüngliches Vorzeichen behalten hat. Mithin ist die neue positive Indikatrix von J_1' mit der ursprünglichen identisch.

Hiermit ist bewiesen:

Satz 3. *Eine Jordansche Mannigfaltigkeit ist eine zweiseitige Mannigfaltigkeit.*

§ 4.

Die Ordnung der Jordanschen Gebiete.

Wir wählen sowohl für J wie für den R_n einen positiven Sinn der Indikatrix, schlagen um einen nicht in J liegenden Punkt A eine Kugel $\varkappa_A$, welche wir zur Bestimmung ihrer positiven Indikatrix als Grenze ihres Innern betrachten, und bilden J durch Projektion aus A eindeutig und stetig auf $\varkappa_A$ ab. Den Grad dieser Abbildung nennen wir die *Ordnung von A in bezug auf J.* Sie ist für hinreichend großes ν identisch mit der Differenz der Anzahlen der positiven und der negativen Kreuzungen mit J_ν eines aus dem Unendlichen nach A gezogenen Weges, von dem kein Eckpunkt und keine Teilstrecke in J_ν liegt und keine Seite eine $(n-2)$-dimensionale Grundseite von J_ν trifft. Sie kann sich bei stetiger Bewegung von A in endlicher Entfernung von J nicht ändern, sodaß wir von einer *Ordnung des Innengebietes* und von einer *Ordnung des Außengebietes in bezug auf J* sprechen können*).

Liegt A im Innengebiete, so können wir nach A aus dem Unendlichen einen Weg ziehen, der von J nur $J'-j$ trifft. Dieser Weg liefert für die Ordnung des Innengebietes sofort ± 1.

---

*) Vgl. J. Hadamard in dem auf S. 97 und 305 dieses Annalenbandes zitierten Buche. [[4]]

Im Außengebiete aber kann man A leicht in solcher Weise bestimmen, daß die Bildmenge von J *nicht überall dicht* in $\varkappa_A$ liegt. Die Ordnung des Außengebietes ist somit gleich Null, und wir sprechen aus:

Satz 4. *In bezug auf eine Jordansche Mannigfaltigkeit hat das Innengebiet die Ordnung* ± 1, *das Außengebiet die Ordnung Null* *).

§ 5.

Die verallgemeinerte Indikatrix.

[[5]] Sei α eine eineindeutige und stetige Abbildung einer geschlossenen zweiseitigen n-dimensionalen Mannigfaltigkeit μ_1 auf eine geschlossene zweiseitige n-dimensionale Mannigfaltigkeit μ_2, c der Grad dieser Abbildung, S_1 ein solches Grundsimplex von μ_1, dessen Bild S_2 innerhalb des Innensimplexes (vgl. Math. Ann. 71, S. 102) eines Elementes $\mathfrak{s}$ von μ_2 liegt, J_1 der Umfang von S_1, J_2 das Bild von J_1, α_ν eine simpliziale Approximierung von α, $J_{2\nu}$ die entsprechende simpliziale Approximierung von J_2, P_1 ein zum Innern von S_1 gehöriger Punkt, P_2 das Bild von P_1.

Wenn dann α durch α_ν mit einem hinreichenden Grade der Genauigkeit approximiert wird, so wird für einen aus P_2 nach der Grenze von $\mathfrak{s}$ gezogenen Weg, von dem kein Eckpunkt und keine Teilstrecke in $J_{2\nu}$ liegt und keine Seite eine $(n-2)$-dimensionale Grundseite von $J_{2\nu}$ trifft, die Differenz der Anzahlen der positiven und der negativen Kreuzungen mit $J_{2\nu}$ gleich $\pm c$ sein. Dieselbe Zahl ist aber nach § 4 gleich ± 1, sodaß sich herausstellt:

Satz 5. *Eine eineindeutige und stetige Beziehung zwischen zwei geschlossenen zweiseitigen n-dimensionalen Mannigfaltigkeiten besitzt entweder den Grad* $+1$ *oder den Grad* -1 **).

Wir bezeichnen nunmehr mit J_1 eine willkürliche in einem Elemente von μ_1 enthaltene und sich innerhalb eines Elementes $\mathfrak{s}$ von μ_2 abbildende Jordansche Mannigfaltigkeit, konstruieren zu ihr dem vorstehenden Aufsatze gemäß ein j_1 und ein γ_1, wählen im Innengebiete von $\gamma_1 + J_1''$ ein

---

[[4]] *) Dieser Satz wird bei Hadamard l. c. ohne Beweis ausgesprochen.

[[6]] **) Dieses Theorem wird von Herrn Lebesgue bei der Herleitung seines Existenzsatzes der „variétés enlacées" (C. R., 27 mars 1911) stillschweigend vorausgesetzt.

Der Begriff des Abbildungsgrades c läßt sich unmittelbar auf eineindeutige und stetige Beziehungen zwischen einem zweiseitigen n-dimensionalen Gebiete μ_1 und einem n-dimensionalen Gebiete μ_2 übertragen. Auch hier kann c nur die Werte
[[5]] $+1$ und -1 besitzen. Falls aber μ_2 einseitig wäre, müßte c nach S. 106 dieses Annalenbandes gleich Null sein, sodaß hiermit die *Invarianz der Zwei- resp. Einseitigkeit* bewiesen ist.

solches Simplex S_1, von dem eine $(n-1)$-dimensionale Seite h_1 in γ_1 liegt, bezeichnen die von den übrigen $(n-1)$-dimensionalen Seiten von S_1 gebildete Punktmenge mit k_1, den vollen Umfang von S_1 mit s_1, und wählen im Innern von S_1 einen Punkt P_1.

Wir wählen die positive Indikatrix von J_1 in solcher Weise, daß die Ordnung von P_1 in bezug auf J_1 gleich $+1$ wird, bestimmen die positive Indikatrix von $\gamma_1 + J_1'$ mittels der von J_1', die von $\gamma_1 + J_1''$ mittels der von J_1'', die von s_1 mittels der von $\gamma_1 + J_1''$, die von $\gamma_1 - h_1 + k_1 + J_1''$ mittels der von J_1''.

Die Ordnung eines Punktes in bezug auf $\gamma_1 + J_1'$ läßt sich mittels der $\mathfrak{F}_{1\nu} + J_{1\nu}'$, und die Ordnung in bezug auf $\gamma_1 - h_1 + k_1 + J_1''$ in analoger Weise definieren. In bezug auf s_1 ist alsdann die Ordnung von P_1 gleich $+1$, in bezug auf $\gamma_1 + J_1'$ resp. $\gamma_1 - h_1 + k_1 + J_1''$ gleich Null.

Die Bilder von P_1, J_1, γ_1, $\mathfrak{F}_{1\nu}$, $J_{1\nu}'$, $J_{1\nu}''$, h_1, k_1, s_1 bezeichnen wir mit P_2, J_2, γ_2, $\mathfrak{F}_{2\nu}$, $J_{2\nu}'$, $J_{2\nu}''$, h_2, k_2, s_2; als positive Indikatrizen von s_2, $\gamma_2 + J_2'$, $\gamma_2 - h_2 + k_2 + J_2''$ wählen wir die Bilder der positiven Indikatrizen von s_1, $\gamma_1 + J_1'$, $\gamma_1 - h_1 + k_1 + J_1''$; die Ordnung in bezug auf $\gamma_2 + J_2'$ resp. $\gamma_2 - h_2 + k_2 + J_2''$ definieren wir mittels der simplizialen Approximierungen von $\mathfrak{F}_{2\nu} + J_{2\nu}'$ resp. $\mathfrak{F}_{2\nu} - h_2 + k_2 + J_{2\nu}''$.

Wenn nun α den Grad $+1$ resp. -1 besitzt, so ist die Ordnung von P_2 in bezug auf s_2 ebenfalls gleich $+1$ resp. -1; in bezug auf $\gamma_2 + J_2'$ resp. $\gamma_2 - h_2 + k_2 + J_2''$ ist aber die Ordnung von P_2 gleich Null, denn vom Umfange von $\mathfrak{s}$ läßt sich ein $\gamma_2 + J_2'$ resp. $\gamma_2 - h_2 + k_2 + J_2''$ nicht treffender Weg nach P_2 ziehen. Mithin ist die Ordnung von P_2 in bezug auf J_2 gleich $+1 + 0 + 0 = +1$ resp. gleich $-1 + 0 + 0 = -1$, und wir haben folgende Eigenschaft bewiesen:

Satz 5a. *Bei einer eineindeutigen und stetigen Beziehung zwischen zwei geschlossenen zweiseitigen n-dimensionalen Mannigfaltigkeiten sind in bezug auf einander entsprechende hinreichend kleine Jordansche Mannigfaltigkeiten die Ordnungen der entsprechenden Innengebiete einander gleich oder entgegengesetzt, je nachdem der Abbildungsgrad $+1$ oder -1 beträgt.*

Wenn wir eine hinreichend kleine Jordansche Mannigfaltigkeit, in bezug auf die das Innengebiet die Ordnung $+1$ besitzt, als „Indikatrix" bezeichnen, können wir also im ersten Falle von einer Beziehung *mit invarianter Indikatrix*, im zweiten Falle von einer Beziehung *mit Umkehrung der Indikatrix* sprechen.

Mithin können wir den Fixpunktsatz der eineindeutigen und stetigen Kugeltransformationen*) jetzt in folgender Form aussprechen:

*) Vgl. Math. Ann. 71, S. 114. Der hier aus der Betrachtung eines Vektorfeldes gewonnene Satz 3 läßt sich auch in unmittelbarem Anschluß an S. 105 begründen [[5]] [[7]]

Jede eineindeutige und stetige Transformation einer Kugel gerader Dimensionenzahl mit invarianter Indikatrix oder einer Kugel ungerader Dimensionenzahl mit Umkehrung der Indikatrix besitzt sicher einen Fixpunkt.

§ 6.

Die Invarianz des Abbildungsgrades.

Seien μ_1, μ_2, μ_3 geschlossene zweiseitige n-dimensionale Mannigfaltigkeiten, α_{12} eine eindeutige und stetige Abbildung c_{12}^{ten} Grades von μ_1 auf μ_2, α_{23} eine eindeutige und stetige Abbildung c_{23}^{ten} Grades von μ_2 auf μ_3, α_{13} die durch α_{12} und α_{23} bestimmte Abbildung von μ_1 auf μ_3.

Wir betrachten die Grundsimplexe von μ_2 als ihre Elemente, und wählen die simplizialen Zerlegungen von μ_1 und die Innensimplexe von μ_2 und μ_3 in solcher Weise, daß in einem gewissen Innensimplexe von μ_3 ein nur von Innensimplexen von μ_2 bedecktes Gebiet existiert. In diesem Gebiete ist dann eine durch gewisse simpliziale Approximierungen von α_{12} und α_{23} bestimmte Bildmenge von μ_1 mit einer durch eine gewisse modifizierte simpliziale Approximierung von α_{13} bestimmten Bildmenge von μ_1 identisch. Mithin ist $c_{13} = c_{12} \times c_{23}$, d. h. *für eine Folge von Abbildungen ist der Grad des Produktes gleich dem Produkte der Grade.*

Jetzt läßt sich die auf S. 106 dieses Annalenbandes offen gelassene Frage nach der Invarianz des Abbildungsgrades, d. h. nach der Unabhängigkeit des Abbildungsgrades von der Wahl der Elemente und Messungsskalen, mit Hilfe des Resultates des § 5 erledigen.

Eine zweiseitige Mannigfaltigkeit *in anderer Weise messen* kommt nämlich darauf hinaus, daß zwischen dieser und *einer anderen* Mannigfaltigkeit eine eineindeutige und stetige Beziehung hergestellt wird. Diese Beziehung besitzt aber nach § 5 entweder den Grad $+1$ oder den Grad -1.

Sei nun α_{12} eine eindeutige und stetige Abbildung c_{12}^{ten} Grades einer geschlossenen zweiseitigen n-dimensionalen Mannigfaltigkeit μ_1 auf eine geschlossene zweiseitige n-dimensionale Mannigfaltigkeit μ_2, und sei c'_{12} der Grad von α_{12} für eine *neue* Messungsart von μ_1 und μ_2. Diese neue Messungsart bestimmt erstens eine Mannigfaltigkeit μ_3 und eine Abbildung

---

mittels der Bemerkung, daß eine Transformation ohne Fixpunkt sich in die antipodische Transformation der Identität stetig überführen läßt, nämlich durch Bewegung jedes Punktes auf dem kürzesten seine beiden Lagen verbindenden Großkreisbogen.

[[8]] Ich benutze diese Gelegenheit, um darauf hinzuweisen, daß P. Bohl den durch Folgerung 2 ausgedrückten besonderen Fall von Satz 3 (S. 114) schon in Bd. 127 des J. f. Math. ausgesprochen hat, ohne freilich den Beweis auszuführen.

c_{31} ten Grades von μ_3 auf μ_1 ($c_{31} = \pm 1$), zweitens eine Mannigfaltigkeit μ_4 und eine Abbildung c_{24} ten Grades von μ_2 auf μ_4 ($c_{24} = \pm 1$), und wir haben dem Obenstehenden gemäß:

$$c'_{12} = c_{34} = c_{31} \times c_{12} \times c_{24} = \pm c_{12}.$$

Mithin gilt:

Satz 6. *Der Abbildungsgrad ist eine Invariante der Analysis Situs.*

In diesem Aufsatze wird die Begründung der Invarianz des Abbildungsgrades auf den Jordanschen Satz gestützt. Sie ist aber davon unabhängig, weil die im § 6 bewiesene Eigenschaft, dass für eine Folge von Abbildungen der Grad des Produktes gleich dem Produkte der Grade ist, zugleich als Folgerung Satz 5 enthält. [[9]]

Bezeichnen wir nämlich den Grad der in Satz 5 gemeinten eineindeutigen und stetigen Beziehung mit c, den Grad der inversen Beziehung mit c', so besitzt das Produkt dieser beiden Beziehungen den Grad cc'. Dieses Produkt ist aber die Identität, sodass cc' gleich 1, und sowohl c wie c' gleich $\pm$ 1 sein muss.

NOTES

[[1]] The concepts of *Erreichbarkeit* and *Unbewalltheit* in the two-dimensional case come from A. Schoenflies (see 1908 A).

[[2]] Brouwer 1911 F.

[[3]] Corrected in Brouwer 1931.

[[4]] J. Hadamard 1910.

[[5]] Brouwer 1911 D.

[[6]] H. Lebesgue 1911 B.

[[7]] A version of this sentence was affixed at the end of 1911 D.

[[8]] P. Bohl 1904.

[[9]] This strip, affixed to Brouwer's own copy (and probably to all, or at least to the greater part, of the reprints), seems to be a private reprint of Brouwer 1911 N.

MA 71, p. 598

1911 N

Bemerkung

zu dem Aufsatze von L. E. J. Brouwer: „Über Jordansche Mannigfaltigkeiten".
Math. Ann. 71, S. 320—327.

In diesem Aufsatze wird die Begründung der Invarianz des Abbildungsgrades auf den Jordanschen Satz gestützt. Sie ist aber davon unabhängig, weil die im § 6 bewiesene Eigenschaft, daß für eine Folge von Abbildungen der Grad des Produktes gleich dem Produkte der Grade ist, zugleich als Folgerung Satz 5 enthält.

Bezeichnen wir nämlich den Grad der in Satz 5 gemeinten eineindeutigen und stetigen Beziehung mit c, den Grad der inversen Beziehung mit c', so besitzt das Produkt dieser beiden Beziehungen den Grad cc'. Dieses Produkt ist aber die Identität, sodaß cc' gleich 1, und sowohl c wie c' gleich ± 1 sein muß.

⟦See 1911 G⟧

Mathematics. — *Ueber freie Umschliessungen im Raume.* Von L. E. J. BROUWER. 1931

(Communicated at the meeting of January 31, 1931).

In meiner Arbeit „Ueber Jordansche Mannigfaltigkeiten" (Math. Ann. 71 (1911), S. 320—327) habe ich bewiesen, dass im n-dimensionalen Cartesischen Raume jeder Punkt einer Jordanschen Mannigfaltigkeit sowohl für das innere wie für das äussere Gebiet erreichbar ist. Im Anschluss an diesen Satz habe ich die evidente, aber im dortigen Zusammenhange trotzdem zu erwähnende Tatsache hervorgehoben, dass eine freie Umschliessung im n-dimensionalen Cartesischen Raume (d.h. eine geschränkte abgeschlossene Punktmenge, welche im R_n zwei Gebiete bestimmt, und von der jeder Punkt für jedes dieser Gebiete erreichbar ist) nicht notwendig eine Jordansche Mannigfaltigkeit, nicht einmal notwendig zusammenhängend im kleinen ist. Anschaulich formuliert, erhält man nämlich im dreidimensionalen Euklidischen Raume sofort eine nicht-Jordansche freie Umschliessung, wenn man von zwei sich gegenüberliegenden parallelen Kanten eines Kubus die Mittelstrecken zusammennäht, und eine nicht im kleinen zusammenhängende freie Umschliessung, wenn man am horizontalen oberen Grenzquadrat des Einheitskubus längs einer Diagonale eine gegen eine vertikale Kante konvergierende Fundamentalreihe von vertikalen hinreichend engen Einstülpungen der Tiefe $\frac{1}{2}$ vornimmt. [[1]]

Im a.a.O. S. 321 gegebenen formelhaften Beispiel, das diese Tatsache analytisch beschreiben soll, hat sich aber ein Versehen eingeschlichen: es muss das in den Formeln auftretende $\cos\varphi$, um eine nicht-Jordansche freie Umschliessung zu erhalten, durch $\cos 2\varphi$, und um eine nicht im kleinen zusammenhängende freie Umschliessung zu erhalten, durch $\cos 2\varphi \,.\, \operatorname{sgn} \sin\left(\varphi + \frac{1}{4}\pi\right)$ ersetzt werden.

Den Anlass zur Publikation dieser auf der Hand liegenden Richtigstellung bildet eine neuerdings in den Comptes-Rendus vom 5.1.1931 (Bd. 192) veröffentlichte Note von Herrn REY PASTOR, die mein oben zitiertes Beispiel zum Gegenstand hat, und in welcher der Versuch, das Beispiel durch ein besseres zu ersetzen, nicht gelungen ist; zur daselbst definierten angeblichen nicht-Jordanschen freien Umschliessung gehören nämlich die Punkte $\varrho = 0$, $1 < z \leqq \sqrt{2}$, welche von unten nicht erreichbar sind, so dass keine freie Umschliessung vorliegt. Um auf dem Wege, der [[2]]

Herrn REY PASTOR vorgeschwebt hat, das Beispiel in Ordnung zu bringen, muss in den zitierten Formeln meiner Arbeit das $cos\ \varphi$ durch $cos\ \varphi \,.\, s_2 (\varphi)$ bzw. durch $cos\ \varphi \,.\, s_4 (\varphi)$ ersetzt werden; hierin stellt $s_2 (\varphi)$ bzw. $s_4 (\varphi)$ die Winkelfunktion vor, die für $\varphi \neq n\pi$ bzw. für $\varphi \neq \frac{n}{2}\pi$ den absoluten Wert 1 besitzt, für $\varphi = 1$ positiv ist, und für $\varphi = n\pi$ bzw. für $\varphi = \frac{n}{2}\pi$ einerseits das Zeichen wechselt, andererseits daselbst alle Werte des geschlossenen Intervalls $(-1, +1)$ annimmt.

NOTES

[[1]] Brouwer 1911 G.
[[2]] Rey Pastor 1931.

Zur Invarianz des n-dimensionalen Gebiets. 1912 C

Von

L. E. J. Brouwer in Amsterdam.

Mein in Bd. 71 der Mathematischen Annalen veröffentlichter Beweis der Invarianz des n-dimensionalen Gebiets stützt sich auf die ihm vorausgeschickte Erledigung eines Hauptbestandteils des n-dimensionalen Jordanschen Satzes. Im folgenden gelangen wir auf viel direkterem Wege zum Ziele, nämlich in unmittelbarem Anschluß an die Invarianz der Dimensionenzahl, unter Heranziehung vom Begriffe des Abbildungsgrades. [[1]]

Die Invarianz der Dimensionenzahl wurde gegründet auf folgenden Satz*):

Im n-dimensionalen Raume R'_n besitzt das eineindeutige und stetige Bild G' eines n-dimensionalen Gebiets G in beliebiger Nähe eines beliebigen seiner Punkte ein Gebiet.

Sei γ' ein solches zu G' gehöriges Gebiet, γ die in γ' abgebildete Punktmenge von G. Alsdann ist γ ein *Teilgebiet* von G, denn zu jedem Punkte von γ existiert in G eine gewisse Umgebung, deren Bild ganz in γ' enthalten ist, welche mithin zu γ gehört.

Sei M ein willkürlicher Punkt von G, M' sein Bild: *zur Begründung der Invarianz des n-dimensionalen Gebiets haben wir zu zeigen, daß G' in R'_n eine volle Umgebung von M' enthält.*

Dazu beschreiben wir in G um M eine kleine, mit ihrem Innengebiete I in G enthaltene $(n-1)$-dimensionale Kugel K, bezeichnen das Bild von I mit I', das Bild von K mit K', und dasjenige von K' in R'_n bestimmte Gebiet, welches M' enthält, mit $\mathfrak{G}'$. Die gegebene Abbildung von G auf R'_n bestimmt dann eine Abbildung von I auf $\mathfrak{G}'$, welche einen gewissen *Grad* c besitzt.

Zu I gehört nun sicher ein Gebiet γ, dem als Bild ein Teilgebiet γ' von $\mathfrak{G}'$ entspricht, und der Grad dieser Abbildung von γ auf γ' ist gleich ± 1.**)

*) Math. Ann. 70, S. 165. [[2]]

**) ibid. 71, S. 598. [[3]]

Dann aber muß auch der Grad c gleich ± 1 sein, denn zu einem willkürlichen Punkte P' von γ' gibt es eine simpliziale Approximierung der Bildmenge von γ und eine simpliziale Approximierung der Bildmenge von I, welche in einer gewissen Umgebung von P' miteinander identisch sind.

Wenn aber c gleich ± 1 ist, *so muß I' überall dicht in $\mathfrak{G}'$ liegen**), muß also in R_n' eine volle Umgebung von M' enthalten.

W. z. b. w.

⟦4⟧ *) ibid. 71, S. 106.

NOTES

⟦1⟧ Brouwer 1911 E.
⟦2⟧ Brouwer 1911 C.
⟦3⟧ Brouwer 1911 N, affixed to 1911 G.
⟦4⟧ Brouwer 1911 D.

Mathematics. — "*On looping coefficients.*" By Dr. L. E. J. BROUWER. (Communicated by Prof. D. J. KORTEWEG.) 1912 E2

[1]

(Communicated in the meeting of February 24, 1912).

Let us suppose in Sp_3 two non-intersecting simple closed curves k_1 and k_2 furnished with a sense of circuit. Then k_1 possesses with respect to k_2 a *looping coefficient* answering to the intuitive notion of the *number of times that* k_1 *circulates around* k_2, and generally defined as $\frac{1}{4\pi} \times$ the variation corresponding to a circuit of k_1 of the solid angle projecting k_2 out of a variable point of k_1.

A first objection to this definition is, that without further agreement it can be applied only to special categories of simple closed curves. For, as soon as e.g. a simple closed curve k intersects of a sheaf s all the rays contained in a certain finite solid angle, the solid angle projecting k out of the vertex of s, has no more a definite value.

A second objection to the definition is, that it cannot be generalized to a notion of *"looping coefficient in Sp_n of a two-sided closed Sp_h with respect to a two-sided closed Sp_{n-h-1} not intersected by Sp_h."*

In the following we shall give a definition for which these two objections have been annulled.

§ 1.

On each of the two curves k_1 and k_2 we construct *a scale of measurement*[1]), and we consider the set R of pairs of points consisting of a point of k_1 and a point of k_2. A part of R determined by an element[2]) of k_1 and an element of k_2 we shall call a *paralleloelement.* It appears as a continuous one-one image of a parallelogram. Each of these image parallelograms can be divided into four triangles with a common vertex inside the parallelogram and with their bases in the sides of the parallelogram. Accordingly we can divide each paralleloelement of R into four two-dimensional elements, and with this we attain that the whole set R is divided into two-dimensional elements which by their mode of being joined cause R to appear as a *closed two-dimensional space.*[3])

Let p be a paralleloelement of R, d_1 resp. d_2 the corresponding element of k_1 resp. k_2, A_1 resp. B_1 the negative resp. positive endpoint of d_1, A_2 resp. B_2 the negative resp. positive endpoint of d_2, we then define the row of pairs of points $(A_2 A_1)$, $(A_2 B_1)$, $(B_2 B_1)$ as a positive indicatrix of the *partitional simplex*[4]) of p determined by those pairs of points, and with the aid of it we fix the positive indicatrix of the four elements of R belonging to p[5]). In this way we determine of all elements of R the positive indicatrix, where for

[[2]] 1) Mathem. Annalen 71, p. 98–100.

2) ibid., p. 97.

3) ibid., p. 98.

4) ibid., p. 100.

5) ibid., p. 101.

two arbitrary elements having a side in common these indicatrices satisfy the relation prescribed for *two-sided* spaces [1]).

So R' is a *closed two-sided two-dimensional space.*

The set of the vector directions of Sp_3 forms likewise a closed two-sided two-dimensional space (of the connection of the sphere) which we shall represent by B. The positive indicatrix of the spheres of Sp_3 (and with it at the same time the positive indicatrix of B) we determine by regarding them as boundary of their inner domain [2]).

If we conjugate to each pair of points consisting of a point of k_1 and a point of k_2 the direction of the vector connecting the two points, we determine a *continuous one-one representation* α *of* R *on* B. To this representation belongs a finite integer c independent of the mode of measurement of R, and therefore also of the mode of measurement of k_1 and k_2, which is called the *degree* of the representation, and possesses the property that the image of R covers positively each partitional domain of B in toto c times [3]).

It is this degree of representation which we define as the *looping coefficient of* k_1 *with respect to* k_2.

By exchange of k_1 and k_2 we find that on one hand the indicatrix of R changes its sign, but on the other hand each image point on B is replaced by its opposite point. *So the looping coefficient of* k_2 *with respect to* k_1 *is equal to the looping coefficient of* k_1 *with respect to* k_2.

We shall now show that for rectifiable curves the looping coefficient of k_1 with respect to k_2 can be expressed by the formula:

$$\frac{1}{4\pi}\int Vol.\ prod.\ (ds_1, ds_2, r^{-2}) \quad . \quad . \quad . \quad . \quad . \quad . \quad (1)$$

This integral namely can be interpreted for rectifiable curves as follows: We construct in k_1 resp. k_2 a simplicial division [4]) z_1 resp. z_2. To this corresponds a simplicial division z of R, whose base simplexes [5]) are determined in connection with the base arcs [6]) of

[1]) ibid., p. 101.

[2]) ibid., p. 108. [[2]]

[3]) ibid., p. 106.

[4]) ibid., p. 101.

[5]) That here the base simplexes are found by division of a parallelo-element, not as l. c. by division of an element, has of course no influence on our reasoning. Moreover, after HADAMARD (comp. J. TANNERY, "Introduction à la théorie des fonctions d'une variable", Vol. II, p. 463) a simplicial division of the parallelo-elements can be subdivided to a simplicial division of the elements. [[3]]

[6]) i. e. one-dimensional base simplexes.

k_1 and k_2 in the same way as we have determined above the elements of R in connection with the elements of k_1 and k_2. Each base arc of z_1 resp. z_2 we replace by the corresponding "chord", i. e. by the straight line segment with the same endpoints. Let $\varkappa_1$ be the chord corresponding to the base arc β_1 of k_1, $\varkappa_2$ the chord corresponding to the base arc β_2 of k_2, r the distance of their midpoints, then $\varkappa_1$ and $\varkappa_2$ regarded as vectors determine together with a vector of size r^{-2} in the direction of the straight line connecting their midpoints, a certain volume product. Of the volume products appearing in this way for the different pairs $(\varkappa_1, \varkappa_2)$ we take the sum S; our integral is to be regarded as $\frac{1}{4\pi} \times$ the limit of S for infinite condensation of z_1 and z_2.

Let us on the other hand represent each pair of points consisting of a point of a chord of k_1 and a point of a chord of k_2, by the endpoint of a vector with fixed origin O, and having the size and direction of the vector connecting the corresponding pair of points. Then for infinite condensation of z_1 and z_2 the ratio of the element of S corresponding to $\varkappa_1$ and $\varkappa_2$ to the value of the solid angle projecting out of O the parallelogram representing the chords $\varkappa_1$ and $\varkappa_2$, approaches indefinitely to unity, and so does the ratio of the element of S corresponding to $\varkappa_1$ and $\varkappa_2$ to the part of B covered for the simplicial approximation [1]) of α corresponding to z, by the "base parallelogram" resulting from β_1 and β_2.

As farthermore on account of the rectifiability of k_1 and k_2 the sum of the absolute values of the elements of S for infinite condensation of z_1 and z_2 cannot exceed a certain finite value, $\frac{S}{4\pi}$ converges indeed to the looping coefficient defined as the degree of the representation α.

On the other hand for rectifiable curves holds also the defintion of the looping coefficient as a variation of a solid angle mentioned in the beginning, and we easily see also this definition to be equivalent to the expressión (1).

§ 2.

Let now in Sp_n be given a two-sided closed h-dimensional space ϱ_1 and a two-sided closed $(n-h-1)$-dimensional space ϱ_2 not cutting ϱ_1, each provided with a positive indicatrix. We make ϱ_1 as well

[2] [1]) Mathem. Annalen 71, p. 102.

as ϱ_2 measurable [1]), and we consider the pairs of points consisting of a point of ϱ_1 and a point of ϱ_2. A part of R determined by an element of ϱ_1 and an element of ϱ_2 we shall call a *paralleloelement.* It appears as a continuous one-one image of a $(h, n-h-1)$-*simplotope* [2]). Let us call a division of a simplotope π into simplexes with one common vertex inside π, whilst the remaining vertices lie in the boundary of π, a "canonic division", then we can bring about such a canonic division by first executing it for the two-dimensional limits, then for each three-dimensional limit by projecting the divisions of its two-dimensional limits out of an arbitrary inner point, then for each four-dimensional limit by projecting the divisions of its three-dimensional limits out of an arbitrary inner point, and so on. Accordingly we can divide the paralleloelements of R into $(n-1)$-dimensional elements in such a way, that by their mode of being joined they cause R to appear as a *closed* $(n-1)$-*dimensional space.*

Let p be a paralleloelement of R, d_1 resp. d_2 the corresponding element of ϱ_1 resp. ϱ_2, $A_1 A'_1 \ldots . A_1^{(h)}$ a positive indicatrix of d_1, $A_2 A'_2 \ldots . A_2^{(n-h-1)}$ a positive indicatrix of d_2, we then define the row of pairs of points $(A_1 A_2)$, $(A'_1 A_2)$, ... $(A_1^{(h)} A_2)$, $(A_1^{(h)} A'_2)$, ... $(A_1^{(h)} A_2^{(n-h-1)})$ as a positive indicatrix of the partitional simplex of p determined by those pairs of points, and with the aid of it we fix the positive indicatrix of the elements of R belonging to p. In this way we determine of all elements of R the positive indicatrix, where for two arbitrary elements having an $(n-2)$-dimensional limit in common these indicatrices satisfy the relation prescribed for *two-sided* spaces.

So R is a *closed two-sided* $(n-1)$-*dimensional space.*

The set of the vector directions of Sp_n forms likewise a closed two-sided $(n-1)$-dimensional space (of the connection of the $(n-1)$-dimensional sphere) which we shall represent by B. The positive indicatrix of the spheres of Sp_n (and with it at the same time the positive indicatrix of B) we determine by regarding them as boundary of their inner domain.

If we conjugate to each pair of points consisting of a point of ϱ_1 and a point of ϱ_2 the direction of the vector connecting the two

[1]) ibid., p. 98—100. [[2]]

[2]) Let in Sp_{n-1} be given a plane h-dimensional space v and a plane $(n-h-1)$-dimensional space w. Let S_v be a simplex in v, S_w a simplex in w. The set of those points of R_{n-1} which in the direction of w project themselves on v in S_v, and in the direction of v project themselves on w in S_w, form by definition an $(h, n-h-1)$-*simplotope.* Of a simplotope the limiting spaces of any number of dimensions are likewise simplotopes. (Comp. P. H. Schoute, "Mehrdimensionale Geometrie", Vol. II, p. 45). [[4]]

points, we determine a *continuous one-one representation* α *of* R *on* B. To this representation belongs a finite integer c independent of the mode of measurement of R, and therefore also of the mode of measurement of ϱ_1 and ϱ_2, which is called the *degree* of the representation, and has the property that the image of R covers positively each partitional domain of B in toto c times.

It is this degree of representation which we define as the *looping coefficient of* ϱ_1 *with respect to* ϱ_2.

Exchange of ϱ_1 and ϱ_2 has only this consequence that the indicatrix of R changes its sign in some cases, and that each image point on B is replaced by its opposite point. *So the looping coefficient of* ϱ_1 *with respect to* ϱ_2 *and the looping coefficient of* ϱ_2 *with respect to* ϱ_1 *are either equal or opposite.*

We shall now show that if ϱ_1 and ϱ_2 are *evaluable*, i. e. if they have a definite finite h-dimensional resp. $(n-h-1)$-dimensional volume, the looping coefficient of ϱ_1 with respect to ϱ_2 can be expressed by the formula:

$$\frac{1}{k_n}\int Vol.\ prod.\ (di_1, di_2, r^{1-n}). \quad . \quad . \quad . \quad . \quad . \quad (2)$$

where k_n represents the $(n-1)$-dimensional volume of an $(n-1)$-dimensional sphere described with a radius 1 in the Euclidean Sp_n.

If namely ϱ_1 and ϱ_2 are evaluable, this integral can be interpreted as follows: We construct in ϱ_1 resp. ϱ_2 a simplicial division z_1 resp. z_2. To this corresponds a simplicial division z of R, whose base simplexes are determined in connection with the base simplexes of ϱ_1 and ϱ_2 in the same way as we have determined above the elements of R in connection with the elements of ϱ_1 and ϱ_2. Each base simplex of z_1 resp. z_2 we replace by the plane simplex with the same vertices. Let $\varkappa_1$ be the plane simplex corresponding to the base simplex β_1 of ϱ_1, $\varkappa_2$ the plane simplex corresponding to the base simplex β_2 of ϱ_2, r the distance of their centres of gravity, then $\varkappa_1$ and $\varkappa_2$, the former regarded as an h-dimensional, the second as an $(n-h-1)$-dimensional vector, determine together with a line-vector of size r^{1-n} in the direction of the straight line connecting their centres of gravity, a certain volume product [1]). Of the volume

[1]) The sign of this volume product we determine as follows: After having formed in the manner described above out of $(-1)^h \times$ the positive indicatrix of $\varkappa_1$ and the positive indicatrix of $\varkappa_2$ an indicatrix of a simplotope s parallel to $\varkappa_1$ and $\varkappa_2$, we add to the latter indicatrix the endpoint of a linevector described out of a point of s in the direction of the straight line connecting the centres of gravity of $\varkappa_2$ and $\varkappa_1$. The sign of the n-dimensional indicatrix found in this way determines he sign of our volume product.

products appearing in this way for the different pairs $(\varkappa_1, \varkappa_2)$ we take the sum S, our integral is to be regarded as $\frac{1}{k_n} \times$ the limit of S for infinite condensation of z_1 and z_2.

Let us on the other hand represent each pair of points consisting of a point of a plane simplex determined by z_1 and a point of a plane simplex determined by z_2, by the endpoint of a vector with fixed origin O, and having the size and direction of the vector connecting the corresponding pair of points. Then for infinite condensation of z_1 and z_2 the ratio of the element of S corresponding to $\varkappa_1$ and $\varkappa_2$ to the value of the solid angle projecting out of O the simplotope representing the simplexes $\varkappa_1$ and $\varkappa_2$, approaches indefinitely to unity, and so does the ratio of the element of S corresponding to $\varkappa_1$ and $\varkappa_2$ to the part of B, filled for the simplicial approximation of α corresponding to z, by the "base simplotope" resulting from β_1 ond β_2.

As farthermore on account of the evaluability of ϱ_1 and ϱ_2 the sum of the absolute values of the elements of S for infinite condensation of z_1 and z_2 cannot exceed a certain finite value, $\frac{S}{k_n}$ converges indeed to the looping coefficient defined as the degree of the representation α.

§ 3.

Let us now consider in Sp_n two sets of points ϱ'_1 and ϱ'_2 which have no point in common and are successively a continuous one-one image of an h-dimensional two-sided closed space ϱ_1 and of an $(n-h-1)$-dimensional two-sided closed space ϱ_2, then for these all the considerations of the former § remain of force. Let farthermore ϱ_1'' be a second continuous one-one image of ϱ_1, and let ϱ_2'' be a second continuous one-one image of ϱ_2, then there exists a quantity η with the property that if the distance of two corresponding points of ϱ_1' and ϱ_1'' as well as the distance of two corresponding points of ϱ_2' and ϱ_2'' is smaller than η, the looping coefficient of ϱ_1' with respect to ϱ_2'' is equal to the looping coefficient of ϱ_1' with respect to ϱ_2'.

From this ensues that in Sp_n the looping coefficient of an h-dimensional two-sided closed space ϱ_1 with respect to an $(n-h-1)$-dimensional two-sided closed space ϱ_2 not intersecting ϱ_1 is equal to the value of the integral

$$\frac{1}{k_n} \int Vol.\ prod.\ (di_1\,,\, di_2\,,\, r^{1-n})$$

for an arbitrary *simplicial approximation* [1]) $a(\varrho_1)$ of ϱ_1 and an arbitrary simplicial approximation $a(\varrho_2)$ of ϱ_2.

Let K_1 and K_2 be in Sp_n two spheres lying outside each other, $a(\varrho_1)$ a simplicial image of ϱ_1 lying inside K_1, $a(\varrho_2)$ a simplicial image of ϱ_2 lying inside K_2. The looping coefficient of $a(\varrho_1)$ with respect to $a(\varrho_2)$ is then *zero*; for, by transferring K_2 with $a(\varrho_2)$ outside K_1 to infinity, we can vary this looping coefficient only continuously, thus not at all.

We can now transform $a(\varrho_2)$ continuously into $a(\varrho_2)$ by causing the base points [2]) of $a(\varrho_2)$ to describe continuous paths, and we can choose for these base point paths such broken lines that in none of the intermediary positions of $a(\varrho_2)$ an $(h-1)$-dimensional element limit of $a(\varrho_1)$ has a point in common with $a(\varrho_2)$, neither an $(n-h-2)$-dimensional element limit of $a(\varrho_2)$ has a point in common with $a(\varrho_1)$, whilst those intermediary positions of $a(\varrho_2)$ which correspond to the angles of the base point paths, have no point in common with $a(\varrho_1)$.

Then for this variation of $a(\varrho_2)$ the looping coefficient of $a(\varrho_1)$ with respect to $a(\varrho_2)$ increases by a unit as often as an element e_1 of $a(\varrho_1)$ is traversed by an element η_2 of $a(\varrho_2)$ *positively*, i.e. in such a way that the volume product of e_1, η_2, and the direction of motion of the traversing point is positive according to the above definition.

If on the other hand we understand by $na(\varrho_2)$ resp. $na(\varrho_2)$ a two-sided $(n-h)$-dimensional *net fragment* [3]), limited by $a(\varrho_2)$ resp. $a(\varrho_2)$ and crossing $a(\varrho_1)$ only in a finite number of points, belonging neither to an $(h-1)$-dimensional base limit of $a(\varrho_1)$, nor to an inner $(n-h-1)$-dimensional base limit of $na(\varrho_2)$ resp. $na(\varrho_2)$, whilst such a crossing is called positive, if in the crossing point the n-dimensional indicatrix composed of $(-1)^h \times$ the positive indicatrix of $a(\varrho_1)$ and the positive indicatrix of $na(\varrho_2)$ resp. $na(\varrho_2)$ is positive, then for the above-mentioned variation of $a(\varrho_2)$ the algebraical sum of the number of positive and the number of negative crossings of $a(\varrho_1)$ and $na(\varrho_2)$ increases likewise by a unit each time that $a(\varrho_1)$ is traversed by $a(\varrho_2)$ positively.

From this ensues that the looping coefficient of ϱ_1 with respect to ϱ_2 can also be defined as the algebraical sum $\omega\{a(\varrho_1), na(\varrho_2)\}$ of the number of positive and the number of negative crossings of an arbitrary simplicial approximation $a(\varrho_1)$ of ϱ_1 and an arbitrary $(n-h)$-dimen-

---

[[2]][[5]] 1) Mathem. Annalen 71, p. 102 and p. 316.

[[5]] 2) ibid., p. 317.

[[5]] 3) ibid., p. 316.

[[518]]

sional net fragment $na(\varrho_2)$, *limited by an arbitrary simplicial approximation* $a(\varrho_2)$ *of* ϱ_2.

That this algebraical sum is unequivocally determined by ϱ_1 and ϱ_2, can also be shown by a direct proof.

If, namely, we have two different net fragments $na(\varrho_2)$ and $n'a(\varrho_2)$, limited by the same simplicial approximation $a(\varrho_2)$, and if we represent the net fragment obtained out of $n'a(\varrho_2)$ by inversion of the indicatrix, by $n''a(\varrho_2)$, then $na(\varrho_2)$ and $n''a(\varrho_2)$ form together a *two-sided closed net* [1]), so that $\omega\{a(\varrho_1), na(\varrho_2) + n''a(\varrho_2)\}$ must be equal to zero, thus $\omega\{a(\varrho_1), n'a(\varrho_2)\} = \omega\{a(\varrho_1), na(\varrho_2)\}$.

If farthermore we have two different simplicial approximations $a(\varrho_1)$ and $a'(\varrho_1)$ corresponding to one and the same mode of measurement of ϱ_1, two different simplicial approximations $a(\varrho_2)$ and $a'(\varrho_2)$ corresponding to one and the same mode of measurement of ϱ_2, and two two-sided net fragments $na(\varrho_2)$ and $na'(\varrho_2)$, which, leaving their rims out of consideration, have the same base points, then for continuous transformation of $a'(\varrho_1)$ into $a(\varrho_1)$ we have:

$$\omega\{a'(\varrho_1), na(\varrho_2)\} = \omega\{a(\varrho_1), na(\varrho_2)\},$$

and for continous transformation of $a'(\varrho_2)$ into $a(\varrho_2)$:

$$\omega\{a'(\varrho_1), na'(\varrho_2)\} = \omega\{a'(\varrho_1), na(\varrho_2)\}.$$

If finally we have two different modes of measurement μ_1 and μ'_1 with corresponding indicatrices of ϱ_1, and two different modes of measurement μ_2 and μ'_2 with corresponding indicatrices of ϱ_2, then on account of the theorem, that a continuous one-one correspondence between two closed spaces possesses the degree ± 1 [2]), there exists a simplicial approximation $a'(\varrho_1)$ corresponding to μ'_1, covering a simplicial approximation $a(\varrho_1)$ corresponding to μ_1 with the degree *one*, and a simplicial approximation $a'(\varrho_2)$ corresponding to μ'_2, covering a simplicial approximation $a(\varrho_2)$ corresponding to μ_2 with the degree *one*, from which ensues immediately:

$$\omega\{a'(\varrho_1), na'(\varrho_2)\} = \omega\{a(\varrho_1), na(\varrho_2)\},$$

with which the proof that the abovementioned algebraical sum depends exclusively on ϱ_1 and ϱ_2, is completed.

In close connection with the looping coefficient is the notion of *enlaced spaces* recently introduced by LEBESGUE [3]). Two spaces enlaced [7]

[1]) ibid., p. 316. [5]

[2]) ibid., p. 324 and p. 598. [5][6]

[3]) C. R., 27 mars 1911.

according to LEBESGUE, possess in our terminology with respect to each other an *odd* looping coefficient. But to justify his definition LEBESGUE has neglected to prove that the being enlaced or not of two spaces is independent of the manner in which they are measured, which property is established only by the above reasonings.

The developments joined by LEBESGUE to his definition can meanwhile be made entirely rigorous by replacing the notion *"enlaced"* by: *"enlaced for a definite mode of measurement."*

NOTES

⟦1⟧ The present paper is related to Lebesgue 1911 B where the concept of *enlacement* (mod 2) was introduced. It seems that Brouwer discovered this concept independently (see the sketch of a letter to R. Baire in 1911 C ⟦1⟧). Lebesgue counts crossings of the polyhedra if one of them is shrunk into a point. Brouwer used the degree of the mapping of the cartesian product into the field of the normed joining vectors – both inefficient devices in comparison with the homological tools which were developed earlier by H. Poincaré 1895 but had to wait for J. W. Alexander 1922 to be combined with Brouwer's methods. This neglect of Brouwer's inevitably led to clumsiness of exposition. For the same reason the present paper failed to exert a direct influence. After J. W. Alexander, 1922, a unified duality theory of intersecting and linking coefficient was constructed by L. Pontrjagin 1931.
⟦2⟧ Brouwer 1911 D.
⟦3⟧ J. Hadamard 1910.
⟦4⟧ Schoute 1905.
⟦5⟧ Brouwer 1911 F.
⟦6⟧ Brouwer 1911 N.
⟦7⟧ Lebesgue 1911 B.

(June 25, 1912).

GÉOMÉTRIE. — *Sur l'invariance de la courbe fermée.* Note de M. **L.-E.-J. Brouwer.** 1912 F ⟦1⟧

D'après la définition de M. Schœnflies, on entend par une *courbe fermée* un ensemble plan, borné, parfait, d'un seul tenant, déterminant dans le plan deux régions, et identique à la frontière commune de ces deux régions.

La propriété d'être une courbe fermée est un *invariant de l'analysis situs*, autrement dit toute image plane, biunivoque et continue d'une courbe fermée est encore une courbe fermée, généralisation du théorème de M. Jordan affirmant que toute image plane, biunivoque et continue du cercle est une courbe fermée.

Le théorème de l'invariance de la courbe fermée a été énoncé par M. Schœnflies, mais il n'en existait pas encore de démonstration, lacune que j'ai comblée moyennant le raisonnement suivant, dont je me borne à indiquer les principes :

Un ensemble fini de points, arrangé dans un ordre cyclique, sera appelé une *chaîne*. Une chaîne pour laquelle le maximum des distances de deux points successifs est inférieur à ε, sera appelée une *chaîne*-ε.

Par une *modification*-η d'une chaîne j'entendrai : 1° un déplacement $< \eta$ d'un de ses points, transformant la chaîne en une chaîne-η; 2° l'intercalation d'un nouveau point, transformant la chaîne en une chaîne-η.

Soit h un nombre fini et positif, e un ensemble borné, parfait, d'un seul tenant et admettant la propriété suivante : pour chaque ε il existe une telle quantité η s'évanouissant avec ε et h telles *chaînes fondamentales* situées dans e, qu'on peut transformer toute chaîne-ε située dans e, moyennant un nombre fini de modifications-η et sans la faire quitter l'ensemble e, en une *chaîne canonique*, se composant d'un nombre fini de chaînes fondamentales. Nous dirons que l'ensemble *e admet une base h-uple de cyclose*. Évidemment la propriété dont il s'agit ici est un *invariant de l'analysis situs*.

Un ensemble n'admettant pas de base $(h - 1)$-uple, mais admettant une base h-uple de cyclose, sera dit *admettre une base réduite h-uple de cyclose*.

On démontre la proposition suivante :

Si un ensemble borné, parfait, d'un seul tenant, détermine dans le plan un nombre fini h + 1 de régions, il admet une base réduite h-uple de cyclose. S'il détermine dans le plan une infinité de régions, il n'admet pas de base finie de cyclose.

Il s'ensuit que *deux ensembles de régions, déterminés dans le plan l'un par un ensemble borné, parfait, d'un seul tenant e, l'autre par une image biunivoque et continue de e, possèdent le même nombre cardinal.*

L'invariance de la courbe fermée forme un cas spécial de ce dernier théorème.

(*Comptes rendus*, t. 154, p. 862, séance du 1er avril 1912.)

NOTE

⟦1⟧ Sketch of 1912 L.

Beweis der Invarianz der geschlossenen Kurve. 1912 L

Von

L. E. J. Brouwer in Amsterdam.

Zweck der vorliegenden Arbeit ist, die in der Schoenfliesschen Theorie der ebenen gestaltlichen Invarianten noch vorhandene Lücke, auf welche ich bei einer früheren Gelegenheit hingewiesen habe*), auszufüllen. [[1]]

§ 1.

Sei $\mathfrak{g}$ ein geschränktes, $(h+1)$-fach zusammenhängendes, ebenes Gebiet, P ein willkürlicher Punkt von $\mathfrak{g}$. Alsdann können wir h, als *Fundamentalkurven* zu bezeichnende, einfache geschlossene Kurven $c_1, c_2, \cdots, c_h$ durch P legen, welche sich nur in P treffen und die Eigenschaft besitzen, daß jede in $\mathfrak{g}$ verlaufende geschlossene stetige Kurve σ sich mittels stetiger Abänderung innerhalb $\mathfrak{g}$ in eine aus einer endlichen Zahl von Kurven c_ν zusammengesetzte „*kanonische stetige Kurve*“ φ überführen läßt.**)

Unter einer *Kette* verstehen wir im folgenden eine zyklisch geordnete, endliche Punktmenge, unter einer ε-*Kette* eine solche Kette, für welche die Entfernung je zweier aufeinanderfolgender Punkte kleiner als ε ist.

Unter einer ε'-*Abänderung* einer Kette $\varkappa$ verstehen wir erstens eine solche Verrückung $< \varepsilon'$ eines Punktes von $\varkappa$, durch welche $\varkappa$ in eine ε'-Kette übergeht, zweitens eine solche Zwischenfügung eines neuen Punktes zwischen zwei aufeinanderfolgende Punkte von $\varkappa$, durch welche $\varkappa$ in eine ε'-Kette übergeht.

*) Vgl. Math. Ann. 68, S. 434, 444. [[2]]

**) Diese stetige Abänderung ist dahin zu verstehen, daß ein von zwei konzentrischen Kreisen k_1 und k_2 begrenztes ebenes Ringgebiet sich derart eindeutig und stetig auf $\mathfrak{g}$ abbilden läßt, daß einerseits σ und k_1, andererseits φ und k_2 einander entsprechen.

Wenn zwei aufeinanderfolgende Punkte einer Kette zusammenfallen, so werden sie als ein einziger Punkt betrachtet. Bei einer ε'-Abänderung bleibt mithin die Anzahl der Punkte der Kette entweder ungeändert, oder sie nimmt um eine Einheit zu, oder sie nimmt um eine Einheit ab.

§ 2.

Sei π eine geschränkte, zusammenhängende, perfekte Punktmenge, welche in der Ebene eine endliche Zahl $h+1$ von Gebieten bestimmt. In jedem dieser Gebiete konstruieren wir das die Punktmenge π im Abstande ε approximierende Polygon*), bezeichnen das von diesen Polygonen begrenzte $(h+1)$-fach zusammenhängende Gebiet mit $\mathfrak{g}_\varepsilon$, das Maximum der Entfernungen, welche die Punkte von $\mathfrak{g}_\varepsilon$ von π besitzen, mit ε_2, wählen in $\mathfrak{g}_\varepsilon$ einen Punkt P beliebig aus, konstruieren durch P in $\mathfrak{g}_\varepsilon$ ein System von h Fundamentalkurven $c_1, c_2, \cdots, c_h$, bestimmen eine Größe $\varepsilon_1 < \varepsilon$ mit der Eigenschaft, daß je zwei auf derselben Kurve c_ν liegende ε_1-Ketten durch eine Folge von ε-Abänderungen auf dieser c_ν ineinander übergeführt werden können, konstruieren auf jeder c_ν eine den Punkt P enthaltende, als *Fundamentalkette* zu bezeichnende ε_1-Kette $\varkappa_\nu$ und nennen eine aus einer endlichen Zahl von Ketten $\varkappa_\nu$, deren je zwei aufeinanderfolgende im Punkte P zusammenhängen, zusammengesetzte ε_1-Kette eine *kanonische Kette*.

§ 3.

Sei $\varkappa$ eine willkürliche in π liegende ε-Kette. Indem wir je zwei aufeinanderfolgende Punkte von $\varkappa$ durch ein Geradensegment verbinden, konstruieren wir einen in $\mathfrak{g}_\varepsilon$ enthaltenen geschlossenen Streckenzug σ, den wir mittels einer stetigen Abänderung α innerhalb $\mathfrak{g}_\varepsilon$ in eine kanonische stetige Kurve φ überführen. Diejenigen Punkte von σ, welche durch α in den Punkt P übergehen, fügen wir in die Kette $\varkappa$ ein. Sodann fügen wir so viele weitere Punkte von σ in die Kette $\varkappa$ ein, daß die entstehende Kette während des ganzen Verlaufes der stetigen Abänderung α eine ε_1-Kette bleibt.

Die vorstehende Konstruktion zeigt, daß h „*Fundamentalketten*" in $\mathfrak{g}_\varepsilon$ existieren mit der Eigenschaft, daß es zu einer willkürlichen in π liegenden ε-Kette $\varkappa$ eine endliche Kettenfolge $\varkappa, \varkappa', \varkappa'', \cdots, \varkappa^{(n)}$ gibt, deren letztes Element eine kanonische Kette ist, während je zwei aufeinanderfolgende Elemente durch eine ε-Abänderung ineinander übergehen.

[[3]] *) Vgl. Schoenflies, Bericht über die Lehre von den Punktmannigfaltigkeiten, II, S. 114.

[[524]]

§ 4.

Wir bezeichnen $\varepsilon + 2\varepsilon_2$ mit ε' und ersetzen jeden Punkt der Ketten $\varkappa', \varkappa'', \cdots, \varkappa^{(n)}$ durch einen ihm möglichst nahe liegenden Punkt von π. In dieser Weise entsteht eine Folge $\varkappa_i, \varkappa_i', \varkappa_i'', \cdots, \varkappa_i^{(n)}$ von ε'-Ketten, in welcher je zwei aufeinanderfolgende Elemente durch eine ε'-Abänderung ineinander übergehen.

Mithin lassen sich zu jedem ε eine solche mit ε unter jede Grenze herabsinkende Größe ε' und h solche in π liegende „*Fundamentalketten*" mit einem gemeinsamen Punkte P_i bestimmen, daß eine willkürliche in π liegende ε-Kette, mittels einer endlichen Folge von ε'-Abänderungen auf π, in eine aus einer endlichen Zahl von Fundamentalketten, deren je zwei aufeinanderfolgende im Punkte P_i zusammenhängen, zusammengesetzte *kanonische Kette* übergeführt werden kann.

Wir werden kurz sagen: „*Die Punktmenge π besitzt eine h-fache Basis der Zyklosis*". ⟦4⟧

§ 5.

Sei h eine endliche Zahl und ϱ eine geschränkte, zusammenhängende, perfekte Punktmenge, welche in der Ebene *mehr als* $h + 1$ Gebiete bestimmt. *Wir behaupten, daß ϱ keine h-fache Basis der Zyklosis besitzen kann.*

Wählen wir nämlich $h + 1$ von ϱ in der Ebene bestimmte *endliche* Gebiete $g_1, g_2, \cdots, g_{h+1}$ und in jedem der g_ν einen Punkt P_ν beliebig aus, so bleibt für hinreichend kleines ε' die *Ordnung**) der P_ν in bezug auf den zu einer in ϱ liegenden ε'-Kette gehörigen geschlossenen Streckenzug bei einer willkürlichen ε'-Abänderung dieser Kette auf ϱ ungeändert. Die zu einem hinreichend kleinen ε gehörigen Fundamentalketten müssen mithin durch ihre Komposition für die Ordnungen der P_ν *alle möglichen Systeme von $h + 1$ ganzen Zahlen* liefern können, wozu es wenigstens $h + 1$ Fundamentalketten geben muß.

Hiermit sind wir nun zu folgendem Ergebnis gelangt:

Eine geschränkte, zusammenhängende, perfekte Punktmenge, welche in der Ebene eine endliche Zahl $h + 1$ von Gebieten bestimmt, besitzt eine h-fache, nicht aber eine $(h - 1)$-fache Basis der Zyklosis.

Eine geschränkte, zusammenhängende, perfekte Punktmenge, welche in der Ebene unendlichviele Gebiete bestimmt, besitzt keine endliche Basis der Zyklosis.

---

*) Vgl. z. B. J. Tannery, „Introduction à la théorie des fonctions d'une variable", II, S. 438. ⟦5⟧

§ 6.

Wenn eine geschränkte, zusammenhängende, perfekte Punktmenge π eine h-fache Basis der Zyklosis besitzt, so bleibt diese Eigenschaft, wie aus ihrer Definition sofort hervorgeht, für ein willkürliches eineindeutiges und stetiges Bild von π bestehen. Aus den am Schlusse des § 5 formulierten Theoremen folgt somit unmittelbar folgender

Satz. *Die Gebietsmengen, die von zwei ebenen, geschränkten, zusammenhängenden, perfekten, einander eineindeutig und stetig entsprechenden Punktmengen in der Ebene bestimmt sind, besitzen dieselbe Kardinalzahl.*

In diesem Satze ist die Invarianz der geschlossenen Kurve enthalten.

NOTES

⟦1⟧ A. Schoenflies 1908 B, p. 119 defined a closed curve as a closed bounded subset of the plane, which divides the plane into two domains in such a way that each of its points is a boundary point of both domains. In 1910 C Brouwer had posed the problem of proving the topological invariance of this concept. See also Y15 where it is mentioned. Here Brouwer proves even more – the invariance of the number of domains determined by a bounded connected closed planar point set.

This work was continued by P. Alexandroff 1926, and from the concept of Zyklosis, though nearer to homotopy, sprang L. Vietoris', 1927, 1929, definition of Betti numbers of all dimensions for compact spaces, which was further developed by P. Alexandroff 1928 A.

⟦2⟧ Brouwer 1910 C.

⟦3⟧ A. Schoenflies 1908 A.

⟦4⟧ The term 'Zyklosis' probably stems from J. B. Listing 1847.

⟦5⟧ J. Hadamard 1910.

Mathematics. "*Continuous one-one transformations of surfaces in themselves*". (5th communication [1])). By Prof. L. E. J. BROUWER. 1912 K2 ⟦1⟧

In CRELLE's Journal, vol. 127, p. 186 Prof. P. BOHL has enunciated without proof the following theorem proved by me (as a particular case of a more general theorem) in vol. 71 of the Mathematische Annalen (compare there page 114): ⟦3⟧ ⟦4⟧

"*Werden die Punkte einer Kugeloberfläche wieder in Punkte der Kugeloberfläche übergeführt und geschieht diese Ueberführung durch stetige Bewegung, welche den Mittelpunkt nicht berührt, so kehrt mindestens ein Punkt in seine frühere Lage zurück. Unter einer stetigen Bewegung ist hier eine Bewegung verstanden, bei welcher die rechtwinkligen Koordinaten stetige Funktionen der Zeit und der Anfangswerte sind.*"

Now I shall show here in the first place that the theorem enunciated and proved in the first communication on this subject [2]), i. e. that each continuous one-one transformation with invariant indicatrix of a sphere in itself possesses at least one invariant point, may be considered as a particular case of the quoted theorem of BOHL [3]). To that end I shall establish the following theorem: ⟦7⟧

"*Any continuous one-one transformation α with invariant indicatrix of a sphere in itself can be transformed by a continuous modification* [4]) *into identity*" [5]).

In order to prove this property we choose in the sphere two opposite points P_1 and P_2 determining a net of circles of longitude and latitude and passing by α into Q_1 and Q_2. By means of a continuous series τ of conform transformations of the sphere in itself we can transform Q_1 and Q_2 into P_1 and P_2. Let c be an arbitrary circle of latitude, described in such a sense that P_1 possesses with respect to c the *order* [6]) $+1$, and c' the image of c for $\alpha\tau$, then P_1 possesses also with respect to c' the order $+1$.

---

1) Compare these Proceedings XI, p. 788; XII, p. 286; XIII, p. 767; XIV, p 300 (1909—1911). ⟦2⟧

2) These Proceedings XI (1909), p. 797. ⟦5⟧

3) This I indicated already shortly Mathem. Ann. 71 (1911), p. 325, footnote *). ⟦6⟧

4) Under a continuous modification of a univalent continuous transformation we understand in the following always the construction of a continuous series of univalent continuous transformations, i. e. a series of transformations depending in such a manner on a parameter, that the position of an arbitrary point is a continuous function of its initial position and the parameter.

5) That this theorem wants a proof is shown by the fact that e.g. for a torus it does not hold.

6) Compare e. g. J. TANNERY, "*Introduction à la théorie des fonctions d'une variable*", vol. II, p. 438. ⟦8⟧

Let P be an arbitrary point coinciding neither with P_1 nor with P_2 and passing by $\alpha\tau$ into R, and let Q be the point corresponding in latitude with P and in longitude with R. Then by transforming the different points R continuously and uniformly along circles of longitude into the corresponding points Q we define a continuous series ϱ of univalent continuous transformations of the sphere in itself with the property that of none of the points R the path passes through P_1 or P_2. So an arbitrary curve c' is transformed by ϱ into a curve c'', with respect to which P_1 possesses likewise the order $+1$, so that c'' covers the corresponding circle of latitude c with the *degree*[1]) $+1$.

From this ensues that an arc of a circle of latitude connecting an arbitrary point P with the corresponding point Q defines unequivocally for any point P an arc of circle of latitude PQ whose variation with P is uniformly continuous, so that it is possible to construct a continuous series ϱ' of univalent continuous transformations of the sphere in itself, transforming each point Q into the corresponding point P, and thereby the transformation $\alpha\tau\varrho$ into identity. But then $\tau\varrho\varrho'$ is the looked out for continuous series of transformations, transforming α into identity.

⟦9⟧ We shall say that two transformations *belong to the same class*, if they can be transformed continuously into each other. We then can state the theorem proved just now in the following form:

THEOREM 1. *All continuous one-one transformations with invariant indicatrix of a sphere in itself belong to the same class.*

As the continuous one-one transformations with invariant indicatrix form a special case of the univalent continuous transformations of degree $+1$[2]), the question arises whether perhaps theorem 1 is a special case of the more general property that all the univalent continuous transformations of the same degree of a sphere in itself belong to the same class. We shall see that this is indeed the case; we shall namely show that any univalent continuous representation of degree zero of a sphere μ on a sphere μ' can be transformed by continuous modification into a representation of μ in a single point of μ', and that any univalent continuous representation of degree $n \gtrless 0$ of a sphere μ on a sphere μ' can be transformed by continuous modification into a *canonical representation of degree n*, i. e. into a representation for which $n-1$ non intersecting simple closed curves of μ are each represented in a single point of μ', whilst the n

⟦4⟧ [1]) Mathem. Ann. 71 (1911), p. 106.

[2]) Mathem. Ann. 71 (1911), p. 106 and 324.

domains determined by these curves are each submitted to a continuous *one-one* representation on μ', and that either all with degree $+1$ or all with degree -1. By means of an indefinitely small modification a canonical representation can be transformed into a simply ramified *Riemann representation,* i.e. into a representation which in the sense of analysis situs is identical to a simply ramified representation of a Riemann surface with n sheets and of genus zero on the complex plane. That all simply ramified Riemann representations belong to the same class, follows, according to a remark made by KLEIN [1]), out of a known theorem of LÜROTH—CLEBSCH.

In order to transform an arbitrarily given univalent continuous representation α of μ on μ' into a representation in a single point, resp. into a canonical representation, we first modify it continuously into a *simplicial approximation* [2]) α', to which we have imparted, by means of eventual subdivisions of the corresponding simplicial divisions of μ and μ', the property that any base triangle of μ covers in μ' either a single base triangle, or a single base side, or a single base point; we then investigate the possibility of finding two base triangles of μ, one positively and the other negatively represented, allowing that we pass from the one to the other by transversing exclusively base sides of μ *not* represented in a single point. If this be the case, μ will possess a positively represented base triangle t_1 and a negatively represented one t_n both represented in the *same* fundamental triangle t' of μ', allowing us to pass from the one to the other by transversing exclusively such base sides of μ, as are represented in the *same* side s_1 of t'. The base triangles $t_2, t_3, \ldots, t_{n-1}$ of μ crossed on this way leading from t_1 to t_n are then also represented entirely in s_1.

Let s_2 and s_3 be the other two sides of t'; by a continuous modification of α' and a suitable farther subdivision of $t_1, t_2, \ldots, t_{n-1}, t_n$, we can generate a representation α'' for which all the triangles $t_1, t_2, \ldots, t_{n-1}, t_n$ are represented entirely in s_2 and s_3, and which possesses still the same property as α', viz. that any base triangle of μ covers in μ' either a single base triangle, or a single base side, or a single base point.

In the same manner as we transformed α' into α'', we transform α'' if possible into α''', and we continue this process until after a

---

1) Compare: "*Ueber Riemann's Theorie der algebraischen Funktionen und ihrer Integrale*". Leipzig, 1882. [[10]]

2) Mathem. Ann. 71 (1911), p. 102. [[4]]

finite number of steps we have reached a representation $a^{(p)}$ no more allowing a suchlike modification.

We now construct on μ all those polygons formed by base sides belonging to $a^{(p)}$ which are represented by $a^{(p)}$ *in a single point.* These polygons divide μ into a finite number of domains $g_1, g_2, \ldots, g_k$. Each domain g_ν, which by $a^{(p)}$ is not represented nowhere dense, admits the property that there is no polygon lying entirely within it or partly within it and partly on its boundary, which is represented by $a^{(p)}$ in a single point [1]). Any two base triangles belonging to the same domain g_ν can be connected within g_ν by a path transversing only base sides *not* represented in a single point, so that of the base triangles of g_ν either no one is represented negatively or no one positively.

As each coherent part of the boundary of g_ν is represented on μ' by a single point, μ' is covered by the image of g_ν with a certain *degree* which we will suppose to be positive. Then there are no negative image triangles, but there are in general singular image triangles with two coinciding vertices.

By considering each coherent part $\gamma_{\nu\tau}$ of the boundary of g_ν as a single point $P_{\nu\tau}$, g_ν is transformed into a sphere sp_ν, and we can deduce a simplicial division of sp_ν from the simplicial division of g_ν belonging to $a^{(p)}$, by bisecting all those base sides of g_ν which touch the boundary but do not lie in the boundary, dividing by means of these bisecting points each base triangle one side of which lies in the boundary, into a triangle and a trapezium to be considered as a base triangle of sp_ν, and dividing those of the remaining base triangles of which sides have been bisected, into new base triangles corresponding to those bisecting points. The simplicial representation $a^{(p)}$ of g_ν on μ' is then at the same time a simplicial representation of sp_ν on μ', whilst by suitable subdivisions of the simplicial divisions

---

[1]) For, as this property holds for polygons formed by base sides, any base triangle of g_ν possesses at most one base side represented in a single point. Therefore each broken line, lying in a single base triangle and not in a single base side, which is represented in a single point, must necessarily lie entirely in a straight line segment connecting two points of the circumference not coinciding with vertices. So a polygon represented in a single point must either consist exclusively of base sides, or it can transverse only such base sides as are represented in one and the same base side of μ'. In the latter case however the series of the base triangles of μ crossed in this way would have to be represented in that selfsame base side of μ', so that each of the two limiting polygons of this series (of which at most one can be illusory) would be a polygon formed by base sides and represented in a single point of μ'.

of sp_ν and μ' we can effectuate that any base triangle of sp_ν covers in μ' either a single base triangle, or a single base side.

By choosing one of the base sides of sp_ν represented by $a^{(\nu)}$ in a single point, and considering it as a single point and accordingly the two base triangles adjacent to it as line segments, sp_ν passes into an other sphere sp_ν' represented likewise simplicially by $a^{(\nu)}$. In the same way we deduce from sp_ν' an other sphere sp_ν'' if this be possible, and we continue this process until after a finite number of steps we obtain a sphere $sp_\nu^{(r)}$ no more possessing for $a^{(\nu)}$ any singular image triangle.

Let us denote by B and D the two base points of $sp_\nu^{(m-1)}$ identified for $sp_\nu^{(m)}$ and by a and c the two base triangles of $sp_\nu^{(m-1)}$ contracted into line segments for $sp_\nu^{(m)}$. Then the triangles a and c have either only the side BD in common, or moreover a second side, which we may assume to contain the vertex B.

In the first case we represent the third vertex of a, resp. c, by A, resp. C, and the domain covered by a and c together, by d. At least one of the base points B and D, say D, *does not* coincide with a point $P_{\nu\tau}$. We then connect in $sp_\nu^{(m-1)}$ outside d the points A and C by an arc of simple curve β situated in the vicinity of the broken line ADC, and we represent the domain included between β and the broken line ADC, by d'. By means of a continuous series of continuous one-one transformations leaving the points of β invariant and transforming each point of AB and BC into points coinciding with it on $sp_\nu^{(m)}$, we can reduce the domain $d+d'$ with its boundary continuously into the domain d' with its boundary. If we represent by $a_\nu^{(m)}$ an arbitrary univalent continuous representation of $sp_\nu^{(m)}$ on μ', then to the continuous reduction of $d+d'$ to d' corresponds a continuous series of univalent continuous representations of $sp_\nu^{(m-1)}$ on $sp_\nu^{(m)}$ transforming the representation obtained by the identification of B and D, into a continuous one-one correspondence ${}_m a_{m-1}$ in which the points $P_{\nu\tau}$ correspond to themselves, thus also a continuous series of univalent continuous representations of $sp_\nu^{(m-1)}$ on μ', leaving invariant the images of the points $P_{\nu\tau}$, and transforming $a_\nu^{(m)}$ considered as a representation of $sp_\nu^{(m-1)}$ on μ', into that representation $a_\nu^{(m-1)}$ of $sp_\nu^{(m-1)}$ on μ', which follows from $a_\nu^{(m)}$ by means of ${}_m a_{m-1}$.

In the second case we represent the third vertex of a and c by F, choose on the side DF of a, the side DF of c, and the common side BF successively three such points A, C, and G, as in passing from $sp_\nu^{(m-1)}$ to $sp_\nu^{(m)}$ are brought to coincidence, connect A within a rectilinearly with B and G, C within c rectilinearly with B and G, and apply the operation of the first case to the pairs of fundamental triangles ABD and CBD; BCG and FCG; BAG and FAG successively [1]).

By applying this operation successively to $sp_\nu^{(r)}$, $sp_\nu^{(r-1)}, \ldots, sp''_\nu$ and sp'_ν we experience that the representation $\alpha^{(\nu)}$ of sp_ν on μ' can be transformed by a continuous modification leaving the images of the points $P_{\nu\tau}$ invariant, into a representation α_h of sp_ν on μ', which follows from $\alpha^{(\nu)}$ by means of a continuous one-one correspondence between $sp_\nu^{(r)}$ and sp_ν. As $sp_\nu^{(r)}$ can be divided into elements each of which is submitted for $\alpha^{(\nu)}$ to a one-one representation of degree $+1$ on a base triangle of μ', it is clear that sp_ν can be divided into elements each of which is submitted for α_h to a one-one representation of degree $+1$ on a base triangle of μ'. The representation α_h of sp_ν on μ' is therefore a *Riemann representation*, and eventually it may be transformed by an indefinitely small modification leaving the images of the points $P_{\nu\tau}$ invariant, into a simply ramified Riemann representation.

By executing this process of modification for all the values of ν for which it is applicable we arrive at a representation α_e being for any of the spheres $sp_1, sp_2, \ldots, sp_k$ either a simply ramified, positive or negative Riemann representation, or a representation nowhere dense.

In each domain g_ν we approximate the boundary parts $\gamma_{\nu\tau}$ by simple closed curves $\varkappa_{\nu\tau}$ not intersecting each other. Each $\varkappa_{\nu\tau}$ includes with the corresponding $\gamma_{\nu\tau}$ a domain $g'_{\nu\tau}$, and the $\varkappa_{\nu\tau}$ situated in the same domain g_ν include together a domain g'_ν. The domains $g'_{\nu\tau}$ belonging to the same τ form together a domain g''_τ. By means of a continuous series of univalent continuous representations of g_ν on sp_ν we can transform identity into a representation which for g'_ν with the exclusion of its boundaries is a continuous one-one representation on sp_ν, whilst $\varkappa_{\nu\tau}$ and $g'_{\nu\tau}$ are represented in $P_{\nu\tau}$. By doing this for all values of ν we transform α_e into a representation α_l being for each of the domains g'_ν and g''_τ after contraction of its rims into points either a simply ramified, positive or negative Riemann representation, or a representation nowhere dense.

The domains g'_ν and g''_τ, which will be represented henceforth by $\mathfrak{g}_1, \mathfrak{g}_2, \ldots, \mathfrak{g}_u$, are determined on μ by a finite number of simple closed curves not intersecting each other.

---

1) If we dropped the condition of the invariancy of the images of the points $P_{\nu\tau}$ (introduced only for the sake of clearness), this second case might have been treated of course in the same manner as the first.

We choose an arbitrary domain g_ν, and suppose in the first place that α_l is for the sphere σ_ν into which g_ν is transformed by contraction of its rims into points, a simply ramified Riemann representation. We then draw on μ' a system of ramification sections belonging to this representation and corresponding to a system of simple closed "ramification curves" on σ_ν. By first leaving the ramification sections on μ' invariant and varying eventually continuously the ramification curves on σ_ν in such a manner that after that they contain no more a point corresponding to a rim of g_ν, and then leaving the ramification curves on σ_ν invariant and contracting the ramification sections on μ' continuously into points, we can transform the representation of σ_ν on μ' determined by α_l continuously into a canonical representation. During this continuous modification the points representing the rims of g_ν vary also in general. Let i_τ be such a rim and $g_{\nu\tau}$ the residual domain of g_ν on μ determined by i_τ. We then can follow the continuous variation of the image point of i_τ by a continuous series of continuous one-one transformations of μ' in itself to which corresponds a continuous modification of the representation of $g_{\nu\tau}$ on μ' determined by α_l. By applying this modification to the representations of all the residual domains of g_ν we generate a representation α'_l of μ on μ' into which α_l can be transformed continuously, and which is a canonical representation for σ.

In the second place we suppose α_l to be for σ_ν a representation nowhere dense. Then we can modify the representation of σ_ν on μ' determined by α_l into a representation in a single point. The variation of the image points of the rims of g_ν implied by this modification, can be followed once more in the way described above by a continuous modification of the representation of the residual domains of g_ν, furnishing us with a representation α'_l of μ on μ' into which α_l can be transformed continuously, and which represents σ_ν in a single point.

By executing this operation for all values of ν successively, we get a representation $\alpha_l^{(+)}$ of μ on μ', into which α_l can be transformed continuously, and which represents each of the domains $g_1, g_2, \ldots g_u$ either after contraction of the rims into points canonically, or in a single point. The sphere μ is now divided by a finite number of non intersecting simple closed curves into a finite number of domains $d_1, d_2, \ldots, d_w$ in such a way that for $\alpha_l^{(+)}$ each of these domains is submitted either after contraction of the rims into points to a continuous one-one representation, or to a representation in a single point. Thus the degree of these representations is 0, +1, or

— 1, according to which we distinguish domains of the first, the second, and the third kind.

If for the representation $\alpha_l^{(\psi)}$, which may be denoted henceforth by α_f, all domains d_ν are of the first kind, we have attained our aim ; for then we have transformed α continuously into a representation of μ in a single point of μ'. So we further confine ourselves to the case that among the d_ν there are domains of the second or of the third kind, and we will suppose that there occur moreover domains of the first kind. Then there is certainly a domain d_η of the first kind adjacent to a domain d_ϑ of the second or third kind. The domain formed by d_η and d_ϑ together, may be indicated by $d_{\eta\vartheta}$, the sphere deduced from $d_{\eta\vartheta}$ by contraction of its rims into points, by $\delta_{\eta\vartheta}$. We then can modify the univalent continuous representation of $\delta_{\eta\vartheta}$ on μ' determined by α_f continuously into a continuous one-one representation of $\delta_{\eta\vartheta}$ on μ'. The variation of the image points of those rims of $d_{\eta\vartheta}$ which originate from d_η necessarily implied by this modification, can once more be followed in the manner described above by a continuous modification of the representation determined by α_f of those residual domains of $d_{\eta\vartheta}$ which originate from d_η, furnishing us with a representation α'_f distinguishing itself thereby from α_f that a domain of the first kind and a domain of the second (resp. third) kind have been united into a single domain of the second (resp. third) kind.

By repeating this operation as many times as possible we arrive after a finite number of steps at a representation $\alpha_f^{(\tau)}$, distinguishing itself thereby from α_f that all the domains of the first kind have been absorbed by domains of the second and of the third kind.

If there are for the representation $\alpha_f^{(\tau)}$, which may be denoted henceforth by α_q domains of the second as well as of the third kind, we consider a domain d_π of the second kind separated by a simple closed curve $i_{\pi\rho}$ from a domain d_ρ of the third kind, and we represent the domain formed by d_π and d_ρ together, by $d_{\pi\rho}$, and the sphere deduced from $d_{\pi\rho}$ by contraction of its rims into points, by $\delta_{\pi\rho}$. Moreover we represent by P_1 the image point of $i_{\pi\rho}$ for α_q, by P_2 the opposite point of P_1 on μ', and we modify the representation of $\delta_{\pi\rho}$ determined by α_q into a representation of $\delta_{\pi\rho}$ in the single point P_2, by diminishing the polar distances measured from P_2 continuously and proportionally to each other to zero. The variation of the image points of the rims of $d_{\pi\rho}$ necessarily implied by this

[[534]]

modification, can be followed in the manner described above by a continuous modification of the representation of the residual domains of $d_{\pi\rho}$ determined by a_q, furnishing us with a representation $a_{q'}$ distinguishing itself thereby from a_q that a domain of the second and one of the third kind have been united into a single domain of the first kind; this domain however, if it does not occupy the whole sphere μ, can be absorbed in the manner described above by an adjacent domain of the second or of the third kind, by which process $a_{q'}$ passes continuously into a representation a'_q, distinguishing itself thereby from a_q that a domain of the second and one of the third kind have been absorbed together by a domain of the second resp. of the third kind.

By repeating this operation as many times as possible we arrive after a finite number of steps at a representation $a_q^{(\zeta)}$ for which the domains d_ρ are either all of the second or all of the third kind. So this representation is a *canonical* one, and we have proved:

THEOREM 2. *All univalent continuous transformations of the same degree of a sphere in itself belong to the same class.* ⟦11⟧

A proof of the inverse theorem has been given Mathem. Ann. 71, p. 105.

In carrying out the ideas sketched in the second communication on this subject [1]) I experienced that in some points of the course of demonstration indicated there, still a tacit part is played by the Schoenfliesian theory of domain boundaries criticized by me [2]), so that the theorems 1 and 2 formulated p. 295 and likewise the "general translation theorem" founded upon them and enunciated without proof Mathem. Ann. 69, p. 178 and 179, cannot be considered as proved [3]), and a question of the highest importance is still to be decided here. ⟦12⟧

The "plane translation theorem" stated at the end of the second communication (p. 297) and likewise Mathem. Ann. 69, p. 179 and 180, has meanwhile been proved rigorously by an other method.[4]) ⟦13⟧ ⟦15⟧

⟦13⟧

1) These Proceedings XII (1909), p. 286—297.

2) Compare Mathem. Ann. 68 (1910), p. 422—434. ⟦14⟧

3) Already the property of p. 288 that the transformation domain constructed in the way indicated there determines at most two residual domains, vanishes for some domains incompatible with the Schoenfliesian theory. ⟦13⟧

4) Compare Mathem. Ann. 72 (1912), p. 37—54. ⟦16⟧

[[17]] [1]) Vgl. diese Proceedings XI, S. 788; XII, S. 286; XIII, S. 767; XIV, S. 300; XV, S. 352; XXII, S. 811. Hinsichtlich der fünften dieser Mitteilungen kann bemerkt werden, dass der dortige Beweis auch unabhängig vom LÜROTH-CLEBSCH-schen Theorem geführt werden kann, nämlich so: a.a.O. S. 357 Z. 21 wählen wir auf μ' eine solche einfache geschlossene Kurve k, welche ein alle Bildpunkte von Rändern und Verzweigungspunkten der g_ν sowie alle nirgends dichten Bilder von g_ν enthaltendes Gebiet g begrenzt, und ziehen $k+g$ stetig zusammen in einen Punkt P von g. Die durch diese stetige Kontraktion von $k+g$ bestimmte stetige Aenderung von α_e führt zu einer „*primitiven Abbildung* α_l *von* μ *auf* μ'", für welche die ausserhalb voneinander gelegenen Innengebiete G_ν einer endlichen Zahl einander nicht treffender einfacher geschlossener Kurven von μ je eineindeutig mit dem Grade ± 1 auf die punktierte Fläche μ' und der Rest von μ auf den Punkt P abgebildet wird. Sodann führen wir mittels wiederholter stetiger Verschmelzungen, jedesmal von einem durch α_l mit dem Grade $+1$ und einem durch α_l mit dem Grade -1 abgebildeten G_ν, α_l in eine „*homogen-primitive Abbildung* α_q *von* μ *auf* μ'" über, deren Gebiete G_ν entweder alle mit dem Grade $+1$ oder alle mit dem Grade -1 abgebildet werden Dass alle homogen-primitiven Abbildungen
[[18]] n-ten Grades von μ auf μ' zur selben Klasse gehören, leuchtet unmittelbar ein.
In der ~~sechsten~~ siebenten Mitteilung ist S. 814 Fussnote [2]) *statt* Math. Annalen 81 *zu lesen* Math. Annalen 82.

NOTES

[[1]] The Dutch version was communicated at the meeting of 28 Sept. 1912. The English version appeared in the Proceedings of the meeting of 28 Sept. 1912. This paper has been incorporated into the present chapter because of the occurrence of the mapping degree and because it shows for two dimensions that the mapping degree characterises the homotopy class of mappings of spheres. The restriction to two dimensions indicates that as to method the paper does not properly fit into the present chapter. It is concerned with the topology of surfaces and as such is akin to the first four papers with the same title and still more closely related to those which follow and some more papers on the topology of surfaces.

[[2]] Brouwer 1909 F2, 1909 H2, 1911 H2, 1911 J2.

[[3]] P. Bohl 1904.

[[4]] Brouwer 1911 D.

[[5]] Brouwer 1909 F2.

[[6]] Brouwer 1911 G.

[[7]] Generalisation to higher dimensions by H. Hopf 1926 A.

[[8]] J. Hadamard 1910.

[[9]] The first occurrence of the term 'class' in the sense of homotopy class of mappings.

⟦10⟧ F. Klein 1882 B. See also a footnote to Brouwer 1920 G2, reprinted below where Lüroth–Clebsch is avoided.
⟦11⟧ The n-dimensional generalisation of this theorem was proved by H. Hopf 1926 A. Generalisations to the case of mappings of n-dimensional polyhedra into the n-sphere, H. Hopf 1933, to mappings of the 3-dimensional into the 2-dimensional sphere, H. Hopf 1931 A.
⟦12⟧ These remarks on 1909 H2 were already mentioned in 1909 H2 ⟦2⟧.
⟦13⟧ Brouwer 1909 H2.
⟦14⟧ Brouwer 1910 C.
⟦15⟧ Brouwer 1910 F.
⟦16⟧ Brouwer 1912 B.
⟦17⟧ This is a footnote to Brouwer 1920 G2.
⟦18⟧ Correction 'siebenten' in Brouwer's hand.

1912M

SUR LA NOTION DE "CLASSE" DE TRANSFORMATIONS D'UNE MULTIPLICITÉ

[[1]] PAR L. E. J. BROUWER.

Soient μ et μ' des multiplicités fermées: nous dirons de deux représentations univoques et continues de μ sur μ' qu'elles appartiennent à la même *classe,* si nous pouvons arriver de l'une à l'autre au moyen d'une modification continue. Les représentations de la même classe possèdent toutes le même *degré* (voir mon travail
[[2]] "Ueber Abbildung von Mannigfaltigkeiten," *Mathem. Annalen,* Bd. 71), mais ce qui est remarquable, c'est qu'en un grand nombre de cas l'inverse de ce théorème est encore vrai, propriété dont j'indiquerai brièvement la démonstration pour le cas que μ et μ' sont toutes les deux des sphères.

Nous commencerons par démontrer la propriété pour certaines catégories spéciales de représentations, d'où nous nous élèverons par étapes aux représentations les plus générales.

Considérons d'abord les *représentations riemanniennes,* c'est-à-dire les représentations qui au sens de l'analysis situs sont identiques aux représentations des surfaces de Riemann de genre zéro sur le plan complexe. D'une part M. Klein a demontré que toutes les représentations de degré n à ramifications simples peuvent être transformées d'une manière continue les unes en les autres en faisant mouvoir convenablement les points de ramification (propriété découlant du théorème de Lüroth-Clebsch sur l'arrangement canonique des coupures de ramification), d'autre part on peut, pour les représentations à ramifications multiples, séparer ces ramifications en un nombre équivalent de ramifications simples, de sorte que pour les représentations riemanniennes notre théorème se trouve démontré.

Considérons ensuite les *représentations canoniques,* pour lesquelles $n-1$ courbes simples fermées, situées dans μ et sans points communs, sont représentées chacune en un seul point de μ', tandis que les n régions déterminées par elles sont représentées biunivoquement, et cela ou toutes positivement, ou bien toutes négativement. Une représentation canonique pouvant être transformée, au moyen d'une modification continue arbitrairement petite, en une représentation riemannienne, notre théorème est encore démontré pour les représentations canoniques.

Passons aux *représentations normales,* ne satisfaisant qu'aux conditions suivantes: (1) elles sont *sans contraction cyclique,* c'est-à-dire il n'existe sur μ aucune courbe simple fermée représentée en un seul point de μ'; (2) elles sont *simpliciales*; (3) elles sont *sans plis**, c'est-à-dire de chaque élément de μ, dont l'image est biunivoque, l'indicatrice de l'image a le même signe. Pour les représentations normales nous démontrons notre théorème en faisant subir aux arcs de courbe de μ' images d'éléments de μ, une sorte d'extension continue les rendant images biunivoques et transformant la représentation normale en une représentation riemannienne.

Envisageons enfin les *représentations simpliciales sans plis*, pour lesquelles il existe sur μ un nombre fini d'ensembles polygonaux d'un seul tenant, représentés chacun en un seul point de μ' et divisant μ en un nombre fini de régions, pour chacune desquelles la représentation est ou normale ou nulle part dense. Pour ces dernières représentations nous démontrons notre théorème en réduisant d'abord les ensembles polygonaux à des courbes simples fermées sans points communs—ce qui nous donne une représentation simpliciale sans plis « à contraction simple »—et réduisant ensuite la représentation de chacune des régions partielles, dont l'image n'est pas nulle part dense, successivement à une représentation riemannienne et à une représentation canonique, ce qui nous donne une telle représentation simpliciale sans plis à contraction simple, pour laquelle les régions partielles de μ se divisent en trois sortes, représentées sur μ' successivement avec le degré 0, avec le degré $+1$, et avec le degré -1. Or, au moyen de nouvelles transformations continues de la représentation, on peut d'une part remplacer l'ensemble d'une région de 2de et une région adjacente de 3me sorte par une seule région de 1re sorte, d'autre part faire absorber une région de 1re sorte par une région adjacente de 2de ou de 3me sorte, procédé qui finit par nous donner une représentation canonique.

Maintenant les *représentations générales* ne présentent plus de difficulté, puisqu'en commençant par les rendre simpliciales et détruisant ensuite les plis morceau par morceau, on peut les transformer d'une manière continue en représentations simpliciales sans plis.

* Par un "*plis*" nous entendons ici une suite finie d'éléments de μ, dont chaque élément, excepté le dernier, possède en commun avec son suivant un "côté de contact," tous ces côtés de contact étant représentés sur un même arc de courbe de μ', et le premier et le dernier élément de la suite étant représentés avec des indicatrices opposées.

NOTES

[1] For an elaboration of this sketch, see the preceding 1912 K2.
[2] Brouwer 1911 D.

1913 A

Über den natürlichen Dimensionsbegriff.

[[1]] Von Herrn *L. E. J. Brouwer* in Amsterdam.

Auf Grund der Invarianz der Dimensionenzahl*) läßt sich die Dimensionenzahl einer Mannigfaltigkeit**) definieren als die Anzahl der Parameter, durch welche sich die Mannigfaltigkeit in der Umgebung eines beliebigen ihrer Punkte eineindeutig und stetig darstellen läßt. Diese „arithmetische“ Definition trägt aber nach *Poincaré****) unserer intuitiven Raumanschauung ungenügend Rechnung. *Poincaré* erhebt deshalb die Forderung einer rekurrenten Definition von etwa folgender Form†):

„*Ein Kontinuum heiße n-dimensional, wenn man es durch ein oder mehrere* $(n-1)$*-dimensionale Kontinua in getrennte Stücke zerlegen kann.*“

Obgleich der n-dimensionale *Jordan*sche Satz††) auf die Möglichkeit einer derartigen Definition deutet, so läßt sich diese in der zitierten Form dennoch nicht aufrecht erhalten.

Zunächst bemerken wir, daß das Wort „*Kontinuum*“ hier sicher nicht etwa im Sinne von „*Mannigfaltigkeit*“ aufgefaßt werden darf; in diesem Falle würde nämlich die Definition erst brauchbar werden, nachdem eine von der Parameterdarstellung unabhängige Charakterisierung der Mannigfaltigkeiten unter den abstrakten Mengen gelungen sein würde. Weil dies aber bis jetzt nicht der Fall ist, so müßte der *Poincaré*schen Defini-

[[2]] *) Vgl. meinen Beweis in Math. Annalen 70, S. 161—165 und die daran anknüpfenden Entwicklungen von *Lebesgue* in C. R. de l'Acad. des sciences, Paris, 27 mars 1911.

[[3]] **) Für die Definition des Begriffes „Mannigfaltigkeit“ vgl. Math. Annalen 71, S. 97.

[[4]] ***) Revue de métaphysique et de morale, 1912, S. 486, 487.

[[5]] †) a. a. O., S. 488.

††) Vgl. den teilweise von *Lebesgue*, teilweise von mir erbrachten Beweis in
[[6]] C. R. de l'Acad. des sciences, Paris, 27 mars 1911, und Math. Annalen 71, S. 305—319.

[[540]]

tion irgendeine allgemeinere abstrakte Charakterisierung des Kontinuums vorausgeschickt werden, z. B. diese: „*Eine Normalmenge* (im *Fréchet*schen [[7]]
Sinne) π *heiße ein Kontinuum, wenn es für je zwei ihrer Elemente* m_1 *und* m_2 *eine zusammenhängende, abgeschlossene**) *Menge gibt, welche Teilmenge von* π *ist und* m_1 *und* m_2 *enthält.*“ **) Für solche allgemeinere Kontinua, welche keine Mannigfaltigkeiten sind, würde aber unsere Definition zu Schwierigkeiten führen; z. B. würde man einem Kegel des Cartesischen Raumes, der sich ja durch einen Punkt zerlegen läßt, nur *eine* Dimension zusprechen dürfen. [[9a]]

Auch die Worte „*ein oder mehrere*“ könnten nicht unverändert beibehalten werden, weil mehrere m-dimensionale Mannigfaltigkeiten zusammen eine $(m+p)$-dimensionale Mannigfaltigkeit bilden können.

Alle diese Mängel lassen sich nun beseitigen, indem wir zunächst die *Poincaré*sche rekurrente Definition wie folgt abändern:

Es sei π irgendeine Normalmenge***), π_1, ϱ und ϱ' drei Teilmengen von π, welche innerhalb π abgeschlossen†) sind und keine gemeinsamen [[12]]
Punkte besitzen. Alsdann heißen ϱ und ϱ' *in* π *durch* π_1 *getrennt*, wenn jede zusammenhängende, abgeschlossene Teilmenge von π, welche sowohl mit ϱ wie mit ϱ' Punkte gemeinsam hat, auch von π_1 mindestens einen Punkt enthält. Der Ausdruck: „π *besitzt den allgemeinen Dimensionsgrad* n“, in welchem n eine beliebige natürliche Zahl bezeichnet, soll nun besagen, daß für jede Wahl von ϱ und ϱ' eine trennende Menge π_1 existiert, welche den allgemeinen Dimensionsgrad $n-1$ besitzt, daß aber nicht für jede Wahl von ϱ und ϱ' eine trennende Menge π_1 existiert, welche einen geringeren allgemeinen Dimensionsgrad als $n-1$ besitzt. Weiter soll der Ausdruck: „π *besitzt den allgemeinen Dimensionsgrad Null bzw. einen unendlichen allgemeinen Dimensionsgrad*“ bedeuten, daß π kein Kontinuum als Teil ent-

---

*) Unter einer *abgeschlossenen* Menge verstehen wir hier eine ihre Grenzelemente enthaltende Menge, in welcher jede unendliche Folge von Elementen mindestens ein [[8]]
Grenzelement aufweist.

**) Diese Definition ist der von *Schoenflies* für die Kontinua des n-dimensionalen [[9]]
Raumes gegebenen nachgebildet (vgl. Bericht über die Lehre von den Punktmannigfaltigkeiten, Bd. II, S. 117).

***) Inwieweit die Definition des Textes auch für Mengen allgemeinerer Art [[10]]
einen naturgemäßen Sinn behält, soll hier unerörtert bleiben.

†) Dieser Ausdruck besagt, daß π_1, ϱ und ϱ' alle ihre in π gelegenen Grenz- [[11]]
punkte enthalten.

hält, bzw. daß zu π weder die Null noch irgendeine natürliche Zahl als ihr allgemeiner Dimensionsgrad gefunden werden kann*).

Dieser Definition läßt sich leicht eine von der Rekurrenz unabhängige Form geben. Dazu denken wir uns die Menge π von zwei Personen A und B der „*Dimensionsoperation*" unterzogen, worunter wir folgendes verstehen: A wählt in π zwei innerhalb π abgeschlossene Teilmengen ϱ und ϱ' beliebig aus, worauf B ϱ und ϱ' in π trennt durch eine innerhalb π abgeschlossene Menge π_1. Sodann wählt A in π_1 zwei innerhalb π_1 abgeschlossene Teilmengen ϱ_1 und ϱ_1' beliebig aus, worauf B ϱ_1 und ϱ_1' in π_1 trennt durch eine innerhalb π_1 abgeschlossene Menge π_2. Dieser Prozeß wird unbeschränkt wiederholt, bis eventuell eine Menge π_h auftritt, welche kein Kontinuum mehr als Teil enthält. Wenn einerseits B unabhängig von den Wahlen der ϱ_ν und ϱ_ν' dafür sorgen kann, daß eine Menge π_h auftritt, deren $h \leq n$, und andererseits A unabhängig von den Wahlen der π_ν dafür sorgen kann, daß *keine* Menge π_h auftritt, deren $h < n$, so werden wir sagen, daß π *den allgemeinen Dimensionsgrad n besitzt.* Wenn dagegen keine natürliche Zahl n existiert mit der Eigenschaft, daß B unabhängig von den Wahlen der ϱ_ν und ϱ_ν' dafür sorgen kann, daß eine Menge π_h auftritt, deren $h < n$, so werden wir sagen, daß π *einen unendlichen allgemeinen Dimensionsgrad besitzt.*

Wenn zu einem Punkte P von π Umgebungen, welche den allgemeinen Dimensionsgrad m, aber keine Umgebungen, welche einen geringeren allgemeinen Dimensionsgrad besitzen, existieren, so werden wir sagen, daß *π in P den Dimensionsgrad m besitzt.* In verschiedenen Punkten kann eine Menge verschiedene Dimensionsgrade besitzen; keiner von diesen kann indes den allgemeinen Dimensionsgrad der Menge übersteigen. Falls in jedem Punkte der Menge der Dimensionsgrad dem allgemeinen Dimensionsgrade der Menge gleich ist, so werden wir sagen, daß die Menge *einen homogenen Dimensionsgrad besitzt.*

Auf Grund der vorstehenden Definitionen soll nun die *Poincaré*sche Forderung vollständig erfüllt werden durch die Begründung von folgendem

Dimensionssatz. *Eine n-dimensionale Mannigfaltigkeit besitzt den homogenen Dimensionsgrad n***).

[[13]] *) Nach dieser Definition wird sowohl für den *Hilbert*schen wie für den *Fréchet*schen R_ω ein unendlicher allgemeiner Dimensionsgrad gefunden.

[[14]] **) Weil der Dimensionsgrad offenbar eine Invariante der Analysis Situs ist, so ist im Dimensionssatz die Invarianz der Dimensionenzahl enthalten.

[[542]]

Zum Beweise dieses Satzes zeigen wir zunächst, daß B bei der Dimensionsoperation dafür sorgen kann, daß $h < n$. Dazu konstruiert B, nachdem A die Mengen ϱ und ϱ' bestimmt hat, eine simpliziale Zerlegung*) ζ von π, und zwar in solcher Weise, daß, wenn wir unter einem ${}_\pi s_\varrho$ bzw. ${}_\pi s_{\varrho'}$ ein entweder in seinem Inneren oder auf seiner Grenze Punkte von ϱ bzw. ϱ' enthaltendes Grundsimplex von ζ verstehen, kein ${}_\pi s_\varrho$ mit einem ${}_\pi s_{\varrho'}$ identisch ist und kein ${}_\pi s_\varrho$ an ein ${}_\pi s_{\varrho'}$ grenzt. Alsdann bilden diejenigen $(n-1)$-dimensionalen Seiten der ${}_\pi s_\varrho$, welche weder in ihrem Inneren noch auf ihrer Grenze Punkte von ϱ enthalten, ein System von zweiseitigen $(n-1)$-dimensionalen Pseudomannigfaltigkeiten**), in welchem übrigens mehrere Elemente oder Elementseiten zusammenfallen können. Die von diesen Pseudomannigfaltigkeiten gebildete Punktmenge wählt B als π_1. Falls darauf A die Mengen ϱ_1 und ϱ'_1 in demselben Teilkontinuum von π_1 wählt, so konstruiert B eine solche simpliziale Zerlegung von π_1, daß kein ${}_{\pi_1} s_{\varrho_1}$ mit einem ${}_{\pi_1} s_{\varrho'_1}$ identisch ist und kein ${}_{\pi_1} s_{\varrho_1}$ an ein ${}_{\pi_1} s_{\varrho'_1}$ grenzt. Alsdann bilden diejenigen $(n-2)$-dimensionalen Seiten der ${}_{\pi_1} s_{\varrho_1}$, welche weder in ihrem Inneren noch auf ihrer Grenze Punkte von ϱ_1 enthalten, ein System von zweiseitigen $(n-2)$-dimensionalen Pseudomannigfaltigkeiten, in welchem übrigens wieder mehrere Elemente oder Elementseiten zusammenfallen können. Die von diesen Pseudomannigfaltigkeiten gebildete Punktmenge wählt B als π_2. In dieser Weise fortfahrend, gelangt B schließlich zu einer Menge π_n, welche kein Kontinuum mehr als Teil enthält, es sei denn, daß der Prozeß schon früher dadurch beendet wurde, daß A einmal ϱ_ν und ϱ'_ν nicht in demselben Teilkontinuum von π_ν wählte.

Wir zeigen zweitens, daß A bei der Dimensionsoperation dafür sorgen kann, daß h nicht kleiner als n ausfällt. Dazu wählt A in π von einem n-dimensionalen Elemente $E_1 E_2 \ldots E_{n+1}$ den Punkt E_1 als ϱ und die $(n-1)$-dimensionale Seite $E_2 \ldots E_{n+1}$ als ϱ'; den zur Elementseite $E_1 E_2$ bzw. $E_1 E_3 \ldots E_{n+1}$ gehörigen Teil von π_1 als ϱ_1 bzw. ϱ'_1; den zur Elementseite $E_1 E_2 E_3$ bzw. $E_1 E_2 E_4 \ldots E_{n+1}$ gehörigen Teil von π_2 als ϱ_2 bzw. ϱ'_2; usw. Um zu beweisen, daß von den Punktmengen $\pi_1, \pi_2, \ldots \pi_n$ keine in Fortfall kommen kann, bezeichnen wir mit τ das Ausgangselement $E_1 E_2 \ldots E_{n+1}$, mit τ_1 die Grenze des von π_1 in τ bestimmten, an den Punkt E_1 grenzenden

*) Math. Annalen 71, S. 101. [[15]]

**) a. a. O., S. 305. [[16]]

Gebiets g, mit τ_2 die Grenze der von π_2 in τ_1 bestimmten, an die Kante $E_1 E_2$ grenzenden Gebietsmenge*) g_1, mit τ_3 die Grenze der von π_3 in τ_2 bestimmten, an die zweidimensionale Seite $E_1 E_2 E_3$ grenzenden Gebietsmenge g_2, usw., konstruieren in τ eine simpliziale Zerlegung von der Dichte ε**), bezeichnen mit γ das n-dimensionale Fragment***), welches von den mitsamt ihrer Grenze zu g gehörigen Grundsimplexen gebildet wird, mit σ_1 den *innerhalb* τ gelegenen Teil der gleichfalls simplizial zerlegt vorliegenden Grenze von γ, mit ε_1 das Maximum der Abstände, welche die Punkte von σ_1 von τ_1 besitzen, mit γ_1 das $(n-1)$-dimensionale Fragment, welches von denjenigen Grundsimplexen von σ_1, die von g_1 einen Abstand $\leqq \varepsilon_1$ besitzen, gebildet wird, mit σ_2 den *innerhalb* σ_1 gelegenen Teil der Grenze von γ_1, mit ε_2 das Maximum der Abstände, welche die Punkte von σ_2 von τ_2 besitzen, und fahren so fort. Alsdann konvergieren $\varepsilon_1, \varepsilon_2, \ldots \varepsilon_n$ mit ε gegen Null, so daß die eventuelle Existenz von $\sigma_1, \sigma_2, \ldots \sigma_n$ diejenige von $\tau_1, \tau_2, \ldots \tau_n$, mithin auch diejenige von $\pi_1, \pi_2, \ldots \pi_n$, in denen ja der Reihe nach $\tau_1, \tau_2, \ldots \tau_n$ als Teilmengen enthalten sind, nach sich ziehen wird.

Hiermit ist der Dimensionssatz zurückgeführt auf folgenden

Hilfssatz. Es sei σ ein simplizial zerlegtes n-dimensionales Element mit den Eckpunkten $E_1, E_2, \ldots E_{n+1}$; γ ein aus Grundsimplexen von σ gebildetes Fragment, das alle an E_1, aber kein an $E_2 E_3 \ldots E_{n+1}$ grenzendes Grundsimplex von σ enthält; σ_1 der innerhalb σ liegende Teil der Grenze von γ; γ_1 ein aus Grundsimplexen von σ_1 gebildetes Fragment, das alle an $E_1 E_2$, aber kein an $E_1 E_3 \ldots E_{n+1}$ grenzendes Grundsimplex von σ_1 enthält; σ_2 der innerhalb σ_1 liegende Teil der Grenze von γ_1; γ_2 ein aus Grundsimplexen von σ_2 gebildetes Fragment, das alle an $E_1 E_2 E_3$, aber kein an $E_1 E_2 E_4 \ldots E_{n+1}$ grenzendes Grundsimplex von σ_2 enthält; σ_3 der innerhalb σ_2 liegende Teil der Grenze von γ_2; usw. Alsdann kann von den Punktmengen $\sigma_1, \sigma_2, \sigma_3, \ldots \sigma_n$ keine verschwinden.

[[17]] *) Unter einer in τ_ν gelegenen Gebietsmenge verstehen wir eine in τ_ν gelegene Punktmenge, von der kein Punkt Grenzpunkt der durch sie in τ_ν bestimmten Komplementärmenge ist.

[[18]] **) Math. Annalen 71. S. 101.

[[19]] ***) a. a. O., S. 306.

[[544]]

Dieser Hilfssatz, auf den schon *Lebesgue* in Math. Annalen 70 die [[20]] Invarianz der Dimensionenzahl zurückgeführt hat, dessen Beweis daselbst aber eine wesentliche Lücke aufweist*), leuchtet unmittelbar ein, wenn wir den von mir in Math. Annalen 71 eingeführten Begriff des *Abbildungsgrades***) heranziehen.

---

*) Vgl. daselbst S. 167, wo derselbe Hilfssatz in folgender Form auftritt: [[21]] *In einem n-dimensionalen Würfel I, der im R_n von den ebenen Räumen $x_i = x_i^0$ und $x_i = x_i^0 + 2l$ $(i = 1, 2, \ldots n)$ begrenzt ist, bestimme man eine Gebietsmenge γ, deren innerhalb I gelegene Grenze σ_1 von einer endlichen Zahl von den $(n-1)$-dimensionalen Seiten von I parallelen $(n-1)$-dimensionalen Raumstücken gebildet wird, welche einen endlichen Abstand besitzt vom Raume $x_1 = x_1^0 + 2l$, und deren Komplementärmenge einen endlichen Abstand besitzt vom Raume $x_1 = x_1^0$. In σ_1 bestimme man eine Gebietsmenge γ_1, deren innerhalb I gelegene Grenze σ_2 von einer endlichen Zahl von den $(n-2)$-dimensionalen Seiten von I parallelen $(n-2)$-dimensionalen Raumstücken gebildet wird, welche einen endlichen Abstand besitzt vom Raume $x_2 = x_2^0 + 2l$, und deren Komplementärmenge einen endlichen Abstand besitzt vom Raume $x_2 = x_2^0$. In dieser Weise fahre man fort. Alsdann kann von den Punktmengen $\sigma_1, \sigma_2, \ldots \sigma_n$ keine verschwinden.* Zum Beweise dieses Satzes wird nun a. a. O. folgende geometrische Schlußweise angewandt: „In σ_1 ist ein Teilkontinuum I_1 enthalten, das sich von $x_i = x_i^0$ bis $x_i = x_i^0 + 2l$ erstreckt für $i = 2, 3, \ldots n$. In der Grenze des zu I_1 gehörigen Teiles von γ_1 ist dann ein Teilkontinuum I_2 enthalten, das sich von $x_i = x_i^0$ bis $x_i = x_i^0 + 2l$ erstreckt für $i = 3, 4, \ldots n$. Ebenso ist in der Grenze des zu I_2 gehörigen Teiles von γ_2 ein Teilkontinuum I_3 enthalten, das sich von $x_i = x_i^0$ bis $x_i = x_i^0 + 2l$ erstreckt für $i = 4, 5, \ldots n$. In dieser Weise werden der Reihe nach $I_1, I_2, \ldots I_n$ definiert, und zeigt sich gleichzeitig die Existenz aller σ_ν, in denen ja die entsprechenden I_ν als Teilmengen enthalten sind.“ Daß indes eine derartige Schlußweise, welche aus der Eigenschaft, daß $I_{\nu-1}$ sich von $x_i = x_i^0$ bis $x_i = x_i^0 + 2l$ erstreckt für $i = \nu, \ldots n$, die Existenz folgert eines I_ν, das sich von $x_i = x_i^0$ bis $x_i = x_i^0 + 2l$ erstreckt für $i = \nu + 1, \ldots n$, nicht ohne weiteres erlaubt ist, erhellt aus folgendem Beispiel: Man wähle im Falle $n = 3$ für γ die Gesamtheit der durch folgende Formeln [[22]] bestimmten Punktmengen:

$$1^0.\ x_1^0 < x_1 < x_1^0 + \tfrac{1}{4} l.\ 2^0.\ x_1^0 + \tfrac{1}{2} l < x_1 < x_1^0 + \tfrac{3}{4} l.$$
$$3^0.\ x_1^0 + \tfrac{1}{4} l \leq x_1 < x_1^0 + \tfrac{1}{2} l,\ x_i^0 + \tfrac{1}{2} l < x_i < x_i^0 + \tfrac{3}{2} l \ (i = 2, 3);$$

für I_1 dasjenige der beiden Teilkontinua, in welche σ_1 zerfällt, das der Würfelseite $x_1 = x_1^0$ am nächsten liegt: für γ_1 diejenigen Punkte von σ_1, für welche $x_2 \leq x_2^0 + \frac{1}{2} l$, unter Ausschluß der Grenzpunkte der zugehörigen Komplementärmenge. Alsdann enthält die Grenze des zu I_1 gehörigen Teiles von γ_1 *kein* Teilkontinuum, das sich von $x_3 = x_3^0$ bis $x_3 = x_3^0 + 2l$ erstreckt.

Inzwischen hat Herr *Lebesgue* mir brieflich einen neuen geometrischen Beweis des Hilfssatzes mitgeteilt, den er demnächst im Bull. de la Soc. Math. de France zu veröffentlichen beabsichtigt.

**) Vgl. daselbst S. 105. [[23]]

[[24]] Die Eigenschaft, daß die Projektion von σ_ν aus der Elementseite $E_1 E_2 \ldots E_\nu$ die Elementseite $E_{\nu+1} E_{\nu+2} \ldots E_{n+1}$ mit dem Grade 1 bedeckt, läßt sich nämlich von ν auf $\nu + 1$ ausdehnen, indem wir zunächst aus ihr folgern, daß die Projektion des in der Elementseite $E_1 \ldots E_\nu E_{\nu+2} \ldots E_{n+1}$ liegenden Teiles der Grenze von σ_ν aus der Elementseite $E_1 \ldots E_\nu$ oder aus der Elementseite $E_1 \ldots E_{\nu+1}$ die Elementseite $E_{\nu+2} \ldots E_{n+1}$ mit dem Grade 1 bedeckt, und sodann σ_ν, indem wir jedesmal ein einziges seiner Grundsimplexe tilgen, stückweise auf γ_ν reduzieren, wobei der in der Elementseite $E_1 \ldots E_\nu E_{\nu+2} \ldots E_{n+1}$ liegende Teil der Grenze von σ_ν schrittweise in $\sigma_{\nu+1}$ übergeht und der entsprechende Projektionsgrad auf $E_{\nu+2} \ldots E_{n+1}$ sich nicht ändern kann. Weil mithin jedes σ_ν $(\nu = 1, 2, 3, \ldots n)$ sich mit dem Grade 1 auf eine $(n - \nu)$-dimensionale Seite von σ projiziert, so kann keines der σ_ν sich auf Null reduzieren. W. z. b. w.

[[546]]

[[Corrections introduced in 1923 D2:

Replace 'p. 147, l.15–20 enthält.' by]]

Es sei π irgendeine Normalmenge[9]), π_1, ϱ und ϱ' drei Teilmengen von π, welche innerhalb π abgeschlossen[10]) sind und keine gemeinsamen Punkte besitzen. Alsdann heissen ϱ und ϱ' *in π durch π_1 getrennt,* wenn π_1 in π eine ϱ enthaltende, aber ϱ' nicht enthaltende Gebietsmenge g bestimmt.[11])

[10]) Dieser Ausdruck besagt, dass π_1, ϱ und ϱ' alle ihre in π gelegenen Grenzpunkte enthalten.

[11]) Diesen der Gebietsmenge g auferlegten Bedingungen können natürlich mehrere Gebietsmengen von π genügen. Im in [1]) zitierten Original hat sich an dieser Stelle eine andere, mit dem übrigen Inhalte des Aufsatzes in keinem Zusammenhang stehende Trennungsdefinition eingeschlichen. Dass die obige (übliche) Definition die in der vorliegenden Abhandlung in Wirklichkeit gebrauchte ist, geht aus dem Zusammenhang hervor, insbesondere aus Fussnote[16]) und dem zugehörigen Passus des [[25]]
Textes. Die daselbst eingeführte, von π_2 in π_1 bestimmte, an die Kante $E_1 E_2$ grenzende Gebietsmenge kann nämlich keinen anderen Sinn haben, als den des Durchschnittes einer schon vorhandenen von π_2 in π_1 bestimmten, an $E_1 E_2$ grenzenden, an $E_1 E_3 \ldots E_{n+1}$ jedoch nicht grenzenden Gebietsmenge mit π_1. Auf die Berichtigung, welche hier anzubringen war, bin ich von Herrn P. Urysohn in [[26]]
Moskau aufmerksam gemacht worden. [[27]]

[[Replace footnote *) on p. 151 by the following]]

[19]) Die „faits bien évidents", welche dieser Beweis (auf S. 167) voraussetzt, sind [[28]]
nämlich unrichtig, und bilden, wenn sie in eine richtige Form gebracht werden, eine Eigenschaft, welche tiefer liegt, als der Hilfssatz selbst. Nachdem Herr Lebesgue (in 1911) auf dieses Versehen hingewiesen worden war, teilte er mir seine Absicht mit, binnen kurzem im Bull. de la Soc. Math. de France einen neuen Beweis des Hilfssatzes zu bringen, von dem er mir gleichzeitig die Hauptzüge auseinandersetzte. Obgleich diese Auseinandersetzungen mich nicht befriedigten, meinte ich dennoch im in [1]) zitierten Original auf die von Herrn Lebesgue zugesagte Veröffent- [[29]]
lichung hinweisen zu müssen. Dieselbe ist indes ausgeblieben und erst in Fundamenta Mathematicae, Bd. 2 (1921), S. 256–285 ist Herr Lebesgue auf den [[30]]
Gegenstand zurückgekommen und hat er einen stichhaltigen Beweis des Hilfssatzes gegeben, der, was den Kern betrifft, mit meinem obigen Beweise von 1913 übereinstimmt, davon aber durch eine unnötig verwickelte Darstellung der Einzelheiten abweicht.

NOTES

⟦1⟧ An exercise book and fragments from the beginning of 1910 (see Y16 ⟦1⟧) show Brouwer trying to prove the invariance of dimension along two different lines. The first and most natural, suggested by ideas of Poincaré, did not immediately lead to success. The second, by blunt use of the mapping degree, was fully successful and led to the proof of the invariance of dimension, Brouwer 1911 C.

Poincaré 1903 had proposed an inductive definition of dimension, a more precise and more elaborate version of which is found in Poincaré 1912 A [1]): a continuum has dimension n if it can be decomposed by cuts (coupures) of dimension $n-1$. This was also the leading idea of Brouwer's first approach. The unusual Dutch word 'coupuur' in those fragments betrays Poincaré's influence, although, rather than through Poincaré 1903, this influence is likely to have reached Brouwer personally at his stay in Paris during the Christmas holidays 1909/10. There are no indications whether after the success along the second route Brouwer still continued attempts along the first route, up to the publication of Lebesgue 1911 A (see Brouwer 1911 C⟦1⟧). Of course when Brouwer saw this paper of Lebesgue, he immediately identified Lebesgue's paving principle as the missing link he could use in his first approach. According to his own testimony he succeeded within a few days in proving Lebesgue's principle (see Y4), which yielded a new proof of the invariance of dimension.

The present paper contains this proof in the setting of an inductive definition of dimension, that is, Brouwer's general dimension concept. As we explained above, it goes back to the spring of 1911. Though it had been promised to Blumenthal early in 1912, Brouwer published it not in *Mathematische Annalen* but in *Crelle*. Probably because of the strange place of publication, neither P. Urysohn nor K. Menger were aware of this paper when they started their own independent researches on dimension. It seems to me that if they had known it from the start and consequently been subjected to its influence, their acknowledgements would have precluded the 1928 quarrel between Brouwer and Menger as well as the attribution of dimension theory to Urysohn and Menger or, for that matter, Menger and Urysohn, rather than to Brouwer.

1924 M contains a correction of 1913 A.

1923 D2 is a reprint of 1913 A with this correction introduced into the text,

[1]) There is much confusion about Poincaré's papers of 1903 and 1912 A. In historical notes on dimension theory it is a habit to quote some paragraphs from Poincaré 1903 (according to 'La valeur de la science') along with a reference to Poincaré 1912 A (according to 'Dernières Pensées'). Since Dernières Pensées was published posthumously, authors stress that Poincaré conceived his ideas on dimension in the last year of his life, which adds the melodramatic feature that by a lucky chance these ideas were preserved in order to be elaborated by Brouwer. F. Riesz 1909 (written in 1906) witnesses a much earlier impact of Poincaré's ideas on dimension. As we noticed Brouwer, too, was influenced by Poincaré's *ideas* though not by the *publication* of 1912 A.

a footnote (11) added to explain the correction, and a footnote (19) relating to Lebesgue.

In the following the text of 1913 A is reprinted with the correction and other relevant pieces from 1923 D2 added at the end. In our notes the numbering of the footnotes of 1913 A according to 1923 D is indicated.

1924 J2 contains supplementary remarks to 1913 A and 1923 D2.

1924 L is virtually identical with part of 1924 J2. It is not republished here.

⟦2⟧ Brouwer 1911 C, Lebesgue 1911 B. Brouwer does not mention 1911 A. Footnote 2 in 1923 D2.

⟦3⟧ Brouwer 1911 D. Footnote 3 in 1923 D2.

⟦4⟧ H. Poincaré 1912 A. Footnote 4 in 1923 D2.

⟦5⟧ H. Poincaré 1912 A. Footnote 5 in 1923 D2.

⟦6⟧ Lebesgue 1911 B. Brouwer 1911 F. Footnote 6 in 1923 D2.

⟦7⟧ Normalmenge im Fréchetschen Sinne: M. Fréchet 1906.

⟦8⟧ Footnote 7 in 1923 D2. At the next opportunity Brouwer uses 'closed' in the usual sense. Indeed, in that period 'closed' often means 'closed and bounded' (in cartesian spaces). In Schoenflies 1908 A it is sometimes not clear what the author means.

⟦9⟧ A. Schoenflies 1908 A. Footnote 8 in 1923 D2.

⟦9a⟧ Brouwer's criticism was anticipated by F. Riesz 1906/07, p. 313.

⟦10⟧ Footnote 9 in 1923 D2.

⟦11⟧ This is the usual meaning of 'closed'. Footnote 10 in 1923 D2.

⟦12⟧ In this paragraph Brouwer's own copy contains the following additions and corrections:

Ink note in the margin, dating, according to the handwriting, from about 1923: im Lennes'schen Sinne zusammenhängend.

The word 'zusammenhängend' in line 18 is underlined with a short pen stroke.

There is a hardly visible pencil correction, which in the history of Brouwer's style of writing must date from much before 1923: the word 'abgeschlossene' in line 18 is deleted and a line with an arrow is drawn from this word to the margin, where one reads 'zu streichen in Übereinstimmung mit S. 150 Fussnote *)'. In a letter to H. Hahn of 4 August 1929 Brouwer asserted that he corrected the text of his own copy as early as March 1913, which certainly means the pencil rather than the ink correction. This is confirmed in the most unexpected way by a note in the proof sheets of A. Schoenflies 1913 (Brouwer read carefully the proofs of Schoenflies' book and advised its author in the most efficient way (see 1910 C ⟦9⟧. and A. Schoenflies 1913, VII, Vorwort)). On p. 382 of these proofsheets he elaborated footnote [2]) by adding

'. / ⟦deleting the period⟧; ebenso die Untersuchungen Brouwers in Math. Ann. 70, S. 161–165 und Journ. f. Math. 142, 146–152 (an letzterer Stelle ist übrigens nach einer Mitteilung Brouwers auf S. 147, Z. 18 das Wort "abgeschlossen" zu streichen)'.

⟦encircled:⟧

durch die Aufnahme dieser Verbesserung im Bericht kommt sie besser unter die Augen des Publikums als durch eine Berichtigung in Crelle. Br.

The last is of course meant as a communication to Schoenflies.

As indicated by the final text of A. Schoenflies 1913, Schoenflies did not take over this addition.

About the dating of this footnote there cannot be the slightest doubt. It is written with the same pen, the same ink, and within the same short period, i.e. August 1913, as the other notes in the proofreadings of A. Schoenflies 1913.

In his publications on this question, Brouwer never made allusions to such 'private' evidence. In a letter to H. Hahn of August 1929 he formulated such behaviour as a principle while mentioning the present case as a particular instance.

Deleting 'abgeschlossen' makes the definition equivalent to that which is introduced as a correction in 1923 D2 and 1924 M. Brouwer 1923 D2, Footnote 11, claimed that the occurrence of 'abgeschlossene' had been unintentional, Brouwer 1924 J2, Footnote 6, called it a clerical mistake.

We shall return to this point, which plays a rôle in the Brouwer–Menger controversy, Brouwer 1928 C.

⟦13⟧ Footnote 12 in 1923 D2.

⟦14⟧ Footnote 13 in 1923 D2.

⟦15⟧ Brouwer 1911 D. Footnote 14 in 1923 D2.

⟦16⟧ Brouwer 1911 E. Footnote 15 in 1923 D2.

⟦17⟧ Footnote 16 in 1923 D2. Formally *Gebietsmenge* is Brouwer's term for relative open set. Its historical origin, however, is the division of the plane or euclidean space by a closed subset into a number of domains; any union of such domains is called a *Gebietsmenge*.

The margin contains a pencil note from the same period as that indicated in ⟦12⟧: 'Im Text wird gemeint der Durchschnitt von τ_1 mit der von π_2 in π_1 bestimmten an die Kante $E_1 E_2$ grenzenden (d.h. ρ_1 enthaltenden) Gebietsmenge'.

⟦18⟧ Brouwer 1911 D. Footnote 17 in 1923 D2.

⟦19⟧ Brouwer 1911 E. Footnote 18 in 1923 D2.

⟦20⟧ H. Lebesgue 1911 A. The *Hilfsatz* is essentially Lebesgue's paving theorem. See Brouwer 1911 C ⟦1⟧.

⟦21⟧ This footnote is in 1923 D2 replaced by a shorter one, numbered 19.

⟦22⟧ An earlier version of this counterexample from Brouwer's papers: Y1–2.

⟦23⟧ Brouwer 1911 D. Footnote 20 in 1923 D2.

⟦24⟧ An earlier version of this proof from Brouwer's papers: Y3. P. Alexandroff 1928 B, p. 622, considers the topological use of the projection σ as a remarkable novelty.

⟦25⟧ Brouwer 1913 A, p. 150 footnote *).

⟦26⟧ See Brouwer 1924 J2 footnote 6 and 7.

⟦27⟧ In 1911 E ⟦12⟧ we touched on the history of the concept of connectedness.

For closed sets it goes back to C. Jordan 1893, p. 25: not divisible into two non-empty closed subsets. Lennes 1911 and – probably independently – Brouwer 1911 E extended it by switching to *relative* closedness. Earlier, starting from the Weierstrass domain concept, another extension had been attempted: a subset of a cartesian space is connected if every pair of points can be connected by closed (bounded) connected sets. But this definition had never been operational. Brouwer 1928 C distinguishes in topological spaces these two concepts as *weak* (Lennes) and *strong* connectedness (Schoenflies). Applied to the complement of a closed set π_1 in a space π, they lead to *strong* and *weak* separation, respectively, of two sets ρ and ρ' by π_1, and finally to a *weak* and a *strong* concept of *dimension*. Formally Brouwer 1913 A uses weak separation, and consequently arrives at the strong dimension concept. In the present place Brouwer asserts that he always meant not weak, but strong separation. Indeed, as Brouwer points out, the proof on p. 150 is based on the use of weak connectedness. One could add the remark that from 1911 onwards Brouwer knew of weak connectedness, and that neither he – nor for that matter, anybody else – had ever used strong connectedness. Moreover the – according to Brouwer – unintentional *abgeschlossene* occurs *after* by footnote 7 the meaning of 'abgeschlossene' had been changed from compact into closed, whereas the word *abgeschlossene*, if it was intended in the sense of Schoenflies *strong connectedness* could only mean compact. It is perhaps no exaggeration if Brouwer 1924 J2 footnote 6 claims it was a clerical mistake: inserting two footnotes related to different uses of *abgeschlossen* may have caused the confusion in which the word slipped into a place where it did not belong. If this is true there remains one puzzling question: why did Brouwer start the inductive definition of dimension with the negation of 'Schoenflies' continua (rather than the negation of connected sets) as 0-dimensional sets, or for that matter, with the empty set as (-1)-dimensional set. Though Brouwer never said so, I would guess that even here he originally meant the negation of connected sets. There is, however, no means of checking it, since it does not affect the proof which is about the (compact) n-dimensional simplex.

Such is the public evidence on what Brouwer called a clerical mistake.. One has to study this evidence carefully in order to become convinced. Brouwer's presentation of his point, in particular the footnote 6 of 1924 J2 in which Lebesgue is made responsible for the delay of the correction, lacks psychological appeal. The fine distinction between weak and strong connexion may appear as an a posteriori implantation – such subtleties in the stone age of topology! Yet Brouwer was this subtle.

What surprised me was the evidence in the anecdota. I would have guessed that Brouwer discovered the 'clerical error' sometime between 1913 and 1923, because in no publication did he claim any date for this discovery. It appears from the Hahn correspondence and from the Schoenflies proofreadings mentioned in Brouwer 1913 A ⟦12⟧ that Brouwer recognised the mistake very early, certainly in 1913, perhaps as early as March 1913, and it becomes more and more probable that he indeed considered it too obvious to need immediate correction. Perhaps he had not even checked whether Schoenflies had accepted his footnote.

⟦28⟧ H. Lebesgue 1911 A.

⟦29⟧ The footnote of 1923 D2 that refers to 1913 A.

⟦30⟧ H. Lebesgue 1921.

Berichtigung zur Abhandlung „Über den natürlichen Dimensionsbegriff", dieses Journal Bd. 142 (1913), S. 146—152.

1924 M [[1]]

Von Herrn *L. E. J. Brouwer* in Amsterdam.

Nach einer mir von Herrn *P. Urysohn* in Moskau gemachten Bemerkung sind auf S. 147 der obigen Abhandlung, Z. 17—20 die Worte:

> „wenn jede zusammenhängende abgeschlossene Teilmenge von π, welche sowohl mit ϱ wie mit ϱ' Punkte gemeinsam hat, auch von π_1 mindestens einen Punkt enthält"

zu ersetzen durch:

> „wenn π_1 in π eine ϱ enthaltende, aber ϱ' nicht enthaltende Gebietsmenge bestimmt".

In der Tat geht aus dem Zusammenhang, insbesondere aus Fußnote [1]) auf S. 150 und dem zugehörigen Passus des Textes hervor, daß es der letztere Trennungsbegriff ist, der den Entwickelungen der obigen Abhandlung zugrunde liegt.

NOTE

[[1]] Brouwer 1913 A.

1924 J2 **Mathematics.** — "*Bemerkungen zum natürlichen Dimensionsbegriff.*"

[[1]] By Prof. L. E. J. Brouwer.

(Communicated at the meeting of June 28, 1924).

§ 1.

[[2]] Wir beschränken uns im folgenden auf *kondensierte Spezies,* d.h. abgeschlossene kompakte metrische Spezies oder Vereinigungen von divergenten Fundamentalreihen abgeschlossener kompakter metrischer Spezies, und betrachten für diese Spezies den vor einiger Zeit von Menger[1]) und Urysohn[2]) eingeführten Dimensionsbegriff, den wir den *MU-Dimensionsbegriff* nennen werden. Die Definition dieses Begriffes hat für kondensierte Spezies folgende Form:

"*Eine Spezies* π *hat eine MU-Dimension* $\leq n$, *wenn ein beliebiger Punkt* m *von* π *für beliebiges positives* ε *eine Umgebung* $U(m) < \varepsilon$ *besitzt, deren Grenze eine MU-Dimension* $\leq n-1$ *besitzt. Eine Spezies* π *hat die MU-Dimension* 0, *wenn sie kein Kontinuum als Teil enthält*".

Mittels vollständiger Induktion in bezug auf n zeigt man leicht, dass die Vereinigung einer endlichen Anzahl oder einer divergenten Fundamentalreihe von Spezies der *MU*-Dimension $\leq n$ wiederum eine Spezies der *MU*-Dimension $\leq n$ darstellt.

Wir betrachten nun in einer Spezies π der *MU*-Dimension $\leq n$ zwei innerhalb π abgeschlossene Teilspezies ϱ und ϱ' von π, welche keine gemeinsamen Punkte besitzen. Dann folgt aus der Verallgemeinerung des Heine-Borelschen Theorems, dass es eine Umgebung $U(\varrho)$ von ϱ gibt, welche eine Grenze π_1 der *MU*-Dimension $\leq n-1$ besitzt und von der ϱ' weder einen Punkt noch einen Grenzpunkt enthält. M. a. W. ϱ *und* ϱ' *können in* π *getrennt werden*[3]) durch eine innerhalb π abgeschlossene Teilspezies π_1 von π der *MU*-Dimension $\leq n-1$.

Wenn wir also einer Spezies π eine *N*-Dimension $\leq n$ zusprechen, wenn sie nach der Terminologie des in [3]) zitierten Aufsatzes einen *allgemeinen Dimensionsgrad* $\leq n$ besitzt, so haben wir:

[[3]] 1) Monatshefte für Mathematik und Physik, Bd. 33, S. 157—160 (nach in 1921 bei der Wiener Akademie hinterlegten Definitionen und Resultaten).

[[4]] 2) Comptes Rendus, t. 175 (1922), S. 440.

[[5]] 3) Diese Proceedings 26, S. 796. Hinsichtlich einer hier in bezug auf das 1913

[[6]] im Journ. f. Math. 142 erschienene Original vorgenommenen Berichtigung vgl. näher § 2, sowie meine in Bd. 21 der Mathem. Zeitschr. erscheinende Note:

[[7]] „*Zum natürlichen Dimensionsbegriff*".

Eine Spezies der MU-Dimension $\leq n$ besitzt auch eine N-Dimension $\leq n$. Weil mittels vollständiger Induktion in bezug auf n auch die Umkehrung dieses Satzes leicht bewiesen werden kann, so ergeben für kondensierte Spezies die Definitionen der *MU*-Dimension und der *N*-Dimension den gleichen Dimensionsbegriff[4]). Die Bezeichnung als "*natürlicher*" *Dimensionsbegriff* scheint mir für diesen Begriff in erster Linie auf Grund der ihm anhaftenden Qualität, welche durch den *N*-Dimensionsbegriff zum Ausdruck kommt, berechtigt zu sein. [[8]]

§ 2.

Dem natürlichen Dimensionsbegriff liegt folgende, der Anschauung entnommene „*natürliche Trennungsdefinition*" zu grunde:

a) *Wenn π_1, ϱ und ϱ' innerhalb π abgeschlossene Teilspezies von ohne gemeinsame Punkte sind, so heissen ϱ und ϱ' in π durch π_1 getrennt, wenn jede zusammenhängende Teilspezies von π, welche sowohl mit ϱ wie mit ϱ' Punkte gemeinsam hat, auch von π_1 mindestens einen Punkt enthält*[5]).

In Journ. f. Math. 142, S. 147 hatte sich in diese Definition ausser jeder Beziehung zum übrigen Inhalte der Abhandlung das Wort „abgeschlossene" eingeschlichen[6]), wodurch die Trennungsdefinition folgende Form erhielt: [[6]]

b) *Wenn π_1, ϱ und ϱ' innerhalb π abgeschlossene Teilspezies von*

---

[4]) Diese Aequivalenz wurde von Herrn URYSOHN schon im Herbst 1923 erwähnt.

[5]) Im in Fussnote [3]) zitierten Aufsatz befindet sich diese Trennungsdefinition in der folgenden, mit der obigen äquivalenten, weniger anschaulichen, aber üblicheren und sich an die Beweisführung enger anschliessenden Form:

c) *Wenn π_1, ρ und ρ' innerhalb π abgeschlossene Teilspezies von π ohne gemeinsame Punkte sind, so heissen ρ und ρ' in π durch π_1 getrennt, wenn π_1 in π eine ρ enthaltende, aber ρ' nicht enthaltende Gebietsmenge bestimmt.*

[6]) Die Veröffentlichung der (übrigens für den Leser auf Grund des Zusammenhanges auf der Hand liegenden) Berichtigung dieses Schreibfehlers scheint vor elf Jahren dadurch unterblieben zu sein, dass ich im Anschluss an die damals in Aussicht gestellten (a. a. O. S. 151 in einer Fussnote beiläufig berührten) näheren Ausführungen LEBESGUES auf die Dimensionstheorie zurückzukommen und dann gleichzeitig die betreffende Korrektur auszuführen beabsichtigte, dass aber, als diese Ausführungen LEBESGUES Jahr aus Jahr ein auf sich warten liessen und meine Aufmerksamkeit von anderen Gegenständen in Anspruch genommen wurde, die Angelegenheit in Vergessenheit geraten ist. Erst anlässlich eines im vorigen Jahre auf der Jahresversammlung der Deutschen Mathematiker-Vereinigung in Marburg von Herrn URYSOHN gehaltenen Vortrags, in welchem auf die a. a. O. zu hebende Unstimmigkeit hingewiesen wurde, hat neuerdings im in [3]) zitierten Aufsatz, sowie m Journ. f. Math. 153 die Richtigstellung stattgefunden. [[9]] [[5]] [[10]]

π ohne gemeinsame Punkte sind, so heissen ϱ und ϱ' in π durch π_1
⟦11⟧ *getrennt, wenn jede zusammenhängende, abgeschlossene Teilspezies von π, welche sowohl mit ϱ wie mit ϱ' Punkte gemeinsam hat, auch von π_1 mindestens einen Punkt enthält.*

Falls man, was ebenfalls möglich ist, die abstrakte Dimensions-
⟦5⟧ theorie auf die letztere Definition gründen will, so ist vom in [5]) zitierten
⟦12⟧ Aufsatz der zwischen S. 798 Z. 4 v. u. und S. 799 Z. 14 v. u. enthaltene Passus wie folgt zu ändern [7]):

"... bezeichnen wir mit τ das Ausgangselement $E_1 E_2 E_3 \ldots E_{n+1}$, für jedes ν mit τ_ν den zu τ gehörigen Teil von π_ν, konstruieren in τ eine simpliziale Zerlegung ζ von der Dichte ε, setzen $\gamma_0 = \tau$, und wählen für jedes ν, nachdem $\gamma_0, \gamma_1, \gamma_2, \ldots, \gamma_\nu$ bestimmt sind, für $\gamma_{\nu+1}$ eine solche Vereinigung von zu γ_ν gehörigen Grundsimplexen von ζ, dass *erstens* die zu den Elementseiten $E_1 E_2 \ldots E_{\nu+1}$ und $E_1 \ldots E_\nu E_{\nu+2} \ldots E_{n+1}$ gehörigen Teile von γ_ν in γ_ν durch $\gamma_{\nu+1}$ getrennt sind, *zweitens* falls $\tau_{\nu+1}$ existiert, $\gamma_{\nu+1}$ für passend gewähltes positives $\varepsilon_{\nu+1}$ besteht aus denjenigen zu γ_ν gehörigen Grundsimplexen von ζ, welche einen Abstand $\leq \varepsilon_{\nu+1}$ von $\tau_{\nu+1}$ besitzen. *Alsdann können wir $\varepsilon_1, \varepsilon_2, \varepsilon_3, \ldots$ alle mit ε gegen Null konvergieren lassen.*

Nehmen wir nämlich einen Augenblick an, dass wir $\varepsilon, \varepsilon_1, \varepsilon_2, \ldots, \varepsilon_\nu$ zusammen gegen Null konvergieren lassen können, während für $\varepsilon, \varepsilon_1, \varepsilon_2, \ldots, \varepsilon_\nu, \varepsilon_{\nu+1}$ dasselbe unmöglich ist, so lässt sich eine positive Grösse a angeben mit der Eigenschaft, dass für jedes positive η in τ_ν in einer Entfernung $> a$ von $\tau_{\nu+1}$ eine η-Kette existiert, von deren
⟦14⟧ Endpunkten das eine in einer Entfernung $< \eta$ von $E_1 E_2 \ldots E_\nu E_{\nu+1}$ und das andere in einer Entfernung $< \eta$ von $E_1 \ldots E_\nu E_{\nu+2} \ldots E_{n+1}$ gelegen ist. Dann aber müsste in τ_ν in einer Entfernung $\geq a$ von $\tau_{\nu+1}$ ein sowohl $E_1 \ldots E_\nu E_{\nu+1}$ wie $E_1 \ldots E_\nu E_{\nu+2} \ldots E_{n+1}$ treffendes abgeschlossenes Kontinuum existieren, was ungereimt ist.

---

7) Dass in dieser Weise die Tragweite des Textes nicht nur formal, sondern auch materiell beeinflusst wird, erläutert folgendes von Herrn Urysohn herrührendes Beispiel: Es sei π die Euklidische Ebene; $E_1 = (0, -1)$, $E_2 = (3, 2)$, $E_3 = (-3, 2)$; π_1 die Vereinigung der folgenden sechs Punktmengen: 1 . $x = 0$. $0 \leq y \leq 1$; 2 . $y = 0$, $-1 \leq x \leq 0$; 3 . $x = -1$, $-2 \leq y \leq 0$; 4 . $y = -2$, $-1 \leq x \leq 1$; 5 . $x = 1$, $-2 \leq y < 0$; 6 . $y = \sin^2 \frac{\pi}{x}$, $0 < x \leq 1$; π_2 die der Trennungsdefinition *b*), *nicht* aber der Trennungsdefinition *a*) genügende (vgl. in diesem Zusammenhang
⟦13⟧ Hausdorff, Grundzüge der Mengenlehre, S. 458) Vereinigung der beiden Punkte $(0, -2)$ und $(0, 1)$. Die durch die alte Wendung des Textes a. a. O. S. 798—799 geforderte, in τ_1 von π_2 bestimmte, an $E_1 E_2$ grenzende, an $E_1 E_3$ jedoch nicht grenzende Gebietsmenge g_1 (durch welche das im Hilfssatz auftretende, aus Grundsimplexen von σ_1 gebildete Fragment γ_1, das alle an $E_1 E_2$, aber kein an $E_1 E_3$ grenzendes Grundsimplex von σ_1 enthält, bedingt ist) *existiert in diesem Falle nicht.*

Weil wir also $\varepsilon, \varepsilon_1, \varepsilon_2, \varepsilon_3, \ldots$ in der Tat alle zusammen gegen Null konvergieren lassen können, so können wir, falls τ_{ν_1} verschwindet, dafür sorgen, dass auch γ_{ν_1} verschwindet. Der Dimensionssatz ist hiermit zurückgeführt auf den Nachweis der Eigenschaft, dass γ_{ν_1} für $\nu_1 \leqq n$ unmöglich verschwinden kann, und hiermit auf den Beweis von folgendem

H i l f s s a t z. Es sei σ ein simplizial zerlegtes n-dimensionales Element mit den Eckpunkten $E_1, E_2, \ldots, E_{n+1}; \ldots$"

NOTES

⟦1⟧ Brouwer here discusses what in 1928 C2 he will call strong dimension, that is based on a strong connectedness concept (see 1913 A ⟦27⟧) instead of weak dimension based on weak connectedness as meant in Brouwer 1913 A, according to the correction in Brouwer 1923 D2. See also the correction 1924 M. A partial reprint of the present paper is 1924 L.

⟦2⟧ *Spezies* instead of *Menge* is intuitionistic terminology.

⟦3⟧ K. Menger 1923. The 1921 draft was afterwards published in K. Menger 1929 A.

⟦4⟧ P. Urysohn 1922. See also its elaboration: P. Urysohn 1925.

⟦5⟧ Brouwer 1923 D2. See 1913 A, 1923 D2.

⟦6⟧ Brouwer 1913 A.

⟦7⟧ Brouwer 1924 L. See also 1913 A, 1923 D2.

⟦8⟧ The local version of the *MU*-dimension is more subtle than the global one, which was only considered by Brouwer. Because of the diverging start of induction the *N*-dimension can be larger than the *MU*-dimension in non-compact spaces. There is a third deviation if 'strong' instead of 'weak' dimension is admitted (see 1913 A, 1923 D2). On the equivalence in compact spaces see P. Urysohn 1926, W. Hurewicz 1927, L. Tumarkin 1928, p. 653.

⟦9⟧ See the story on the Brouwer–Lebesgue quarrel in 1911 C ⟦1⟧. I venture to guess the culpit was Schoenflies rather than Lebesgue. Perhaps Brouwer had never checked whether A. Schoenflies 1913 had adopted his addition to a footnote (see 1913 A ⟦12⟧), until the 1923 talk with Urysohn, when he discovered that Schoenflies had not done so. Of course this would not have been a reason for attacking Schoenflies.

⟦10⟧ Brouwer 1924 M.

⟦11⟧ This is the 'weak separation' (1913 A ⟦27⟧).

⟦12⟧ The corresponding text in 1913 A is p. 149, 2 f.b. 'bezeichnen' upto p. 150, 10 f.b. 'E_{n+1};'.

⟦13⟧ F. Hausdorff 1914.

⟦14⟧ On p. 637 in line 5 and 4 f.b. 'das' must be changed into 'der'. It is almost incredible that Brouwer who wrote and spoke German excellently used 'Punkt' as a neutral noun as it is in Dutch. Perhaps he had originally written 'Enden' instead of 'Endpunkten' and preserved the article when he changed it. Strangely enough this mistake was not even corrected by the editors of *Mathematische Zeitschrift* in 1924 L.

Mathematics. — *Zur Geschichtschreibung der Dimensionstheorie.* 1928 C2
By Prof. L. E. J. BROUWER. [[1]]

(Communicated at the meeting of September 29, 1928).

§ 1.

Die Dimensionstheorie habe ich in einem 1913 erschienenen Aufsatz
„Ueber den natuerlichen Dimensionsbegriff" (Journ. f. Math. 142, S. [[2]]
146—152) fuer diejenigen Spezies, fuer welche die Dimension einen
natuerlichen Sinn besitzt, naemlich fuer die kondensierten metrischen [[3]]
Spezies, begruendet, indem ich einen d i m e n s i o n e l l e n R e c h t f e r t i-
g u n g s s a t z brachte, d.h. einen im Bereich der kondensierten metrischen
Spezies sinnvollen Dimensionsbegriff definierte, fuer den ich zeigen konnte,
dass die n-dimensionalen Mannigfaltigkeiten die Dimension n besitzen.

Fuer diesen Dimensionsbegriff boten sich naturgemaess zwei Varianten
dar, welche den beiden damals fuer kondensierte metrische Spezies
gangbaren Zusammenhangsbegriffen entsprachen. Es fuehrt naemlich in
gleicher Weise der schwache Zusammenhang, d.h. die Unzerlegbarkeit
in zwei Relativgebietspezies (LENNES 1911), zur starken Trennung (LEN- [[4]]
NES 1911), zur schwachen Dimension und zum schwachen Rechtferti-
gungssatz, wie der starke Zusammenhang, d.h. die Unzerlegbarkeit in
zwei Stueckspezies (SCHOENFLIES 1908), zur schwachen Trennung, zur [[5]]
starken Dimension und zum starken Rechtfertigungssatz. Der schwache
Rechtfertigungssatz ist eine unmittelbare Folge des starken, aber nicht
umgekehrt. Ein Beweis des (im unten besprochenen MENGERschen Buche [[6]]
nicht erwaehnten) starken Rechtfertigungssatzes wurde erst 1924 veroeffent-
licht und zwar in meiner Note „Bemerkungen zum natuerlichen Dimen-
sionsbegriff" (diese Proceedings 27, S. 635—638). [[7]]

In meinem zitierten Aufsatz vom Jahre 1913 wird zwar schon der [[2]]
starke Rechtfertigungssatz dem Wortlaute nach formuliert, der darauf
folgende Beweis bezieht sich aber aufs deutlichste auf den schwachen
Rechtfertigungssatz, so dass ersichtlich nur dieser gemeint ist, und der in
der Formulierung des Satzes enthaltene Schreibfehler jedem mitdenkenden
Leser sofort offenbar wird. Dass dieser Schreibfehler lange Jahre hindurch
keine gedruckte Verbesserung erfuhr, hatte erstens die allgemeine Ursache
der Fahrlaessigkeit muehelos einzusehenden Schreibfehlern gegenueber,
zweitens einen in diesen Proceedings 27, S. 136 mitgeteilten speziellen [[7]]
Grund.

In meinem zitierten Aufsatz von 1913 habe ich mich auf die Be- [[2]]
gruendung der Dimensionstheorie beschraenkt und auf die Veroeffent-
lichung weiterer dimensionstheoretischer Entwicklungen verzichtet,

einerseits weil mit dem Beweis des Rechtfertigungssatzes das gestellte erkenntnistheoretische Ziel erreicht war, andererseits weil [1]) fuer die anschliessenden Betrachtungen (in erster Linie diejenigen welche sich um den Summensatz und den Zerlegungssatz gruppieren) nicht wie fuer den Beweis der Rechtfertigungssatzes [2]) eine intuitionistische Verwirklichung nahe lag.

Die Aufstellung des schwachen Dimensionsbegriffes und der Beweis
des schwachen Rechtfertigungssatzes, d. h. die Begruendung der Dimen-
⟦2⟧ sionstheorie, ist mit meiner zitierten Arbeit von 1913, und nicht etwa
mit der ersten oeffentlichen Formulierung des in derselben enthaltenen
offensichtlichen Versehens, ebenso unanfechtbar zu datieren, wie der Beweis
⟦10⟧ der Invarianz des Gebietes mit meiner betreffenden Arbeit von 1911, und
nicht mit der ersten oeffentlichen Ausfuellung der in der ihr zugrunde
liegenden Definition der Pseudomannigfaltigkeit befindlichen offensicht-
⟦11⟧ lichen Luecke, oder auch die mengentheoretische Begruendung der ebenen
⟦13⟧ Geometrie mit HILBERT's bezueglichem Originalaufsatz von 1906, und
⟦14⟧ nicht mit dem Wiederabdruck dieses Aufsatzes, in welchem meinen
brieflichen Angaben vom Oktober 1909 zufolge ein offensichtliches Ver-
sehen in den zugrunde liegenden Axiomen richtiggestellt wurde.

§ 2.

Auf meine im § 1 besprochene Begruendung der Dimensionstheorie folgen URYSOHN, der 1922, von einer lokalen Form der schwachen Dimensionsdefinition [3]) ausgehend, den Summensatz und den Zerlegungssatz fand [4]) und im Zusammenhang mit diesen Saetzen in den Jahren 1922—1924 zahlreiche dimensionstheoretische Fragen zur Klaerung brachte [5]), und MENGER und ALEXANDROFF, die, ersterer seit 1924, letzterer seit 1925, die Dimensionstheorie weiter ausbauten und vertieften. Weil indessen die scharfsinnigen Untersuchungen von URYSOHN, MENGER und ALEXANDROFF grossenteils einen bloss formalistischen Charakter besitzen, reicht der intuitionistische (d.h. sinnvolle) Bestand der Dimensionstheorie heute kaum ueber denjenigen von 1913 hinaus.

§ 3.

⟦6⟧ Im neulich erschienenen MENGERschen Buche "Dimensionstheorie"
wird eine historische Darstellung der Entstehung der Dimensionstheorie

---

⟦8⟧ [1]) Vgl. meine einschlaegige Aeusserung in „Intuitionistische Mengenlehre", diese Pro-
ceedings 23 (1920), S. 950.

⟦6⟧ [2]) Vgl. meine (im unten besprochenen MENGERschen Buche nicht erwaehnte) Note
⟦9⟧ „Intuitionistische Einfuehrung des Dimensionsbegriffes", diese Proceedings 29 (1926),
S. 855—863.

⟦6⟧ [3]) Auf die (im unten besprochenen MENGERschen Buche nicht explizite erwaehnte)
Aequivalenz der URYSOHNschen Form mit der meinigen haben URYSOHN fuer kompakte
⟦15⟧ Spezies und ich fuer kondensierte Spezies unabhaengig voneinander hingewiesen (vgl. Fund.
Math. 8, S. 328; diese Proceedings 27 (1924), S. 636).

⟦16⟧ [4]) Comptes Rendus 175, S. 440.

⟦17⟧ [5]) Fund. Math. 7, 8.

gegeben, welcher gemaess MENGER als ihr Begruender, ich (mit EUKLID und einigen anderen) als Vorlaeufer und URYSOHN als Nachfolger zu gelten haetten. Diese Darstellung widerspricht dem oben in § 1 und § 2 dargelegten wirklichen Sachverhalt, ist denn auch falsch und bedeutet ein Antasten der Autorschaft eines Lehrers und eines verstorbenen Kommilitonen.

§ 4.

Meine Autorschaft wird im MENGERschen Buche angetastet dadurch,
dass zu verstehen gegeben wird, dass die BROUWERsche Note aus dem [[2]]
Jahre 1913 eine Definition der schwachen Dimension (jedenfalls nach
dem damaligen Stande der Wissenschaft) weder explizite noch implizite
enthalte, dass also derjenige, der spaeter eine solche Definition explizite
niedergeschrieben habe, berechtigt sei, sich selbst als Entdecker des
Dimensionsbegriffes und als Begruender der Dimensionstheorie zu qualifizieren. Diese These wird erhaertet mit zwei falschen Zitaten:

E r s t e n s wird (auf S. 85) meine Arbeit von 1913, die in den bisheri- [[2]]
gen Veroeffentlichungen MENGERs regelmaessig regulaer mit der schwachen [[18]]
Dimensionsdefinition, d.h. unter Anbringung der aus dem anschliessenden
Text folgenden, zum Ueberflusse auch formal und explizite in der Literatur
befindlichen Richtigstellung, zitiert worden war, hier auf einmal mit der
starken Dimensionsdefinition angefuehrt, wobei also die MENGER einerseits als sich unmittelbar darbietend, andererseits als in der Literatur
vorhanden bekannte Richtigstellung in einer den Leser irrefuehrenden
und den Sachverhalt entstellenden Weise totgeschwiegen wird. [[19]]

Z w e i t e n s wird (auf S. 86) im Anschluss an das vorstehende Zitat
behauptet, dass die der daselbst dem Wortlaute nach enthaltenen Form
der starken Definition dem Wortlaute nach aehnliche Form der schwachen
Definition auf dem „i n z w i s c h e n e r s t e n t s t a n d e n e n" modernen
Zusammenhangsbegriff beruhe (eine Behauptung, der man kaum einen
anderen Sinn beilegen kann, als den einer Anfechtung der frueheren —
sowohl durch mich wie durch MENGER erfolgten — Hinstellung des [[20]]
betreffenden Versehens als Schreibfehler), waehrend in Wirklichkeit die
Definition des schwachen Zusammenhangs (wie diejenige der starken
Trennung) in einer von MENGER selbst in seiner ersten Veroeffentlichung [[21]]
ordnungsgemaess zitierten Abhandlung von LENNES aus dem Jahre 1911 [[4]]
vorzufinden ist.

§ 5.

Die URYSOHNsche Autorschaft in bezug auf die lokale Form der
schwachen Dimensionsdefinition wird im MENGERschen Buche angetastet [[6]]
mit der Anfuehrung (daselbst S. 83 u. 84) von zwei Akten, welche der
zitierten URYSOHNschen Comptes-Rendus-Note gegenueber eine Prioritaet
besitzen sollen, naemlich:

1. Eine Hinterlegung bei der Wiener Akademie der Wissenschaften aus dem Jahre 1921.

[[22]] Diese (mir vorliegende, uebrigens in diesen Proceedings 29, S. 1122—1123 wörtlich abgedruckte) Hinterlegung enthaelt aber nur vage Andeutungen ueber die Dimensionalitaet „gewisser Kontinua" der euklidischen Raeume, denen kaum mehr als die Hoffnung, dieselbe spaeter einmal durch eine rekursive Definition festzulegen, zu entnehmen ist (Schon um die Dimension 2 zu definieren wird der dazu notwendige Begriff der Dimensionalitaet 1 fuer kompakte Mengen welche keine Kontinua sind, als „aus diskreten Kurven bestehend" erklaert, mit der nicht weiter ausgefuehrten Bemerkung, dass „dieses diskret sich praezisieren lasse").

2. Eine angeblich im Februar 1922 bei den Monatsh. f. Math. u. Phys. eingereichte Note, in welcher die schwache Dimensionsdefinition fuer Teilmengen von euklidischen Raeumen ausgesprochen sein soll.

In Wirklichkeit ist eine derartige Note nicht erschienen und es hat
[[23]] sich davon auch kein Manuskript vorgefunden. Es existiert nur ein (mir vorliegender) Brief von MENGER an HAHN vom 15. Febr. 1922, der folgenden Passus enthaelt: *„Ich hatte die kleine Arbeit, die Sie Herr Professor zu lesen die grosse Guete hatten, mit einer Definition der n-dimensionalen Mengen beschlossen*[6]*), die korrekterweise, wie ich nun glaube*[6]*), so haette lauten sollen:"* (Es folgt die lokale Form der schwachen Dimensionsdefinition fuer Teilmengen von euklidischen Raeumen). *„..... Ich vermute*[6]*) nun, dass u.a. folgender Satz gilt: n-dimensional im R_n sind die und nur die Mengen mit einem innern Punkt im R_n."*

Es handelt sich also nur um eine Vermutung (nicht einmal um eine unbewiesene Behauptung) des schwachen Rechtfertigungssatzes bzw. der Brauchbarkeit der lokalen Form des schwachen Dimensionsbegriffes fuer Teilmengen des R_n, also keineswegs um eine Begruendung, geschweige denn um die erste Begruendung, der Dimensionstheorie.

Ebensowenig wie die beiden obigen Akten gibt MENGER's erste Publi-
[[24]] kation „Ueber die Dimensionalitaet von Punktmengen, Erster Teil" (der Druckerei ueberreicht 12. 12. 1923, druckfertig erklaert 28. 1. 1924, erschienen 12. 4. 1924, auf dem Umschlage der Monatsh. f. Math. u. Phys. 33 und im MENGERschen Buche „Dimensionstheorie" unrichtig datiert mit dem Erscheinungsjahre 1923) ihm einen Anteil an der Begruendung oder dem Ausbau der Dimensionstheorie, weil in dieser Publikation nichts steht, was am Datum ihrer Einreichung nicht laengst (naemlich in meiner
[[2]] [[6]] Arbeit von 1913 und in der URYSOHNschen Comptes-Rendus-Note von 1922) veroeffentlicht und allgemein zugaenglich waere.

[[26]] Die fuer Prioritaetsfragen in Betracht kommenden MENGERschen Veroeffentlichungen zur Dimensionstheorie setzen erst ein mit seinen

6) Sperrung von mir.

Aufsaetzen „Ueber die Dimension von Punktmengen" (diese Pro- [[27]]
ceedings 27, S. 639—643, 1924) und „Ueber die Dimension von [[28]]
Punktmengen, II. Teil" (Monatsh. f. Math. u. Phys. 34, S. 137—161, der
Redaktion eingereicht 6. 10. 1924, der Druckerei ueberreicht 15. 10. 1924,
druckfertig erklaert 30. 12. 1924, erschienen 20. 9. 1926, auf dem Um- [[29]]
schlage der Monatsh. f. Math. u. Phys. 34 und im MENGERschen Buche [[6]]
„Dimensionstheorie" unrichtig datiert mit dem Erscheinungsjahre 1924).
Bei MENGER's Erscheinen auf der dimensionstheoretischen Buehne lagen
mithin in der Literatur bereits vor die Definition der schwachen und der
starken Dimension (inklusive des Aequivalenzbeweises der URYSOHNschen
Form mit der meinigen fuer die schwache Dimension), die Formulierung
und der Beweis des schwachen und des starken Rechtfertigungssatzes
(inklusive der — in MENGER's Debutarbeit ordnungsgemaess zitierten — [[30]]
oeffentlichen formalen Richtigstellung des oben in § 1 und § 4 erwaehnten
Schreibfehlers) und die Formulierung der Hauptsaetze fuer die schwache
Dimension.

§ 6.

Die URYSOHNsche Autorschaft in bezug auf die Beweise des Summen-
satzes und des Zerlegungssatzes wird im MENGERschen Buche angetastet [[6]]
dadurch, dass daselbst (S. 94 u. 157) die MENGERschen Beweise dieser
Saetze mit 1924 und die URYSOHNschen [7]) mit 1926 datiert werden.

Der Sachverhalt ist aber dieser, dass die (im Fruehling 1923 eingereichte) einschlaegige URYSOHNsche Abhandlung sicher nicht spaeter als mit URYSOHN's Todestag am 17.8.1924 und die einschlaegige MENGERsche Abhandlung nach dem vorigen § sicher nicht frueher als mit ihrem Eingangstag am 6. 10. 1924 zu datieren ist, so dass auch in bezug auf die Beweise des Summensatzes und des Zerlegungssatzes URYSOHN's Prioritaet ausser Zweifel steht.

Ueberdies wird die URYSOHNsche Autorschaft in bezug auf die For-
mulierung und den Beweis der Umkehrung des Zerlegungssatzes im
MENGERschen Buche verdunkelt dadurch, dass der Zerlegungssatz und [[6]]
seine Umkehrung daselbst (S. 156 u. 157) nicht als selbstaendige Saetze,
sondern als Teile von durch Kombination mit Ergaenzungen geringerer
Tragweite entstehenden erweiterten Saetzen eingefuehrt werden.

---

7) Die URYSOHNsche Formulierung des Summensatzes befindet sich nicht, wie MENGER [[31]]
angibt, auf S. 260, sondern auf S. 337 von Fund. Math. 8.

KNAW Proc. 32, p. 1022

In Prof. L. E. J. BROUWER's article *Zur Geschichtschreibung der* [[32]]
Dimensionstheorie (these Proceedings Vol. 31) p. 957, line 5—6 the words:

auf dem Umschlage der Monatsh. f. Math. u. Phys. 34 und

are to be struck out.

NOTES

⟦1⟧ Not until the twenties were Brouwer's topological concepts appreciated and assimilated by other mathematicians. In this respect his work on dimensionality of topological spaces hardly differed from that on manifolds, mappings, Jordan theorems and so on. The proper difference was in the place of publication: the paper on dimension was published in *Crelle* rather than in *Mathematische Annalen* as had been the bulk of his topological work. So it looked like an erratic block in Brouwer's work, which it certainly was not. In 1911 C ⟦1⟧ I ventured to guess what the course of history would have been if Brouwer 1913 A had also been published in Mathematische Annalen. Now H. Hahn, for example, overlooked it, and P. Urysohn and K. Menger did not notice it when they started their own work on dimension. Their start was also mutually independent. In one direction this is obvious since Urysohn died in 1924 before he could have learned about Menger's advances in dimension theory. In the other direction Menger's activity in dimension theory prior to Urysohn's first note in 1922 was witnessed by H. Hahn and O. Schreier and is proven by *a posteriori* publications. In the term Menger–Urysohn-dimension the authors are recorded in alphabetic as opposed to strictly historical order, which would give Urysohn a slight edge.

In 1923 or 1924 Menger had got in touch with Brouwer, in 1925–26 he was part of a group of young topologists who gathered around Brouwer in Holland (see P. Alexandroff 1969 A). Brouwer communicated papers of Menger to the Royal Netherlands Academy, Menger dedicated the reprints of a systematic and historical survey on dimension theory, K. Menger 1926 A, to Brouwer. Correspondence up to 1928 shows Menger attached to Brouwer. In 1928 Menger's book on dimension theory, K. Menger 1928, appeared. Brouwer 1928 C2 reacted with a sharp reproof to Menger's report on Brouwer's and Urysohn's part in founding dimension theory. A mammoth correspondence ensued between Brouwer and H. Hahn, Menger's teacher, editor (with W. Wirtinger) of the Monatshefte für Mathematik und Physik, a friend of both of them and mediator in this thorny affair. The whole story is paradigmatic for the logic and psychology of scholarly controversies, a rich source for students interested in this field. In former notes I have shed light on the mathematical and historical background of this controversy – here a few supplementary remarks will suffice.

From the beginning Brouwer agreed in principle with the draft of Menger's defence, which he offered to communicate to the Royal Netherlands Academy for publication. He made only a few slight reservations with respect to some formulations he wanted to be changed. To a few of them Menger agreed, to others he did not. Many new versions were proposed. Of course as it happens in such events, the superficial reader would not have been able to distinguish the one formulation from the other, and the more expert ones did not need any explanation. Brouwer's position gradually stiffened, and finally after the publication of Menger's anecdota,

Menger 1929 A, in *Monatshefte*, and a few other incidents, the affair concluded with the publication of Menger's defence, K. Menger 1930, in Monatshefte. Brouwer did not react publicly to this paper. See, however, the end of the present note.

Reviewing the controversy after so many years, with no relevant new data available, and reluctant to resume it, I should stress one thing, namely that if Menger got into trouble, he had asked for it (which as usual may mean, through imprudence rather than malice). In his effort to show his independence of Urysohn (which nobody had doubted) Menger, by arranging the evidence, went as far as suggesting that he was even simultaneous with Urysohn, which he certainly was not. With respect to Brouwer there are three points worth mentioning. Firstly his statement on the invariance of dimension (p. 243: 'Wäre die Dimension nicht topologisch invariant, so würde dies höchstens der Wichtigkeit der topologischen Abbildungen Abbruch tun'), which suggests belittling Brouwer's work; secondly, calling a living great mathematician, along with a few dead ones, a *Vorläufer* (p. 84), which would be in bad taste even if it were true, and finally the report on Brouwer 1913 A and 1923 D2: Menger (p. 85–86) reproduces Brouwer's 1913 A definition in full, without correcting or even mentioning what Brouwer had called a clerical mistake, and he cites the 1923 D2 correction not as a correction but as 'andere Trennungsdefinitionen' (strong separation). This meant ignoring Brouwer's own account. On top of this he asserted that the definition of separation intended in Brouwer 1913 A according to Brouwer 1923 D2 'sich . . . auf den inzwischen entstandenen modernen Zusammenhangsbegriff beruft' which meant that in 1913 Brouwer could not possibly have aimed at strong separation (and, consequently, weak dimension), since the basic notion of weak connectedness had not yet been invented at that time. If this were true, it would mean that Menger had caught Brouwer in a lie. Of course neither was it true, nor did Menger intend to accuse Brouwer of lying.

Though Brouwer never put it this way, it is clear that this was the reason for Brouwer's sharp reaction. In his defence, Menger 1929 B, retracted the argument on the dating of weak connectedness and excused it as a '*fehlerhafte Bemerkung*' though apparently he continued to doubt Brouwer's credibility. He explained his error as having been caused by Brouwer 1913 A's failure to cite Lennes 1911, which might have led him to attribute weak connection to Hausdorff 1914 – a bad explanation since in former publications, as well as in his book, Menger had always correctly ascribed weak connectedness to Lennes 1911.

After having finished the present manuscript, I got acquainted with K. Menger 1932, a private print with the typographical appearance of the *Monatshefte*, indicating what seem to be posthumous alterations of references to Brouwer's work in P. Urysohn 1925. Though not intended as such, the paper reads like a grave accusation. It seems that neither Brouwer nor other people concerned ever knew about it. Since this private print is virtually inaccessible, it does not make much sense to discuss it without quoting it in full extent, along with a few docu-

ments from Brouwer's estate, which would mean resuming the whole affair. The main point in Menger 1932 is a divergence in the text of a footnote referring to Brouwer 1913 A in P. Urysohn 1925, pp. 37–38, such as it occurs in a few preprints, compared with the final publication (preprints are an old habit of Fundamenta Mathematicae):
in the preprints,

(1) Il se propose de résoudre cette question.

in the final publication,

(2) Il résout cette question.

The first formulation suggests that Brouwer's proof is wrong, which it in fact is without the correction 1923 D2 of the 'clerical mistake'.

It is in itself quite improbable that Urysohn after having met Brouwer and been befriended by him, would have chosen such an impolite formulation as is (1), even if he had believed Brouwer to be wrong.

Before reaching the preprint version, the footnote had been modified at least twice by Urysohn himself (which is stated by K. Menger 1932). The changes consisted mainly in deleting an explicit statement of Brouwer's proof being wrong. Along with this alteration the version (1) was changed into (2), yet later back into (1). Obviously the latter change was caused by a failure to repeat the second time the full first step modification. The only essential change in the footnote, made after Urysohn's death, was adding a reference to Menger's work. This change and a few others had been agreed upon when Alexandroff and Urysohn visited Brouwer, shortly before Urysohn's death. Urysohn's work was edited by Alexandroff, though because of obstructions to the postal intercourse between USSR and Poland, Brouwer mediated between Alexandroff and the editors of Fundamenta Mathematicae.

⟦2⟧ Brouwer 1913 A. See also Brouwer 1923 D2.

⟦3⟧ In fact it was formulated for complete, rather than compact spaces.

⟦4⟧ J. Lennes 1911.

⟦5⟧ A. Schoenflies 1908 A.

⟦6⟧ K. Menger 1928.

⟦7⟧ Brouwer 1924 J2. See also Brouwer 1913 A and 1923 D2.

⟦8⟧ Brouwer 1919 D1 in volume 1.

⟦9⟧ Brouwer 1926 B2 in volume 1.

⟦10⟧ Brouwer 1911 E.

⟦11⟧ Brouwer 1921 F.

⟦13⟧ Brouwer can only mean Hilbert 1902.

⟦14⟧ Hilbert 1909. See also Brouwer's letter to Hilbert, Y8. See Y8 ⟦2⟧.

⟦15⟧ P. Urysohn 1926. Brouwer 1924 J2. See also 1924 J2 ⟦8⟧.

⟦16⟧ P. Urysohn 1922.

⟦17⟧ P. Urysohn 1925, 1926.

⟦18⟧ For example K. Menger 1924 B, footnote 6, mentioned explicitly Brouwer's *Schreibfehler*.

⟦19⟧ Menger mentioned 1923 D, yet not as a correction.
⟦20⟧ Brouwer 1924 J2, K. Menger 1924 B, footnote 6.
⟦21⟧ K. Menger 1923, footnote 3.
⟦22⟧ K. Menger 1926 B.
⟦23⟧ This manuscript could indeed not be found in 1926, as is witnessed by Menger 1926 B in a most subtle manner, so in 1928 Brouwer was correct in his belief that it no longer existed. Menger 1926 B is a first print of a manuscript and a letter of Menger of 1921; two other manuscripts of 1921 and 1922 are mentioned in the same context. When Brouwer accepted Menger 1926 B for publication in KNAW Proceedings, he insisted on distinguishing between documents which still existed, and others. This was done in such a way that some documents were marked as 'noch vorhanden', whereas others are mentioned without this mark. It is hard to understand why the *correction of a paper* of 1921 is published in 1926 *without the paper itself*, and only a careful reader of Menger 1926 B will discover that the mark 'noch vorhanden' is to be associated with the correction, and not with the paper itself, which, indeed, was not available when Brouwer asked for it in 1926.

In the meantime the manuscript seems to have been recovered. It was printed in Menger 1929 A. Here and in Menger 1926 B the correction to this paper lacks the final sentence from 'Ich vermute nun' onwards, which is quoted by Brouwer.
⟦24⟧ K. Menger 1923.
⟦26⟧ See also K. Menger 1929 A, a publication of anecdota from 1921 to 1923. In this period Menger was in possession of the dimension definition, the sum theorem for closed sets and the theorem on the dimension of n-space. The publication also makes it likely that up to the end of 1923 he did not know P. Urysohn 1922, which contains the greater part of P. Urysohn's results in dimension theory, though without proofs.
⟦27⟧ K. Menger 1924 A.
⟦28⟧ K. Menger 1924 B.
⟦29⟧ Correction, see below.
⟦30⟧ K. Menger 1924 B, footnote 6.
⟦31⟧ P. Urysohn 1926.
⟦32⟧ Correction, published in KNAW Proceedings 32, p. 1022.

1912 H

2. Über den Kontinuitätsbeweis für das Fundamentaltheorem der automorphen Funktionen im Grenzkreisfalle.

⟦1⟧

Von L. E. J. Brouwer in Amsterdam.

Die Menge der Systeme von Grenzkreispolygoninvarianten gegebener Signatur $(p, n, l_1, \ldots, l_n)$ läßt sich nach Fricke[1]) eineindeutig und stetig auf das Innere eines $(2n+6p-6)$-dimensionalen Würfels abbilden. In bezug auf den Rand dieses Würfels läßt sich ein zuerst von Poincaré[2]) bewiesener Satz wie folgt formulieren:

Hilfssatz. „*Wenn die einer Folge von Grenzkreisgruppen der Signatur $(p, n, l_1, \ldots, l_n)$ entsprechenden Systeme von Polygoninvarianten gleichmäßig gegen den Würfelrand konvergieren, so konvergiert unter den entsprechenden Polygonen eine Teilfolge gegen ein Grenzkreispolygon von veränderter Signatur, und unter den zugehörigen mit n verschiedenen Punkten signierten Riemannschen Flächen eine Teilfolge konstanter, endlicher Blätterzahl k gegen eine Fläche von veränderter Signatur.*"

Auf Grund dieses Satzes führt Poincaré[3]) den Kontinuitätsbeweis der Existenz linear-polymorpher Funktionen auf Riemannschen Flächen unter den folgenden beiden stillschweigenden Voraussetzungen:

Theorem 1. *Die Klassen der Riemannschen Flächen vom Geschlechte p bilden eine singularitätenfreie $(6p-6)$-dimensionale Mannigfaltigkeit.*[4])

⟦2⟧ 1) Fricke-Klein, Theorie der automorphen Funktionen, Bd. 2, S. 305.

⟦3⟧ 2) Acta Mathematica 4, S. 250—276. Vgl. auch Fricke-Klein, a. a. O. S. 333, 363, 404.

⟦4⟧ 3) A. a. O. S. 276—278.

⟦5⟧ 4) Für die Definition der (singularitätenfreien) q-dimensionalen Mannigfaltigkeit vgl. Brouwer, Math. Ann. 71, S. 117.

Theorem 2. *In einer q-dimensionalen Mannigfaltigkeit bildet das eineindeutige und stetige Bild eines q-dimensionalen Gebiets*[1]) *wiederum ein Gebiet.*

Die Anwendung des Theorems 1 aber läßt sich umgehen, indem wir den Kontinuitätsbeweis in folgender Weise ein wenig abändern:

Wir wählen $m > 2p - 2$[2]) und betrachten auf der einen Seite die Menge M_π der zu den Systemen von Polygoninvarianten der Signatur $(p, n, l_1, \ldots, l_n)$ gehörigen automorphen Funktionen mit ausschließlich einfachen Verzweigungspunkten und mit m einfachen, außerhalb der singulären Stellen der Gruppe liegenden Polen im Fundamentalbereich[3]), auf der anderen Seite die Menge M_f der aus m numerierten Blättern zusammengesetzten, mit n verschiedenen im Endlichen gelegenen Punkten signierten über die Ebene ausgebreiteten Riemannschen Flächen vom Geschlechte p[3]) mit $2m + 2p - 2$ numerierten einfachen, im Endlichen gelegenen Verzweigungspunkten; die Numerierung, sowohl der Blätter wie der Verzweigungspunkte, soll dabei der kanonischen Anordnung im Sinne von Lüroth-Clebsch entsprechen.[4]) ⟦6⟧

Zu jedem Punkte von M_π gibt es eine durch $4m + 2n + 4p - 8$ reelle Parameter eineindeutig und stetig darstellbare Umgebung; diese Parameter sind die $2n + 6p - 6$ Polygoninvarianten, die m komplexen Stellen der Pole in einem Fundamentalbereich, und die $m - p - 1$ komplexen Verhältnisse von $m - p$ willkürlich anzunehmenden Polresiduen.

Die Menge M_f bildet ein Kontinuum und besitzt zu jeder ihr angehörigen Fläche eine durch $4m + 2n + 4p - 8$ reelle Parameter eineindeutig und stetig darstellbare Umgebung.

Die Menge derjenigen Punkte von M_f, welche den Punkten von M_π entsprechen, bezeichnen wir mit B_f. Sie enthält zu jedem ihrer

---

1) Als q-dimensionales Gebiet wird eine solche Teilmenge einer q-dimensionalen Mannigfaltigkeit bezeichnet, welche zu jedem ihrer Punkte eine volle Umgebung und zu je zwei ihrer Punkte einen sie verbindenden einfachen Kurvenbogen enthält.

2) Um die Gedanken zu fixieren, setzen wir im folgenden $p > 1$ voraus.

3) Automorphe Funktionen bzw. über die Ebene ausgebreitete Riemannsche Flächen, welche sich nur um additive und multiplikative Konstanten unterscheiden, betrachten wir im folgenden als identisch.

4) Es ist dies dieselbe Menge M_f, mittels welcher Klein in seiner Schrift über Riemann (1882) zuerst gezeigt hat, daß die Klassen der Riemannschen Flächen vom Geschlechte p ein *Kontinuum* bilden, und welche ebenfalls direkt zu beweisen gestattet, daß in der Umgebung *jeder* Riemannschen Fläche vom Geschlechte p ohne Ausnahme die Klassen der *zerschnittenen* Riemannschen Flächen vom Geschlechte p eine *singularitätenfreie* $(6p - 6)$-dimensionale Mannigfaltigkeit bilden (vgl. Brouwer, Gött. Nachr. 1912). ⟦6⟧ ⟦7⟧

Punkte das eineindeutige und stetige Bild eines Gebietes von M_π als Umgebung, muß mithin auf Grund des obenstehenden Theorems 2 in M_f eine Gebietsmenge erfüllen. Um zu zeigen, daß B_f mit M_f identisch ist, brauchen wir nur noch den Satz zu beweisen, daß jeder zu M_f gehörige Grenzpunkt von B_f notwendig in B_f enthalten ist.[1])

Zu diesem Zwecke nehmen wir einen Augenblick an, daß ein zu M_f gehöriger Grenzpunkt f_ω von B_f *nicht* in B_f enthalten ist, betrachten eine gegen f_ω konvergierende Folge $f_1, f_2, f_3, \ldots$ von Punkten von B_f und unterscheiden zwei Fälle:

1. *Die entsprechenden Systeme von Polygoninvarianten konvergieren gleichmäßig gegen den Würfelrand.* Alsdann existierte nach dem obenstehenden Satze von Poincaré zu jedem f_p eine solche derselben Klasse angehörige k-blättrige Fläche φ_p, daß eine Teilfolge τ der φ_ν als Limeselement eine Riemannsche Fläche *von veränderter Signatur* bestimmte. Andererseits aber muß in τ eine Teilfolge τ' existieren, deren Pole auf den entsprechenden f_ν gegen bestimmte Stellen konvergieren, und in τ' eine Teilfolge τ'', welche gegen eine einzige Limesfläche *von derselben Signatur* als f_ω konvergiert, was ein Widerspruch ist.

2. *Die entsprechenden Systeme von Polygoninvarianten besitzen einen Limespunkt innerhalb des Würfels.* Alsdann bestimmen die den f_ν entsprechenden Elemente von M_π ein ebenfalls zu M_π gehöriges Limeselement. Die dieser automorphen Limesfunktion entsprechende über die Ebene ausgebreitete Riemannsche Fläche f_ω müßte mithin zu B_f gehören, was wieder ein Widerspruch ist.

Somit ist B_f notwendig mit M_f identisch.

Im obenstehenden ist der Kontinuitätsbeweis zurückgeführt auf die Begründung des Theorems 2, d. h. des Theorems von der Invarianz des Gebiets, welche Begründung ich in einer demnächst in den Mathematischen Annalen erscheinenden Arbeit ausgeführt habe.[2])

Diskussion.

Im Anschluß an den Vortrag des Herrn Brouwer teilt Herr Koebe mit, daß ihm neuerdings in Anknüpfung an Kleins allgemeinen Ansatz die Durchführung des Kontinuitätsbeweises für *alle* Kleinschen Funda-

1) Mittels der von Herrn Koebe in der nachstehenden Diskussion angedeuteten Methode läßt sich dieses Theorem auf den Fall des *allgemeinen* Kleinschen Fundamentaltheorems ausdehnen. Damit wird dann zugleich die vorliegende Form des Kontinuitätsbeweises auf diesen allgemeinen Fall übertragbar. Vgl. meinen
[[8]] am 13. Januar 1912 in der Gött. Ges. der Wiss. vorgelegten Brief an R. Fricke

[[9]] 2) Die Arbeit ist inzwischen in Bd. 71, S. 305—313 erschienen. Vgl. auch
[[10]] Bd. 72, S. 55 und die Bd. 71, S. 314 zitierten Abhandlungen von Baire und Lebesgue.

mentaltheoreme und weitere Probleme der konformen Abbildung gelungen sei. Er erläutert dies kurz an dem Beispiel der Uniformisierung durch automorphe Funktionen des Schottkyschen Typus, ein Fall, welcher der Poincaréschen Methode bereits unzugänglich sei.[1])

P. Koebe.

Bei der Erwiderung hebt Herr Brouwer hervor, daß die Schwierigkeiten des Kontinuitätsbeweises in zwei Gruppen zerfallen, von denen die erste, *topologische*, zu welcher die Singularitätenfreiheit der Modulmannigfaltigkeit und die Invarianz des Gebiets gehören, welche durch seinen eben gehaltenen Vortrag gleichzeitig für *alle* Kleinschen Fundamentaltheoreme erledigt worden sei[2]), durch die Mitteilung von Herrn Koebe weder überwunden noch umgangen werde. Diese Mitteilung beziehe sich ausschließlich auf die zweite (für den Grenzkreisfall nicht in Betracht kommende) Gruppe, reiche aber *in Vereinigung mit dem Brouwerschen Vortrage* aus, um dessen Resultat auf das *allgemeine Kleinsche Fundamentaltheorem* auszudehnen.

L. E. J. Brouwer.

1) Vgl. hierzu auch den Schlußabschnitt des nachfolgenden Referates von Koebe. P. Koebe. ⟦11⟧ ⟦12⟧

2) Vgl. Brouwer, Gött. Nachr. 1912.

⟦13⟧

Die obige neben dem von Herrn KOEBE redigierten Diskussionstext allzuwenig motiviert aussehende „Erwiderung" bezieht sich auf die von Herrn KOEBE nach meinem Vortrage ausgesprochene Behauptung, dass bij Anwendung des KOEBE'schen Verzerrungssatzes der Kontinuitätsbeweis für das allgemeine KLEIN'sche Fundamentaltheorem in solcher Weise geführt werden könne, dass die topologischen Schwierigkeiten unter Heranziehung POINCARÉ'scher Reihen umgangen werden. Dass der Schlussabschnitt des nachstehenden KOEBE'schen Referates zu dieser Behauptung im Widerspruch steht, lässt sich dadurch erklären, dass *im von Herrn* KOEBE *in Karlsruhe gehaltenen Vortrage* von dem Kontinuitätsbeweis überhaupt nicht die Rede gewesen, sodass insbesondere die auf S. 162 befindliche Anmerkung [1]) vollständig irreführend ist. Hierzu ist weiter zu bemerken, dass sowohl das nachstehende KOEBE'sche Referat, wie auch die in den Göttinger Nachrichten vom Jahre 1912 erschienene KOEBE'sche Note über den Kontinuitätsbeweis erst im Sommer 1912 gedruckt worden sind, und dass Herr KOEBE schon Neujahr 1912 im Besitze einer Abschrift meines über den Kontinuitätsbeweis handelnden, in den Göttinger Nachrichten vom Jahre 1912 im Auszug abgedruckten Briefes an R. FRICKE war. ⟦14⟧ ⟦11⟧ ⟦11⟧ ⟦11⟧ ⟦15⟧ ⟦16⟧ ⟦8⟧

Das auf der zweiten Seite dieses Briefabdruckes befindliche Zitat auf Herrn KOEBE ist in der Druckerei von mir unbekannter Seite, ohne meine Mitwirkung oder Vorkenntnis eingefügt worden; in der That sind die bezüglichen KOEBE'schen Noten mir erst nach ihrem Erscheinen bekannt geworden. ⟦17⟧ ⟦18⟧

AMSTERDAM, 1. Mai 1913.

NOTES

⟦1⟧ The first attempts, in the 1880's, at uniformisation of algebraic functions proceeded by means of automorphic functions (Klein 1882 A, Poincaré 1884): The automorphic function of a discrete group of fractional linear mappings maps the fundamental domain onto a cut up Riemann surface. In order to invert this procedure one must know whether all Riemann surfaces can be obtained in this way. The clue is to count parameters, of Riemann surfaces of a given ramification type on the one hand, of discrete groups on the other, or equivalently, of canonically cut up Riemann surfaces and fundamental domains, which in fact depend on the same number of parameters. The second method, due to F. Klein 1882 A involves comparing open varieties, the first, Poincaré's, is that of closed varieties. (Compare the exposition in P. Koebe 1914, Klein 1923 A.) By constructing automorphic functions the variety of groups or of fundamental domains is mapped into that of Riemann surfaces or cut up Riemann surfaces. It is a continuous, and under suitable norming, a one-to-one mapping of one variety into another of the same dimension. Nothwithstanding Cantor's 1877 discovery of the equivalence of cartesian spaces of all dimensions under one-to-one mappings, in the 1880's it was still quite natural to make sure by counting parameters that the image variety is filled up by the images of the original one. This then was the 'continuity proof' of uniformisation. As a matter of fact, Klein's use of open varieties was grossly insufficient, whereas Poincaré's had still to be completed by proving the invariance of the concept of n-dimensional domain under topological mappings. Moreover Poincaré's closing the variety of Riemann surfaces by a variety of codimension 2 rested on the tacit assumption that a variety could not be separated by a subvariety of codimension $\geqq 2$, but this gap was not noticed as early as this.

At present, if a Riemann surface is to be uniformised, it is wrapped up with, rather than cut up into, a simply connected surface, which is then conformally mapped upon a circular disk (or the plane, or the plane closed at infinity). In this framework automorphic functions are an *a posteriori* bonus: the automorphisms of the universal wrapping with respect to the given Riemann surface are reflected by a group of fractional linear mappings of the circular disk, and by inverting the uniformisation one gets an automorphic function of that group. This lucid idea goes back to H. A. Schwarz (an oral communication of 11 April 1882, mentioned in Correspondance 1923), but for a quarter of a century it could not extricate uniformisation from the dead end of automorphic functions, extensively cultivated in the school of F. Klein. P. Koebe 1907 and H. Poincaré 1907 finally proved uniformisation in Schwarz's spirit, and though this proof virtually included solving the open problems of automorphic functions, the dominance of automorphic functions still subsisted for quite a few years, to some extent even in P. Koebe's work. The correspondence between Klein and Koebe in 1910 (letters of Koebe to Klein of 11 April 1910, 10 October 1910, 13 October 1910 Univ. Bibl. Göttingen) still centers around the continuity method in uniformisation.

It is one of the ironies of history that shortly after uniformisation had been extricated from the dead end of automorphic functions, this dead end was broken through: Brouwer 1911 E proved the invariance of n-dimensional domain. No doubt this was the reason why Brouwer was invited to participate, along with F. Klein, P. Koebe, L. Bieberbach and E. Hilb, in what we would now call a symposium on automorphic functions, at the Deutsche Naturforscherversammlung and meeting of the Deutsche Mathematiker-Vereinigung in Karlsruhe, on 27 September 1911. The present paper is part of the report on this symposium. The contemporary echo of Brouwer's achievement in automorphic functions is heard in Fricke–Klein 1912 II where much attention is paid to it in the preface and the main part.

Op 19 June 1911 in a letter to Blumenthal (Y1) Brouwer had asked which topological theorem Blumenthal had orally indicated to him as indispensable for the continuity proof of the existence of polymorphic functions on Riemann surfaces; Brouwer doubted whether this was really Jordan's theorem. In a letter of 26 August 1911 (but in the mean time more letters may have been exchanged) Blumenthal recommended Brouwer to get in touch with Fricke about automorphic functions. He doubted whether even the Jordan theorem without some converse would suffice to justify the continuity argument. (We have observed that Poincaré's approach used that varieties are not separated by subvarieties of codimension 2.) Fragments, probably from the same period, show Brouwer intensely busy with automorphic functions. In a short time he succeeded in mastering this field, such as cultivated by Poincaré and the Klein school. Later on, in a letter of 24 February 1912 to Hilbert (Univ. Bibl. Göttingen), Brouwer characterised automorphic functions as '· · · den betreffenden Gegenstand, der ja ohnehin meinen Interessen ziemlich fern liegt und mit dem ich mich nur beiläufig auf Wunsch von Klein beschäftigt hatte · · ·'. Klein may have extended this wish as soon as he learned about Brouwer's proof of the invariance of *dimension*. When Brouwer wrote to Blumenthal (Y1), his manuscript on invariance of domain was finished. It is probable that at this date he was already in possession of the results he would communicate at Karlsruhe – what he asked Blumenthal for may have been additional information. It is not clear whether Brouwer contacted Fricke before Karlsruhe. Fricke did not attend this meeting. A contact in November is testified by Brouwer's letter to Hilbert of 31 May 1912 (Univ. Bibl. Göttingen).

As well as in the Karlsruhe lecture the subject is dealt with in 1912 D and 1912 G, which appeared at an earlier date than that one. 1919 L2 is also related to automorphic functions.

In the present paper the summary of Brouwer's lecture is followed by a discussion remark of Koebe and an unrelated rejoinder of Brouwer. This discrepancy is explained in a note, dated 1 May 1913, stuck into Brouwer's own reprint, probably a private print. (See also another version of this note in Brouwer 1919 L2, footnote 2.) It is clear from these data that Karlsruhe had been the scene of a clash between Brouwer and Koebe. This is confirmed by a paragraph of a letter of Blumenthal to Brouwer of 8 October 1911: 'Es hat mir sehr leid getan, dass ich Sie in Karlsruhe nicht mehr getroffen habe. Bernstein sagte mir, Sie hätten mit Koebe über den Kontinuitätsbeweis gestritten. Es ist nicht angenehm, mit ihm zu diskutieren, aber er hat bisher noch nie einen Fehler gemacht, daher bin ich geneigt, ihm auch in dieser Sache Kredit zu geben (ebenso wie seinerzeit Lebesgue, also vielleicht auch mit Unrecht), besonders da ich so ungefähr zu sehen glaube, was er mit dem Verzerrungssatz machen kann. Bernstein freilich schreibt ihm Ansichten zu, die wohl sehr anfechtbar wären. Aber ich halte das für ein Missverständnis Bernsteins.'

According to Brouwer's report in a letter to Hilbert of 24 February 1912 (Univ. Bibl. Göttingen), Koebe 'sagte ⟦· · ·⟧ am Ende des Vortrags folgendes:

"Weil auf Grund meines Verzerrungssatzes bei stetiger Änderung der Modulen *nichts passieren kann*, sind die von Herrn Brouwer erbrachten L⟦ösu⟧ngen der Schwierigkeiten der Invarianz des Gebiets und der Singularitätenfreiheit der Modulmannigfaltigkeit in meinem Gedankengange entbehrlich".

Worauf ich nachdrücklich erwiderte:

"Der Verzerrungssatz kann nur das von Poincaré für den Grenzkreisfall erreichte Resultat, und damit zugleich meinen Kontinuitätsbeweis auf das allgemeinste Fundamentaltheorem ausdehnen; meine Beiträge bleiben bei dieser Ausdehnung aber nach wie vor in ihrem vollem Umfang erforderlich."

Darauf sprach Koebe die unsinnigen Worte:

"Was Herr Brouwer gebracht hat, das mache ich mit Poincaré'schen Reihen," und dann schloss Klein die Diskussion.'

After many more annoying discussions at Karlsruhe, Brouwer had decided not to publish his work on automorphic functions, lest he got into more trouble with Koebe. But he afterwards changed his mind under the pressure of several people. According to a letter to Hilbert of 31 May 1912 he had felt the need of making sure of what Koebe really knew in order to protect himself against possible accusations of triviality or plagiarism. So he had got in touch with Koebe, but did not get any relevant information. However, both had agreed to exchange the manuscripts of forthcoming publications. On 2 January 1912 Brouwer sent Koebe the manuscript of 1912 D but Koebe refused to reciprocate. A long correspondence with Hilbert, Klein, Koebe and probably others developed. Y20 is the most illuminating among those pieces. See also 1912 D ⟦22⟧.

In papers of a later date Koebe used the *Verzerrungssatz* to eliminate Poincaré's closure trick, albeit only in special cases. At the Karlsruhe meeting Koebe may have believed the *Verzerrungssatz* could do more, that is, replace the invariance of domain argument. In any event when he edited his discussion remark and his own Karlsruhe lecture, Koebe 1912 A, he knew this was wrong. The only extant letter of Koebe to Brouwer (of early February 1912, Univ. Bibl. Göttingen) is uncompromising or even insulting although at about the same time a letter of Koebe to Klein of 30 March 1912 (Univ. Bibl. Göttingen) reveals little less than face saving surrender. However, Koebe 1912A, and still more Koebe 1914, in particular p. 51, is a candid acknowledgement of Brouwer's view. See also the impartial account in F. Klein 1923.

Brouwer's note of 1 May 1913 makes an allusion to another incident. Brouwer 1912 D had appeared with a change in the main text and in a footnote (see 1912 D ⟦7⟧). The change implies a more positive acknowledgment of some claims of Koebe though not of those related to the continuity proof. This *unauthorised* change was signalled by Brouwer; see ⟦17⟧ .

Oral tradition tells a cloak-and-dagger story about this footnote: On some dark afternoon in March an unidentified person waring a large hat, a turned up collar, and blue glasses called at Dieterick'sche Univ.-Buchdruckerei W. Fr. Kaestner in Göttingen, the printing office of *Göttinger Nachrichten*, and asked for the printer's proof of the next issue. He got it, and after a while he gave it back and left. The identity of this person has never been determined, nor is it known whether he made any change in Brouwer's reading proof, which of course disappeared after printing.

I do not know how much of this story is true. To a trustworthy friend of mine who years later asked him about this incident, Koebe explained it as a trick somebody had played on him. Though the revised edition of the footnote gives information which at that time was not publicly available, the hypothesis that it was a practical joke cannot at all be excluded in the Göttingen ambiance. Koebe was a picturesque character whose honesty and frankness forbade him to disguise his greatness as a mathematician: he travelled incognito, he often said that in his birthplace Luckenwalde the streetboys shouted after him 'Da geht der grosse Funktionentheoretiker' (which earned him the nickname of 'der grösste Luckenwalder Funktionentheoretiker'), and one of his best known sayings was 'Europa spricht davon – Koebe versendet Separate'. That such a character should be the target of a practical joke, in a place where practical jokes were not unusual, is not a farfetched hypothesis. In a letter to Klein of 25 May 1914 (Univ. Bibl. Göttingen) Koebe complained about gossip spread about him in Göttingen, related to the sticker in Brouwer 1912 H. It seems incredible that for nearly two years Koebe had been unaware of the whole affair, but this does not prove hypocrisy. Gossip, if unjustified, goes a long way before it finally reaches the ears of the ones incriminated.

From Brouwer's correspondence with Klein (Univ. Bibl. Göttingen) one can infer that Brouwer discovered the unauthorised change in 1912 D in the last week of June 1912. Koebe persevered in withholding the manuscripts of his publications even after they had been referred to in Brouwer 1912 D without Brouwer's knowledge. It seems that even Klein did not succeed in changing Koebe's mind (Letter to Klein of 1 July 1912, Univ. Bibl. Göttingen).

The final and an intermediate edition of the present paper are extant, both of them typewritten. The final edition contains several notes which were afterwards deleted. One of them is related to the discussion with Koebe; it does not provide new information. A not completely identical copy of the final edition is an enclosure of a letter to Hilbert of 27 February 1912 (Univ. Bibl. Göttingen).

⟦2⟧ Fricke–Klein 1912.

⟦3⟧ H. Poincaré 1884, Fricke–Klein 1912.

⟦4⟧ H. Poincaré 1884.

⟦5⟧ Brouwer means 1911 D, p. 97.

⟦6⟧ F. Klein 1882 A. J. Lüroth 1871. A. Clebsch 1873.

⟦7⟧ Brouwer 1912 G.

⟦8⟧ Brouwer 1912 D.

⟦9⟧ Brouwer 1911 E.

⟦10⟧ Brouwer 1911 F. R. Baire 1907 B. H. Lebesgue 1911 B.

⟦11⟧ P. Koebe 1912 A. The text of this intervention is incompatible with Brouwer's rejoinder (see ⟦1⟧).

⟦12⟧ Brouwer 1912 D, 1912 G.

⟦13⟧ This note stuck into Brouwer's own copies and probably into all reprints, is also mentioned by O. Blumenthal in a letter to F. Klein (Univ. Bibl. Göttingen) of 5 June 1914, and in a letter of P. Koebe to Klein (Univ. Bibl. Göttingen) of 25 May 1914. It seems to be a private print, made in Holland. Another version of this note, in Brouwer 1919 L2, footnote 2.

⟦14⟧ Read 'bei' instead of 'bij'. The error along with other evidence betrays the Dutch compositor.

⟦15⟧ P. Koebe 1912 B.

⟦16⟧ Among Brouwer's papers there is a copy of a typewritten precursor manuscript of Brouwer 1912 G (rather than 1912 D) which bears Koebe's note: 'Ich bin gerne bereit, nach den Weihnachtsferien behufs Abfassung einer Gött. Nachr. Note mit Ihnen in Verbindung zu treten. Vorher bin ich anderweitig zu sehr in Anspruch genommen. Mit bestem Gruss P. Koebe.' The manuscript and the note are undated; in any case they antedate Christmas 1911.

⟦17⟧ See ⟦1⟧ and Brouwer 1912 D ⟦7⟧.

⟦18⟧ P. Koebe 1912 B, 1912 C.

Über die topologischen Schwierigkeiten des Kontinuitätsbeweises der Existenztheoreme eindeutig umkehrbarer polymorpher Funktionen auf Riemannschen Flächen.

1912 D

Von

L. E. J. Brouwer in Amsterdam. ⟦1⟧

(Auszug aus einem Briefe an R. Fricke). ⟦2⟧

Vorgelegt von Herrn F. Klein in der Sitzung am 13. Januar 1912 durch den vorsitzenden Sekretär.

Sei $\varkappa$ eine Klasse von diskontinuierlichen linearen Gruppen mit einer bestimmten charakteristischen Signatur; es gilt in bezug auf diese Klasse das „allgemeine Kleinsche Fundamentaltheorem", wenn zu jeder kanonisch zerschnittenen mit n Punkten signierten Riemannschen Fläche vom Geschlechte p ein und nur ein kanonisches System von Fundamentalsubstitutionen einer Gruppe der Klasse $\varkappa$ gehört. ⟦3⟧

Bei der Kontinuitätsmethode, deren Klein sich in Bd. 21 der Mathematischen Annalen und Poincaré in Acta math. Bd. 4 zur Begründung des Fundamentaltheorems bedient, kommt es auf den Beweis folgender sechs Sätze an. ⟦4⟧ ⟦5⟧

1. Die Klasse $\varkappa$ enthält zu jedem ihr angehörigen kanonischen System von Fundamentalsubstitutionen ohne Ausnahme eine durch $6p-6+2n$ reelle Parameter eineindeutig und stetig darstellbare Umgebung.

2. Bei stetiger Änderung der Fundamentalsubstitutionen innerhalb der Klasse $\varkappa$ ändert sich die entsprechende kanonisch zerschnittene Riemannsche Fläche ebenfalls stetig.

3. Zwei verschiedene kanonische Systeme von Fundamentalsubstitutionen der Klasse $\varkappa$ können nicht derselben zerschnittenen Riemannschen Fläche entsprechen.

4. Wenn eine Folge von kanonisch zerschnittenen mit n Punkten signierten Riemannschen Flächen vom Geschlechte p gegen eine kanonisch zerschnittene mit n Punkten signierte Riemannsche Fläche vom Geschlechte p konvergiert und wenn jeder Fläche der Folge ein kanonisches System von Fundamentalsubstitutionen der Klasse $\varkappa$ entspricht, so entspricht der Limesfläche ebenfalls ein kanonisches System von Fundamentalsubstitutionen der Klasse $\varkappa$.

5. Die kanonisch zerschnittenen mit n Punkten signierten Riemannschen Flächen vom Geschlechte p bilden bei Zugrundelegung des Riemannschen Klassenbegriffs eine stetige Mannigfaltigkeit, welche zu jeder ihr angehörigen Fläche ohne Ausnahme eine durch $6p-6+2n$ reelle Parameter eineindeutig und stetig darstellbare Umgebung enthält.

6. Im r-dimensionalen Raume ist das eineindeutige und stetige Bild eines r-dimensionalen Gebiets wiederum ein Gebiet.

Die Sätze 1, 2, 3, 4 lasse ich hier außer Betracht. Für den
[[5]] Grenzkreisfall sind sie schon von Poincaré in Bd. 4 der Acta Mathematica erledigt worden [1]), während für den allgemeinsten Fall nur die Sätze 3 und 4 noch des erschöpfenden Beweises harren;
[[7]] indes ist es Herrn Koebe gelungen, auch diese Lücke vollständig auszufüllen [2]).

Die Sätze 5 und 6 bilden die von Ihnen hervorgehobenen [3]) topologischen Schwierigkeiten des Kontinuitätsbeweises. Von diesen ist Satz 6 durch meine unlängst erschienene Arbeit „Beweis der Invarianz des n-dimensionalen Gebiets" [4]) erledigt worden, und die Anwendung von Satz 5 wird am einfachsten umgangen, indem wir den Kontinuitätsbeweis in folgender, abgeänderter Form führen:

---

1) Der Poincaré'sche Beweis des Satzes 4 (l. c. pag. 250—276) bedarf vom
[[6]] jetzigen Standpunkte der Mengenlehre einiger geringfügiger Abänderungen, auf welche ich in einem auf der Naturforscherversammlung in Karlsruhe gehaltenen Vortrage hingewiesen habe.

2) Vgl. dessen vorläufige Note vom 13. Januar 1912 in den Gött. Nachr. „Begründung der Kontinuitätsmethode im Gebiete der konformen Abbildung und
[[8]] Uniformisierung" und „zur Begründung der Kontinuitätsmethode" in den Leipziger Berichten.

[[9]] 3) Fricke-Klein, Theorie der automorphen Funktionen, Bd. 2, S. 413.

[[10]] 4) Mathem. Annalen 71, S. 305—313. Vgl. auch die daselbst S. 314 zitierten Abhandlungen von Baire und Lebesgue, und eine demnächst in Mathem.
[[11]] Annalen 72 erscheinende Arbeit, in welcher der Satz fast unmittelbar aus der Invarianz der Dimensionenzahl gefolgert wird.

Satz 6 ist in der allgemeinen Form des Textes natürlich entbehrlich, wenn
[[12]] man beweisen kann, daß die Beziehung zwischen den Systemen von Fundamentalsubstitutionen und den entsprechenden zerschnittenen Riemannschen Flächen eine analytische ist.

[[604]]

Wir wählen $m > 2p - 2$ [1]) und betrachten auf der einen Seite die Menge M_g der zur Klasse $\varkappa$ gehörigen automorphen Funktionen mit ausschließlich einfachen Verzweigungspunkten und mit m einfachen, außerhalb der singulären Stellen der Gruppe liegenden Polen im Fundamentalbereich [2]), auf der anderen Seite die Menge M_f der aus m numerierten Blättern zusammengesetzten, über die Ebene ausgebreiteten, mit n im Endlichen gelegenen Punkten signierten Riemannschen Flächen vom Geschlechte p mit $2m + 2p - 2$ numerierten einfachen, im Endlichen gelegenen Verzweigungspunkten [2]); die Numerierung sowohl der Blätter wie der Verzweigungspunkte soll dabei der kanonischen Anordnung im Sinne von Lüroth-Clebsch entsprechen. ⟦13⟧

Die Menge M_f bildet ein Kontinuum und besitzt zu jeder ihr angehörigen Fläche ohne Ausnahme eine durch $4p - 8 + 2n + 4m$ reelle Parameter eineindeutig und stetig darstellbare Umgebung.

Zu einer willkürlichen der Menge M_g angehörigen automorphen Funktion φ existiert in M_g eine durch $4p - 8 + 2n + 4m$ reelle Parameter bestimmbare Umgebung u_φ; diese Parameter sind die m komplexen Stellen der Pole im Fundamentalbereich, die $m - p - 1$ komplexen Verhältnisse von $m - p$ willkürlich anzunehmenden Polresiduen und die $6p - 6 + 2n$ Parameter der kanonischen Systeme von Fundamentalsubstitutionen. Das Wertegebiet der zu u_φ gehörigen Parameter bildet ein $(4p - 8 + 2n + 4m)$-dimensionales Gebiet w_φ.

Der Funktion φ entspricht eine diskrete Menge von zu M_f gehörigen Flächen. Weiter folgern wir aus den Sätzen 1, 2, 3 und aus der Bemerkung, daß eventuelle birationale Transformationen einer Riemannschen Fläche in sich (nicht nur für die einzelne Riemannsche Fläche, sondern auch für die Gesamtheit der zu u_φ gehörigen Riemannschen Flächen) nicht beliebig klein werden können [3]),

---

1) Um die Gedanken zu fixieren, setzen wir im Folgenden $p > 1$ voraus.

2) Automorphe Funktionen, bez. über die Ebene ausgebreitete Riemannsche Flächen, welche sich nur um additive und multiplikative Konstanten unterscheiden, betrachten wir als identisch.

3) Birationale Transformationen einer Riemannschen Fläche vom Geschlechte $p > 1$ in sich können nämlich nach Hurwitz (Math. Annalen 41, S. 428) unmöglich ein kanonisches Schnittsystem in ein äquivalentes kanonisches Schnittsystem überführen. (Zwei kanonische Schnittsysteme heißen äquivalent, wenn sie sich durch stetige Variierung auf der als fest gedachten Fläche ineinander überführen lassen). Dasselbe gilt übrigens ganz allgemein von jeder periodischen, eineindeutigen und stetigen Transformation einer Fläche vom Geschlechte $p > 1$ in sich, weil die Überlagerungsfläche, welche sich um die Menge der Fixpunkte aperiodisch herum- ⟦14⟧ ⟦15⟧

daß einem hinreichend klein gewählten w_φ eine Menge von in M_f gelegenen eineindeutigen und stetigen Bildern, mithin auf Grund von Satz 6 eine in M_f gelegene Gebietsmenge entspricht. Dann aber entspricht auch der ganzen Menge M_g in M_f eine Gebietsmenge G_f.

Wir formulieren nun Satz 4 in folgender Form:

„Wenn eine Folge von kanonisch zerschnittenen Flächen von M_f gegen eine kanonisch zerschnittene Fläche von M_f konvergiert und wenn jeder Fläche der Folge ein kanonisches System von Fundamentalsubstitutionen der Klasse $\varkappa$ entspricht, so entspricht der Limesfläche ebenfalls ein kanonisches System von Fundamentalsubstitutionen der Klasse $\varkappa$" [1]).

Diese Eigenschaft zieht unmittelbar nach sich, daß die Gebietsmenge G_f in M_f keine Grenze besitzen kann, mithin die ganze Mannigfaltigkeit M_f ausfüllen muß. Hiermit ist das allgemeine Fundamentaltheorem bewiesen für jede in geeigneter Weise kanonisch zerschnittene, mit n Punkten signierte Riemannsche Fläche vom Geschlechte p, auf welcher algebraische Funktionen mit m außerhalb der Stigmata liegenden einfachen Polen und mit ausschließlich einfachen Verzweigungspunkten existieren, d. h. (für hinreichend großes m) für jede in geeigneter Weise kanonisch zerschnittene, mit n Punkten signierte Riemannsche Fläche vom Geschlechte p überhaupt, dann aber auf Grund Ihres Würfelsatzes [2])
⟦20⟧
zugleich für jede irgendwie kanonisch zerschnittene, mit n Punkten signierte Riemannsche Fläche vom Geschlechte p.
⟦21⟧ ⟦22⟧

windet, sonst einer periodischen, eineindeutigen und stetigen Transformation ohne Fixpunkte unterliegen würde.

Der im Texte benutzte Satz läßt sich seinerseits auch auf Grund der Uniformisierung beweisen, z. B. im von Poincaré auf Grund eines Schreibens von
⟦16⟧ Klein in Acta Math. 7, S. 15—19 entwickelten Gedankengange.

⟦17⟧ 1) In meinem Karlsruher Vortrage habe ich gezeigt, wie diese Form des
⟦18⟧ ⟦19⟧ Satzes sich für den Grenzkreisfall aus den Acta Math. 4, S. 250—276 enthaltenen Entwicklungen Poincaré's ablesen läßt.

⟦9⟧ 2) Fricke-Klein, Theorie der automorphen Funktionen, Bd. 2, S. 305.

⟦606⟧

NOTES

⟦1⟧ The reprint pagination is completed with the Journal pagination.

Several handwritten manuscripts, the final and a 'hyperfinal' one, which are typewritten, are extant. This last manuscript contains corrections introduced after the first proofreading, and a great number of unedited notes, part of which are included in the present edition.

Printer's proofs are extant, dated 8 February 1912, 23 February 1912, and 28 February 1912. The second is dated 26 Februari 1912 by Brouwer, the last is signed for press by Brouwer on 2 March 1912, though in a letter to Hilbert on 9 March 1912 (Univ. Bibl. Göttingen) he asked Hilbert 'meine Note in den Göttinger Nachrichten nun drucken zu lassen'. Major changes were introduced in the first printer's proof, among which are the change of title, which originally was 'Über den Kontinuitätsbeweis der Existenztheoreme ⟦· · ·⟧', the addition of footnote 1 on p. 604, and footnote 2 on p. 604 (though not the unauthorised text which was finally printed), an elaboration of footnote 4 on p. 604 and of footnote 3 on p. 605, a changing of footnote 1 on p. 606 and two final paragraphs, which, however, do not appear in the second proof. Minor changes were made in the second printer's proof.

⟦2⟧ I was greatly surprised to discover that two handwritten drafts of this 'Letter to Fricke' – dated Amsterdam 22 December 1911 and starting 'Im Anschluss an unsere letzte Unterhaltung · · ·' – were addressed to 'Hochgeehrter Herr Geheimrat' which I thought could only mean Felix Klein. How could the addressee of the letter have changed from 22 December 1911 to 13 January 1912, when Klein communicated it to *K. Gesellschaft der Wissenschaften zu Göttingen* through their secretary?, I asked. I found the explanation in another draft letter among Brouwer's papers (here edited as Y6; see Y6 ⟦1⟧) which was certainly addressed to Klein. I guessed that Klein had changed the name of the addressee into Fricke. The reason must have been Klein's illness – mentioned in Y6 – which prevented him from looking more closely into Brouwer's note, or the argument that the honour of being addressed by Brouwer was better deserved by Fricke with whom Brouwer must have corresponded for some time on the subject of the note.

Meanwhile I discovered that at that time Fricke was *Geheimrat* as well (properly, *Geheimer Hofrat*) so he could have been the addressee of this letter right from the start. Brouwer's letter to Fricke would have been sent to Klein (by Brouwer or Fricke) in order to be communicated to the *K. Gesellschaft der Wissenschaften.* Klein would have proposed that Brouwer introduce some changes, which was accepted by Brouwer Y6.

I have explained this in detail as a warning against misapprehensions regarding the destination of this letter.

⟦3⟧ 'allgemeine' added in compliance with Klein's wishes. See 6.

⟦4⟧ F. Klein 1882 A.

⟦5⟧ H. Poincaré 1884.

⟦6⟧ Added in the first proof.

⟦7⟧ The text of the manuscript

> 'indes soll in demnächst erscheinenden Arbeiten von Herrn Koebe auch diese Lücke ausgefüllt werden'

was changed in the first proof to

> 'indes hat Herr Koebe mir mitgeteilt, dass er in demnächst erscheinenden Arbeiten auch diese Lücke vollständig ausgefüllt hat [2]).
>
> [2]) Über Satz 3 findet sich schon eine vorläufige Bemerkung in einer am 20. Mai 1910 vorgelegten Note der Göttinger Nachrichten.'

This version appears in the second and third proof. The version actually printed was the result of an unauthorised change (see 1912 H ⟦1⟧). It differs from the authorised one by the more positive character of acknowledging Koebe's claimed results and by mentioning forthcoming papers of Koebe which Brouwer could not have known at that time. The change does not, however, involve Koebe's claims over which Brouwer and Koebe had clashed in Karlsruhe.

⟦8⟧ P. Koebe 1912 B, 1912 C. These are the papers which Brouwer complains Koebe had promised him and afterwards withheld (see 1912 H ⟦1⟧). The paper cited in the authentic version is P. Koebe 1910.

⟦9⟧ Fricke–Klein 1912.

⟦10⟧ Brouwer 1911 E. R. Baire 1907 B. Lebesgue 1911 B.

⟦11⟧ Brouwer 1912 C. The reference was added in the first proof.

⟦12⟧ This addition was promised in the letter to Klein Y6.

⟦13⟧ J. Lüroth 1871. A. Clebsch 1873.

⟦14⟧ The footnote got its present form in the first proof. See also Y6.

⟦15⟧ A. Hurwitz 1892. See also Brouwer 1919 L2, Y22, Y23.

⟦16⟧ H. Poincaré 1885 B.

⟦17⟧ In the manuscript footnote 1 on p. 606 was: 'Wie Herr Koebe auf der Naturforscherversammlung in Karlsruhe auseinandergesetzt hat, lässt sich dieser Satz in elegantester Weise folgern aus dem Verzerrungssatze.' This was deleted in the first proof. The present note 1 appears in the second proof although it was not indicated in the first proofreading. The reference to Koebe may have been deleted because meanwhile Koebe had recognised that he could not furnish the proof.

⟦18⟧ H. Poincaré 1884.

⟦19⟧ Add 'in' after 'aus' in the second line of footnote 1.

⟦20⟧ The 'hyperfinal manuscript' contains a strange Dutch note relating to 'Würfelsatzes'. Here follows a translation of this note:

> *Detailed proof.* Let P be a point of the cube for which the uniformisation according to κ is granted, Q a point for which it has to be proved. Draw within the cube the joining line segment PQ. For every point R of it there is certainly a point of M_f such that if for R the uniformisation is granted we can transfer it to all points of a line segment s_R of PQ including R, while moreover if for an arbitrary point of s_R the uniformisation is granted we can transfer it to R.

– So while preserving the uniformisation, we can proceed in the direction PQ, from P to P', from P' to P'', from P'' to P''', and so on. Let the point sequence $P^{(\alpha)}$ have a limit point R, then there is a point $P^{(k)}$ belonging to s_R, from which we can transfer the uniformisation to R. Then we go from R to R', R'', and so on up to a limit point S, and so on.

⟦21⟧ To the first proof Brouwer added two more paragraphs which, however, do not appear in the second proof. Here follows their text:

Andererseits haben Sie bewiesen (footnote: Fricke-Kkein, Theorie der automorphen Funktionen, Bd. 2, S. 305), dass die kanonischen Zerschneidungen derjenigen Flächen von M_f, welche sich durch ein zu einer bestimmten Klasse von *Grenzkreisgruppen* gehöriges *System* von Fundamentalsubstitutionen uniformisieren lassen, ein Kontinuum bilden.

Mithin bilden nun nicht nur die unzerschnittenen, sondern auch die kanonisch zerschnittenen Flächen von M_f ein Kontinuum, sodass das Fundamentaltheorem auch für jede *irgendwie kanonisch zerschnittene* Riemannsche Fläche vom Geschlechte p bewiesen ist.

⟦22⟧ The 'hyperfinal manuscript' contains a few more notes (translated as far as they are in Dutch):

Perforce publish the Karlsruhe lecture (in Jahresberichte ?) with the remark that I explained to Koebe that I worked with the polygons instead of the groups before he had told me a word about his ideas.

Indicate to Koebe the later changes in this lecture; the original text counted only simple poles and counted as different a pole residu 0 if conceived as attached to two different points.

I really intended to 'circumvent' the continuity of the 'set of the cut up surfaces of genus p' – this follows from the fact that I communicated the proof of the non-singularity of that set earlier to Fricke, so that circumventing *only this* would not have been meaningful.

Den vollständigen Inhalt der vorstehenden Note habe ich in den Tagen 27.–29. Sept. in Karlsruhe Herrn Koebe und mehreren anderen Fachgenossen (u.a. den Herren Bieberbach, Bernstein und Rosenthal) mündlich mitgeteilt, nachdem zuvor in der Sitzung der Deutschen Mathematiker-Vereinigung ich den vollständigen Kontinuitätsbeweis für den Grenzkreisfall vorgetragen hatte, und Herrn Koebe darauf hingewiesen hatte, dass Satz 4 sich für den allgemeinsten Kontinuitätsbeweis aus seinem Verzerrungssatze folgern lasse.

Mit dem Worte 'Mannigfaltigkeit' wird immer eine stetige ⟦that is connected⟧ Punktmenge gemeint, und zwar eine solche, welche in der Umgebung eines Punktes allgemeiner Lage die Struktur eines Zahlenraumes gewisser Dimensionszahl besitzt. Auf diese Stetigkeit habe ich deshalb nicht nachdrücklicher hingewiesen, weil sie in *meinem* Beweisgange keine Rolle spielt, und am Schluss von selbst als Folgerung herauskommt.

In der ersten Redaktion setzte das S. 4, Z. 1 befindliche "*Hiermit*" den Frickeschen Würfelsatz stillschweigend voraus.

Y 6 Letter to F. Klein, draft or copy

⟦1⟧ Es hat mir sehr leid getan zu hören, wie Sie Sich durch Ihre rastlosen und selbst-
losen Bemühungen im Interesse der Wissenschaft und des Gemeinwohls ganz
überarbeitet haben. Möchten Sie Sich in Zukunft doch etwas mehr schonen:
wir brauchen Sie ja noch so lange als unser ⟦sic⟧ Führer und Meister.

⟦2⟧ Bei der Korrektur werde ich gerne Ihren Bemerkungen Rechnung tragen;
ich werde den betreffenden Satz bezeichnen als Ihr "*allgemeines* Fundamental-
theorem" und ich werde darauf hinweisen, dass mein Satz 6 überflüssig wird,
sobald die analytische Beziehung der beiden Mannigfaltigkeiten nachgewiesen
ist (leider hat mir dieser Nachweis bis jetzt nicht gelingen wollen). Ueber Ihren
⟦3⟧ von Poincaré in Acta 7 wiedergegebenen Beweis der Endlichkeit der bir⟦ationa-
len⟧. Transform⟦ationen⟧. bei $p > 1$ möchte ich mir die Bemerkung erlauben,
dass dieser Beweis die Uniformisierung voraussetzt, also in meinem Gedanken-
gang nicht angewandt werden darf. Ueberdies brauche ich nicht nur die Endlich-
keit der Transformationen einer bestimmten Fläche, sondern von *allen* Flächen
vom Geschlechte p. (Es wäre ja a priori möglich, dass zwar die kleinste bir⟦atio-
nale⟧. Transf⟦ormation⟧. einer gegebenen Fläche eine endliche Maximalver-
rückung $\mathcal{F}$ besässe, dass aber dieses $\mathcal{F}$ bei Variierung der Fläche beliebig klein
werden könnte. Noether berichtete mir unlängst, dass auch ihm vom letzten
⟦4⟧ Satze kein Beweis unabh⟦ängig⟧. vom Hurwitz'schen bekannt ist. Eine aus-
führliche Redaktion der Arbeit würde ich sehr gern für die Annalen redigieren.

Mit den besten Wünschen für Ihre baldige und vollkommene Wiederherstellung und mit herzlichem Gruss

Ihr verehrender

NOTES

⟦1⟧ This is a one page draft or copy. The whole page is crossed through. It is clear from the details that the only possible addressee is Felix Klein, but no such letter has been found in Klein's papers, which contain quite a few of Brouwer's letters. However, some letter of this kind must have been written by Brouwer around New Year 1912 to Klein. The letter is related to 1912 D, which Klein submitted to the *K. Gesellschaft der Wissenschaften* on 13 January 1912. Since Brouwer promises changes to be made in the printer's proof, this letter cannot have been written much earlier than this date. The lack of New Year's wishes indicates an even later date.

⟦2⟧ Brouwer means his note 1912 D. The announced changes were indeed made during proofreading.

⟦3⟧ F. Klein 1882 A, H. Poincaré 1885 B.

⟦4⟧ See Brouwer 1912 D ⟦15⟧, Brouwer 1919 L2, Y22, Y23. A. Hurwitz 1892.

Letter to Koebe

Y 20

⟦Univ. Bibl. Göttingen⟧

Amsterdam, 14.2.12 ⟦1⟧

Overtoom 565

Abschrift ⟦1⟧

Geehrter Herr Koebe

Glücklicherweise bin ich noch im Besitze des abgekürzten Textes meines Karlsruher Vortrags, den ich hier beilege, damit Sie nicht länger behaupten können, als habe ich in Karlsruhe im Vortrage oder im Gespräche etwa die „*geschlossenen*" ⟦2⟧ Mannigfaltigkeiten Poincaré's verwendet.

Dass Sie eine solche Behauptung äussern konnten, beweist übrigens nur, dass die moderne Mengenlehre Ihnen absolut fremd sein muss. Sind ja die mit den angeblich „geschlossenen⟦"⟧ Mannigfaltigkeiten arbeitenden Entwicklungen Poincaré's der reinste Blödsinn, und nur dadurch zu entschuldigen, dass es zur Zeit ihrer Abfassung noch gar keine Mengenlehre gab.

Dass sich dennoch auf Grund der übrigen Poincaré'schen Entwicklungen der Beweis des „Weierstrass'schen Satzes" (in der Terminologie Klein's) und damit ⟦3⟧ der Kontinuitätsbeweis für den Grenzkreisfall durchführen lässt, das war eben der Inhalt meines Karlsruher Vortrag.

Durch Ihre Mitteilungen habe ich nur die weitere Einsicht erlangt, dass mittels Ihres Verzerrungssatzes meine Methode sich auf das allgemeinste Fundamentaltheorem übertragen lässt.

Was Sie Sich aus meinem Vortrage oder aus unseren Gesprächen über von mir benutzte „geschlossene Mannigfaltigkeiten" erinnern, bezieht sich auf folgendes: Dadurch, dass ich im beiliegenden Texte automorphe Funktionen, welche sich nur um additive und multiplikative Konstanten unterscheiden, als identisch ⟦4⟧ betrachte, erreiche ich, dass zu jedem inneren Punkte des Würfels eine *geschlossene* Mannigfaltigkeit von m-poligen Funktionen gehört. Nur hiermit wird das S. 3 Z. 19 des beiliegenden Textes befindliche „*Alsdann*" gerechtfertigt, denn nur auf ⟦5⟧ Grund dieser Geschlossenheit erlangt man die Sicherheit, dass eine Punktfolge von M_π immer dann einen zu M_π gehörigen Grenzpunkt besitzt, wenn die zugehörige Punktfolge des Würfels einen Grenzpunkt innerhalb des Würfels besitzt.

Ich bitte Sie, mir den beiliegenden Text nach einigen Tagen zurückzusenden.

Besten Gruss

L E J Brouwer

Warum schicken Sie mir nicht eine Abschrift Ihres Manuskriptes, wie ich es ⟦6⟧ Ihnen getan, und wie Sie es mir versprochen haben?

NOTES

[[1]] Copy, appended to a letter of Brouwer to Hilbert, 24 February 1912. – See Brouwer 1912 H [[1]].

[[2]] H. Poincaré 1884. See also 1912 H [[1]].

[[3]] F. Klein 1882 A.

[[4]] Brouwer 1912 H, p. 155, footnote [3]).

[[5]] Brouwer 1912 H, p. 156, l. 20.

[[6]] See Brouwer 1912 H [[1]].

Über die Singularitätenfreiheit der Modulmannigfaltigkeit. 1912G

Von

L. E. J. Brouwer in Amsterdam. ⟦1⟧

Vorgelegt in der Sitzung am 22. Juni 1912 durch Herrn D. Hilbert.

In den Klein-Poincaré'schen Kontinuitätsbeweisen der Klein'schen Fundamentaltheoreme spielt eine wichtige Rolle die Eigenschaft, daß bei Zugrundelegung des Riemann'schen Klassenbegriffs die nach irgend einem Modell kanonisch zerschnittenen bzw. die unzerschnittenen Riemannschen Flächen vom Geschlechte p eine stetige Mannigfaltigkeit bilden, welche in der Umgebung von jedem ihrer Punkte ohne Ausnahme durch $6p-6$ reelle Parameter eineindeutig und stetig darstellbar ist. Obwohl nun die Anwendung dieser Eigenschaft, wie ich in einem auf der Naturforscherversammlung in Karlsruhe gehaltenen Vortrage, sowie in einem am 13. Januar 1912 in der Gött. Ges. der ⟦2⟧
Wiss. vorgelegten Briefe an R. Fricke gezeigt habe, bei dem Kontinuitätsbeweise umgangen werden kann, so möchte ich andererseits hervorheben, daß wenigstens für die nach irgend einem Modell kanonisch zerschnittenen Riemannschen Flächen die Nichtexistenz von Singularitäten in einfacher Weise direkt[1]) hergeleitet werden kann und zwar wieder auf Grund des Hurwitz'schen Theorems, daß eine birationale Transformation einer Riemannschen Fläche in sich unmöglich ihre Differentiale erster Gattung invariant lassen kann[2]).

---

1) d. h. unabhängig von den Fundamentaltheoremen, welche den betreffenden Satz natürlich auch als Folge nach sich ziehen.

2) Mathem. Ann. 41, S. 428. ⟦3⟧

Um dies darzutun, betrachten wir die Menge μ der kanonisch zerschnittenen, über die Ebene ausgebreiteten, höchstens m-blättrigen Riemannschen Flächen vom Geschlechte $p\,(m > 2p-2)$, wobei wir solche Flächen, welche sich nur um additive und multiplikative Konstanten unterscheiden, als identisch betrachten. Diese Menge μ besteht aus einer die nur einfache, im Endlichen gelegene Verzweigungspunkte aufweisenden m-blättrigen Flächen[1]) enthaltenden Menge M, welche zu jedem ihrer Elemente eine durch $4m+4p-8$ reelle Parameter darstellbare Umgebung besitzt, und einer Menge γ, welche von einem Teile der Grenzelemente von M gebildet wird. Für hinreichend großes m existiert dann zu jedem Elemente von μ ein zur selben Riemann'schen Klasse gehöriges Element von M, sodaß wir nur zu zeigen haben, daß in der Umgebung eines willkürlichen Elementes von M die Klassen der kanonisch zerschnittenen Flächen vom Geschlechte p eineindeutig und stetig durch $6p-6$ reelle Parameter darstellbar sind.

Sei f_ω eine willkürliche zu M gehörige Riemannsche Fläche und $f_1, f_2, f_3, \ldots$ eine gegen f_ω konvergierende Folge von zu M gehörigen Riemannschen Flächen. Es existiert dann nach Ritter[2]) eine Folge von algebraischen Kurven $\varphi_1(x,y)=0$, $\varphi_2(x,y)=0$, $\varphi_3(x,y)=0, \ldots$, welche der Reihe nach zu $f_1, f_2, f_3, \ldots$ gehören und gegen eine einzige, zu f_ω gehörige algebraische Limeskurve $\varphi_\omega(x,y)=0$ konvergieren. Von diesen Kurven dürfen wir annehmen, daß sie alle denselben Grad $2n$ besitzen und nur mehrfache Punkte mit getrennten Tangenten aufweisen, daß die unendlich fernen Punkte der X- und der Y-Axe für sie alle als n-fache Punkte auftreten, daß die im Endlichen gelegenen mehrfachen Punkte von $\varphi_\omega(x,y)=0$ mit Ausnahme eines einzigen n-fachen Punktes P gewöhnliche Doppelpunkte D_α sind und daß von den im Endlichen gelegenen mehrfachen Punkten der $\varphi_\nu(x,y)=0$ gegen jeden Punkt D_α je ein gewöhnlicher Doppelpunkt und gegen P der Rest konvergiert.

Zu $\varphi_\omega(x,y)=0$ bzw. zu jeder Kurve $\varphi_\nu(x,y)=0$ betrachten wir nun die Menge ϱ_ω bzw. ϱ_ν der höchstens m Pole aufweisenden rationalen Funktionen von x und y; diese Funktionen setzen sich (abgesehen von der additiven Konstante) linear und homogen zusammen aus $m-p$ Elementarfunktionen, in welche die m Pole (x', y'), (x'', y''), $\ldots$, $(x^{(m)}, y^{(m)})$ algebraisch als Parameter ein-

---

⟦4⟧ 1) Die Blätter und die Verzweigungspunkte dieser Flächen denken wir der kanonischen Anordnung im Sinne von Lüroth-Clebsch entsprechend numeriert.

⟦5⟧ 2) Mathem. Ann. 46, S. 227.

⟦804⟧

⟦588⟧

gehen, während die zu den ϱ_ν gehörigen Elementarfunktionen gegen die zu ϱ_ω gehörigen Elementarfunktionen konvergieren. Die Systeme von Verzweigungspunkten $\mathfrak{F}_\omega$ bzw. $\mathfrak{F}_\nu$ der von ϱ_ω bzw. von den ϱ_ν in μ bestimmten Flächen bilden somit $(4m-2p-2)$-dimensionale, von $2m-p-1$ komplexen Parametern abhängende algebraische Gebilde. Die $\mathfrak{F}_\nu$ konvergieren gegen $\mathfrak{F}_\omega$, und zwar gleichmäßig, denn die Konvergenz besteht für jede Wahl der gegen f_ω konvergierenden Folge $f_1, f_2, f_3, \ldots$

Sei ε eine bestimmte, beliebig klein zu wählende Größe und g ein solches Wertegebiet der $2m-p-1$ komplexen Parameter, daß das entsprechende Teilgebiet h_ω von $\mathfrak{F}_\omega$ in M enthalten ist, sich regulär verhält und nur unterhalb ε liegende Schwankungen der Differentialquotienten erster Ordnung in bezug auf die Parameter aufweist. Alsdann existiert in M eine solche Umgebung B von f_ω, daß für jede zu B gehörige Fläche f_α das dem Parametergebiete g entsprechende Teilgebiet h_α von $\mathfrak{F}_\alpha$ in M enthalten ist, sich regulär verhält und nur unterhalb 2ε liegende Schwankungen der Differentialquotienten erster Ordnung in bezug auf die Parameter aufweist, während h_α mit f_α stetig variiert und für f_ω in h_ω übergeht. Wenn wir unter den h_α je zwei miteinander zusammenhängende als identisch betrachten, so kann die Menge der h_α auf ein $(4m+4p-8)-(4m-2p-2)=(6p-6)$-dimensionales Gebiet eineindeutig und stetig abgebildet werden.

Wir behaupten weiter, daß je zwei dieser h_α verschiedenen Gebilden $\mathfrak{F}_\alpha$ angehören müssen. Im entgegengesetzten Falle würde nämlich eine solche gegen $\mathfrak{F}_\omega$ konvergierende Folge $\mathfrak{F}_1, \mathfrak{F}_2, \ldots$ existieren, deren jedes $\mathfrak{F}_\nu$ wenigstens zwei der h_α, welche wir mit h_ν und h'_ν bezeichnen wollen, enthielte. Sei p_ω eine zu h_ω gehörige Fläche, p_ν bzw. p'_ν eine zu h_ν bzw. h'_ν gehörige Fläche, welche für unbegrenzt wachsendes ν gegen p_ω konvergiert, so kann die p_ν und p'_ν verbindende birationale Transformation für unbegrenzt wachsendes ν nicht unbegrenzt klein werden, sodaß sie gegen eine birationale Transformation von p_ω in sich konvergieren muß. Dies hieße aber, daß die kanonisch zerschnittene Fläche p_ω eine das Schnittsystem in ein äquivalentes Schnittsystem überführende birationale Transformation in sich zuließe, was gegen das oben zitierte Hurwitz'sche Theorem verstoßen würde.

Hieraus folgt unmittelbar, daß die Menge der $\mathfrak{F}_\alpha$ sich eineindeutig und stetig auf ein $(6p-6)$-dimensionales Gebiet abbilden läßt, sodaß die Klassenmenge der kanonisch zerschnittenen Riemannschen Flächen vom Geschlechte p in der Umgebung von f_ω, mithin wegen der willkürlichen Wahl von f_ω in der

Umgebung von jedem ihrer Punkte sich wie eine singularitätenfreie $(6p-6)$-dimensionale Mannigfaltigkeit verhält.

Der vorstehende Beweis gilt für die nach irgend einem dem allgemeinen Klein'schen Fundamentaltheorem entsprechenden Modell kanonisch zerschnittenen, nicht aber für die unzerschnittenen Riemannschen Flächen. In der Tat läßt er die Frage unentschieden, ob in der Klassenmenge der unzerschnittenen Flächen die Klassen der regulären Flächen nicht eine im Sinne der Analysis Situs singuläre Rolle spielen.

Amsterdam, 12. Juni 1912.

NOTES

⟦1⟧ The reprint pagination is completed with the journal pagination. – See 1912 H, 1912 F. A printer's proof dated 8 July 1912 and signed for press, 11 July 1912, is extant.
⟦2⟧ Brouwer 1912 D.
⟦3⟧ Hurwitz 1892. See also Brouwer 1919 L2, Y22, Y23.
⟦4⟧ J. Lüroth 1871. A. Clebsch 1873.
⟦5⟧ Ritter 1895.

Über die Erweiterung des Definitionsbereichs einer stetigen Funktion. 1918C

Von

L. E. J. Brouwer in Amsterdam. [[1]]

Es handelt sich um den Beweis der folgenden Eigenschaft:

Ist A eine abgeschlossene Punktmenge des n-dimensionalen Raumes und f eine beschränkte Funktion, die auf A definiert und in jedem Häufungspunkte von A stetig ist, so kann man eine im ganzen Raum stetige, beschränkte Funktion F finden, die in jedem Punkte von A gleich f ist.

Von dieser Eigenschaft existieren explizite Beweise von De la Vallée Poussin*) und Bohr**), die beide die Funktion F mittels eines unendlichen Prozesses (ersterer mittels einer unendlichen Reihe, letzterer mittels eines Integrals) konstruieren. Eine elementarere Hilfsmittel benutzende Herstellungsart der Funktion F läßt sich aus dem § 4 meines Beweises der Invarianz des n-dimensionalen Gebiets (Math. Ann. 71, S. 309—311), dessen Gegenstand von einer allgemeineren Aufgabe gebildet wird, wie folgt herausschälen. [[4]]

Wir nehmen mit dem n-dimensionalen Raum eine solche Fundamentalreihe $\zeta_1, \zeta_2, \zeta_3, \cdots$ von simplizialen Zerlegungen (vgl. Math. Ann. 71, S. 101) vor, daß $\zeta_{\nu+1}$ eine Unterteilung von ζ_ν ist, und daß die Größe der zu ζ_ν gehörigen Grundsimplexe für unbeschränkt wachsendes ν gleichmäßig gegen Null konvergiert. Zu jedem ζ_ν kann man eine solche von Grundsimplexen gebildete in der Komplementärmenge A' von A enthaltene abgeschlossene Gebietsmenge G_ν bestimmen, deren Grenze für unbeschränkt wachsendes ν gleichmäßig gegen A konvergiert. Wir setzen $G_1 = g_1$, bezeichnen die von denjenigen Grundsimplexen von $G_{\nu+1}$, welche nicht in G_ν enthalten sind, gebildete Punktmenge mit $g_{\nu+1}$ und verstehen unter einem σ_ν ein zu g_ν gehöriges Grundsimplex von G_ν. Wir dürfen annehmen, daß für

*) „Intégrales de Lebesgue, Fonctions d'Ensemble, Classes de Baire", § 125. [[2]]

**) Mitgeteilt von Carathéodory in „Vorlesungen über reelle Funktionen", § 541, 542. [[3]]

jedes ν die Grenzen von G_ν und $G_{\nu+1}$ keinen gemeinschaftlichen Punkt besitzen, und kein σ_ν gleichzeitig die Grenze von $G_{\nu-1}$ und die Grenze von G_ν berührt.

Von einem σ_ν, welches die Grenze von G_ν berührt, können für $1 \leqq p \leqq n-1$ eine oder mehrere p-dimensionale Seiten nach $\zeta_{\nu+1}$ zerlegt sein. Von den derartigen σ_ν bestimmen wir in folgender Weise eine „aus ζ_ν und $\zeta_{\nu+1}$ gemischte" simpliziale Zerlegung: Wir gehen aus von den schon vorhandenen simplizialen Zerlegungen der Kanten. Sodann bestimmen wir von jeder noch nicht zerlegten zweidimensionalen Seite, welche zerlegte Kanten besitzt, eine „gemischte" simpliziale Zerlegung, indem wir aus einem in ihrem Innern willkürlich gewählten Punkt die schon vorhandene simpliziale Zerlegung ihrer Kanten projizieren; die übrigen zweidimensionalen Seiten bleiben unzerlegt bzw. behalten ihre bisherige Zerlegung bei. In dieser Weise fortfahrend, bestimmen wir, nachdem die Zerlegungen der p-dimensionalen Seiten festgelegt sind, von jeder noch nicht zerlegten $(p+1)$-dimensionalen Seite, welche zerlegte p-dimensionale Seiten besitzt, eine „gemischte" simpliziale Zerlegung, indem wir aus einem in ihrem Innern willkürlich gewählten Punkt die schon vorhandene simpliziale Zerlegung ihrer p-dimensionalen Seiten projizieren; die übrigen $(p+1)$-dimensionalen Seiten bleiben unzerlegt bzw. behalten ihre bisherige Zerlegung bei.

Wenn wir unter den $\mathfrak{s}_\nu$ einerseits diejenigen σ_ν, welche die Grenze von G_ν nicht berühren, andererseits die Grundsimplexe der aus ζ_ν und $\zeta_{\nu+1}$ gemischten simplizialen Zerlegungen derjenigen σ_ν, welche die Grenze von G_ν berühren, verstehen, so können wir die Punktmenge A' als eine offene n-dimensionale Mannigfaltigkeit (vgl. Math. Ann. 71, S. 98) und die zu den verschiedenen Werten von ν gehörigen $\mathfrak{s}_\nu$ als die Grundsimplexe einer simplizialen Zerlegung ζ dieser offenen n-dimensionalen Mannigfaltigkeit auffassen.

Nunmehr weisen wir jedem Grundpunkte H von ζ einen solchen Punkt H_1 von A zu, welcher von ihm eine das Doppelte der kleinstmöglichen nicht übersteigende Entfernung besitzt, und setzen den Wert der zu bestimmenden Funktion F in H gleich ihrem schon bekannten Werte in H_1. Um weiter den Wert von F in einem willkürlichen Punkte P von A', welches dem Innern oder der Grenze des Grundsimplexes $\mathfrak{s}$ von ζ angehört, zu bestimmen, verfahren wir folgendermaßen: P ist Schwerpunkt von gewissen in den Eckpunkten $E_1, \cdots, E_{n+1}$ von $\mathfrak{s}$ konzentrierten positiven Massen $\mu_1, \cdots, \mu_{n+1}$. Wenn die dem obigen gemäß in $E_1, \cdots, E_{n+1}$ bestimmten Werte von F der Reihe nach gleich $\varphi_1, \cdots, \varphi_{n+1}$ sind, so setzen wir den Wert von F in P gleich $\frac{\varphi_1 \mu_1 + \cdots + \varphi_{n+1} \mu_{n+1}}{\mu_1 + \cdots + \mu_{n+1}}$. Hiermit ist der Wert von F in jedem Punkte sowohl von A wie von A', mithin in jedem Punkte des Raumes festgelegt.

Der Stetigkeitsbeweis braucht nur für einen willkürlichen Punkt K der Begrenzung von A geführt zu werden Zu diesem Zwecke wählen wir eine beliebig kleine positive Zahl ε_1. Alsdann existiert eine solche mit ε_1 gegen Null konvergierende positive Zahl ε_2, daß in jedem in einer Entfernung $\leqq \varepsilon_2$ von K liegenden, zu A gehörigen Punkte der Wert von f um einen absoluten Betrag $\leqq \varepsilon_1$ von $f(K)$ verschieden ist, und eine solche mit ε_2 gegen Null konvergierende positive Zahl ε_3, daß jedes $\mathfrak{s}_\nu$, welches in seinem Innern oder auf seiner Grenze einen in einer Entfernung $\leqq \varepsilon_3$ von K liegenden Punkt enthält, ganz innerhalb einer Entfernung $\leqq \frac{1}{4}\varepsilon_2$ von K gelegen ist, mithin nur solche Werte von F, welche von $F(K)$ um einen absoluten Betrag $\leqq \varepsilon_1$ verschieden sind, aufweisen kann, so daß *in jedem in einer Entfernung* $\leqq \varepsilon_3$ *von* K *liegenden Punkte des Raumes der Wert von* F *um einen absoluten Betrag* $\leqq \varepsilon_1$ *von* $F(K)$ *verschieden ist.*

MA 79, p. 403 1919 E

Nachträgliche Bemerkung über die Erweiterung des Definitionsbereichs einer stetigen Funktion.

Von

L. E. J. Brouwer in Amsterdam. [[5]]

Die Erweiterung des Definitionsbereichs einer auf einer abgeschlossenen Punktmenge definierten stetigen Funktion, welche, wie ich S. 209—211 dieses Annalenbandes hervorgehoben habe, durch Spezialisierung einer Math. Ann. 71, S 309—311 von mir definierten Konstruktion erhalten wird, ist seitdem ebenfalls, nicht nur, wie angegeben, von De la Vallée Poussin und Bohr, sondern auch von Tietze im Journ. f. Math. 145, S. 10—11 [[6]] erzielt worden, und zwar steht Tietzes Methode der meinigen sehr nahe und besitzt den gleichen elementaren Charakter: es werden nur, statt Simplexe, Kuben gebraucht.

NOTES

⟦1⟧ See 1919 E.

⟦2⟧ De la Vallée Poussin 1916.

⟦3⟧ C. Carathéodory 1918.

⟦4⟧ Brouwer 1911 E.

⟦5⟧ See Brouwer 1918 C.

⟦6⟧ Tietze 1914. There now are many more proofs.

Lebesguesches Maß und Analysis Situs. 1918 D

Von

L. E. J. Brouwer in Amsterdam.

Zweck des vorliegenden Aufsatzes ist, für das Lebesguesche Maß die vollständige Abwesenheit von für die Analysis Situs invarianten Eigenschaften darzutun.*) Die Beweisführung, welche für den linearen Fall noch keinerlei Schwierigkeiten bietet**), beschränkt sich im folgenden auf ebene Punktmengen; doch läßt sich, wie ich glaube, die hier angewandte Methode ohne prinzipielle Änderungen auf Punktmengen des n-dimensionalen Raumes übertragen.

§ 1. Ableitung eines Hilfssatzes.

Wir denken in der Ebene ein rechtwinkliges Koordinatensystem, verstehen unter einer Breitelinie $b_{a_1 a_2 a_3 \cdots}$ (bzw. unter einer Längelinie $l_{a_1 a_2 a_3 \cdots}$) eine gerade Linie, welche der X-Achse (bzw. der Y-Achse) parallel ist und von ihr eine im ternalen System gemessene positive Entfernung $0, a_1 a_2 a_3 \cdots$ besitzt, und betrachten eine nirgends dichte abgeschlossene Punktmenge C, welche ganz zum von den Längelinien l_0 und l_3 und den Breitelinien b_0 und b_3 begrenzten abgeschlossenen Quadrate k gehört, aber weder von l_0 noch von l_3 Teilsegmente enthält. Wir wählen eine solche Fundamentalreihe $\eta_1, \eta_2, \cdots$ von positiven Zahlen < 1, daß das unendliche Produkt

*) Man vergleiche folgende Äußerung Borels, welche mir auf eine entgegengesetzte Vermutung hinzuweisen scheint: „Les ensembles qui ne sont pas de mesure nulle sont formés d'une matière simple, avec des ensembles continus positifs ou négatifs; ils sont hétérogènes au continu; les ensembles de mesure nulle peuvent être, au contraire, sensiblement homogènes au continu, c'est-à-dire identiques à eux-mêmes dans des intervalles aussi petits que l'on veut . . . la notion d'ensemble de mesure nulle est primordiale" (Bull. Soc. Math. de France 41 (1913), S. 17). [[1]]

**) Beispiele topologischer Äquivalenz von linearen Nullmengen und Komplementärmengen von Nullmengen finden sich u. a. in einem Schreiben von mir an Herrn Blumenthal aus dem Jahre 1913; doch waren solche damals auch anderen Mathematikern, z. B. Herrn Bohl, bekannt; das erste in der Literatur auftretende Beispiel dürfte sich finden bei Carathéodory, Vorlesungen über reelle Funktionen, § 337. [[2]]

$\prod_\nu (1 - \eta_\nu)$ *nicht* verschwindet, und das unendliche Produkt $\prod_\nu (1 + \eta_\nu)$ einen endlichen Wert besitzt, und nehmen mit dem Quadrate k der Reihe nach eine Fundamentalreihe $\tau_1, \tau_2, \cdots$ von eineindeutigen und stetigen Transformationen vor, welche alle Breitelinien nebst l_0 und l_3 in sich überführen, und im folgenden näher definiert werden sollen.

Definition von τ_1. Wir bezeichnen mit s_0, s_1 und s_2 die Streifen, in welche k durch b_1 und b_2 zerlegt wird, und bestimmen eine endliche Folge $q_1, q_2, q_3, \cdots$ von mit k homothetischen und keinen Punkt von C enthaltenden Quadraten, welche abwechselnd zwischen b_0 und b_1 und zwischen b_2 und b_3 liegen, während ihre Mittelpunkte wachsende Abszissen besitzen, keine Längelinie zwei q_ν trifft, der erste Mittelpunkt auf l_0 und der letzte auf l_3 gelegen ist, und für jedes ν der Abstand der Schnittpunkte der Längelinien durch die Mittelpunkte von q_ν und $q_{\nu+1}$ mit einer willkürlichen Breitelinie weniger als ε_1 beträgt. Der Minimalwert der Seitenlängen der q_ν sei ζ_1. Die Schnittpunkte derjenigen beiden Längelinien, welche $q_{2\nu-1}$ in drei kongruente Rechtecke zerlegen, mit b_0 bzw. b_1 bezeichnen wir mit $A_{2\nu-1}$ und $B_{2\nu-1}$ bzw. $C_{2\nu-1}$ und $D_{2\nu-1}$. Die Schnittpunkte derjenigen beiden Längelinien, welche $q_{2\nu}$ in drei kongruente Rechtecke zerlegen, mit b_2 bzw. b_3 bezeichnen wir mit $E_{2\nu}$ und $F_{2\nu}$ bzw. $G_{2\nu}$ und $H_{2\nu}$. Sodann verstehen wir unter den *Hauptlinien zweiter Ordnung* die Streckenzüge $B_{2\nu-1}\, D_{2\nu-1}\, E_{2\nu}\, G_{2\nu}$ und $H_{2\nu}\, F_{2\nu}\, C_{2\nu+1}\, A_{2\nu+1}$ nebst den in l_0 und l_3 enthaltenen Seiten von k. Zerlegen wir die zwischen zwei bestimmten aufeinanderfolgenden Hauptlinien zweiter Ordnung enthaltenen Teilstrecken von Breitelinien in je zwei Segmente, welche auf jeder Breitelinie dasselbe Verhältnis aufweisen, so nennen wir den geometrischen Ort der Teilpunkte eine *Längelinie zweiter Ordnung*. Diejenige Längelinie zweiter Ordnung, welche b_0 im selben Punkte wie $l_{a_1 a_2 a_3 \cdots}$ trifft, bezeichnen wir mit ${}_2l_{a_1 a_2 a_3 \cdots}$ und bestimmen τ_1 durch die Eigenschaft, daß sie das in k liegende Segment von $l_{a_1 a_2 a_3 \cdots}$ für jede Wahl der a_ν in ${}_2l_{a_1 a_2 a_3 \cdots}$ überführt. Durch geeignete Wahl von ε_1 sorgen wir dafür, daß für jede Wahl von a_1 die zwischen ${}_2l_{a_1}$ und ${}_2l_{a_1+1}$ enthaltene Teilstrecke einer willkürlichen Breitelinie zwischen $\frac{1}{3}(1 - \eta_1)$ und $\frac{1}{3}(1 + \eta_1)$ liegt.

Definition von τ_2. Sei m_2 eine solche positive ganze Zahl, daß für jede Wahl von $a_1, a_2, \cdots, a_{m_2}$ die zwischen ${}_2l_{a_1 \cdots a_{m_2}}$ und ${}_2l_{a_1 \cdots a_{m_2-1}(a_{m_2}+1)}$ enthaltene Teilstrecke einer willkürlichen Breitelinie $< \frac{1}{9}$ ist. Wir bezeichnen mit $s_{\mu 0}, s_{\mu 1}, s_{\mu 2}$ die Streifen, in welche s_μ durch $b_{\mu 1}$ und $b_{\mu 2}$ zerlegt wird, und bestimmen in jedem s_μ eine endliche Folge ${}_\mu q_1, {}_\mu q_2, {}_\mu q_3, \cdots$ von mit k homothetischen und keinen Punkt von C enthaltenden Quadraten, welche abwechselnd zwischen $b_{\mu 0}$ und $b_{\mu 1}$ und zwischen $b_{\mu 2}$

[596]

und $b_{\mu+1}$ liegen, während die Längelinien zweiter Ordnung durch ihre Mittelpunkte b_μ in Punkten mit wachsenden Abszissen treffen, keine Längelinie zweiter Ordnung zwei ${}_\mu q_\nu$ trifft, der erste Mittelpunkt auf l_0 und der letzte auf l_3 gelegen ist, und für jedes ν der Abstand der Schnittpunkte der Längelinien zweiter Ordnung durch die Mittelpunkte von ${}_\mu q_\nu$ und ${}_\mu q_{\nu+1}$ mit einer willkürlichen Breitelinie weniger als ε_2 beträgt. Der Minimalwert der Seitenlängen der ${}_\mu q_\nu$, wenn wir sowohl μ wie ν variieren lassen, sei ζ_2. Die Schnittpunkte derjenigen beiden Längelinien zweiter Ordnung, welche die der X-Achse parallele Mittellinie von ${}_\mu q_{2\nu-1}$ in drei gleiche Stücke zerlegen, mit $b_{\mu 0}$ bzw. $b_{\mu 1}$ bezeichnen wir mit ${}_\mu A_{2\nu-1}$ und ${}_\mu B_{2\nu-1}$ bzw. ${}_\mu C_{2\nu-1}$ und ${}_\mu D_{2\nu-1}$. Die Schnittpunkte derjenigen beiden Längelinien zweiter Ordnung, welche die der X-Achse parallele Mittellinie von ${}_\mu q_{2\nu}$ in drei gleiche Stücke zerlegen, mit $b_{\mu 2}$ bzw. $b_{\mu+1}$ bezeichnen wir mit ${}_\mu E_{2\nu}$ und ${}_\mu F_{2\nu}$ bzw. ${}_\mu G_{2\nu}$ und ${}_\mu H_{2\nu}$. Sodann verstehen wir unter den *Teilhauptlinien dritter Ordnung von* s_μ die Streckenzüge ${}_\mu B_{2\nu-1}$ ${}_\mu D_{2\nu-1}$ ${}_\mu E_{2\nu}$ ${}_\mu G_{2\nu}$ und ${}_\mu H_{2\nu}$ ${}_\mu F_{2\nu}$ ${}_\mu C_{2\nu+1}$ ${}_\mu A_{2\nu+1}$ nebst den in l_0 und l_3 enthaltenen Seiten von s_μ. Zerlegen wir die zwischen zwei bestimmten aufeinanderfolgenden Teilhauptlinien dritter Ordnung von s_μ enthaltenen Teilstrecken von Breitelinien in je zwei Segmente, welche auf jeder Breitelinie dasselbe Verhältnis aufweisen, so nennen wir den geometrischen Ort der Teilpunkte eine *Teilllängelinie dritter Ordnung von* s_μ. Unter einer *Längelinie dritter Ordnung* verstehen wir einen aus Teillängelinien dritter Ordnung der verschiedenen s_μ zusammengesetzten, b_0 und b_1 verbindenden Streckenzug. Diejenige Längelinie dritter Ordnung, welche b_0 im selben Punkte wie $l_{a_1 a_2 a_3 \cdots}$ trifft, bezeichnen wir mit ${}_3 l_{a_1 a_2 a_3 \cdots}$ und bestimmen τ_3 durch die Eigenschaft, daß sie ${}_2 l_{a_1 a_2 a_3 \cdots}$ für jede Wahl der a_ν in ${}_3 l_{a_1 a_2 a_3 \cdots}$ überführt. Wir wählen $\varepsilon_2 < \frac{1}{27}\zeta_1$, und sorgen weiter durch geeignete Wahl von ε_2 dafür, daß für jede Wahl von $a_1, a_2, \cdots, a_{m_2}$ das Verhältnis von der zwischen ${}_3 l_{a_1 \cdots a_{m_2-1} a_{m_2}}$ und ${}_3 l_{a_1 \cdots a_{m_2-1} (a_{m_2}+1)}$ enthaltenen Teilstrecke einer willkürlichen Breitelinie und der zwischen ${}_2 l_{a_1 \cdots a_{m_2-1} a_{m_2}}$ und ${}_2 l_{a_1 \cdots a_{m_2-1} (a_{m_2}+1)}$ enthaltenen Teilstrecke derselben Breitelinie zwischen $1-\eta_2$ und $1+\eta_2$ liegt.

Definition von τ_n. Sei m_n eine solche positive ganze Zahl $> m_{n-1}$, daß für jede Wahl von $a_1, a_2, \cdots, a_{m_n}$ die zwischen ${}_n l_{a_1 \cdots a_{m_n-1} a_{m_n}}$ und ${}_n l_{a_1 \cdots a_{m_n-1} (a_{m_n}+1)}$ enthaltene Teilstrecke einer willkürlichen Breitelinie $< \frac{1}{3^n}$ ist. Wir bezeichnen mit $s_{\mu_1 \cdots \mu_{n-1} 0}$, $s_{\mu_1 \cdots \mu_{n-1} 1}$, $s_{\mu_1 \cdots \mu_{n-1} 2}$ die Streifen, in welche $s_{\mu_1 \cdots \mu_{n-1}}$ durch $b_{\mu_1 \cdots \mu_{n-1} 1}$ und $b_{\mu_1 \cdots \mu_{n-1} 2}$ zerlegt wird, und bestimmen in jedem $s_{\mu_1 \cdots \mu_{n-1}}$ eine endliche Folge ${}_{\mu_1 \cdots \mu_{n-1}} q_1$, ${}_{\mu_1 \cdots \mu_{n-1}} q_2$,

${}_{\mu_1\cdots\mu_{n-1}}q_3, \cdots$ von mit k homothetischen und keinen Punkt von C enthaltenden Quadraten, welche abwechselnd zwischen $b_{\mu_1\cdots\mu_{n-1}0}$ und $b_{\mu_1\cdots\mu_{n-1}1}$ und zwischen $b_{\mu_1\cdots\mu_{n-1}2}$ und $b_{\mu_1\cdots\mu_{n-1}3}$ liegen, während die Längelinien n^{ter} Ordnung durch ihre Mittelpunkte $b_{\mu_1\cdots\mu_{n-1}}$ in Punkten mit wachsenden Abszissen treffen, keine Längelinie n^{ter} Ordnung zwei ${}_{\mu_1\cdots\mu_{n-1}}q_\nu$ trifft, der erste Mittelpunkt auf l_0 und der letzte auf l_3 gelegen ist, und für jedes ν der Abstand der Schnittpunkte der Längelinien n^{ter} Ordnung durch die Mittelpunkte von ${}_{\mu_1\cdots\mu_{n-1}}q_\nu$ und ${}_{\mu_1\cdots\mu_{n-1}}q_{\nu+1}$ mit einer willkürlichen Breitelinie weniger als ε_n beträgt. Der Minimalwert der Seitenlängen der ${}_{\mu_1\cdots\mu_{n-1}}q_\nu$, wenn wir sowohl ν wie die μ_ν variieren lassen, sei ζ_n. Die Schnittpunkte derjenigen beiden Längelinien n^{ter} Ordnung, welche die der X-Achse parallele Mittellinie von ${}_{\mu_1\cdots\mu_{n-1}}q_{2\nu-1}$ in drei gleiche Stücke zerlegen, mit $b_{\mu_1\cdots\mu_{n-1}0}$ bzw. $b_{\mu_1\cdots\mu_{n-1}1}$ bezeichnen wir mit ${}_{\mu_1\cdots\mu_{n-1}}A_{2\nu-1}$ und ${}_{\mu_1\cdots\mu_{n-1}}B_{2\nu-1}$ bzw. ${}_{\mu_1\cdots\mu_{n-1}}C_{2\nu-1}$ und ${}_{\mu_1\cdots\mu_{n-1}}D_{2\nu-1}$. Die Schnittpunkte derjenigen beiden Längelinien n^{ter} Ordnung, welche die der X-Achse parallele Mittellinie von ${}_{\mu_1\cdots\mu_{n-1}}q_{2\nu}$ in drei gleiche Stücke zerlegen, mit $b_{\mu_1\cdots\mu_{n-1}2}$ bzw. $b_{\mu_1\cdots\mu_{n-1}3}$ bezeichnen wir mit ${}_{\mu_1\cdots\mu_{n-1}}E_{2\nu}$ und ${}_{\mu_1\cdots\mu_{n-1}}F_{2\nu}$ bzw. ${}_{\mu_1\cdots\mu_{n-1}}G_{2\nu}$ und ${}_{\mu_1\cdots\mu_{n-1}}H_{2\nu}$. Sodann verstehen wir unter den *Teilhauptlinien* $(n+1)^{ter}$ *Ordnung von* $s_{\mu_1\cdots\mu_{n-1}}$ die Streckenzüge ${}_{\mu_1\cdots\mu_{n-1}}B_{2\nu-1}$ ${}_{\mu_1\cdots\mu_{n-1}}D_{2\nu-1}$, ${}_{\mu_1\cdots\mu_{n-1}}E_{2\nu}$ ${}_{\mu_1\cdots\mu_{n-1}}G_{2\nu}$ und ${}_{\mu_1\cdots\mu_{n-1}}H_{2\nu}$ ${}_{\mu_1\cdots\mu_{n-1}}F_{2\nu}$ ${}_{\mu_1\cdots\mu_{n-1}}C_{2\nu+1}$ ${}_{\mu_1\cdots\mu_{n-1}}A_{2\nu+1}$ nebst den in l_0 und l_3 enthaltenen Seiten von $s_{\mu_1\cdots\mu_{n-1}}$. Zerlegen wir die zwischen zwei bestimmten aufeinanderfolgenden Teilhauptlinien $(n+1)^{\text{ter}}$ Ordnung von $s_{\mu_1\cdots\mu_{n-1}}$ enthaltenen Teilstrecken von Breitelinien in je zwei Segmente, welche auf jeder Breitelinie dasselbe Verhältnis aufweisen, so nennen wir den geometrischen Ort der Teilpunkte eine *Teillängelinie* $(n+1)^{ter}$ *Ordnung von* $s_{\mu_1\cdots\mu_{n-1}}$. Unter einer *Längelinie* $(n+1)^{ter}$ *Ordnung* verstehen wir einen aus Teillängelinien $(n+1)^{\text{ter}}$ Ordnung der verschiedenen $s_{\mu_1\cdots\mu_{n-1}}$ zusammengesetzten, b_0 und b_1 verbindenden Streckenzug. Diejenige Längelinie $(n+1)^{\text{ter}}$ Ordnung, welche b_0 im selben Punkte wie $l_{a_1 a_2 a_3\cdots}$ trifft, bezeichnen wir mit ${}_{n+1}l_{a_1 a_2 a_3\cdots}$ und bestimmen τ_n durch die Eigenschaft, daß sie ${}_n l_{a_1 a_2 a_3\cdots}$ für jede Wahl der a_ν in ${}_{n+1}l_{a_1 a_2 a_3\cdots}$ überführt. Wir wählen

$$\varepsilon_n < \frac{1}{27}\zeta_{n-1}\left(< \frac{1}{243}\zeta_{n-2} < \cdots < \frac{1}{3^{2n-3}}\zeta_2 < \frac{1}{3^{2n-1}}\zeta_1\right)$$

und sorgen weiter durch geeignete Wahl von ε_n dafür, daß für jede Wahl von $a_1, a_2, \cdots, a_{m_n}$ das Verhältnis von der zwischen ${}_{n+1}l_{a_1\cdots a_{m_n-1} a_{m_n}}$ und ${}_{n+1}l_{a_1\cdots a_{m_n-1}(a_{m_n}+1)}$ enthaltenen Teilstrecke einer willkürlichen Breitelinie

[[598]]

und der zwischen ${}_n l_{a_1 \cdots a_{m_n - 1} a_{m_n}}$ und ${}_n l_{a_1 \cdots a_{m_n - 1} (a_{m_n} + 1)}$ enthaltenen Teilstrecke derselben Breitelinie zwischen $1 - \eta_n$ und $1 + \eta_n$ liegt.

Die Limestransformation von $\tau_1, \tau_1 \cdot \tau_2, \tau_1 \cdot \tau_2 \cdot \tau_3, \cdots$ ist eine eineindeutige und stetige Transformation τ des Quadrates k in sich, welche das in k liegende Segment von $l_{a_1 a_2 a_3 \ldots}$ für jede Wahl der a_ν in einen solchen b_0 und b verbindenden und jede Breitelinie nur in einem Punkte treffenden einfachen Kurvenbogen ${}_\omega l_{a_1 a_2 a_3 \cdots}$ überführt, daß für jede Wahl von n und den μ_ν innerhalb des Streifens $s_{\mu_1 \cdots \mu_n}$ zum Innern eines Quadrates ${}_{\mu_1 \cdots \mu_n} q_\sigma$ gehörige, also außerhalb C liegende Punkte dieses Kurvenbogens liegen. Mithin enthält C von keinem Kurvenbogen ${}_\omega l_{a_1 a_2 a_3 \cdots}$ einen Teilbogen, sodaß die zu τ reziproke Transformation die Menge C in eine nirgends dichte abgeschlossene Punktmenge C' überführt, welche von keiner Längelinie $l_{a_1 a_2 a_3 \ldots}$ ein Segment enthält. Mit anderen Worten, wir haben bewiesen:

Hilfssatz: *Sei $PQRS$ ein abgeschlossenes Quadrat, und C eine in diesem Quadrat enthaltene nirgends dichte abgeschlossene Punktmenge, zu welcher kein Segment von QR oder PS gehört; es existiert eine eineindeutige und stetige Transformation des Quadrates in sich, welche jede Quadratseite und jeden der Seite PQ parallelen Querschnitt des Quadrates invariant läßt, und C in eine Punktmenge C' überführt, welche kein der Quadratseite QR paralleles gerades Liniensegment enthält.*

§ 2. Der Inhalt der abgeschlossenen Punktmengen.

Wir denken wieder im abgeschlossenen Quadrate k mit den Eckpunkten P, Q, R, S (von denen Q im Koordinatenanfangspunkt, P auf der X-Achse und R auf der Y-Achse liegt) eine nirgends dichte abgeschlossene Punktmenge C, zu welcher kein Segment von QR oder PS gehört und bestimmen nach § 1 eine eineindeutige und stetige Transformation τ' von k in sich, welche alle geraden Liniensegmente $y = c$ nebst den Quadratseiten $x = 0$ und $x = 1$ und allen Punkten der Quadratseite $y = 0$ invariant läßt, und C in eine Punktmenge C' überführt, welche kein gerades Liniensegment $x = c$ enthält. Die Entfernung, welche ein willkürlicher Punkt von k von C' besitzt, bezeichnen wir mit ϱ. Alsdann ist ϱ eine sich über k erstreckende stetige, nirgends negative Funktion von x und y, welche in allen Punkten von C', aber sonst nirgends verschwindet. Wir setzen

$$\frac{\int_0^y \varrho(xy)\,dy}{\int_0^1 \varrho(xy)\,dy} = \eta(xy).$$

Die Formeln
$$x' = x$$
$$y' = \eta(xy)$$
definieren eine eineindeutige und stetige Transformation τ'' von k in sich, welche C' in eine solche nirgends dichte abgeschlossene Punktmenge C'' überführt, daß der lineare Inhalt der auf einer willkürlichen Geraden $x = c$ liegenden Teilmenge von C'' verschwindet, mithin auch der ebene Inhalt von C'' selbst verschwindet.

Sei t die durch $\tau''^{-1} \cdot \tau'^{-1}$ bestimmte, die Eckpunkte invariant lassende Transformation der Quadratseite QR in sich, M der Mittelpunkt von k, und Q_n bzw. R_n der Schnittpunkt von MQ bzw. MR mit der Geraden $x = 2^{-n-1}$. Durch Projektion aus M bestimmt dann t gleichzeitig eine Transformation des geraden Liniensegmentes $Q_n R_n$ in sich, welche einen willkürlichen Punkt W_n dieses Liniensegmentes in W_n' überführt. Unter ${}_aR_n$ verstehen wir einen solchen Punkt von $Q_n R_n$, daß der Schnittpunkt der geraden Linie $M\,{}_aR_n$ mit der Y-Achse eine Ordinate $\frac{a}{2^n}$ besitzt, und betrachten ein Trapezium ${}_aR_n\,{}_{a+1}R_n\,{}_{2a}R_{n+1}\,{}_{2a+2}R_{n+1}$. Sei Z die Mitte von ${}_aR_n\,{}_{a+1}R_n$ und U die Mitte von ${}_aR_n'\,{}_{a+1}R_n'$, und sei ${}_a\tau_n$ die eineindeutige und stetige Transformation, welche das Trapezium ${}_aR_n\,{}_{a+1}R_n\,{}_{2a}R_{n+1}\,{}_{2a+2}R_{n+1}$ in solcher Weise in das Trapezium ${}_aR_n'\,{}_{a+1}R_n'\,{}_{2a}R_{n+1}'\,{}_{2a+2}R_{n+1}'$ überführt, daß für jedes α die gerade Linie, welche die Strecken ${}_aR_n\,{}_{2a}R_{n+1}$ und ${}_{a+1}R_n\,{}_{2a+2}R_{n+1}$ bzw. die Strecken ${}_aR_n Z$ und ${}_{2a}R_{n+1}\,{}_{2a+1}R_{n+1}$ bzw. die Strecken $Z\,{}_{a+1}R_n$ und ${}_{2a+1}R_{n+1}\,{}_{2a+2}R_{n+1}$ nach dem Verhältnis α zerlegt, in die gerade Linie, welche die Strecken ${}_aR_n'\,{}_{2a}R_{n+1}'$ und ${}_{a+1}R_n'\,{}_{2a+2}R_{n+1}'$ bzw. die Strecken ${}_aR_n' U$ und ${}_{2a}R_{n+1}'\,{}_{2a+1}R_{n+1}'$ bzw. die Strecken $U\,{}_{a+1}R_n'$ und ${}_{2a+1}R_{n+1}'\,{}_{2a+2}R_{n+2}'$ nach dem Verhältnis α zerlegt, übergeht. Bei Variierung von n und a bestimmen die verschiedenen ${}_a\tau_n$ eine eineindeutige und stetige Transformation τ_{QR} der abgeschlossenen Dreiecksfläche MQR in sich, bei welcher die Quadratseite QR der Transformation t unterliegt, und die zur Dreiecksfläche gehörige Teilmenge von C'' in eine nirgends dichte abgeschlossene Punktmenge vom Inhalt Null übergeht. Wenn wir in analoger Weise τ_{RS} und τ_{PS} definieren, so gelangen wir zu einer eineindeutigen und stetigen Transformation τ''' von k in sich, bei welcher der Quadratumfang derselben Transformation wie bei $\tau''^{-1} \cdot \tau'^{-1}$ unterliegt, und C'' in eine nirgends dichte abgeschlossene Punktmenge C''' vom Inhalt Null übergeht.

Betrachten wir nun die Transformation $\tau' \cdot \tau'' \cdot \tau'''$, so stellt sich folgende Eigenschaft heraus:

Satz 1: *Sei $PQRS$ ein abgeschlossenes Quadrat, und C eine in diesem Quadrat enthaltene nirgends dichte abgeschlossene Punktmenge, zu welcher*

[[600]]

kein Segment von QR oder PS gehört; es existiert eine eineindeutige und stetige Transformation des Quadrates in sich, welche jeden Punkt des Quadratumfanges invariant läßt und C in eine Punktmenge vom Inhalte Null überführt.

Sei nun C_1 eine völlig willkürliche in k enthaltene nirgends dichte abgeschlossene Punktmenge. Dann existiert ein Ordnungstypus $^*\omega + \omega$ von der Y-Achse parallelen, links gegen QR und rechts gegen PS konvergierenden geraden Querschnitten

$$\cdots, K_{-2}L_{-2}, \quad K_{-1}L_{-1}, \quad K_0L_0, \quad K_1L_1, \quad K_2L_2, \cdots$$

von k, welche keine zu C_1 gehörige Segmente enthalten.*) Auf Grund von Satz 1 gibt es für jedes ν eine eineindeutige und stetige Transformation t_ν des abgeschlossenen Reckteckes $K_\nu L_\nu L_{\nu+1} K_{\nu+1}$ in sich, welche jeden Punkt seines Umfanges invariant läßt, und die zu ihm gehörige Teilmenge ${}_\nu C_1$ von C_1 in eine Punktmenge ${}_\nu C_1'$ vom Inhalt Null überführt. Aus der Betrachtung der Vereinigung der verschiedenen Transformationen t_ν ergibt sich sodann folgendes Resultat:

Satz 2: *Sei k ein abgeschlossenes Quadrat und C_1 eine in k enthaltene nirgends dichte abgeschlossene Punktmenge; es existiert eine eineindeutige und stetige Transformation von k in sich, welche jeden Punkt des Quadratumfanges invariant läßt und C_1 in eine Punktmenge vom Inhalt Null überführt.*

Wir nehmen nunmehr an, daß die dem Innern von k angehörige Teilmenge von C_1 nicht-abzählbar ist. Alsdann liegt im Innern von k eine perfekte punkthafte Teilmenge C_2 von C_1, und kann man im Innern von k eine einfache geschlossene Kurve γ konstruieren, welche C_2 als Bestandteil enthält.**) Sei ε eine beliebige zwischen 0 und 1 liegende positive Zahl, so kann man im Innern von k eine einfache geschlossene „nichtquadrierbare" Kurve γ' konstruieren, deren Inhalt $> 1 - \frac{1}{2}\varepsilon$ ist***), und auf γ' eine solche punkhafte perfekte Punktmenge C' bestimmen, daß das Maß der Komplementärmenge von C' auf $\gamma' < \frac{1}{2}\varepsilon$, mithin der Inhalt von $C' > 1 - \varepsilon$ ist. Zwischen C_2 und C' läßt sich eine eineindeutige und stetige Beziehung herstellen; diese läßt sich zu einer eineindeutigen und stetigen Beziehung zwischen γ und γ', und letztere sich zu einer einein-

*) Sei nämlich $a_1, a_2, a_3, \cdots$ eine Folge von gegen Null konvergierenden positiven Zahlen. Diejenigen Werte von x, zu denen in C_1 enthaltene, der Y-Achse parallele gerade Liniensegmente $\geq a_\nu$ gehören, bilden für jedes ν eine nirgends dichte abgeschlossene Menge σ_ν, und die Komplementärmenge von $\mathfrak{S}\{\sigma_1, \sigma_2, \cdots\}$ ist überall dicht zwischen 0 und 1.

**) Schoenflies, Bericht über die Mengenlehre II, S. 258, 259. [[3]]

***) Ibid. S. 257. [[3]]

deutigen und stetigen Transformation von k in sich erweitern*), bei welcher jeder Punkt des Quadratumfanges invariant bleibt. D. h. wir haben bewiesen:

Satz 3: *Sei k ein abgeschlossenes Quadrat der Seitenlänge* 1, *und C_1 eine in k enthaltene, im Innengebiete von k nicht-abzählbare nirgends dichte abgeschlossene Punktmenge; es existiert eine eineindeutige und stetige Transformation von k in sich, welche jeden Punkt des Quadratumfanges invariant läßt und C_1 in eine Punktmenge, deren Inhalt dem Werte* 1 *beliebig nahe kommt, überführt.*

Wir fassen das Ergebnis dieses Paragraphen wie folgt zusammen:

Theorem 1: *Sei k ein abgeschlossenes Quadrat der Seitenlänge* 1, *und C eine in k enthaltene, im Innengebiete von k nicht-abzählbare abgeschlossene Punktmenge; bei den eineindeutigen und stetigen Transformationen von k in sich schwankt der Inhalt der Punktmengen, in welche C übergeht, zwischen* 0 *(inklusive) und* 1 *(exklusive), wenn C nirgends dicht ist; zwischen* 0 *(exklusive) und* 1 *(exklusive), wenn C weder nirgends dicht noch überall dicht ist.*

§ 3. Das Maß der inneren und äußeren Grenzmengen.

Wir denken das abgeschlossene Quadrat k zerlegt in eine überall nicht-abzählbare äußere Grenzmenge C und eine innere Grenzmenge I. Sei $C = \mathfrak{S}(C_1, C_2, C_3, \cdots)$, wobei die abgeschlossene Punktmenge C_ν für jedes ν in der abgeschlossenen Punktmenge $C_{\nu+1}$ enthalten ist, und sei I_ν die Komplementärmenge von C_ν in k.

Wir zerlegen k in vier kongruente homothetische Teilquadrate, mit jedem der letzteren nehmen wir die gleiche Zerlegung vor, und fahren in dieser Weise unbeschränkt fort. Alle so entstehende Teilquadrate von k werden wir als Quadrate $\varkappa$ bezeichnen. Wir definieren nun eine Fundamentalreihe $\tau_1, \tau_2, \cdots$ von eineindeutigen und stetigen Transformationen von k wie folgt:

Definition von τ_1. Sei C_{α_1} eine im Innengebiete von k nicht-abzählbare Punktmenge. Auf Grund des Satzes 3 von § 2 definieren wir τ_1 als eine eineindeutige und stetige Transformation von k in sich, welche jeden Punkt des Umfanges von k invariant läßt und C_{α_1} in eine Punktmenge mit Inhalt $> \frac{1}{2}$ überführt. Die Punktmenge, in welche C_ν bzw. I_ν durch τ_1 übergeführt wird, bezeichnen wir mit C_ν' bzw. I_ν'.

Definition von τ_2. Wir zerlegen I'_{α_1} in solcher Weise in als $\varkappa_2$ zu bezeichnende Quadrate $\varkappa$, deren Größe bei Konvergenz gegen C'_{α_1} aber auch

[[3]] *) Schoenflies, Bericht über die Mengenlehre II, S. 209—212. Vgl. auch meine
[[4]] Bemerkung dazu in Math. Ann. 68, S. 427, 428.

[[602]]

nur dabei unbeschränkt abnimmt, daß zwei willkürliche Punkte von k, deren Entfernung $\geqq \frac{1}{2}$ ist, durch τ_1 in zwei *nicht* zum selben $\varkappa_2$ gehörige Punkte übergeführt werden. Sodann wählen wir eine endliche Menge von als $\varkappa_2'$ zu bezeichnenden Quadraten $\varkappa_2$ aus, deren Gesamtinhalt mehr als $\frac{3}{4}$ des Maßes von I'_{α_1} beträgt. Sei C'_{α_2} eine im Innengebiete jedes $\varkappa_2'$ nicht-abzählbare Punktmenge. Auf Grund des Satzes 3 von § 2 definieren wir τ_2 als eine eineindeutige und stetige Transformation von k in sich, welche jeden Punkt der Umfänge der $\varkappa_2'$ und jeden außerhalb der $\varkappa_2'$ liegenden Punkt invariant läßt und C'_{α_2} in eine Punktmenge überführt, deren Inhalt innerhalb jedes $\varkappa_2'$ mehr als $\frac{2}{3}$ des Inhaltes dieses $\varkappa_2'$ beträgt. Die Punktmenge, in welche C_ν' bzw. I_ν' durch τ_2 übergeführt wird, bezeichnen wir mit C_ν'' bzw. I_ν''.

Definition von τ_n. Wir zerlegen $I^{(n-1)}_{\alpha_{n-1}}$ in solcher Weise in als $\varkappa_n$ zu bezeichnende Quadrate $\varkappa$, deren Größe bei Konvergenz gegen $C^{(n-1)}_{\alpha_{n-1}}$ aber auch nur dabei unbeschränkt abnimmt, daß jedes $\varkappa_n$ in einem $\varkappa_{n-1}$ enthalten ist und zwei willkürliche Punkte von k, deren Entfernung $\geqq \frac{1}{2^{n-1}}$ ist, durch $\tau_1 \cdot \tau_2 \cdots \tau_{n-1}$ in zwei *nicht* zum selben $\varkappa_n$ gehörige Punkte übergeführt werden. Sodann wählen wir eine endliche Menge von als $\varkappa_n'$ zu bezeichnenden Quadraten $\varkappa_n$ aus, deren Gesamtinhalt mehr als $\frac{3}{4}$ des Maßes von $I^{(n-1)}_{\alpha_{n-1}}$ beträgt. Sei $C^{(n-1)}_{\alpha_n}$ eine im Innengebiete jedes $\varkappa_n'$ nicht-abzählbare Punktmenge. Auf Grund des Satzes 3 von § 2 definieren wir τ_n als eine eineindeutige und stetige Transformation von k in sich, welche jeden Punkt der Umfänge der $\varkappa_n'$ und jeden außerhalb der $\varkappa_n'$ liegenden Punkt invariant läßt und $C^{(n-1)}_{\alpha_n}$ in eine Punktmenge überführt, deren Inhalt innerhalb jedes $\varkappa_n'$ mehr als $\frac{2}{3}$ des Inhaltes dieses $\varkappa_n'$ beträgt. Die Punktmenge, in welche $C^{(n-1)}_\nu$ bzw. $I^{(n-1)}_\nu$ durch τ_n übergeführt wird, bezeichnen wir mit $C^{(n)}_\nu$ bzw. $I^{(n)}_\nu$.

Die Limestransformation von τ_1, $\tau_1 \cdot \tau_2$, $\tau_1 \cdot \tau_2 \cdot \tau_3$, $\cdots$ ist eine eineindeutige und stetige Transformation τ des Quadrates k in sich, welche C in eine äußere Grenzmenge vom Maß 1 und I in eine innere Grenzmenge vom Maß 0 überführt. Wir formulieren mithin:

Satz 4: *Sei k ein abgeschlossenes Quadrat der Seitenlänge 1, welches in eine überall nicht-abzählbare äußere Grenzmenge C und eine innere Grenzmenge I zerlegt ist; es existiert eine eineindeutige und stetige Transformation von k in sich, welche C in eine Punktmenge vom Maß 1 und I in eine Punktmenge vom Maß 0 überführt.*

Wir denken nunmehr k zerlegt in eine äußere Grenzmenge C und eine überall dichte innere Grenzmenge I. Sei wieder $C = \mathfrak{S}(C_1, C_2, C_3, \cdots)$, wobei die abgeschlossene Punktmenge C_ν für jedes ν in der abgeschlossenen Punktmenge $C_{\nu+1}$ enthalten ist, so ist jedes C_ν nirgends dicht. Sei wieder I_ν die Komplementärmenge von C_ν in k.

Wir definieren in folgender Weise eine Fundamentalreihe $\tau_1, \tau_2, \tau_3, \cdots$ von eineindeutigen und stetigen Transformationen von k:

Definition von τ_1. Auf Grund des Satzes 2 von § 2 definieren wir τ_1 als eine eineindeutige und stetige Transformation von k in sich, welche jeden Punkt des Umfanges von k invariant läßt und C_1 in eine Punktmenge vom Inhalt Null überführt. Die Punktmenge, in welche C_ν bzw. I_ν durch τ_1 übergeführt wird, bezeichnen wir mit C'_ν bzw. I'_ν.

Definition von τ_2. Wir zerlegen I_1' in solcher Weise in als $\varkappa_2$ zu bezeichnende Quadrate $\varkappa$, deren Größe bei Konvergenz gegen C_1' aber auch nur dabei unbeschränkt abnimmt, daß zwei willkürliche Punkte von k, deren Entfernung $\geqq \frac{1}{2}$ ist, durch τ_1 in zwei *nicht* zum selben $\varkappa_2$ gehörige Punkte übergeführt werden. Auf Grund des Satzes 2 von § 2 definieren wir τ_2 als eine eineindeutige und stetige Transformation von k in sich, welche jeden Punkt von C_1' und jeden Punkt der Umfänge der $\varkappa_2$ invariant läßt und C_2' in eine Punktmenge überführt, welche innerhalb jedes $\varkappa_2$ vom Inhalt Null ist. Die Punktmenge, in welche C_ν' bzw. I_ν' durch τ_2 übergeführt wird, bezeichnen wir mit C_ν'' bzw. I_ν''.

Definition von τ_n. Wir zerlegen $I_{n-1}^{(n-1)}$ in solcher Weise in als $\varkappa_n$ zu bezeichnende Quadrate $\varkappa$, deren Größe bei Konvergenz gegen $C_{n-1}^{(n-1)}$ aber auch nur dabei unbeschränkt abnimmt, daß jedes $\varkappa_n$ in einem $\varkappa_{n-1}$ enthalten ist und zwei willkürliche Punkte von k, deren Entfernung $\geqq \frac{1}{2^{n-1}}$ ist, durch $\tau_1 \cdot \tau_2 \cdots \tau_{n-1}$ in zwei *nicht* zum selben $\varkappa_n$ gehörige Punkte übergeführt werden. Auf Grund des Satzes 2 von § 2 definieren wir τ_n als eine eineindeutige und stetige Transformation von k in sich, welche jeden Punkt von $C_{n-1}^{(n-1)}$ und jeden Punkt der Umfänge der $\varkappa_n$ invariant läßt und $C_n^{(n-1)}$ in eine Punktmenge überführt, welche innerhalb jedes $\varkappa_n$ vom Inhalt Null ist. Die Punktmenge, in welche $C_\nu^{(n-1)}$ bzw. $I_\nu^{(n-1)}$ durch τ_n übergeführt wird, bezeichnen wir mit $C_\nu^{(n)}$ bzw. $I_\nu^{(n)}$.

Die Limestransformation von $\tau_1, \tau_1 \cdot \tau_2, \tau_1 \cdot \tau_2 \cdot \tau_3, \cdots$ ist eine eineindeutige und stetige Transformation τ des Quadrates k in sich, welche C in eine äußere Grenzmenge vom Maß 0 und I in eine innere Grenzmenge vom Maß 1 überführt. Wir formulieren mithin:

Satz 5: *Sei k ein abgeschlossenes Quadrat der Seitenlänge 1, welches in eine äußere Grenzmenge C und eine überall dichte innere Grenzmenge I*

zerlegt ist; es existiert eine eineindeutige und stetige Transformation von k in sich, welche C in eine Punktmenge vom Maß 0 *und I in eine Punktmenge vom Maß* 1 *überführt.*

Das Ergebnis dieses Paragraphen kann wie folgt zusammengefaßt werden:

Theorem 2: *Sei k ein abgeschlossenes Quadrat der Seitenlänge* 1, *welches in eine im Innern von k nicht-abzählbare äußere Grenzmenge C und eine im Innern von k nicht-abzählbare innere Grenzmenge I zerlegt ist; bei den eineindeutigen und stetigen Transformationen von k in sich schwankt daß Maß der Punktmengen, in welche I übergeht, zwischen* 0 *(inklusive) und* 1 *(inklusive), wenn I überall dicht und C überall nicht-abzählbar ist; zwischen* 0 *(inklusive) und* 1 *(exklusive), wenn I nicht überall dicht, aber C überall nicht-abzählbar ist; zwischen* 0 *(exklusive) und* 1 *(inklusive), wenn I überall dicht, aber C nicht überall nicht-abzählbar ist; zwischen* 0 *(exklusive) und* 1 *(exklusive), wenn weder I überall dicht noch C überall nicht-abzählbar ist.*

NOTES

⟦1⟧ E. Borel 1913.
⟦2⟧ C. Carathéodory 1918.
⟦3⟧ A. Schoenflies 1908 A.
⟦4⟧ Brouwer 1910 C.

1924 K2

Mathematics. — "*On the n-dimensional simplex star in* R_n." By Prof. L. E. J. BROUWER.

(Communicated at the meeting of December 27, 1924).

[[1]] Each of the two following §§ contains another proof of Mr. WILSON's theorem II deduced in the preceding communication

§ 1.

Let α be a p-dimensional face of an n-dimensional simplex S in an R_n. The flat p-dimensional spaces of R_n parallel to α and having at least one point in common with S form an $(n-p)$-dimensional simplex in the R_{n-p} of *all* flat p-dimensional spaces of R_n which are parallel to α.

For let $P_1, P_2, \ldots, P_{n+1}$ be the vertices of S and $P_1, P_2, \ldots, P_{p+1}$ those of α. Let $x_1, x_2, \ldots, x_{n+1}$ be barycentric coordinates with respect to S. Then an arbitrary flat p-dimensional space of R_n parallel to α is given by $(x_1+x_2+\ldots+x_{p+1}) : x_{p+2} : \ldots : x_{n+1} = a_{p+1} : a_{p+2} : \ldots : a_{n+1}$, the necessary and sufficient condition that it should have at least one point in common with S laying in the absence of an a_μ and an a_ν of opposite signs. We may regard these $a_{p+1}, a_{p+2}, \ldots, a_{n+1}$ as homogeneous coordinates of the p-dimensional spaces considered.

Keeping now the vertices $P_1, P_2, \ldots, P_{p+1}$ unchanged, we replace $P_{p+2}, \ldots, P_{n+1}$ by $P'_{p+2}, \ldots, P'_{n+1}$, and represent the barycentric coordinates with respect to the new simplex S' by $x'_1, x'_2, \ldots, x'_{n+1}$. Then the following formulae of transformation hold:

$$\left.\begin{array}{l} x'_1 : x'_2 : \ldots : x'_{p+1} : x'_{p+2} : \ldots : x'_{n+1} = (x_1 + {}_1a_{p+2}x_{p+2} + \ldots + {}_1a_{n+1}x_{n+1}) : \\ : \ldots : (x_{p+1} + {}_{p+1}a_{p+2}\, x_{p+2} + \ldots + {}_{p+1}\, a_{n+1}\, x_{n+1}) : \\ : ({}_{p+2}a_{p+2}\, x_{p+2} + \ldots + {}_{p+2}a_{n+1}\, x_{n+1}) : \ldots : ({}_{n+1}a_{p+2}\, x_{p+2} + \ldots + {}_{n+1}a_{n+1}x_{n+1}) \end{array}\right\} \quad (I)$$

By means of S' in the same way as by means of S, we introduce homogeneous coordinates for the flat p-dimensional spaces parallel to α and call these coordinates $a'_{p+1}, a'_{p+2}, \ldots, a'_{n+1}$; they are evidently homogeneous linear functions of $a_{p+1}, a_{p+2}, \ldots, a_{n+1}$ by the formulae (I).

Thus if in R_n we have an n-dimensional simplex star of centre P_1, having the p-dimensional face $P_1\, P_2 \ldots P_{p+1}$, then the flat p-dimensional spaces of R_n parallel to α form an R_{n-p}, and those of them having at least one point in common with any simplex of the star of which $P_1 P_2 \ldots P_{p+1}$ is a face, form in that R_{n-p} an $(n-p)$-dimensional simplex star. For the $(n-p)$-dimensional simplexes derived from the n-dimensional simplexes $P_1 \ldots P_{p+1} P'_{p+2} \ldots P'_{n+1}$ and $P_1 \ldots P_{p+1} P''_{p+2} \ldots P''_{n+1}$ of the star have always a face in common of as many dimensions as there are indices ϱ for which $P'_\varrho = P''_\varrho$.

We assume now that for $\nu < n$, a ν-dimensional simplex star in R_ν of centre C fills a neighbourhood of C in R_ν, and consider a point A of an n-dimensional simplex star σ in R_n, which belongs neither to the boundary nor to a $(p-1)$-dimensional face of σ, but to the p-dimensional face α of the star, with the vertices $P_1, P_2, \ldots, P_{p+1}$ (P_1 is the centre of the star). Of the set of the flat p-dimensional spaces of R_n parallel to α any neighbourhood of α fills a neighbourhood of A in R_n.

Consequently the set of those flat p-dimensional spaces of R_n parallel to α which cut a simplex of σ of which α is a face, contain a neighbourhood of A in R_n and since the parts of these flat p-dimensional spaces which lie outside σ have a non-vanishing minimum distance from A, σ also fills a neighbourhood of A in R_n.

With this property Mr. WILSON's theorem is proved. For, if the *closed* set σ did not fill a neighbourhood of P_1 in R_n, there would be a point A of σ different from P_1, not belonging to the boundary of σ and being a limiting-point of points of R_n not belonging to σ.

§ 2.

Theorem 1. *Given in R_n a finite number of flat $(n-2)$-dimensional spaces $l_1, l_2, \ldots, l_p$ and two points P' and P'' not in any l_ν and in a distance a from each other, then P' and P'' can be joined by a polygonal path of span a containing no point of the l_ν.*

Proof. Let v'_ν be the flat $(n-1)$-dimensional space through P' and l_ν, and v''_ν that through P'' and l_ν. Further let P be a point of R_n neither in a v'_ν nor in a v''_ν, but inside the $(n-1)$-dimensional sphere of diameter $P'P''$. Then the segments $P'P$ and PP'' constitute a polygonal path satisfying the required conditions.

Theorem 2. *Let S be an n-dimensional simplex star in R_n, b the intrinsic boundary of S (i. e. the set of points formed by those faces of S not having the centre of S as vertex), K a point of S not belonging to b, G the domain complementary to S in R_n, and g the boundary (obviously belonging to S) of G in R_n. Then K can not belong to g.*

Proof. A point of S belonging neither to b nor to an $(n-2)$-dimensional face of S has always a neighbourhood (with respect to R_n) laying entirely in S. Consequently g is the sum (union) of b and a closed set of points c entirely in the $(n-2)$-dimensional faces of S. Let d be the distance (measured in R_n) of K from b.

Assume now that K belongs to g. In that case at a distance $< \frac{1}{4} d$ from K, there could be assigned on the one hand a point P' of G and on the other a point P'' belonging to S but not to an $(n-2)$-dimensional face of S. By theorem 1, P' and P'' could then be joined by a polygonal path π of span $< \frac{1}{2} d$ not meeting c, which, just because its span is $< \frac{1}{2} d$, also would not meet b and thus would have no point in common with g. But this result is contradictory, since π joins the point P' of G to the point P'' not in G. The assumption that K belongs to g is thus reduced to an absurdity proving theorem 2.

NOTE

⟦1⟧ W. Wilson 1924. Wilson's Theorem II is the following: In any R_n any n-dimensional simplex star of centre A_0 contains an n-dimensional spherical region of centre A_0.

CHAPTER 7

Topology of surfaces

Mathematics. — "*Ueber eineindeutige, stetige Transformationen von Flächen in sich*" (sechste Mitteilung[1])). By Prof. L. E. J. BROUWER. 1919 L2 [[1]]

(Communicated in the meeting of November 30, 1918).

§ 1. In einem in 1912 in den Göttinger Nachrichten (S. 603—606[2])) in Auszug abgedruckten Briefe an R. FRICKE habe ich (S. 605, Fussnote[3])) kurz skizziert, wie das von HURWITZ herrührende analytische Theorem, dass *birationale Transformationen einer Riemannschen Fläche vom Geschlechte $p > 1$ in sich unmöglich ein vollständiges kanonisches Schnittsystem der Fläche in ein äquivalentes kanonisches Schnittsystem überführen können*, mittels der Analysis Situs bewiesen werden kann, wobei sich seine Gültigkeit herausstellt für *alle periodischen, eineindeutigen und stetigen Transformationen.* Die damalige Andeutung wird im folgenden näher präzisiert und gerechtfertigt werden. [[2]] [[3]]

Sei O die gegebene zweiseitige Fläche, I die Menge der bei der n-periodischen, eineindeutigen und stetigen, die Ränder invariant lassenden Transformation t von O invarianten Punkte. Wir nehmen an, dass jedes von I in O bestimmte Gebiet von t in sich transformiert wird (was, wenn t die Indikatrix von O invariant lässt, stets der Fall ist), und unterziehen die Wirkung von t auf eines dieser Gebiete, welches wir mit ω bezeichnen werden, einer näheren Betrachtung. Dabei ziehen wir im Falle, dass t die Indikatrix von O umkehrt, jeden eventuellen für t *nicht* invarianten Rand von ω

[1]) Vgl. diese Proceedings. XI, S. 788; XII, S. 286; XIII, S. 767; XIV, S. 300; XV, S. 352.

[2]) Das daselbst S. 604 auf künftige Publikationen von P. KOEBE (der Neujahr 1912 im Besitze einer Abschrift meines Briefes an R. FRICKE war) hinweisende Zitat ist nach der Erledigung der Korrekturen von einer mir unbekannten Hand, ohne meine Mitwirkung oder Vorkenntnis eingefügt worden; die bezüglichen Noten sind mir erst nach ihrem Erscheinen bekannt geworden. [[4]]

In engem Zusammenhang mit dem Inhalte meines (Anfang März 1912 gedruckten) Briefes an R. FRICKE stehen die Karlsruher Verhandlungen über automorphe Funktionen vom Jahre 1911; der über dieselben erstattete Bericht (Jahresber. d. D. M. V. XXI), ist, ebenso wie die in Gött. Nachr. 1912 erschienene KOEBE'sche Mitteilung über den Kontinuitätsbeweis, im Sommer 1912 gedruckt worden. Das in diesem Berichte enthaltene Referat über den Vortrag von P. KOEBE (insbesondere die auf S. 162 befindliche Anmerkung [1])) ist insofern irreführend, dass im wirklichen KOEBE'schen Vortrage, nach der ihm vorangegangenen, S. 156—157 wiedergegebenen kurzen Diskussion mit mir, vom Kontinuitätsbeweise überhaupt nicht wieder die Rede gewesen ist. [[2]] [[5]] [[6]] [[7]] [[5]]

in einen Punkt zusammen und fügen diesen Punkt zu ω hinzu. Sei R die Menge der übrig bleibenden (für t invarianten) Ränder r_α von ω und seien (a_1, b_1), (a_2, b_2), (a_3, b_3), ... die Rückkehrschnittpaare einer *vollständigen* kanonischen Zerschneidung von ω. Wir nehmen an, dass innerhalb jedes dieser Rückkehrschnittpaare ein a_ν so gewählt werden kann, dass nicht nur a_ν selbst, sondern auch seine beiden Seiten von t ungeachtet der Ränder äquivalent abgebildet werden (was, wenn t jeden Rückkehrschnitt der kanonischen Zerschneidung ungeachtet der Ränder äquivalent abbildet und die Indikatrix von O umkehrt, stets der Fall ist). Weiter schliessen wir den Fall aus, dass entweder kein Rückkehrschnittpaar (a_ν, b_ν) und höchstens zwei Ränder r_α, oder kein Rand r_α und nur ein einziges Rückkehrschnittpaar (a_ν, b_ν) existiert.

§ 2. Im Falle, dass in ω wenigstens zwei Rückkehrschnittpaare (a_ν, b_ν) existieren, verstehen wir unter Ω diejenige (kein Rückkehrschnittpaar mehr aufweisende) Schottkysche Ueberlagerungsfläche von ω, welche die a_ν als blättertrennende Ufer besitzt, unter L die Menge derjenigen Ränder l_α von Ω, welche durch eine unendliche Zahl von Ueberschreitungen der a_ν auf ω erzeugt werden, unter ${}_1a_\nu$, ${}_2a_\nu$, ... die Ueberlagerungsbilder von a_ν auf Ω, unter a'_ν das durch t auf ω bestimmte Bild von a_ν, unter ${}_1a'_\nu$, ${}_2a'_\nu$, ... die Ueberlagerungsbilder von a'_ν auf Ω. Wenn wir jedem ${}_\mu a_\nu$ dasjenige ${}_\lambda a'_\nu$ zuordnen, dessen Umlaufkoeffizienten zwischen den l_α die gleichen absoluten Werte besitzen, wie die entsprechenden Umlaufkoeffizienten von ${}_\mu a_\nu$, so ist in Anschluss daran eine durch die Ueberlagerung von Ω über ω in t übergehende, eineindeutige und stetige Transformation t' von Ω in sich bestimmt, welche, ebenso wie t, n-periodisch sein muss und jeden Rand l_α invariant lässt. Ob *alle* Ränder von Ω für t' invariant sind, lassen wir dahingestellt, ziehen aber jeden eventuellen für t' *nicht* invarianten Rand in einen Punkt zusammen und fügen diesen Punkt zu Ω hinzu.

Im Falle, dass in ω nur ein einziges Rückkehrschnittpaar (a, b) existiert, verstehen wir unter Ω diejenige (kein Rückkehrschnittpaar mehr aufweisende) Schottkysche Ueberlagerungsfläche von ω, welche a als blättertrennendes Ufer besitzt, unter l_1 und l_2 diejenigen Ränder von Ω, welche durch eine unendliche Zahl von Ueberschreitungen von a auf ω erzeugt werden, unter ${}_1r$, ${}_2r$, ... die Ueberlagerungsbilder auf Ω eines (für t der Annahme gemäss invarianten) Randes r von ω. Alsdann existiert eine durch die Ueberlagerung von Ω über ω in t übergehende, eineindeutige und stetige Transformation t' von Ω in sich, welche ${}_1r$ (mithin auch ${}_2r$, ${}_3r$, ...) invariant lässt, so dass sie, ebenso wie t, n-periodisch sein muss. Ob *alle* Ränder von Ω für t' invariant sind, lassen wir wieder dahingestellt und ziehen jeden eventuellen für t' *nicht* invarianten Rand in einen Punkt zusammen, den wir zu Ω hinzufügen.

Im Falle, dass in ω kein Rückkehrschnittpaar (a_ν, b_ν) existiert, verstehen wir unter Ω die Fläche ω selbst und unter t' die Transformation t selbst.

In jedem der drei Fälle besitzt die Fläche Ω *wenigstens drei* Ränder und ist sie eineindeutiges stetiges Bild eines Teilgebietes der Kugel, während t' eine n-periodische, eineindeutige und stetige Transformation von Ω in sich darstellt, welche jeden Rand, aber keinen Punkt invariant lässt.

Wir dürfen annehmen, dass von den Rändern von Ω *wenigstens einer isoliert* ist. Im entgegengesetzten Falle können wir nämlich in Ω ein Gebiet g bestimmen, zu dessen Grenze kein Grenzpunkt von Rändern von Ω gehört, und welches, ebenso wie seine Komplementärmenge, Ränder von Ω enthält. Die Vereinigung von g und seinen Bildern für $t', t'^2, \ldots t'^{n-1}$ bildet eine Fläche, welche, ebenso wie Ω, wenigstens drei Ränder besitzt und von t' mit invarianten Rändern und ohne invariante Punkte in sich transformiert wird, überdies aber einen isolierten Rand besitzt.

§ 3. Seien $P_1, P_2, \ldots P_{n-1}$ die Punkte, in welche ein in Ω willkürlich gewählter Punkt P von $t', t'^2, \ldots t'^{n-1}$ der Reihe nach übergeführt wird. Ein P und P_1 verbindender stetiger Kurvenbogen j_P bildet mit seinen von $t', t'^2, \ldots t'^{n-1}$ bestimmten Bildern eine geschlossene stetige Kurve k_P. Die Menge R' der Ränder r'_α von Ω lässt sich in solcher Weise in eine endliche Zahl ≥ 3 von voneinander isolierten Teilmengen $R'_1(k_P)$, $R'_2(k_P)$, $\ldots R'_m(k_P)$ zerlegen, dass der Umlaufkoeffizient von k_P zwischen einem zu $R'_\nu(k_P)$ und einem zu $R'_\lambda(k_P)$ gehörigen Rand modulo n gleich einem durch ν und λ bestimmten, für $\nu = \lambda$ fortfallenden Wert $c_{\nu\lambda}(k_P) \geq 0$ und $< n$ ist. Diese $R'_\nu(k_P)$ behalten ihre Brauchbarkeit und die zugehörigen $c_{\nu\lambda}(k_P)$ ihre Gültigkeit bei, wenn für festes P der Bogen j_P diskontinuierlich geändert wird. Mittels gleichzeitiger stetiger Variierung von P und j_P sieht man weiter ein, dass auch durch Aenderung des Punktes P die Rolle der $R'_\nu(k_P)$ und $c_{\nu\lambda}(k_P)$ nicht gestört wird. Indem wir aber P in hinreichender Nähe von R'_h und j_P in passender Weise wählen, können wir dafür sorgen, dass der Umlaufkoeffizient von k_P zwischen einem zu $R'_\nu(k_P)$ und einem zu $R'_\lambda(k_P)$ gehörigen Rand, für ν und λ beide von h verschieden, fortfällt und

[[613]]

hieraus folgern wir unmittelbar, dass $c_{\nu\lambda}(k_P) = 0$ für jedes r, jedes λ, jedes P und jedes j_P.

§ 4. Wir bezeichnen einen willkürlich gewählten isolierten Rand von Ω mit ϱ, die Menge der übrigen Ränder von Ω mit R'_0 und betrachten die sich um R' aperiodisch herumwindende Ueberlagerungsfläche S von Ω. Diese Fläche S ist eineindeutiges und stetiges Bild der Cartesischen Ebene; ihr Rand R'' enthält einen aus ϱ hervorgegangenen Teil R_1'' und einen aus R'_0 hervorgegangenen Teil R_2''; diese beiden Teile sind voneinander isoliert.

Sei P ein Punkt von Ω in der Umgebung von ϱ, P_1 sein von t' bestimmtes Bild. Wir verbinden P und P_1 in der Umgebung von ϱ durch einen solchen stetigen Kurvenbogen j_P, welche mit seinen von $t', t'^2, \ldots . t'^{n-1}$ bestimmten Bildern eine geschlossene stetige Kurve k_P erzeugt, deren Umlaufkoeffizient zwischen zwei willkürlichen Rändern von Ω fortfällt. Die Möglichkeit, einen derartigen Kurvenbogen j_P herzustellen, folgt aus § 3. Sei P_m der Anfangspunkt, P_{1m} der Endpunkt eines Ueberlagerungsbildes von j_P auf S, so existiert eine durch die Ueberlagerung von S über Ω in t' übergehende, eineindeutige und stetige Transformation t'' von S in sich, welche P_m in P_{1m} überführt. Alsdann lässt die Transformation t''^n den Punkt P_m invariant, so dass t'', ebenso wie t', n-periodisch sein muss. Hiermit sind wir aber zu einem Widerspruch gelangt, weil eine periodische, eineindeutige und stetige Transformation der Cartesischen Ebene in sich ohne invariante Punkte nicht existieren kann.

§ 5. Im Falle, dass t die Indikatrix von O invariant lässt und von einer vollständigen kanonischen Zerschneidung von O jeden Rückkehrschnitt samt seinen beiden Seiten ungeachtet der Ränder äquivalent abbildet, besitzt t dieselbe Eigenschaft in bezug auf ω (was unmittelbar wie folgt eingesehen werden kann: Sei s ein Rückkehrschnitt von ω, der ω nicht zerlegt, so entspricht einer stetigen Variierung von s in O, wenn die Ränder von ω je in einen zu ω hinzuzufügenden Punkt zusammengezogen werden, eine stetige Variierung von s in ω). Wäre nun ω samt seiner Grenze nicht identisch mit O, so besässe ω einen *durch eine zusammenhängende perfekte Menge von für t invarianten Punkten abgeschlossenen Rand* und würde auf Grund davon die in den §§ 3 und 4 hinsichtlich der Transformation t' von Ω angestellte, auf einen Widerspruch führende Ueberlegung *auch im Falle, dass in ω kein Rückkehrschnittpaar (a_ν, b_ν) und nur zwei Ränder existieren*, in Kraft bleiben. Mithin ist in diesem Falle ω *samt seiner Grenze mit O identisch und O entweder eine Kugel, oder ein Zylinder, oder eine Cartesische Ebene, oder ein Torus.*

NOTES

⟦1⟧ Communication VI is not related to Communications I–V (1909 F2, 1909 H2, 1911 H2, 1911 J2, 1912 K2). It starts a series of investigations on mappings of surfaces which considerably differ in character from the earlier ones.
⟦2⟧ Brouwer 1912 D. A clearer exposition in Y22. In the present exposition. conformal mappings have been eliminated.
⟦3⟧ A Hurwitz 1892. Though Hurwitz' theorem is easily accessible by modern techniques, the editor knows of no publication of a simple proof.
⟦4⟧ Similar remarks on the strip attached to 1912 H.
⟦5⟧ Brouwer 1912 H.
⟦6⟧ P. Koebe 1912 B.
⟦7⟧ P. Koebe 1912 A.

Y 22 Letter to Hurwitz

⟦Univ. Bibl. Göttingen⟧

Amsterdam, 30.12.1911
Overtoom 565

⟦1⟧ Hochgeehrter Herr Professor

Entschuldigen Sie, dass ich die Freiheit nehme, mich mit einer Frage an Sie zu wenden. Ich brauche nämlich folgenden Satz:

„Eine birationale Transformation τ einer Riemannschen Fläche vom Geschlechte $p > 1$ in sich kann niemals ein (aus p in einem Punkte C zusammenhängenden Rückkehrschnittpaaren bestehendes) kanonisches Schnittsystem s in ein *äquivalentes* kanonisches Schnittsystem s' überführen." (s und s' heissen äquivalent, wenn sie sich durch stetige Bewegung auf der Fläche ineinander überführen lassen)

Von der Richtigkeit dieses Stazes habe ich mich in folgender Weise überzeugt:

„Die Transformation τ habe n Fixpunkte. Wir konstruieren ein erweitertes kanonisches Schnittsystem S, welches sich aus s und n aus C nach den Fixpunkten führenden Einschnitten zusammensetzt, und welches von der Transformation τ in S' übergeführt wird. Wäre nun s' mit s äquivalent, so müsste *auf Grund der Periodizität von* τ auch S' mit S äquivalent sein. (S und S' heissen äquivalent, wenn sie sich durch stetige Bewegung auf der Fläche *ohne Verrückung der n Fixpunkte* ineinander überführen lassen)

Wir konstruieren nun zur Riemannschen Fläche die in der Theorie der automorphen Funktionen übliche einfach zusammenhängende „Ueberlagerungsfläche", welche sich um die n Fixpunkte aperiodisch herumwindet, und welche sich konform auf ein Kreisinneres abbilden lässt, wobei die n Fixpunkte und ihre Reproduktionen auf den Kreisumfang rücken. Der Transformation τ entspricht sodann eine konforme Transformation *ohne Fixpunkte* dieses Kreisinneren in sich, welche auf Grund der Äquivalenz von S und S' *periodisch* sein müsste, was ein Widerspruch ist, weil eine periodische konforme Transformation eines Kreisinneren in sich immer einen Fixpunkt aufweist."

Es scheint mir nun aber sehr wahrscheinlich, dass der fragliche Satz sich vom
⟦2⟧ Standpunkte der kombinatorischen Konstruktion der „regulären" Riemannschen
⟦3⟧ Flächen viel einfacher wird einsehen lassen, und sogar von Ihren früheren diesbezüglichen Untersuchungen eine unmittelbare Folge sein muss.

Habe ich mit dieser Vermutung recht? Und ist der Satz irgendwo schon ausgesprochen? Für eine gütige Nachricht darüber wäre ich Ihnen zu grossem Dank verpflichtet.

Mit grösster Hochachtung
Ihr ergebenster
L E J Brouwer.

NOTES

⟦1⟧ This letter is related to Brouwer 1912 D, p. 605, footnote 3, and Brouwer 1919 L2. See also 1912 D ⟦15⟧, Y6, Y23, 1919 S ⟦4⟧. It expresses the main idea more clearly than do 1912 D and 1919 L2. However, in 1919 L2 the use of the conformal mapping is eliminated. It seems that this is what Brouwer had aimed at in the present letter.

⟦2⟧ The term 'regular Riemann surfaces' for Riemann surfaces admitting non-trivial automorphisms is ascribed to F. Klein. However F. Klein 1878 speaks of 'regulär eingeteilte Riemannsche Flächen'. See W. Dyck 1880 who speaks of regular Riemann surfaces.

Y 23 Letter to Hurwitz

⟦Univ. Bibl. Göttingen⟧

Amsterdam, 5.1.12
Overtoom 565

⟦1⟧ Hochgeehrter Herr Professor

Für Ihre schleunige Antwort bringe ich Ihnen meinen verbindlichsten Dank. Entschuldigen Sie, dass ich die Freiheit nehme, nochmal auf den Gegenstand zurückzukommen; vielleicht habe ich mich in meinem ersten Schreiben etwas undeutlich
⟦2⟧ ausgedrückt. Meine Frage betraf die von Ihnen in Mathem. Annalen Bd. 32 und 41 studierten „regulären" Riemannschen Flächen, d.h. Flächen mit konformen Transformationen in sich. Sei F eine solche Fläche, p ihr Geschlecht, $I_1, I_2, \cdots, I_p$ linear-unabhängige, zu F gehörige Integrale erster Gattung. Zu jeder konformen Transformation der Fläche F in sich gehört ein System von p linearen Transformationsgleichungen

$$\begin{cases} I'_1 = a_{11}I_1 + a_{12}I_2 + \text{———} + a_{1p}I_p. \\ \text{———————————} \\ \text{— — — — — — ————— — —} \\ I'_p = a_{p1}I_1 + a_{p2}I_2 + \text{————} + a_{pp}I_p. \end{cases}$$

Mein Satz kommt nun darauf hinaus, dass dieses Gleichungssystem sich niemals auf:

$$\begin{cases} I'_1 = I_1 \\ \text{------} \\ \text{-----} \\ \\ I'_p = I_p \end{cases}$$

reduzieren kann.

Von diesem Satze, den ich vollkommen streng bewiesen zu haben glaube, würde ich meinen, er müsse sich doch auf einem viel direkteren Wege einsehen lassen.
⟦3⟧ Als ich aber in einem auf der Naturforscherversammlung in Karlsruhe gehaltenen Vortrage den Satz zur Lösung eines Uniformisierungsproblems benutzte, fand ich unter den vielen sachkundigen Anwesenden keinen, der den Satz kannte, und man verwies mich an Sie als den bei weitem kompetentesten in diesen Fragestellungen.

Andererseits scheint mir Ihre in Annalen 41 enthaltene Abhandlung die geometrische Konstruktion der regulären Flächen so erschöpfend darzustellen, dass ich mir immer wieder sage der Satz müsse doch irgendwie implizite da drin stecken. Wenn Sie gelegentlich dies einmal gütigst nachsehen wollten, würden Sie mich zu grossem Dank verpflichten. ⟦4⟧ ⟦5⟧

Mit ausgezeichneter Hochachtung
Ihr ergebenster
L E J Brouwer.

NOTES

⟦1⟧ Compare Y22 ⟦1⟧. Brouwer explains the problem anew in terms of integrals of the first kind.
⟦2⟧ A. Hurwitz 1888, 1892.
⟦3⟧ Brouwer 1912 H.
⟦4⟧ A. Hurwitz 1892.
⟦5⟧ Hurwitz answered:

Zürich, 6. Januar 1912.

Hochgeehrter Herr College,
Soeben erhalte ich Ihren Brief und beeile ich mich, Ihnen darauf zu antworten, da in der neuen Form des Satzes mir die Sache nun ganz klar ist. Sie finden den von Ihnen gemeinten Satz in sehr einfacher Weise bewiesen in meiner Arbeit „Über algebraische Gebilde mit eindeutigen Transformationen in sich" (Math. Annalen Bd 41 p. 403 ff.) auf pag. 428. Die von Ihnen in Ihrem Briefe angeschriebenen linearen Transformationsgl. der Integrale 1 G. sollten wie auf p. 428 oben meiner Arbeit noch je eine additive Constante erhalten, welche beim Übergang zu den Differentialen indessen wegfällt. – Mit ergebenstem Grusse Ihr A. Hurwitz

Of course this does not answer Brouwer's question, but in his second letter Brouwer did not add the proviso that the theorem should be proved independently of uniformisation. To the reference Brouwer added 'Vgl. Appell-Goursat p. 387.', which means Appell & Goursat 1895.

1919 S

Über die periodischen Transformationen der Kugel.

Von

L. E. J. Brouwer in Amsterdam.

⟦1⟧ Die Resultate der vorstehenden Arbeit des Herrn v. Kerékjártó waren mir seit mehreren Jahren bekannt; u. a. habe ich im Jahre 1911 Herrn F. Bernstein die beiden folgenden Sätze mitgeteilt:

1. *Jede periodische, eineindeutige und stetige Transformation der Kugel mit invarianter Indikatrix ist einer euklidischen Drehung topologisch äquivalent.*

⟦2⟧ 2. *Jede involutorische, eineindeutige und stetige Transformation der Kugel mit umkehrender Indikatrix ist einer euklidischen Spiegelung um den Mittelpunkt oder um eine Mittelebene topologisch äquivalent.*

Meinen damaligen Beweis, der mir zwar weniger einfach, dafür aber leichter verallgemeinerungsfähig als der vorstehende scheint, teile ich hier mit. Er beruht auf folgender Spezialisierung des allgemeineren in Fuß-
⟦3⟧ note [3]) von S. 605 der Göttinger Nachrichten 1912 enthaltenen Theorems:

Wenn eine periodische, eineindeutige und stetige Transformation einer schlichtartigen Fläche F alle Ränder invariant läßt und keinen invarianten Punkt aufweist, so ist die Anzahl der Ränder von F entweder 2 oder 0.

⟦4⟧ Die zu diesem Satze führende, a. a. O. angegebene Schlußweise liefert nämlich gleichzeitig den Zusatz, daß *im Falle einer periodischen, eineindeutigen und stetigen Transformation der Kugel, deren Fixpunkte die Kugel nicht in Teilgebiete zerlegen, die Anzahl der Fixpunkte entweder 2 oder 0 ist.* An diesen Zusatz anknüpfend, betrachten wir der Reihe nach:

1. *Die n-periodische Transformation t mit invarianter Indikatrix der Kugel K.* Indem wir in K die beiden Fixpunkte tilgen, erhalten wir eine Zylinderfläche Z. Indem wir je zwei durch eine Potenz von t ineinander übergehende Punkte von Z identifizieren, erhalten wir eine als *Modulfläche von t in bezug auf Z* zu bezeichnende zweiseitige Fläche M. Das auf M bestimmte Bild eines nicht zusammenziehbaren Rückkehrschnittes von Z ist ebenfalls nicht zusammenziehbar, so daß M nicht einfach zusammen-
⟦5⟧ hängend sein kann. Andererseits kann M keine mehrfache Basis der Zyklosis

⟦620⟧

besitzen, weil in dem Falle auch Z eine mehrfache Basis der Zyklosis besitzen müßte. Mithin ist M, ebenso wie Z, eine Zylinderfläche, deren n-periodische Überlagerungsfläche die Fläche Z zurückgibt, und hieraus folgt, daß *t einer euklidischen Drehung über einen Winkel $\frac{2\pi}{n}$ topologisch äquivalent ist*. Gleichzeitig ersehen wir, daß t auch in bezug auf K eine Modulfläche, und zwar eine vom Zusammenhange der Kugel, besitzt.

2. *Die involutorische Transformation t mit umkehrender Indikatrix und höchstens zwei Fixpunkten der Kugel K.* Die nach eventueller Tilgung der Fixpunkte restierende Fläche bezeichnen wir wieder mit Z und die Modulfläche von t in bezug auf Z, welche in diesem Falle einseitig ist, mit M. Wenn Z eine Zylinderfläche wäre und j das auf Z bestimmte Bild eines einseitigen Rückkehrschnittes i von M, so könnte i die Fläche M nicht zerlegen, mithin könnte j die Menge der Paare von durch t ineinander übergehenden Punkten von Z nicht zerlegen, was unmöglich ist. Wir sehen also, daß Z mit K identisch und M eine geschlossene Fläche ist. Weiter entspricht jeder nicht zusammenziehbare Rückkehrschnitt von M notwendig einer Verbindungskurve zweier durch t ineinander übergehender Punkte von K. Weil aber je zwei dieser Verbindungskurven sich stetig ineinander überführen lassen, so lassen auch je zwei nicht zusammenziehbare Rückkehrschnitte der einseitigen Fläche M sich stetig ineinander überführen, d. h. M besitzt den Zusammenhang der projektiven Ebene. Weil weiter die zweiseitige Verdoppelung von M die Fläche K zurückgibt, so *ist t einer euklidischen Spiegelung um den Mittelpunkt topologisch äquivalent*.

3. *Die involutorische Transformation t mit umkehrender Indikatrix und mehr als zwei Fixpunkten der Kugel K.* In diesem Falle kann von den durch die Menge ϱ der Fixpunkte bestimmten Restgebieten keines in sich transformiert werden, so daß unter diesen Restgebieten zwei existieren, welche durch t vertauscht werden. Mithin liegen in K auch zwei einfach zusammenhängende Gebiete g und g', deren Grenze in ϱ enthalten ist und welche durch t vertauscht werden. Diese Gebiete g und g' müssen notwendig eine gemeinsame Grenze besitzen, welche zusammen mit g und g' die Kugel K erschöpft und sowohl für g wie für g' unbewallt ist, mithin eine einfache geschlossene Kurve darstellt, *womit t sich als einer euklidischen Spiegelung um eine Mittelebene topologisch äquivalent erwiesen hat.*

4. *Die $2n$-periodische ($n > 1$) Transformation t mit umkehrender Indikatrix der Kugel K.* Wir bezeichnen die für t^2 auftretenden Fixpunkte mit F_1 und F_2, die Modulfläche von t^2 in bezug auf K mit M, die durch die Abbildung von K auf M aus t hervorgehende involutorische Transformation mit umkehrender Indikatrix von M mit s. Wenn s eine F_1 und

F_2 enthaltende einfache geschlossene Kurve j von Fixpunkten besäße, so würde die für t invariante Bildmenge von j auf K aus $2n$ F_1 und F_2 verbindenden, einander außer in F_1 und F_2 nicht treffenden einfachen Kurvenbogen bestehen, deren Reihenfolge von t umgekehrt werden müßte von denen aber jede durch eine geeignete Potenz von t in jede andere überführbar wäre, was für $n > 1$ unmöglich ist. Wenn also s eine Kurve j von Fixpunkten besitzt, so werden F_1 und F_2 durch j getrennt. Das Bild von j auf K ist eine F_1 und F_2 trennende einfache geschlossene Kurve k, so daß *t sich im topologischen Sinne als eine (die Kurve k n-periodisch in sich transformierende) Drehspiegelung erweist.* Wenn aber s keinen Fixpunkt besitzt, so bezeichnen wir die Modulfläche von s in bezug auf M mit N und wählen in N einen willkürlichen den Bildpunkt von F_1 und F_2 nicht enthaltenden einseitigen Rückkehrschnitt i. Das Bild von i auf M ist eine F_1 und F_2 trennende einfache geschlossene Kurve j, das Bild von j auf K ist eine einfache geschlossene Kurve k und *t ist im topologischen Sinne eine (die Kurve k $2n$-periodisch in sich transformierende) Drehspiegelung.*

Unter den Folgerungen aus den obigen Theoremen hebe ich als eine der bemerkenswertesten diese hervor: Wenn wir unter einer *topologischen Involution n^{ter} Ordnung* einer Fläche F eine solche Zerlegung von F in Systeme von *höchstens n* Punkten verstehen, daß die Menge dieser Systeme sich eineindeutig und stetig auf eine als *Modulfläche der Involution* zu bezeichnende Fläche M abbilden läßt (wie leicht zu beweisen, *ist alsdann F eine n-blättrig über M ausgebreitete Riemannsche Fläche*), so gilt der Satz, daß *jede Gruppe von n eineindeutigen und stetigen Transformationen mit invarianter Indikatrix einer geschlossenen zweiseitigen Fläche F eine Involution n^{ter} Ordnung von F darstellt.**)

*) Vgl. eine der Amsterdamer Akademie d. W. am 29. März 1919 vorzulegende Mitteilung. [6]

NOTES

[1] B. v. Kérékjárto 1919 C.
[2] *involutorisch* means periodic.
[3] Brouwer 1912 D.
[4] If 'a.a.O.' means Brouwer 1912 D, then it should be said that its footnote betrays little of the *Schlussweise*. It is strange that Brouwer did not cite Brouwer 1919 L2, which provides a complete proof of Hurwitz' theorem.
[5] *Zyklosis* means the first Betti group (or the fundamental group). See Brouwer 1912 L.
[6] Brouwer 1919 N2.

Mathematics. — "*Ueber topologische Involutionen.*" By Prof. L. E. J. BROUWER. 1919 N2 [[1]]

(Communicated in the meeting of March 29, 1919).

Unter einer *topologischen Involution n-ter Ordnung* einer h-dimensionalen Mannigfaltigkeit F verstehen wir eine solche Zerlegung von F in Systeme von *höchstens n* Punkten, dass die Menge dieser Systeme sich eineindeutig und stetig auf eine als *Modulmannigfaltigkeit der Involution* zu bezeichnende h-dimensionale Mannigfaltigkeit M abbilden lässt.

§ 1. *Involutionen von Linien.*

Sei P ein Punkt der Modullinie M, der ein solches Segment s von M begrenzt, von dem jeder Punkt n verschiedene Bildpunkte auf der Linie F besitzt, während fur P selbst $m > 1$ dieser Bildpunkte in einem Punkte Q von F zusammenfallen. Wenn wir einen Punkt C von s in hinreichender Nähe von P wählen, so zerfallen die weder P noch C entsprechenden, in der Nähe von Q gelegenen Punkte von F in drei und nur drei Kategorien: ein solcher Punkt ist nämlich *entweder* Bildpunkt eines zwischen C und P liegenden Punktes von M und lässt sich alsdann ohne Berührung der in der Nähe von Q liegenden Bildpunkte $D_1, D_2, \ldots D_m$ von C mit Q verbinden, *oder* Bildpunkt eines durch C von P getrennten Punktes von M, in welchem Falle er sich ohne Berührung von $Q, D_1, D_2, \ldots D_m$ aus der Nähe von Q entfernen lässt, *oder* schliesslich Bildpunkt eines durch P von C getrennten Punktes von M, in welchem Falle er sich sowohl ohne Berührung von $D_1, D_2, \ldots D_m$ mit Q verbinden, wie ohne Berührung von $Q, D_1, D_2, \ldots D_m$ aus der Nähe von Q entfernen lässt. Hiermit sind wir aber für $m > 1$ zu einem Widerspruch gelangt, so dass *jeder Punkt von M notwendig n verschiedene Bildpunkte auf F besitzt.* Hieraus folgt unmittelbar, *erstens,* dass F eine *geschlossene* Linie ist, *zweitens,* dass *die Involution n-ter Ordnung von F einer n-periodischen Rotationsgruppe topologisch äquivalent ist.*

§ 2. *Involutionen von Flächen.*

Sei β ein Gebiet der Modulfläche M, von dem jeder Punkt n

verschiedene Bildpunkte auf der Fläche F besitzt, P ein solcher erreichbarer Punkt der Grenze von β; dass auf einem gewissen aus β nach P führenden Wege m der n Bildpunkte gegen den Bildpunkt Q von P konvergieren. Sei j eine auf M in der Nähe von P um P gezogene einfache geschlossene Kurve, k ihr auf F in der Nähe von Q gelegenes Bild. Wenn j hinreichend klein gewählt wird, so ist ein keinem Punkte von j entsprechender, in der Nähe von Q gelegener Punkt von F *entweder* Bildpunkt eines innerhalb j liegenden Punktes von M und lässt sich alsdann ohne Berührung von k mit Q verbinden, *oder* Bildpunkt eines ausserhalb j liegenden Punktes von M, in welchem Falle er sich ohne Berührung von k aus der Nähe von Q entfernen lässt. Weil mithin k auf F *zwei und nur zwei Gebiete* bestimmt, so ist k eine einfache geschlossene Kurve, auf welcher die gegebene Involution von F eine Involution mit j als Modullinie bestimmt, deren Ordnung notwendig $= m$ sein muss, so dass sie einer m-periodischen Rotationsgruppe topologisch äquivalent ist. Hieraus folgt unmittelbar, *erstens*, dass die Bildpunkte der erreichbaren Punkte der Grenze von β, mithin auch diese erreichbaren Punkte selbst isoliert sind, dass also diejenigen Punkte von M, welche weniger als n Bildpunkte besitzen, ebenso wie die entsprechenden Bildpunkte selbst, isoliert sind, *zweitens*, dass *die Fläche F eine über die Modulfläche M n-blättrig ausgebreitete Riemannsche Fläche darstellt.*

§ 3. *Endliche Gruppen von Linien.*

Wir betrachten eine endliche Gruppe G von n eineindeutigen und stetigen Transformationen mit invarianter Indikatrix einer Linie F. Weil jede Transformation von G periodisch ist, so muss F notwendig geschlossen sein und kann für keine Transformation von G ein invarianter Punkt existieren. Sei nun s ein Segment von F, von dem der eine Endpunkt S_1 durch die Transformation t von G in den anderen Endpunkt S_2 übergeht, das aber übrigens kein Paar für G äquivalenter Punkte enthält. Seien $S_2, S_3, \ldots S_m, S_{m+1} = S_1$ die Punkte von F, in welche S_1 durch die sukzessiven Potenzen der m-periodischen Transformation t übergeht. Alsdann kann auf keinem Segmente $S_h S_{h+1}$ ausser dem Endpunktepaar ein Paar für G äquivalenter Punkte existieren, so dass zwei Punkte von F nur dann für G äquivalent sind, wenn sie durch eine Potenz von t ineinander übergehen. Weil mithin die Gruppe G ausschliesslich die Potenzen von t enthält, so ist sie einer n-periodischen Rotationsgruppe topologisch äquivalent, d. h. sie ist eine *Involution n-ter Ordnung.*

[[624]]

§ 4. *Endliche Gruppen von Flächen.*

Wir betrachten eine endliche Gruppe G von n eineindeutigen und stetigen Transformationen mit invarianter Indikatrix einer geschlossenen zweiseitigen Fläche F. Sei P ein für eine Untergruppe γ von G invarianter Punkt, τ eine (offenbar periodische) Transformation von γ. Der Transformation τ von F entspricht eine ebenfalls periodische, einen Bildpunkt von P invariant lassende Transformation der einfach zusammenhängenden Ueberlagerungsfläche von F. Hieraus folgt nach dem *Rotationssatze* von KERÉKJÁRTÓ[1]), dass P auf F eine von für τ invarianten Punkten freie volle Umgebung besitzt, so dass ebenfalls eine volle Umgebung von P auf F existiert, innerhalb deren *keine* Transformation von γ einen Punkt invariant lässt. Wenn wir also in hinreichender Nähe von P eine P in ihrem Innern enthaltende einfache geschlossene Kurve konstruieren, so bestimmt dieselbe zusammen mit ihren von γ erzeugten Bildern eine gleichfalls eine einfache geschlossene Kurve darstellende *äussere Grenze* k, welche von γ in solcher Weise in sich transformiert wird, dass für keine Transformation von γ ein invarianter Punkt auftreten kann. Hieraus folgt unmittelbar, dass γ *von einer einzigen periodischen Transformation* t *erzeugt wird.* Wenn wir nun das System der für G mit P äquivalenten Punkte mit π bezeichnen, so können wir, indem wir t in obiger Weise auf die einfach zusammenhängende Ueberlagerungsfläche von F ausdehnen, mittels des Rotationssatzes weiter folgern, dass in der Menge der Systeme von für G äquivalenten Punkten von F eine volle Umgebung von π existiert, welche sich inklusive π eineindeutig und stetig auf ein Flächenstück abbilden lässt. Hiermit hat sich herausgestellt, dass die Gruppe G eine (übrigens spezielle) *Involution n-ter Ordnung* darstellt.

[1]) Mathem. Annalen 80, S. 36—41. ⟦2⟧

NOTES

⟦1⟧ The paper is closely related to 1919 L2 and 1919 S.

⟦2⟧ B. v. Kérékjárto 1919 C. It should be: pp. 36–38.

1919 P2 **Mathematics.** — "*Aufzählung der periodischen Transformationen*

⟦1⟧ *des Torus*". By Prof. L. E. J. BROUWER.

(Communicated in the meeting of May 3, 1919).

§ 1. *Transformationen mit invarianter Indikatrix und invarianter*

⟦2⟧ *Zyklosis.*

Wir werden sagen, dass eine Transformation *die Zyklosis invariant lässt*, wenn sie *jeden Zykel* (d. h. jede nicht zusammenziehbare einfache geschlossene Kurve) *inklusive des Umlaufssinnes äquivalent transformiert.* Alsdann kann eine n-periodische Transformation t mit invarianter Indikatrix und invarianter Zyklosis des Torus R keinen Punkt invariant lassen [1]), so dass ihre Modulfläche [2]) nach der HURWITZschen Formel [3]) ein Torus T sein muss. Die Flächen R und T besitzen eine gemeinsame einfach zusammenhängende Ueberlagerungsfläche S, welche sich in solcher Weise topologisch (d. h. eineindeutig und stetig) auf eine Cartesische Ebene C abbilden lässt, dass die Translation von C

$$(\tau) \quad . \quad . \quad . \quad \begin{cases} x' = x + a \, (a \text{ ganz}) \\ y' = y \end{cases}$$

der Transformation t und die von den Translationen

$$\begin{cases} x' = x + 1 \\ y' = y \end{cases} \quad \text{und} \quad \begin{cases} x' = x \\ y' = y + 1 \end{cases}$$

erzeugte Gruppe derjenigen Gruppe von S, welche alle Punkte von T invariant lässt, entspricht. Zwei Punkte P_1 und P_2 von C entsprechen nun dann und nur dann demselben Punkte von T, wenn sie *entweder* demselben Punkte von R entsprechen, *oder* P_1 durch eine Potenz von τ in einen Punkt P_3, der demselben Punkte von R entspricht wie P_2, übergeführt wird. Hieraus folgt *erstens*, dass eine Translation von C der Form

$$\begin{cases} x' = x + b \;\; (b \text{ ganz}) \\ y' = y + 1 \end{cases}$$

angegeben werden kann, welche alle Punkte von R invariant lässt,

---

⟦3⟧ 1) Diese Proceedings XXI, S. 707—710.

⟦4⟧ 2) Ibid., S. 1143.

⟦5⟧ 3) Math. Annalen 41, S. 404.

zweitens, dass, wenn n der Minimalwert ist, für den die Translation von C

$$\begin{cases} x' = x + n \\ y' = y \end{cases}$$

alle Punkte von R invariant lässt, die Zahlen a und n relativ prim sind. Für ein geeignetes torisches Koordinatensystem auf R wird somit die Transformation t durch folgende Formeln dargestellt:

$$\begin{cases} \varphi' = \varphi \\ \psi' = \psi + \dfrac{2\pi a}{n} \quad (a \text{ und } n \text{ ganz und relativ prim}). \end{cases}$$

§ 2. *Transformationen mit invarianter Indikatrix und variabler Zyklosis.*

Sei t die betrachtete n-periodische Transformation des Torus R, t' eine mit t korrespondierende Transformation der einfach zusammenhängenden Ueberlagerungsfläche S von R. Alsdann geht t' durch eine geeignete topologische Abbildung von S auf eine Cartesische Ebene C in eine Transformation t'' von C über, welche von einer periodischen, ganzzahligen, homogen-linearen Transformation der Determinante $+1$ nur beschränkte Abweichungen aufweist, mithin sich, ebenso wie die letztere Transformation, periodisch, eineindeutig und stetig auf den unendlichfernen Kreis von C ausdehnen lässt, wobei dieser Kreis einen invarianten Umlaufssinn, mithin keinen invarianten Punkt besitzt, so dass t'' im endlichen von C und deshalb auch t auf R einen invarianten Punkt besitzen muss. Weil somit die Modulfläche B von t von R *nicht* unverzweigt überdeckt wird, so bezitzt B nach der obigen Hurwitzschen Formel den Zusammenhang der Kugel und liegt R nach der Art einer regulären Riemannschen Fläche vom Geslecht 1 n-blättrig und mit einem n-blättrigen Verzweigungspunkte im eben ermittelten für t invarianten Punkte über B ausgebreitet. Hieraus folgt, dass nur vier Fälle möglich sind [1]: [[5]]

I. $n = 2$; R überdeckt B mit vier zweiblättrigen Verzweigungspunkten.

II. $n = 6$; R überdeckt B mit einem zweiblättrigen, einem dreiblättrigen und einem sechsblättrigen Verzweigungspunkt.

III. $n = 4$; R überdeckt B mit zwei vierblättrigen und einem zweiblättrigen Verzweigungspunkt.

IV. $n = 3$; R überdeckt B mit drei dreiblättrigen Verzweigungspunkten.

---

[1]) Appell et Goursat, *Théorie des fonctions algébriques*, Paris, Gauthier-Villars, 1895, S. 241. [[6]]

In jedem dieser vier Fälle ist die Transformation t durch die angegebene Struktur der entsprechenden Riemannschen Fläche vollständig charakterisiert.

§ 3. *Involutorische Transformationen mit umkehrender Indikatrix.*

Wir bezeichnen die Menge der für die betrachtete Transformation t des Torus R invarianten Punkte mit I. Alsdann sind drei Fälle zu unterscheiden:

α. Die Komplementärmenge von I in R besteht aus mehr als einem Gebiete. Weil keines dieser Gebiete von t in sich mit invarianten Randpunkten transformiert werden kann, so werden dieselben von t paarweise verwechselt. Zwei derartige von t verwechselte Gebiete g_1 und g_2 müssen aber eine gemeinsame Grenze besitzen, welche sowohl für g_1 wie für g_2 unbewallt [1]) ist, was nur möglich ist, wenn sie aus zwei äquivalenten Zykeln besteht. Dann aber wird t für ein geeignetes torisches Koordinatensystem wie folgt dargestellt:

$$\begin{cases} \varphi' = -\varphi \\ \psi' = \psi. \end{cases}$$

β. Die Komplementärmenge von I in R besteht aus einem einzigen, nicht mit R identischen Gebiete g. Wenn g nicht schlichtartig wäre, so würde mit t eine involutorische Transformation mit umkehrender Indikatrix der einfach zusammenhängenden Ueberlagerungsfläche S von R, deren invariante Punkte nur ein einziges Gebiet begrenzten, korrespondieren, was unmöglich ist. Weil mithin g schlichtartig ist, so werden seine Ränder von t paarweise verwechselt [2]). Zwei derartige von t verwechselte Ränder r_1 und r_2 sind aber beide für g unbewallt, so dass sie in einem Zykel k von R zusammenfallen und das Gebiet $g+k$ nicht mehr schlichtartig ist. Eventuelle Ränder des Gebietes $g+k$ müssten einerseits für t invariant sein, andererseits von t paarweise verwechselt werden, können mithin nicht existieren, so dass $g+k$ mit R identisch ist und t für ein geeignetes torisches Koordinatensystem wie folgt dargestellt wird:

$$\begin{cases} \varphi' = -\varphi \\ \psi' = \psi + \varphi. \end{cases}$$

γ. Die Menge I fällt fort. In diesem Falle besitzt t eine geschlossene, einseitige Modulfläche M, welche R als ihre zweiseitige Verdoppelung besitzt, mithin notwendig eine einseitige Ringfläche sein muss. Wir können also auf M zwei einander nicht treffende einseitige Zykeln r_1 und r_2 wählen. Weil r_1 und r_2 die nicht zu ihnen gehörenden

[7] [1]) Math. Annalen 71, S. 321.

[8] [2]) Ibid. 80, S. 36—41.

Punkte von M nicht in getrennte Gebiete zerlegen, so können auch die Bilder s_1 und s_2 auf R von r_1 und r_2 die nicht zu ihnen gehörenden Paare von für t äquivalenten Punkten von R nicht in getrennte Gebiete zerlegen, d. h. s_1 und s_2 zerlegen R in zwei für t äquivalente Gebiete und t wird für ein geeignetes torisches Koordinatensystem wie folgt dargestellt:

$$\begin{cases} \varphi' = -\varphi \\ \psi' = \psi + \pi. \end{cases}$$

§ 4. *Nicht-involutorische Transformationen mit umkehrender Indikatrix.*

Sei t die betrachtete $2n$-periodische Transformation des Torus R. Weil für $n > 1$ keine $2n$-periodische, ganzzahlige, homogen-lineare Transformation der Determinante -1 in zwei Variabeln existiert, so lässt t^2 die Zyklosis von R invariant, besitzt mithin als Modulfläche einen Torus T, auf welchem das Bild von t eine involutorische Transformation t_1 mit umkehrender Indikatrix darstellt. Wir führen auf T die nach § 3 zu t_1 gehörigen torischen Koordinaten φ und ψ ein, betrachten zwei durch die Gleichungen $\varphi = a$ und $\varphi = -a$ bestimmte Zykeln r_1 und r_2 von T und bezeichnen die Bildzykeln von r_v auf R mit ${}_1s_v, {}_2s_v, \ldots {}_ms_v$. Weil alle Zykeln ${}_\mu s_v$ für t äquivalent sind und ihre zyklische Reihenfolge von t umgekehrt wird, so kann ihre Anzahl nur *zwei* betragen und muss $m = 1$ sein. Mithin wird die alle Punkte von R invariant lassende Transformationsgruppe der einfach zusammenhängenden Ueberlagerungsfläche S von R und T durch die Translationen

$$\begin{cases} \varphi' = \varphi \\ \psi' = \psi + 2n\pi \end{cases} \quad \text{und} \quad \begin{cases} \varphi' = \varphi + 2\pi \\ \psi' = \psi + 2h\pi \end{cases}$$

erzeugt. Wenn wir nunmehr φ und $\psi - h\varphi$ auf T als neue torische Koordinaten (φ, ψ) einführen, so werden diese Translationen durch

$$\begin{cases} \varphi' = \varphi \\ \psi' = \psi + 2n\pi \end{cases} \quad \text{und} \quad \begin{cases} \varphi' = \varphi + 2\pi \\ \psi' = \psi \end{cases}$$

und t_1 durch

$$\begin{cases} \varphi' = -\varphi \\ \psi' = \psi + k\varphi \end{cases} \quad \text{bzw.} \quad \begin{cases} \varphi' = -\varphi \\ \psi' = \psi + k\varphi + \pi, \end{cases}$$

mithin t in denselben Koordinaten durch

$$\begin{cases} \varphi' = -\varphi \\ \psi' = \psi + k\varphi + 2q\pi \end{cases} \quad \text{bzw.} \quad \begin{cases} \varphi' = -\varphi \\ \psi' = \psi + k\varphi + (2q+1)\pi, \end{cases}$$

also auf jeden Fall durch

$$\begin{cases} \varphi' = -\varphi \\ \psi' = \psi + k\varphi + g\pi \end{cases}$$

und in korrespondierenden torischen Koordinaten *von R* durch

$$\begin{cases} \varphi' = -\varphi \\ \psi' = \psi + \frac{k}{n}\varphi + \frac{g}{n}\pi \end{cases}$$

dargestellt. Auf Grund der Eindeutigkeit von t auf R muss $\frac{k}{n} =$ einer ganzen Zahl e sein. Indem wir aber für geeignetes b, φ und $\psi - b\varphi$ auf R als neue torische Koordinaten (φ, ψ) einführen, können wir erreichen, dass e entweder $= 0$ oder $= 1$ wird, so dass wir schliesslich für t zu den folgenden Formeln gelangt sind:

$$\begin{cases} \varphi' = -\varphi \\ \psi' = \psi + \frac{g}{n}\pi \end{cases} \quad \text{bzw.} \quad \begin{cases} \varphi' = -\varphi \\ \psi' = \psi + \varphi + \frac{g}{n}\pi. \end{cases}$$

wo g und n relativ prim sind.

NOTES

⟦1⟧ The paper largely overlaps Brouwer 1919 G, 1919 H though the approaches are different.
⟦2⟧ 'mit invarianter Zyklosis' means invariance of every single element of the fundamental group.
⟦3⟧ Brouwer 1919 L2.
⟦4⟧ Brouwer 1919 N2. It should be p. 1144.
⟦5⟧ A. Hurwitz 1892.
⟦6⟧ Appell & Goursat 1895.
⟦7⟧ Brouwer 1911 G.
⟦8⟧ Brouwer 1919 S. It should be pp. 39–41.

GÉOMÉTRIE. — *Énumération des surfaces de Riemann régulières de genre un.* Note de **M. L.-E.-J. Brouwer.** 1919G [[1]]

Soient S une surface régulière de genre un; a_1, a_2, ..., a_q ses points de ramification; si au point a_i les feuillets de la surface se partagent en cycles de r_i feuillets, quatre cas sont à distinguer ([1]) :

I.	$q = 4$,	$r_1 = r_2 = r_3 = r_4 = 2$.		
II.	$q = 3$,	$r_1 = 2$,	$r_2 = 3$,	$r_3 = 6$.
III.	$q = 3$,	$r_1 = 4$,	$r_2 = 2$,	$r_3 = 4$.
IV.	$q = 3$,	$r_1 = r_2 = r_3 = 3$.		

Nous traçons une ligne de passage joignant a_1 et a_4 et passant par a_2 et a_3 au cas I; joignant a_1 et a_3 et passant par a_2 aux cas II, III et IV.

Soit R la surface à connexion simple *superposée* à S et se composant par conséquent d'une infinité de feuillets. Représentons R sur le plan euclidien P, la division de R en feuillets par les lignes de passage peut être représentée topologiquement par une division régulière de P en polygones fondamentaux. Ces polygones fondamentaux sont :

Au cas I, des rectangles présentant *deux* orientations différentes et dont la hauteur est à la largeur comme 2 à 1;

Au cas II, des triangles équilatéraux présentant *six* orientations différentes;

Au cas III, des carrés présentant *quatre* orientations différentes;

Au cas IV, des rhombes présentant *trois* orientations différentes et dont les angles sont égaux à $\frac{1}{3}\pi$ et $\frac{2}{3}\pi$ respectivement.

En prenant un polygone fondamental de chaque orientation et en joignant ces polygones, on forme un carré C aux cas I et III, un hexagone régulier H aux cas II et IV.

([1]) Voir, par exemple, Appell et Goursat, *Théorie des fonctions algébriques*, Paris, Gauthier-Villars, 1895, p. 241. [[2]]

S'il existait deux polygones fondamentaux de P, différemment orientés et représentant le même feuillet de S, il existerait une rotation de P laissant invariants tous les points de S, ce qui évidemment est impossible. Donc, le groupe des transformations de P laissant invariants tous les points de S et *déterminant complètement la surface* S, est un *groupe de translations t.*

En choisissant sur P les axes des coordonnées perpendiculaires à deux côtés de C ou de H respectivement, et l'unité linéaire égale à la distance de deux côtés opposés de C ou de H respectivement, le groupe *t*, engendré par les deux transformations

$$(1)\qquad \begin{cases} x' = x + n, \\ y' = y \end{cases}$$

et

$$(2)\qquad \begin{cases} x' = x + m, \\ y' = y + p \end{cases}$$

(m, n et p représentant des entiers), est assujetti à la seule condition d'être invariant pour une rotation de $\frac{1}{2}\pi$ au cas III, de $\frac{1}{3}\pi$ aux cas II et IV, ce qui entraîne pour m, n et p les conditions suivantes :

Au cas III, m et n divisibles par p et $\left(\frac{m}{p}\right)^2 + 1$ divisible par $\frac{n}{p}$;

Aux cas II et IV, m et n divisibles par p et $\left(\frac{m}{p}\right)^2 + \frac{m}{p} + 1$ divisible par $\frac{n}{p}$;

⟦3⟧ Au cas I ~~non seulement~~ les entiers m, n et p sont complètement arbitraires~~, mais on doit encore ajouter la possibilité que *t* est engendré par la *seule* translation (2)~~.

(*Comptes rendus*, t. 168, p. 677, séance du 31 mars 1919.)

ERRATA.

(Séance du 31 mars 1919.)

Note de M. *L.-E.-J. Brouwer*, Énumération des surfaces de Riemann régulières de genre un :

Page 678, la dernière phrase doit être réduite ainsi : Au cas I, les entiers m, n et p sont complètement arbitraires.

NOTES

⟦1⟧ Overlaps 1919 P2. Riemann surfaces of a genus > 1 which allow non-trivial automorphisms were called regular by F. Klein 1878 (Cp. Y22 ⟦2⟧). Brouwer extends this terminology to surfaces of genus 1. Cf. Errata. The offprint pagination is completed with the journal pagination.

⟦2⟧ Appell & Goursat 1895.

⟦3⟧ Deletion of 'non seulement' et 'mais ⋯ translation (2)' in Brouwer's hand. Cf. Errata.

1919 H GÉOMETRIE. — *Enumération des groupes finis de transformations*
⟦1⟧ *topologiques du tore.* Note de M. L.-E.-J. Brouwer.

Dans une Communication à l'Académie royale d'Amsterdam (voir
⟦2⟧ Procès-verbal de la séance du 29 mars 1919), j'ai démontré que les sys-
⟦3⟧ tèmes de points équivalents pour un groupe fini g de transformations topo-
logiques (c'est-à-dire biunivoques et continues) à indicatrice invariante
d'une surface fermée et bilatérale S, forment une nouvelle surface fermée
et bilatérale M, que j'appelle la *surface modulaire de g par rapport à* S, et
sur laquelle S est étendue à la manière d'une surface de Riemann au sens
⟦4⟧ large. Dans le cas que S est un tore, il s'ensuit de la formule de M. Hurwitz
(voir *Mathem. Annalen*, t. 41, p. 404) que le genre de M ne peut être que
⟦5⟧ zéro ou un. Au premier cas, S est une surface de Riemann régulière de
⟦6⟧ genre ~~zéro~~ et l'analyse du groupe g correspondant est contenue dans ma
⟦7⟧ Communication du 31 mars 1919 (voir p. 677-678 de ce tome). Au second
cas, S est étendu sur M sans ramification, de sorte qu'une même surface à
connexion simple R est *superposée* à S et à M. Représentons R sur un plan
cartésien C de manière que le groupe des transformations de R laissant
invariants tous les points de M corresponde au groupe engendré sur C par
les deux transformations $x' = x + 1$, $y' = y$ et $x' = x$, $y' = y + 1$. Alors le
groupe t des transformations de R laissant invariants tous les points de S
et *déterminant complètement le groupe g*, correspond au groupe engendré
sur C par les deux transformations $x' = x + n$, $y' = y$ et $x' = x + m$,
$y' = y + p$ (m, n et p désignant des entiers arbitraires).

Passons aux groupes G *à indicatrice variable*, les transformations à indicatrice invariante de G forment un sous-groupe g de G rentrant dans le cas précédent, donc donnant lieu à une surface modulaire M, sur laquelle l'image de G est engendrée par une transformation topologique *et involutive* à indicatrice renversée i.

Si M *est de genre un*, on démontre que, pour une représentation convenable de R sur C, la transformation j de C correspondant à i s'obtient sous l'une des trois formes suivantes :

(I) $x'=-x,\ y'=y$; (II) $x'=-x,\ y'=y+\frac{1}{2}$; (III) $x'=-x,\ y=y+x$;

en même temps que le groupe de C correspondant au groupe des transformations de R laissant invariants tous les points de M, est engendré par les deux transformations $x'=x+1, y'=y$ et $x'=x, y'=y+1$.

Le groupe t étant assujetti à la condition d'être invariant pour la trans- ⟦8⟧
formation j, les entiers m, n et p doivent satisfaire aux conditions suivantes :

Aux cas I et II : $2m$ divisible par n;

Au cas III : m et n divisibles par p; $\left(\frac{m}{p}\right)^2+2\left(\frac{m}{p}\right)$ divisible par $\frac{n}{p}$.

Si M *est de genre zéro*, d'après le théorème de Kerékjártó, la transformation i est topologiquement équivalente à l'une des deux transformations suivantes de la sphère :

(α) $\varphi=\varphi \quad \psi'=-\psi$; (β) $\varphi'=\varphi+\pi, \quad \psi'=-\psi$,

φ et ψ désignant la longitude et la latitude dans un système de coordonnées sphériques.

Nous traiterons séparément les quatre cas I, II, III et IV, distingués dans ma Communication du 31 mars :

Le *cas* I se divise en quatre sous-cas, dont le premier présente pour i le cas (β), les trois autres le cas (α).

a. i échangeant d'une part a_1 et a_3, d'autre part a_2 et a_4, nous traçons sur M une courbe de Jordan fermée et invariante pour i, sur laquelle les deux couples de points (a_1, a_3) et (a_2, a_4) se séparent et dans laquelle nous choisissons la ligne de passage.

b. i échangeant d'une part a_2 et a_3, d'autre part a_1 et a_4, a_1 et a_2 étant situés du même côté de la ligne λ, c'est-à-dire du lieu des points invariants pour i, nous choisissons la ligne de passage de manière qu'elle ne rencontre λ qu'en un seul point, qui la divise en deux arcs équivalents pour i.

c. i échangeant a_1 et a_3, tandis que a_2 et a_4 sont situés sur λ, nous choisissons la ligne de passage de manière qu'elle ne rencontre λ qu'en les seuls points a_2 et a_4 et que son arc $a_1 a_3$ est divisé par a_2 en deux parties équivalentes pour i.

d. Les deux couples de points (a_1, a_3) et (a_2, a_4) se séparant sur λ, nous choisissons la ligne de passage dans λ.

En précisant convenablement la représentation de R sur le plan euclidien P, une transformation j de P correspondant à i s'obtient dans les quatre sous-cas sous les formes :

[[9]] $$\begin{cases} 2x' = 1 - 2x, \\ 2y' = 1 + 2y, \end{cases} \qquad \begin{cases} 2x' = 1 - 2x, \\ y' = y, \end{cases} \qquad \begin{cases} x' = y, \\ y' = x, \end{cases} \qquad \begin{cases} x' = x, \\ y' = 1 - y, \end{cases}$$

respectivement (l'origine des coordonnées étant choisie en un point représentant de a_2), dont résultent pour le groupe t les conditions suivantes :

Aux sous-cas a, b et d : $2m$ divisible par n.

Au sous-cas c : m et n divisibles par p; $\left(\frac{m}{p}\right)^2 - 1$ divisible par $\frac{n}{p}$.

Le *cas II* présente pour i nécessairement le cas (α), tandis que a_1, a_2 et a_3 sont situés sur la ligne λ, dans laquelle nous choisissons la ligne de passage. En précisant convenablement la représentation de R sur P, j s'obtient sous la forme

$$x' = -y, \qquad y' = -x$$

(l'origine des coordonnées représentant a_3). Il s'ensuit pour le groupe t, qu'ou $\frac{n}{p} = 1$, ou bien $\frac{n}{p} = 3$ et $\frac{m}{p} = 3h + 1$, h désignant un entier arbitraire.

Le *cas III* présente pour i nécessairement le cas (α) et se divise en deux sous-cas :

a. i échangeant a_1 et a_3, tandis que a_2 est situé sur λ, nous choisissons la ligne de passage de manière qu'elle ne rencontre λ qu'en le seul point a_2, qui la divise en deux arcs équivalents pour i ;

b. a_1, a_2 et a_3 étant situés tous les trois sur λ, nous choisissons la ligne de passage dans λ.

En précisant convenablement la représentation de R sur P, j s'obtient dans les deux sous-cas sous les formes

$$x' = -y, \qquad y' = -x; \qquad x' = y, \qquad y' = x$$

respectivement (l'origine représentant a_2 au premier sous-cas, a_1 au second sous-cas).

Dans les deux sous-cas, on trouve $\frac{n}{p} = 1$, si $\frac{m}{p}$ est pair ; $\frac{n}{p} = 1$ ou 2, si $\frac{m}{p}$ est impair.

Le *cas IV* présente encore pour i nécessairement le cas (α) et se divise en deux sous-cas :

a. i échangeant a_1 et a_3, tandis que a_2 est situé sur λ, nous choisissons la ligne de passage de manière qu'elle ne rencontre λ qu'en le seul point a_2, qui la divise en deux arcs équivalents pour i;

b. a_1, a_2 et a_3 étant situés tous les trois sur λ, nous choisissons la ligne de passage dans λ.

En précisant convenablement la représentation de R sur P, j s'obtient dans les deux sous-cas sous les formes

$$x' = -y, \quad y' = -x; \qquad x' = y, \quad y' = x$$

respectivement (l'origine représentant a_2 au premier sous-cas, a_1 au second sous-cas). [[10]]

Dans les deux sous-cas, il s'ensuit pour le groupe t, qu'ou $\frac{n}{p} = 1$, ou bien $\frac{n}{p} = 3$ et $\frac{m}{p} = 3h + 1$, h désignant un entier arbitraire.

Je termine par la remarque que le problème analogue pour la sphère se résout par la même méthode, mais beaucoup plus simplement.

(*Comptes rendus*, t. 168, p. 845, séance du 28 avril 1919.)

ERRATA.

—

(Séance du 28 avril 1919.)

Note de M. *L.-E.-J. Brouwer*, Énumération des groupes finis de transformations topologiques du tore :

Page 845, ligne 8 en remontant, *au lieu de* genre zéro, *lire* genre un.

ERRATA.

—

(Séance du 28 avril 1919.)

Note de M. *L.-E.-J. Brouwer*, Énumération des groupes finis de transformations topologiques du tore :

Page 846, ligne 16, *au lieu de* invariant, pour, *lire* invariant pour.
Page 847, ligne 15, *au lieu de* $2y' = 1 - 2y$, *lire* $2y' = 1 + 2y$.
Même ligne, *au lieu de* $y = 1 - y$, *lire* $y' = 1 - y$.
Page 848, ligne 17, *au lieu de* représentent, *lire* représentant

NOTES

⟦1⟧ The reprint pagination is completed with the journal pagination. Overlaps 1919 P2, cf. Errata.
⟦2⟧ Brouwer 1919 N2.
⟦3⟧ This seems historically the first use of the term ‘topological transformation’. In 1919 L2 and 1919 S Brouwer still spoke about ‘one-to-one continuous transformations’. 1919 P2 is later than the present paper. Terms such as ‘topologically equivalent’ are older.
⟦4⟧ A. Hurwitz 1892.
⟦5⟧ Regular Riemann surface see Y22 ⟦2⟧.
⟦6⟧ Correction ‘un’ for ‘zéro’ in Brouwer’s hand. Cf. Errata.
⟦7⟧ Brouwer 1919 G.
⟦8⟧ Deletion of the comma in Brouwer’s hand. Cf. Errata.
⟦9⟧ Corrections ‘+’ for ‘−’ and ‘y'’ for ‘y’ in Brouwer’s hand. Cf. Errata.
⟦10⟧ Correction ‘représentant’ for ‘représentent’ in Brouwer’s hand. Cf. Errata.

 [[1]]

GÉOMÉTRIE. — *Sur les points invariants des transformations topologiques des surfaces.* Note de M. L.-E.-J. BROUWER. 1919 J

J'ai démontré en 1909 que toute transformation topologique (c'est-à-dire biuniforme et continue) d'une surface bilatérale fermée de genre zéro à indicatrice invariante laisse au moins un point invariant ([1]). Il importe de remarquer que ce théorème n'est valable que pour le genre zéro. En effet, pour chaque surface bilatérale fermée S de genre p supérieur à zéro, on peut construire d'une manière très simple des transformations (et même des transformations périodiques) topologiques, à indicatrice invariante et ne laissant aucun point invariant. [[2]]

PREMIER CAS : $p = 1$. — Soient φ et ψ des coordonnées bicirculaires sur le tore S. La transformation

$$\varphi' = \varphi,$$
$$\psi' = \psi + \frac{2\pi}{n}$$

possède les propriétés requises.

SECOND CAS : $p = 2$. — Soient T et T′ deux tores congruents, φ et ψ des coordonnées bicirculaires sur T. La transformation t de T définie par les formules [[6]]

$$\varphi' = \varphi + \psi,$$
$$\psi' = -\varphi$$

possède un seul point invariant P. Au voisinage de P traçons une courbe simple fermée k, invariante pour t et divisant T en une région f contenant P et une région g. Soient t', k', f' et g' les images de t, k, f et g sur T′.

([1]) Ma plus simple démonstration de cette propriété a été communiquée par M. HADAMARD dans sa Note sur l'indice de KRONECKER, insérée dans la seconde édition de l'*Introduction à la théorie des fonctions* de TANNERY. La même démonstration se trouve dans un Mémoire récent de M. BIRKHOFF (*American Transactions*, t. 18, p. 289) où aussi une extension du théorème est obtenue par la même méthode. [[3]] [[4]] [[5]]

En identifiant sur les courbes congruentes k et k' les points correspondants, nous formons une surface bilatérale fermée S de genre 2, se composant de k, g et g' et sur laquelle t et t' définissent une transformation possédant les propriétés requises.

Troisième cas : $p > 2$. — Soient φ et ψ des coordonnées bicirculaires sur le tore T, t la transformation de T, définie par les formules

$$\varphi' = \varphi,$$
$$\psi' = \psi + \frac{2\pi}{p-1}.$$

Attachons au tore $p-1$ anses telles que les $p-1$ paires de surfaces de contact ne se touchent pas et se correspondent pour les diverses puissances de t. Détruisons ensuite les surfaces de contact, nous obtenons une surface bilatérale fermée S de genre p, sur laquelle t définit une transformation possédant les propriétés requises.

Passons aux transformations à indicatrice renversée. Nous construirons pour chaque surface bilatérale fermée S de genre p une transformation (même une transformation involutive) topologique, à indicatrice renversée et ne laissant aucun point invariant.

Premier cas : $p = 0$. — Soit φ la longitude, ψ la latitude dans un système de coordonnées géographiques sur la sphère S. La transformation

$$\varphi' = \varphi + \pi,$$
$$\psi' = -\psi$$

possède les propriétés requises.

Deuxième cas : $p = 1$. — Soient φ et ψ des coordonnées bicirculaires sur le tore S. La transformation

$$\varphi' = -\varphi,$$
$$\psi' = \psi + \pi$$

possède les propriétés requises.

Troisième cas : $p = 2q$. — Soit T une sphère, t une transformation de T construite d'après le premier cas. Attachons à la sphère p anses telles que les p paires de surfaces de contact ne se touchent pas et se correspondent deux à deux pour t. Détruisons les surfaces de contact, nous obtenons une surface bilaterale fermée S de genre p, sur laquelle t définit une transformation possedant les propriétés requises.

QUATRIÈME CAS : $p = 2q + 1$. — Soit T un tore, t une transformation de T construite d'après le second cas. Attachons au tore $p - 1$ anses telles que les $p - 1$ paires de surfaces de contact ne se touchent pas et se correspondent deux à deux pour t. Détruisons les surfaces de contact, nous obtenons une surface bilatérale fermée S de genre p, sur laquelle t définit une transformation possédant les propriétés requises.

(*Comptes rendus*, t. 168, p. 1042, séance du 26 mai 1919.)

ERRATA.

(Séance du 26 mai 1919.)

Note de M. *L. E.-J. Brouwer*, Sur les points invariants des transformations topologiques des surfaces :

Page 1042, ligne 9 en remontant, *au lieu de* La transformation de T, *lire* La transformation t de T.

Même page, ligne 5 de la note, *au lieu de* p. 2089, *lire* p. 289.

NOTES

⟦1⟧ The reprint pagination is completed with the journal pagination. Cf. Errata.
⟦2⟧ Brouwer 1909 F2.
⟦3⟧ J. Hadamard 1910.
⟦4⟧ G. Birkhoff 1917.
⟦5⟧ Correction 289 in Brouwer's hand.
⟦6⟧ Addition of *t* in Brouwer's hand. Cf. Errata.

Mathematics. — *"Ueber die Struktur der perfekten Punktmengen"* (dritte Mitteilung [1])). By Prof. L. E. J. BROUWER. 1920 A2

(Communicated in the meeting of October 25, 1919).

§ 1.

Ein *Flächensystem* bzw. eine *Fläche* ist im folgenden definiert mittels eines solchen zweidimensionalen Fragmentes [2]) bzw. zusammenhängenden zweidimensionalen Fragmentes, in welchem die in einem willkürlich gewählten Elementeckpunkte mündenden Elementseiten entweder alle oder alle bis auf zwei je zwei Elementen gemeinsam sind. Im ersteren Falle sprechen wir von einem *gewöhnlichen*, im letzteren von einem *aussergewöhnlichen* Elementeckpunkt, während wir eine Elementseite gewöhnlich oder aussergewöhnlich nennen, je nachdem sie zu zwei oder zu einem einzigen Elemente gehört. Das Flächensystem bzw., die Fläche besteht aus dem Fragmente *mit Ausnahme* der aussergewöhnlichen Elementseiten und Elementeckpunkte, welche zusammen die *Grenze* des Flächensystems bzw. der Fläche bilden.

Ein Flächensystem bzw. eine Fläche, deren Elemente Grundsimplexe einer simplizialen Zerlegung des Flächensystems oder der Fläche ω sind, wird ein *Teilflächensystem* bzw. eine *Teilfläche* von ω genannt werden.

Sei α ein auf der Fläche ω nicht-negativer Charakteristik gelegenes und auf ω kein elementares [3]) Restgebiet bestimmendes, abgeschlossenes Kontinuum, $\alpha' = \omega, \alpha'', \alpha''', \ldots$ eine α approximierende Folge von Teilflächen von ω, durch welche also α als $\mathfrak{D}(\alpha', \alpha'', \ldots)$ bestimmt ist, so dürfen wir für jedes ν annehmen, dass $\alpha^{(\nu)}$ auf ω kein elementares Restgebiet bestimmt, dass $\alpha^{(\nu+1)}$ eine Teilfläche von $\alpha^{(\nu)}$ ist und dass die Grenzen von $\alpha^{(\nu)}$ und $\alpha^{(\nu+1)}$ keinen Punkt gemeinsam haben. Sei $k^{(\nu)}$ die Charakteristik von $\alpha^{(\nu)}$, $m^{(\nu)}$ die maximale Anzahl von einander nicht treffenden, zusammen nicht zerstückelnden einfachen geschlossenen Kurven von $\alpha^{(\nu)}$, so kann für wachsendes ν [3]

[1]) Für die erste und zweite Mitteilung vgl. diese Proceedings XII, S. 785 XIV, S. 137. [1]

[2]) Vgl. Math. Annalen 71, S. 306. [2]

[3]) Ein Teilgebiet von ω heisst elementar, wenn es nur auf ω zusammenziehbare einfache geschlossene Kurven enthält.

weder $k^{(\nu)}$ noch $m^{(\nu)}$ zunehmen, so dass eine endliche positive Zahl g existiert mit der Eigenschaft, dass $k^{(g+\mu)} = k^{(g)}$ und $m^{(g+\mu)} = m^{(g)}$ für jedes nicht negative μ. Hieraus folgt, dass die von $\alpha^{(g+\mu+1)}$ in $\alpha^{(g+\mu)}$ bestimmten Restflächen alle Zylinderflächen sind, dass die topologische Gestalt von $\alpha^{(g+\mu)}$ für jedes μ gleich derjenigen von $\alpha^{(g)}$ ist und dass $\alpha^{(g+\mu+1)}$ aus $\alpha^{(g+\mu)}$ durch Zurücknahme der Ränder nach Innen entsteht, so dass zwischen den $\alpha^{(g+\mu)}$ und α die folgenden Beziehungen existieren:

1. Zu jeder Kombination $(\varepsilon, \varepsilon_1, \mu)$ gibt es ein solches nur von ε und μ abhängendes und für festes μ mit ε gegen 0 konvergierendes ε_2, dass jede ε-Kette [1]) von $\alpha^{(g+\mu)}$ mittels einer endlichen Folge von ε_2-Abänderungen [1]) innerhalb $\alpha^{(g+\mu)}$ in eine ε_1-Kette von α übergeführt werden kann.

2. Zu jedem ε gibt es ein solches mit ε gegen 0 konvergierendes ε_3, dass jede ε-Kette eines $\alpha^{(g+\mu)}$ mittels einer endlichen Folge von ε_3-Abänderungen innerhalb $\alpha^{(g+\mu)}$ in eine Kette von α übergeführt werden kann.

3. Jedes $\alpha^{(g+\mu)}$ besitzt die gleiche minimale Multiplizität der Basis
[[5]] der Zyklosis [2]), wie α.

4. Zu jeder Kombination eines hinreichend kleinen ε', eines ε'_1 und eines μ gibt es ein solches nur von ε' und μ abhängendes und für festes μ mit ε' gegen 0 konvergierendes ε_2, dass jedes System von $(\varepsilon, \varepsilon')$-Fundamentalketten [2]) von $\alpha^{(g+\mu)}$ mittels einer endlichen Folge von ε_2-Abänderungen innerhalb $\alpha^{(g+\mu)}$ in ein System von $(\varepsilon_1, \varepsilon'_1)$-Fundamentalketten von α übergeführt werden kann.

5. Zu jedem ε' gibt es ein solches mit ε' gegen 0 konvergierendes ε'', dass jedes System von $(\varepsilon, \varepsilon')$-Fundamentalketten eines $\alpha^{(g+\mu)}$ mittels einer endlichen Folge von ε''-Abänderungen innerhalb $\alpha^{(g+\mu)}$ in ein System von $(\varepsilon, \varepsilon'')$-Fundamentalketten von α übergeführt werden kann.

6. Zu jedem ε gibt es ein solches ε^0, dass jedes System von $(\varepsilon, \varepsilon')$-Fundamentalketten von α gleichzeitig ein System von $(\varepsilon^0, \varepsilon')$-Fundamentalketten eines jeden $\alpha^{(g+\mu)}$ ist.

Wir bringen diese Beziehungen zum Ausdruck, indem wir α als *zyklomatisches Extrakt* von $\alpha^{(g)}$ bezeichnen.

Sei nunmehr α ein willkürliches auf ω gelegenes abgeschlossenes Kontinuum, α_1 die Vereinigung von α und den elementaren Restgebieten von α, $\alpha_1^{(g_1)}$ eine α_1 als zyklomatisches Extrakt enthaltende Teilfläche von ω. Alsdann gibt es zu jeder Kombination $(\varepsilon, \varepsilon_1)$ ein

[[4]] 1) Vgl. Math. Annalen 72, S. 422.

[[4]] 2) Ibid., S. 424.

solches mit ε gegen 0 konvergierendes ε_2, dass jede ε-Kette von $\alpha_1^{(g_1)}$ mittels einer endlichen Folge von ε_2-Abänderungen innerhalb ω in eine ε_1-Kette von α übergeführt werden kann. *Wir werden auch α ein zyklomatisches Extrakt von $\alpha_1^{(g_1)}$ nennen.* nur von ε ab-hängendes und

Insbesondere werden wir ein zyklomatisches Extrakt eines Elementes als *Elementarkontinuum*, ein zyklomatisches Extrakt einer Zylinderfläche als *Zylinderkontinuum* bezeichnen.

§ 2.

Sei α eine auf der Fläche ω nicht-negativer Charakteristik gelegene, auf ω kein elementares Restgebiet bestimmende, abgeschlossene Punktmenge, $\alpha' = \omega, \alpha'', \alpha''', \ldots$ eine α approximierende Folge von Teilflächensystemen von ω, so dürfen wir für jedes ν annehmen, dass $\alpha^{(\nu)}$ auf ω kein elementares Restgebiet bestimmt, dass $\alpha^{(\nu+1)}$ ein Teilflächensystem von $\alpha^{(\nu)}$ ist und dass die Grenzen von $\alpha^{(\nu)}$ und $\alpha^{(\nu+1)}$ keinen Punkt gemeinsam haben. Sei $\beta^{(\nu)}$ das aus $\alpha^{(\nu)}$ durch Tilgung derjenigen Stücke, welche die topologische Gestalt eines Elementes oder eines Zylinders besitzen, hervorgehende Flächensystem, S ein willkürliches Stück von $\beta^{(\nu)}$, $k(S)$ die Charakteristik von S, $m(S)$ die maximale Anzahl von einander nicht treffenden, zusammen nicht zerstückelnden einfachen geschlossenen Kurven von S, so sind die Zahlen $k(S)$ und $m(S)$ beide nicht-negativ und nicht beide gleich Null. Mithin existieren zwei ganze nicht-negative Zahlen g und h mit der Eigenschaft, dass jedes $\beta^{(g+\mu)}$ sich aus h Stücken $\beta_1^{(g+\mu)}$, $\beta_2^{(g+\mu)}, \ldots \beta_h^{(g+\mu)}$ zusammensetzt, dass die topologische Gestalt von $\beta_s^{(g+\mu)}$ für jedes μ und jedes s gleich derjenigen von $\beta_s^{(g)}$ ist, dass $\beta_s^{(g+\mu+1)}$ aus $\beta_s^{(g+\mu)}$ durch Zurücknahme der Ränder nach Innen entsteht und dass das als $\mathfrak{D}(\beta_s^{(g)}, \beta_s^{(g+1)}, \ldots)$ bestimmte Stück von α ein zyklomatisches Extrakt von $\beta_s^{(g)}$ ist.

Das hiermit erhaltene Resultat kann in der folgenden Form, welche, wie eine triviale Ueberlegung zeigt, für eine Fläche ω negativer Charakteristik ihre Gültigkeit behält, ausgedrückt werden:

Zu jeder auf einer Fläche ω gelegenen abgeschlossenen Punktmenge α existiert ein Teilflächensystem ψ von ω mit der Eigenschaft, dass α sich aus erstens von den Stücken von ψ je einem zyklomatischen Extrakt, zweitens einer Menge von Elementarkontinuen und Zylinderkontinuen zusammensetzt.

Ist insbesondere ω eine Kugel, so sind alle Stücke von α Elementarkontinua.

Ist ω eine projektive Ebene, so sind entweder alle Stücke von α

Elementarkontinua, oder alle bis auf eines, das ein zyklomatisches Extrakt von ω ist.

Ist ω ein Torus, so sind entweder alle Stücke von α Elementarkontinua, oder alle bis auf eines, das ein zyklomatisches Ektrakt von ω ist, oder endlich alle bis auf eine abgeschlossene zyklisch geordnete Menge von Zylinderkontinuen; die Teilzylinder von ω, von denen diese Zylinderkontinua zyklomatische Extrakte sind, sind alle auf ω stetig ineinander überführbar.

⟦7⟧ ERRATUM.

In Professor BROUWER's paper: "*Ueber die Struktur der perfekten Punktmengen* (dritte Mitteilung)", p. 473 of this Vol., l. 1 from the top

for: mit ε gegen 0 konvergierendes
read: nur von ε abhängendes und mit ε gegen 0 konvergierendes

NOTES

⟦1⟧ The first two communications 1910 B2, 1911 B2 are not related to the third. A French summary of the present paper: Brouwer 1919 K. Written corrections added in the Dutch edition 1920 A1 have been accounted for in the English edition and in a correction ⟦7⟧.
⟦2⟧ Brouwer 1911 D.
⟦3⟧ The gothic D indicates intersection (a notation of G. Cantor).
⟦4⟧ Brouwer 1912 L.
⟦5⟧ Here *Zyklosis* means fundamental group.
⟦6⟧ Insertion in l. 1 after solches 'nur von ε abhängendes und' according to Brouwer's correction and an Erratum, see ⟦7⟧.
⟦7⟧ Erratum from the same volume: KNAW Proc. 22 (1920), p. 974.

1920 B2

Mathematics. — "*Ueber eineindeutige, stetige Transformationen von Flächen in sich*". (~~Sechste~~ Siebente Mitteilung [1])). By Prof. L. E. J. BROUWER. [[1]]

(Communicated at the meeting of March 27, 1920).

Die in der fünften Mitteilung über diesen Gegenstand für die Kugel ausgeführte *Aufzählung aller Transformationsklassen* wird hier für die projektive Ebene erbracht werden.

Sei t eine eindeutige stetige Transformation der projektiven Ebene π in sich, k eine einseitige einfache geschlossene Kurve von π, h die Verdoppelung von k, G das in π von h umschlossene zweiseitige Gebiet, k' das Bild von k für t. Wir werden t *erster* oder *zweiter Art* nennen, je nachdem k' einseitig oder zweiseitig ist. Eine Transformation erster und eine zweiter Art können offenbar niemals derselben Klasse angehören.

§ 1. *Die Transformationsklassen erster Art.*

Sei T eine der beiden durch t bestimmten Abbildungen von $G+h$ auf die zweiseitige Verdoppelung β von π, G' bzw. h' das Bild von G bzw. h für T, I der Inhalt von β für eine bestimmte elliptische Massbestimmung in π. Alsdann ist, wenn wir G und β mit passenden Indikatrizen versehen, der Inhalt einer willkürlichen simplizialen Approximierung von G' gleich $\frac{2n+1}{2} I$, wo n eine nichtnegative ganze Zahl ist, welche wir den *Grad* von t nennen werden. Alle Transformationen erster Art, welche derselben Klasse angehören, besitzen offenbar denselben Grad.

Um auch die umgekehrte Eigenschaft zu beweisen, werden wir zwei Methoden angeben, von denen die erste vom Resultate der fünften Mitteilung über diesen Gegenstand Gebrauch macht, die zweite dem Beweisgange dieser Mitteilung parallel läuft.

Erste Methode. Wir konstruieren in G eine einfache geschlossene Kurve r_2 und eine innerhalb r_2 gelegene einfache geschlossene Kurve r_1, und bezeichnen das Innengebiet von r_1 met G_1, das Zwischengebiet von r_1 und r_2 mit G_2, das Zwischengebiet von r_2 und h mit

[1]) Vgl. diese Proceedings XI, S. 788; XII, S. 286; XIII, S. 767; XIV, S. 300 XV, S. 352 (1909—1912). [[2]]

G_3. Wir werden t eine *gegen P reduzierte Transformation n^{ten} Grades* nennen, wenn T die Kurven h und r_1 je eineindeutig und beide mit dem gleichen Umlaufssinn auf den (β in zwei der Reihe nach mit den Graden n und $n+1$ überdeckte Hälften β_1 und β_2 zerlegenden) Grosskreis m, die Kurve r_2 auf den in β_2 gelegenen Pol P von m und die Gebiete G_2 und G_3 je eineindeutig auf das von P und m begrenzte Gebiet abbildet. Nach dem Resultate der fünften Mitteilung über diesen Gegenstand können wir zwei willkürliche gegen P reduzierte Transformationen n^{ten} Grades *unter Invarianz der durch dieselben bestimmten Abbildungen von r_3 und h* stetig ineinander überführen.

Hiermit ist aber unser Ziel erreicht: eine beliebige Transformation erster Art lässt sich nämlich durch stetige Modifizierung in eine gegen P reduzierte Transformation überführen, indem wir zunächst der Kurve h' die erforderliche Gestalt erteilen und sodann unter Invarianz aller Punkte von h' den Prozess zu Ende führen.

Zweite Methode. Wir werden t eine *normalisierte Transformation n^{ten} Grades* nennen, wenn *erstens* h' eine einfache geschlossene Kurve und eineindeutiges Bild von h ist (durch welches also β in zwei der Reihe nach mit den Graden n und $n+1$ überdeckte Hälften β_1 und β_2 zerlegt wird) und *zweitens* T eine einfach verzweigte Riemannsche Abbildung ist, deren Verzweigungspunkte alle in β_2 gelegen sind. In diesem Falle können wir in β_2 nach der Lüroth-Clebschschen Methode ein solches System von Verzweigungsschnitten mit dazu gehöriger Blätteranordnung anbringen, dass h' eine ganz im ersten Blatt gelegene Kurve wird. Aus dieser Bemerkung folgt unmittelbar, dass *alle normalisierten Transformationen n^{ten} Grades zur selben Klasse gehören.*

Wir werden t eine *kanonische Transformation n^{ten} Grades* nennen, wenn *erstens* h' ein Grosskreis und eineindeutiges Bild von h ist und *zweitens* n in G gelegene einander nicht treffende einfache geschlossene Kurven von T in solcher Weise in je einen einzigen Punkt von β übergeführt werden, dass die von diesen Kurven bestimmten Teilgebiete von G alle mit dem Grade $+1$ eineindeutig und stetig abgebildet werden, und zwar die nicht an h grenzenden auf die einfach oder mehrfach punktierte Kugel β, das an h grenzende auf eine von h' umschlossene, im allgemeinen ebenfalls punktierte Halbkugel. In diesem Falle können wir t *zunächst* mittels einer beliebig kleinen, alle Punkte von h' invariant lassenden stetigen Modifizierung in solcher Weise umformen, dass T eine einfach verzweigte Riemannsche Abbildung mit lauter, nicht nur in β, sondern auch in π, verschiedenen Verzweigungspunkten wird, und sodann mittels einer weiteren stetigen Abänderung in eine normalisierte Transformation überführen. *Mithin gehören auch alle kanonischen Transformationen n^{ten} Grades zur selben Klasse.*

Eine beliebige Transformation erster Art lässt sich aber durch stetige Modifizierung in die kanonische Form bringen: um dies zu bewerkstelligen, formen wir sie zunächst so um, dass h' ein Grosskreis und eineindeutiges Bild von h wird und wenden sodann unter Invarianz aller Punkte von h' die in der fünften Mitteilung über diesen Gegenstand erörterte Abänderungsmethode an, welche hier nur dahin zu ergänzen ist, dass a. a. O. S. 355 oben unter den ⟦3⟧ Gebieten g_ν auch ein durch h begrenztes Gebiet g_{ν_h} auftritt, das für $a^{(p)}$ nicht nirgends dicht abgebildet wird, während wir mittels einer beliebig kleinen stetigen Modifizierung von $a^{(p)}$ erreichen können, dass kein weiterer Teil der Grenze von g_{ν_h} mit h zusammenhängt und dass das Bild von g_{ν_h} keine auf h' gelegenen Verzweigungspunkte aufweist; weiter tritt nebst den a.a.O. S. 359 und 360 unterschiedenen Gebieten erster, zweiter ⟦3⟧ und dritter Art noch *ein einziges Gebiet vierter Art* auf, das eine der von h' umschlossenen Halbkugeln eineindeutig und stetig, entweder positiv oder negativ, überdeckt, während das a. a. O. im vierten Absatz von S. 359 angegebene Verfahren eventuell auch zu ⟦3⟧ verwenden ist, um ein Gebiet zweiter bzw. dritter Art mit einem angrenzenden negativen bzw. positiven Gebiete vierter Art zu einem positiven bzw. negativen Gebiete vierter Art zu vereinigen. *Mithin gehören alle Transformationen erster Art n^{ten} Grades zur selben Klasse.*

§ 2. *Die Transformationsklassen zweiter Art.*

Sei wieder T eine der beiden durch t bestimmten Abbildungen von $G+h$ auf die zweiseitige Verdoppelung β von π und G' bzw. h' das Bild von G bzw. h für T, so wird β von einer willkürlichen simplizialen Approximierung von G' entweder überall mit einem geraden oder überall mit einem ungeraden Grade überdeckt. Im ersteren Falle werden wir t eine *gerade*, im letzteren Falle eine *ungerade Transformation zweiter Art* nennen. Die Transformationen einer Transformationsklasse zweiter Art sind offenbar entweder alle gerade, oder alle ungerade.

Sei θ die Fläche vom Zusammenhange der Kugel, welche aus π durch Identifizierung aller Punkte von k erhalten wird. Wir werden t eine *in k kontrahierte Transformation* nennen, wenn k' sich auf einen einzigen Punkt reduziert und zwar insbesondere eine *einfache in k kontrahierte Transformation,* wenn β für T von θ entweder mit dem Grade 0 oder mit dem Grade 1 überdeckt wird. Alsdann

⟦649⟧

folgt aus dem Resultate der fünften Mitteilung über diesen Gegenstand unmittelbar, dass *alle einfachen in k kontrahierten Transformationen derselben Parität zur selben Klasse gehören.*

Eine beliebige Transformation t zweiter Art lässt sich aber durch stetige Modifizierung in die Form einer einfachen in k kontrahierten Transformation bringen: um dies zu bewerkstelligen, formen wir sie zunächst in eine in k kontrahierte Transformation um, wobei also h' sich auf einen einzigen Punkt P reduziert und β für T' von θ mit einem gewissen Grade m überdeckt wird, und modifizieren sodann t in solcher Weise weiter, dass h' der Reihe nach alle Lagen von zweimal durchlaufenen, durch P und den Gegenpunkt Q von P als Pole bestimmten Breitekreisen erhält, und sich schliesslich in Q zusammenzieht. In diesem Augenblicke wird β für T' von θ entweder mit dem Grade $m+2$ oder mit dem Grade $m-2$ überdeckt: durch geeignete Einrichtung des Verfahrens können wir dafür sorgen, dass ein beliebig gewählter dieser beiden Werte erreicht wird. Hieraus folgt, dass wir durch passende Wiederholung desselben Prozesses t in eine einfache in k kontrahierte Transformation überführen können. *Mithin gehören alle Transformationen zweiter Art derselben Parität zur selben Klasse.*

§ 3. *Die Minimalzahlen der Fixpunkte.*

Weil einer eindeutigen stetigen Transformation von π in sich zwei eindeutige stetige Transformationen von β in sich entsprechen, welche nicht beide den Grad -1 besitzen, mithin nicht beide fixpunktfrei sein können [1]), so *besitzt eine eindeutige stetige Transformation der projektiven Ebene π in sich wenigstens einen Fixpunkt.*

Dass andrerseits *für keine Transformationsklasse von π die Minimal-*
[[5]] *zahl der Fixpunkte mehr als 1 beträgt* [2]), erhellt aus der folgenden Transformation erster Art n^{ten} Grades:

$$\begin{cases} tg\,\psi' = tg\,\psi + \cos\varphi \\ \varphi' = (2n+1)\,\varphi, \end{cases}$$

wo mit φ und ψ Länge und Breite auf β bezeichnet werden, und aus der folgenden geraden bzw. ungeraden Transformation zweiter Art:

$$\begin{cases} \varphi' = \varphi \\ \omega' = 0 \text{ bzw. } \omega' = 2\omega, \end{cases}$$

wo mit φ und ω Länge und Polabstand auf β bezeichnet werden.

[[4]] 1) Vgl. Math. Annalen 71, S. 114.

2) Wegen der Beantwortung der analogen Frage für die Kugel und die beiden Ringflächen vgl. meine demnächst im Anschluss an einen Aufsatz von J. Nielsen
[[6]] in Math. Annalen 81 erscheinende Notiz: „*Ueber die Minimalzahl der Fixpunkte bei den Klassen von eindeutigen stetigen Transformationen der Ringflächen*".

ERRATUM. [[7]]

In Vol. XXII of these Proceedings, on page 811 in the title of the paper: *"Ueber eineindeutige, stetige Transformationen von Flächen in sich"*, the words "Sechste Mitteilung" are to be replaced by "Siebente Mitteilung". [[8]]

On page 814 of the same paper, the example given of a transformation of the first kind of degree n with a single invariant point, is irrelevant. [[9]] Moreover, from footnote [2]) on page 865 of Vol. XXIX of these Proceedings, follows that by transformations of the first kind of positive degree, certainly at least two invariant points appear. [[10]]

In Vol. XXIII of these Proceedings, on page 232 in the title of the paper: *"Ueber eineindeutige, stetige Transformationen von Flächen in sich"*, the words "siebente Mitteilung" are to be replaced by "Achte Mitteilung". [[11]]

NOTES

[[1]] Correction of 'Sechste' into 'Siebente' in Brouwer's hand. See also the Erratum [[8]]. Not reprinted: a French summary, 1920 C (where in the title 'L.-E.-Z. Brouwer' is to be changed into 'L. E. J. Brouwer').
[[2]] Brouwer 1909 F2, 1909 H2, 1911 H2, 1911 J2, 1912 K2. Brouwer overlooked the sixth communication of this series, viz. 1919 L2; this was corrected in the above Erratum. The present communication is only related to 1912 K2.
[[3]] Brouwer 1912 K2.
[[4]] Brouwer 1911 D.
[[5]] This is wrong. See [[9]].
[[6]] J. Nielsen 1920 B. Brouwer 1920 E. Read: Math. Annalen 82.
[[7]] From KNAW Proceedings, 29 (1926), 1133.
[[8]] See [[1]].
[[9]] See [[5]].
[[10]] Brouwer 1926 D2.
[[11]] Brouwer 1920 G2.

1920 E

Über die Minimalzahl der Fixpunkte bei den Klassen von eindeutigen stetigen Transformationen der Ringflächen.

Von

L. E. J. Brouwer in Amsterdam.

[[1]] Die von Herrn Nielsen im vorstehenden Aufsatz für die topologischen Transformationen der Ringflächen erzielten Resultate werden im folgenden auf die eindeutigen stetigen Transformationen ausgedehnt. Dabei werden zwei derartige Transformationen zur selben *Klasse* gerechnet, wenn sie sich (unter Erhaltung der Eindeutigkeit und der Stetigkeit) stetig ineinander überführen lassen [1]).

§ 1.

Die zweiseitige Ringfläche.

Sei (A, B) eine Basis der Fundamentalgruppe des Torus, so gehört zu jeder eindeutigen stetigen Transformation dieser Fläche in sich ein Formelsystem

$$A' = A^m B^n$$

$$B' = A^p B^q,$$

wo m, n, p, q keinerlei Beschränkung unterliegende ganze Zahlen bezeichnen. Die Menge der Transformationsklassen des Torus und die Menge der Zahlengruppen (m, n, p, q) entsprechen sich eineindeutig. Durch Anwendung der Nielsenschen Methode findet man mühelos, daß *jede Trans-*

[[2]] 1) Vgl. Proc. V. Int. Congr. of Math. (Cambridge 1912) **2**, S. 9; Amsterd. Ber., holl. Ausgabe, **21**, S. 300, engl. Ausgabe **15**, S. 352 (1912). Im Anschluß an das Resultat dieser Arbeiten ersieht man mittels Betrachtung der Transformation der Cartesischen Ebene

$$\operatorname{arc}\cot x' = n \operatorname{arc}\cot x \qquad y' = y + 1$$

[[3]] im Zusammenhang mit dem in Math. Ann. **71**, S. 114 formulierten Satz 3, daß *im Falle der Kugel* die Fixpunktminimalzahl bei der Klasse der Transformationen vom Grade -1 gleich 0, bei jeder anderen Klasse gleich 1 ist.

formation der Klasse (m, n, p, q) *mindestens* $|mq - np + m - q + 1|$ *Fixpunkte aufweist,* und durch Betrachtung passender linearer Transformationen ersieht man sofort, daß diese Minimalzahl bei jeder Klasse auch wirklich vorkommt.

§ 2.

Die einseitige Ringfläche.

Sei (A, C) die von Nielsen benutzte Basis der Fundamentalgruppe der einseitigen Ringfläche E, so gehört zu einer eindeutigen stetigen Transformation t von E in sich ein Formelsystem

$$A = A^m C^n$$

$$C' = A^p C^q.$$

Hierin ist, weil das Bild der amphidromen Flächenkurve A wieder amphidrom sein muß, $n = 0$; weiter folgt aus der Relation $A'C'A'C'^{-1} = 1$, daß q ungerade ist. Wir erhalten also das zur Transformation t gehörige Formelsystem in der Gestalt

$$A' = A^m$$

$$C' = A^p C^{2r+1}.$$

Je nachdem p gerade oder ungerade ist, werden wir die Transformation t sowie ihre Klasse *gerade* oder *ungerade* nennen. Alsdann enthält jede gerade bzw. ungerade Klasse eine Transformation mit dem Formelsystem

$$(1) \qquad \begin{array}{l} A' = A^m \\ C' = C^{2r+1} \end{array} \quad \text{bzw.} \quad \begin{array}{l} A' = A^m \\ C' = AC^{2r+1} \end{array},$$

wo m nicht-negativ ist. Die Menge der Zahlengruppen (m, r) (m eine nicht-negative, r eine beliebige ganze Zahl) entspricht sowohl der Menge der geraden wie der Menge der ungeraden Transformationsklassen von E eineindeutig.

Wenn man in der von Nielsen angegebenen Weise aus (A, C) eine Basis (A, B) der Fundamentalgruppe des E doppelt überdeckenden Torus R herleitet, so entsprechen einer Transformation t von E, welche einer der beiden Klassen (m, r) angehört, zwei der Reihe nach den Klassen $(m, 0, 0, 2r+1)$ und $(-m, 0, 0, 2r+1)$ angehörende Transformationen t_1 und t_2 von R; wenn f, f_1, f_2 der Reihe nach die Anzahlen der Fixpunkte von t, t_1, t_2 bezeichnen, so ist $f = \frac{1}{2}(f_1 + f_2)$.

[[653]]

Nun haben wir aber nach § 1:

$$f_1 \geqq |2r(m-1)|$$

$$f_2 \geqq |2r(m+1)|,$$

mithin für $m \neq 0$:

$$f \geqq |2mr|$$

und für $m = 0$:

$$f \geqq |2r|,$$

während man wieder innerhalb jeder Klasse (m, r) leicht eine lineare Transformation bestimmt, welche *nicht mehr* als diese Minimalzahl von Fixpunkten aufweist.

(Eingegangen am 20. Januar 1920.)

NOTES

⟦1⟧ J. Nielsen 1920 B.
⟦2⟧ Brouwer 1912 M, 1912 K2.
⟦3⟧ Brouwer 1911 D.

Aufzählung der Abbildungsklassen endlichfach zusammenhängender Flächen.[1]

1921 D

Von

L. E. J. Brouwer in Amsterdam.

Sei O ein Punkt der endlichfach zusammenhängenden Fläche μ, F die Gruppe der geschlossenen stetigen Kurven von μ durch O (welche in bezug auf F nur dann als verschieden betrachtet werden, wenn sie sich nicht mittels stetiger Modifizierung unter Festhaltung von O ineinander überführen lassen), N eine aus den, falls μ zweiseitig ist, der Fundamentalrelation

$$a_1 a_2 a_1^{-1} a_2^{-1} \ldots a_{n-1} a_n a_{n-1}^{-1} a_n^{-1} a_{n+1} \ldots a_{n+m} = 1$$

und, falls μ einseitig ist, der Fundamentalrelation

$$a_1^2 a_2^2 \ldots a_{n-1}^2 a_n^2 a_{n+1} \ldots a_{n+m} = 1$$

genügenden Kurven $a_1, a_2, \ldots, a_{n+m}$ (welche wir in dieser Reihenfolge geordnet denken) bestehende „*Normalbasis*“ von F, O' ein Punkt der endlichfach zusammenhängenden Fläche μ', G die Gruppe der geschlossenen stetigen Kurven von μ' durch O', M eine aus den Kurven $b_1, b_2, \ldots, b_r$ bestehende Normalbasis von G.

Zu einer eindeutigen stetigen Abbildung σ von μ auf μ', welche O in O', mithin jedes a_ν in eine Kurve a'_ν durch O' überführt, gehört ein „*Transformationsformelsystem*“

$$a'_\nu = \varphi_\nu(b_1 \ldots b_r) \qquad (\nu = 1, 2, \ldots, n+m), \tag{1}$$

wo die φ Produkte darstellen.

Sei $\chi(b_1 \ldots b_r)$ ein willkürliches Element von G, so nennen wir das Formelsystem

$$a'_\nu = \chi \varphi_\nu \chi^{-1} \qquad (\nu = 1, 2, \ldots, n+m) \tag{2}$$

[1]) Vgl. ein der Amsterdamer Akad. d. Wiss. am 26. Juni 1920 vorgelegtes und ein der Pariser Akad. d. Wiss. am 12. Juli 1920 vorzulegendes Referat. [[1]]

zu (1) *ähnlich.* Innerhalb der Klasse von σ bestimmt man leicht O in O' überführende Abbildungen von μ auf μ', zu denen (2) als Transformationsformelsystem gehört. Umgekehrt gehört zu jeder O in O' überführenden Abbildung von μ auf μ', welche zur Klasse von σ gehört, ein zu (1) ähnliches Transformationsformelsystem.

Somit bestimmt jede Klasse von Abbildungen von μ auf μ' (zu welcher ja immer O in O' überführende Abbildungen gehören) eine vollständige Menge untereinander ähnlicher Transformationsformelsysteme. Wir bezeichnen diese Menge als das *formale Bild* der Klasse und stellen das Problem, *die Bedingungen zu ermitteln, unter denen zwei dasselbe formale Bild B besitzende* [1]) *Abbildungsklassen* $\varkappa_1$ *und* $\varkappa_2$ *von* μ *auf* μ' *identisch sind.*

Den Ausgangspunkt der Lösung dieses „*Klassenproblems*" bildet das Theorem [2]), daß *zwei eindeutige stetige Abbildungen einer Kugel* μ *auf eine Kugel* μ' *sich dann und nur dann stetig ineinander überführen lassen, wenn sie denselben Grad besitzen.* Wir stellen diesem Theorem zunächst das Korollar an die Seite, daß *zwei eindeutige stetige Abbildungen* σ_1 *und* σ_2 *einer mit einer Indikatrix versehenen Kreisfläche k mit der Grenze g auf eine mit einer Indikatrix versehene Euklidische Kugel* μ', *welche für g dasselbe Bild g' bestimmen, sich dann und nur dann unter Festhaltung aller Punkte von g' stetig ineinander überführen lassen, wenn je zwei entsprechende simpliziale Approximierungen von* σ_1 *und* σ_2 *denselben Inhalt besitzen.*

Sodann nennen wir zwei Segmente von g', welche entweder zusammenfallen und dabei in bezug auf den Umlaufssinn von g' entgegengesetzt gerichtet sind oder zueinander antipodisch gelegen und dabei in bezug auf den Umlaufssinn von g' gleich gerichtet sind, ein *Segmentenpaar erster Art von* g'. Dagegen bezeichnen wir zwei Segmente von g', welche entweder zusammenfallen und dabei in bezug auf den Umlaufssinn von g' gleich gerichtet sind oder zueinander antipodisch gelegen und dabei in bezug auf den Umlaufssinn von g' entgegengesetzt gerichtet sind, als *Segmentenpaar zweiter Art von* g'.

Wenn g' Segmentenpaare erster Art enthält, welche unter Erhaltung ihrer charakteristischen Eigenschaft und unter Invarianz ihrer Endpunkte nach Belieben stetig variiert werden können, im übrigen aber unveränderlich ist, so werden wir sagen, daß g' eine *Variabilität erster Art* besitzt.

Wenn g' nebst eventuellen Segmentenpaaren erster Art solche zweiter Art enthält, welche unter Erhaltung ihrer charakteristischen Eigenschaft

---

[1]) Wie man unmittelbar einsieht, ist diese Eigenschaft von der Wahl der Punkte O und O' unabhängig.

⟦2⟧ [2]) Amsterd. Berichte, holl. Ausgabe **21**, S. 309; engl. Ausgabe **15**, S. 360 (1912).

und unter Invarianz ihrer Endpunkte nach Belieben stetig variiert werden können, im übrigen aber unveränderlich ist, so werden wir sagen, daß g' eine *Variabilität zweiter Art* besitzt.

Wenn g' einzelne Segmente enthält, welche unter Festhaltung ihrer Endpunkte nach Belieben stetig variiert werden können, so werden wir sagen, daß g' eine *Variabilität dritter Art* besitzt.

Diese Definitionen ermöglichen uns, aus unserem Korollar das nachstehende Theorem zu folgern:

Hauptsatz. *Zwei eindeutige stetige Abbildungen σ_1 und σ_2 einer mit einer Indikatrix versehenen Kreisfläche k mit der Grenze g auf eine mit einer Indikatrix versehene Euklidische Kugel des Flächeninhaltes* 1, *welche für g dasselbe Bild g' bestimmen, lassen sich, wenn für g' eine der drei obigen Variabilitäten vorliegt, dann und nur dann stetig ineinander überführen:*

1. *wenn g' eine Variabilität erster Art besitzt und je zwei entsprechende simpliziale Approximierungen von σ_1 und σ_2 denselben Inhalt haben;*

2. *wenn g' eine Variabilität zweiter Art besitzt und die Inhalte je zweier entsprechender simplizialer Approximierungen von σ_1 und σ_2* modulo 2 *gleich sind;*

3. *wenn g' eine Variabilität dritter Art besitzt.*

Wir konstruieren nunmehr auf μ ein der Normalbasis N entsprechendes, in O zusammenhängendes *kanonisches Rückkehrschnittsystem R*, durch welches also μ in eine schlichte Fläche $\mathcal{F}$, deren Grenze g in R liegt, dabei übrigens einzelne Segmente von R der Fundamentalrelation entsprechend zweimal durchlaufen kann, und m, je von einem Flächenrande r_ϱ und einer zu R gehörigen, r_ϱ umschließenden „Randschlinge“ s_ϱ begrenzten, Zylinderflächen ζ_ϱ $(\varrho = 1, 2, \ldots, m)$ zerlegt wird. Alsdann können wir $\varkappa_1$ und $\varkappa_2$ durch zwei solche Abbildungen σ_1 und σ_2 repräsentieren, welche für O dasselbe Bild O', für R dasselbe Bild R' und für jedes ζ_ϱ dasselbe auf Grund einer eindeutigen stetigen Abbildung α_ϱ von $\zeta_\varrho + \sigma_\varrho$ auf σ_ϱ erklärte, mithin kurvenartige, Bild ζ_ϱ' bestimmen. Den Abbildungen σ_1 und σ_2 lassen wir zwei solche Abbildungen τ_1 und τ_2 von $\mathcal{F} + g$ auf die *universelle Überlagerungsfläche μ''* von μ' (durch welche σ_1 und σ_2 mittels der α_ϱ vollständig bestimmt sind) entsprechen, welche für g dasselbe geschlossene Bild g'' aufweisen.

Nach diesen Vorbereitungen sind wir imstande, das Klassenproblem der Reihe nach für die folgenden vier Hauptfälle zu erledigen.

Erster Hauptfall: μ ist geschlossen und μ' hat den Zusammenhang der Kugel.

Alsdann ist μ'' mit μ' identisch. Alle Abbildungsklassen von μ auf μ' besitzen dasselbe formale Bild. Wir legen auf μ' eine elliptische Maßbestimmung fest, auf Grund deren diese Fläche den Flächeninhalt 1 bekommt. Für σ_1 und σ_2 wählen wir *Normalabbildungen,* für welche R' sich auf den Punkt O' reduziert. Wenn σ_1 und σ_2 sich stetig ineinander überführen lassen, so lassen sie sich auch *unter Invarianz von* O' *als Bildpunkt von* O stetig ineinander überführen. Wir suchen mithin nach den Bedingungen, unter denen die letztere Überführbarkeit besteht, und unterscheiden zwei Unterfälle:

Erster Unterfall: μ *ist zweiseitig.* Weil einer außer der Invarianz von O' als Bildpunkt von O keinerlei Beschränkung unterliegenden stetigen Änderung von σ_ν eine stetige Änderung von τ_ν mit *Variabilität erster Art* für g'' entspricht, so ergibt der Hauptsatz als *sekundären Klassencharakter* den Inhalt zugehöriger Normalabbildungen, d. h. den *Grad beliebiger zugehöriger Abbildungen.*

Zweiter Unterfall: μ *ist einseitig.* Weil einer außer der Invarianz von O' als Bildpunkt von O keinerlei Beschränkung unterliegenden stetigen Änderung von σ_ν eine stetige Änderung von τ_ν mit *Variabilität zweiter Art* für g'' entspricht, so ergibt der Hauptsatz als *sekundären Klassencharakter* die Inhaltsparität zugehöriger Normalabbildungen, d. h. die *Parität beliebiger zugehöriger Abbildungen* [1]).

Zweiter Hauptfall: μ ist geschlossen und μ' hat den Zusammenhang der projektiven Ebene.

Alsdann besitzt μ'' den Zusammenhang der Kugel. Wir legen auf μ' eine elliptische Maßbestimmung fest, auf Grund deren μ'' den Flächeninhalt 1 bekommt. Dem Punkte O' von μ' entsprechen die zueinander antipodisch gelegenen Punkte O'' und O''' von μ''. Wir wählen auf μ' eine gerade Linie l durch O', der auf μ'' ein Großkreis h durch O'' und O''' entspricht, auf l einen Umlaufssinn λ, und für σ_1 und σ_2 *Normalabbildungen,* für welche jeder zu R gehörige Rückkehrschnitt entweder in O', oder eineindeutig in l, und zwar das erste Mal, daß er in g auftritt, mit dem Umlaufssinne λ abgebildet wird. Wenn σ_1 und σ_2 sich stetig

---

[1]) Unter der *Parität* einer eindeutigen stetigen Abbildung σ einer *geschlossenen* Fläche μ auf eine beliebige Fläche μ' verstehen wir die Parität der Multiplizität, mit der eine beliebige simpliziale Approximierung von σ einen beliebigen Punkt von μ', der nicht dem Bilde einer Grundseite angehört, überdeckt.

[[658]]

ineinander überführen lassen, so geht dabei τ_1 stetig entweder in τ_2 über oder in die zu τ_2 antipodische Abbildung. Somit läßt in diesem Falle τ_1 sich *mittels stetiger Änderung von* σ_1 *unter Invarianz von* O' *als Bildpunkt von* O stetig überführen entweder in τ_2, oder in die durch Spiegelung um h aus τ_2 entstehende, mithin für g dasselbe Bild g'', wie τ_1 und τ_2, bestimmende Abbildung τ_3. Wir suchen somit nach den Bedingungen, unter denen eine der beiden letzteren Überführbarkeiten besteht, und unterscheiden fünf Unterfälle:

Erster Unterfall: *μ ist zweiseitig und alle Rückkehrschnitte von μ werden zweiseitig abgebildet.* Weil einer außer der Invarianz von O' als Bildpunkt von O keinerlei Beschränkung unterliegenden stetigen Änderung von σ_ν eine stetige Änderung von τ_ν mit *Variabilität erster Art* für g'' entspricht, so ergibt der Hauptsatz als *sekundären Klassencharakter* den absoluten Wert der auf μ'' gemessenen Inhalte zugehöriger Normalabbildungen, d. h. den *absoluten Wert der auf μ'' gemessenen Grade beliebiger zugehöriger Abbildungen.*

Zweiter Unterfall: *μ ist einseitig und alle einseitigen Rückkehrschnitte von μ werden einseitig abgebildet.* In diesem Falle hat entweder das Bild von $\mathcal{F}$ für τ_2 oder das Bild von $\mathcal{F}$ für τ_3 den Inhalt $\frac{p}{2}$, wo p ganz und nicht-negativ ist, die Parität von n besitzt, und gleichzeitig den absoluten Wert des Grades der von σ_2 bestimmten Abbildung der zweiseitigen Verdoppelung von μ auf μ'' darstellt. Weil weiter einer außer der Invarianz von O' als Bildpunkt von O keinerlei Beschränkung unterliegenden stetigen Änderung von σ_ν eine stetige Änderung von τ_ν mit *Variabilität erster Art* für g'' entspricht, so ergibt der Hauptsatz als *sekundären Klassencharakter* den absoluten Wert der auf μ'' gemessenen Inhalte zugehöriger Normalabbildungen, oder auch den *absoluten Wert der Grade beliebiger zugehöriger Abbildungen der zweiseitigen Verdoppelung von μ auf μ''.*

Dritter Unterfall: *μ ist zweiseitig und wenigstens ein Rückkehrschnitt von μ wird einseitig abgebildet.* In diesem Falle wird auch wenigstens ein zu R gehöriger Rückkehrschnitt einseitig abgebildet. Weil mithin einer außer der Invarianz von O' als Bildpunkt von O keinerlei Beschränkung unterliegenden stetigen Änderung von σ_ν eine stetige Änderung von τ_ν mit *Variabilität zweiter Art* für g'' entspricht, so ergibt der Hauptsatz als *sekundären Klassencharakter* die Parität der auf μ'' gemessenen Inhalte zugehöriger Normalabbildungen, d. h. die *Parität der auf μ'' gemessenen Grade zugehöriger Normalabbildungen.*

Vierter Unterfall: *μ ist einseitig, wenigstens ein einseitiger Rückkehrschnitt von μ wird zweiseitig abgebildet, und die zu B gehörigen Ab-*

[[659]]

bildungen sind gerade. In diesem Falle wird auch wenigstens ein zu R gehöriger Rückkehrschnitt zweiseitig abgebildet und bedeckt eine beliebige zu B gehörige Abbildung von μ auf μ'' je zwei antipodische Punkte von μ'' mit derselben Parität, so daß die Anzahl der auf μ' einseitig abgebildeten Rückkehrschnitte von R *gerade* ist. Aus der letzteren Eigenschaft folgt unmittelbar, daß die Inhalte der Bilder von $\mathfrak{F}$ für τ_2 und τ_3 entweder beide ganz und gerade oder beide ganz und ungerade sind. Weil weiter einer außer der Invarianz von O' als Bildpunkt von O keinerlei Beschränkung unterliegenden stetigen Änderung von σ_ν eine stetige Änderung von τ_ν mit *Variabilität zweiter Art* für g'' entspricht, so ergibt der Hauptsatz als *sekundären Klassencharakter* die *Parität der auf μ'' gemessenen Inhalte zugehöriger Normalabbildungen.*

Fünfter Unterfall: *μ ist einseitig, wenigstens ein einseitiger Rückkehrschnitt von μ wird zweiseitig abgebildet, und die zu B gehörigen Abbildungen sind ungerade.* In diesem Falle wird auch wenigstens ein zu R gehöriger Rückkehrschnitt zweiseitig abgebildet und bedeckt eine beliebige zu B gehörige Abbildung von μ auf μ'' je zwei antipodische Punkte von μ'' mit verschiedener Parität, so daß die Anzahl der auf μ' einseitig abgebildeten Rückkehrschnitte von R *ungerade* ist. Aus der letzteren Eigenschaft folgt unmittelbar, daß entweder der Inhalt des Bildes von $\mathfrak{F}$ für τ_2 oder der Inhalt des Bildes von $\mathfrak{F}$ für τ_3 einen Wert $q + \frac{1}{2}$, wo q ganz und gerade, besitzt. Weil weiter einer außer der Invarianz von O' als Bildpunkt von O keinerlei Beschränkung unterliegenden stetigen Änderung von σ_ν eine stetige Änderung von τ_ν mit *Variabilität zweiter Art* für g'' entspricht, so folgt aus dem Hauptsatz, daß der *sekundäre Klassencharakter in Fortfall kommt,* m. a. W., daß *zu jedem in diesem Falle befindlichen formalen Bilde nur eine einzige Klasse gehört.*

Dritter Hauptfall: μ ist offen und μ'' ist geschlossen.

Alsdann besitzt μ'' den Zusammenhang der Kugel und μ' den Zusammenhang der Kugel oder der projektiven Ebene. Weil schon der stetigen Veränderlichkeit von σ_ν *unter Erhaltung ihres Charakters und der α_0 sowie unter Invarianz von O' als Bildpunkt von O* eine stetige Veränderlichkeit von τ_ν mit *Variabilität dritter Art* für g'' entspricht, so folgt aus dem Hauptsatz, daß der *sekundäre Klassencharakter in Fortfall kommt,* m. a. W., daß *zu jedem im dritten Hauptfall befindlichen formalen Bilde nur eine einzige Klasse gehört.*

Vierter Hauptfall: μ'' ist offen.

Aldann bestimmen wir auf μ'' eine Euklidische bzw. hyperbolische Metrik, verstehen unter η eine variable reelle Zahl zwischen 1 und 2 und unter τ_η eine solche Abbildung von $\mathfrak{F} + g$ auf μ'', daß die Bilder P''_1, P''_2, P''_η eines beliebigen Punktes P für τ_1, τ_2, τ_η kollinear sind, ihre gegenseitigen Abstände der Relation

$$P''_1 P''_\eta : P''_\eta P''_2 = (\eta - 1) : (2 - \eta)$$

genügen und P''_η zwischen P''_1 und P''_2 gelegen ist. Eine stetige Variierung von η mit dem Anfangswerte 1 und dem Endwerte 2 bedingt einen stetigen Übergang von τ_1 in τ_2 und dementsprechend von σ_1 in σ_2. Mithin kommt auch hier der *sekundäre Klassencharakter in Fortfall*, d. h. *zu jedem im vierten Hauptfall befindlichen formalen Bilde gehört nur eine einzige Klasse.*

(Eingegangen am 3. 7. 1920.)

NOTES

⟦1⟧ Brouwer 1920 G2, 1920 D are summaries of the present paper. Reprinted from 1920 G2 only a footnote related to Brouwer 1912 K2. In 1920 G2 p. 232 line 2 change 'siebente' into 'achte', in footnote [1]) 'sechsten' into 'siebenten'.

Not reprinted 1920 D. Here p. 89, l. 11 f.b. 'χ^{-2}' is to be changed into 'χ^{-1}', p. 89, l. 8 f.b. 'sont de' is to be changed into 'peuvent être choisies dans'.

The subject was resumed by H. Hopf, 1931 B.

⟦2⟧ Brouwer 1912 K1, K2.

CHAPTER 8

Mechanics

The motion of a particle on the bottom of a rotating vessel under the influence of the gravitational force

1918 E

by

L. E. J. Brouwer [[1]]

(Amsterdam)

We imagine a surface Ω with a horizontal tangential plane at O and rotating with a constant angular velocity ω around a vertical axis through O, and a heavy particle subjected to gliding friction, moving on Ω, which consequently has an equilibrium state at O. We will make a few remarks with respect to the motions of the particle; its mass will be assumed to be the unit.

§ 1. *The equations of motion*

We take a rectangular coordinate system, the XZ- and the YZ-plane of which coincide with the principal normal sections of Ω at O, the positive Z-axis being directed upwards. The surface then rotates with this coordinate system around the Z-*axis* with a constant angular velocity ω; we assume the rotation to take place from the positive X-axis towards the positive Y-axis. If we choose x and y as necessary coordinates of the mass point, then there exists for its motion a potential function $V = gz$ and a kinetic energy $T = \frac{1}{2}[(\dot{x}-y\omega)^2+(\dot{y}+x\omega)^2+\dot{z}^2]$ thus a kinetic potential

$$L = \tfrac{1}{2}[(\dot{x}-y\omega)^2+(\dot{y}+x\omega)^2+\dot{z}^2]-gz,$$

where z is the function of x and y arising from the shape of Ω.

Thus we get:

$$\frac{\partial L}{\partial \dot{x}} = \dot{x}-y\omega+\frac{\partial z}{\partial x}\dot{z}$$

$$\frac{\mathrm{d}}{\mathrm{d}t}\left(\frac{\partial L}{\partial \dot{x}}\right) = \ddot{x}-\dot{y}\omega+\frac{\partial z}{\partial x}\ddot{z}+\frac{\partial^2 z}{\partial x^2}\dot{x}\dot{z}+\frac{\partial^2 z}{\partial x \partial y}\dot{y}\dot{z}$$

$$\frac{\partial L}{\partial x} = (\dot{y}+x\omega)\omega+\frac{\partial^2 z}{\partial x^2}\dot{x}\dot{z}+\frac{\partial^2 z}{\partial x \partial y}\dot{y}\dot{z}-g\frac{\partial z}{\partial x}$$

$$\frac{\partial L}{\partial \dot{y}} = \dot{y}+x\omega+\frac{\partial z}{\partial y}\dot{z}$$

$$\frac{\mathrm{d}}{\mathrm{d}t}\left(\frac{\partial L}{\partial \dot{y}}\right) = \ddot{y}+\dot{x}\omega+\frac{\partial z}{\partial y}\ddot{z}+\frac{\partial^2 z}{\partial x \partial y}\dot{x}\dot{z}+\frac{\partial^2 z}{\partial y^2}\dot{y}\dot{z}$$

$$\frac{\partial L}{\partial y} = -(\dot{x}-y\omega)\omega + \frac{\partial^2 z}{\partial x \partial y}\dot{x}\dot{z} + \frac{\partial^2 z}{\partial y^2}\dot{y}\dot{z} - g\frac{\partial z}{\partial y}.$$

If we denote the normal pressure exerted by the surface on the mass point by N and the coefficient of friction by μ, then the equations of motion are:

$$\ddot{x} - 2\omega\dot{y} - \omega^2 x + (g+\ddot{z})\frac{\partial z}{\partial x} + \frac{\mu N}{\sqrt{(\dot{x}^2+\dot{y}^2+\dot{z}^2)}}\left[\dot{x}\left\{1+\left(\frac{\partial z}{\partial x}\right)^2\right\} + \dot{y}\frac{\partial z}{\partial x}\cdot\frac{\partial z}{\partial y}\right] = 0$$

(1)

⟦2⟧

$$\ddot{y} + 2\omega\dot{x} - \omega^2 y + (g+\ddot{z})\frac{\partial z}{\partial y} + \frac{\mu N}{\sqrt{(\dot{x}^2+\dot{y}^2+\dot{z}^2)}}\left[\dot{x}\frac{\partial z}{\partial x}\cdot\frac{\partial z}{\partial y} + \dot{y}\left\{1+\left(\frac{\partial z}{\partial y}\right)^2\right\}\right] = 0$$

where N is still unknown, whereas

$$\ddot{z} = \frac{\partial^2 z}{\partial x^2}\dot{x}^2 + 2\frac{\partial^2 z}{\partial x \partial y}\dot{x}\dot{y} + \frac{\partial^2 z}{\partial y^2}\dot{y}^2 + \frac{\partial z}{\partial x}\ddot{x} + \frac{\partial z}{\partial y}\ddot{y}.$$

Besides this let us consider the equations of frictionless motion

$$\ddot{x} - 2\omega\dot{y} - \omega^2 x + (g+\ddot{z})\frac{\partial z}{\partial x} = 0$$

(2)

$$\ddot{y} + 2\omega\dot{x} - \omega^2 y + (g+\ddot{z})\frac{\partial z}{\partial y} = 0$$

where likewise

$$\ddot{z} = \frac{\partial^2 z}{\partial x^2}\dot{x}^2 + 2\frac{\partial^2 z}{\partial x \partial y}\dot{x}\dot{y} + \frac{\partial^2 z}{\partial y^2}\dot{y}^2 + \frac{\partial z}{\partial x}\ddot{x} + \frac{\partial z}{\partial y}\ddot{y},$$

and let $\ddot{x}'$ and $\ddot{y}'$ be the values of $\ddot{x}$ and $\ddot{y}$ following from (2), expressed by x, y, $\dot{x}$, $\dot{y}$. Furthermore we put

$$\frac{\partial^2 z}{\partial x^2}\dot{x}^2 + 2\frac{\partial^2 z}{\partial x \partial y}\dot{x}\dot{y} + \frac{\partial^2 z}{\partial y^2}\dot{y}^2 + \frac{\partial z}{\partial x}\ddot{x}' + \frac{\partial z}{\partial y}\ddot{y}' = \ddot{z}'$$

$$\ddot{x} - \ddot{x}' = \ddot{x}''$$

$$\ddot{y} - \ddot{y}' = \ddot{y}''$$

$$\ddot{z} - \ddot{z}' = \ddot{z}''$$

so that the following relation holds:

$$\ddot{z}'' = \frac{\partial z}{\partial x}\ddot{x}'' + \frac{\partial z}{\partial y}\ddot{y}'' \qquad (3)$$

From (2) it follows

⟦666⟧

$$D\ddot{x}' = -E\frac{\partial z}{\partial x} + \left[1+\left(\frac{\partial z}{\partial y}\right)^2\right](2\omega\dot{y}+\omega^2 x) - \frac{\partial z}{\partial x}\cdot\frac{\partial z}{\partial y}(-2\omega\dot{x}+\omega^2 y)$$

$$D\ddot{y}' = -E\frac{\partial z}{\partial y} + \left[1+\left(\frac{\partial z}{\partial x}\right)^2\right](-2\omega\dot{x}+\omega^2 y) - \frac{\partial z}{\partial x}\cdot\frac{\partial z}{\partial y}(2\omega\dot{y}+\omega^2 x),$$

hence

$$D\ddot{z}' = E - Dg + \frac{\partial z}{\partial x}(2\omega\dot{y}+\omega^2 x) + \frac{\partial z}{\partial y}(-2\omega\dot{x}+\omega^2 y),$$

where

$$D = 1+\left(\frac{\partial z}{\partial x}\right)^2+\left(\frac{\partial z}{\partial y}\right)^2 \quad \text{and} \quad E = g+\frac{\partial^2 z}{\partial x^2}\dot{x}^2+2\frac{\partial^2 z}{\partial x\partial y}\dot{x}\dot{y}+\frac{\partial^2 z}{\partial y^2}\dot{y}^2.$$

Observing (3) we now find easily from (1):

$$\ddot{x}'' = \frac{-\mu N\dot{x}}{\sqrt{(\dot{x}^2+\dot{y}^2+\dot{z}^2)}}$$

$$\ddot{y}'' = \frac{-\mu N\dot{y}}{\sqrt{(\dot{x}^2+\dot{y}^2+\dot{z}^2)}},$$

hence

$$\ddot{z}'' = \frac{-\mu N\dot{z}}{\sqrt{(\dot{x}^2+\dot{y}^2+\dot{z}^2)}}.$$

In order finally to determine N, we note that the total force exerted on the particle equals $\underline{\ddot{x}-2\omega\dot{y}-\omega^2 x}_x+\underline{\ddot{y}+2\omega\dot{x}-\omega^2 y}_y+\underline{\ddot{z}}_z$. Thus the total force exerted by the surface upon the particle equals

$$(\underline{\ddot{x}'-2\omega\dot{y}-\omega^2 x}_x+\underline{\ddot{y}'+2\omega\dot{x}-\omega^2 y}_y+\underline{\ddot{z}'+g}_z)+(\underline{\ddot{x}''}_x+\underline{\ddot{y}''}_y+\underline{\ddot{z}''}_z) =$$

$$= \left[\underline{-(g+\ddot{z}')\frac{\partial z}{\partial x}}_x+\underline{-(g+\ddot{z}')\frac{\partial z}{\partial y}}_y+\underline{g+\ddot{z}'}_z\right]+[\underline{\ddot{x}''}_x+\underline{\ddot{y}''}_y+\underline{\ddot{z}''}_z], \qquad [[2]]$$

where the first expression between the brackets represents the normal component, the second the tangential component. Thus:

$$N = (g+\ddot{z}')\sqrt{\left[1+\left(\frac{\partial z}{\partial x}\right)^2+\left(\frac{\partial z}{\partial y}\right)^2\right]}$$

or, after substituting the expression found above for $\ddot{z}'$:

$$N\sqrt{D} = E + \frac{\partial z}{\partial x}(2\omega\dot{y} + \omega^2 x) + \frac{\partial z}{\partial y}(-2\omega\dot{x} + \omega^2 y).$$ [1])

§ 2. *The two cases of formal stability*

If we investigate only motions with small deviations from equilibrium, that is with small values of x, y, $\dot{x}$, $\dot{y}$, $\ddot{x}$, $\ddot{y}$, if further we assume that the fraction is very small, and if we neglect the influence of both the frictional forces and of the terms of the equations of higher than the first degree x, y, $\dot{x}$, $\dot{y}$, $\ddot{x}$, $\ddot{y}$, then denoting the radii of curvature of the intersection of Ω with the XZ-plane and the YZ-plane by r_1 and r_2 respectively, and suppose $1/r_1 \geqq 1/r_2$, the following approximate equations of motion hold:

$$\begin{aligned} \ddot{x} - 2\omega\dot{y} + \left(\frac{g}{r_1} - \omega^2\right)x &= 0 \\ \ddot{y} + 2\omega\dot{x} + \left(\frac{g}{r_2} - \omega^2\right)y &= 0 \end{aligned} \tag{1}$$

If, in order to solve these differential equations, we put $x = ae^{st}$, $y = be^{st}$, we have the following relations between a, b, and s:

$$\begin{aligned} as^2 + 2\omega bs + a\left(\frac{g}{r_1} - \omega^2\right) &= 0 \\ bs^2 + 2\omega as + b\left(\frac{g}{r_2} - \omega^2\right) &= 0 \end{aligned} \tag{2}$$

such that s must satisfy the equation:

$$\begin{vmatrix} \dfrac{g}{r_1} - \omega^2 + s^2 & -2\omega s \\ 2\omega s & \dfrac{g}{r_2} - \omega^2 + s^2 \end{vmatrix} = 0$$

or, after developing

[1]) This value of N can also be found as that of the normal component of $\underline{K} + \underline{T}$, where $\underline{K}$ represents the total force exerted by the surface upon the mass point and $\underline{T}$ an arbitrary tangential force, which we choose as

$$\underline{-\ddot{x}}\,x + \underline{-\ddot{y}}\,y + \underline{-\ddot{x}\frac{\partial z}{\partial x} - \ddot{y}\frac{\partial z}{\partial y}}\,z,$$

such that

$$\underline{K} + \underline{T} = \underline{-2\omega\dot{y} - \omega^2 x}\,x + \underline{2\omega\dot{x} - \omega^2 y}\,y + \underline{E}\,z\,.$$

The normal component of the force $\underline{K} + \underline{T}$, multiplied by $\sqrt{D}$ then appears immediately in the form given in the text for $N\sqrt{D}$.

$$\left(\frac{g}{r_1}-\omega^2\right)\left(\frac{g}{r_2}-\omega^2\right)+s^2\left(\frac{g}{r_1}+\frac{g}{r_2}+2\omega^2\right)+s^4=0. \tag{3}$$

The necessary and sufficient condition for formal stability, that is for stability of the motion satisfying the approximate equations of motion therefore consists in that the equation (3) has two negative roots in s^2.

If we assume this to be the case and put the roots equal to $-\lambda_1^2$ and $-\lambda_2^2$ (λ_1 and λ_2 being positive), then the 'approximate' motion is additively composed of *two principal vibrations S_1 and S_2*, where the principal vibration S_ν takes place according to the equations

$$\begin{aligned} x &= p\sin(\lambda_\nu t+\alpha) \\ y &= k_\nu p\sin(\lambda_\nu t+\alpha+h_\nu), \end{aligned} \tag{4}$$

α and p being arbitrary parameters and h_ν and k_ν fixed constants. Further by substitution in (1), it appears that $h_\nu=\frac{1}{2}\pi$ and k_ν can be found from

$$-\lambda_\nu^2+2\omega k_\nu\lambda_\nu+\left(\frac{g}{r_1}-\omega^2\right)=0.$$

The above mentioned condition for formal stability postulates that the following three inequalities are satisfied:

1. $\left(\frac{g}{r_1}-\omega^2\right)\left(\frac{g}{r_2}-\omega^2\right)>0,$

2. $\frac{g}{r_1}+\frac{g}{r_2}+2\omega^2>0,$

3. $\left(\frac{g}{r_1}-\frac{g}{r_2}\right)^2+8\omega^2\left(\frac{g}{r_1}+\frac{g}{r_2}\right)>0.$

According to the way in which inequality 1. is satisfied, we distinguish two cases:

First case: Stability for small angular velocities, when $(g/r_2)-\omega^2>0$, so that inequalities 2. and 3. are automatically satisfied. This can only happen if both r_1 and r_2 are positive, that is, if the surface is biconcave at 0 in upward direction, therefore possessing a shape which brings the particle necessarily into stable equilibrium even without rotation.

Second case: Stability for large angular velocities, when $(g/r_2)-\omega^2<0$; as in the previous case, this will never happen if both r_1 and r_2 are negative; negative values of r_1 and r_2 in inequality 2. would require that

$$8\omega^2\left(-\frac{g}{r_1}-\frac{g}{r_2}\right)>4\left(-\frac{g}{r_1}-\frac{g}{r_2}\right)^2,$$

therefore a fortiori that

$$8\omega^2\left(-\frac{g}{r_1}-\frac{g}{r_2}\right)>\left(\frac{g}{r_1}-\frac{g}{r_2}\right)^2$$

so that inequality 3. could not possibly be satisfied. Apart from the case that both r_1 and r_2 are positive, so that the inequalities 2. and 3. are automatically satisfied, *stability for large angular velocities can also take place if r_1 is positive and r_2 negative, that is, in the case of a rotating saddle, in which case there would not be any stability without rotation.*

Indeed, if we consider such a rotating saddle $g/r_1 = \alpha$ and $g/r_2 = -\beta$, then α and β are positive magnitudes and the three inequalities of the stability condition become

I. $\omega^2 > \alpha$

II. $2\omega^2 > \beta - \alpha$

III. $(\beta+\alpha)^2 > 8\omega^2(\beta-\alpha)$.

As long as $\beta \leqq \alpha$ (that is, if the principal curvature in O, concave in the upward direction, is not exceeded by the principal curvature in the downward direction) the inequalities II and III are automatically satisfied; formal stability is therefore assured as soon as inequality I is satisfied, i.e. as soon as the rotational velocity ω exceeds a certain limit.

If, however, $\beta > \alpha$, then equality III yields a maximum M and inequalities I and II yield a minimum m for the rotational velocity ω. There will be rotational velocities for which formal stability takes place, provided $M > m$, in other words, if both inequalities $\dfrac{(\beta+\alpha)^2}{\beta-\alpha} > 8\alpha$ and $\dfrac{(\beta+\alpha)^2}{\beta-\alpha} > 4(\beta-\alpha)$ are satisfied, that is if $\beta < 3\alpha$ (since the third inequality requires only $\beta \neq 3\alpha$). Therefore as long as in O the principal curvature, concave in an upward direction, is less than three times the one concave in a downward direction, there are rotation velocities for which the motion of the particle on the rotating saddle yields formal stability.

§ 3. *The absolute stability for small angular velocities*

If $\omega^2 < g/r_2$, then besides the formal stability shown above there is also absolute stability for an arbitrary coefficient of friction[1]).

To prove this we consider the function

$$H = -L + \dot{x}\frac{\partial L}{\partial \dot{x}} + \dot{y}\frac{\partial L}{\partial \dot{y}} = \tfrac{1}{2}(\dot{x}^2+\dot{y}^2+\dot{z}^2-\omega^2x^2-\omega^2y^2)+gz.$$

This function H is an integral of the differential equations (2) of § 1, that is, we have the following relation

[1]) This also holds for other than gliding frictional forces, as can be shown by an argument similar to the one used in this paragraph.

$$\frac{\partial H}{\partial \dot{x}}\ddot{x}' + \frac{\partial H}{\partial \dot{y}}\ddot{y}' + \frac{\partial H}{\partial \dot{z}}\ddot{z}' + \frac{\partial H}{\partial x}\dot{x} + \frac{\partial H}{\partial y}\dot{y} + \frac{\partial H}{\partial z}\dot{z} = 0,$$

and the value of $\mathrm{d}H/\mathrm{d}t$ following from the equations of motion (1) of § 1 equals

$$\frac{\partial H}{\partial x}\ddot{x}'' + \frac{\partial H}{\partial \dot{y}}\ddot{y}'' + \frac{\partial H}{\partial \dot{z}}\ddot{z}'' = \dot{x}\ddot{x}'' + \dot{y}\ddot{y}'' + \dot{z}\ddot{z}'' = -\mu N\sqrt{(\dot{x}^2 + \dot{y}^2 + \dot{z}^2)} = -\mu N v,$$

so that during the actual motion of the particle the value of H continually decreases.

Now for small deviations from equilibrium

$$2H = (1+\varepsilon)\left[\dot{x}^2 + \dot{y}^2 + \left(\frac{g}{r_1} - \omega^2\right)x^2 + \left(\frac{g}{r_2} - \omega^2\right)y^2\right] = 2H_1(1+\varepsilon),$$

where ε represents a very small magnitude, and H_1 is an essentially positive homogeneous quadratic expression in $\dot{x}$, $\dot{y}$, x, y, such that the variety $H_1 = c$ in the Cartesian space R of the x, y, $\dot{x}$, $\dot{y}$ encloses an interior domain that for an indefinitely decreasing value of c converges to the origin. Now suppose that the state of motion at $t = 0$ corresponds to a point P of R for which $H = \eta$ (η being very small), then everywhere within the variety $H_1 = 2\eta$ the relation $H = h_1(1+\varepsilon)$ holds, where ε means a very small magnitude. If now among values of t, increasing from 0 onwards, t_1 is the first time that in R a point of the variety $H_1 = 2\eta$ is reached, then the orbit in R between $t = 0$ and $t = t_1$ would lie entirely within the variety $H_1 = 2\eta$; under such circumstances H would necessarily assume values $> \eta$, which contradicts the continual decrease of the value of H we proved above. *Thus the orbit in R cannot reach the variety $H_1 = 2\eta$ for any positive value of t whatsoever*, so that the deviations of the particle from equilibrium remain small for all positive values of t and the motion of the particle possesses absolute stability.

§ 4. *Coming to rest in the case of small angular velocities*

From the formula $\mathrm{d}H/\mathrm{d}t = -\mu N v$ derived in § 3 it follows that

$$\frac{\mathrm{d}H}{\mathrm{d}s} = \frac{\mathrm{d}H/\mathrm{d}t}{\mathrm{d}s/\mathrm{d}t} = \frac{\mathrm{d}H/\mathrm{d}t}{v} = -\mu N = -\mu g(1+\varepsilon),$$

so that the motion of the particle converges to a single endpoint E, which is reached after a finite path.

If $V(x, y)$ represents the scalar value of the vector $\underline{V}(x, y)$, that is, of the vector $\underline{\ddot{x}}'_x + \underline{\ddot{y}}'_y + \underline{\ddot{z}}'_z$ in the point (x, y) for $\dot{x} = \dot{y} = 0$, and $S(x, y)$ the value of N in the point (x, y) for $\dot{x} = \dot{y} = 0$, then it is clear that *E cannot lie outside the closed curve τ given by the equation $V(x, y) = \mu S(x, y)$.*

We will denote the vector $\underline{\ddot{x}}_x + \underline{\ddot{y}}_y + \underline{\ddot{z}}_z$ by $\underline{A}$, the vector $\underline{\ddot{x}}'_x + \underline{\ddot{y}}'_y + \underline{\ddot{z}}'_z$ by $\underline{B}$, the vector $\underline{\ddot{x}}''_x + \underline{\ddot{y}}''_y + \underline{\ddot{z}}''_z$ by $\underline{W}$. Then we have $\underline{B} = \underline{V} + \underline{U} + \underline{Z}$, where

$$D\underline{V} = \underbrace{-g\frac{\partial z}{\partial x} + \left[1+\left(\frac{\partial z}{\partial y}\right)^2\right]\omega^2 x - \frac{\partial z}{\partial x}\cdot\frac{\partial z}{\partial y}\cdot\omega^2 y}_{x} +$$

$$+ \underbrace{-g\frac{\partial z}{\partial y} - \frac{\partial z}{\partial x}\cdot\frac{\partial z}{\partial y}\cdot\omega^2 x + \left[1+\left(\frac{\partial z}{\partial x}\right)^2\right]\omega^2 y}_{y} +$$

$$+ \underbrace{-g\left[\left(\frac{\partial z}{\partial x}\right)^2 + \left(\frac{\partial z}{\partial y}\right)^2\right] + \frac{\partial z}{\partial x}\omega^2 x + \frac{\partial z}{\partial y}\omega^2 y}_{z}$$

$$D\underline{U} = \underbrace{-\frac{\partial z}{\partial x}\cdot\frac{\partial z}{\partial y}(-2\omega\dot{x}) + \left[1+\left(\frac{\partial z}{\partial y}\right)^2\right]2\omega\dot{y}}_{x} +$$

$$+ \underbrace{\left[1+\left(\frac{\partial z}{\partial x}\right)^2\right](-2\omega\dot{x}) - \frac{\partial z}{\partial x}\cdot\frac{\partial z}{\partial y}\cdot 2\omega\dot{y}}_{y} + \underbrace{\frac{\partial z}{\partial y}(-2\omega\dot{x}) + \frac{\partial z}{\partial x}\cdot 2\omega\dot{y}}_{z}$$

$$D\underline{Z} = \underbrace{(g-E)\frac{\partial z}{\partial x}}_{x} + \underbrace{(g-E)\frac{\partial z}{\partial y}}_{y} + \underbrace{g-E}_{z},$$

We denote the scalar values of $\underline{A}$, $\underline{W}$, $\underline{B}$, $\underline{U}$, $\underline{Z}$ by A, W, B, U, Z, respectively. The vectors $\underline{V}$ and $\underline{U}$ are lying in the tangential plane of Ω, moreover the vector $\underline{U}$ is orthogonal to the orbit of the particle in Ω, the vector $\underline{Z}$ is directed according to the normal to Ω.

Let θ be the angle between the vector $\underline{V}$ and the velocity vector $\underline{v}$ of the particle (taken as positive if the rotation of $\underline{V}$ towards $\underline{v}$ takes place in the same sense as that from the positive X-axis to the positive Y-axis), then the component of $\underline{A}$ in the direction of the orbit equals $V\cos\theta - W$, the component of $\underline{A}$ in the tangential plane orthogonal to the orbit (taken as positive if the rotation of this component towards $\underline{v}$ takes place in the same sense as that from the positive X-axis to the positive Y-axis) equals $V\sin\theta + U$.

If E does not coincide with O, then as the particle approaches E, the angle θ must converge either to 0 *or to* π. Indeed if we suppose that for given $\varepsilon_1 > 0$ in an indefinite vicinity of E there are points P of the orbit such that $\sin^2\theta > \varepsilon_1$, then at points P, sufficiently near to E, the component of A orthogonal to the orbit will be approximately equal to $V\sin\theta$, and approximately lie in the tangential plane. Therefore, at these points the vector $\underline{v}$ is going to turn towards the vector $\underline{V}$, and this happens with an angular velocity with respect to which one may neglect the turning velocity of the vector $\underline{V}$ belonging to the particle moving in the orbit, so that θ^2 *is decreasing*. We can, therefore, be sure that as the particle is approaching *E, in any case θ converges to a single limiting value, but if this value were different* *from* 0 *or* π, *one should be able to indicate in the orbit of the particle a final segment* FE *for which the time integral of the acceleration would have a direction different from the velocity in* F.[1])

If E is lying within the curve τ or if E happens to lie on τ such that $\lim \theta = \pi$, then the scalar deceleration of the motion of the particle converges to a finite limit, *therefore the terminal* E *is reached in a finite time*. We will now show that E *cannot lie on* τ *in such a way that* $\lim \theta = 0$, from which it follows immediately that *in each and every case the particle comes to rest after a finite time.*

Let us assume that E is lying on τ and that $\lim \theta = 0$. Let F be a point of the orbit converging to E, and s the length of the arc FE of the orbit. Then the scalar deceleration at F is

$$\frac{d^2 s}{dt^2} = (W_F - W_E) + (V_E - V_F \cos\theta)$$

or, if by W'_F we denote the value which W would possess at F if the particle would come to rest there,

$$\frac{d^2 s}{dt^2} = (W_F - W'_F) + (W'_F - W_E) + (V_E - V_F \cos\theta).$$

If the vectors $\underline{V}$ and $\underline{Q} = \underline{\frac{\partial z}{\partial x}\,x} + \underline{\frac{\partial z}{\partial y}\,y} + \underline{\left(\frac{\partial z}{\partial x}\right)^2 + \left(\frac{\partial z}{\partial y}\right)^2 z}$ do *not* coincide at E, $W_F - W'_F$ equals $(1+\varepsilon)hv$, where v is the scalar velocity at F, h a constant, and ε may mean any magnitude with no other property than converging to 0 as F approaches E.

If, however, the vectors $\underline{V}$ and $\underline{Q}$ coincide at E and we denote the angles at F between the vector Q and the vectors $\underline{V}$ and $\underline{v}$ by θ_1 and θ_2 respectively, such that $\theta_2 = \theta_1 + \theta$, then $W_F - W'_F = (1+\varepsilon)q\theta_2 v + (1+\varepsilon)kv^2$, where q is a constant and k is a positive constant, while ε in both occurrences has the above meaning. Therefore also

$$W_F - W'_F = (1+\varepsilon)q\theta_1 v + (1+\varepsilon)q\theta v + (1+\varepsilon)kv^2 = \varepsilon s + (1+\varepsilon)q\theta v + (1+\varepsilon)kv^2$$

(since θ_1 is in a non-infinite limit ratio to s).

[1]) If we make use of the property that E cannot lie on τ so that $\lim \theta = 0$ – to be shown below – then the relation between θ and v near E can be expressed by the approximate equation $d \sin \theta / dv = a + b \sin \theta / v$, where a and b are positive for $\lim \theta = 0$ and negative for $\lim \theta = \pi$. If we choose $\sin \theta$ and v as plane Cartesian coordinates, then this approximate equation determines the orbits of a one-parameter projective group, which leaves invariant the origin, the infinite point of the Y-axis and a second infinite point; for postive b the point of collision of these orbits is at infinity, for negative b at the origin; in this connection it is easily shown that the case $\lim \theta = \pi$ is an *exceptional case* among the orbits of the particle whereas *in general* these orbits yield $\lim \theta = 0$. [3]

Further $W_F' - W_E = \eta s$, where η is a magnitude which converges to 0 with the coordinates x_E and y_E of E.

Finally $V_E - V_F \cos\theta = (1+\varepsilon)p\theta^2 - (1+\varepsilon)r_1 s$, where p and r_1 are positive constants, whereas ε has the above meaning in both occurrences, and r_1 does *not* converge to 0 with the coordinates x_E and y_E.

Consequently for the orbital segment either the relation

$$\frac{d^2s}{dt^2} = (1+\varepsilon)hv + (1+\varepsilon)p\theta^2 - (1+\varepsilon)rs,$$

or the relation

$$\frac{d^2s}{dt^2} = (1+\varepsilon)q\theta v + (1+\varepsilon)kv^2 + (1+\varepsilon)p\theta^2 - (1+\varepsilon)rs,$$

holds, where h and η are constants, and p, r and k are positive constants.

If we put $U = uv$, then in the orbital segment FE the scalar value of the component of $\underline{A}$, orthogonal to the orbit and within the tangent plane, equals $(1+\varepsilon)V_E\theta + (1+\varepsilon)u_E v$. If we now assume that for a given $\varepsilon_1 > 0$ in an indefinite vicinity of E there exist points P of the orbit for which $V_E\theta = (-1 \pm \varepsilon_1)u_E v$, then in points P sufficiently near E the vector $\underline{v}$ keeps turning in a sense approximately directed towards a vector at P in the tangent plane to Ω which makes an angle of $-u_E \cdot v/V_E$ with $\underline{V}$, whereas compared with the angular velocity of the rotation of the vector $\underline{v}$, which approximately equals $u_E \cdot \varepsilon_1$, one may neglect both the rotational velocity of the vector $\underline{V}$ of the particle moving along the orbit and the flux of v. From this it follows that $\zeta = \dfrac{V_E \cdot \theta}{u_E \cdot v} + 1$ is decreasing, which contradicts the assumption that in any vicinity of E there exist points P; consequently there is a terminal segment HE of the orbit where ζ is either permanently $> \varepsilon_1$ or permanently $< \varepsilon_1$. Further, since corresponding to a given ε_1 we can indicate a terminal segment GE of the orbit with the property that as long as $\zeta \geqq \varepsilon_1$, according to their absolute values the logarithmic flux of ζ is larger than the logarithmic flux of v, the value ε_1 cannot permanently be exceeded by ζ within an arbitrary terminal segment of the segment GE. *Thus we can be sure that in an indefinite approach of the particle to E there is convergence of ζ to 0 and of $\dfrac{V_E \cdot \theta}{u_E \cdot v}$ to -1*, so that one of the following two cases must apply:

First case: On the orbital segment FE the relation

$$\frac{d^2s}{dt^2} = (1+\varepsilon)hv - (1+\varepsilon)rs$$

$$\frac{dv}{ds} = (1+\varepsilon)a - (1+\varepsilon)b\frac{s}{v}, \tag{1}$$

holds where a is a constant converging to 0 with the coordinates x_E and y_E of E, and b a positive constant not converging to 0 with the coordinates x_E and y_E of E, such that $a^2 < 4b$ may be assumed, therefore for the integral curves of the differential equation in the vs-plane within a certain neighborhood of the origin the angle between the tangent and the radius vector cannot decrease below a certain minimum, and none of these integral curves can possess a terminal segment lying in the first quadrant, terminating at the origin, which on the other hand would be a necessary consequence of an orbital curve of the particle in the first case.

Second case: On the orbital segment FE the relation

$$\frac{\mathrm{d}^2 s}{\mathrm{d}t^2} = mv^2 - rs + \varepsilon v^2 + \varepsilon s,$$

holds, or if we put $v^2 = w$:

$$\frac{\mathrm{d}w}{\mathrm{d}s} = cw - ds + \varepsilon w + \varepsilon s, \tag{2}$$

where c is a constant and d a positive constant. An orbital curve of the particle in this case would necessarily lead to an integral curve of (2) *within the first quadrant of the ws-plane* with a terminal segment ending at the origin. On this curve approaching the origin $\mathrm{d}w/\mathrm{d}s$ would converge to 0 and also w/s would converge to 0; then because of (2) $\mathrm{d}w/\mathrm{d}s$ would be permanently negative along a certain terminal segment and *would therefore lie outside the first quadrant.*

From this contradiction it follows that also this second kind of orbital curves of the particle is impossible; that E, therefore, cannot lie on τ in such a way that $\lim \theta = 0$, which finishes the proof that the state of rest is reached after a finite lapse of time.

NOTES

⟦1⟧ This paper, published in Dutch (under the title *Beweging van een materieel punt op de bodem eener draaiende vaas onder den invloed der zwaartekracht*) and translated for the present edition, has never had the attention it deserved.

Its publication was preceded by a correspondence between Brouwer, O. Blumenthal, and G. Hamel, which for historical reasons is reproduced after 1918 E. After Blumenthal's letter one would expect a letter (probably lost) in which Brouwer sketched the contents of the present paper.

The present paper has been discussed by O. Bottema 1951, dedicated to Brouwer on his 70th birthday; this paper also contains remarks on an analogous problem in four dimensions.

The following remarks are due to a letter of W. T. Koiter and O. Bottema:

There is an analogy between Brouwer's problem and that of the critical speed of a vertical shaft with unequal flexural rigidities, bearing a mass and turning rapidly. This problem is described by the equations

$$\begin{aligned} \ddot{x} - 2\omega\dot{y} + (\omega_1^2 - \omega^2)x &= 0 \\ \ddot{y} + 2\omega\dot{x} + (\omega_2^2 - \omega^2)y &= 0 \end{aligned} \qquad \text{(A)}$$

(neglecting friction). Thus, in contrast with Brouwer's $g/r_1, g/r_2, \omega_1^2, \omega_2^2$ are always positive, though different from each other. For $\omega_1^2 < \omega_2^2$ stability prevails if $\omega^2 < \omega_1^2$ or $> \omega_2^2$, whereas for $\omega_1^2 < \omega^2 < \omega_2^2$ it is unstable. It seems that the first treatise on this subject is A. Föppl 1895 (also in A. Föppl 1910), under the title 'Schwingungen von schnell umlaufenden Hängespindeln'. It is astonishing that Hamel did not mention it in his letter 3 to Blumenthal, especially since Th. v. Kármán, who was told about Brouwer's problem (letter 2) had discussed Föppl's work in Kármán 1910.

This relation with Brouwer's work is important because of the practical significance of the problem of the critical speed. Brouwer's work covers a broader field since he stated stability under certain conditions even for $\omega_1^2 < 0$ and $\omega_2^2 > 0$.

It is also strange that Brouwer's results are not found in the current textbooks. The reason is probably that they were published in Dutch, in a Dutch periodical. There seems to be no other literature (except O. Bottema 1951) on the problem of the particle in a rotating vessel. There is, however, much about the problem of the critical speed, described by (A), which is also dealt with by A. Stodola 1924.

⟦2⟧ A few formulae have been corrected.

⟦3⟧ Point of collision – see Brouwer 1915 B, p. ⟦320⟧.

Blaricum, 5. 5. 14.

Lieber Herr Blumenthal, ⟦1⟧

Darf ich Sie einen Moment belästigen? Bei meinen elementarmechanischen Uebungen stosse ich auf folgendes höchst einfache Problem, das ich aber nicht lösen kann:

'Ein schwerer materieller Punkt befindet sich in Gleichgewichtslage auf dem Boden im Inneren einer Vase (also im Punkte, wo die Tangentialebene horizontal ist). Die beiden Hauptkrümmungsradien im Gleichgewichtspunkte seien R_1 und R_2. Die Masse des beweglichen Punktes sei m, der Reibungskoeffizient k. Die Vase wird nun einer gleichförmigen Rotation um eine vertikale Axe durch den Bodenpunkt und mit einer angulären Geschwindigkeit ω unterworfen. Es fragt sich, ob dabei das Gleichgewicht im Bodenpunkte *stabil* bleibt, wenigstens für *sehr kleine* Werte von k.'

Wählen wir die Normale im Bodenpunkt als Z-axe, die Tangenten an den Hauptnormalschnitten daselbst als X- und Y-axe, so werden die Diff.gleichungen des Problems:

$$\ddot{x} - 2\omega\dot{y} + kmg\frac{\dot{x}}{\sqrt{\dot{x}^2+\dot{y}^2}} + x\left(\frac{mg}{R_1} - \omega^2\right) = 0$$

$$\ddot{y} + 2\omega\dot{x} + kmg\frac{\dot{y}}{\sqrt{\dot{x}^2+\dot{y}^2}} + y\left(\frac{mg}{R_2} - \omega^2\right) = 0$$

Für $\omega^2 < mg/R_1$ und $< mg/R_2$ (also für die kleineren Rotationsgeschwindigkeiten) sieht man leicht, ohne die Diff.gleichungen zu lösen, dass die Stabilität erhalten bleibt. Wie steht es aber für grössere Werte von ω? Die Frage ist ja physisch so furchtbar einfach, dass ich überzeugt bin, dass ihre Lösung (wenigstens für $R_1 = R_2$, d.h. für den Fall, dass die Vase ein Rotationskörper ist) sich entweder leicht machen lassen oder sich doch in der Literatur vorfinden müsse.

Sind Sie dem Problem vielleicht einmal begegnet? Oder sehen Sie irgend einen Ansatz zur Lösung der Diff.gleichungen? (welche inzwischen zur Beantwortung der Stabilitätsfrage wahrscheinlich nicht nötig ist) – Oder wollen Sie vielleicht in meinem Namen einmal bei Hamel anfragen, der als eingefleischter Mechaniker in ⟦2⟧ diesen Dingen mehr bewandert sein muss, und den ich zwar persönlich kenne, aber nicht genügend, um mich direkt an ihn wenden zu können. *Auf keinen Fall sollten Sie bei der Sache viel Zeit verlieren*, das ist sie nicht wert.

Im voraus vielen Dank und herzlichen Gruss. Mit Juel bin ich seit einiger Zeit ⟦3⟧ in regem Briefwechsel; ich hoffe Ihnen diesbezüglich demnächst meine definitive

Nachricht zu schicken. Dies wird übrigens vielleicht besser mündlich gehen, in welchem Falle ich einmal an einem schönen Frühlingstag nach Aachen komme.

Ihr
[[4]] Brouwer

NOTES

[[1]] Correspondence related to 1918 E. See also Y98.
[[2]] G. Hamel.
[[3]] There is some correspondence relating to Juel among Brouwer's papers.
[[4]] O. Blumenthal answered on the third page of Brouwer's original letter (no date mentioned):

Lieber Herr Brouwer!
Ich habe Ihre Frage auf einem Spaziergang Hamel und Karman vorgelegt. Die Antwort, die Sie erhalten, rührt von Hamel her. Es ergiebt sich, dass das Gleichgewicht immer *stabil* ist. Beweis:

1). Es giebt für jeden Reibungskoeffizienten k einen endlichen Bereich B um den tiefsten Punkt des Gefässes, in der Art, dass ein in irgend einem Punkte dieses Bereiches ohne Geschwindigkeit hingebrachter Massenpunkt in Ruhe bleibt.

Folgt daraus, dass die Zentrifugalkraft und Coriolis-Kraft mit abnehmendem Abstand von dem tiefsten Punkte abnimmt, dagegen die Reibung aus der Ruhelage einen festen Wert hat.

Demnach ist die Gleichgewichtslage stabil gegenüber einer reinen *Lagen*veränderung.

2). Sie ist auch stabil gegenüber einer Geschwindigkeitsänderung. Denn wir können die Anfangsgeschwindigkeit, und die Anfangsverschiebung so klein wählen, dass der Massenpunkt sicher im Inneren des Bereiches B die Geschwindigkeit Null erhält, worauf dann 1) anwendbar wird.

Denn: die von der Centrifugalkraft + Schwerkraft herrührende potentielle Energie wächst quadratisch mit der Entfernung vom tiefsten Punkt. Dagegen verzehrt die Reibung Energie proportional mit dem zurückgelegten Wege, also mindestens proportional mit dem Abstand von der Ausgangslage. Daher nimmt zunächst die Energie sicher ab, und wenn die Anfangsenergie genügend klein ist, so wird sie in einem Punkte des Bereiches B Null.

Ich hoffe, Sie sind mit dem Beweis zufrieden. Es ist übrigens wahrscheinlich, dass das Problem schon behandelt ist, bei Regulatoren.

Wie steht es mit Edinburgh? Juel hat mir gestern eine Arbeit aus der Kopenhagener Akademie über Kurven 3. und 4. Ordnung geschickt. Ich hoffe demnach, dass die Betrachtungen richtig sind.

Beste Grüsse!
Ihr
O. Blumenthal,

Brouwer probably answered in a letter in which he gave a full account of his own results as published later on in 1918 E. The contents of this letter can be retraced by means of G. Hamel's letter of 9 June 1914 to O. Blumenthal, which the latter forwarded to Brouwer:

Aachen, den 9. Juni 1914.

Lieber Herr Blumenthal!

Mit grösztem Vergnügen habe ich den Brouwer'schen Brief und seine schönen und merkwürdigen Resultate gelesen. Sie sind alle richtig und wahrscheinlich neu.

Man wuszte bis jetzt folgendes:

Wenn man die Gleichungen hat

$$\ddot{x}-2\omega\dot{y}+ax = 0$$

$$\ddot{y}+2\omega\dot{x}+by = 0,$$

so herrscht Stabilität 1) wenn $a>0$ und $b>0$, 2) wenn sowohl $a<0$ als auch $b<0$, aber ω^2 hinreichend grosz ist. (Das steht auch in meinem Buch.) Das Neue entsteht bei Brouwer dadurch, dass die a und b das ω^2 in der Form $a = \frac{g}{R_1} - \omega^2$ und $b = \frac{g}{R_2} - \omega^2$ enthalten.

Was nun Widerstände angeht, so ist auch bekannt (Satz von W. Thomson, steht auch in meinem Buch), dass bei Dämpfung und bei negativen a, b Stabilisierung durch ω unmöglich ist. Ich will Ihnen nun jetzt Folgendes beweisen: *Die Gleichgewichtslage $x=0$, $y=a$ ist bei Hinzutreten von Gleitreibung stabil, was auch immer a und b sein mögen.* Das heisst: werde der Punkt zu Anfang an irgendeine Stelle $x_0 = r_0 \cos\varphi_0$ und $y_0 = r_0 \sin\varphi_0$ gesetzt und ihm die Geschwindigkeit v_0 erteilt, so bleibt auf jeden Fall der Punkt in einem beliebig kleinen Bereich, wenn man r_0 und v_0 hinreichend klein nimmt.

Beweis: Sei die Reibungskraft $\frac{1}{2}k$, so lautet der Energiesatz

$$\dot{x}^2+\dot{y}^2+ax^2+by^2+ks = h \equiv v_0^2+(a\cos^2\varphi_0+b\sin^2\varphi_0)r_0^2,$$

wo s der Weg ist.

Nun sei r eine erreichte Entfernung $>r_0$. Dann ist

$$k(r-r_0) < ks < h-r^2(a\cos^2\varphi+b\sin^2\varphi)$$

$(x = r\cos\varphi,\ y = r\sin\varphi)$.

Oder wenn m das Minimum von $a\cos^2\varphi+b\sin^2\varphi$ bedeutet (oder a oder b selber),

$$k(r-r_0) < k-mr^2 \tag{1}$$

Das sagt, dass die Grade $y = k(r-r_0)$ stets unterhalb der Parabel $y = k-mr^2$ bleibt. [1])

Beachten wir, dass

$$h = v_0^2+r_0^2(a\cos^2\varphi_0+b\sin^2\varphi_0) > mr_0^2$$

ist, dass also (1), was selbstverständlich ist, für $r = r_0$ erfüllt ist.

Wenn dann m positiv ist, sagt (1) aus, dass für r ein Spielraum besteht, der mit h beliebig klein wird. (Fig. 1).

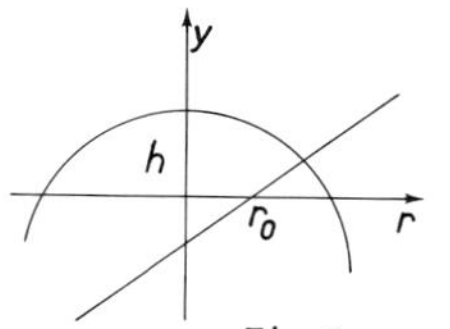

Fig. 1

Wenn aber m negativ ist, können zwei Fälle eintreten, entweder

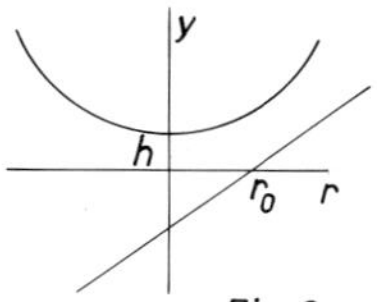

Fig. 2

oder

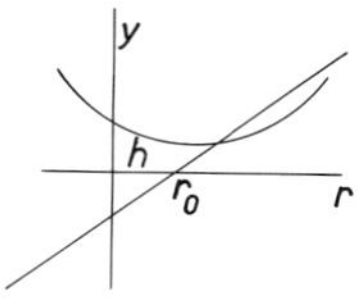

Fig. 3

Im zweiten Falle ergiebt (1) eine Beschränkung für r, im ersten nicht.

Nun kann ich aber bei gegebenen m und k r_0 und h noch beliebig klein machen, und die einzige Bedingung $h > mr_0^2$ (wobei $m < 0$!) ist zu erfüllen, oder wenn Stabilität für alle φ_0 da sein soll, $h > Mr_0^2$, wo M das Maximum von $a\cos^2\varphi_0+b\sin^2\varphi_0$ ist. Auf jeden Fall kann ich r_0 und h so klein wählen, dass Fall 2 (Fig. 3) vorliegt, womit der Satz bewiesen ist.

Das ist aber *nicht das*, was Herr Brouwer wissen will. Mein Beweis sagt nur: ich kann mit Vorsicht wirklich einen Punkt an die Stelle $x = 0$, $y = 0$ setzen, gleichgültig wie die Fläche aussieht und wie sie rotiert. Die Vorsicht muss nun um so gröszer sein, je kleiner k, je glatter die Fläche ist.

[1]) Hamel means that only those r are possible for which the straight line stays below the parabola.

Herr Brouwer aber möchte wissen, ob er seine Resultate auf einer genügend glatten Fläche experimentell verifizieren kann, d.h. ob bei hinreichend kleinem k die Bahnkurven hinreichend benachbart an die von ihm errechneten bleiben.[1]) Wenigstens für eine Zeit, die grosz genug ist, um das Charakteristische zu zeigen. Und das möchte ich jetzt als richtig beweisen.

Genauer beweise ich den Satz: (exakte Formulierung am Schluss) Durch die Reibung erleiden die Integrationskonstanten C der reibungslosen Bewegung eine Veränderung ΔC, für welche

$$|\Delta C| \leqq Mkt \quad \text{(im stabilen Falle so richtig)}$$

ist, wobei M eine konstante Zahl bedeutet.

Daraus folgt: 1) Instabile Bewegungen (Exponentialbewegungen) bleiben instabil.

2) Stabile Bewegungen können in folgende drei Typen übergehen: a) dauernd oscillatorische b) solche, die gegen eine Ruhelage konvergieren c) solche, die langsamer ins Unendliche gehen wie $kt \sin \alpha t$. Ich vermute, dass a) gar nicht vorkommt, dass b) dann eintritt, wenn die Konstanten $a, b > 0$ sind, c) dann wenn beide <0 sind.

Aber jedenfalls genügt dieser Satz, auch im Falle 2c), um die Brouwer'schen Sätze experimentell zu bestätigen. Denn die Instabilität ist eine langsame, noch langsamer, als die Instabilität eines aufrechtstehenden Kreisels bei Luftwiderstand. So wie ein rotierender Kreisel stabil genug steht, um die charakteristische Stabilisierung zu zeigen, auch wenn er langsam umfällt, so wird es möglich sein, einen Punkt auf einem rotierenden Sattel lang genug zu halten, wenn man alles glatt genug macht. Er wird nur langsam entgleiten.

Den *Beweis* will ich nur kurz skizzieren. Die Gleichungen lauten

$$\ddot{x} - 2\omega\dot{y} + ax + \tfrac{1}{2}k \cos\psi = 0$$

$$\ddot{y} + 2\omega\dot{x} + by + \tfrac{1}{2}k \sin\psi = 0$$

wo ψ den Richtungswinkel ($\operatorname{tg}\psi = \dot{y}/\dot{x}$ angiebt.

Ich setze

$$x = x_1 + \xi, \qquad y = y_1 + \eta,$$

wo $x_1 y_1$ den reibungslosen Gleichungen genügen. Dann bekomme ich

$$\ddot{\xi} - 2\omega\dot{\eta} + a\xi + \tfrac{1}{2}k \cos\psi = 0$$

$$\ddot{\eta} + 2\omega\dot{\xi} + b\eta + \tfrac{1}{2}k \sin\psi = 0.$$

[1]) Here Hamel deleted part of the text.

Ich integriere mittels Variation der Konstanten. Seien

$$\xi_\lambda = e^{\omega_\lambda t} \qquad \eta_\lambda = b_\lambda e^{\omega_\lambda t} \qquad (\lambda = 1, 2, \cdots 4)$$

die 4 Partikularlösungen der reibungsfreien Gleichungen, so bekomme ich mit dem Ansatz

$$\xi = \sum c_\lambda \xi_\lambda; \qquad \eta = \sum c_\lambda b_\lambda \xi_\lambda$$

für die c die Gleichungen

$$\sum \dot{c}_\lambda \xi_\lambda = 0, \qquad \sum \dot{c}_\lambda b_\lambda \xi_\lambda = 0$$

$$\sum \dot{c}_\lambda \omega_\lambda \xi_\lambda = -\tfrac{1}{2}k \cos\psi; \qquad \sum \dot{c}_\lambda \omega_\lambda b_\lambda \xi_\lambda = -\tfrac{1}{2}k \sin\psi.$$

Nun ist die Determinante, die beim Auflösen in den Nenner kommt, eine Konstante mal $\xi_1 \xi_2 \xi_3 \xi_4 = e^{(\omega_1+\omega_2+\omega_3+\omega_4)t} = 1$, also konstant (die charakteristische Gleichung ist biquadratisch, daher $\omega_1+\omega_2+\omega_3+\omega_4 = 0$). Die Zählerdeterminanten aber für $\dot{c}_\lambda$ enthalten das Produkt der drei anderen ξ, sind also eine Konstante mal $\xi_\lambda^{-1} = \text{const}\, e^{-\omega_\lambda t}$.

Also ist

$$\dot{c}_\lambda = \int_0^t C_\lambda k e^{-\omega_\lambda t} dt, \quad \text{wo} \quad |C_\lambda| < M_\lambda \qquad ^{1})$$

und M_λ eine feste Zahl. (Wegen $|\cos\psi| \leqq 1$, $|\sin\psi| \leqq 1$.)

Haben wir nun den instabilen Fall (alles reell), so folgt

$$c_\lambda = \overline{C_\lambda} k \int_0^t e^{-\omega_\lambda t} dt = \overline{C_\lambda} k (1 - e^{-\omega_\lambda t})$$

und also

$$\xi = \sum k\overline{C_\lambda}(e^{\omega_\lambda t} - 1),$$

$$x = \sum (\text{const} + k\overline{C_\lambda}) e^{\omega_\lambda t} + F,$$

wo $|F| <$ feste Zahl, und auch $\overline{C_\lambda}$, der Mittelwert [1]) von C_λ, $<$ als feste Zahl M ist.

Sind aber die ω komplex, so kommen wir auf Integrale der Art

$$c_\lambda = \int_0^t C_\lambda k \begin{matrix}\cos\\ \sin\end{matrix} \alpha_\lambda t \, dt, \quad \text{wo} \quad |C_\lambda| < M_\lambda$$

und daraus folgt sofort

$$|C_\lambda| < kM_\lambda t$$

wie behauptet.

Ich habe beim Beweise Verschiedenheit der 4 Wurzeln der charakteristischen Gleichung angenommen, die Grenzfälle sind aber auch nicht von Bedeutung.

[1]) Note that the C_λ depend on $\cos\psi$ and $\sin\psi$.

Also genau heiszt mein Satz so:

Bei Reibung erleiden die Integrationskonstanten Änderungen ΔC, *wobei im stabilen Falle*

$$|\Delta C| \leqq Mkt$$

ist und M eine feste Zahl, im instabilen Falle

$$|\Delta C| \leqq Mk(1-e^{-\omega_\lambda t}),$$

M ebenfalls eine feste Zahl.

Es ist mir ganz recht, wenn Sie diesen Satz Herrn Brouwer mitteilen wollen; hoffentlich habe ich jetzt das, was er möchte.

Mit bestem Grusz
Ihr
Hamel

Y 98 Letter to G. Hamel, Brouwer's copy

Amsterdam, 20.6.14.

[[1]] Sehr geehrter Herr Hamel,

Blumenthal hat Ihren Brief vom 9.6. an mich weitergesandt; erlauben Sie, dass ich mich nunmehr direkt an Sie wende, um Ihnen für Ihr Interesse und Ihre freundliche Hilfe herzlich zu danken. Ihr Gedanke, die praktische Stabilität auf eine 'langsame' Entfernung von der Gleichgewichtslage zurückzuführen, hat mich sehr überrascht, scheint mir aber zutreffend, und ich teile Ihre Ueberzeugung, dass nunmehr die Stabilität auf dem glatten rotierenden Sattel sich experimentell nachweisen lassen werden [[sic!]] muss.

Den Beweis glaube ich noch kürzer führen zu können, als es in Ihrem Briefe geschehen ist. Die allgemeine Lösung der reibungslosen Bewegungsgleichungen sei:

$$\begin{aligned} x &= c_1 e^{\lambda_1 t} + c_2 e^{-\lambda_1 t} + c_3 e^{\lambda_2 t} + c_4 e^{-\lambda_2 t} \\ y &= k_1 c_1 e^{\lambda_1 t} - k_1 c_2 e^{-\lambda_1 t} + k_2 c_3 e^{\lambda_2 t} - k_2 c_4 e^{-\lambda_2 t} \\ \dot{x} &= \lambda_1 c_1 e^{\lambda_1 t} - \lambda_1 c_2 e^{-\lambda_1 t} + \lambda_2 c_3 e^{\lambda_2 t} - \lambda_2 c_4 e^{-\lambda_2 t} \\ \dot{y} &= k_1 \lambda_1 c_1 e^{\lambda_1 t} + k_1 \lambda_1 c_2 e^{-\lambda_1 t} + k_2 \lambda_2 c_3 e^{\lambda_2 t} + k_2 \lambda_2 c_4 e^{-\lambda_2 t} \end{aligned} \tag{I}$$

Der Bewegungszustand des materiellen Punktes lässt sich einerseits definieren durch die Werte von x, y, $\dot{x}$, $\dot{y}$, andererseits durch die vier Grössen $\gamma_1, \gamma_2, \gamma_3, \gamma_4$, worunter wir verstehen die zugehörigen Werte von c_1, c_2, c_3, c_4, falls man dem gegebenen Bewegungszustande den Nullpunkt der Zeit entsprechen lässt. Die Wertsysteme $(x, y, \dot{x}, \dot{y})$ und $(\gamma_1, \gamma_2, \gamma_3, \gamma_4)$ hangen eineindeutig und homogenlinear zusammen. Die einfach unendliche Mannigfaltigkeit der zu einer Bahnkurve (worunter wir im f[[olgenden]]. eine reelle Bahnk[[urve]]. verstehen) gehörigen Wertsysteme $(\gamma_1, \gamma_2, \gamma_3, \gamma_4)$ hat im stabilen Falle die Eigenschaft, dass jeder $(\mathrm{mod.}\,\gamma_\nu)$, mithin auch $\sqrt{\Sigma(\mathrm{mod.}\,\gamma_\nu)^2}$ konstant ist. Man hat nämlich $\gamma_1(t) = e^{\lambda_1 t}\gamma_1(0)$, usw. Ebenso ist für zwei (auch was die Zeit betrifft) benachbarte Bahnkurven jeder $(\mathrm{mod.}\,\mathrm{d}\gamma_\nu)$, mithin auch $\sqrt{\Sigma(\mathrm{mod}\,\mathrm{d}\gamma_\nu)^2}$ konstant. Man hat nämlich wieder $\mathrm{d}\gamma_1(t) = e^{\lambda_1 t}\mathrm{d}\gamma_1(0)$, usw.

Aus den Gleichungen (I) folgt nun weiter, dass das Verhältnis von

$$\sqrt{\mathrm{d}x^2 + \mathrm{d}y^2 + \mathrm{d}\dot{x}^2 + \mathrm{d}\dot{y}^2}$$

zum entsprechenden Werte von $\sqrt{\Sigma(\mathrm{mod.}\,\mathrm{d}\gamma_\nu)^2}$ zwischen zwei festen Grenzen schwankt, welche beide von 0 und ∞ verschieden sind. Den Maximalwert von $\sqrt{\mathrm{d}x^2 + \mathrm{d}y^2 + \mathrm{d}\dot{x}^2 + \mathrm{d}\dot{y}^2}$, gemessen von jedem Punkte der einen bis zum 'nächsten'

(d.h. ein minimales $dx^2+dy^2+d\dot{x}^2+d\dot{y}^2$ aufweisenden) Punkte der anderen Bahnkurve nenne ich die Wegdistanz der beiden benachbarten Bahnkurven; sei h der auf Grund des vorhergehenden sicher existierende Maximalwert im ganzen Bahnkurvensystem des Verhältnisses der Maximal- und Minimalwerte für irgend zwei benachbarte Bahnkurven des in derartiger Weise gewonnenen Ausdrucks $\sqrt{dx^2+dy^2+d\dot{x}^2+d\dot{y}^2}$.
Alsdann ist die im Zeitelement dt auf Grund der Gleitreibung durchlaufene Wegdistanz $< hk\,dt$. Die Zunahme der 'Verrückung' (d.h. des Minimums von $\sqrt{x^2+y^2+\dot{x}^2+\dot{y}^2}$ für die Bahnkurve) sind [[sic]] mithin ebenfalls $< hk\,dt$ und die Zunahme der 'Verrückung' zwischen den Zeitpunkten t_0 und t ist $< hk(t-t_0)$, womit Ihr Satz bewiesen ist.

Noch eine Bemerkung möchte ich mir erlauben. Sie schreiben in Ihrem Briefe die Bewegungsgleichungen in folgender Form an:

$$\ddot{x}-2\omega\dot{y}+ax = 0$$

$$\ddot{y}+2\omega\dot{x}+by = 0,$$

und sprechen die *Vermutung* aus, dass im Falle $a, b > 0$ die Reibung Konvergenz gegen eine Ruhelage verursacht. Dass die Ruhe sogar schon nach einer endlichen Zeit erreicht wird, folgt aus der Existenz des Energieintegrals der reibungslosen Bewegung:

$$H \equiv \tfrac{1}{2}(\dot{x}^2+\dot{y}^2+ax^2+by^2) = c.$$

Für die durch die Reibung verursachten Änderungen von H gilt nämlich:

$$\frac{dH}{dt} = -k\sqrt{\dot{x}^2+\dot{y}^2}.$$

Die positiv-definite Form H kann mithin nur abnehmen, womit die Stabilität der Bewegung mit Reibung erwiesen ist. Weiter muss die totale Bahnlänge endlich sein, weil $dH/dt = -k$. Liegt nun der Endpunkt der Bahn in endlicher Entfernung von der Gleichgewichtslage, so treten in der Nähe dieses Endpunktes nur unbeschränkt abnehmende Geschwindigkeiten auf, deren Richtung sich infolge der anziehenden Kraft ($X = -ax$, $Y = -by$) und der Reibungskraft der Richtung der anziehenden Kraft nähern muss, und diese Richtung nicht überschreiten kann, andererseits aber erreichen muss, weil sonst die Beschleunigung gegen einen endlichen Limes konvergieren würde, dessen Richtung von der Limesrichtung der Geschwindigkeit verschieden wäre, sodass die Geschwindigkeit nicht vom Integrale der Beschleunigung erschöpft werden könnte. Die Geschwindigkeit hat mithin im Endpunkte die Richtung des Vektors $-(a\boldsymbol{x}_x+b\boldsymbol{y}_y)$, sodass der Punkt sich während des Endes der Bahn der Gleichgewichtslage ($x = 0$, $y = 0$) *nähert*. Der Limeswert der Anziehungskraft kann somit nicht der Reibungskraft gleich sein, weil sonst in der Nähe des Bahnendes die Komponenten von Geschwindigkeit und Beschleunigung in der Richtung der Limestangente das gleiche Vorzeichen besässen!

Mithin dominiert in der Nähe des Endpunktes die Reibungskraft mit einem endlichen Ueberschuss gegenüber der Anziehungskraft, sodass nachdem eine Entfernung ε vom Endpunkte erreicht ist, der Endpunkt selbst nach einem Zeitelemente der Ordnung ε erreicht wird.

Liegt aber der Endpunkt der Bahn in der Gleichgewichtslage selbst, so werden x, y, $\dot{x}$, $\dot{y}$ schliesslich alle verschwindend klein. In dem Augenblick aber dominiert die Reibungskraft, und die Ruhelage ist nach einer verschwindend kleinen Zeit erreicht.

Mit nochmaligem Dank, in grosser Hochachtung,

Ihr ergebenster
L E J Brouwer.

NOTE

⟦1⟧ Brouwer's answer to G. Hamel (see Y97 ⟦4⟧). The paper is historically interesting because of its direct approach to problems of differential topology, which is unusual for that period.

ABBREVIATIONS

Abh. Akad. Berlin	Abhandlungen der Königlich Preussischen Akademie der Wissenschaften zu Berlin
Abh. Göttingen	Abhandlungen der Königlichen Gesellschaft der Wissenschaften zu Göttingen
Abh. Math. Sem. Hamburg	Abhandlungen aus dem Mathematischen Seminar der Hamburgischen Universität
Acta Math.	Acta Mathematica
Advances in Math.	Advances in Mathematics
Amer. J. Math.	American Journal of Mathematics
Annales Ecole norm. sup.	Annales Scientifiques de l'Ecole Normale Supérieure
Annals of Math.	Annals of Mathematics
Arch. for Math. og Naturvid.	Archiv for Mathematik og Naturvidenskab
Atti IV Congr. Intern. Mat. Roma	Atti del IV Congresso Internazionale dei Matematici
Berichte Leipzig	Berichte über die Verhandlungen der Königlich Sächsischen Gesellschaft der Wissenschaften zu Leipzig, Mathematisch-Physische Klasse
Bull. A. M. S.	Bulletin of the American Mathematical Society
Bull. Sci. Math.	Bulletin des Sciences Mathématiques
Bull. Soc. Math. Belg.	Bulletin de la Société Mathématique de Belgique
Bull. Soc. Math. France	Bulletin de la Société Mathématique de France
Canad. J. Math.	Canadian Journal of Mathematics
Christiaan Huygens	Christiaan Huygens, Internationaal Mathematisch Tijdschrift
Colloques CNRS	Colloques Internationaux du Centre National de la Recherche Scientifique
Comm. Math. Helvet.	Commentarii Mathematici Helvetici
Comp. Math.	Compositio Mathematica
C. R. Paris	Comptes rendus hebdomadaires des séances de l'Académie des Sciences, Paris
Enseignement Math.	L'Enseignement Mathématique
Euclides	Euclides, Tijdschrift voor de didactiek van de exacte wetenschappen
Fund. Math.	Fundamenta Mathematicae
Indag. Math.	Indagationes Mathematicae ex actis quibus titulus Proceedings
J. ber. D. M. V.	Jahresbericht der Deutschen Mathematiker-Vereinigung
Journ. de Math.	Journal de mathématiques pures et appliquées
Journ. Ecole Polytechnique	Journal de l'Ecole Polytechnique
Journ. r. u. angew. Math.	Journal für die reine und angewandte Mathematik
KNAW Jaarboek	Koninklijke Nederlandse Akademie van Wetenschappen, Jaarboek
KNAW Proc.	Koninklijke Nederlandse Akademie van Wetenschappen, Proceedings
KNAW Verh.	Koninklijke Nederlandse Akademie van Wetenschappen, Verhandelingen
KNAW Versl.	Koninklijke Nederlandse Akademie van Wetenschappen, Verslag
MA	Mathematische Annalen

Matemat. Sbornik	Matematičeskij Sbornik
Meded. Intern. Inst. Wijsbegeerte	Mededeelingen van het Internationaal Instituut voor Wijsbegeerte te Amsterdam
Memorie Accad. Sci. Bologna	Memorie della Reale Accademia delle Scienze dell'Istituto di Bologna
Messenger of Math.	The Messenger of Mathematics
Monatsber. Akad. Berlin	Monatsberichte der Königlich Preussischen Akademie der Wissenschaften zu Berlin
Monatshefte Math. u. Phys.	Monatshefte für Mathematik und Physik
MZ	Mathematische Zeitschrift
Nachr. Göttingen	Nachrichten von der Koeniglichen Gesellschaft der Wissenschaften zu Göttingen und der Georg-Augustus Universität ⟦up to 1896⟧. Mathematisch-Physikalische Klasse ⟦from 1897⟧
Nieuw Arch. Wisk.	Nieuw Archief voor Wiskunde
Phil. Mag.	The London, Edinburgh and Dublin Philosophical Magazine and Journal of Science
Phil. Trans.	Philosophical Transactions of the Royal Society of London
Proc. V Intern. Congr. Math. Cambridge	Proceedings of the Fifth International Congress of Mathematicians
Proc. London Math. Soc.	Proceedings of the London Mathematical Society
Proc. Nat. Acad. Sci.	Proceedings of the National Academy of Sciences of the United States of America
Proc. Roy. Soc. London	Proceedings of the Royal Society of London
Quarterly Journ. of Math.	The Quarterly Journal of Pure and Applied Mathematics
Rend. Acc. Linc. Roma	Atti della Reale Accademia dei Lincei. Rendiconti. Classe di scienze fisiche, matematiche et naturali, Roma
Rend. Circ. Mat. Palermo	Rendiconti del Circulo Matematico di Palermo
Rivista di Mat.	Rivista di Matematica
Sitzber. Akad. Berlin	Sitzungsberichte der Königlich Preussischen Akademie der Wissenschaften zu Berlin Sitzungsberichte der Preussischen Akademie der Wissenschaften, Physikalisch-Mathematische Klasse, Berlin
Sitzber. Erlangen	Sitzungsberichte der Physikalisch-Medizinischen Sozietät zu Erlangen
South African J. Sci.	South African Journal of Sciences
Tôhoku Math. Jour.	The Tôhoku Mathematical Journal
Trans. A. M. S.	Transactions of the American Mathematical Society
Trans. R. Irish Acad. Dublin	The Transactions of the Royal Irish Academy Dublin
Vierteljahrsschrift Zürich	Vierteljahrsschrift der Naturforschenden Gesellschaft in Zürich

LITERATURE

1915 J. W. ALEXANDER. II. A proof of invariance of certain constants of Analysis Situs. Trans. A. M. S. 16, pp. 148–154.

1922 – A proof and extension of the Jordan-Brouwer separation theorem. Trans. A. M. S. 23, pp. 333–349.

1923 – Invariant points of a surface transformation of given class. Trans. A. M. S. 25, pp. 173–184.

1924 – An example of a simply connected surface bounding a region which is not simply connected. Proc. Nat. Acad. Sci. 10, pp. 8–10.

1926 – Combinatorial Analysis Situs. Trans. A. M. S. 28, pp. 301–329.

1926 P. ALEXANDROFF. Über kombinatorische Eigenschaften allgemeiner Kurven. MA 96, pp. 512–554, in particular pp. 519, 527.

1928 A – Untersuchungen über Gestalt und Lage abgeschlossener Mengen beliebiger Dimension. Annals of Math. (2) 30, pp. 101–187.

1928 B – Über den allgemeinen Dimensionsbegriff und seine Beziehungen zur elementaren geometrischen Anschauung. MA 98, pp. 617–636, in particular pp. 621, 622.

1932 – Dimensionstheorie. Ein Beitrag zur Geometrie der abgeschlossenen Mengen. MA 106, pp. 161–238.

1969 A – Die Topologie in und um Holland in den Jahren 1920–1930. Nieuw Arch. Wisk. (3), 17, pp. 109–127.

1969 B – (ed.) Problemy Gil'berta, Moscow 1969.

1935 P. ALEXANDROFF and H. HOPF. Topologie I. Berlin, Springer.

1937 P. ALEXANDROFF, H. HOPF and L. PONTRJAGIN. Über den Brouwerschen Dimensionsbegriff. Comp. Math. 4, pp. 239–255.

1928 P. ALEXANDROFF and P. URYSOHN. Über nulldimensionale Punktmengen. MA 98, pp. 89–106, in particular footnotes 16a, 20.

1973 A. A. ANDRONOV, E. A. LEONTOVICH, I. I. GORDON and A. G. MAIER. Qualitative theory of second order dynamic systems. Translated from Russian by D. Louvish, New York, J. Wiley.

1921 L. ANTOINE. Sur l'homéomorphie de deux figures et de leurs voisinages. Thesis Strasbourg = Journ. de Math. (8) 4, pp. 221–325.

1895 P.APPELL and E.GOURSAT. Théorie des fonctions algébriques et de leurs intégrales. Paris, Gauthier-Villars.

1889 C.ARZELÀ. Funzioni di linee. Rend. Acc. Linc. Roma (4) 5_1, pp. 342–348.

1895 – Sulle funzioni di linee. Memorie Accad. Sci. Bologna (5) 5, pp. 225–244.

1896 – Sull'essistenza degl'integrali nelle equazioni differenziali ordinarie. Memorie Acad. Sci. Bologna (5) 6, pp. 131–140.

1907 A R.BAIRE. Sur la non-applicabilité de deux continus à n et $n+p$ dimensions. C. R. Paris 144, pp. 318–321.

1907 B – Sur la non-applicabilité de deux continus à n et $n+p$ dimensions. Bull. Sci. Math. 31, pp. 94–99.

1883 I.BENDIXSON. Quelques théorèmes de la théorie des ensembles de points. (Extrait d'une lettre adressée à M. Cantor à Halle.) Acta Math. 2, pp. 415–429.

1901 – Sur les courbes définies par des équations différentielles. Acta Math. 24, pp. 1–88.

1972 R.BENNET. Countable dense homogeneous spaces. Fund. Math. 74, pp. 189–194, in particular Theorem 3.

1900 F.BERNSTEIN. Ueber einen Schoenflies'schen Satz der Theorie der stetigen Functionen zweier reeller Veränderlichen. Nachr. Göttingen 1900, pp. 98–102.

1970 G.BESSAGA and A.PELCZYŃSKI. The estimated extension theorem, homogeneous collections and skeletons, and their applications to the topological classification of linear metric spaces and convex sets. Fund. Math. 69, pp. 153–190, in particular prop. 6.9 and 4.3.

1898 L.BIANCHI. Sull'applicabilità di due spazi colla medesima curvatura di Riemann costante. Rend Acc. Linc. Roma (5) 7_2, pp. 147–155 = Opere IX (1958), pp. 7–15.

1917 G.BIRKHOFF. Dynamical systems with two degrees of freedom. Trans A. M. S. 18, pp. 199–300, in particular p. 289 = Collected Mathematical Works II (1950), pp. 1–102, in particular p. 91.

1904 P. Bohl. Über die Bewegung eines mechanischen Systems in der Nähe der Gleichgewichtslage. Journ. r. u. angew. Math. 127, pp. 179–276, in particular p. 186.

1898 E. Borel. Leçons sur la théorie des fonctions. Paris, Gauthier-Villars, p. 44.
1913 – Les ensembles de mesure nulle. Bull. Soc. Math. France 41, pp. 1–19.
1922 – Méthodes et problèmes de théorie des fonctions. (Coll. de monographies sur la théorie des fonctions.) Paris, Gauthier-Villars.

1933 K. Borsuk. Drei Sätze über die n-dimensionale euklidische Sphäre. Fund. Math. 20, pp. 177–190.

1951 O. Bottema. On the stabilization of equilibrium by rotation. KNAW Proc. 54, pp. 61–65.

1877 G. Cantor. Ein Beitrag zur Mannigfaltigkeitslehre. Journ. r. u. angew. Math. 84, pp. 242–258 = Gesammelte Abhandlungen 1932, pp. 119–133.
1879 – Ueber einen Satz aus der Theorie der stetigen Mannigfaltigkeiten. Nachr. Göttingen 1879, pp. 127–135 = Gesammelte Abhandlungen 1932, pp. 134–138.
1883 – Ueber unendliche, lineare Punktmannichfaltigkeiten. MA 21, pp. 545–591, in particular 576 = Gesammelte Abhandlungen 1932, in particular p. 194.
1884 – Ueber unendliche, lineare Punktmannichfaltigkeiten. Fortsetzung. MA 23, pp. 453–488 = Gesammelte Abhandlungen 1932, pp. 210–246.
1885 – Über verschiedene Theoreme aus der Theorie der Punctmengen in einem n-fach ausgedehnten stetigen Raume G_n. II. Acta Math. 7, pp. 105–124 = Gesammelte Abhandlungen 1932, pp. 261–277.
1895 – Beiträge zur Begründung der transfiniten Mengenlehre. I. MA 46, pp. 481–512 = Gesammelte Abhandlungen 1932, pp. 282–311.

1918 C. Carathéodory. Vorlesungen über reelle Funktionen. Leipzig-Berlin, Teubner. 2. Aufl. 1927.

1930 E. Cartan. La théorie des groupes finis et continus et l'analysis situs. Mémorial des Sciences Mathématiques 42, in particular p. 17 = Oeuvres Complètes (1952) I_2, pp. 1165–1224, in particular p. 1181.

1883 F. CASPARY. Zur Theorie der Thetafunctionen mit zwei Argumenten. Journ. r. u. angew. Math. 94, pp. 74–86.

1840 A. CAUCHY. Mémoire sur l'intégration des équations différentielles. Exercices d'analyse I (1840), pp. 328–331 = Oeuvres Complètes (2) 11, pp. 399–465, in particular pp. 400–404.

1845 A. CAYLEY. On certain results relating to quaternions. Phil. Mag. (3) 26, pp. 141–145 = Collected Math. Papers I (1889), pp. 123–126.

1855 – Recherches ultérieures sur les déterminants gauches. Journ. r. u. angew. Math. 50, pp. 299–313, in particular 312 = Collected Math. Papers II (1889), pp. 202–215, in particular p. 215.

1859 – A sixth Memoir upon quantics. Phil. Trans. 149, pp. 61–90 = Collected Math. Papers II (1889), pp. 561–592.

1869 A E. B. CHRISTOFFEL. Ueber die Transformation der homogenen Differentialausdrücke zweiten Grades. Journ. r. u. angew. Math. 70, pp. 46–70.

1869 B – Ueber ein die Transformation homogener Differentialausdrücke zweiten Grades betreffendes Theorem. Journ. r. u. angew. Math. 70, pp. 241–245.

1873 A. CLEBSCH. Zur Theorie der Riemann'schen Fläche. MA 6, pp. 216–230.

1873 W. K. CLIFFORD. Preliminary sketch of biquaternions. Proc. London Math. Soc. (1) 4, pp. 381–395, in particular 389–390 = Mathematical Papers 1882, pp. 181–200, in particular pp. 192–193.

1910 G. COMBEBIAC. Pour une théorie de la mesure. Enseignement math. 12, pp. 89–97, 200–209.

1911 – Note complémentaire sur les fonctions de mesure. Enseignement math. 13, pp. 381–391.

1923 CORRESPONDANCE d'Henri Poincaré et de Felix Klein. Acta Math. 39, pp. 94–132 = F. Klein, Gesammelte Mathematische Abhandlungen III, pp. 587–621.

1930 D. VAN DANTZIG. Ueber topologisch homogene Kontinua. Fund. Math. 15, pp. 102–125.

1936 – Zur topologischen Algebra. III. Brouwersche und Cantorsche Gruppen. Comp. Math. 3, pp. 408–426, in particular 408.

1907 M.Dehn, P.Hegaard. Analysis Situs. Encyklopädie der mathematischen Wissenschaften III 1, 1, pp. 153–220, in particular p. 160.

1910 A.Denjoy. Continu et discontinu. C. R. Paris 151, pp. 138–140.

1892 U.Dini. Grundlagen für eine Theorie der Functionen einer veränderlichen reellen Grösse. ⟦Translation of U. Dini, Fondamenti per la teorica delle funzioni di variabili reali, Pisa 1878, by J. Lüroth and A. Schepp.⟧ Leipzig, Teubner.

1880 W.Dyck. Ueber Aufstellung und Untersuchung von Gruppen und Irrationalität regulärer Riemann'scher Flächen. MA 17, pp. 473–509.

1911 F.Engel. ⟦Review of Brouwer 1909 B.⟧ Jahrbuch über die Fortschritte der Mathematik 40, p. 194.

1913 – ⟦Review of Brouwer 1910 H.⟧ Jahrbuch über die Fortschritte der Mathematik 41, pp. 181–182.

1926 G.Feigl. Fixpunktsätze. J. ber. D. M. V. 34_{II}, pp. 124–130.

1928 – Fixpunktsätze für spezielle n-dimensionale Mannigfaltigkeiten. MA 98, pp. 355–398.

1895 A.Föppl. Ueber den ruhigen Gang von schnell umlaufenden Hängespindeln. Civilingenieur 41, pp. 621–632.

1910 – Vorlesungen über Technische Mechanik, VI, Die wichtigsten Lehren der höheren Dynamik, Leipzig, Teubner, in particular pp. 66–72.

1874 W.Frahm. Ueber gewisse Differentialgleichungen. MA 8, pp. 35–44.

1906 M.Fréchet. Sur quelques points du calcul fonctionnel. Rendiconti Palermo 22, pp. 1–74 = Thèse Paris, Gauthier-Villars.

1910 – Les dimensions d'un ensemble abstrait. MA 68, pp. 145–168, in particular 159.

1928 – Les espaces abstraits et leur théorie considérée comme introduction à l'analyse générale. Paris, Gauthier-Villars.

1873 Fresdorf. Diss. Göttingen. ⟦No more details available.⟧

1955 H.Freudenthal. Poincaré et les fonctions automorphes. Le Livre du Centenaire de la naissance de Henri Poincaré (1854–1954). Paris, Gauthier-Villars, 1955, pp. 212–219.

1965 – Lie groups in the foundations of geometry. Advances in Math. 1, pp. 145–190, in particular pp. 148–157.

1968 – L'algèbre topologique – en particulier les groupes topologiques et de Lie. 12e Congrès International d'Histoire des Sciences, Paris, 1968. Colloques, pp. 223–243 = Revue de Synthèse (3), Nos 49-52.

1967 H. Freudenthal and A. Heyting. Levensbericht van L. E. J. Brouwer. KNAW Jaarboek 1966–67, pp. 335–340. ⟦Translation in the present volume.⟧

1897 R. Fricke and F. Klein. Vorlesungen über die Theorie der automorphen Functionen. I. Leipzig, Teubner.

1912 R. Fricke and F. Klein. Vorlesungen über die Theorie der automorphen Functionen. II. Leipzig, Teubner.

1840 C. F. Gauss. Allgemeine Lehrsätze in Beziehung auf die im verkehrten Verhältnissen des Quadrats der Entfernung wirkenden Anziehungs- und Abstossungs-Kräfte = Werke V (1867), pp. 195–242.

1952 A. M. Gleason. Groups without small subgroups. Annals of Math. 56, pp. 193–212.

1919 B. P. Haalmeijer. Note on linear homogeneous sets of points. KNAW Proc. 22, 681–683.

1923 – Über lineare homogene Punktmengen. MZ 16, pp. 92–102.

1910 J. Hadamard. Sur quelques applications de l'indice de Kronecker. In: J. Tannéry, Introduction à la théorie des fonctions, 2e ed.; tome II, pp. 437–477.

1912 G. Hamel. Elementare Mechanik. Leipzig–Berlin, Teubner.

1848 W. R. Hamilton. Researches respecting quaternions. First series. (Read 13 November 1843.) Trans. R. Irish Acad. Dublin 21, pp. 199–296, in particular pp. 256–259 = The Mathematical Papers III, 1967, pp. 159–226, in particular pp. 197–199.

1903 G. H. Hardy. The cardinal number of a closed set of points. Messenger of Math. (2) 33, pp. 67–69.

1914 F. Hausdorff. Grundzüge der Mengenlehre. Leipzig, Veit.

1919 – Über halbstetige Funktionen und deren Verallgemeinerung. MZ 5, pp. 292–309, in particular pp. 296–297.

1937 A. VAN HEEMERT. Topologische Gruppen und unzerlegbare Kontinua. Comp. Math. 5, pp. 319–326.

1943 – De R_n-adische voortbrenging van algemeen-topologische ruimten met toepassingen op de theorie van niet splitsbare continua. Thesis Groningen.

1908 K. HENSEL. Theorie der algebraischen Zahlen I. Leipzig, Teubner.

1905 G. HESSENBERG. Begründung der elliptischen Geometrie. MA 61, pp. 173–184.

1900 D. HILBERT. Mathematische Probleme, Nachr. Göttingen 1900, pp. 253–297 = Gesammelte Abhandlungen III, 1935, pp. 290–329.

1902 – Ueber die Grundlagen der Geometrie. MA 56, pp. 381–422 = Grundlagen der Geometrie, Anhang IV.

1909 – Grundlagen der Geometrie. Leipzig-Berlin, Teubner. 3. Auflage.

1904 E. W. HOBSON. On inner limiting sets of points in a linear interval. Proc. London Math. Soc. (2) 2, pp. 316–326.

1925 H. HOPF. Über die Curvatura integra geschlossener Hyperflächen. MA 95, pp. 340–367.

1926 A – Abbildungsklassen n-dimensionaler Mannigfaltigkeiten. MA 96, pp. 209–224.

1926 B – Vektorfelder in n-dimensionalen Mannigfaltigkeiten. MA 96, pp. 225–250.

1928 A – Eine Verallgemeinerung der Euler-Poincaréschen Formel. Nachr. Göttingen 1928, pp. 127–136.

1928 B – Zur Topologie der Abbildungen von Mannigfaltigkeiten. I. Neue Darstellung der Theorie des Abbildungsgrades für topologische Mannigfaltigkeiten. MA 100, pp. 579–608.

1929 – Über die algebraische Anzahl von Fixpunkten. MZ 29, pp. 493–524.

1930 – Über wesentliche und unwesentliche Abbildungen von Komplexen. Matemat. Sbornik 37, pp. 53–62.

1931 A – Über die Abbildungen der dreidimensionalen Sphäre auf die Kugelfläche. MA 104, pp. 637–665.

1931 B – Beiträge zur Klassifizierung der Flächenabbildungen. Journ. r. u. angew. Math. 165, pp. 225–236.

1933 – Die Klassen der Abbildungen der n-dimensionalen Polyeder auf die n-dimensionale Sphäre. Comm. Math. Helvet. 5, pp. 39–54.

1966 – Einige persönliche Erinnerungen aus der Vorgeschichte der heutigen Topologie. Colloque de Topologie CBRM Louvain, pp. 9–20 = (approximately) Ein Abschnitt aus der Entwicklung der Topologie. J. ber. D. M. V. 68, pp. 182–192.

1927 W. HUREWICZ. Normalbereiche und Dimensionstheorie. MA 96, pp. 736–764, in particular p. 763.

1929 – Über ein topologisches Theorem. MA 101, pp. 210–218.

1948 W. HUREWICZ and H. WALLMAN. Dimension theory. (Princeton Mathematical Series.) Princeton Univ. Press.

1888 A. HURWITZ. Ueber diejenigen algebraischen Gebilde, welche eindeutige Transformationen in sich zulassen. MA 32, pp. 290–308 = Nachr. Göttingen 1887, pp. 85–107 = Mathematische Werke I (1932), pp. 241–259.

1892 – Ueber algebraische Gebilde mit eindeutigen Transformationen in sich. MA 41, pp. 403–442 = Mathematische Werke I (1932), pp. 391–430.

1896 E. JAHNKE. Ueber ein allgemeines aus Thetafunctionen von zwei Argumenten gebildetes Orthogonalsystem und seine Verwendung in der Mechanik. Sitzber. Akad. Berlin 1896, pp. 1023–1030.

1897 – Ueber einen Zusammenhang zwischen den Elementen orthogonaler Neuner- und Sechzehnersysteme. Journ. r. u. angew. Math. 118, pp. 224–233.

1902 – Über Drehungen im vierdimensionalen Raum. J. ber. D. M. V. 11, pp. 178–182.

1904 – Bemerkung zu der am 27. Februar 1904 vorgelegten Notiz von Herrn Brouwer "Over een splitsing van de continue beweging om een punt O van R_4 in twee continue bewegingen om O van R_3's". KNAW Versl. 12, pp. 940–941.

1914 JAHRBUCH über die Fortschritte der Mathematik 42, Jahrgang 1911, in particular p. 1032.

1910 S. JANISZEWSKI. Sur la géométrie de lignes cantoriennes. C. R. Paris 151, pp. 198–201.

1893 C. JORDAN. Cours d'Analyse de l'Ecole Polytechnique I, 2nd ed. pp. 25, 91–99, 3rd ed. 1909, pp. 25, 91–99.

1879 E. JÜRGENS. Allgemeine Sätze über Systeme von zwei eindeutigen und stetigen reellen Functionen von zwei reellen Veränderlichen. Leipzig, Teubner.

1899 – Der Begriff der n-fachen stetigen Mannigfaltigkeit. J. ber. D. M. V. 7_{I}, pp. 50–55.

1910 TH. V. KÁRMÁN. Festigkeitsprobleme im Maschinenbau. In: Encyklopädie der mathematischen Wissenschaften IV, 4, pp. 311–385, in particular pp. 383–385.

1909 ED. KASNER. Tautochrones and brachistochrones. Bull. A. M. S. 15, pp. 475–483.

1919 B. V. KERÉKJÁRTÓ, Über die Brouwerschen Fixpunktsätze. MA 80, pp. 29–32.

1919 – Über Transformationen des ebenen Kreisringes. MA 80, pp. 33–35.

1919 – Über die periodischen Transformationen der Kreisscheibe und der Kugelfläche. MA 80, pp. 36–38.

1885 W. KILLING. Die Mechanik in den Nicht-*Euklidischen* Raumformen. Journ. r. u. angew. Math. 98, pp. 1–48.

1892 – Ueber die Grundlagen der Geometrie. Journ. r. u. angew. Math. 109, pp. 121–186.

1893 – Einführung in die Grundlagen der Geometrie, I. Paderborn, Schöningh, 1893.

1873 F. KLEIN. Ueber die sogenannte Nicht-Euklidische Geometrie (Zweiter Aufsatz) MA 6, pp. 112–145 = Gesammelte Mathematische Abhandlungen I (1921), pp. 311–343.

1878 – Ueber die Transformation siebenter Ordnung der elliptischen Functionen. MA 14, pp. 428–471, in particular p. 458 = Gesammelte Mathematische Abhandlungen III (1923), pp. 90–135, in particular p. 121.

1882 A – Neue Beiträge zur Riemann'schen Functionentheorie (mit zwei lithographierten Tafeln). MA 21, pp. 141–218 = Gesammelte Mathematische Abhandlungen III (1923), pp. 630–710.

1882 B – Ueber Riemann's Theorie der algebraischen Functionen und ihrer Integrale. Leipzig, Teubner.

1890 – Zur Nicht-Euklidischen Geometrie. MA 37, pp. 544–572 = Gesammelte Mathematische Abhandlungen I (1921), pp. 353–383.

1894 – Autographierte Vorlesungshefte. MA 45, pp. 140–152, in particular p. 141 = (partially) Gesammelte Mathematische Abhandlungen I (1921), pp. 498–502.

1923 A – Zum Kontinuitätsbeweise des Fundamentaltheorems. Gesammelte Mathematische Abhandlungen III, pp. 731–741. (Mit E. Bessel-Hagen.)

1923 B – Verschiedene Verzeichnisse. Anhang of Gesammelte Mathematische Abhandlungen III, in particular p. 29, 1.13–14.

1928 H.KNESER. Glättung von Flächenabbildungen. MA 100, pp. 609–617.

1907 P.KOEBE. Über die Uniformisierung beliebiger analytischer Kurven. Nachr. Göttingen 1907, pp. 191–210.

1910 – Über die Uniformisierung der algebraischen Funktionen mit imaginärer Substitutionsgruppe (Fortsetzung und Schluss). Nachr. Göttingen 1910, pp. 180–189.

1912 A – Referat über automorphe Funktionen und Uniformisierung. J. ber. D. M. V. 21, pp. 157–163.

1912 B – Begründung der Kontinuitätsmethode im Gebiete der konformen Abbildung und Uniformisierung (Voranzeige). Nachr. Göttingen 1912, pp. 879–886.

1912 C – Zur Begründung der Kontinuitätsmethode. Berichte Leipzig 64, pp.59–62.

1914 – Über die Uniformisierung der algebraischen Kurven. IV. MA 75, pp. 42–129.

1891 F.KÖTTER. Ueber die Bewegung eines festen Körpers in einer Flüssigkeit. Sitzber. Akad. Berlin 1891, pp. 47–56.

1895 – Ueber eine Darstellung der Richtungscosinus zweier orthogonaler Coordinatensysteme durch Thetafunctionen zweier Argumente, welche die Lösungen mehrerer Probleme der Mechanik als Spezialfälle umfasst. Sitzber. Akad. Berlin 1895, pp. 807–814.

1969 G.KREISEL and M.H.A.NEWMAN. Luitzen Egbertus Jan Brouwer 1881–1966. Elected For. Mem. R. S. 1948. Biographical Memoirs of Fellows of the Royal Society, pp. 38–68.

1869 L.KRONECKER. Über Systeme von Functionen mehrer Variabeln. Monatsber. Akad. Berlin 1869, pp. 159–193, 688–698 = Werke I (1895), pp. 175–226.

1922 C.KURATOWSKI. Théorie des continus irréductibles entre deux points I. Fund. Math. 3, pp. 200–231, in particular Bibliographie pp. 230–231.

1927 – Idem II. Fund. Math. 10, pp. 225–275, in particular Bibliographie pp. 274–275.

1928 – Sur la structure des frontières communes à deux régions. Fund. Math. 12, pp. 20–42, in particular Bibliographie pp. 41–42.

1952 – Topologie I, 3rd ed.; II, 2nd ed. (Monografie Matematyczne 20–21.) Warszawa.

1911 A H. Lebesgue. Sur la non-applicabilité de deux domaines appartenant respectivement à des espaces à n et $n+p$ dimensions. (Extrait d'une lettre à M. O. Blumenthal) MA 70, pp. 166–168 = Oeuvres scientifiques IV (1973), pp. 170–172.

1911 B – Sur l'invariance du nombre de dimensions d'un espace et sur le théorème de M. Jordan relatif aux variétés fermées. C. R. Paris 152, pp. 841–843 = Oeuvres scientifiques IV (1973), pp. 173–175.

1921 – Sur les correspondances entre les points de deux espaces. Fund. Math. 2, pp. 256–285 = Oeuvres scientifiques IV (1973), pp. 177–206.

1922 – Notice sur les travaux scientifiques de M. Henri Lebesgue. Toulouse. 89 p., in particular pp. 79–81 = Oeuvres scientifiques I (1972), pp. 97–175, in particular pp. 165–167.

1924 – Sur le théorème de Schoenflies. Fund. Math. 6, pp. 96–99 = Oeuvres scientifiques IV (1973), pp. 207–210.

1911 N. J. Lennes. Curves in non-metrical analysis situs with an application in the calculus of variations. Amer. J. Math. 33, pp. 287–326.

1885 S. Lie. Untersuchungen über Transformationsgruppen. II. Arch. for Math. og Naturvid. 10, pp. 353–413 = Gesammelte Abhandlungen V (1924), pp. 507–551.

1890 – Ueber die Grundlagen der Geometrie I. II. Berichte Leipzig 42, pp. 284–321, 355–418 = Gesammelte Abhandlungen II_1 (1935), pp. 380–468.

1893 – Theorie der Transformationsgruppen III, Leipzig, Teubner.

1905 E. Lindelöf. Remarques sur un théorème fondamental de la théorie des ensembles. Acta Math. 29, pp. 183–190.

1869 R. Lipschitz. Untersuchungen in Betreff der ganzen homogenen Functionen von n Differentialen. Journ. r.u. angew. Math. 70, pp. 71–102. Entwicklung einiger Eigenschaften der quadratischen Formen von n Differentialen. I. II. Journ. r.u. angew. Math. 71, pp. 274–287, 288–295.

1870 B – Fortgesetzte Untersuchungen in Betreff der ganzen homogenen Functionen von n Differentialen. Journ. r. u. angew. Math. 72, pp. 1–56.

1876 – Sur la possibilité d'intégrer complètement un système donné d'équations différentielles. Bull. Sci. Math. 10, pp. 149–159.

1847 J. B. Listing. Vorstudien zur Topologie, Göttinger Studien 1847, Göttingen 1848.

1904 H.A.LORENTZ. Maxwells elektromagnetische Theorie. Encyklopädie der mathematischen Wissenschaften. V, 2, pp. 63–144, in particular pp. 120–121, Footnote 61.

1871 J.LÜROTH. Note über Verzeigungsschnitte und Querschnitte in einer Riemann'schen Fläche. MA 4, pp. 181–184.

1878 – Ueber gegenseitig eindeutige und stetige Abbildung von Mannigfaltigkeiten verschiedener Dimensionen aufeinander. ⟦I.⟧ Sitzber. Erlangen 10, pp. 190–195.

1899 – Über gegenseitig eindeutige und stetige Abbildung von Mannigfaltigkeiten verschiedener Dimensionen aufeinander. II. Sitzber. Erlangen 31, pp. 87–91.

1906 – Über Abbildung von Mannigfaltigkeiten. MA 63, pp. 222–238.

1910 S.MAZURKIEWICZ. Sur la théorie des ensembles. C. R. Paris 151, pp. 296–298.

1920 – Sur un ensemble G_δ ponctiforme qui n'est homéomorphe à aucun ensemble linéaire. Fund. Math. 1, pp. 61–81.

1923 K.MENGER. Über die Dimensionalität von Punktmengen. I. Monatshefte Math. u. Physik 33, pp. 148–160.

1924 A – Ueber die Dimension von Punktmengen. KNAW Proc. 27, pp. 639–643.

1924 B – Über die Dimension von Punktmengen. II. Monatshefte für Math. u. Phys. 34, pp. 137–161.

1926 A – Bericht über die Dimensionstheorie. J. ber. D. M. V. 35, pp. 113–150.

1926 B – Zur Entstehung meiner Arbeiten über Dimensions- und Kurventheorie. KNAW Proc. 29, pp. 1122–1124.

1928 – Dimensionstheorie. Leipzig, Teubner.

1929 A – Zur Dimensions- und Kurventheorie. Monatshefte Math. u. Phys. 36, pp. 411–432.

1930 – Antwort auf eine Note von Brouwer. Monatshefte Math. u. Phys. 37, pp. 175–182.

1932 – Über die Hinweise auf Brouwer in Urysohns Mémoire. 5p. Wien, Selbstverlag.

1892 R.MILESI. Sulla impossibile coesistenza della univocità e della continuità nella corrispondenza che si può stabilire fra due spazi continui ad un numero differente di dimensioni. Rivista di Mat. 2, pp. 103–106.

1839 F.Minding. Wie sich entscheiden lässt, ob zwei gegebene krumme Flächen aufeinander abwickelbar sind oder nicht .... Journ. r. u. angew. Math. 19, pp. 370–387.

1844 F.N.M.Moigno. Leçons de calcul différentiel et de calcul intégral, d'après les méthodes, et les ouvrages publiés ou inédits de M. A.-L. Cauchy. II, in particular, Leçons 26, 27, 28, 33.

1952 D.Montgomery and L.Zippin. Small subgroups of finite dimensional groups. Annals of Math. (2) 56, pp. 213–241.

1955 D.Montgomery and L.Zippin. Topological transformation groups. New York, Interscience Publishers.

1878 E.Netto. Beitrag zur Mannigfaltigkeitslehre. Journ. r. u. angew. Math. 86, pp. 263–268.

1927 J.v.Neumann. Zur Theorie der Darstellungen kontinuierlicher Gruppen. Sitzber. Akad. Berlin 1927, pp. 76–90 = Collected Works I (1961), pp. 134–148.

1933 – Die Einführung analytischer Parameter in topologischen Gruppen. Annals of Math. (2) 34, pp. 170–190 = Collected Works II (1961), pp. 366–386.

1920 A J.Nielsen. Über fixpunktfreie topologische Abbildungen geschlossener Flächen. MA 81, pp. 94–96.

1920 B – Über die Minimal zahl der Fixpunkte bei den Abbildungstypen der Ringflächen. MA 82, pp. 83–93.

1927/1929/1933 – Untersuchungen zur Topologie der geschlossenen zweiseitigen Flächen. I. Acta Math. 50, pp. 189–358, II. Acta Math. 53, pp. 1–76, III. Acta Math. 58, pp. 87–167.

1942 – Abbildungsklassen endlicher Ordnung. Acta Math. 75, pp. 23–115.

1881 P.Opitz. Einige Sätze über die Anziehung in mehrfach ausgedehnten Gaussischen und Riemannschen Räumen. Diss. Göttingen.

1900 W.F.Osgood. Ueber einen Satz des Herrn Schoenflies aus der Theorie Functionen zweier reeller Veränderlichen. Nachr. Göttingen 1900, pp. 94–97.

1890 A G.Peano. Sur une courbe, qui remplit toute une aire plane. MA 36, pp. 157–160.

1890 B – Démonstration de l'intégrabilité des équations différentielles ordinaires. MA 37, pp. 182–228.

1885 E. PHRAGMÉN. Über die Begrenzung von Continua. Acta Math. 7, pp. 43–48.

1881 H. POINCARÉ. Mémoire sur les courbes définies par une équation différentielle. [[I.]] Journ. de Math. (3) 7, pp. 375–422 = Oeuvres I (1928), pp. 3–44.

1882 – Mémoire sur les courbes définies par une équation différentielle. [[II.]] Journ. de Math. (3) 8, pp. 251–296 = Oeuvres I (1928), pp. 44–84.

1884 – Sur les groupes des équations linéaires. Acta Math. 4, pp. 201–311 = Oeuvres II (1916), pp. 300–401.

1885 A – Mémoire sur les courbes définies par une équation différentielle. III. Journ. de Math. (4) 1, pp. 167–244 = Oeuvres I (1928), pp. 90–158.

1885 B – Sur un théorème de M. Fuchs. Acta Math. 7, pp. 1–32 = Oeuvres III (1934), pp. 4–31.

1895 – Analysis situs. Journ. Ecole Polytechnique (2) 1, pp. 1–123 = Oeuvres VI (1953), pp. 193–288.

1899 – Les méthodes nouvelles de la mécanique céleste. III. Paris, Gauthier-Villars.

1903 – L'espace et ses trois dimensions. Revue de Métaphysique et de morale 11, pp. 281–301, 407–429, in particular pp. 299–301, 407–408. Also in: La valeur de la science. Paris, Flammarion 1905(?), pp. 59–136, in particular, pp. 92–99.

1904 – Rapport sur les travaux de M. Hilbert. Izvestija fiziko-matematičeskogo obščestva pri imperatorskom kazanskom universitete (2) 14, pp. 10–48, in particular pp. 44–46; almost identical with a review of D. Hilbert 1902 in Bull. Sci. Math. (2) 26_{I} (1902), pp. 249–272, in particular pp. 270–271.

1907 – Sur l'uniformisation des fonctions analytiques. Acta Math. 31, pp. 1–64 = Oeuvres IV (1950) pp. 70–139.

1912 A – Pourquoi l'espace a trois dimensions. Revue de Métaphysique et de Morale 20, pp. 484–504, in particular pp. 486–493. Also in Dernières Pensées, Paris, Flammarion 1913, pp. 57–97, in particular 63–70.

1912 B – Sur un théorème de géométrie. Rend Circ. Mat. Palermo 33, pp. 375–407 = Oeuvres VI (1953), pp. 499–538.

1913 – Dernières Pensées. Paris, Flammarion.

1905 D. POMPÉIU. Sur la continuité des fonctions de variables complexes. Annales Toulouse (2) 7, pp. 264–315 = Thèse. Paris, Gauthiers-Villars.

1931 L. Pontrjagin. Über den algebraischen Inhalt topologischer Dualitätssätze, MA 105, pp. 165–205.

1931 J. Rey Pastor. Une propriété caractéristique des variétés de Jordan. C. R. Paris 192, pp. 27–29.

1868 B. Riemann. Ueber die Hypothesen, welche der Geometrie zu Grunde liegen. Habilitationsschrift Göttingen 1854. Abh. Göttingen 1868, pp. 1–20 = Gesammelte Mathematische Werke 1876, pp. 254–269.

1905 F. Riesz. Über mehrfache Ordnungstypen I. MA 61, pp. 406–421.
1906/07 – A térfogalom genesise. Mathematikai és physikai lapok 15 (1906), pp.97–122; 16 (1907), pp. 145–161 = (translated) Die Genesis des Raumbegriffs. Mathematische und naturwissenschaftliche Berichte aus Ungarn 24, pp. 309–353 = Gesammelte Arbeiten I (1960), pp. 67–109, 110–161.

1895 E. Ritter. Die Stetigkeit der automorphen Functionen bei stetiger Abänderung des Fundamentalbereiches. II. MA 46, pp. 200–248.

1870 E. Schering. Die Schwerkraft im Gaussischen Raum. Nachr. Göttingen 1870, pp. 311–321.
1873 – Die Schwerkraft in mehrfach ausgedehnten Gauss'schen und Riemann'schen Räumen. Nachr. Göttingen 1873, pp. 149–159.

1925 W. Scherrer. Translationen über einfach zusammenhängende Gebiete. Vierteljahrsschrift Zürich 70, pp. 77–84.

1923 E. Schmidt. Über den Jordanschen Kurvensatz. Sitzber. Akad. Berlin 1923, pp. 318–329.

1899 A. Schoenflies. Ueber einen Satz aus der Analysis situs. Nachr. Göttingen, 1899, pp. 282–290.
1900 – Die Entwickelung der Lehre von den Punktmannigfaltigkeiten. J. ber. D. M. V. 8_2, pp. 1–250.
1903 A – Über den Beweis eines Haupttheorems aus der Theorie der Punktmengen. Nachr. Göttingen, 1903, pp. 21–31.
1903 B – Beiträge zur Theorie der Punktmengen. I. MA 58, pp. 195–234.
1904 – Beiträge zur Theorie der Punktmengen. II. MA 59, pp. 129–160.
1906 A – Beiträge zur Theorie der Punktmengen. III. MA 62, pp. 286–328.
1906 B – Die Beziehungen der Mengenlehre zur Geometrie und Funktionentheorie. J. ber. D. M. V. 15, pp. 557–576.

1908 A – Die Entwicklung der Lehre von den Punktmannigfaltigkeiten. II. Leipzig, Teubner = Jber. D. M. V., 2. Ergänzungsband.

1908 B – Bemerkung zu meinem zweiten Beitrag zur Theorie der Punktmengen. MA 65, pp. 431–432.

1910 – Bemerkung zu dem vorstehenden Aufsatz des Herrn L. E. J. Brouwer. MA 68, pp. 435–444.

1913 – Entwickelung der Mengenlehre und ihrer Anwendungen. I, 1. Hälfte. [[2. Aufl. Leipzig, Berlin, Teubner; von Die Entwicklung der Lehre von den Punktmannigfaltigkeiten. J. Ber. D. M. V. 8 (1900).]]

1905 P.H.SCHOUTE. Mehrdimensionale Geometrie. II. (Sammlung Schubert 36.) Leipzig, G. J. Göschen.

1886 F.SCHUR. Ueber den Zusammenhang der Räume constanten Riemann'schen Krümmungsmaasses mit den projectiven Räumen. MA 27, pp. 537–567.

1928 E.SPERNER. Neuer Beweis für die Invarianz der Dimensionszahl und des Gebietes. Abh. Math. Sem. Hamburg 6, pp. 265–272.

1934 – Über die fixpunktfreien Abbildungen der Ebene. Abh. Math. Sem. Hamburg 10, pp. 1–48.

1924 A.STODOLA. Dampf- und Gasturbinen, 6. Aufl., Berlin, J. Spinger, in particular pp. 931–932.

1928 S.STOÏLOW. Sur les transformations continues et la topologie des fonctions analytiques. Ann. Ecole norm. sup. (3) 45, pp. 347–382.

1902 E.STUDY. Über Nicht-Euklidische und Liniengeometrie. Jber. D. M. V. 11, pp. 313–340.

1903 – Geometrie der Dynamen. Leipzig, Teubner, in particular p. 248.

1878 TAGEBLATT der Naturforscherversammlung zu Cassel 1878.

1878 J.THOMAE. Sätze aus der Functionentheorie. Nachr. Göttingen 1878, pp. 466–468.

1914 H.TIETZE. Über Funktionen, die auf einer abgeschlossenen Menge stetig sind. Journ. r. u. angew. Math. 145, pp. 9–14.

1931 H.TIETZE and L.VIETORIS. Beziehungen zwischen den verschiedenen Zweigen der Topologie. Enzyklopädie der mathematischen Wissenschaften $III_{1\,II}$, pp. 141–237 ital. (III AB 13), in particular p. 155, footnote 20.

1953 A J.TITS. Etude de certains espaces métriques. Bull. Soc. Math. Belg. 1952, pp. 44–52.
1953 B – Sur un article précédent .... Bull. Soc. Math. Belg. 1953, pp. 124–125.

1928 L.TUMARKIN. Über die Dimension nicht abgeschlossener Mengen. MA 98, pp. 637–656, in particular p. 653.

1922 P.URYSOHN. Les multiplicités Cantoriennes. C. R. Paris 175, pp. 440–442.
1923 – Sur une fonction analytique partout continue. Fund. Math. 4, pp. 144–150.
1925 – Mémoire sur les multiplicités Cantoriennes. Fund. Math. 7, pp. 30–137.
1926 – Mémoire sur les multiplicités Cantoriennes. Suite. Fund. Math. 8, pp. 225–359.

1916 G.DE LA VALLÉE POUSSIN. Intégrales de Lebesgue, Fonctions d'ensemble, Classes de Baire. Paris, Gauthier-Villars.

1905 O.VEBLEN. Theory on plane curves in non-metrical analysis situs. Trans. A. M. S. 6, pp. 83–98.

1927 L.VIETORIS. Über den höheren Zusammenhang kompakter Räume und eine Klasse von zusammenhangstreuen Abbildungen. MA 97, pp. 454–472.
1929 – Zum höheren Zusammenhang der kompakten Räume. MA 101, pp. 219–225.

1880 C.WEIERSTRASS. Zur Functionenlehre. Monatsber. Akad. Berlin 1880, pp. 719–743, in particular 721 = Gesammelte Werke 2 (1895), pp. 201–223, in particular p. 203.

1924 W.WILSON. The equivalence in R_n of the n-dimensional simplex star and the spherical neighborhood. KNAW Proc. 27, pp. 772–777.
1928 – Representation of manifolds. MA 100, pp. 552–578.

1898 W.A.WIJTHOFF. De biquaternion als bewerking in de ruimte van vier afmetingen ⟦the biquaternion as an operation in the fourdimensional space⟧. Thesis Amsterdam.

1953 H.YAMABE. A generalization of a theorem of Gleason. Annals of Math. 58, pp. 351–385.

1917 K. YONEYAMA. Theory of continuous set of points. Tôhoku Math. Journ. 12, pp. 43–158, in particular pp. 60–62.

1903 A W.H. YOUNG. Zur Lehre der nicht abgeschlossenen Punktmengen. Berichte Leipzig 1903, pp. 287–293.

1903 B – On the analysis of linear sets of points. Quarterly Journ. of Math. 35, pp. 102–116.

1904 A – On closed sets of points and Cantor's numbers. Proc. London Math. Soc. (2) 1, pp. 230–246.

1904 B – On sequences of sets of intervals containing a given set of points. Proc. London Math. Soc. (2) 1, pp. 262–266.

1905 – Ordinary inner limiting sets in the plane or higher space. Proc. London Math. Soc. (2) 3, pp. 371–380.

1909 L. ZORETTI. La notion de ligne. Annales Ecole norm. sup. (3) 26, pp. 485–497.

1910 A – Sur la notion de ligne. C. R. Paris 151, pp. 201–203.

1910 B – Remarque au sujet de la Remarque de M. Brouwer. Annales Ecole norm. sup. (3) 27, p. 567.

1911 L. Z⟦ORETTI⟧. ⟦Review of A. Schoenflies 1908 A.⟧. Bull. sci. math. (2) 35_{I}, pp. 283–288, in particular p. 287.